D1801263

Guides to Information Sources

Information Sources in
Engineering

Guides to Information Sources

A series under the General Editorship of
Ia C. McIlwaine,
M.W. Hill and
Nancy J. Williamson

This series was known previously as 'Butterworths Guides to Information Sources'.

Other titles available include:

Information Sources in Architecture and Construction (Second edition)
  edited by Valerie J. Nurcombe
Information Sources in Chemistry (Fourth edition)
  edited by R.T. Bottle and J. F. B. Rowland
Information Sources in Physics (Third edition)
  edited by Dennis F. Shaw
Information Sources in Finance and Banking
  by Ray Lester
Information Sources in Environmental Protection
  edited by Selwyn Eagle and Judith Deschamps
Information Sources in Grey Literature (Third edition)
  by C.P. Auger
Information Sources in Music
  edited by Lewis Foreman
Information Sources in the Life Sciences (Fourth edition)
  edited by H.V. Wyatt
Information Sources in Sport and Leisure
  edited by Michele Shoebridge
Information Sources in Patents
  edited by C.P. Auger
Information Sources for the Press and Broadcast Media
  edited by Selwyn Eagle
Information Sources in Information Technology
  edited by David Haynes
Information Sources in Pharmaceuticals
  edited by W.R. Pickering
Information Sources in Metallic Materials
  edited by M.N. Patten
Information Sources in the Earth Sciences (Second edition)
  edited by David N. Wood, Joan E. Hardy and Anthony P. Harvey
Information Sources in Cartography
  edited by C.R. Perkins and R.B. Barry
Information Sources in Polymers and Plastics
  edited by R.T. Adkins
Information Sources in Economics (Second edition)
  edited by John Fletcher

Guides to Information Sources

Information Sources in
Engineering
Third Edition

Editors
K.W. Mildren
P.J. Hicks

London · Melbourne · Munich · New Jersey

© 1996 Bowker-Saur, a division of Reed Elsevier (UK) Ltd.

All rights reserved. No part of this publication may be reproduced or transmitted in any form or by any means (including photocopying and recording) without the written permission of the copyright holder except in accordance with the provisions of the Copyright, Designs and Patents Act 1988 or under the terms of a licence issued by the Copyright Licensing Agency, 90 Tottenham Court Road, London W1P 9HE. The written permission of the copyright holder must also be obtained before any part of this publication is stored in a retrieval system of any nature. Applications for the copyright holder's written permission to reproduce, transmit or store in a retrieval system any part of this publication should be addressed to the publisher.

Warning: The doing of any unauthorized act in relation to a copyright work may result in both a civil claim for damages and criminal prosecution.

British Library Cataloguing in Publication Data
A catalogue record for this title is available from the British Library

Library of Congress Cataloging-in-Publication Data
A catalog record for this book is available from the Library of Congress

Published by Bowker-Saur, Maypole House, Maypole Road,
East Grinstead, West Sussex RH19 1HU, UK
Tel: +44(0)1342 330100 Fax: +44(0)1342 330191
E-mail: lis@bowker-saur.co.uk
Internet Website: http://www.bowker-saur.co.uk/service/
Bowker-Saur is part of REED REFERENCE PUBLISHING

ISBN 1-85739-057-1

Cover design by Calverts Press
Typesetting by Wyvern Typesetting
Printed on acid-free paper
Printed and bound in Great Britain by Bell & Bain Ltd, Glasgow

Series editors' foreword

The second half of the 20th century has been characterized by the recognition that our style of life depends on acquiring and using information effectively. It has always been so, but only in the information society has the extent of the dependence been recognized and the development of technologies for handling information become a priority. These modern technologies enable us to store more information, to select and process parts of the store more skilfully and transmit the product more rapidly than we would have dreamt possible only 40 years ago. Yet the irony still exists that, while we are able to do all this and are assailed from all sides by great masses of information, ensuring that one has what one needs just when one wants it is frequently just as difficult as ever. Knowledge may, as Johnson said in the well known quotation, be of two kinds, but information, in contrast, is of many kinds and most of it is, for each individual, knowable only after much patient searching.

The aim of each Guide in this series is simple. It is to reduce the time which needs to be spent on that patient searching; to recommend the best starting point and sources mostly likely to yield the desired information. Like all subject guides, the sources discussed have had to be selected, and the criteria for selection will be given by the individual editors and will differ from subject to subject. However, the overall objective is constant; that of providing a way into a subject to those new to the field or to identify major new or possibly unexplored sources to those already familiar with it.

The great increase in new sources of information and the overwhelming input of new information from the media, advertising, meetings and conferences, letters, internal reports, office memoranda, magazines, junk mail, electronic mail, fax, bulletin boards etc. inevitably tend to make one reluctant to add to the load on the mind and memory by consulting books and journals. Yet they, and the other traditional types of printed material, remain for many purposes the most reliable sources of information. Despite all the information that is instantly accessible via the new technologies one still has to look things up in databooks, monographs, journals, patent specifications, standards, reports both official and commercial, and on maps and in atlases. Permanent recording of facts, theories and opinions is still carried out primarily by publishing in printed form. Musicians still work from printed scores even though they are helped by sound recordings. Sailors still use printed charts and tide tables even though they have radar and sonar equipment.

However, thanks to computerized indexes, online and CD-ROM, searching the huge bulk of technical literature to draw up a list of references can be undertaken reasonably quickly. The result, all too often, can still be a formidably long list, of which a knowledge of the nature and structure of information sources in that field can be used to put in order of likely value.

It is rarely necessary to consult everything that has been published on the topic of a search. When attempting to prove that an invention is genuinely novel, a complete search may seem necessary, but even then it is common to search only obvious sources and leave it to anyone wishing to oppose the grant of a patent to bear the cost of hunting for a prior disclosure in some obscure journal. Usually, much proves to be irrelevant to the particular aspect of our interest and whatever is relevant may be unsound. Some publications are sadly lacking in important detail and present broad generalizations flimsily bridged with arches of waffle. In any academic field there is a 'pecking order' of journals so that articles in one journal may be assumed to be of a higher or lower calibre than those in another. Those experienced in the field know these things. The research scientist soon learns, as it is part of his training, the degree of reliance he can place on information from co-workers elsewhere, on reports of research by new and (to him) unknown researchers, on data compilations and on manufacturers of equipment. The information worker, particularly when working in a field other than his own, faces very serious problems as he tries to compile, probably from several sources, a report on which his client may base important actions. Even the librarian, faced only with recommending two or three books or journal articles, meets the same problem though less acutely.

In the Bowker-Saur Guides to Information Sources we aim to bring you the knowledge and experience of specialists in the field. Each author regularly uses the information sources and services described and any tricks of the trade that the author has learnt are passed on.

Nowadays, two major problems face those who are embarking upon research or who are in charge of collections of information of every kind. One is the increasingly specialized knowledge of the user and the concomitant ignorance of other potentially useful disciplines. The second problem is the trend towards cross-disciplinary studies. This has led to a great mixing of academic programmes – and a number of imprecisely defined fields of study. Courses are offered in Environmental Studies, Women's Studies, Communication Studies or Area Studies, and these are the forcing ground for research. The editors are only too aware of the difficulties raised by those requiring information from such hybrid subject fields and this approach, too, is being handled in the series alongside the traditional 'hard disciplines'.

Guides to the literature have a long and honoured history. Marion Spicer of SRIS recently drew to our attention a guide written in 1891 for engineers. No doubt there are even earlier ones. Nowadays, with the information and even the publishing fields changing quite frequently, it is necessary to update guides every few years and this we do in this present Series.

Michael Hill
Ia McIlwaine
Nancy Williamson

About the contributors

Bill Addis studied at the University of Cambridge and gained his engineering training with the aeroengine division of Rolls Royce. He is currently a lecturer at the University of Reading where his interests are mainly in the fields of building engineering and design. He is the author of *Structural engineering: the nature of theory and design* (1990) and *The art of the structural engineer* (1994).

Ken Anderson is currently Senior Lecturer in Environmental Engineering at the University of Newcastle upon Tyne, and is also Executive Director of Environmental Technology Consultants Ltd. He received his PhD in 1964 and since then has a wide academic and professional experience in Environmental Engineering, both in the UK and overseas. His specialization is in the field of aerobic and anaerobic biological treatment of industrial wastewaters and the management of solid and hazardous wastes. He is the author of over 100 technical papers and is currently regional editor for developing country papers submitted to Water Research.

C.P. Auger is a freelance writer and information consultant. He has a number of publications to his credit, including works on grey literature and patents.

Chris Baile is currently the Product Development Manager for a new commercial service from The British Library. Before that he was the Information Officer (Engineering) at The British Library for five years. He has also worked in information posts for British Gas and Shell Expro. Amongst other works, he has contributed to *Engineers' guide to product information: sources and use* (1991).

David Brimage obtained his doctorate in chemistry at the University of Leicester and then trained as a Librarian. After working in

x   About the Contributors

Birmingham and Durham he spent 10 years at the Polytechnic of the South Bank as Faculty Librarian for Science and Engineering. After a spell in industry with Ferranti he is now a subject librarian at the University of Portsmouth.

Peter Bonsall is Reader in Transport Planning Methods and Director of the Transport Planning and Engineering Masters course at Leeds University's Institute for Transport Studies. He has been active in transport research since 1974 and has initiated or directed projects in a number of areas including simulation of car-pooling schemes, the use of new technologies in data collection, project appraisal software and the impact of telecommunications on travel demand. He has published widely and has acted as consultant to several governmental bodies.

Brian Boyle obtained his first degree in Computer Systems and Robotics from the Department of Electrical and Electronic Engineering of Liverpool University. He is currently employed as a Research Assistant within the Department of Marine Technology of Cranfield University, where he is currently concluding his PhD research. The subject of his research is the development of a system for computer control of an underwater manipulator, utilising real time simulation and modelling, and three dimensional graphical feedback.

Ron Brewster studied Electrical Engineering at South East Technical College while working as an apprentice with Kelvin Hughes Division of Smiths Industries, where he later worked with the marine radar development team. He has also worked for Standard Telecommunications Laboratories, working for several years on the development of microwave strip-line techniques and equalization techniques for data modems. He gained his MSc degree at Aston University in 1967 and his doctorate in 1972, where he is now Senior Lecturer in Telecommunications.

Michael Bussell graduated in Civil Engineering at Bristol University in 1966, and since then he has worked for Ove Arup & Partners, a major international engineering consultancy. After 20 years' general design experience, he joined the firm's Research & Development group where he now provides in-house specialist technical advice, as well as writing external technical guides under contract. He is particularly interested in engineering aspects of building conservation.

John Corbett is Professor at the School of Industrial Manufacturing and Science, Cranfield University. His research interests are in Precision Engineering and Design for Manufacture, in which he has conducted funded research for government and industry. He has undertaken consultancies and presented seminars all over the world. His publications include contributions for the IMechE, IProdE and the BSI, and he was recently joint author/editor of the book *Design for Manufacture* (Addison-Wesley).

Neville Dean is a Senior Lecturer in Mathematics on the Cambridge campus of Anglia Polytechnic University. He holds an MA degree from St. John's College, Cambridge, and a DPhil degree from the University of Sussex. He is also a Chartered Mathematician and a Chartered Physicist. His industrial experience includes mathematical modelling aimed at improving consumer products, and leading research projects on electrical contact materials. He has a special interest in the application of mathematical modelling in engineering and science.

Jonathan Dell is a lecturer in Electronics at York University, specializing in digital electronics, microprocessor applications and industrial measurement systems. His current research centres on parallel computing architectures for application in real-time reconstruction of the images obtained from industrial tomographic systems. He previously worked at Sheffield Halam University, Tektronix International, Amsterdam, and Texas Instruments Ltd., Bedford.

Gerald Druce read Mechanical Engineering at Imperial College, London and worked on the design and development of mechanisms for International Computers Ltd. before joining the teaching staff of the University of Surrey. He gained his PhD in 1976 through research into the improvement of cam design and manufacture. He has published more than 20 papers and is currently Vice-Chairman of the ESDU Mechanisms Committee. In retirement he has been appointed a Visiting Research Fellow at the University of Surrey where he continues to investigate means of enhancing cam performance.

David Elliott is a Chartered Civil Engineer currently lecturing in Civil and Environmental Engineering at the University of Newcastle upon Tyne, and is an Executive Director of Environmental Engineering Consultants Ltd. His specialized interests are in the modelling dispersion of pollutants in aquatic and atmospheric environments, and in the impact of civil engineering activities on the environment. He has wide academic and professional experience working on consultancy projects for WHO, the European Union, and the Asian Development Bank.

Denis E. Filer was appointed Director-General of The Engineering Council in 1988, and retired from this post in 1995. Before that he was director of engineering in ICI, with whom he had worked since 1955. After graduating from Manchester University, he served with the Royal Electrical and Mechanical Engineers. He later joined the Territorial Army, in which he became a colonel, and was awarded the Territorial Decoration. He was awarded the CBE in January 1992.

Terry Hanson wrote his chapter when working at the University of

xii *About the Contributors*

Connecticut Library as Head of the Research and Information Services Department. Since then he has returned to his former post of Sub-Librarian (Electronic Information Services) at the University of Portsmouth. He has written widely on many aspects of library service.

John Harris graduated in physics from the University of Birmingham in 1956. After working at UKAEA Sellafield on Research and Development for the early nuclear power programme, he joined the Engineering Department of the University of Manchester in 1960, where, ever since, he has taught, researched and published many papers on nuclear engineering. He has interests in plant safety, reliability and maintenance, has co-authored two textbooks, edits a periodical and consults in these areas.

Peter Hicks read Engineering at Oxford University and has spent 10 years working in the areas of design, manufacture, materials and quality control in industry. He is now at the University of Portsmouth and responsible for the Design and Computer-aided Engineering Group in the Department of Mechanical Engineering. He is a member of the Institution of Mechanical Engineers Degree Accreditation Panel, Faculty Head of Quality Assurance and Senior Research Associate of an EPSRC Teaching Company. He has published papers and refereed national and international conferences in the areas of pedagogy and design research.

Steve C. Hughes is Director of the Biomedical Engineering Group at the University of Surrey. After graduating in nuclear physics he has worked for the Atomic Energy Authority, and in Medical Biophysics at the University of Dundee, becoming a founding member of the School of Biomedical Engineering Sciences there. Since 1983 he has been involved in leading a postgraduate group with interests in gait analysis, human shape, tissue mechanics, rehabilitation engineering and bioelectric phenomena.

Chengi Kuo is an engineering graduate of the University of Glasgow, where he gained both his BSc and his PhD. After working on computer applications to ship building he spent two years in the USA, and became a professor at the University of Strathclyde in 1972. He has gained recognition and an international reputation in safety of marine systems, computer applications to ship technology, surface support for subsea work and novel educational methods. He works closely with industry and has acted as a consultant to more than 30 international companies and organizations. He is the author of four books and editor of a further nine, and has published over a hundred papers and articles.

David Langford is Professor of Construction at the Department of Civil Engineering, University of Strathclyde. Before that he was Director of Postgraduate Studies, School of Architecture and Building Engineering, Bath University. He was commissioned by the BRE to contribute to the reports, *Faster building for industry* and *Faster building for commerce*, and in 1990 he was given an annual research award by the CIOB to research construction marketing in the 1990s, which is now published in a report.

R.S. McMaster is Director of the Automation and Robotics Programme, Cranfield University, and runs the Advanced Automation and Design MSc course. He is also responsible for managing various commercial and research contracts and providing consultancy in automation and robotics. He has also held a number of positions in industry, including the aerospace and mechanical power transmission industries. He is a registered Chartered Engineer and a council member of the British Robot Association.

John T. McMullan is Professor of Physics and Director of the Energy Research Centre of the University of Ulster. He has considerable research and consultancy experience in the energy field, and specialises in CFC replacement in the refrigeration and air-conditioning industries, and in advanced combustion and fuel conversion technologies for the utilisation of fossil and biomass fuels, including energy crops. He has published many books and research papers and is editor of the *International Journal of Energy Research*.

Ken Mildren graduated in physics at the University of Liverpool. He worked in Liverpool City Libraries prior to taking a postgraduate course in librarianship at the College of Librarianship, Aberystwyth. He was appointed as Engineering Librarian at Portsmouth Polytechnic and is also now Academic Services Librarian at the University of Portsmouth.

Paul O'Callaghan holds the Chair of Energy Management at Cranfield University. Having graduated from the University of Wales, and after a period in industry, he joined the University and established the well-known MSc course, Energy Conservation and the Environment, which has now run for 23 years. His teaching and research interests are in the areas of integrated environmental management and multimedia applications, and he is Fellow of the Institution of Mechanical Engineers. He acts as consultant to many international companies and institutions and is the author of numerous books and technical publications.

Charles Oppenheim is Professor of Electronic Library Research, De Montfort University, Leicester. Before this he has been Professor of Information Science at Strathclyde University, and Business Development Manager at Reuters Ltd. He is Vice President of ASLIB, and was President of the Institute of Information Scientists, 1994–95. A frequent contributor to the professional literature and a well known public speaker, he is on the editorial board of a number of journals and is one of the UK's representatives on the European Commission's legal advisory board.

Anthony Parker began his career with the Central Electricity Generating Board before gaining a BSc in Electrical Engineering and an MSc from research into the thermal characteristics of power equipment, at Southampton University. He was awarded a PhD in 1974 before joining the Department of Electrical and Electronic Engineering at the University of Adelaide, South Australia, where he is now Head of Department.

Martin John Pitt is a Chartered Chemical Engineer and Chartered Chemist who has worked in both roles in industry. He has published books on process instrumentation and waste disposal, and is a compiler for *Bretherick's handbook of reactive chemical hazards*. He is currently a Senior Lecturer in Process Engineering at the University of Sheffield, with research interests in two-phase extraction and reaction, and chemical process safety.

David Radcliffe is a Senior Lecturer in the Department of Mechanical Engineering at the University of Queensland, Australia, working in the areas of engineering design, manufacturing systems and biomedical engineering. Previously he worked on several large industrial design projects and has taught at the Universities of Melbourne, Adelaide and Stanford.

Arkady Retik is a Senior Lecturer in Construction Management in the Department of Civil Engineering, University of Strathclyde, Glasgow. After graduating in 1981, he worked initially as a structural engineer and later as a project manager. In 1985 he obtained his MSc and DSc in the area of Construction Engineering and Management. Before joining Strathclyde in 1992, he spent 18 months as a research fellow in the Department of Quantity Surveying at Salford University. He is a well known author and researcher in the field of construction information technology.

Bob Rhodes is currently Sub-librarian (Information) at Loughborough University of Technology. He qualified initially as a Metallurgist before moving to a research association, advising industry on the use of steel. He then studied Library and Information Science at Sheffield University before working in the library serving the materials and engineering faculties. He has authored a number of papers and articles, and has also written a Unesco manual on training scientists and technologists in using libraries.

Peter Richards was originally a Zoologist from Aberdeen University. He worked first in industry in Cell Culture, and then completed a PhD

About the Contributors xv

on Polar Pycnogonids, a marine arthropod group, before moving into Biomedical Engineering at the University of Surrey 18 years ago. His research has involved microscope techniques, heart-lung bypass, heart valves, soft tissue properties and tissue/electrical interfaces.

Phil Ruston graduated in 1988 from University College London with an MA degree in Library and Information Studies. He has worked for The British Library as a cataloguer, and in 1991 joined its Business Information Service (BIS) as Information Officer. He talks regularly on Production Information at BIS seminars on Information Sources.

Philip Sargent is currently Senior Software Manager at Quillion Systems Ltd. and is working on engineering data management issues. He was formerly a Senior Consultant in telecommunications strategy modelling at Analysis Ltd. He has a materials science PhD degree and spent five years at Carnegie Mellon University, Pittsburgh, and Cambridge University researching materials information use in the product development process.

Tariq Pervez Sattar completed a PhD in Adaptive Control Systems in 1987. He is now a senior lecturer at the South Bank University, London and leads the teaching and research work in Control, Robotics and Real Time Systems.

C.T. Shaw is Professor of Mining and Dean of the Royal School of Mines, Imperial College of Science, Technology and Medicine, London, and he has been an Independent Consulting Mining Engineer since 1977. He is the Secretary General of the Society of Mining Professors-Societät der Berbaukunde and the editor of *Mining Resources Engineering*.

John Smart graduated from Imperial College, London, in 1967 and was awarded a PhD degree in 1972. He then spent eight years in industry, first with Stone Manganese Marine, who manufactured ship's propellers, then with Powdrex, who made high speed steel powder and components. He has been at Manchester University since 1980 where his research interests are in numerical stress analysis, particularly in probabilistic design methods for brittle materials, and in the analysis of mechanical joints.

Michael Smith graduated in mineral sciences from the University of Leeds before undertaking research in hydrometallurgy. He was employed by British Steel in research and development and later in the consultancy company. In 1981 he joined the staff of Imperial College as a lecturer in mineral technology.

Peter Stanley graduated from Cambridge University in 1954. After three years with Rolls-Royce Ltd., Aero-engine Division, he went to Nottingham University and developed research interests in experimental stress analysis, pressure vessel analysis and behaviour and probabilistic design methods for brittle materials. He has been at Manchester University since 1979, where his principal research interest is the development and application of thermoelastic stress analysis techniques.

John Watton began his engineering career in industry before graduating from University College Cardiff, obtaining a PhD degree in 1969. Continuing research includes digital control, fault diagnosis, the application of computational fluid dynamics to component design and the application of artificial networks to system design. His design contribution has resulted in machines now in commercial operation. He is currently Reader in Mechanical Engineering (Fluid Power) in the Cardiff School of Engineering, University of Wales, Cardiff, and has published a number of papers and books.

Contents

Series editors' foreword   v
About the Contributors   ix
Preface   xxi

Introduction

1   Information and the engineer
     D.E. Filer   1

Primary information sources

2   Reports
     C. P. Auger   3
3   Standards
     C.P. Auger   15
4   Patents and patent information
     C. Oppenheim   31
5   Journals, conferences and theses
     C.W. Baile   63
6   Product information
     P.J. Ruston   81

Secondary information sources

7   Abstracting and indexing services, bibliographies and reviews
     R.G. Rhodes   97
8   Electronic sources of engineering information
     T.A. Hanson   107

9 Standard reference sources
     D.R.G. Brimage   121

Specialized subject fields

10 Stress analysis
     P. Stanley and J. Smart   133
11 Kinematics, dynamics and machines
     G. Druce   151
12 Thermodynamics and thermal systems
     P.W. O'Callaghan   199
13 Energy technology
     J.T. McMullan   223
14 Nuclear engineering
     J. Harris   259
15 Chemical engineering
     M.J. Pitt   269
16 Fluid mechanics and fluid power systems
     J. Watton   285
17 Materials information sources for engineers
     P. Sargent   301
18 Safety engineering
     M.J. Pitt   315
19 Design and ergonomics
     P.J. Hicks   329
20 Manufacturing techniques
     J. Corbett   353
21 Robotics and automated manufacture
     R.S. McMaster and B.G. Boyle   377
22 Control engineering
     T.P. Sattar   389
23 Electronics
     J. Dell   411
24 Electric power systems and machines
     A.M. Parker   425
25 Communications engineering
     R.L. Brewster   441
26 Mathematics and statistics for engineers
     N. Dean   451
27 Marine technology
     C. Kuo   467
28 Environmental control engineering
     G.K. Anderson and D.J. Elliott   485
29 Transportation and traffic planning and engineering
     P.W. Bonsall   495

30	Construction engineering W. Addis and M. Bussell	519
31	Structural engineering M. Bussell and W. Addis	547
32	Construction management D.A. Langford and A. Retik	569
33	Mineral process engineering M.R. Smith	589
34	Mining C.T. Shaw	607
35	Biomedical engineering S.C. Hughes and P.R. Richards	623
36	Concurrent engineering D.F. Radcliffe	649
Index		661

Preface to the third edition

The major change in this third edition is that all the subject chapters have been written by specialists in each subject area, although in some cases they may have sought assistance from their local librarian or information officer. Material included in the chapters has been chosen by the contributors on a selective basis and some chapters are more comprehensive than others. Any variations in the style of reference citation, caused by the long gestation period of the book, have been kept to a minimum.

The chapters have not been subjected to restrictive editing, thereby retaining the individual style of the contributors. Most of the chapters were initially written in late 1993 and 1994, with a few received in 1995; although many have been revised subsequently.

We would like to thank the contributors, the series editors, the sub-editor, numerous colleagues who have assisted with various queries and the editorial staff at Bowker-Saur. Particular thanks are due to our Heads of Department and the Dean of Engineering at the University of Portsmouth for their encouragement and support.

P.J. Hicks
K.W. Mildren

CHAPTER ONE

Information and the engineer

D. E. FILER

Engineering has always been an information-rich activity. It has a part to play in almost every aspect of modern civilization. In areas such as communications, transport, medicine and energy production, engineering is the source of the creation of real wealth. It is therefore not surprising that engineers use a wide range of information, as is illustrated by the contents page of this book.

The analysis and application of information have always been important parts of the engineer's function. These requirements have become increasingly important as the pressure of competition now demands ever more frequent product changes and more complex and flexible production processes which need to operate at an optimum level of quality and performance. In addition, no engineering project can take place in isolation from its commercial context or from wider social and ethical considerations; for example, as part of the product design process, engineers need to obtain information about relevant patent or copyright legislation. Similarly, safety requirements need to be investigated and environmental considerations are an increasingly important element in project design and manufacture.

Engineering provides the means to turn scientific and technical knowledge into useful and saleable products to satisfy customer requirements. Given that such requirements are dynamic, it follows that engineers always need to be aware of new or different ways of producing competitive products. Tracking down and making use of available information are important parts of this process. These include, for example, finding out about modern materials and processing in order to explore the scope for taking advantage of the properties of such materials to produce products which function more effectively and are thereby more attractive to the customer.

2 Information and the engineer

In recent years, there have been fundamental changes in the role of engineers in manufacturing. Engineers now design and develop products and, at the same time, concern themselves with manufacture and often marketing. The distinctions between specific engineering disciplines are becoming blurred as engineers work in teams on complex designs and processes which call for knowledge and information beyond a particular discipline and often beyond the technical aspects of engineering itself; for example, engineers now need not only to be technically proficient but also to have management skills and business acumen.

The nature of engineering in the modern world puts a premium on engineers being flexible, adaptable and always ready for change in the tools and techniques they employ in their professional lives. The pace of technological change is developing at an ever increasing rate. Every year, new technologies come into play, new materials and processes are introduced and more up-to-date methods of managing the manufacturing process also bring about new approaches. Engineers are at the centre of this technological change. Therefore they need to know where to find information that will keep their knowledge and skills relevant to contemporary requirements and how to access such information in a structured and methodical manner using printed and computerized sources of information. This book is designed to help engineers to identify and access the information they need to do their job in a business environment in which the pace of technological change is accelerating, the legislative structure is more complex, safety requirements are more stringent and the time scales within which engineering projects must be carried out are ever tighter.

CHAPTER TWO

Reports

C. P. AUGER

Introduction

The term 'report', which originally meant 'to carry back', today has a number of connotations, including common talk; an account given or an opinion formally expressed after investigation or consideration or collation of information; a description or epitome or reproduction of a scene or speech or court case, especially for a newspaper; and a periodical statement on a pupil's work or conduct.

In the context of engineering information, the report is most appropriately identified by reference to the purpose for which it is written and the circulation allocated to it. The main areas of activity are research, development, testing and evaluation. Typically, engineering reports will contain experimental details and production data which are then consolidated into definitive information such as statistics, materials properties and processed test results. Subsequently, the information is developed to the stage where it can be imparted to a wider audience, either within the originating organization itself or externally into the public domain. This point is a crucial one, for it is here that technology transfer takes place. In particular, the report will contain the summaries, conclusions and recommendations which make it valuable to a wider readership. The format is different from that adopted in other types of speedy technical communication, such as papers to conferences and articles in journals, in that little or no editing or refereeing takes place. The report is simply given a wider distribution under its original identity. In cases where it is absorbed into a report-handling agency's system or special collection, it may acquire an accession number within a given series which will afford it universal recognition. Essentially, however, the text remains in its original state.

4 Reports

Other types of report encountered in the engineering literature include status reports, project summaries, marketing and market research reports, and project evaluations. Of these, marketing and market research reports often present difficulties of access for the would-be reader, since, because of their high cost and short life, they are not acquired on a comprehensive basis by libraries and other agencies, yet they may contain information crucial for the success of a technical project.

The report as a communications medium

Although it is possible to identify reports series, particularly in aeronautics, which had their origins in the early years of the century, the development of the report as a major means of communication dates back only to about the year 1941. It was at that time that the US Office of Scientific Research and Development was set up, and its task was to serve as a centre for mobilizing the scientific resources of the nation and applying the results of research to national defence. Today, reports are still the prime means of communicating within the defence industry, but their virtues have also been recognized by the R&D community in general.

The special characteristic of a report as a communications medium is that it is a response document, usually written with a specific readership in mind. Many engineering companies and establishments regard the report as the prime means of technical communication, supported where appropriate by memoranda, lectures and presentations. Often, the format is well established to ensure uniformity of layout and content, and, in the case of research and development reports, specific guidance is available in the form of BS 4811: 1986 (British Standards Institution). An advantage of the technical report is that the information it contains can be made available with a minimum of delay. Even documents which are deliberately bulky in order to include all the details essential to a thorough understanding of the work described can be produced quickly and cheaply in small quantities. Indeed, thoroughness is a further characteristic of reports, since they are investigative accounts in the sense of detailing failed or partially successful projects as well as programmes of work where all the stages went according to plan.

Once prepared and issued for the specified requirements of the parent organization, reports can, if it is felt appropriate, be made available to outside bodies without further modification. This availability is, of course, subject to certain considerations and the wider audience has to accept that the contents will not have been reworked or edited in a way which might make the document more readable or less parochial. The currency of reports both within the originating organization and in the public domain tends to vary with the subject matter and its treatment. Some reports may be quickly superseded by more formal types of communications, particularly the journal article; others may become standard works in their own right and fulfil a small but steady demand over the years. The general experience, as might be expected with documents produced to meet an immediate need, is that readership drops off dramatically after a couple of years of a report's availability within an organization or its inclusion in an outside collection.

Typical series

The bestowal of an identification to any report starts with its originating organization, where numbers or combinations of letters and numbers are allocated to each document as it is written. Since the number of reports-issuing agencies and bodies around the world runs into many thousands, there is unfortunately a corresponding multiplicity of coding systems.

Attempts have been made to collate and list numbers allocated to reports and in this respect a major work of reference is Aronson's codes dictionary (Aronson, 1986), the purpose of which is to identify or provide associations for most of the codes that have been applied to reports. Its coverage is very wide, and not confined to documents issued in the USA, although these do form the majority. The dictionary also includes notes which give guidance on ambiguous codes and the major reports-issuing agencies. It is the outstanding work in its field for, while other compilations have tackled the question of report number identification, notably the British Library (1992), none has approached the comprehensiveness of the dictionary.

As reports flow into reports-handling agencies, they receive fresh identities in the form of accession numbers, and it is these numbers which form the basis of several widely known series. Those most likely to be encountered in engineering include the following, which have become universally recognized for identification and ordering purposes:

- AD: here stands for Astia Document, with Astia in turn originally meaning the Armed Services Technical Information Agency, a now defunct US organization. The National Technical Information Service (NTIS), discussed below, is currently responsible for issuing AD reports for public use;
- PB: here stands for Publications Board, another example of initials remaining in use long after the body itself has disappeared. The NTIS is also responsible for issuing PB reports;

6 Reports

- N: here signifies a document processed by the National Aeronautics and Space Administration (NASA), also discussed below.

Announcement services

The most effective way of keeping a reports collection in harmony with the needs of a library and its users is to select reports considered to be relevant as and when they are announced and made available. Such a selection policy can be implemented by consulting the various announcement journals and services provided by agencies serving the major subject areas. Most of the key titles are published in the USA, although their coverage is worldwide.

National Technical Information Service

The major reports announcement journal is undoubtedly *Government Reports Announcements and Index* (*GRA&I*), issued every two weeks by NTIS, which provides full bibliographical report entries arranged by subject category and sub-category.

GRA&I is supplemented by a keyword index, a personal author index, a corporate author index, a contract/grant number index and an NTIS order/report number index. The scope of the publication reflects the role of NTIS, an agency of the US Department of Commerce, as the central research development and engineering body.

The NTIS information collection increases by approximately 70 000 new works annually, many of which contain foreign technology or marketing information. All are permanently available for sale either directly from the many thousands of titles in shelf stock or from the microfiche masters of titles less in demand.

The categorization scheme for the subjects covered in *GRA&I* is one based on a system endorsed in 1964 by a body in existence at that time (but now disbanded), namely the Committee on Scientific and Technical Information (COSATI) of the Federal Council for Science and Technology. Details are contained in an NTIS report, order number AD 612 200. From 1987 onwards, *GRA&I* began to use the NTIS subject category and sub-category version of the COSATI scheme, which is suitable for online searching and is described in the booklet *Subject category descriptions* (PR 832, free from NTIS).

GRA&I is of great value to the staffs of libraries and information centres because of its detailed and comprehensive coverage of the reports literature. Users, as for example practising engineers whose immediate interests are confined to a particular discipline, may prefer a more selective approach and choose to consult *NTIS Alerts* (formerly

NTIS *Abstract Newsletters*), a series of bulletins published twice a month and intended to focus on nearly 200 highly specialized topics. Reports available from NTIS have various price codes and can be purchased direct from NTIS headquarters at Springfield, Virginia, or from one of the worldwide network of official distributors. In the case of the UK, the official distributor, known as the Official UK Managing Dealer, is Microinfo Limited of Alton, Hampshire. Access to the NTIS bibliographical database is available via seven online hosts and covers the period 1964 to date.

Reports issued by NTIS also form part of the collections of major libraries and can, for example, be borrowed from the British Library Document Supply Centre.

National Aeronautics and Space Administration

The announcement journal *Scientific and Technical Aerospace Reports* (*STAR*) is a major component of NASA's information system, covering aeronautics, space and supporting disciplines. *STAR* is published twice a month and announces current publications of the following types: NASA, NASA contractor and NASA grantee reports; reports issued by other US government agencies, domestic and foreign institutions, universities and private firms; translations in report form; NASA-owned patents and patent applications, and certain dissertations and theses. *STAR* is complemented by *International Aerospace Abstracts* (*IAA*), published by the American Institute of Aeronautics and Astronautics, which concentrates on journal, book and conference literature.

The subject scope of *STAR*, which is arranged into major subject divisions split into further specific subject categories and one general category division, includes all aspects of aeronautics and space research and development. The section devoted to engineering covers general engineering, communications, electronics and electrical engineering, fluid mechanics and heat transfer, instrumentation and photography, lasers and meters, mechanical engineering, quality assurance and reliability, and structural mechanics. Citations and abstracts are drawn from NASA's own resources and also from other publications, notably *GRA&I*. NASA files since 1962 are available online through the host ESA/IRS, but access is restricted to member states of the European Space Agency.

United States Department of Energy

The third major US reports announcement service which deals with many aspects of engineering is *Energy Research Abstracts* (*ERA*), a journal providing abstracting and indexing coverage of all scientific and technical reports, conference papers and proceedings, theses and monographs originated by the US Department of Energy, its laboratories, energy centres and contractors. *ERA* also covers other energy information prepared in report form by federal and state government organizations, foreign governments, and domestic and foreign universities and research organizations.

ERA, which may be regarded as the successor to the long-established *Nuclear Science Abstracts*, is comprehensive in its scope, encompassing the US Department of Energy's research, development, demonstrations and technical programmes, resulting from its broad charter for energy sources, supplies, safety, environmental impacts and regulation.

The complete subject category scheme, with scope notes, is available as *DOE/TIC-4584 Energy data base: subject categories and scope*. Abstracts are arranged according to subject categories, with five indexes: corporate author; personal author; subject index; contract number, and report number index. In addition, *ERA* provides details of reports collections and availability.

Technical reports in Great Britain

At one time, it was the responsibility of the Department of Industry's Technology Reports Centre (TRC) to handle technical reports from government research establishments, government-sponsored projects and other British and overseas government sources. The openly available paper copy reports accessions to TRC were announced in the journal *R&D Abstracts*.

However, as the collection developed, it became apparent that there was an increasing overlap between the reports activities of TRC and those of the British Library Lending Division (BLLD), now the British Library Document Supply Centre (BLDSC). It was decided that TRC should withdraw from reports-handling activities and that instead reliance should be placed on the British Library, with its far greater resources and economies of scale.

Consequently, TRC's stock of T-numbered reports for the period 1972-1981 was transferred to Boston Spa, where it is now available to registered users. The decision to end TRC's reports-handling activities meant the final issue of *R&D Abstracts* appeared on 15 December 1981. The file retains a reference value since *R&D Abstracts* was comprehensively indexed and the abstracts themselves were very full and informative.

Technical reports of interest to Britain's defence community are now handled by the Ministry of Defence's Defence Research Information Centre (DRIC) in Glasgow, the primary function of which is to disseminate scientific and technical information to the Ministry itself, the Defence Research Agency (DRA), and the UK firms and organizations working on UK MoD contracts. DRIC concentrates in particular on information contained in scientific and technical reports of a research nature and publishes a monthly announcement service, *DRIC Abstracts*, which appears in two editions: one classified UK CONFIDENTIAL and one classified UK RESTRICTED.

British Library Document Supply Centre

The main announcement journal for openly available British reports is now *British Reports, Translations and Theses* (*BRTT*), a bibliography published monthly (with indexes which cumulate annually) by BLDSC. The coverage of *BRTT* includes reports, theses and translations which can be difficult to identify and locate. *BRTT*, which is based on the collection at BLDSC, is intended to increase awareness of what has been issued and help to promote its wider use.

A keyterm index is issued with each monthly part, and the arrangement of the individual entries follows a version of the COSATI subject category list noted above, but now called the SIGLE Classification Scheme (details of SIGLE are given below). The entries in *BRTT* differ from other major reports announcement services in that they are confined to bibliographic citations without abstracts, but with carefully chosen explanatory terms.

International Atomic Energy Agency

At the international level, an important source of information on engineering reports is the announcement journal *INIS Atomindex* (INIS is the International Nuclear Information System of IAEA in Vienna). INIS is a cooperative, decentralized information system set up by IAEA and its member states, and its purpose is to construct a database identifying publications relating to nuclear science and its peaceful applications. Member states and cooperating international bodies scan the scientific and technical literature published within their boundaries or by them and select from it those items which fall within the subject scope of INIS, and process them according to agreed standards and rules.

The arrangement of *INIS Atomindex* is such that, within each subject category, the references are arranged by type of literature: non-conventional (technical reports, pre-conference papers, patents) followed by conventional publications (books, journal articles and so on). As with the US announcement services, the indexing is very thorough and includes personal authors, corporate entries, subjects, conferences and report numbers.

Forschungsberichte

In Germany, an important announcement journal is *Forschungsberichte*, published quarterly by VCH Verlag, Weinheim. The service began in 1976, and entries are grouped according to the SIGLE classification, and within it specific subject categories are arranged by titles rendered into English, with the original German titles provided after the author's name and affiliation. No abstracts are included, but many of the items cited are available from the Technische Informationsbibliothek, Hanover (TIB, noted below).

Availability and supply of reports

In the past, the availability and supply of reports have presented many problems to would-be users, notably because of difficulties with identification and acquisition, either on loan or by purchase. The reports-handling agencies noted above have, through their highly organized announcement service, made great progress in bringing a degree of order to the scene, but there are still areas which may be regarded as needing further effort. This requirement has been recognized by the International Federation of Library Associations, which has instituted a programme of Universal Availability of Publications (UAP), on the premise that the widest possible availability of published material (that is, recorded knowledge issued for public use) for intending users, wherever and whenever they need it, is an essential element in economic, scientific, technical, social, educational and personal development. UAP provides a useful, broad framework within which problems and solutions relating to areas such as reports literature can be set. Centralized systems for collecting and distributing reports are one demonstration of the concept of UAP in action. At the national level, NTIS represents a good example, while at the international level the System for Information on Grey Literature in Europe (SIGLE) is making some headway.

Grey literature has been defined as that literature which is not readily available through normal bookselling channels and is therefore difficult to identify and obtain. In addition to reports, which constitute the bulk of grey literature, the expression is commonly applied to technical specifications, trade literature, conference proceedings and preprints, official publications, theses and dissertations, translations and locally issued documents.

SIGLE is coordinated by the European Association for Grey Literature in Europe (EAGLE) and input is drawn from a number of national SIGLE centres around Europe, with the BLDSC representing the UK. File data extend back to 1980 and are made up of reports (40 per cent), theses (25 per cent) and other non-conventional literature (35 per cent). The SIGLE database can be accessed via the hosts STN International and BLAISE-LINE. It is possible to form the impression that the reports literature is much larger than it really is, particularly if a judgement is made purely on the numbers of items appearing in individual issues of announcement journals. In fact, there is a considerable overlap, and an examination of any issue of *GRA&I* will show that it contains many entries for items from the US Department of Energy and NASA which have already been cited respectively in *ERA* and *STAR*. A similar duplication occurs when *ERA* and *STAR* quote PB and AD reports originating with NTIS and so cited in *GRA&I*.

Special collections of reports

In the United Kingdom the main collection of British reports is housed at the BLDSC, Boston Spa, where the acquisition of such material has been a constant activity since the 1960s. In many cases, the special arrangements entered into with issuing bodies have over the years proved highly successful in obtaining reports from organizations such as the United Kingdom Atomic Energy Authority, the Science and Engineering Research Council, the Ministry of Defence and so on.

BLDSC also maintains a collection of overseas reports, especially those from the US agencies which are collectively responsible for the bulk of the many thousands of reports issued annually (not just in science and technology), namely NTIS, NASA, the US Department of Energy and the Educational Resource Information Center (ERIC). This enormous intake of overseas material is normally received in microform (as distinct from the paper copies available for sale through officially appointed agents); in 1993, BLDSC reported an annual intake of 110 000 report documents in microform format.

In Europe, a notable reports collection is that maintained by the Technische Informationsbibliothek (TIB) at the University of Hanover, where a service called TIBQUICK offers access to 3.8 million volumes and microforms, including some 900 000 dissertations and reports from the USA, 83 000 unpublished German research reports and many thousands of reports from the former Soviet Union.

Formats

Reports come in all shapes and sizes – indeed, the variety of format is one of the characteristics of the genre. Quality of production and, more importantly, reproduction varies considerably. Reports can range from conventionally printed documents indistinguishable in presentation and appearance from ordinary books and pamphlets, to typescripts about which the issuing agencies are sometimes obliged to give warnings such as 'portions of this document are not fully legible'.

For ease of storage, duplication and transmission, microforms, in particular the microfiche, have been widely adopted in the dissemination of reports literature. The standard adopted is a microfiche in the form of a 105 by 148 mm transparency with a capacity of 98 frames, 96 of which are normally available for textual matter at one page per frame. Since a very high proportion of reports is usually far less than 100 pages in length, the microfiche enables reports to be disseminated as single transparencies. Trailer fiche are available where reports do exceed this length. The great disadvantage of the microfiche from the user's point of view is that the document does not have the immediacy of a paper copy because a viewing machine of some sort is required. In certain cases, the poor quality of the original does not permit the adequate reproduction of paper copies from fiche, so that the reader has no option but to use a viewer.

Treatment of reports in libraries

It follows that, because of the originators' absolute control, at least initially, over the content and distribution of reports, accessibility to the body of knowledge contained within the reports literature can be difficult, arbitrary and, at times, even frustrating. It is one of the reasons that a consistent reports acquisitions policy becomes so hard to implement – if it is decided to collect all items in a given series, gaps are likely to show up which, on investigation, prove to be due to reports not being available without special permission and the completion of 'need to know forms'.

Many libraries and engineering organizations acquire two sorts of reports collections. Firstly, they act as the archival repository for their own reports. Usually, the documents in such a collection are termed 'internal reports' and may be subject to restrictions of availability to certain grades of staff or to special departments. Secondly, libraries will collect reports from other organizations, quite often on an exchange basis, or sometimes as the result of an automatic distribution agreement with a major issuing agency. Automatic distribution is fine in principle, but in practice can lead to the receipt of hundreds of unwanted documents which then cost time and effort to store and eventually discard.

Routines

Many of the larger libraries and engineering concerns usually have a reports section, with specially trained staff whose main responsibility is to deal with all aspects of the reports literature. This is particularly the case with engineering organizations which issue large numbers of reports themselves, in addition to acquiring many titles from outside sources.

The degree to which special handling routines will need to be introduced into a research centre, library or engineering establishment will depend greatly on the quantity of reports being handled which fall into restricted availability category, usually because of commercial or security considerations. Such routines may involve the introduction of limited access zones or even the provision of strong-rooms. One effect of restrictions of this nature is that booking-out and return procedures can become elaborate and tedious, and so discourage reports as a source of information.

In smaller libraries and establishments, reports can be handled on an *ad hoc* basis and treated in a manner similar to that for other documents of like format, such as trade literature, standards and patents. A split will necessarily occur between paper copies, which in themselves present shelving and storage problems due to features such as plastic spines or strip bindings, and microfiche, which need to be housed in special wallets or filing cabinets. The variety of formats, which effectively precludes any form of browsing, emphasizes the need for adequate catalogues and indexes, and the practices followed by the main announcement journals can often be adopted with profit.

Cataloguing reports

Reports can be treated for the purpose of cataloguing just like any other material added to the library, and so subject to the cataloguing rules which apply to the collection as a whole. In practice, however, reports tend to be treated somewhat differently.

It is true that some very pertinent help is available in the form of the *Guidelines for descriptive cataloguing of reports* (*Guidelines*, 1986), originally prepared in 1978 by the oddly-named Committee on Information Hang-ups (an indication perhaps that reports are regarded as difficult material?). The Guidelines, which were revised in 1986, provide rules to ensure uniformity of descriptive cataloguing among agencies, libraries and information centres which exchange bibliographical information in the reports literature.

Despite the existence of these and similar guidelines, many organizations prefer to devote the minimum of cataloguing effort to their reports collections and rely instead on published indexes and bibliographies. Where the quantities are sufficient to justify the practice, reports are best filed in numerical order, even though gaps will occur representing the reports of no interest to a particular library or organization.

Outlook

Despite its acknowledged shortcomings as a mainstream means of disseminating information, the report is a permanent feature of the literature of engineering, and its use is likely to grow as the alternatives to conventional publishing routes increase through the application of new technologies, especially desk-top publishing. For the engineer wishing to acquaint a circle of peer readers with the latest developments in his or her particular field of expertise, the report as a disseminating medium has no equal; by the same token, the engineer wishing to find out the latest happenings as announced by others working in the same area cannot afford to ignore the reports literature. Finally, it should be mentioned that a much fuller treatment of all aspects of reports as part of the grey literature is contained in a companion volume in the present series (Auger, 1994).

References

Aronson, E. J. (ed.) (1986) *Report series codes dictionary*. 3rd ed. Detroit: Gale Research.

Auger, C. P. (1994) *Information sources in grey literature*. 3rd ed. London: Bowker-Saur.

British Library (1992) *Alphanumeric reports publications index (ARPI)*. Boston Spa: BLDSC.

British Standards Institution (1986) *BS 4811: specification for the presentation of research and development reports*. London: BSI.

Guidelines (1986) for descriptive cataloging of reports: a revision of the COSATI standard for descriptive cataloging of government scientific and technical reports. (PB 86-112349). Springfield, VA: National Technical Information Service.

CHAPTER THREE

Standards

C. P. AUGER

Introduction

Standards are officially approved specifications applicable in various sectors of trade and industry, and they cover such topics as methods of testing, terminology, performance and construction requirements, and codes of practice. Usually standards are prepared by mutual agreement among the interested parties concerned, and are subsequently used to simplify and rationalize production and distribution; to ensure uniformity, reliability and safety; and to eliminate wasteful variety. Standards may also be considered as restraints which sometimes hinder the development of new and improved ideas, and so act as a brake to scientific and technical progress.

On balance however standards must be regarded as vital to the success of any advanced industrial society, and the various collections and series available at national and international level are ample evidence of the important contribution standards make to the manufacturing and commercial aspects of everyday life. In this chapter it is possible only to indicate the main trends of standardization, to name some of the most quoted series, and to show the major areas of use as they may apply to engineering. The approach has been to move from the general to the specific, taking in activities at the international, regional, national and local levels.

The average standard specification is not a particularly lengthy document, often in fact a pamphlet of just a few pages in length, typically containing details of methods, measurements, definitions, properties and processes. The standard document invariably has an identifying alpha-numeric code which in many cases can acquire a widely and instantly recognized significance. Many different bodies issue standards, including national and international standards organizations, trade associations, and government departments.

In the world of engineering it is readily evident from an examination of any established reference work or handbook that standards constitute a common technical language. For example Green (1992) covers a wide range of topics, such as *American Standard Drafting Room Practice*, ANSI Y14.I, and *British Standard. Hexagon socket screws and wrench keys: metric series*, BS 4168. Adler (1993) makes reference to SI units (ISO 31 and ISO 1000); engineering temperature measurement (*VDE/VDI Guideline* 3511); permanent magnet materials (DIN 17410); salt spray (fog) testing (ASTM B 117); and vehicle bodies for passenger cars (SAE J 826b), to name just a few.

International standards

Two of the most important international agencies for standardization are the International Organization for Standardization (ISO) and the International Electrotechnical Commission (IEC), both of which are considered below. There is in addition a number of other international agencies concerned with standards, and typical of them are the International Civil Aviation Organization (ICAO) and the International Maritime Organization (IMO). ICAO, which has its headquarters in Montreal, Canada, sets international technical standards and recommended practices for all areas of civil aviation, including airworthiness, air navigation, traffic control and pilot licensing; IMO, which is based in London, fosters inter-governmental cooperation in technical matters relating to international shipping, especially with regard to safety at sea. It is also charged with preventing and controlling marine pollution caused by shipping.

Whereas the prospect of an air disaster or the chance of a major marine pollution incident may be a great stimulus to participating countries to reach agreements on standards within a reasonable time scale, in other areas there appears to be less urgency in the schedule allowed from proposal to final promulgation. The need for faster technological integration in general has been made by many commentators, as for example Ralph (1992), whilst specific problems, such as the case for and the implications of a harmonized European plug and socket, have been aired by Dossett (1994).

An insight into the involved process of reaching agreement in committee has been provided by Schultz (1984), who gives a flowchart on the progress of an international/regional standard from conception to adoption. Harmony can be difficult to achieve in areas where technology is still advancing, as has proved the case in the field of electronic data interchange (EDI).

Finally, it should be noted that many standards are international in all but name – this is especially so of some of the standards prepared by the major American technical societies, such as the American Society of Mechanical Engineers (ASME) and the Society of Automotive Engineers (SAE).

International Organization for Standardization

ISO (the International Organization for Standardization) has its headquarters in Geneva and is a worldwide federation of national standards bodies, at present comprising 110 members, one in each country. The object of ISO is to promote the development of standardization and related activities in the world with a view to facilitating international exchange of goods and services, and to develop cooperation in the spheres of intellectual, scientific, technological and economic activity. The results of ISO technical work are published as International Standards.

The scope of ISO covers standardization in all fields except electrical and electronic engineering standards, which are the responsibility of the International Electrotechnical Commission (IEC – see below).

ISO work is carried out through 2832 technical bodies. More than 30 000 experts from all parts of the world participate each year in ISO technical work, which by the end of 1994 had resulted in 9652 ISO standards.

International standardization originally began in the electrotechnical field over 90 years ago. While some attempts were made in the 1930s to develop international standards in other technical fields, it was not until ISO was created that an international organization devoted to standardization as a whole came into existence. Following a meeting in London in 1946, delegates from 25 countries decided to create a new international organization the object of which would be to facilitate the international coordination and unification of industrial standards. The new organization began to function officially on 23 February 1947.

Because the name of the organization would have different abbreviations in different languages (IOS in English, OIN in French), it was decided to use a word derived from the Greek *isos*, meaning equal. Therefore the short form of the Organization's name is always ISO. The technical work of ISO is carried out through technical committees (TC). Each technical committee or sub-committee has a secretariat assigned to an ISO member body, of which there were 81 in 1995.

By the end of 1994 there were 185 technical committees, 636 sub-committees, 1975 working groups, and 36 *ad hoc* study groups.

An international standard is the result of an agreement between the member bodies of ISO, and may be used as such, or may be implemented through incorporation in national standards of different countries. A first important step is the committee draft (CD) – a document circulated for study within the technical committee or subcommittee. When agreement is finally reached within the committee, the draft is sent to the Central Secretariat for registration as a draft international standard (DIS); the DIS is then circulated to all member bodies. If 75 per cent of the votes cast are in favour of the DIS, it is accepted for publication as an International Standard. Most standards require periodic revision. Several factors combine to render a standard out of date: technological evolution, new methods and materials, new safety and quality requirements. To take account of these factors, ISO has established the general rule that all ISO standards should be reviewed at intervals of not more than five years. Sometimes it is necessary to revise a standard earlier. A list of all ISO standards is given in the *ISO catalogue*. Other ISO publications include *ISO membership*; *ISO memento* (which lists secretariats for each technical committee); *ISO liaisons*; *ISO technical programme*; and the *ISO Bulletin* (issued monthly).

Technical information on standards is coordinated through ISONET, an organization of 63 national standards information centres providing rapid access to more than half a million standards, technical regulations and other standards-type documents.

The ISO/IEC Information Centre at the ISO Central Secretariat has data on 15 800 international standards from ISO, IEC and 26 other international organizations. Details are contained in the *KWIC index of international standards*.

The range of subjects covered by ISO standards is listed in the *ISO catalogue* in a structure based on the International Classification for Standards (ICS) and extends from 01 – Generalities, terminology, standardization, documentation to 97 – Housekeeping, entertainment, sports.

International Electrotechnical Commission

The International Electrotechnical Commission (IEC), which is also based in Geneva, came into being in 1906 and is thus senior to the ISO by some 40 years. ISO and IEC have established a formal agreement which spells out clearly the relationship between the two organizations. In accordance with the terms of the agreement, ISO and IEC together constitute the specialized system for worldwide standardization.

Questions and issues on matters relating specifically to the development of standardization in the electrical and electronic engineering fields are regarded as the concern of IEC. Other subject areas are the responsibility of ISO. In matters of international standardization not related to any particular technology, ISO undertakes in consultation with IEC to safeguard any electrotechnical interests which may be involved. To ensure the necessary technical coordination, ISO and IEC have established a joint ISO/IEC Technical Programming Committee.

The members of the IEC are the national committees, one for each country, which are required to be as representative as possible of all the electrotechnical interests in the country concerned: users, governmental authorities, and teaching and professional bodies. National committees obtain a large measure of support from industry and are mostly recognized by their governments. Full information on the IEC and its national committees can be found in the *IEC directory*.

Other IEC publications include the *IEC catalogue*, available in separate English and French versions; the *IEC Bulletin*, with separate English, French and Spanish editions; and the *Technical Guide* series, in which 38 ISO and ISO/IEC titles have so far been published.

Regional standards

Many standards are published which cover the needs of regions or special areas, most notably in the European Communities, where standardization is regarded as a very important factor in the internal market. The view is taken that standards have for a long time been misused as technical barriers to trade within Europe. The Commission in Brussels does not take responsibility for establishing the desired standards – instead it leaves industry itself to look for voluntary solutions, and the necessary agreements are reached through the European Committee for Standardization (CEN), which also covers the European Free Trade Association (EFTA). CENELEC (European Committee for Electrotechnical Standardization) is the relevant body for electrotechnical standards, and CEE (International Commission for Conformity Certification of Electrical Equipment) is the organization for certification.

The European standards series EN covers the decision of CEN and CENELEC, and, once adopted by each member state, the standards are printed as national standards. The EN standards are usually listed in the national standards catalogues, as for example BS EN ISO 9000: 1994 *Quality management and quality assurance standards*, a standard which provides a framework for management systems, procedures and work instructions. The standard was previously identified as BS 5750: 1987, but the name was changed to reflect better its international status when a revision took place in 1994.

By following European standardization and conformity assessment procedures, manufacturers are able to affix the CE Mark to their industrial products, thus indicating that they conform to Community requirements. The CE Mark has come to be regarded as a technical passport and is designed to cover specific categories of products – the *EU*

Directive on machinery safety, for instance, came into force in January 1995 and thereafter any machinery built is required to have a CE Mark on it. The importance of the *Directive* has been described by Laidler (1995), and Hall (1995) highlights the many different standards which now need to be consulted in building a machine.

In 1994 the Community published *Common standards for enterprises* (CO-83–94–410-EN-C) which gives a complete listing of the technical harmonization laws falling under the new approach. For a recent appraisal of the harmonization law, see the study by Farr (1992). A further guide is *Standards for access to the EC markets*, which explains EU standardization policy in detail and is available from CEN/CENELEC, Brussels. European standards are often coordinated by the relevant specialist sectors of industry, as for example the *Data book for tyres and rims*, published by the European Tyre and Rim Technical Organization (ETRTO), Brussels.

The European Coal and Steel Community (ECSC) is responsible for the *EURONORM* series, which has been published since 1955, and whose titles are available (in Dutch, English, French, German and Italian) from the Office of Official Publications of the European Communities in Luxembourg. In 1989 the American National Standards Institute (ANSI) set up a Brussels office to strengthen ties within the standards community in Europe because many in the United States were concerned about the possible approach of an economic 'Fortress Europe'. These fears appear to have disappeared and the office was closed in 1995.

Other areas of regional standardization include the Nordic countries – Denmark, Finland, Norway and Sweden – where the responsible body is Internordisk Standerisering (INSTA). In the Americas the Comision Panamericana de Normas Tecnicas (COPANT) operates from Buenos Aires, with standards issued in Spanish.

National standards

Many countries have established national standards systems and space limitations preclude mention of all but a representative few of them. However, it is worth recalling that national standards systems were developed in the nineteenth century at least in part as a result of two quite different influences. Firstly, there was the increasing need for compatibility within ranges of basic engineering components, and nowhere is this more clearly illustrated than in the search for a system of uniform screw threads – see for example the papers by Whitworth (1841), Briggs (1865) and Lowenherz (1889) covering respectively developments in England, the United States and Germany.

Secondly, there were the expressions of public outrage and alarm at engineering and engineers in general as a series of disasters caused great loss of life and enormous material damage – see for example the notice published in the *Manchester Guardian* of 16 August 1854 on concern over the frequent explosions of steam boilers. The notice gave impetus to the setting up of the Association for the Prevention of Steam Boiler Explosions, later to become the Manchester Steam Users' Association, one of many bodies dedicated to setting and raising performance and safety standards.

United Kingdom – British Standards Institution

In the United Kingdom, national standards are the responsibility of the British Standards Institution (BSI), the first national standards body in the world. BSI was established in 1901 as the Engineering Standards Committee and took the name British Standards Institution in 1929 on receiving a Royal Charter. Today the major function of BSI is to help British industry compete effectively in world markets. The work of BSI in standards, testing, quality assurance and expert guidance is geared to enable British companies to meet the quality needs of buyers at home and abroad.

Bibliographical details of all British Standards are set out in the *BSI standards catalogue* (formerly the *British standards yearbook*) which lists over 13 000 BSI publications. Each year more than 1200 new or revised standards are issued. Standards are drawn up by all those who have a particular interest in the subject – manufacturers, users, research organizations, government departments and consumers. All standards are made available for public comment before they are published in their final form. The *Catalogue* is updated by twelve monthly supplements.

Most British Standards are published in the *General Series*, identified by the prefix BS, initials which have spread far beyond the bounds of engineering literature and are now referred to in all types of publications as indicators of quality and safety. For example, *Good Housekeeping* (June 1995), in an article on outdoor lighting for the garden, urges readers to check that the products they buy are marked as conforming to BS 4533: 1986–1990; BS EN 60598: 1992–; BS 3535: 1990; or EN 60742: 1990. BSI publications can be purchased from sales outlets around the country and they can also be consulted at a limited number of public, university and other libraries listed in the *Catalogue*.

BSI also issues a number of specialized series of standards, including *Codes of Practice*, *Automobile Series*, *Marine Series*, *Aerospace Series*, and *BSI Handbooks*. Although BSI is a national standards body, it is international in outlook and offers for sale over 100 000 different foreign and international standards, including:

ISO/IEC complete range

DIN complete set of official English translations
ASME complete set of Boiler and Pressure Vessel Codes
Underwriters' Laboratory complete set.

BSI also holds extensive collections of US Military Specifications and Federal regulations.

The *BSI standards catalogue* includes an index to corresponding international/British Standards which lists ISO, IEC, CISPR (International Special Committee on Radio Interference) and CIE (International Commission on Illumination) documents. The 1994 *Catalogue* also has a section called *BSI best sellers* which gives details of titles from major technical publishers to complement BSI's own range of publications.

The main categories under which the titles are grouped are:

Environment/health and safety
Building and construction
Engineering
Information technology.

A similar compilation is provided by AFNOR (see below) in its *Catalogue des ouvrages*, enabling the reader to place the subject of standardization in a broader perspective.

A *Microfiche index to British Standards* is published by Technical Indexes Limited in collaboration with BSI. The service covers microfilm and microfiche and provides a guide to locating full texts on film, plus a subject index. In the United Kingdom the microform products are only available to subscribing members of BSI.

Another alternative to the traditional paper format for standards is the compact disc, and PERINORM provides standards information on CD-ROM as the result of a joint project of the national standards bodies of Germany, France and Great Britain. Since its introduction in 1989, PERINORM has been continually developed and improved, and since 1991 the three partners involved in the production of the disc have offered all CEN/CENELEC members the opportunity of having their national standards databases included on PERINORM. Austria, Switzerland and The Netherlands have already done so, and other standards bodies are set to follow. A further development is NORMIMAGE, a series of CD-ROM databases utilizing the same retrieval software as PERINORM to search and display in facsimile format the complete texts of standards relating to specific technical fields. The discs are designed as databases of the relevant national standards, supplemented by European and international standards. The first two discs in the series are No 1 *Quality*, and No 2 *Fasteners*.

United Kingdom – Other Standards Activities

The government department responsible for general policy on standards, quality and design is the Department of Trade and Industry

(DTI), which oversees a number of executive agencies, including the National Physical Laboratory and the National Weights and Measures Laboratory (discussed below).

In matters of defence there is a long tradition of standards preparation to cater for the specialized and often exacting demands of the armed services. The main series are *Defence standards* (DEF/STAN), *Defence specifications* (DEF/SPEC), and the aerospace specifications originally produced by the Directorate of Technical Development (DTD) of the Ministry of Aviation and its successor departments. The current trend is to consolidate all defence specifications into one series DEF/STAN, and to use national and international standards wherever appropriate. Information on military and defence standards of whatever origin is available from the Defence Research Information Centre (DRIC), part of the Ministry of Defence Procurement Executive. The role of DRIC is to disseminate scientific and technical information to the UK defence community, which is defined as Ministry of Defence branches and establishments, armed service units, military colleges and British firms and organizations working on Government defence contracts.

Such contracts frequently cite military standards and require them to be followed in executing the work the Ministry requests to be done. DRIC issues a helpful and informative booklet *A guide to services*, available from its Glasgow office.

National standards – France

In France the official body responsible for issuing national standards is the Association Française de Normalisation (AFNOR), created in 1926 and based in Paris. AFNOR issues *Normes Françaises* (NF) divided into four broad categories:

1 Fundamental standards (concerned with terminology, metrology and similar topics);
2 Standards for methods of testing;
3 Standard specifications (which establish the characteristics of a product);
4 Standards for organization and service (which describe the functions of an enterprise and the activities of the service).

AFNOR is the French branch of the Comité Européen de Normalisation (CEN) and administers 19 000 national and international standards currently in force.

Details of AFNOR specifications and related ISO and CEN documents are contained in the *AFNOR catalogue des normes françaises*, with information on translated standards provided in the *Catalogue of English translations*. Collections of standards arranged by broad subject and varying in size from a single volume of several hundred pages (e.g. *Conteneurs pour le transport des marchandises*, 1992) to multi-volume sets (e.g. *Caoutchouc*, 1992) can also be purchased. Full details are given in the *Catalogue des recueils*.

National standards – Germany

The official standards organization in Germany is the Deutsches Institut für Normung eV (DIN), a widely influential body based in Berlin. Publications issued by DIN are all published by Beuth Verlag, a publishing house whose profits are employed to finance the work undertaken by DIN for the public good.

DIN is not a government agency but a registered association. DIN standards are technical rules framed to promote rationalization, quality management, safety, environmental protection and communication in industry, technology, science, government and the public domain. In DIN standards work is carried out by 44 000 external experts organized into 4300 technical committees. Draft standards are made available for public comment, and published standards are reviewed for continuing relevance every five years at least.

In preparing its standards, DIN ensures that they can be cited in legislation and legal transactions as a description of the current state of technology.

The *DIN Katalog für technische Regeln* contains the bibliographic data for all DIN standards and draft standards, as well as data for more than 200 other collections of technical rules. In all, the *Katalog* provides references to more than 100 000 German, foreign and international technical rules; it is published annually and is regularly updated by supplements issued during the course of the year.

DIN standards which have been translated in full into English (of which there are more than 4000) are indicated by the letters *En*. The Deutsche Informationszentrum für Technische Regeln (DITR) is the German information centre for standards, rules and regulations; it was founded in 1979 and is housed in the same building as DIN. DITR has a major standards library with a total stock of over 550 000 items, and offers a comprehensive enquiry service, particularly as an aid to promoting economic connections across national frontiers.

DITR also maintains a bibliographical database, DITR DATENBANK, which stores not the texts of the standards themselves, but the most important reference data relating to standards and technical rules.

The first ten years of the activities of DITR, and incidentally the increasing complexity of the European standards scene, have been described by Marschall (1989).

Germany has many other standards issuing bodies in addition to DIN; for example, the Verband Deutscher Elektrotechniker (VDE, the Association of German Electrical Engineers) issues standards on vehicle ignition interference suppression (VDE 0879), and the Verband der Automobilindustrie (VDA, the German Automotive Industry Association) has developed a VDA unit module for luggage compartment capacity which is more commonly used than DIN-ISO 3832.

National standards – United States

In the United States there are very many bodies which are active in the preparation of standards, and although there is a central organization, the New York based American National Standards Institute (ANSI), it differs from the national standards organizations of other countries in that its main function is to coordinate and approve the standards developed by qualified technical and professional societies, trade associations and other groups, which voluntarily submit their specifications to ANSI for approval. ANSI was at one time known as the American Standards Association (ASA) and the initials still persist in the designation of film speeds. The total number of standards producing bodies runs to many hundreds and the National Institute of Standards and Technology (NIST) publishes from time to time *Standards activities organisations in the United States* (*Special Publication* 681), which is a directory identifying organizations which develop, publish and revise standards in the public and private sectors. Another guide to bodies concerned with standards is the *Standards and Specifications Information Bulletin*, published since 1964 by the National Standards Association and covering US government and industry standards and specifications. The compilation provides details of issuing organization, Federal Supply Classification Code, adoption, ANSI designation, and suppliers of products conforming to standards included. The *Bulletin* has grown to a database of over 125 000 records.

ANSI itself has published its *Catalogue of American national standards* on an annual basis since 1977. The work is updated by the journal *ANSI Reporter* which gives news of national and international standardization activities, and by the periodical *ANSI Standards Action*, which lists final approval actions, new standards and proposals for future work. An informational brochure from ANSI became available in June 1995. Many American technical societies are major publishers of voluntary standards, whilst other societies comment and contribute on a smaller scale. In particular the American Society for Testing and Materials (ASTM) has developed an entire library of standards of its own in the *Annual book of ASTM standards* which runs to many volumes covering broad technical topics such as iron and steel products, rubber, electronics, test methods and magnetic properties. The work of ASTM is updated by the monthly *ASTM Standardization News*.

The American Society of Mechanical Engineers (ASME) produces a widely recognized set of standards in the form of the ASME *Boiler and pressure vessel code*, which embraces standards, codes of practice, case reports, inspection and test methods affecting the safety of boilers, pressure vessels and nuclear plant components. The Society of Automotive Engineers (SAE) issues the multi-volume *SAE handbook* on a yearly basis. The handbook contains many hundreds of standards, recommended practices and information reports covering all aspects of the automobile industry. One of the most universally recognized groups of SAE standards is that covering engine oil viscosity, references to which appear in practically every car owner's instruction manual.

In the field of electrical engineering the National Electrical Manufacturers Association (NEMA) has standards in two classes:

1 NEMA Standard, which relates to a product commercially standardized and subject to repetitive manufacture; the standard has to receive approval by at least 90 per cent of the members of the appropriate NEMA sub-committee;
2 NEMA Suggested Standard for Future Design, which may not have been regularly applied to a commercial product but which suggests a sound engineering approach to future developments; the future standard has to receive approval by at least two-thirds of the appropriate NEMA sub-committee.

In aerospace the American Institute of Aeronautics and Astronautics (AIAA) both issues standards of its own (e.g. *AIAA standards for aerodynamic decelerator and parachute drawings*, approved by ANSI and adopted by the Department of Defense, 1991), and takes part in the general discussion on standardization, as for example in its report *The future of aerospace standards* (1990) which emphasizes the importance of international consensus standards.

Standards in engineering materials and manufacturing processes have proliferated to the extent that no engineering library can be considered complete without one or more guides to equivalent standards. For example, the American Society for Metals (ASM) publishes the *Worldwide guide to equivalent irons and steels* (3rd edition 1992), with a companion compilation for non-ferrous metals and alloys. In the field of manufacturing some 5000 standards from 16 countries are recorded in *International Standards Index – welding and related processes* (1991, Woodhead Publishing).

The Department of Defense (DoD) plays a major role in the preparation and dissemination of military standards covering the needs of the United States armed forces. DoD standards and specifications which are not subject to security restrictions on a need-to-know basis are published through the United States Government Printing Office and are listed in the *US Department of Defense index of specifications and standards*, which has appeared regularly since 1952. Information on DoD standards and matters concerning military standardization is summarized in abstracts appearing in the twice-monthly *Government Report Announcements and Index (GRA&I)*.

National standards – other countries

Many other industrialized countries have set up national standards bodies and regularly publish catalogues and announcement journals. The following are just a representative sample: Canadian Standards Association, Toronto – *Catalogue* and *Info Update*; Dansk Standardiseringsraad (Danish Standards Association), Hellerup – *Dansk Standard*; Japanese Standards Association, Tokyo – *JIS Yearbook*; Standardiseringkommission i Sverige (SIS – Swedish Standards Institution), Stockholm – *Maanadens Standard*; Nederlands Normalisatie-Instituut (Netherlands Standards Institute), Delft – *Normalisatie Nieuws*; South Africa Bureau of Standards, Pretoria – *Bulletin*; Bureau of Indian Standards, New Delhi – *Standards India*; and Asociación Española y Certificación (Spanish Standards and Certification Association), Madrid – *UNE*.

Standards laboratories

A considerable proportion of the data used in the preparation and drafting of engineering standards is based on work carried out in the many laboratories scattered around the world which are responsible for metrology and precise measurements techniques. At the international level the Bureau International des Poids et Mesures (BIPM), which is based in Sèvres in France and came into being after the signing in Paris in 1875 of the Convention du Mètre, conducts scientific work into:

Mass and related quantities
Frequency and time scales
Length, including stabilized lasers
Electricity
Photometry and radiometry
Ionizing radiations
Temperature.

The BIPM operates under the exclusive supervision of the Comité International des Poids et Mesures (CIPM), which itself comes under the authority of the Conférence Générale des Poids et Mesures (CGPM). The CIPM and the BIPM work in close cooperation with other governmental and international organizations; in particular there are close links with the International Organization for Legal Metrology (OIML), founded in 1955.

Publications of the BIPM include the official reports of the General

Conferences (*Comptes Rendus des Séances de la Conférence Générale des Poids et Mesures*), of the annual meetings of the CIPM (*Procès-Verbaux des Séances du Comité International des Poids et Mesures*), and of the meetings of the Consultative Committees (*Sessions des Comités Consultatifs*). One of the most widely used publications of the BIPM is *Le Système International d'Unités* (5th edition 1985) which describes the modern system of units (SI units) used in the measurement of all physical quantities, the principal units being the metre, kilogram, second, ampere, kelvin, candela and mole. The journal *Metrologia* is edited from the offices of BIPM.

In Great Britain the National Physical Laboratory (NPL) is one of the world's leading centres for the precise measurement techniques and standards which are essential in engineering sectors making products with high added value. NPL is an Executive Agency operating under the aegis of the Department of Trade and Industry (DTI). The DTI also has responsibility for the National Weights and Measures Laboratory, the focus for legal metrology in the United Kingdom.

In Germany matters concerning units are covered in DIN 1301 – *Units*, and in DIN 1304 – *General Formula Symbols*. The whole field of legal units has been reviewed by Haeder and Gärtner (1980).

In the United States the National Institute of Standards and Technology (NIST), formerly the National Bureau of Standards (NBS) conducts research and provides a national basis for American physical measurement systems. The NIST is a prolific publisher, and titles include the *NIST handbook*; *NIST monographs*; *NIST special publications*; *NIST technical notes*; and the *NIST Journal of Research*.

Engineering company standards

Many large engineering companies prepare and use their own standards specifications, drawing in part on national and international standards, but always tailoring the final versions to local needs. Thus many firms find it worthwhile to draw up specifications on topics such as health and safety, especially the treatment of hazardous materials; on industrial materials used in the manufacture of products; and on manufacturing processes such as heat treatment or surface finishing. Such activities are normally in the hands of standards engineers who combine specialist subject knowledge with skills in drafting to prepare standards which are clearly written and workable in practice. Often, as in the national and international scene, approval is sought and obtained for drafts circulated to technical committees within the organization.

Such activities can prove time-consuming and expensive and over recent years the trend has been for companies to reduce to an essential minimum the production of their own specifications and to rely more and more on published documents. Nevertheless a considerable body of company standards material has grown up, to the extent that some specifications are quoted as though they were openly published, and in consequence become sought by outsiders.

In Great Britain the interests of the standards engineer and indeed those of any individual concerned with the application of standards are catered for by the British Standards Society (BSS), which is part of the British Standards Institution. The Society promotes the techniques and benefits of standardization, particularly as a basis for quality. Membership is achieved by the payment of an entrance fee and an annual subscription.

References

Adler, U. (ed.) (1993) *Automotive handbook* 3rd ed. Stuttgart: Robert Bosch.
Briggs, R. A. (1865) A uniform system of screw threads. *Journal of the Franklin Institute*, **32**, 111–125.
Dossett, D. (1994) The harmonized European plug and socket. *Executive Engineer*, **2**(1), 8–9; 12.
Farr, S. (1992) *Harmonization of technical standards in the EC*. London: Chancery Law (European Practice Library Series).
Green, R. E. (ed.) (1992) *Machinery's handbook* 24th ed. New York: Industrial Press.
Haeder, W. and Gärtner, E. (1980) *Die gesetzlichen einheiten in der technik*, 5th ed. Berlin: Beuth Verlag.
Hall, D. (1995) Standards. *Engineering*, **236**(5), 48–49.
Laidler, P. (1995) Marked machines. *Engineering*, **236**(5), 46; 48–49.
Lowenherz, L. (1889) Recommendations on the introduction of uniform screw threads. *Zeitschrift für Instrumentenkunde*, (9), 396–418.
Marschall, H. W. (1989) Ten years of DITR – compte rendu for Europe. *DIN – Mitteilungen*, **68**(11), 531–538.
Ralph, W.J. (1992) European standards and 1992. *Professional Engineering*, **5**(1), 33–35.
Schultz, K. P. (1984) Aufbau und arbeitsweise übernationalen normenorganisationen. *DIN – Mitteilungen*, **63**(7), 365–374.
Whitworth, J. (1841) On a uniform system of screw threads. *Institution of Civil Engineers, Minutes of Proceedings*, (1), 157–160.

CHAPTER FOUR

Patents and patent information

C. OPPENHEIM

What is a patent?

A patent is a bargain or contract between a Government or other patent issuing authority, and an inventor. The Government or patent issuing authority gives the inventor a monopoly right preventing others from making, using or selling the invention for a certain time; in return, the inventor must disclose all that he or she knows about the invention in a document, known as a patent specification, which is published, often called laid Open to Public Inspection (OPI) and is available for anyone to read. The monopoly right is a negative right – it is a right to prevent others from doing things without permission; just because someone owns a patent does not automatically give him or her the right to make, use or sell the invention him or herself. This is because a patent may be an improvement on an earlier patent, and the inventor of the improvement will not be able to make, use or sell the improvement without permission from the owner of the earlier patent. This is not unusual in the world of patents.

The word 'patent' derives from the Latin word for 'open'. In the Middle Ages, the monarch would issue important declarations, such as, for example, the Magna Carta, the appointment of an individual to be a judge and so on, using so-called 'Letters Patent' or open letters. These were not sealed, so anyone could read them, but had the King's seal attached so that everyone could see they were authentic and had the monarch's authority. Letters Patent are still issued to this day; for example, when someone is knighted. Among the Letters Patent that were issued in England (the early history of patents is tied to English history) in the Middle Ages were many giving individuals monopoly rights; for example, a monopoly right to be the only person to make salt in the country, or the only person allowed to import senna pods. Over a time, the shorthand word 'patent' came to refer to monopoly rights for inventions only.

The first English patent for such monopolies was granted in 1331 to allow certain foreigners to practise their crafts in England. The patent was not for an invention as such. Over time, an increasing number of patents was issued for monopolies, but few were for inventions. Instead, they were used as a method of raising revenue. The monarch would grant the patent to some wealthy individual in return for badly needed cash. The wealthy individual hoped to more than recoup his initial investment in profits deriving from the monopoly. This annoyed the populace at large, who were frustrated by being unable to undertake normal commercial enterprises due to monopolies granted by the monarch. In 1623, the English Parliament passed the important Statute of Monopolies. This Act made all monopolies illegal, with one exception: namely patents for invention. This was the first Act in the world to protect patents for invention. The wording of the crucial Clause is still clear and unambiguous despite being more than 350 years old: 'Monopolies will be granted for the term of 14 years or under hereafter to be made of the sole working or making of any manner of new manufacture within this realm to the true and first inventor of such manufactures, which others at the time shall not use, so long as also they be not contrary to the law nor mischievous to the State by raising prices of commodities at home or generally inconvenient'.

This Statute of Monopolies has, of course, been superseded many times. The current Patents Act in the UK, for example, is the Patents Act 1977. The basic principles remain, however. These are:

- The patent is issued for a fixed term
- It is issued for a genuinely new invention
- It gives a monopoly right preventing others from using the invention
- It is granted to the true inventor
- It is for 'a manner of manufacture' – some industrial application or other
- The invention must be lawful
- The monopoly right should not be abused to hurt trade

All these principles remain in modern patent law throughout the world, not just in the UK.

Patentability

To be patented, the invention must fulfil certain requirements. In particular, it must:

- be new;
- involve an inventive step;
- have industrial application; and
- not be one of a select group of inventions or concepts which are excluded from patentability.

Incidentally, do note that, perhaps surprisingly, there is NO requirement for the invention to *work*. There is no requirement, for example, to submit a working model of a machine, or to demonstrate the product in any way to the Patent Office. The Patent Office *will* require that in the patent document that is submitted – the *patent specification* – evidence is provided of the efficacy of the invention, but that is quite a different matter from the Patent Office testing the invention for itself.

Novelty

The first requirement, then, is that the invention is new. How does one decide if an invention is new – or *novel*, to use patent jargon? In most countries' patent law (but not in the USA), the key date is the so-called *priority date*. I will not define that now, but let us take that as a clearly defined date in patent law. An invention is novel if, at its priority date, the invention was not known in any way to the public at large. In US law, incidentally, a much less precise date is used: the date when the invention was first thought of. This is a much harder date to prove than a priority date, that is, as will become clear, a clearly legally definable date.

There are two ways an invention could have become known to the public before the priority date. The first is called *prior publication*, the second is called *prior use*, sometimes known as *prior user*.

Prior publication is defined as follows: if, at any time before the priority date, a publication was made available to the public describing the invention in sufficient detail for someone 'skilled in the art' – a notional expert – to understand it, then the invention cannot be patented.

It is for this reason that major scientific and technical libraries in the world record not just the receipt of journals and books, but also date-stamp every few pages those books and journals with the date the item was made available to the library users.

Consider an imaginary example. Let us assume there IS a prior publication, as a description of the invention in question appeared in the *Outer Mongolian Journal of Engineering and Biochemistry*, vol. 25, pp. 125 –140, but the journal has sales of only one copy in the world, the library in which the journal was placed has not had a visitor enter its doors for the past 50 years, the article was written in an obscure form of Outer Mongolian and this language is now extinct.

Is the invention patentable? The law is quite clear. This is prior publication and the invention cannot be patented. It is irrelevant whether any member of the public read the item; what is crucial is that someone in theory could have done so. Therefore, it makes no difference WHERE in the world the prior publication appeared. It makes no difference how obscure the item was. All that matters is the publication was in a place where the public are entitled to visit, and that someone sometime might have read the item before the priority date.

Consider another hypothetical example. Imagine I have invented a machine to travel through time. I apply for a patent. Someone objects because there is a prior publication. This is *The time machine* by H. G. Wells, first published in 1895. Is it a prior publication?

Again, the law is clear. No, this is NOT a prior publication because the text of H. G. Wells' story fails to describe in sufficient detail how the machine works. The prior publication must give enough information for someone skilled in the art to make the invention.

This is not to say that fiction cannot ever be cited as a prior publication. In a famous case about 20 years ago, an invention for a new type of cat flap was rejected on the grounds that a similar type of cat flap had been drawn – not even written, but drawn – in a British children's cartoon magazine *The Beano*. Of course, it is very rare for fiction to be cited as prior publication. Prior publication can, incidentally, be oral as well as written – as in a speech; that is why it is VERY dangerous to make public speeches about an invention before applying for a patent.

The other way to stop an invention being novel is prior use. This is defined as before the priority date; the invention must not have been USED in such a way that the public could see it in action, or see the products of the invention.

Prior public use is a rather more difficult idea than prior publication. For example, imagine a computer being shown for the first time. It incorporates new circuitry that improves its speed. The public view the computer, and can see it is working faster, but do not know what the trick is. That is probably NOT prior public user. On the other hand, imagine a car with a new type of engine that is shown at a motor show. The public can look at the car, lift the bonnet, but cannot take the engine to bits. That IS probably prior user, as an intelligent mechanic may be able to work out the new engine's principles from external visual inspection.

Consider a final example. You are making rivets with a new machine. You sell the rivets. They look no different from normal rivets, but your new machine produces them much more cheaply than other machines. Prior user? Probably, as it could be argued that minute inspection of the rivets would show microscopic differences from traditional rivets, and an expert may be able to deduce from those differences what the key element of your invention was. This approach to prior user is just, otherwise an unscrupulous manufacturer would sell his rivets for as long as possible, and only when it was clear the secret was about to go public would then to apply for a patent to get monopoly rights. Each case is always treated on its merits, but as a rule one should not show the machine, or products from the machine, in public before applying for the patent.

This also explains why commercial organizations are paranoid about letting strangers into their labs. If the company routinely allowed strangers to wander around, then an invention developed in the lab could be invalidated by prior user because a stranger – a member of the public – walking through could have seen the invention and grasped its significance.

The invention must involve an inventive step

This is often called 'obviousness' – the invention must not be obvious. Aspirin cures headaches. Penicillin kills bacteria. What if I invent a pill containing both aspirin and penicillin? Let us imagine there is no prior publication or prior user, and the combined pill both kills bacteria and cures headaches. Should I get a patent? No, because mixing the two chemicals is obvious. There is nothing clever, or inventive, in what I have done. What if the mixture of penicillin and aspirin made left-handed people right-handed? Now that would NOT be obvious, and I could indeed then get a patent for the invention – but only for its unexpected use. I could not stop people using the mixture to cure colds.

Most patent Acts around the world require that on the date of the priority date, and without any benefit of hindsight, the invention shall not be obvious to someone skilled in the art – our notional expert again. To draw an analogy – you have been struggling with a really difficult crossword clue for half an hour. You give up. A friend tells you the answer. 'Of course,' you say, 'how obvious!'. You now have the hindsight and the thing is totally clear. It is very difficult to get an expert to avoid hindsight when he or she is questioned about obviousness. Furthermore, what is obvious to one person is not obvious to another. For these reasons, obviousness is not as precise an idea as novelty. The sorts of questions that have to be asked are:

- Would our expert have assessed the chances of success of the yet unmade invention as high enough to justify trying it out?
- If the invention is so obvious, why hasn't anyone made it before?
- Does the invention satisfy a long felt need? If yes, then why hasn't someone tried the invention before?
- Is the invention the application of a well-known product for a well-known purpose?
- Is the invention merely the addition of features that already exist – such as my aspirin and penicillin example?

The invention must be capable of industrial application

This criterion rarely causes problems. It really is a modern version of the Statute of Monopolies' 'manner of new manufacture'. Virtually all inventions are capable of passing this requirement.

Excluded inventions

There are certain inventions that in most countries may not be patented. These include the following:

- Discoveries, such as a new chemical compound in the bark of a tree in the Amazon jungle, cannot be patented.
- Scientific theories cannot be patented.
- Mathematical techniques cannot be patented.
- Literary, musical and like works cannot be patented as they enjoy protection under copyright instead. The position of computer programs is discussed below.
- Schemes or rules for performing mental acts, such as mental arithmetic methods, cannot be patented.
- Methods of playing games, such as a new chess opening or a new bidding system in bridge, cannot be patented. However, board games as such can be patented, complete with their formal rules. Indeed, the game of 'Monopoly'™ was itself the subject of a patent!
- Computer programs cannot be patented. There is some pressure to change this rule, and I foresee changes in the next few years. In practice, many computer programs are protected by the simple device of trade secrets.
- Any method of treatment of the human or animal body by surgery cannot be patented, presumably because an important surgical technique should not be the subject of sordid commercialism; however, pharmaceuticals CAN be patented, so there is no real logic in this. It is debatable (in ethical terms) whether pharmaceuticals should be the subject of patents. In favour is that it encourages further research. Against is that companies are making profits out of other peoples' illnesses. In practice, all developed countries permit the protection of pharmaceutical compounds and mixtures containing them.
- Inventions likely to induce offensive, immoral or anti-social behaviour cannot be patented. Among devices that would fall under this heading are torture devices, equipment to make electricity meters under-record, and mantraps.

A safe-breaking device WOULD be patentable, because not every use of them would be illegal – you may need to break into your own safe and you have lost the keys, for example. Years ago, condoms could not be patented under this rule, but this is no longer the case. The European Patent Office recently refused a patent for a genetically engineered mouse that develops cancer easily. This was refused a patent on morality grounds because it caused the mouse unnecessary suffering.

Finally, it is worth noting that the whole question of whether genetically engineered living creatures can be patented is currently lively. It is quite probable that they will be explicitly permitted soon. Already, patents for genetically engineered micro-organisms have been granted, and it is difficult to see where the line should be drawn, so, it is argued, no line should be drawn anywhere. Again, this raises significant ethical issues, but these are beyond the scope of this chapter.

Obtaining a patent

The description that follows represents UK law. However, most countries' laws do not differ in significant regards. Any individual, groups of individuals or company can apply for a patent at the Patent Office. The applicant can be the inventor him/herself, or his/her employer if the person made the invention as part of his or her duties as an employee. If, on the other hand, the invention was created by an employee but it was NOT part of his/her duties to invent, then the invention belongs to the employee – although the employer is probably entitled to sack the employee for using company resources.

If the idea of an invention is stolen by a third party (the patent term is *obtained*), then the true inventor can object when the patent is granted, and can either have the patent cancelled or have it transferred to him/herself.

One must apply for a patent on standard forms, answering various basic questions and providing a full description of the invention in the patent specification.

You will recall I mentioned 'priority date' when referring to novelty and to obviousness. The timing of an application to the Patent Office is the key factor both with defining what the 'priority date' is and with the likelihood of commercial success of the product. The earlier one puts in an application, the more likely it is no-one else gets in an application him/herself on the same subject, and the less likely it is that any prior publication or prior user takes place. On the other hand, the earlier one puts in the application, the less developed is the invention and the more likely it is that you have not thought through all its ramifications and implications.

For this reason, it is traditional in the UK to put in an early application, often called a 'provisional' to the Patent Office. All that is required is a brief indication of the scope of the invention and a modest fee. This establishes the 'priority date', i.e. the date that the application was made. The applicant then has twelve months in which to formalize his/her application by submitting the full detailed specification, an abstract, the claims and a much larger fee. By doing this, the applicant has established the earlier date as the priority date, and has had twelve months to work up his/her invention. If, during those twelve months, the applicant decides for one reason or another that the invention is not viable (a not uncommon occurrence), he/she does nothing at the end of the twelve-month period; he/she loses the initial fee and there is no priority date. Should someone then later on apply for a patent for exactly the same invention, the Patent Office will ignore the earlier application that had been applied for.

The applicant does not have to put in the provisional. He/she could go straight in for the full application with claims and abstract, and many applicants, particularly large corporations who understand the patent system fully, do just that. Then the priority date is simply the date of that application.

There is a third way of getting a priority date; this involves the use of the International Convention, also known as the Paris Convention.

The International Convention

The International Convention, or Paris Convention, was first agreed in 1883, and all major countries of the world are parties to it. The basis of the Convention is simple; each country will give the same rights and obligations to foreigners as to their own nationals. The important feature as regards priority dates is as follows: if a person applies for a patent in one country and thereby establishes a priority date, as long as the person then applies for patents within twelve months of that priority date in other countries, then any countries that receive such later applications agree to regard the earlier date as the priority date of that application. These later applications are called *Convention Applications*. To gain this benefit, the applicant must quote the earlier priority date on his/her application. This earlier priority date and country are duly noted in the documentation associated with the later application.

This is helpful for information scientists and librarians trying to link various patents from around the world together, as will become clear later in this chapter.

There is nothing to stop the applicant applying abroad later than the twelve-month period, but then he/she cannot claim that earlier priority date, and the priority date is simply the date when that later application was made. These are known as *Non-Convention Applications*.

Each one of the patent applications is treated individually by the different Patent Offices. Some will have an easy passage and will quickly be published without amendment; some will be amended and be published later on. Some will fail and will never see the light of day. Those that are published are linked together by a common priority date. Patents that are linked by such a common priority date are called a *family of patents*. The first member of a family to appear – often, but not necessarily always the first one to be applied for – is known as the *basic* patent; all others that subsequently appear, claiming the same convention priority date, are known as *equivalents*.

Families of patents are important to librarians and information scientists. Let us imagine you are interested in an invention, and all you are aware of is a Japanese patent. You do not like the idea of translating it. So you check whether there is an equivalent in an easier language. You find, say, that there is a Canadian equivalent. You obtain that and use it instead. Families represent a free translation service.

The application procedure

Having put in the patent application, what happens next? If it just a provisional application, the Patent Office simply files it away. If it is a full application, whether based on earlier priority or not, the Patent Office carries out a literature search. It is looking for prior publications. In theory, the Patent Examiner assigned to the case should look for all prior publications, but in practice they usually just look at earlier patents. Then, eighteen months after the priority date, the Patent Office publishes both the application itself and the results of its literature search. The documents are numbered in simple sequential order and are known as *early published applications*. They have no legal significance at this stage.

Anyone can read and inspect the early published application document and decide whether it represents an important invention and whether the Patent Office's search has been rigorous enough. Any third party can see an early published application and if it realises the Patent Examiner has missed some key prior publication, it can draw the Patent Office's attention to this omission.

The applicant, too, looks at the search report and any comments put in by third parties, and then decides whether it is worthwhile to pursue the application. In practice, more than half the applications fall by the wayside at this stage.

If the applicant does persist, he/she then pays another fee, and the application is then subjected to detailed legal examination of the invention's novelty and obviousness. The applicant may be required to amend the wording of the application or its claims.If the applicant cannot agree with the decisions of the examiner, he/she can go to appeal, all the way to the House of Lords if need be.

Sooner or later, depending on how protracted the negotiations are, the patent is granted. Now a second document appears: the granted patent. As you would imagine, far fewer of these appear than early published applications. This second document has the same numbering as the first, but whereas the first has a number like 2345678A, the granted patent is numbered 2345678**B**. Whereas early published applications are published in simple numeric order, granted patents – the B documents – appear randomly and there is no way of an outsider knowing when an early published application will appear as a B document. This patent immediately takes effect and will last, assuming certain renewal fees are paid, for up to 20 years from the priority date. In practice, only about ten per cent of all patents granted are renewed for their full term; these, of course, represent the most important inventions. The owner of the patent, incidentally, is called the *patentee*. If the gestation of the patent has been particularly long and hard, and the granted patent only appeared, say, ten years after priority date, then the effective life of the patent appears to be just ten years. If the patentee sues for infringement of his patent, he can backdate the legal action to the date the early published application became OPI (18 months after priority date).

Effect of the grant

The effect of getting a patent is simple. You can prevent any third party from making, using, importing or selling your invention. If anyone does any of those acts without your permission, you can sue for damages.

Damages in patent cases are usually very high, and the costs are higher still, so anyone thinking of making, using, selling or importing the invention without permission (*infringing* the patent) does so at severe risk.

The patent documents comprise an abstract, the text of the specification and claims. It may well also contain one or more drawings. In addition, it contains detailed bibliographic information provided by the Patent Examiner. This information includes priority date and details, the name and address of the applicant, and a series of classification symbols identifying the subject matter of the invention. I will discuss patent classification schemes later in this chapter. Helpfully, all the bibliographic information, which always appears on a separate front page, is broken down into headings preceded by a number. This numbering system is common to all Patent Offices, and so even a patent in a foreign language can be inspected for its bibliographic data, always in a standard format. This is particularly helpful when you try to identify equivalents from common priority information.

The claims describe clearly the scope of the monopoly requested by the applicant. Typically, a broad claim 1 is given followed by a series of more specific claims that are sub-sets of the first claim. This is so that if the first claim is knocked out for obviousness, lack of novelty, etc. in some future legal case, the remaining claims still provide some measure of monopoly protection for the patentee.

The specification must include the best way of undertaking the invention known to the applicant, although he/she is perfectly at liberty to surround this best method with several inferior methods, and not to make explicit which one is the best.

There are certain other requirements for the specification, such as *sufficiency*, i.e. it must provide enough information for an expert to repeat the invention. This means the description must be clear and unambiguous, and nothing must be kept back from the specification. Failure to comply could mean the cancellation or amendment of the patent in any future legal case.

Thus, the British patent system is characterized by early publication eighteen months after priority date, and separate examination, and two sets of published documents. Most of the major countries of the world follow this pattern, with the same basic rules. The prime exception is the USA, which does not bother with early published applications, has a rigorous examination, and just one document appears, the final granted patent. Because most countries publish applications 18 months after priority date, one gets a flood of applications published around the world during the same week for a single application, if that has been applied for in many countries. Most countries publish their patent documents weekly.

The patent is a piece of property. It can be sold (called *assignment*), leased (*licensed*), mortgaged, given away and passed on in your will. If you assign your patent, the person you assigned it to then gains all the rights and obligations of the patentee. Assignment is unusual, but not unknown. More common is licensing. This is equivalent to renting out your own house. You give someone permission to make, use, sell or import your invention, but the patent is still ultimately yours. The two most common types of license are *exclusive licences*, where you license the patent to just one third party, and *non-exclusive licences*, where you license the rights to many people. In return, you will be paid cash, typically a lump sum, or a percentage of sales or profits based on the product of the patent.

The idea of licensing is common in much of intellectual property, such as in trade marks, copyright and so on. What is unusual about patent licensing is that there is a provision for *compulsory licences*.

If it can be demonstrated that the patentee has failed to exploit his/her invention, and yet is preventing others from making, using or selling the invention, then any interested third party can demand that the Comptroller of the Patent Office issues a compulsory licence. This means the patentee MUST agree to give a licence to that third party, on terms to be agreed – or, if they cannot agree, then on terms imposed by the courts. Compulsory licences are very rarely applied for, and even more rarely granted, but the provision is important because it ensures patentees do not just 'sit on' their inventions.

Quite separately, the Crown – the Government – has the right to demand compulsory licences of any patentee for any reason. This provision has been used in the past by the National Health Service when obtaining pharmaceuticals. Even more extensive powers regarding patents are given to Government during war time, but they are beyond the scope of this chapter.

Infringement

Infringement, too, has some sort of analogy with property owning, as it is akin to someone trespassing on your property. Infringement is, simply enough, a third party making, importing, offering to supply, using or selling your invention without your permission, whether your invention relates to a patented product or to a product made by a patented process.

What defences has an infringer got if he/she is taken to court accused of infringement? There are many defences, but let me make clear one defence he/she does NOT have. It is no defence to say you did not know about the patent's existence. Ignorance is no excuse at all. The major defences are:

1. It is not infringement because what I am doing does not fall within the claims of the patent. This is one of the most important defences.
2. The person suing is not entitled to sue because he/she does not own the patent in question.
3. The infringement took place at a time that the patent had expired or had lapsed.
4. Statute of limitations – more than six years have passed since the alleged infringement.
5. The patent in question is invalid – this is probably the most regularly used defence. In other words, the patent should never have been granted, because of prior publication, prior user, obviousness, because the invention does not work as is claimed, and so on. For example, I noted earlier that there is no obligation to demonstrate that an invention works; equally, you must not mislead the Patent Office by claiming in writing it does things that it does not do. An applicant can lose a patent because the specification does not disclose the invention clearly and fully enough for it to be performed by someone skilled in the art. If the defence works, the patent in question is struck down. This is known as *revocation*.

Very few infringement actions take place. This is because either the patent is so strong and so clearly valid that no one dares infringe, or it is so weak that people infringe with immunity. A court case only occurs when both parties genuinely believe they have a good chance of winning.

The biggest case in recent years on patent infringement was *Polaroid* v. *Kodak* for the instant camera. Polaroid sued Kodak in the US courts for an instant camera it had launched, claiming the Kodak device infringed Polaroid's patents. Kodak denied it and carried on selling their cameras throughout the court action. Because of this recklessness, when they lost the case, they had to pay billions of dollars in punitive damages.

It is, incidentally, an offence in UK law to issue threats of patent infringement action without carrying out that threat. So it is an offence to issue threatening letters. You must go straight ahead with the action, or do nothing.

It is also, incidentally, an offence to claim you have a patent when you do not. So you must not call your product 'Patented', or quote a patent number, if the product is not patented. You cannot call a product patented if you have just an early published application to your name.

The US patent system

This system, though not as old as the UK one, dates back to the original Constitution of the USA in 1788. The current system dates back to 1836, although of course there have been detailed changes to the law since. The USA was for many years the world's most prolific country for patent documents; it has now been overtaken by Japan. However, the vast majority of Japanese patent documents are low-grade 'pot boiler' type documents, whereas US patents enjoy an unrivalled reputation for their high quality. Some 5 000 000 US patents have been issued, making them a key source of scientific and technical information.

The US system is nowadays very much out of line with most other countries' systems. The most important distinguishing features are as follows:

- The system does not depend on filing dates for establish priority
- A full statement of the prior art is needed
- It is not an early publication, deferred examination system
- Examination is very strict, and many years can pass before the patent is granted
- The patent is in the name of the inventor, not a company
- The patents are highly regarded in terms of presentation
- There is a low proportion of invalid or weak patents issued

- The patent term is seventeen years from date of publication. Only recently have renewal fees become a requirement.

The question of novelty is particularly important. The date on which novelty is assessed is the date the inventor first had the idea, NOT the date that he/she filed the application. This means, if you are going to patent in the USA, you need to keep meticulous notebooks recording your thoughts, your experiments, your results, and have them dated and authenticated. This is usually done using signed and dated laboratory notebook reports. Therefore all large corporations all around the world routinely require that their scientists and engineers fill out these reports, and have them authenticated, for one day a US patent application may be filed, and then the notebooks may have to be provided in evidence.

This system has advantages and disadvantages. One advantage is that the company is not under the same pressure to get a patent application in early, when the invention has not been fully developed. Only worthwhile inventions are applied for. On the other hand, filing dates are clear and cannot really be disputed, whereas laboratory notebooks suffer from ambiguities – was that off-hand remark on page two really the first time he had that idea? – and could potentially be open to falsification.

Looking at the question of a full statement of prior art, the applicant is required to disclose all he/she knows about prior art. Failure to do so is a criminal offence, known as 'fraud on the Patent Office'. Most other patent issuing authorities are much more relaxed about this matter. This means the text of the patent is useful in providing a literature survey of the topic in hand.

Once someone has applied for a US patent, the process of search and examination continue hand in hand. No one can find out what is going on, and there is no early published application or search report. Only when the patent is finally granted does anyone know that invention had been the subject of a patent application. All you know in advance is the title of the application, which is designed not to be meaningful – such as 'chemical process'.

This means there is just one publication, the final granted patent, which appears out of the blue. The examination process is very strict. One aspect of this strictness is that the applicant is often advised to withdraw his/her original application, and re-submit it. This leads to what are known as 'Continuation in Part', with a new filing date. This makes it particularly difficult to link US patents to equivalents elsewhere, as they may have similar subject matter and may have started life with identical priority dates, but, due to continuations in part, now seem to have a different filing date. Skilled patent searchers do not have a problem with this.

Another peculiarity of the US system is the polite fiction that only individuals get patents, not companies. So all patents are in the name of the inventors, although they are free to assign their rights to a company. If you want to check a company's US patents through the US Patent and Trademark Office's (USPTO) manual indexes, you find you have to check each inventor instead in turn. Fortunately, computerised information services, which I will be describing in due course, solve the problem.

It is perhaps worth saying that 25 years ago the US patent system was typical for most countries. Few countries used early published applications, and most just issued one patent, on grant. There are moves to bring the US system into line with the rest of the world.

International treaties

The Patent Cooperation Treaty (PCT) is designed to help developing countries that have problems with their Patent Offices doing literature searches. It provides for a centralized literature search for any Patent Office run by one of several major approved literature search offices, including the USA, Japan, Sweden, the Netherlands, Austria and Australia. The application is made to some country's Patent Office. It passes the application to the PCT central office, WIPO (World Intellectual Property Organization, a UN special agency), based in Geneva, which passes it on to the appropriate search body. The search body runs a full literature search through, at the very minimum, all patents published since 1920, from ten major patent issuing authorities plus 160 leading scientific and technical journals. Each office is supposed to search at least this minimum; if possible, more. The search is aimed at both obviousness and novelty. WIPO charges a modest fee to the national Patent Office for this service.

Eighteen months after priority date, WIPO publishes the application plus the search report. The specification, abstract and search report always appear in English as well as the original language that the applicant used. This is, of course, invaluable to English-speakers. Assuming the applicant wishes to proceed to grant, he/she then pays the national Patent Office a further fee, and the application is then examined. Once again, for a fee, the national Patent Office can farm this stage out to a larger and more experienced Patent Office, not necessarily in the same country as where the search was run. This facility is not as often used as the search service. Ultimately, the national Patent Office decides whether to grant the national patent or not.

So, the PCT results in a sequence of early published applications published by WIPO in Geneva, but thereafter local national patents just result. The PCT is well used by developing countries, and is increasingly popular with many organizations in developed countries as well.

The PCT is therefore a facility to help developing countries to conduct the search and examination. However, the result is still a series of national patents, subject to interpretation under local laws. Perhaps more significant is the trend towards merging national patent systems into regional patent systems; in other words, to merge the laws. The most important of these is the European Patent Convention (EPC). The European Patent Office (EPO), based in Munich, offers applicants the ability in a single examination to get patents in most European countries; not just EEC, but others, such as Scandinavia, Austria, eastern European countries, etc. The application is made to Munich. The procedures are essentially as for British applications, but you specify which countries you want protection in. There is a modest extra fee for each country you specify. There is then an early published European application and search report, numbered sequentially and followed by an A. You then go for substantive examination. The final granted European patent appears, numbered the same but B. That patent is then automatically valid in the countries specified. You get a 'bundle of national patents' automatically from the publishing of a single granted European patent.

So, in the UK, you can have a valid patent that is either a British patent, or a European patent that designates the UK. Both are equally valid, and there are now more EPO patents appearing designating the UK than there are 'true' British patents being published.

The European route is used heavily by larger companies. Its pricing structure – a high up-front fee, but low marginal cost for extra countries – means it is very cost-effective if you want to take out patents in many countries at once. Gradually, over the years, the use of national patents will decline, but there will always be the need for a cheap local system for small firms and private inventors.

The European Patent Convention is not the only regional patent system, but it is by far the largest and most important. There are others in Africa and South America.

The logical final stage of the path from PCT to regional patent systems is unification of patent law across countries, so that you get a transnational patent. With the EPO, you specify which countries you want protection in. With a true regional system, you would have no such choice; you would automatically be covered in all the countries, and national patents would cease to exist. No such system exists, although the European Union has for more than 20 years been trying to develop a Community Patent Convention, which would result in the abolition of national patents within the Union and culminate in a single patent covering the whole European Union. This idea has made very little progress over the years and it may be another ten years at least before it becomes reality.

Patents information

Patents represent a bargain between an inventor and the State. In return for a monopoly right, the State expects the inventor to reveal all that he/she knows about the invention.

Indeed, failure to fully disclose details of the invention, or the best method of doing the invention, are grounds for rejecting a patent application or for revoking the patent after it has been granted. In addition, most developed countries of the world have a deferred examination system, so that there is a flood of early published applications, the majority of which continue no further, and then a smaller number of final patents containing similar subject matter – though not necessarily identical, as the examiner may have insisted on changes. The result is a flood of patents documents appearing world-wide in a variety of languages.

Importance of patents as sources of information

On the face of it, patents are difficult documents to handle, and this is why so many librarians and information officers avoid them if they can. They are not written to the style of a book or a journal article. They are legal documents, written in a peculiar mixture of technical and legal jargon known as 'patentese'. About 1 000 000 new patent documents are published each year in the world, and well over half of these are in the difficult languages of Japanese, Chinese and Russian. Fifty-four per cent of current patent output is in the Japanese language alone. It has been estimated that 25 000 000 patent documents exist at the moment. Then again, not many libraries stock patents, and those that do rarely have adequate indexes to their collections, or staff who are knowledgeable about patents. Often patents are only available on microform, regarded by many people as a user-unfriendly medium. Most countries' patents appear in apparently random number order, and it is therefore difficult to be sure if a particular patent has appeared or not, and, if it has appeared, whether the library has lost it or not. Because the patents – especially B documents – appear in semi-random number order, libraries must keep them loose in box files and cannot bind them into sets.

In addition, many countries are slow to print their specifications, and many patent documents are poor quality photocopies of faint typescript. Also, patents can be expensive to buy.

There are advantages, though, as well. In each country, patents are issued by just one publishing authority: the local Patent Office. They are issued regularly, usually weekly. The Patent Offices issue gazettes and official journals that list all the patents published. Many of these have excellent indexes. Patent documents never go out of print. Even very old patents are available for purchase, although often only as photocopies. There is a world-wide classification system for patents, the *International Patent Classification* (see below) which is widely used throughout the world. Seventy-five Patent Offices use this system to classify their documents, making it easy to monitor current patents in a chosen field, or to do retrospective searches. Very importantly, the format and layout of most patent documents in the world follow internationally agreed standards, and so you can come across a patent document from a new country – say, Estonia – for the first time, and be confident what the bibliographic information is, where the abstract is and so on. There is, also, either no copyright on patents, or copyright is waived, and so you can photocopy. There are also excellent abstracting and indexing services (see below) which cover the literature in a comprehensive and timely manner.

Nevertheless, most people's perceptions of patents as sources of information are negative. The advantages appear to be outweighed by the many disadvantages.

However, patents ARE used a lot and, indeed, the most profitable bibliographic database producers are those that cover the patents literature. Their advantages must outweigh these disadvantages. In short, people have pressing reasons to use patents.

The contents and layout of patent documents

Until the 1960s, the format and layouts of patents reflected the local Patent Office practice. Since then, through the good offices of WIPO, there has been steady progress towards standardization. The standards relate not just to the presentation of the data and text, but also to how the bibliographic information is presented. The universal acceptance of these standards has helped the handling, readability and treatment of patents by information scientists. The most important feature of the modern patent specification is its *title page*. This page is added by the patent issuing authority to the front page of the specification. The title page provides bibliographic information, an abstract, and often a drawing or two. Each unit of the title page is assigned a unique numerical code, the so-called *INID Code* (INID stands for Internationally agreed Numbers for Identification of Data). Among the most important INID codes are the following:

(11) Document number
(13) Kind of document (early published application, granted patent, etc.)
(19) Name of patent issuing authority
(21) Local application number
(22) Local application date

(26) Language of document
(31) Priority application number
(32) Priority date
(33) Priority country
(41) Publication date (if unexamined application)
(42) Publication date (if examined patent)
(51) International Patent Classification symbols
(52) Local patent classification symbols
(54) Title
(57) Abstract
(71) Name of applicant
(72) Name of inventor
(74) Name of patent agent or attorney
(81) PCT designated States
(84) Designated States for a regional patent application, such as EPC

The patent contains a title, bibliographic information and an abstract. In addition, it contains the specification and the claims. The title, written by the applicant, is deliberately vague – 'Improvements in motor vehicles', 'New chemical compounds' and so on. There is no legal requirement for explicit titles.

The bibliographic information identifies the patent document itself and is supplied by Patent Examiners to a consistent standard; it ties the patent to other members of the same patent family – that is to say, by common priority details. The abstract is written by the applicant. Although in theory the Patent Examiner can insist on an abstract being rewritten to make it clearer, in practice the original wording is usually untouched. Frequently, the applicant will attempt to hide the key features of his/her invention from the abstract, as the abstract will be widely used in disseminating information about the patent to the wide world; so it is useful, but cannot be totally relied upon. The specification is a detailed description of the invention, written in patentese, but with a fairly well established structure of, first, describing the problem that is being addressed; second, describing previous attempts to solve the problem and what was wrong with them: these two parts are often known as the *prior art*. The document then summarises the major features and advantages of the current invention. It then describes the invention in detail, perhaps including scientific experimental results, and will cross-refer to any drawings. The applicant should provide as full a description as possible, with a description that would be clear to a skilled expert, not necessarily to the man in the street.

The claims describe precisely the monopoly that is being claimed and are always written in a fairly complex manner. From the legal point of view, the claims are the crucial element of the patent, but from the information point of view it is perhaps less important – unless your search is specifically for legal information. Beginners commonly assume that the claims represent a concise summary of the invention, and can be used as a summary. It does not serve the purpose of being a summary, and therefore the claims should be used with caution. The claims usually involve both broad and narrow claims.

The broad claims cover a range of technology that the inventor wants protected; the narrower claims are there in case the patent is challenged for obviousness or novelty and the broad claim may be knocked out, but narrower claims are allowed to survive. This happens quite often. Then the patentee is left with some protection, not as large as originally hoped for, but enough. Without doubt, one of the narrower claims will precisely encompass the key best bit of the invention, but it is not always possible to work out which claim represents the key element.

When a patent infringement suit is considered, or when the validity of the patent is considered, the scope of protection is determined by the patent claims. The specification provides useful background, but no more. Obviously, therefore, the wording of the claims is crucial. That is why inventors should use the services of a qualified patent agent.

Advantages of patents as sources

There are four advantages of patents as sources of information. Firstly, there are many patents, so it is a rich vein of literature; secondly, patents frequently represent the only source of information on a topic; thirdly, they frequently represent the earliest source of information on a topic; finally, they provide more detail.

As I have already indicated, about 1 000 000 patent documents appear each year. Not all of these represent new inventions, of course. Many are family members from basics that have already appeared. Precise figures are not available, but a good guess would be that 35% of them represent genuinely new items – they are basics – and 65% of them are equivalents.

Therefore, perhaps 350 000 new inventions are described in patents each year. This is similar in size to the number of new articles that appear in the whole of engineering each year. Thus, patents represent a sizeable part of the scientific and technical literature appearing world-wide.

It is sometimes argued that even if all those patents do appear, they are likely to duplicate things in the journal literature, so they can be safely ignored. This is a common misconception. In the majority of cases, *the patent literature is the only source*. There have been many studies carried out on this subject. They demonstrate consistently and convincingly that the vast majority of material in the patent literature is NEVER duplicated in the journal literature. The figures vary considerably according to the study, but in most cases the result is that about 90% of patent documents never get duplicated in the journal literature. Thus, of the 350 000 new patents appearing each year, at least 300 000 of them will never see the light of day in any other medium.

A more subtle argument is that if the invention is *important*, then it will appear also in journal articles. Not many studies have been done on this topic and, indeed, it is arguable about what exactly is, or is not, 'important', but those which have show the same pattern as I have just described; even in important cases, the chances are the patent will be your only source.

Nonetheless, some 50 000 new patents each year ARE duplicated by journal articles. Surely we can at least safely ignore these? No, because usually *patents offer the earliest source of information*. The first thing the inventor MUST do is apply for a patent. Only after he/she has established a priority date can journal articles be considered.

Often, the inventor will not bother with a journal article at all. He/she may have moved on to other interests, or his/her employer may discourage journal articles because they would draw attention to the invention through a more popular medium.

Combine all these factors with the fact that an early published application appears eighteen months after priority date, and that learned journals can easily take one to two years to publish an article after receipt, and it is no surprise that patents appear first. There are innumerable famous cases of patents being the first documents to appear, often by many years. Among notable examples are the punch card, radar, the videodisk, and the keel of the Australian yacht which won the Americas Cup in 1983.

Finally, *patents provide more information*. Journal and book editors require brevity and clarity. Patent law requires maximum full description. The inventor must reveal ALL he knows about the invention. Patents frequently run to hundreds of pages. Indeed, it is arguable that their length makes them harder to use and makes it easier for an inventor to hide the key element among much dross, but, there is no question, a patent is the place to go for the detailed information.

There are other factors worth remembering, too. Patents cover all areas of science and technology; they offer solutions to problems rather than theoretical ideas; they have historical significance and are of great value to historians of science and technology and they have unexpected uses in techno-commercial areas (see below).

For some types of search patents are the only possible source

If a librarian or information scientist has to do a search in the scientific or technical areas, he/she would be foolish – and a poor information professional – to ignore them. In some cases, there is no choice: patents are the ONLY source that can be used. Many organizations need to track patents for legal reasons. The main examples of these sorts of searches are *patentability searches* and *infringement searches*.

Imagine that your employer has developed an invention and wishes to know if anyone has thought of the idea before; in other words, the company wants to know if there have been one or more prior publications. If there were any prior publications, carrying out more R&D work on the invention, and applying for patent protection world-wide, would be a waste of time and money. So it is usual to ask the information scientist at an early stage to undertake a patentability search. This is a search of all the relevant scientific and technical literature. Patent Offices usually search patents as their only, or their prime, source. So you, too, MUST search the patent literature.

It is important to note that you need to obtain 100% recall – you must not miss a single relevant item – in a patentability search. Just doing a rough and ready search (the sort of thing you might do for a general search request) is quite inadequate. Your search must be thorough and comprehensive. If you miss a relevant item, and it only becomes known many months or years later, you have wasted an awful lot of your employer's money!

The second sort of search where you MUST search the patent literature is an infringement search. Imagine that your employer wishes to make, use or sell a particular product. The company needs to be sure that, by doing so, it does not infringe anyone's patent. So, before going into the market, the prudent company asks its librarians or information scientists to undertake an infringement search. This time you need only search through patents in the country or countries in question – those countries where your employer plans to make, use or sell the product. Remember though, if the country is the UK or other European countries, you need to check both British (or other national) patents and European patents that designate the UK or the other countries in question. You only need to search for patents at this stage. You need to get 100% recall again.

If the information professional finds no relevant patents, then he/she can advise accordingly. What if the professional finds one or more relevant patents? Then further follow up activity is required. He/she needs to ask two questions, in this order: is the patent in force, and, if so, is the patent valid?

First, he/she must check the *status* of the patent. Is it still in force? Perhaps all he/she has found so far is an A early published application. The professional needs to find out whether a granted patent was issued, whether the applicant withdrew his/her application in the light of the literature search, or whether it is still undergoing examination by the Patent Office. If it has been granted, he/she needs to find out if the patentee has paid the renewal fees, thereby keeping the patent alive, or whether he/she has failed to do so, allowing the patent to *lapse*, or whether the patent has reached the end of its 20-year life and has *expired*.

Let us assume the professional found a relevant patent, and it is still in force. Does this mean the employer may NOT proceed to make, use or sell the product? If the employer is really keen, it may instruct the professional to do a *validity search* to try and knock the patent out by finding a prior publication or evidence of obviousness. No Patent Office is infallible (although the fact that the USPTO has granted a US patent is strong evidence of a patent's validity). Their searches for the prior art are limited as they tend to search patents only. So just because a patent has been granted does NOT guarantee that it is a valid patent! If the professional finds a relevant prior publication, or clear evidence of obviousness or prior user, the employer can make use or sell the invention notwithstanding the presence of the patent.

If the employer does make, use or sell the product, and is sued by the patentee, its defence will be that the patent is invalid. The case will be argued over the validity of the patent, and if the employer is proved right, then the offending patent will be revoked. Therefore the professional's search and advice can be critical.

There is a common thread in these two types of legal patent searches – the need for 100% recall. Unlike most information searching, no hits is what you want to get, but reporting to your employer that there are no hits when there is one could prove very costly to your employer – far more costly than missing an item in a general scientific or technical search.

For this reason, information retrieval services specializing in patents are designed to maximize recall. High recall implies low precision – in other words, false drops to wade through. So patent searching is very different in style from other searching. Also, patent searches take longer than others, because, if no hits are found, the searcher must persist through as many sources as possible. In short, patent searching requires more patience, more diligence and involves more sifting through items than most types of searches.

The problems involved in patent searching also affect the pricing of databases. Online databases have traditionally been charged on a connect hour plus display charge basis. For most searching, this is a reasonable approach, as you tend to be looking for hits, and the more hits you pull out, the more you pay. However, for patent searchers, zero hits are often good news from the employer's point of view. So a search that ends up with no hits is both cheap for the patent searcher and is a positive result.

However, because patents databases online are not just used for infringement and patentability searches – they are excellent for mainstream scientific and technical searching – there are two groups of searchers searching the same database but with different perceptions of what is value for money. In practice, the pricing of patents databases follows the traditional structure, and so patentability and infringement searching are a bargain.

Patent searching

There are five types of searches one can carry out on the patent literature: legal status searches; author searching, which is called 'name searching' in the patent community; patent family searching; searching for bibliographic information, and subject matter searching.

Legal status searches are to ascertain whether the patent is in force, whether the application has proceeded to grant, whether anyone has opposed the patent and so on.

Each major national Patent Office or patent issuing authority maintains an official register that can be inspected in a Search Room. For a modest fee, you can inspect the Register, take notes and, for a further fee, you can get an official statement of whether the patent is in force or not, or what its status is. For another fee, you can inspect the so-called 'file wrapper', which comprises originals of, or copies of, all the correspondence between the Patent Office and the applicant from the first application to date. None of this is kept confidential, but you can only inspect a file wrapper after the date of early publication.

If you are interested in European patents, you can get information about their legal status either online, by visiting the European Patent Office, or by post.

There are a few commercial online services that offer some legal status information. Although it is fairly authoritative, readers should remember that the information is not always guaranteed to be 100% accurate, and you can get noise on the line. Therefore, if you do need an authoritative statement, you should always go to or contact by letter the Patent Office concerned directly.

Name searching is fairly straightforward. There are some excellent online databases offering patents information (see below), and they can be searched using the name of the inventor(s) or patentee(s) as search terms. In addition, Patent Offices themselves often provide printed name indexes to patents granted or published. Two factors to remember are:

1. There are two sets of authors to most patents: the actual inventors and the employer who owns the patent. The number of patents taken out by private individuals is quite small. Online databases distinguish these two types of author into different searchable fields.
2. Quite often, the name a company patents under is not the same as its well-known name. Frequently, a special subsidiary company is set up just for patenting; for example, British Coal patents under the name 'Coal Industries'.

Patent family searching involves tracing members of the patent family that are linked by a common priority. You may well know the basic patent – the first one to appear – or you may have the priority details, or you may know some other family member. You may want the family details simply to assess the scale of patenting – it is plausible that the bigger the family, the more important the patentee considers the patent to be, and therefore the more important the invention – or you may want an equivalent in a convenient language or you simply want to know the number of the equivalent in some particular country.

The easiest way by far of carrying out patent family searches, and the only way that can be recommended, is either using an online database such as INPADOC or WORLD PATENTS INDEX, or printed/microform derivatives from them. They provide comprehensive family information based on the number of any member of the family or the priority details. Years ago, one used to have to do laborious manual name searches in each country in turn. A family search could easily take a week to do.

One proviso should be mentioned on family searches. Non-convention equivalents are not always picked up by online services. A non-convention equivalent is one that has the same subject matter as another patent, but does not claim the same priority. This can happen if the applicant does not get round to applying in other countries until AFTER the twelve-month grace period.

Although patent information services DO try to identify and then note these where they can, it is not easy for them. So, just because you do not find a family member in a particular country using one of the commercial services, it is no guarantee that the relevant 'equivalent' patent does not exist. It may well do so, but without claiming that common priority. The only way to be sure is to do a comprehensive name or subject search as well in the country of interest. Then you would pick up that non-convention equivalent as a hit.

Searching by *bibliographic information* is easy enough online, but difficult to do using manual services. Assuming you have some piece of data, such as a priority date, priority country, priority number, publication date and so on, most patents CD-ROM and online databases are designed to allow you to use these as search terms.

Subject matter searching

These types of searches are handled in the same way as for general literature; you can search online or using CD-ROMs by words in the text, or by controlled language indexing, or by both. Starting with word searches, some databases offer you abstract words; some offer you full text of the patent specification; some offer you words in the main claim, or in all the claims; some databases offer combinations thereof.

The patent title is usually very vague, so some database producers (notably Derwent Information – see below) add value by making the title more explicit. As far as the abstract goes, some database producers use the abstract as supplied by the applicant, no matter how poorly written. Others prepare their own abstracts.

Searching the full text of a patent is not always a good idea because of the arcane patentese used and because of the large number of false drops that will inevitably result. Most searchers prefer to use abstract words. Nonetheless, full-text searching has benefits for certain types of searching difficult to conduct using title words or controlled language. Mead Data Central's LEXPAT is the leading product in this field.

What about controlled language indexing? There are two types that readers should be aware of. Firstly, there is the coding supplied by certain commercial database producers; secondly, there is the Patent Offices' own national (or international) classification systems.

As far as the commercial database producers are concerned, readers need only be aware of one organization's system and that is Derwent Information (formerly Derwent Publications). Derwent Information dominates the patents information world. About 30 years ago, it developed its own indexing system, which has been developed and refined over the years. It is particularly strong in the area of chemical compounds, but is also useful in all other technologies. Consistently applied by Derwent coders on all new patents, it provides efficient recall and precision through the mass of the patent literature. It is quite a complex system in the chemical areas, but is easy to learn in other technologies.

Patent classification schemes

Patent Offices have for many years been obliged to check the prior art before granting a patent. It was recognized that old patents – which traditionally were all they looked through – must be classified to make searching them easier, so various Patent Offices over the years developed their own classification schemes.

Initially, all the schemes were purely hierarchical, with a broad subject area broken down into smaller subject areas, then even smaller areas, until eventually, corresponding to a long and complex mark, you would get a very detailed heading. Unlike the Dewey Decimal Classification, these schemes did not cover all areas of knowledge, but just practical crafts. In more recent years, they have developed their schemes into a complex mixture of both traditional classification and indexing terms that are independent of the broad classification scheme. Modern patent classifications are a mixture of hierarchical classification and keywords (although the keywords happen to be not English words but codes).

The Patent Offices usually classify the invention as shown in its claims, rather than the description of the invention in the specification: the two are not always identical.

Besides developing their own, non-compatible classifications, the Patent Offices have tried to co-operate by developing a general all purpose classification called the *International Patent Classification* (IPC). Some countries ONLY classify using the IPC. Others classify both with the IPC and with their own national classifications. It is interesting that if you get a patent family, and you look at the IPCs assigned by the various offices to the family members, they rarely are in accord with each other. This is partly because of different depth of knowledge of the IPC, partly because of different styles of classifying and partly different levels of competence.

Patent classifications are arcane and difficult to understand. As a result, expert patent searchers who understand what makes them tick can be found in most developed countries.

How can one search using these classification schemes? Various national Patent Offices issue printed or microform indexes to their patents, sorted by classmark, but it is easier to search using classifications on patent databases online or on CD-ROM rather than laboriously going through the printed indexes. Most online patent databases include the relevant national classifications and/or the IPCs as searchable fields. Guides to the various national classification schemes and to the IPC are issued by the appropriate patent-issuing bodies.

Abstracting and indexing services covering patents

What are the abstracting and indexing services that cover patents? Very broadly, they can be divided into two types: services that happen to cover patents in among other sources, such as journals and reports, and services that only cover patents and no other sources. Those in electronic form can be traced through Bowker-Saur's *World databases in patents*.

As a rule, abstracting and indexing services that happen to cover patents alongside other sources cover patents very poorly. Numerous studies have shown up major deficiencies in their coverage. The reasons for this are various, but include: a basic lack of understanding of the patent literature; the problems of the vast volume of patents that makes them very selective in their coverage, or an unwillingness to recognize the importance of patents as sources.

Only in one sector – the chemicals sector – are patents well covered. Services such as *World Aluminium Abstracts* and *Zinc Abstracts* cover patents in their specialist areas quite well. The biggest service of all, *Chemical Abstracts*, now covers them fairly comprehensively.

So, except for the chemical area, you cannot rely on standard indexing and abstracting services for retrieving patents. Certainly, services such as *Engineering Index* cannot be relied upon for patents coverage.

What about specialist patents abstracting and indexing services? The major ones include: CLAIMS; the various EPO databases; INPADOC; JAPIO; USPA; INPI; CASSIS family; FIRST, and WORLD PATENTS INDEX.

CLAIMS is a series of online patent databases covering US patents only. CLAIMS is of particular interest because it goes back to 1950, and because it adds its own coding for chemical patents. Abstracts are provided from 1971 onwards, and the main claim from 1965. The service is not cheap, but it is the best database for US patents. It is available on DIALOG and is shortly to be loaded on to other hosts.

The various European Patent Office databases provide bibliographic details and legal status for European patent documents, the full text of the IPC as used by the EPO, and classmarks and patent numbers for a comprehensive library of old patents from 20 countries, going back to 1877, which are used by the EPO Examiners when doing novelty searches.

INPADOC, which is based in Vienna, used to be an independent organization, but was acquired by the European Patent Office in the early 1990s. This service provides both microfiche and online databases. It provides a record of patents from 56 patent issuing authorities going back to 1968, and provides titles (in original language) and bibliographic details, including IPCs assigned. It is the best service for name and for patent family searches, because of the large number of countries covered, but for subject matter searching it is very limited, and it is handicapped by one's inability to see an abstract. It is available through ORBIT (now Questel.Orbit), DIALOG and on its own online host. The EPO databases and INPADOC have recently been subsumed under the general name EPIDOS, though it is probable that searchers will continue to use the older, more familiar names for some time to come.

JAPIO, another ORBIT database, as its name suggests, provides online bibliographic data and English language abstracts on Japanese unexamined applications from 1976 to date. It is a useful source for Japanese patents.

USPA provides bibliographic information plus the abstract and claims of US patents since 1970. There is a separate file offering US classmarks and US patent numbers, but nothing else. This latter database covers the period 1790 to date, since the first US patents, and is probably the oldest database in the world for time-span covered. It is available on ORBIT.

INPI is a series of online databases on Questel.Orbit, covering all French patents, and also European patents. The amount of data is limited to bibliographic information plus classmarks.

The CASSIS family is a series of CD-ROM products on US patents, including US classification, keywords, company names, bibliographic data, inventors and abstracts, going back to 1969; the US classification list goes back to 1790. US patents are widely available on CD-ROM; other products of the same type are produced by Research Publications Inc.

FIRST is one of many CD-ROMs produced by EPIDOS. This one is a version of the EPO databases and includes PCT publications. One is able to scan the front page of the patent for viewing. EPIDOS produce a wide range of CD-ROM products, covering EPO, PCT and various national patents, some in full text, some with images.

As has already been indicated, Derwent Information is THE dominant player in the patents information business. It offers a variety of printed, microfilm and database services, with comprehensive and in-depth coverage of patents world-wide. The services include alerting, or SDI, as well as retrieval services. Most major companies in the world subscribe to Derwent's services. Their services fall into two broad groups. There is a series of high priced subscription services, and some generally available online databases. The subscription services, notably *Chemical Patents Index* and *Electrical Patents Index*, offer in-depth coding and retrieval of patents in their respective areas. They are outside the scope of this chapter, but interested readers should contact Derwent Publications for further details.

The best known Derwent service is their publicly available online service WORLD PATENTS INDEX (WPI). This online database, available on Questel.Orbit, STN and DIALOG, provides coverage of all subjects back to 1974, and offers coverage of chemical patents even further back to 1963. Since the lifetime of a patent is 20 years from priority date (or seventeen years from publication date in the USA), there are now *no* in-force patents anywhere from the countries and authorities covered by Derwent that are not covered by the online service.

Essentially, Derwent provides titles, home-grown abstracts, bibliographic information and in-depth coding for every patent taken from 32 patent issuing authorities. The patents are grouped into families, making family searching very simple, although not as extensive as INPADOC's. The great strength of the database is its consistent coding and extremely well written abstracts. This makes subject matter searching easier and more reliable on WPI than on any other service. Although not cheap, WPI online has to always be your database of choice for patent searching.

Derwent has just launched CD-ROM products based on WPI, offering abstracts with drawings. In addition, drawings from its abstracts are available online. It does offer special chemical structure searching software for chemical patent searching, and it also offers a product called PATSTAT for statistical analysis of patents.

One of the strengths of patents as sources is that they could be useful for techno-commercial intelligence. Because of the legal requirement to file a patent application if you want a monopoly, firms are forced to disclose through the patent literature their research interests. Patents are therefore widely used for the analysis of the performance of companies' – and even whole countries' – R&D effort. One can assess how productive a company is, what areas it is moving into and what areas it is moving out of, who the most productive inventors are and so on. This information is widely used for commercial intelligence purposes. By tracking patent applications, you can detect early new research priorities, can identify the productive inventors you want to head hunt, can identify who is likely to be your competition if you want to move into a new subject area and so on.

However, to achieve this aim, you need to be able to manipulate large-scale patent numbers and detect trends within those numbers. This requires the use of statistical analysis softwares, and Derwent has developed a PC software, specifically for use on its own databases, for analysing trends in patenting activity. Searchers run a search – say, on a particular company – and then analyse its patenting activity, produce histograms and pie charts, etc. Alternatively, you can pick a subject area, and break down the patentees by country, say. Much active competitor tracking takes place using this software.

The *OJ*, information brokers and miscellaneous patent information sources

The Official Journal is the official weekly journal of the UK Patent Office. All major Patent Offices issue similar publications. This publication records weekly the applications for patents, the applications published, the patents granted and the legal actions involving specified patents. It also notes all European patents that designate the UK. It is widely used by patent searchers for SDI purposes.

The UK and other Patent Offices will set up SDI profiles for clients, based on its own patent classification, and will routinely send such clients all new patents corresponding to the chosen profile.

The UK Patent Office will run batch searches based on its classification and will send a computer printout of the patent numbers which are hits. This is a fairly archaic system and the Patent Office is working on putting up its own database of its own patents online. In the meantime, it offers a way of searching UK patents only which complements the services Derwent and INPADOC can provide.

The Patent Information Network is funded by The British Library to ensure that certain major central reference libraries maintain good collections of UK and other major countries' patents and have staff who are knowledgeable in patents information.

Finally, there are various information brokers who specialize in running patent searches for clients. The most important of these are the British Library Online Search Service, the Patent Office itself and various independent patent searching organizations. The British Library will simply run straightforward literature and family searches, but both the Patent Office and some of the patent information brokerages will also run patentability searches, infringement searches and status checks.

Further reading and keeping up to date

The only recent informative texts on patents and patent searching are those by Auger (1992) and Van Dulken (1992). Kaye (1992) includes two chapters on patents and is well written. The key journal in the field is *World Patent Information*, published by Pergamon. The Institute of Information Scientists has a Special Interest Group for patent searchers. Its journal, unimaginatively entitled *Searcher*, is useful for keeping up to date with recent developments in patent information. The proceedings of the International Online Information Meeting and of the annual Chemical Information Conference organized by Infonortics usually contain a number of articles on patent information. Journals such as *Online* regularly run columns or articles on aspects of patents searching.

References

Auger, C. P. (1992) *Information sources in patents.* London: Bowker-Saur.
Kaye, D. (1992) *Information and business.* London: Library Association.
Van Dulken, S. (1992) (ed.) *Introduction to patents information.* London: British Library, Science Reference and Information Service.

CHAPTER FIVE

Journals, conferences and theses

C. W. BAILE

We have been living with the problems of the information explosion for some considerable time now. Despite some decrease in the number of publications during the recent recession, there is still a problem in seeking and collating relevant information to satisfy a need.

The three sources covered by this chapter are considered to be the most prolific in number. As this has been the case for some years, the documents themselves and the articles within them have been well controlled both bibliographically and through abstracting and indexing services.

One reason why these sources are held in such esteem is because they are the primary means for scientists and technologists to achieve status and recognition in the scientific and technical community. New information technology has created possible formats for more effective dissemination of ideas other than printed material. These formats are still in a nascent stage and, until they are accepted by the community, they will not achieve the same status as the traditional formats.

Journals

Surveys of engineers' information needs have generally found that use of journals by engineers is wide. Closer inspection usually finds that the type of journal: academic, professional, trade, etc. which is preferred varies according to whether the engineer is in research, production or commerce.

Of the many thousands of journals published, studies show that only about 10 per cent are used regularly and that these supply some 95 per cent of information needs. This group is known as the 'core journals'.

These studies, however, often consider only academic or professional journals. Studies of the use of trade journals are rare, despite the importance of this material for engineers. Anyone wishing to set up a service for a particular set of engineers would be best advised to contact a librarian in a similar field to determine the important literature.

Academic journals

The academic journal is regarded as a medium of great importance. Publication of a paper in such a journal is essential to gain recognition and esteem for both the writer and the writer's organization. In fact, academic journals will be read by only a small number of engineers, these usually being involved in research. Most academic journals cover a very specific subject area and so engineers involved in design or production would need to review regularly a great number of these. The use of articles in these journals is generally stimulated when the article is matched through a search in an abstracting and indexing service.

The contents of an academic journal generally consist of a series of papers, often quite lengthy and with an abstract and accompanying bibliography. The papers will report on research activities or case studies. There may also be a number of shorter papers, possibly interim reports, and letters or some other means for readers to comment on previous papers. Advertisements do not generally form an important part of this type of journal, although there may be calls for papers for conferences.

Publication is often monthly or bi-monthly. Each issue may contain a selection of papers relating to the subject area of the journal, or there may be special issues which focus on a very specific part of the area. Papers will be vetted by an editor or editorial board, so the quality is generally high.

Publishers of academic journals are generally commercial but in many cases the journal will relate to a prestige organization within the subject area. The relationship may be a joint venture or may only invoke the name of the organization through contacts with the editor.

Academic journals are very well covered by abstracting and indexing services so access to the information within should rarely cause a problem.

Possible titles range from the general, *Mechanical Engineering* for example, to the specific, *Journal of Constructional Steel Research*.

Professional journals

Journals published by professional organizations usually fall into two categories. The first is the organization's proceedings or transactions. Generally these have similar properties to those discussed in academic journals above. Their size varies from the small single part journal, for example the *Transactions* of the Institute of Quarrying to the comprehensive series such as the *IEEE Transactions*, the *Proceedings* of the Society of Photo-Optical Engineers and, to a smaller extent, the *Proceedings* of both the Institution of Mechanical Engineers and the Institution of Electrical Engineers.

The second category of journals published by professional organizations could be seen as a current awareness service to members. These journals include such contents as diaries of events, professional matters, current technology, updates on standards and legislation, library acquisitions and letters. There may also be useful articles overviewing areas of concern to members and these are generally written in a less formal style than possible equivalents in the transactions.

These journals are usually monthly publications with 50–100 pages. Advertisements may be included. The price is generally cheaper to members of the organization.

In many cases these journals will be of interest to engineers outside the immediate subject area of the profession. Some of these journals address the engineering profession as a whole, for example *Engineering Management* of the Institution of Electrical Engineers or *Professional Engineering* of the Institution of Mechanical Engineers.

Coverage of professional journals by abstracting and indexing services is patchy, although most major articles will be included.

Trade journals

The trade journal is a type of literature of great importance in trade and industry. There are many thousands of trade journals published throughout the world in a full range of engineering subjects.

An exact description of the contents of a trade journal would be hard to produce. The contents respond to the different kinds of information which is required to service the targeted industry sector. Our definition of a trade journal may also overlap with business journals as they usually contain a wide subject range of articles. Trade journals are usually published commercially and can act as a means of communication within a trade or industry. These journals may include some of the following types of information:

- original technical or scientific articles
- non-original technical or scientific articles
- statistics, market information
- price information
- product and process information
- general news of the trade or industry (e.g. profiles on people, contracts, conference details)
- abstracts or details of relevant articles in related journals or of patents or books
- directory information

- management and legal information
- advertisements for products or jobs
- political and economic information.

The typical trade journal is a monthly publication with 50–100 pages. Of these, more than half may be advertisements. The revenue from advertisers will account for part of the cost of publication. The trade journal may therefore be very cheap.

Technical articles in the journal will usually be reviews of specific technical topics, reviews of machinery, technical evaluations of machines, buyers' guides and specifications or original scientific and technical work. Non-technical articles may include features on management topics, labour relations, individual firms, organizations and personalities, legal and government information and historical material.

An abstract of an article is rarely supplied in a typical trade journal so, unlike academic journals, it is not usually possible to determine immediately the basic content and conclusions of more lengthy articles.

Most trade journals regularly include a 'current news' feature. Indeed, it is just this currency which makes the trade journal such a valuable source of information. Sometimes information of this nature is included in general news columns and technical digests. These sections may include also a vast range of different types of information such as obituaries, retirements, new appointments, price changes, contract announcements, new laws and regulations and the latest technical ideas.

Most articles in a trade journal do not provide bibliographic references but they may note relevant standards, regulations or trade literature for machinery or products from specific companies. This feature is valuable as there are no official listings of manufacturers' catalogues. Unless one is on a company's mailing list or subscribes to an appropriate 'technical package library' it can be difficult to obtain systematically the latest trade literature from a company.

As noted above, over half the contents of a trade journal may be advertising material. There is usually a 'reader's service' provided to put the reader in touch with the advertisers for more information.

The poor coverage of trade journals by abstracting and indexing services will be discussed later in this chapter, but at this stage it should be mentioned that such journals hardly ever produce their own indexes of articles. The exceptions tend to be those journals that are very technical in nature.

The large number of trade journals available means that it is possible in this chapter to give only a hint of what is available. Means of identifying appropriate journals covering specific countries and industries will be discussed later. The list gives a selection of trade journals which are available on the shelves of The British Library Science Reference and Information Service in London. The code in brackets at the end of each reference gives an idea of the country coverage of the journal.

- General
  The Engineer, monthly (international)
  Engineering, monthly (international)
- Chemicals
  Journal für Praktische Chemie, Chemiker-Zeitung, 11 per annum (GER)
- Computing
  Computer Weekly, weekly (UK)
  Computing, weekly (UK)
- Construction
  Building Services Contractor, bi-monthly (USA)
  Construction Equipment, monthly (USA)
  Construction Weekly, weekly (UK)
  HAC: Heating and Air Conditioning Journal, monthly (UK)
  Heating / Piping / Air Conditioning, monthly (USA)
  New Civil Engineer, weekly (UK)
  Roofing, monthly (UK)
  Surveyor, weekly (UK)
  Window Industries, monthly (UK)
- Defence
  Defence Helicopter, monthly (international)
  The Military Engineer, 7 per annum (USA)
- Electrical
  Appliance, monthly (USA)
  Appliance Manufacturer, monthly (USA)
  CEE News, monthly (USA)
  Electrical Wholesaler, monthly (UK)
- Electronic
  Electronic Engineering, monthly (UK)
- Energy
  European Power News, 8 per annum (international)
  LP Gas Review, monthly (UK)
  Pipeline and Gas Journal, monthly (USA)
- Hydraulic engineering
  Power International, monthly (international)
  Promofluid, 10 per annum (France)
- Machinery
  Industrial Equipment News, monthly (UK)
  Machinery Market, weekly (UK)
  Plant Manager's Journal, monthly (UK)
  Schweizer Maschinemarkt, weekly (Switzerland)
  What's New in Industry, 18 per annum (UK)

- Materials
  Ceramic Industry, monthly (USA)
  Ceramics International, 7 per annum (international)
  Coating, monthly (international)
  Concrete, bi-monthly (international)
  Engineering Distributor, monthly (UK)
  Foundry Management and Technology, monthly (USA)
  Interceram, bi-monthly (international)
  Iron and Steel Engineer, monthly (USA)
  Kunststoffe Plastics, monthly (GER)
  Metallurgia, monthly (UK)
  Modern Casting, monthly (USA)
  Modern Metals, monthly (international)
  Modern Plastics International, monthly (international)
  Rubber World, monthly (international)
  Steel Times, monthly (Europe)
  Welding and Metal Fabrication, 10 per annum (UK)
  Wood Technology, monthly (USA)
  Zeitschrift für Metallkunde, monthly (GER)
- Nuclear engineering
  Nuclear Engineering International, monthly (international)
  Nuclear News, monthly (international)
- Offshore engineering
  Offshore Engineer, monthly (international)
- Petroleum engineering
  Petroleum Review, monthly (UK)
  Pipeline and Utilities Construction, fortnightly (international)
- Sanitary engineering
  Aquatechnic International, 10 per annum (UK)
  Wasserwirtschaft, monthly (GER)
  Water Engineering and Management, monthly (USA)
- Transport
  Air Transport World, monthly (international)
  Aviation Week and Space Technology, weekly (international)
  Commercial Motor, weekly (international)
  Flight International, weekly (international)
  Interavia, monthly (international)
  Railway Track and Structures, monthly (USA)
  Service Station, monthly (UK)
  Wings, bi-monthly (North America)

The trade journal is a useful compendium of a variety of information. The information is current and the publication is usually cheap. For these reasons engineers use trade journals frequently and it may be the only form of scanning which they carry out. Information contained in trade journals is generally well researched but readers should be aware of the influence of the advertisers and any bias which may be the result.

Controlled circulation journals

A number of the trade journals listed above are available free of charge to specific groups of people, although they may be bought by anyone. For example, *Construction Europe*, is free to construction managers in EC and EFTA countries. These 'controlled circulation journals' are very common and becoming increasingly so. Controlled circulation journals are driven by advertising. The industrial advertiser needs to reach all those companies operating in industries where its products can be sold, and particularly the individuals within each company who are in a position to specify or purchase.

If the industrial publisher can be certain that his journal reaches the right establishments and the right individuals with quantifiable research or circulation information, and can prove all this to advertisers, then advertising may be able totally to support the journal and obviate the need for subscriptions. This is exactly the state of affairs with controlled circulation journals. It should be stressed that controlled circulation journals are not necessarily of lower quality than paid circulation journals, even though they may carry on average a higher percentage of advertisements. The advertisements are, of course, very important in such journals.

The caveat on trade journals bias above is particularly applicable to controlled circulation journals.

House journals

House journals are produced by industrial organizations, business houses, the public service and similar organizations for the benefit of consumers and/or employees. As with trade journals, the nature of house journals varies enormously from one title to another but they can be classified broadly into three types:

- prestige journals which are equivalent in almost every respect to standard academic journals, often these contain research papers, e.g. *IBM Technical Disclosure Bulletin*
- magazines containing non-technical information and possibly some mention of company products (but no technical details), e.g. *Gaslife*, British Gas Scotland
- periodical catalogues which are similar in all respects to trade literature (i.e. manufacturers' catalogues), e.g. *Elastomers' notebook*, Du Pont de Nemours International S.A.

House journals are generally free of charge or at least very inexpensive, but they are rarely included in journal directories and are infrequently included in abstracting and indexing services.

Newspapers

Several examples of journals in newspaper format are included in the trade journal list above (e.g. *Computer Weekly*), but to a limited extent general newspapers can be a useful source of engineering information. As most newspapers appear daily, new information can be published speedily. One disadvantage, however, is that few newspapers produce an index or are included in general engineering abstracting and indexing services. Tracing information on a specific product or technology can therefore be time-consuming and, more often than not, fruitless.

The *Financial Times* is a good example of a newspaper that covers product information systematically through its regular 'Technology' page. Information can be retrieved from the monthly and annual indexes published by Financial Times Business Information.

Newsletters

Newsletters are small, concise publications on specific subjects. They are designed to be current and to avoid the usual time-lag associated with more prestigious publications.

They are often produced by small organizations, or a unit within a larger organization. They can be very useful for noting new developments or concise summaries. They generally have a small distribution and therefore may be relatively expensive. Publication frequency varies enormously and may be intermittent.

Contents may include overviews, technical articles, reviews of products of interest, a 'stop press' section and a diary of events.

An increasing number of newsletters are available online. Such databases as PTS NEWSLETTER and MCGRAW-HILL ONLINE include a large number of newsletters in full text.

Their coverage by abstracting and indexing services is not thorough and if a reference is found it may be difficult to obtain the issue other than from the database supplier or The British Library Document Supply Centre.

Alternatives to traditional journals

The demands from readers for timely and easily assimilable information has led publishers to explore other formats in which to publish. Of the various approaches to the problem, the two most promising formats are the synopsis journal and the electronic journal.

SYNOPSIS JOURNAL

The idea of a synopsis journal is to prevent the reader from receiving more information than he/she is able to deal with. Two distinct ways to achieve this have arisen: either to send each reader selective papers according to a subject profile from a central collection of all papers; or to send the reader a summary of each paper received and then provide the full text on demand.

Experiments with synopsis journals in engineering have not been particularly successful. At present there are no synopsis journals in engineering. For two years, from 1977, the Institution of Mechanical Engineers published synopsis journals in engineering and production engineering. Although the readers gave the journals considerable support, the venture foundered because few authors were willing to contribute. Very few requests were received for the full text of a paper so the authors did not feel that publication gave them sufficient prestige.

ELECTRONIC JOURNALS

A true electronic journal, produced only in electronic form, has not yet been published in engineering. Indeed it is rare to find one in any branch of science and technology.

A more common occurrence is the publication of a journal in both hard copy and electronic formats. Although the electronic formats are known colloquially as 'full-text' this rarely means that the full journal is available. The advertisements, some of the product lists and buying guides and even illustrations may not be included. This is not helpful for the engineer searching for product information, but this form of searching should not be dismissed as there will be much of interest still remaining, such as comment, reviews and overviews.

All the quality newspapers are published on CD-ROM or are mounted online and most are available through FT PROFILE, the Financial Times host or as part of the TEXTLINE database. TEXTLINE includes also a number of local papers which may include product reviews or local company information. The US host DIALOG and the Swiss host DataStar have a large number of full-text sources available which include many newspapers, from Europe and North America. All the hosts may be accessed directly or the European Space Agency host, ESA-IRS, has set up a link between ESA-IRS and FT PROFILE. ESA-IRS is a very good host for engineering information so this link is a very useful tool for the engineer.

The number of journals which are available online is increasing quickly. Trade journals are not so well represented, but many of the journals of the professional organizations are available as well as a considerable number of newsletters including *Aerospace Daily*, *Byte*, *Chemical Engineering* and *Engineering News Record*. There are far too many to list and the interested searcher should contact the host helpdesks for the latest information.

Most hosts have acknowledged that full-text databases may not be searched by the experienced searcher. In many cases the hosts have set up a different approach to searching these databases so that the searcher is asked questions by a series of 'menus' rather than having to learn the command language of the host.

Sources of information about journals

A number of useful directories exists which compile details of a wide range of journals.

Benn's media (Benn Business Information Services) annual. The directory is produced in three volumes. The latest edition is the 143rd, 1995, and has entries for over 9000 journals, 2000 newspapers and 2000 directories.

The first volume covers the UK and has a number of sections including lists of newspapers, journals and house journals. Extensive details are given for each title listed. The journals section is arranged under broad subject headings. For example, under the heading 'Production engineering' thirteen titles are listed. A wide variety of types of journals are included, but there appears to be a considerable emphasis on technical and trade journals.

The second volume covers European journals and the third international journals. About 250 countries are surveyed in the 141st edition. The sequence of titles varies under each country. Newspapers are listed first but journals are then listed either under subject classifications (for major countries), or in a single alphabetical sequence, or are subdivided according to place and/or frequency of publication. The subject classification, when provided, varies from one country to another, reflecting the subject/industry bias of that country. Norway, for example, has 15 subject headings including such terms as 'Agriculture and forestry', 'Fisheries' and 'Handicrafts and Small industry'.

Standard periodical directory (Oxbridge Communications) annual. The latest edition is the 15th and lists over 70 000 publications in the USA and Canada. Types of journals included are consumer magazines, trade journals, newsletters, government publications, house journals, directories, transactions and proceedings of scientific societies, yearbooks, etc. The entries are listed in classified order and there is also a cross-index to subjects. For example, under the heading 'Metal and Metalworking' there are over 300 titles listed. A reference indicates that further related titles will be found under 'Machinery' and 'Mining and minerals'. The entries give not only full bibliographic details but also brief descriptions of subject coverage. At the end of the directory a title index is provided.

A particularly useful feature of the *Standard periodical directory* is a list of house journals. They are listed in a classified sequence under the term 'House organs' (the equivalent North American term). About 2000 titles are listed. Unfortunately there is no subject access to these titles, nor is there a company name index.

Ulrich's international periodicals directory (R. R. Bowker) annual. The latest edition is the 31st and is published in three volumes. The directory provides bibliographic details for over 112 000 journals of all kinds from most countries of the world.

The main section of the directory is a list of journals in classified subject sequence. A cross-index to the subject headings lists some 1600 keywords. For example, if the term 'Paving' is looked up in the cross-index, the reader is referred to relevant journals under 'Building & construction'. It is also suggested that titles listed under 'Transportation – roads & traffic' may be of use. A title index to the journals is included.

The directory is a very thorough, comprehensive and well laid-out work. Its only disadvantage for the user tracing trade journals on a particular topic is that a relatively small number of the total entries refer to these. This is due chiefly to the controlled circulation and elusive nature of many of this type of journal. The service is also available online through DIALOG, ESA-IRS, on CD-ROM and on microfiche.

Willing's press guide (Reed Information Services) annual. This is a particularly valuable bibliographic reference work for tracing trade journals and newspapers from throughout the world. There is a bias towards the UK, as this is the country of publication. The 1993 issue lists about 2600 UK newspapers and over 10 000 journals and annuals from the UK with many more from the rest of the world.

The first section of *Willing's press guide* has an alphabetical listing of UK publications together with bibliographic details, subscription details and circulation figures, etc. This sequence includes many different types of journals such as newspapers, trade journals, scientific journals and yearbooks. The second section is the 'Overseas section' and provides similar listings for individual countries other than the UK. A 'Classified index' to the publications is included. A list of over 300 subject index terms is provided, giving the titles and countries of relevant journals. For example, in the 1993 issue, 'Process engineering' lists two international titles, two for Australia, seven for Germany, three for the Netherlands, one for New Zealand and four for the USA. Clearly this list is far from comprehensive, but it is very useful nevertheless to have the information brought together in this way.

Willing's press guide also has a newspaper index for the UK, arranged by geographical coverage, and ends with a 'Services and Supplier' directory which lists the organizations under subject headings.

Abstracting and indexing of journals

Publishers of academic and professional journals frequently produce an index for this type of journal. They are also generally well covered by the abstracting and indexing services.

Publishers of trade journals, house journals and newspapers rarely produce their own indexes. It is particularly important, therefore, that the journals should be well covered by abstracting and indexing services so that information may be retrieved readily. Unlike, for example, a typical academic journal in a scientific subject, which might contain a small number of substantial articles that may be abstracted succinctly and indexed, trade journals, house journals and newspapers tend to contain a vast number of minor articles. Indexing and abstracting such articles is an enormous task and may be aggravated even further by the difficulty of finding suitable index terms. It is not surprising therefore that existing abstracting and indexing services are far from ideal. Many titles are not covered at all and others are only partially covered. Titles originating from the USA are, in general, better catered for than European titles, and the rest of the world is covered even less well. House journal coverage is minimal, even when journals contain substantial technical information.

Many abstracting and indexing services are available in electronic format for searching online. This method of searching offers a number of advantages over manual searching:

- online searching is very much quicker than manual searching
- a computer file may be searched more thoroughly than the printed alternative
- complex combinations of concepts may be searched for
- the end result is a neat printed list of references which in many cases includes abstracts.

Of course, online searching is not always preferable to manual searching. Sometimes an appropriate database does not exist, but most of the major abstracting and indexing journals are available online and many services appear in electronic form only.

Ulrich's international periodicals directory includes in its entries the service which abstracts or indexes the journal. Some directories exist which also note the online services that index particular journals, e.g. EBSCO's *Serials Directory* on CD-ROM.

Directory of periodicals online: indexed, abstracted and full-text, science and technology (Federal Document Retrieval) annual. The directory includes the journals covered by over 600 databases. The list of periodicals includes over 19 000 titles. The directory includes journals which are either abstracted or included full-text on the databases.

A large number of abstracting and indexing services exists in engineering. Some of these cover all areas of engineering, others are quite specific in their coverage (see Chapter 7 and later subject chapters).

PTS PROMPT (Predicasts) online. This service has a particular emphasis on trade journals and is therefore valuable for a comprehensive search of literature. The service is available through DataStar, DIALOG, FT PROFILE, Questel (now Questel.Orbit) and STN.

McCarthy Cards (McCarthy Information). Rather than an abstracting or indexing service this is a press cuttings service. Over forty newspapers (including a few weeklies such as *The Economist*) are scanned and all relevant news and comment articles are photocopied and sent to the subscribers. Each article has the publication source information included. When further information is found, a replacement sheet is issued carrying both the current and the earlier information. Services available cover companies in the UK, Europe, North America and Australia. In addition, there is a general 'industry service' which monitors over sixty areas of industrial activity. Information extracted for this service includes trade and commodity statistics and comment on new developments. The service is also available online through FT Profile and on CD-ROM.

Conferences

Engineers are said to be poor users of information but one source which they do enjoy is conferences. This is generally attendance at a conference rather than reading the later proceedings. Conferences provide a useful forum both to listen to and discuss new developments and ideas given in the conference papers. They also allow engineers to meet informally but in a structured context to make contacts with peers and discuss subjects which they themselves are working on.

A conference may be called to highlight a particular area of concern. Out of this a new multi-disciplinary subject may arise to be served by regular meetings and a journal or newsletter.

Generally, information given at a conference is current. A conference is often used as the forum for disclosing new theories or products. Some conferences are accompanied by exhibitions which give the delegates an opportunity to view new products and services related to their work.

Many conferences are regular events, either annual or biennial. There is also a range of one-off conferences or regular conferences with a general title which focus on a specific topic each time. Subject coverage may range from the well-defined to an overview of new developments in a broad discipline. Conferences may also range from the small specialist seminar to the large prestigious event with thousands of delegates.

The organizers of conferences include professional organizations, academic institutions, government agencies and commercial organizations.

Conference proceedings

Although information given at conferences is generally current, it may be some time before the proceedings are published. By that time it is possible that some of the papers may have been published in journals. This paper may either be similar to the conference paper itself or shorter, fuller, more detailed, accompanied by the workings of research or with results from a later stage. This duplication may lead to some confusion when both references are found in an indexing search.

Proceedings may contain just the papers, or sometimes only a selection of papers, given at the conference. In some cases, only abstracts will be available for some or all of the papers. A more complete set of proceedings should also contain summaries or transcripts of any discussion sessions. These sessions may be taped and attached to the proceedings. Some proceedings also contain papers sent to the organizers on topics related to the conference but which are not formally presented by a speaker. These are known as posters.

Proceedings may be published as one-offs, a part of a series of proceedings or, sometimes, included as a special report in a journal.

Sources of information about forthcoming conferences

Many conferences are announced either in advertisements or in the diary sections of professional or academic journals. If there is a call for papers, this may precede the conference by anything up to a year and allow people not presenting a paper to note the date in their diaries.

Several useful services exist to inform people of forthcoming conferences, of these the most comprehensive is the *World Meetings* series.

World Meetings: United States and Canada and *World Meetings: Outside United States and Canada* (Macmillan, Inc.) monthly. The entries are listed in six-month sections for two years after the publication date of the issue. Each entry is given a unique number, which reflects the sequence in which information on conferences is received by the publishers. The entries are then listed in ascending order of this unique number. Each entry gives conference title, date, location, sponsor, contents, deadlines for papers, expected availability of proceedings and, if one exists, details of an accompanying exhibition. There are indexes by conference title, keyword, date, location, publication, deadline for papers and sponsor. The service is relatively comprehensive for major events, particularly in North America and Europe, but naturally can be patchy for smaller events.

Abstracting and indexing of conferences

Many of the standard abstracting and indexing services include conference proceedings as sources to be covered. In particular, the database COMPENDEX*PLUS has good coverage. In addition, several services exist which concentrate solely on conferences.

Conference Papers Index (Cambridge Scientific Abstracts) seven times a year. Conference titles are listed under broad subject headings and then papers are listed below each title. Each issue has a subject index which is cumulated annually. The service is available online through DIALOG, ESA-IRS and STN.

Directory of Published Proceedings, series SEMT (Interdok Corporation) ten times a year. The main list is in accession number order and there is a combined subject and sponsor index and an editor index.

Index of Conference Proceedings Received (British Library) monthly. This service lists the conference holdings of the British Library. As the library places a particular emphasis on this form of material, its holdings are very comprehensive. The index includes conferences only by the title but they are listed under keywords, so one conference may appear in several places. The lists are cumulated annually. This service is also available online through BLAISE-LINE and on CD-ROM.

Index to Scientific & Technical Proceedings (Institute for Scientific Information) monthly. The main list gives the title and details of the conference, including bibliographic, meeting location and sponsorship information. Each entry then lists the papers giving title, authors and their corporate affiliations. The indexes include: author and editor, sponsor, meeting location, corporate affiliation by name and geographic location, and a permuted index of significant words from the titles of both the conferences and the papers. The service is available online through BIDS and part of the service is included in the IST&B Search database on DIMDI.

Theses

Theses, or dissertations as they are called in North America, are results of a student's studies in advanced higher education. They usually contain original research and are often on a very specific subject.

The increasing sponsorship of academic research by industry means that the results are often of considerable importance for practising engineers. The results may be published later in a journal article or book but much is only available in the thesis.

Published theses

The bibliographic control of theses is well covered by Davinson (1977). Access is generally available either through The British Library Document Supply Centre or by applying to the awarding institution. Many series of these are published on microfilm by such organizations as University Microfilms International (UMI).

The thesis should contain all the research results in an extended analysis. There will be an abstract and extended list of references.

Sources of information about theses

Information about which theses are being written may be available from the awarding bodies or such organizations as the Science and Engineering Research Council. Alternatively it may be possible to deduce the general subject areas by finding out what research is being undertaken at an academic organization.

Current research in Britain, physical sciences (Longman Cartermill) annual. The main entries give the titles of the research projects under the academic organization. There is an index to research project titles by keyword in a separate volume.

In addition, Longman Cartermill publish a number of guides to research centres around the world. These do not name individual research projects but give an indication of which subject areas concern the organizations.

ABSTRACTING AND INDEXING OF THESES

There are three important services which abstract or index theses.

British Reports, Translations and Theses (BRTT) (The British Library) monthly. This is primarily an indexing service for grey literature but it includes UK theses received by The British Library Document Supply Centre. The main entries give brief details of the theses and there is a monthly subject and author index which is cumulated annually. The contents are available online on the SIGLE database through BLAISE-LINE.

Dissertation Abstracts (UMI) monthly. Section B is titled 'Science and engineering' but only covers North American universities. The service includes an increasing number of European theses in a separate Section C. The main entries include the abstracts from the thesis and are arranged by subject. There is also an author index. The service is available online through DIALOG, DataStar and STN and on CD-ROM.

Index to Theses Accepted for Higher Degrees by the Universities of Great Britain and Ireland and the Council for National Academic Awards (Aslib) quarterly. Curiously the title does not state that the entries are arranged in broad subject headings with an author and subject index. Abstracts to the theses are usually included.

References

Davinson, D. (1977) *Theses and dissertations as information sources.* London: Clive Bingley.

CHAPTER SIX

Product information

P. J. RUSTON

In BS 4940: 1973 a product is defined as 'any material or item of any complexity from raw material to a complete engineering system, which is offered for sale. The term includes services having no material content.' There is thus a massive amount of product information through which an engineer must sift in order to find relevant data. A need exists for information concerning the components and materials of a firm's own products and perhaps more importantly of its competitors. The engineering firm without easy access to information on products immediately places itself at a disadvantage in the marketplace. It is imperative that a company be fully aware of its competitors' products and of developments in the industry. An engineer requires the capacity to be able to develop a product so as to maintain and improve its market standing. The most efficient way that this can be achieved is through a regular supply of product information.

It is not only in industry that we find a need for product information but also in everyday life. Whether considering which make of computer to buy or where to obtain that elusive spare part for our washing machine, we are constantly evaluating products. However, while recognizing its relevance to us all, in this chapter I shall consider the importance of product information to industry.

Engineers and designers need to be freed from the frustrations, delays and difficulties entailed in finding and selecting product information from an increasing mass of data. With ready access to product information they may be allowed more time to be professional engineers and designers rather than information researchers. In outlining the major sources of product information available I hope to indicate the process by which engineers can best obtain the data required to maintain and improve the position of their firm in the market. When evaluating product information it is possible to identify both primary and secondary sources of data.

Primary sources

The most obvious starting point for anyone interested in product information is with the manufacturer. Primary sources of product information are those pieces of data generated by firms themselves. They cover a broad range of material including the following:

- research documents and test certificates
- catalogues
- brochures
- advertisements
- handbooks
- data sheets
- price lists
- specifications
- annual reports and reviews
- house journals

From the listing above we can see that primary sources of product information encompass more or less all material published by a firm about its own products. Quite often this category of material is referred to as 'trade literature'. Indeed, a fair definition of the term trade literature would be 'primary sources of product information', although in recent years this has been broadened to incorporate the financial details of a firm.

It is apparent that trade literature comes in a wide variety of shapes and forms of material. This can range from the flimsy advertising flyer to the weighty product catalogue running to several hundred pages. Major problems with this type of material are its variety and lack of standardization. There is a British Standard relating to the production of trade literature, specifically in the construction industry, but this is rarely adhered to. Although its aim is to 'help manufacturers, trade associations and research organizations to provide specifiers, contractors, operatives on the site and building owners with the right type of information at the right time and in the right place ...', it is unlikely that many manufacturers are aware of the existence of BS 4940: 1973.

Product catalogues and handbooks are the most extensive sources of product information. Generally they will include precise details of the whole range of a firm's products. A well produced catalogue contains data sheets, range lists, samples and colours, price lists, design data, specifications, repair and servicing, a supply guide and a technical services guide. The loose-leaf design of the catalogue/handbook enables data sheets to be easily added or withdrawn. The data sheet provides concise information about one particular product allowing the prospective purchaser to investigate its suitability.

Primary sources contain detailed technical specifications which allow the engineer a complete insight into the workings of the product. The two exceptions to this rule are annual reports and house journals. These are more likely to advertise the existence of a firm's new products rather than provide detailed technical data. Most annual reports, for example, highlight new products that a firm has launched during the previous year.

The house journal is one of the most important sources of product information available and yet it remains one of the most underused forms of material. This is largely due to a lack of awareness on the part of the user and partly due to a lack of availability. Most firms of any repute publish a house journal although these do vary tremendously in quality. House journals, by their very nature, are intended primarily for internal circulation within the firm. However, as well as being circulated to staff the journal is usually sent to customers and quite often to competitors as well. It may seem strange for one firm to allow competitors access to its own journal but this happens regularly, particularly in areas such as the oil industry. It is unlikely, therefore, that you will find information in a house journal that is confidential. Nevertheless you will discover details of a firm's products before they are published in the trade press. Many house journals, such as the General Electric Company's *GEC Review*, are professionally produced containing detailed technical specifications. Each issue usually contains product news, new contracts, test details, product files and reviews. As with trade journals, many contain business reply cards that can be returned to the firm in order to obtain further product details or a visit from a representative of the firm. Despite being freely available, access to house journals can be a problem. The British Library Business Information Service has a collection of around 2000 of the better quality house journals but there are few other publicly available collections. The British Association of Industrial Editors (BAIE), who coordinate the publishing of house journals in this country, also has a sizeable collection. As with all primary sources of product information, the best way of obtaining copies is to ask the firm if you can be placed on their mailing list for house journals and other relevant documentation.

Secondary sources

In the event that it becomes difficult to obtain information directly from the manufacturer it is necessary to access secondary sources of product

84 Product information

information. A third party, usually a commercial publisher, will report on products via either printed or non-print media.

Printed sources

It is possible to build up a portfolio of product information for a particular firm using the following printed sources:

- trade directories
- trade journals
- exhibition catalogues
- trademarks
- patents.

TRADE DIRECTORIES

In order to identify the manufacturer of a given product it is most effective to consult a trade directory. According to the latest edition of *Directories in print* (1994) (Gale Research) there are over 15 000 directories worldwide. A healthy proportion of these directories cover industry, ranging from those of international and national coverage to those highlighting very specific industry sectors. Trade directories are diverse in shape and format but ideally should contain a list of names and addresses of manufacturers. An increasing proportion of directories provide details of suppliers, distributors and agents as well as manufacturers. The quality of the major directories is continually improving with features such as lists of key personnel and financial data being added to the basic contact record. The financial data are particularly important in considering from whom a product should be purchased. Most directories are compiled through the use of a questionnaire. It is obvious, therefore, that the more detail a firm can provide the greater the possibility a prospective customer will make an initial enquiry concerning the products available.

The most important general trade directories are listed below:

Worldwide. World business directory 3rd ed. (1994) (Gale Research) 4 vols. Produced by the World Trade Centers Association, this directory contains detailed information on more than 100 000 businesses involved in international trade. Volume 4 has very useful product, industry and alphabetic indexes. If you are looking for the manufacturers of stainless steel rods in Guatemala then this is the place to find them.

North America. Thomas register of American manufacturers (Thomas) annual, 26 vols. This major listing of American manufacturers is separated into three parts: volumes 1–16 list manufacturers by products and services; volumes 17–18 contain company profiles; and volumes 19–

26 form a catalogue file. In these latter eight volumes there are 11 000 pages of catalogues from nearly 2000 companies giving detailed product specifications, drawings, photographs, availability and performance data.

Europe. Dun and Bradstreet Europa (Dun and Bradstreet), annual, 4 vols. A pan-European source of data on 60 000 companies with a listing of manufacturers arranged by SIC code.

United Kingdom. Key British enterprises (Dun and Bradstreet) annual, 6 vols. Contains information on the top 50 000 companies in the UK with cross-referencing by town and county as well as by industry.

'*Kompass*' (Reed Information Services) annual, 5 vols. Now in its thirtieth year, *Kompass* continues to provide unrivalled access to about 42 000 UK manufacturers. Vol. 1, products and services, has a very specific classification system enabling the user to highlight manufacturers, distributors, suppliers, importers and exporters of 41 000 different products. Kompass also produces volumes for most other European countries.

Once general trade directories have been exhausted it may be necessary to extend the search further by consulting industry-specific sources. Engineers should be aware of the following specialist trade directories:

Engineer buyers guide (Benn) annual. Benn are responsible for publishing some of the leading UK trade directories of which this is a typical example. Considered to be the definitive guide to products and services, and with a particularly detailed section on machine tools, the *Buyers Guide* is now in its 95th year. The directory also contains valuable information on foreign companies and their UK agents; associations, institutions and societies.

Dial engineering (Reed Information Services) annual. Circulated annually to engineers, buyers and purchasing officers as well as other key purchasing decision makers. The directory combines detailed listings of products and services with a time-saving reader reply section.

EEF directory: the voice of engineering (Guardian Communications) annual. Contains a classified index of Engineering Employers Federation members' products and services.

European electronics directory (1993) (Elsevier Advanced Technology). Along with Benn, Elsevier are leading directory publishers in the UK. This particular example contains details of 10 300 European manufacturers and representatives and has a classified section dealing with 1600 products. It is designed to be useful for managers selling to and buying from the electronics industry in Europe.

DABS compendium (MBR Publications) annual. An A–Z compendium of building products designed to help the busy specifier find first-stage technical information about the materials of the leading building industry suppliers. There is a free reply card service to help contact the manufacturers listed in the detailed classified products section.

RIBA product selector (RIBA Services) annual. This three volume work names approximately 7000 firms with addresses, telephone numbers, trade names and types of product supplied, arranged in 496 product groups. RIBA Services also publish *RIBA product data*, a library of structured product information containing over 2000 pages of product data sheets, cross-referenced from the *Product selector*.

There are numerous other trade directories relating to even more specific aspects of engineering. Examples of these include *Process engineering directory* (Benn) and *Instrument engineer's yearbook* (Institute of Measurement and Control). There are also those directories covering areas linked to engineering such as construction. Indeed the *ASC* (Wilmington Publishing) directory of building materials is a prime example. Now a sizeable three volume work, *ASC* proudly proclaims to be 'the only desk-top, ready-made product literature reference system'. ASC has a comprehensive index of suppliers as well as detailed product listings.

In order to locate the trade directory most likely to satisfy your individual needs it is worth consulting any of the publications listed below:

Trade directories of the world (1990) (Croner) loose-leaf.
Top 3000 directories and annuals 11th ed. (1994) (Dawson).
Directories in print (Gale Research) annual 2 vols. An annotated guide to over 15 000 directories published worldwide.
Current British directories 12th ed. (1993) (CBD Research). Guide to over 14 000 directories published in the British Isles.

Each of these publications has a degree of subject access to the directories listed therein.

TRADE JOURNALS

Most quality trade journals contain a section dedicated to product information. Quite often this takes the form of several pages devoted to new product developments. There is usually a reader reply service within each issue allowing further information to be obtained on particular products if required. Using journals in this way can be a cost-effective means of getting onto a firm's mailing list for trade literature. Many journals publish an annual buyers' guide that often acts as a useful trade directory. Regularly scanning trade journals such as those listed here is to be recommended:

Engineering Distributor
What's New in Industry
Machinery Market
Professional Engineering

What's New in Processing
Turbomachinery International
Industrial Equipment News
Engineering World
Process Equipment News
Manufacturing Engineer

The series of 'What's New ...' journals published by Morgan-Grampian is an excellent source of free trade literature. Each issue is devoted to product information containing a datafile, product locator and details of how to obtain free catalogues. For example, *What's new in industry*, which calls itself the 'guide to products and equipment', has incredibly detailed product finders and issues regular supplements such as 'What's new in materials and finishing'.

This is only a small selection of the many trade journals available. It is imperative that you be familiar with those journals published in the area of your industrial activity. The trade journal is a vital source of current awareness and an invaluable means of obtaining free product information.

By their very nature most trade journals concentrate on a specific industry. There are, however, a few journals that monitor product developments across the whole industry. A prime example here is *World's New Products* published monthly by World Business Publications in London. Each issue highlights new products being brought to the market in a range of sectors. These industries include chemicals and materials, industrial equipment, testing and equipment, electrical, and control and instrumentation. Each entry has a description of the product and the contact details for the manufacturer.

Another example is *New Products* published by the distributing firm of RS Components Ltd in Corby. Each issue reviews new products and considers applications, trends and innovations in particular sectors such as semiconductors. As well as providing a useful means for product sourcing, these journals can be an extremely entertaining read.

EXHIBITION CATALOGUES

Tracking down firms operating in smaller industry sectors can prove a time-consuming, wearisome and often unproductive task. While general trade directories such as *Kompass* are very good, occasionally it is worth consulting exhibition catalogues in order to identify manufacturers of a particular product. The guide to Interbuild, the International Building and Conference Exhibition, for instance, contains an extensive classified index. Using the Interbuild catalogue it is possible to locate manufacturers of everything from edgebanding to anti-vandal fittings. Likewise, specialist exhibitions such as Firesafe, Fire Safety Equipment, produce catalogues with detailed listings of firms operating in the industry. If a firm has gone to the effort of promoting its products at an exhibition, particularly a major trade fair, it is safe to assume that the firm is in a sound financial position and may well be worth conducting business with.

TRADEMARKS

The term trademark is used to cover both symbols and words, although words are often referred to as trade names. It is usually the case that the only reference appearing on a product is the trademark and thus it assumes strategic importance in tracing the manufacturer. Similarly, the manufacturer may choose to register for a trademark in order that a new product may be readily identified by the purchasing public.

In both instances, the prime source of information is provided by the Trademarks Registry. Part of the Patent Office, the Registry conducts searches for registered and pending British trademarks.

An alternative to contacting the Trademarks Registry is to carry out your own search of the trademark directories and registers that are available. The leading reference sources are listed below:

Industrial trade names (Reed Information Services) annual. Includes sections on trade names in the UK, company information, lapsed trade names and agencies.
Trademark register of the US 35th ed. (1993) (Trademark Register). Contains a detailed classification of trademarks in the USA.
Companies and their brands (Gale Research) annual, 2 vols. Covers 47 000 US companies with 255 000 trade names.
International brands and their companies 4th ed. (1994) (Gale Research). Over 97 000 international products and their manufacturers, importers and distributors with addresses.

There are, of course, directories of trade names relating to specific industries. A typical example is *New trade names in the rubber and plastics industries* produced by RAPRA. Many of the trademark directories are now available via online databases, such as TRADEMARKSCAN for US trademarks, and it may well prove more cost-effective to search for trademarks online. Alternatively, it may be worthwhile contacting particular industry libraries or associations for information on trademarks. The Building Centre in London, for example, has details of around 80 000 trade names specific to the construction industry.

PATENTS

Patents are an invaluable and much underused secondary source of product information. There are in excess of 30 million patent specifications and each provides detailed product descriptions. The main purpose of a patent is to provide enough technical information so that anyone can manufacture the product. However, it is possible to use patents as an effective means of detecting manufacturers in certain industry sectors. With the increasing availability of patent literature on CD-ROM this is opening up a whole new avenue for locating product information.

STANDARDS

It is increasingly important for manufacturers to establish that their products are fit for purpose before market launch. In this country products should meet the standards required by the British Standards Institution. If the standard levels are met it gives the manufacturer a stronger competitive edge. It is essential, therefore, that engineers be aware of product standards before any commercial applications are considered. A prospective purchaser will give more credence to a manufacturer whose product has passed the rigours of standards testing and evaluation. Standards form an important part of the product information picture. Again their availability on CD-ROM is bringing heightened awareness of the technical information contained in standards.

Non-print sources

Four categories of non-print sources of product information can be identified. These follow a natural progression from verbal, machine-readable and online sources through to the most recent development of compact disc (CD-ROM). The CD-ROM has revolutionized all areas of the information industry in providing easy access to vast stores of data at a cost-effective rate. Thankfully CD-ROM publishers have not been slow to exploit the area of product information.

VERBAL

Into this category can be placed visits from firms' representatives and attendance at exhibitions and trade fairs. An enquiry to a manufacturer regarding a particular product may well solicit a visit from a representative in order to discuss its merits in greater detail. The expert knowledge the representative brings can save time and investment in a product not entirely suited to the purchasers' needs.

Attendance at exhibitions is a useful way to become aware of the latest products coming onto the market. For instance, at the annual International Building and Construction Exhibition, Interbuild, it is possible to meet buyers, specifiers and manufacturers representing some 12 000 firms from over seventy different countries. This can be an effective means of making contacts within the industry at the same time as keeping abreast of new developments.

Product information

MACHINE-READABLE SERVICES

A large proportion of product information can still be found on machine-readable services. The information usually appears on microfilm cartridges or microfiche and consists primarily of manufacturers' catalogues. These so-called 'package libraries' were introduced in the 1970s to meet the demand for ready access to trade literature. The leading supplier in this field remains Technical Indexes Ltd. TI are currently offering the following product catalogues on microform: Construction and Civil Engineering; Electronic Engineering (including Rapid Update); Engineering Design and Manufacturing; Process Engineering; and Taiwanese Catalogue File. The services provided by TI are on an annual rental and include comprehensive indexing and regular updating. With sales in excess of £11 million and pre-tax profits of £5 million (1991 figures), TI has turned the provision of product information into a very lucrative business.

The initial attraction of 'package libraries' was in the amount of space they saved in contrast to maintaining a library of hard-copy catalogues. Technology is quickly overtaking these services and they now appear cumbersome in comparison to CD-ROM. While a large number of firms still make use of the microfilm cartridges provided by TI it is only a matter of time before they are rendered obsolete by the arrival of the CD-ROM services.

CD-ROM SERVICES

The decline in machine-readable sources of product information has been heralded by the appearance of CD-ROM systems that can retrieve far larger amounts of data in a much quicker time. The huge storage capability of CD-ROMs has been harnessed to bring an unrivalled warehouse of up to 12 000 original scanned A4 pages of text onto each single 5″ disk. The equivalent of 270 000 A4 pages can be stored on one disk and the retrieval of information takes a matter of seconds. The CD-ROM technology is of immeasurable value to engineers as a time-saving device. In the past, hours could be spent sourcing the right component and now the process can take little or no time at all.

Once again it is Technical Indexes who are leading the way with the new technology. At present there are two TI product/supplier catalogues available on CD-ROM: the US VENDOR CATALOGUES/MASTER DIRECTORY and the UK CONSTRUCTION AND CIVIL ENGINEERING INDEX. There are many other related systems entering the market, including BRITISH STANDARDS on CD-ROM, and it is only a matter of time before the complete range of microfilm services is accessible via the new medium. As with TI's microfilm services, the CD-ROMs contain comprehensive indexing and are updated regularly. In addition to retrieving information far more quickly than manual or microfilm sources, the

CD-ROM offers the advantage of being extremely user-friendly. No extensive training is required and there are not the prohibitive pressures of connect-time associated with searching online databases. Finally, the CD-ROM workstation is very neat thus saving premium floorspace.

ONLINE SERVICES

There are many online databases that provide details of manufacturers and suppliers of products. Quite often these equate to hard-copy trade directories, such as KOMPASS ONLINE and the THOMAS REGISTER ON-LINE. Using online databases can be a very speedy and effective means of tracking down the manufacturer of a particular product. The databases can also be used for generating lists for mailshots and marketing exercises. There are, however, few databases that offer trade literature online.

One such database is CODUS, available via Codus Ltd in Sheffield. Aimed at 'freeing you from the tyranny of product lists and data sheets', CODUS contains technical information on approximately 90 000 quality-assured electronic components. The database includes the description, limiting conditions of use, characteristics and post-conditioning limits of the components. Data on all components which comply with CECC, BS EN ISO 9000, and IECQ specifications are supported by comprehensive MIL 217E and HRD5 failure-rate prediction, to give a complete picture in one source. There are several files on the CODUS system. CODUS PLUS is the main database offering access to components by manufacturer's part numbers, generic type numbers, NATO stock numbers and specifications. The Electronic Parts Information Centre, CODUS EPICTM, is the UK MoD's preferred products list of electronic components, available to all companies who hold current MoD contracts. And finally, CODUS SIRE, the Semiconductor Index of Radiation Effects database, which is offered to MoD authorized users only. CODUS provides a vital service to leading aerospace and electronics companies and defence establishments. It is used by many electronic equipment manufacturers in selecting components at the design stage, producing and updating company preferred lists, and identifying alternative sources of supply.

The USA is particularly well served by sources of product information via online databases. The GIDEP database, Government-Industry Data Exchange Program, produced by the US Navy is very similar in content to the CODUS system in the UK. It contains four files of non-classified technical data on design, development, production and operation of all types of equipment and systems from lawn mowers to computers and aircraft. GIDEP consists of four separate files, the Engineering Data File, Reliability – Maintainability Data File, Metrology Data File and Failure Experience Data File. The Engineering Data File has evaluation and qualification test reports, materials specifications and manufacturing process. The data are submitted by users in industry and government agencies, including approximately 1300 military, aerospace and commercial industrial organizations. The service is only available direct from the GIDEP Operations Center in America.

Thomas Online are responsible for two databases essential for those with an interest in product information. The first, THOMAS REGISTER ONLINE, equates roughly to the hard-copy trade directories and holds details on 152 000 US manufacturers. The second, THOMAS NEW INDUSTRIAL PRODUCTS DATABASE, contains descriptive and technical information on more than 77 000 new industrial products and systems worldwide. The data include products and trade name, synonyms, model number, SIC, features, performance specifications, uses and price. Although there is a US bias the database can be used to source new products and systems for all types of industrial applications from adhesives and sealants, electronics, materials handling, metalworking machinery and sensors to wrenches. Both databases are available on the host system DIALOG.

Another invaluable online database for product information, accessible via both DIALOG and DataStar, is PREDICASTS NEW PRODUCT ANNOUNCEMENTS/PLUS. NPA/PLUS contains the complete text of approximately 200 000 press releases from more than 15 000 firms on new products and technologies. The database gives announcements of new products, product modifications, and new technologies and processes from manufacturers, distributors and services in nearly sixty industries. Each record includes product description, specifications and applications; information on trade names, prices, model numbers, availability and licensing agreements. Information may be retrieved by searching product codes based on SIC, product trade name, uses and applications and special feature codes indicating discussions of price or performance specifications. Around 800 records a week are added to NPA/PLUS and, despite its US bias, it is an extremely useful means of keeping up-to-date with the latest technological developments in a given industry.

In the UK, KOMPASS ONLINE remains the leading database for identifying manufacturers of a particular product. Over 160 000 firms are held on the database arranged in 45 000 categories of product in the business-to-business sector. Apart from the *Kompass* trade directories the data are collated from other sources including *Kelly's Directories*, *Directory of Directors*, *Dial Industry*, *British Exports* and *UK Trade Names*. As with many other online databases *Kompass* is also now available on CD-ROM.

Finally, it is not possible to cover online sources without mentioning the COMPENDEX database. COMPENDEX is the single most comprehensive source of data for engineers, covering as it does more than 2.8 million citations from over 4500 journals, books and reports. Produced by Engineering Information Inc. in the USA the database covers all areas of engineering. Accessible via DIALOG and DataStar, COMPENDEX provides essential background information for engineers on a broad range of technical subjects.

Using online databases for product information can seem expensive as most host systems charge both connect-time and a per record cost. However, when set against the amount of research time saved, this type of searching can be very cost-effective. Online databases can be particularly useful for sourcing manufacturers of certain materials or components and can also be used for producing mailing lists as part of a promotional campaign for a firm's product. Searching files such as COMPENDEX and PTS NPA/PLUS on a regular basis is essential for current awareness. This can be an effective means of maintaining a competitive edge over rival firms and is more cost-effective than scanning journals manually.

Trade literature collections

Wall (1986) spoke of the need for a Central Product Information Body. As the doyen of writers on the subject of product information, obviously Wall's views have to be considered seriously. However, the idea of such a body was probably overambitious. It is doubtful if there would ever be the demand to justify the expense of maintaining such a collection. As the national library, the British Library should and does take a leading role in maintaining a collection of trade literature. With over a million businesses in the UK it would be impossible to maintain a comprehensive collection of trade literature. If the British Library can continue to house a representative sample of the UK's output of trade literature then this will be the nearest equivalent to a Central Product Information Body.

British Library Science Reference and Information Service

The BL houses its large collection of trade literature in the business section of the Science Reference and Information Service. The collection consists of product catalogues, brochures, data sheets, annual reports and house journals from in the region of 25 000 firms. The coverage is mainly UK but the selection guidelines have recently been broadened to incorporate the trade literature of US, European and Japanese firms. There is also an archival collection of trade literature containing material from around 7500 firms covering the period 1840–1940. This collection is particularly useful for those attempting to restore old materials or buildings to their former glories. There is a degree of subject access to the collections so that it is possible, for example, to identify manufacturers of miners' lamps in the 1890s without too much difficulty. If it is impractical to visit the collection in London, the Business Information Service, responsible for maintaining the collections, will answer enquiries by mail, fax (0171 412 7453) or telephone (0171 412 7454).

Science Museum Library

The Science Museum Library has a fine collection of archival trade literature, second in size only to that of the British Library. The collection contains material from several thousand UK manufacturing firms. However, the collection is not being extended at the present time. It is to be hoped this situation will soon be rectified.

Business Archives Council

The Business Archives Council coordinates the archiving of trade literature in the UK. In addition to running seminars on business archives and publishing a regular journal and newsletter, the BAC offers advice to firms on how to operate their own libraries. Indeed, most manufacturing firms of a reasonable size maintain their own in-house library of trade literature. These collections can be quite substantial and are often accessible to enquirers from outside the firm. As an example, Taywood Engineering Ltd has an information centre in Southall covering the field of construction and civil engineering. As well as trade literature, Taywood has over 20 000 volumes and 500 journal titles in its library. Similarly, Brown and Root Ltd has a technical information centre in London containing data relating to the energy sector, the oil and gas industry, and engineering and construction. While these centres are primarily for internal use they can be an important source of trade literature for the external enquirer.

PBV Consult

There are several former libraries of the Property Services Agency that contain useful collections of trade literature. One of these is that of PBV Consult, based in Edinburgh, which covers civil engineering, interior design, mechanical and electrical engineering. The former headquarters library of the PSA, based in Croydon, is now operated by the firm TBV. The libraries of Unicorn Construction Services Ltd. and Building Management South and West Ltd., formerly PSA Building Management South and West, are based in Bristol and cover the areas of building design, electrical and mechanical engineering.

Building Centre

Based in London and set up to service the construction industry, the Building Centre houses a wide range of product displays and a comprehensive information department. The Centre houses permanent and temporary displays, including an impressive brick gallery, and product showrooms for the display of building products, materials and components. An increasing number of leading firms are locating technical staff in their showrooms enabling specifiers to receive first hand application and commercial advice. The Information Exchange holds product literature and brochures on over 1500 firms which are free of charge to all visitors. One of the most effective construction industry data search and retrieval systems in the country enables unbiased information on materials, standards and regulations to be given to users of the Centre.

Since May 1993 the Centre has been home to the European Construction Centre. The first of its kind in the European Community, the key objective of the European Construction Centre is to assist commercial organizations of all types in the field of construction and associated industries to exploit the business opportunities which are now emerging in the wider European market. The ECC has a database containing product information as well as commercial, technical and legislative information.

In collaboration with Quantarc/Poulter Communications Plc, the Building Centre has been responsible for developing one of the first CD-ROM packages to contain product information. The BCQ-CONSTRUCTION INDUSTRY INFORMATION SYSTEM contains over 150 000 pages of product literature on compact disc and is maintained and updated daily. BCQ provides immediate access to product literature, building regulations, British Standards and essential technical information for the construction industry.

Obtaining product information

As we have seen, there is a range of sources of trade literature. Most library collections have been developed through material freely acquired from manufacturing firms. In the British Library the collection is developed through the regular mailshotting of firms asking to be placed on their mailing list for all literature. The majority of firms is only too glad to deposit material as this gives them a free advertising opportunity for their products. It is worthwhile drawing up a list of firms operating in relevant industries and asking if they will send details of their products whenever published. In this way it is possible to have a ready-made current awareness service, and at little or no cost. Even recognized competitors are often happy to send material as part of their corporate marketing strategy.

Future trends

The climate for product information in the UK is currently a healthy one. There is a vast range of sources available ranging from hard-copy collections to online services. There are CD-ROM services appearing on the market containing product information, such as those provided by TI and BCQ, and it is reassuring that this is the case, as CD-ROM appears to be the technology of the future. As a user of the BCQ CD-ROM commented, 'it is essential, in our opinion, to provide our clients with the widest possible range of product information and current specifications at the touch of a button. In addition, the facility to print hard copies as required renders the traditional form of library system obsolete.' Engineers can be comforted by the fact that the product information industry, without which they cannot successfully operate, is well fitted as we move towards the 21st century.

References

BS 4940: 1973 Recommendations for the presentation of technical information about products and services in the construction industry. London: British Standards Institution.
Department of the Environment, Property Services Agency (1979) Better trade literature. London: Property Services Agency.
Edmonds, D. (1985) How to find product information. Business Information Review, **2**(1), 5–12.
Newton, D.C. (1991) Trade marks: an introductory guide and bibliography 2nd ed. London: British Library Science Reference and Information Service.
Wall, R. A. (ed.) (1986) Finding and using product information: from trade catalogues to computer systems. Aldershot: Gower.
Wall, R. A. (ed.) (1992) Engineers' guide to product information: sources and use. London: Aslib.
Wall, R. A. (1992) Product information problems: an introduction. Aslib Information, **20**(5), 205–206.

CHAPTER SEVEN

Abstracting and indexing services, bibliographies and reviews

R. G. RHODES

This chapter is mainly concerned with various abstracting and indexing sources in printed form. These are the traditional sources or tools for locating relevant documents for study, research and similar purposes. The emphasis here is on those sources which give some broad coverage in engineering, the specific sources being referred to in the chapters on the relevant subjects.

The printed sources are now complemented by a variety of computerized information sources, an area in which a great deal of change is taking place which can affect the use of abstracting and indexing sources. For example, a particular source may exist in print format and with one or more computerized versions giving the searcher a greater choice.

The use of abstracting and indexing sources in printed form is now less, as a result of the computerized sources. Even so, their role is still important and searchers should bear in mind the need to use the most appropriate source of information and not necessarily the most convenient.

There are areas in which abstracting and indexing sources are particularly suitable:

- for a *comprehensive* search which is likely to involve a mix of printed and computerized sources
- coverage of *early* work which is mainly prior to the 1970s
- when there is *uncertainty* about the area of interest and *browsing* is needed to some extent – examples are looking for ideas for research or new products
- a *preferred* option; printed sources often have better typefaces and allow some assessment of material as details are located, but notes

of potentially useful material should be made. Time taken to locate and digest information is more significant than simply the search time.

Use of abstracting and indexing services and associated sources

Abstracting and indexing sources and their use are not an end in themselves but a part of the means of achieving goals such as study, research and design as undertaken by individuals and teams. Ideally, selection and use of these sources should be easy, effective and efficient. This whole book is intended to assist with selection and can be supplemented by consultation with library and information specialists particularly those with an engineering background or responsibility. Design is often carried out by interdisciplinary teams and the inclusion of a suitable information specialist on the team is worthy of consideration.

Using abstracting and indexing services is not always easy and so they need to be as 'user friendly' as possible. *Chemical Abstracts* is a good model in many aspects.

A critical aspect of use is the choice of keyword and search of index headings. The so-called 'end-user' is in a good position to select terms used to describe the area of interest. With interdisciplinary projects, the information specialist can carry out the search. Use of alternative keywords and headings is important and the producers include 'see' and 'see also' references which are important signposts to relevant material.

Searching for information is a task in its own right and is best carried out at an early stage in the project. Usually, the search needs to be updated particularly for longer duration projects. The temptation to leave information seeking until the information is needed should be avoided for this leads to partial coverage. In research, ignorance of earlier work may lead to needless duplication and then doing it as a secondary task with the associated waste of time and money. In engineering design the early decisions are made in activities which themselves have low costs but relatively high commitment to costs of the final product.

Abstracting and indexing journals can also be used for the updating both of an ongoing project and also for general current awareness.

As to the actual sources, these themselves are varied in origin, coverage for example. An abstracting service provides both source details of the document (bibliographic details) and an abstract or summary. The abstract can be critical, informative or indicative. Critical and informative abstracts can assist in evaluating the usefulness of a document and sometimes remove the need to consult the full document. Producing an abstract can delay the appearance of the details in abstracts for several months after the appearance of the paper. An indexing service does not normally suffer as long a delay but details supplied are limited to the basic bibliographical details. Some confusion can occur from the misleading titles of sources with for example *Engineering Index* including abstracts.

Both abstracting and indexing journals are produced at regular intervals, often monthly. To help use them, there are author and subject indexes produced at intervals, often annually. An exception are the three INSPEC sources (e.g. *Electrical and Electronics Abstracts*) which both have six-month and five-year author and subject indexes.

For an unusual index, there is the Geodex Retrieval System used with *Geotechnical Abstracts* and the *Geodex System/s Structural Information Service*. In both sources, subject cards are matched and aligned numbers (representing abstracts) noted. It looks more difficult than it is to use but it is an effective method of carrying out multiple term searches.

A bibliography is normally a single publication and can be annotated, in that the bibliographical details of documents are supplemented with some indication of the context. A review is also specific in its coverage of a topic. The material is usually evaluated and in context. Some journals publish collections of reviews and those for engineering are listed later.

Because engineering is a progressive field, for example, interest in a car chassis is mostly historical rather than applicable, the useful life of many bibliographies and reviews is limited. Thus, they do not have the same role as in some other fields such as literature. It can be difficult to locate useful bibliographies and reviews. There are listings but their coverage of engineering is often relatively low compared to literature, for example.

Guides to abstracting and indexing services

One choice is the *Index and abstract directory: an international guide to services and serials coverage* now in its third edition (1993) and published by EBSCO. In some ways, this is comparable with the familiar *Ulrich's international periodicals directory*, published annually (with quarterly updates) by Bowker. Both indicate which abstracting and indexing service covers a particular journal and both have lists of abstracting services by subject. The former has them all grouped together, whereas *Ulrich* lists first the journals and then the abstracts and indexing services. Another interesting recent source is Gorman and Mills (1992).

Abstracting and indexing services of general engineering interest

Engineering Index claims to be 'the world's first and most comprehensive collection of time-saving abstracts on engineering developments'. It is published monthly with annual cumulations by Engineering Information Inc. Currently, there are over 110 000 entries per year with good coverage of most engineering fields. US publications are particularly well covered. Conference and meetings papers are included. Despite the title, abstracts are included.

Cambridge Scientific Abstracts have teamed up with Engineering Index to produce nine abstracting services. They are:

BioEngineering Abstracts (formerly Bioengineering and Biotechnology Abstracts)
Civil and Structural Engineering Abstracts
Computer and Information Systems Abstracts
Electronics and Communications Abstracts
Environmental Engineering Abstracts
Manufacturing and Process Engineering Abstracts
Materials Science and Engineering Abstracts
Mechanical Engineering Abstracts (formerly ISMEC Bulletin)
Solid State and Superconductivity Abstracts

Focusing on one of these, namely *Mechanical Engineering Abstracts*, when this was published solely by Cambridge Scientific Abstracts as ISMEC Bulletin, it would have had its own place in this listing with coverage in production engineering as well as mechanical engineering, and resembled the INSPEC services, of which more later. Now the Engineering Index input indicates that searching *Engineering Index* removes the need to also search *Mechanical Engineering Abstracts*. A similar situation also exists for the other eight services for Cambridge Scientific Abstracts.

Of possible broader coverage from the USA is an indexing journal: *Applied Science and Technology Index* which is published by H. W. Wilson Co. Often there is less duplication than might be expected. A UK source comparable in some respects is *Current Technology Index* published by Bowker-Saur and describes itself as '... a subject and author index to British technical periodicals. Coverage includes all branches of engineering ... and various management techniques such as work study, operational research and ergonomics'. This stimulates a comment on a side issue on the coverage of engineering management. Some abstracting and indexing services give little coverage of this area and so it is useful to know of the services that include this aspect. Also published by Bowker-Saur is *Catchword and Trade Name Index* (CATNI) which is particularly useful for civil and structural engineering.

Another UK service is INSPEC which can be searched collectively by computer but as printed services are available as *Physics Abstracts, Electrical and Electronics Abstracts* and *Computer and Control Abstracts*.

Here, we will concentrate on *Electrical and Electronics Abstracts*. The coverage of electrical and electronics is good, INSPEC being both the creation of the Institution of Electrical Engineers and the largest abstracting service in the United Kingdom. *Electrical and Electronics Abstracts* is published monthly with subject and author indexes every six months, some specialist indexes are included with the author indexes. The current annual number of items is 80 000 plus, with abstracts. It may sound obvious but a little care is needed using the indexes to ensure that each six-month period is covered. Five cumulative indexes speed up the searching process

A UK service that also followed this practice is known as FLUIDEX in its computerized form and various names for the printed abstracts. The service has formerly been provided by the BHR Group Limited which developed the British Hydromechanics Research Association's service. Most of the publications in the set were acquired by Elsevier in 1991 and some changes were made including amalgamation of titles and changes of title.

These large services with various sections lead one to the two notable services *Referativnyi Zhurnal* and *Pascal – Bibliographie Internationale* (formerly *Bulletin Signalétique*).

Referativnyi Zhurnal is published by the All-Union Institute of Scientific and Technical Information (VINITI) in Moscow and is believed to be the largest and most comprehensive service for science and technology. The comprehensiveness is thought to vary from section to section, of which over thirty relate to engineering and are published monthly. Some of the sections, particularly those translated into English, may be mentioned in the specialist chapters.

Pascal – Bibliographie Internationale is the comprehensive French-language service published monthly in various sections by the Centre National de la Recherche Scientifique (CNRS) in Nancy. The computerized version is known as PASCAL. The printed service has sixty-five sections covering various fields.

Variations can be considered. There are times when comprehensive coverage is needed or a highly specialized search in which any material is difficult to find. For example, German literature (language and country) might be needed. A personal choice would be to search the subject or industry-based computerized systems such as ZDE (Dokumentation Elektrotechnik). Searching for Japanese technical literature could be demanding. There is *Abstracts of Science and Technology in Japan* which has three parts *Electronics and Communications; Energy Technology; Agriculture, Forestry and Fisheries*. These are published quarterly arranged by the Universal Decimal Classification (UDC) with a keyword index. They are held by the British Library, Science Reference Information Service (SRIS) which operates a Japanese Information Service.

Indexes have not been considered separate to the abstracting services, so two have been noted already, these are *Applied Science and Technology Index* and *Current Technology Index*. A third one is *Science Citation Index* published by the Institute for Scientific Information in Philadelphia.

In spite of the substantial amount of material in *Science Citation Index* the coverage of engineering is weak compared to the life sciences, for example. Even so, it can prove useful in areas like medical electronics and has located items when all else has failed. The computerized versions will often be preferred to the printed indexes (keyword).

A major abstracting service of value but occasionally overlooked is *Applied Mechanics Reviews* which is published monthly by the American Society of Mechanical Engineers. It accurately describes itself as 'an assessment of the world literature in engineering sciences', and perhaps 'engineering sciences' should be highlighted for its value is mostly to researchers in engineering who may find its critical reviews and well written abstracts useful. A good review can save the researcher some time and be a convenient point for reading up a topic.

One or two lesser known indexes may be found useful. An example is the *Five-Year Index to ASTM Technical Papers and Reports* which is published at intervals to supplement the *Fifty Year Index* . . . (1898–1950) by the American Society for Testing and Materials in Philadelphia. Other professional institutions produce indexes to their publications, e.g. Institution of Civil Engineers, American Society of Civil Engineers.

There are important categories in other chapters of which report literature is a good example. The major abstracting journals do often complement or are complemented by the specialist services covered in other chapters, engineering materials being a good example. Materials have very good abstracting and indexing services of their own but additional references are usually to be found in the major services as well. There can be duplication but a personal view is that duplication is not bad news. On the contrary, it may suggest the searcher is achieving good coverage.

For current awareness purposes, there are printed services such as *Current Contents: Engineering, Technology and Applied Sciences* published by the Institute for Scientific Information, for people who like scanning contents pages.

Engineers place some importance on current awareness and there was a survey conducted in the early 1970s which indicated the amount of time typical engineers were prepared to spend on reading for current awareness. Suggestions on the effective use of time were also made. A good mix of sources would seem ideal and the researchers proposed scanning between four and ten journals, one of which should be an abstracting journal. As developments have been made since the survey, alternative secondary sources, probably computer-based, would now be useful. The use of networked lists of contents of recent journals might be popular. Of the ordinary journals, candidates for general coverage could include *Engineer*, *Engineering* or *New Scientist*. A useful source which is widely scanned but which might surprise a few is *Financial Times*. In any event, there could be an agreement for engineers to read one or more newspapers. A journal from the professional institution might be also in the mix.

Bibliographies (including some catalogues)

General bibliographies and catalogues

Some searches will need to include books in their coverage possibly for comprehensiveness or to provide the basic background reading material. Some books can be very specialized, conference proceedings (see also chapter 5), for example. Some abstracting journals cover books but most focus more strongly on papers in journals.

Another preliminary detail, is the form of the source. Most of the printed bibliographies now have their computerized equivalents most of which are paid for by subscription (e.g. GLOBAL BOOKS IN PRINT CD-ROM) rather than cost by use (e.g. online). A personal choice is often the catalogues of large and specialized libraries which can be searched only by computer-based services, an example is WORLDCAT with 25 million items listed (available via OCLC First Search) which is a modern version of the union catalogue. It recently proved useful on an engineering search that was difficult because of the viewpoint.

A printed alternative to WORLDCAT would appear to be *Library of Congress. National Union Catalog: Books* published monthly by the Library of Congress. Although it provides good coverage, it lacks the awareness normally needed in engineering topics. *Cumulative Book Index* published monthly by H. W. Wilson is easier and up-to-date.

The various national bibliographies are like national library catalogues among the tools of the reference librarian. Examples of national bibliographies are the *British National Bibliography* (BNB), published weekly by the British Library in Boston Spa; *Bibliographie Nationale Française* published fortnightly by the Bibliothèque Nationale de France; *Deutsche Nationalbibliographie* published by Buchhändler-Vereinigung in Frankfurt.

Good sources for routine searches with emphasis on up-to-date material are lists of books currently available from publishers. Two English-language sources are *Books in Print* published by Bowker in New York and *Whitaker's Books in Print* published by Whitaker in London. Whitaker also publish associated services such as *Whitaker's Book List*. As Europe becomes a single market, so the equivalent books in print for other European languages become useful and the printed versions are used personally for these. There is a variety of others of this type for African, Canadian, French and German books in print.

For awareness of recent books and ones to be published there is a variety of published sources such as *Bookseller* published by Whitaker. In addition, there are catalogues from publishers and suppliers. Reviews can also be found in many engineering journals.

Specialized bibliographies

This is really an area for those chapters on specific areas of engineering for bibliographies are usually very specialized. On the whole, they do not get the use that bibliographies in the humanities, arts and social sciences do. A general source is *Bibliographic Index*, published half-yearly by H. W. Wilson. A quick scan will show fewer items under headings such as concrete than the author of fiction or current social issues.

Apparent exceptions do sometimes exist and the Macdonald Bibliographic Series published in the early 1970s was quite useful, covering fields such as electronics. But these were really part bibliography and partly a guide to sources.

Reviews

Two stages of information gathering from published sources are identifying documents of possible interest and reading them. Reading can be time-consuming particularly if an intellectual process of analysis and evaluation is taking place. Reviews are useful in both setting a benchmark and date, very convenient for both searching and reading.

Major review serials in engineering in the English language include the following selection:

Advances in Applied Mechanics (Academic Press)
Advances in Automobile Engineering (Pergamon)
Advances in Chemical Engineering (Academic Press)
Advances in Computers (Academic Press)
Advances in Corrosion Science and Technology (Plenum)
Advances in Cryogenic Engineering (Plenum)

Advances in Electrochemistry and Electrochemical Engineering (Wiley)
Advances in Electronics and Electron Physics (Academic Press)
Advances in Engineering Software (Elsevier Applied Science)
Advances in Environmental Science and Technology (Wiley)
Advances in Heat Transfer (Academic Press)
Advances in Materials Research (Wiley)
Advances in Nuclear Science and Technology (Plenum)
Advances in Radio Research (Academic Press)
Advances in Space Science and Technology (Academic Press)
Annual Review of Fluid Mechanics (Annual Reviews)
Annual Review of Materials Science (Annual Reviews)
Annual Reviews of Industrial and Engineering Chemistry (American Chemical Society)
Control and Dynamic Systems – Advances in Theory and Applications (Academic Press)
Progress in Aerospace Sciences (Pergamon)
Progress in Astronautics and Aeronautics (AIAA)
Progress in Combustion Science and Technology (Pergamon)
Progress in Construction Science and Technology (MTP)
Progress in Materials Science (Pergamon)
Recent Advances in Engineering Science (Gordon and Breach)
Reviews in Chemical Engineering (Reidel)

Substantial numbers of reviews appear in primary journals and these can be traced via search tools such as INSPEC which has a classification general review. Some engineering is based on applying science and a good source of reviews in scientific fields is the *Index to Scientific Reviews* published by the Institute for Scientific Information.

In contrast, there are also product reviews which can be useful in engineering design. *Which*, published by the Consumer Association is a well-known journal containing product reviews. One source which is less well-known is *What to Buy for Business* published monthly by Garrard House in Bromley, Kent. It is worth using for training in engineering design in aspects of evaluating products. Its presentation of product features is first class.

Product reviews appear in numerous primary journals and these can be traced by conventional sources such as INSPEC and CTI PLUS. INSPEC uses product reviews as a classification.

References

Gorman, G. E. and Mills, J. J. (1992) *Guide to current indexing and abstracting services in the third world.* London: Zell.

CHAPTER EIGHT

Electronic sources of engineering information

T. A. HANSON

In the second edition of this book the nearest equivalent to this chapter concentrated on the use of commercial online information services. At that time online searching, with its mediated and costed access, was the premium level service for those who could afford it, though there were by then some attempts to open up the process for direct end-user access with services like KNOWLEDGE INDEX from DIALOG and BRS After Dark.

In the decade or so since the second edition the storage and dissemination of scholarly information by electronic means has mushroomed and has become much more complex. The purpose of the chapter for the third edition is to give an overview of electronic sources of engineering information. There are two sections: Bibliographic databases and Engineering information on the Internet.

Bibliographic databases

The traditional online database market has been affected significantly by the growth and market penetration of CD-ROM but continues to thrive nonetheless. There have been many changes in ownership of the main online hosts and many takeovers and mergers. The most significant of these concerns the US giant DIALOG and the market leader in Europe, DataStar. Their merger in 1993 has been followed by the break up of one of the other large multinational players InfoPro, formerly Maxwell Online. InfoPro has been split into three and sold off to different companies; BRS Online to CD-Plus, ORBIT to Questel. Mead Data Central, home of LEXIS/NEXIS, was acquired by Reed Elsevier in 1994. Finally, in January 1995 DIALOG Information was renamed Knight-Ridder Information, though the DIALOG name is being kept for the online service.

In response to the phenomenal growth of CD-ROM the online companies have been surprisingly content and disinclined to major change. CD-ROM has succeeded because it has enabled libraries to offer their users all the power of online searching, through easy and attractive interfaces and with predictable, subscription-based pricing. Though some hosts have introduced major interface improvements, such as FT PROFILE, DataStar and DIALOG, aimed particularly at end-users, there has been little progress on the pricing front. There were changes that shifted the focus of charging away from overall time spent online and towards the 'quality time' spent searching and retrieving information. This trend was led by ESA/IRS and has been continued by recent announcements from DIALOG (mid-1994) of a rearrangement of the pricing structure with less emphasis on connect time. More recently, CDP Technologies (now OVID Technologies), the new name for CD-Plus, made an announcement to this effect soon after their takeover of BRS Online in late 1994.

In a sense it might be said that the impact of CD-ROM has been to reduce the need for online to change because online users now are the hard core who like things just the way they are. They do not need end-user interfaces and subscription pricing; they are happy with fast and efficient command line interfaces and, as expert searchers, the ticking clock is not a problem.

Many of the principal bibliographic databases in engineering are now available both as online services and on CD-ROM. In fact CD-ROM publisher SilverPlatter now offers online (Internet) access to many of its databases as an alternative. Potential subscribers thus often have a choice of formats but it goes even further with options such as:

- *Tape leasing* from the database producer. This is an attractive option for those whose usage levels are high enough to justify the leasing costs. The arrangement involves mounting the tapes on a local computer system and using an interface of choice.
- *Mounting the files on local hard disk.* Some companies, such as SilverPlatter and OVID Technologies, offer network arrangements based on the use of local PC hard disks attached to local area network (LAN) servers. In the case of SilverPlatter the data is delivered on CD-ROM and copied onto the hard disks. OVID Technologies delivers the data by DAT (Digital Audio Tape). The advantage of the hard disk arrangement is the superior networking and speed of access inherent in magnetic storage. At the time of writing the fastest access time of CD-ROM drives is a little under 200ms (milliseconds) whereas the typical magnetic hard disk access time is around 10ms.
- *BIDS and similar cooperative schemes.* A further option in some

parts of the world is the mounting of the database on a wide area network (WAN) serving many institutions. The BIDS (Bath Information and Data Services) service in the UK is an excellent example of this arrangement. Databases are mounted at the University of Bath and accessed by subscribing university sites throughout the country via JANET (the Joint Academic NETwork).

Database listings

GENERAL: SCIENCE AND TECHNOLOGY

Science Citation Index. The SCIENCE CITATION INDEX (SCI) is produced by the Institute for Scientific Information and is one of several subject based citation indexes produced by them. The greatest strengths of SCIENCE CITATION INDEX are the citation approach itself and the subject comprehensiveness across the sciences and technology. The citation approach offers an attractive alternative to the traditional keyword search. It involves searching for articles according to citation relationships. A typical search would have a particular known document (book or journal) as a starting point and the search would result in a list of articles in which the known document had been cited. This approach works particularly well in specifically defined research areas and frequently discovers useful material that would not have emerged through a conventional keyword search. The coverage of SCIENCE CITATION INDEX is currently about 3200 journals and from 1974 to date. As a database SCIENCE CITATION INDEX is available on CD-ROM, direct from ISI or online via several hosts including DataStar, DIALOG, DIMDI and STN. It is also available to the British academic community as part of the BIDS service (see above).

Index to Scientific and Technical Proceedings (ISTP). This file is also from ISI and indexes the published proceedings of scientific and technical conferences worldwide. The indexing is down to the individual conference paper level as well as to the general proceedings. Approximately 4100 published proceedings per year are indexed in ISTP. Coverage is from 1982 to date. The database is available online via DIMDI and BIDS.

NTIS (National Technical Information Service). This database corresponds to the *Government Reports Announcements & Index.* It provides abstracts of US government-sponsored research and development reports. About 250 departments and agencies of government are included. Coverage is from 1964. The NTIS database is available online from OVID Technologies, DataStar, DIALOG, ESA/IRS, ORBIT, Questel (now Questel.Orbit) and STN.

Conferences in Energy, Physics, Mathematics and Chemistry. This is another conference literature database. It is produced by the German company FIZ Karlsruhe and available on the STN online host. It concentrates on scientific and technological conferences with coverage since 1976.

Conference Proceedings Index. This is a general conference proceedings reference database produced by the British Library which covers all subjects since 1964. It is available online via BLAISE and on CD-ROM from the BL, as BOSTON SPA CONFERENCES ON CD-ROM.

ENGINEERING AND APPLIED SCIENCE

Compendex Plus. This the world's most comprehensive general engineering and technology database service. It is the electronic equivalent of the printed abstracting service *Engineering Index*, plus conference records. Both the printed and electronic services are produced by Engineering Information Inc. (Ei) of Hoboken, New Jersey. The coverage of COMPENDEX PLUS includes about 4500 journals, publications of engineering societies and organizations such as the American Institute of Chemical Engineers (AIChE), selected reports, conference papers and books. The database in its online form has been available since 1970 and grows at the rate of about 200 000 records per year. The current size is about 2.8 million records. COMPENDEX PLUS is available online from various hosts including DataStar, DIALOG, Questel.Orbit and STN. It is also available on CD-ROM from KR Information OnDisc.

COMPENDEX is a natural starting point for any search on engineering-related topics. It is strong in all areas, though there are of course specialist databases in some of the engineering sub-disciplines, such as INSPEC for electrical engineering and METADEX for metallurgy. Indeed, as with many other disciplines, there are quality database services at many levels within the general area of engineering. COMPENDEX is a comprehensive database for the widely defined discipline of engineering. At the higher, broader, level of science and technology there is the SCIENCE CITATION INDEX. INSPEC and METADEX are at the next level down from COMPENDEX and at an even more specific level is WORLD ALUMINIUM ABSTRACTS.

Applied Science and Technology Index (ASTI). This is one of the range of popular databases from H.W. Wilson Company. The approach is to index the 'main' titles in particular subject areas and as such offers a good and easy to use service for the undergraduate market. The coverage is from 1983 to date and about 400 journals are indexed, compared to COMPENDEX's 4500. ASTI is available online from Wilsonline, CDP and OCLC, and on CD-ROM from both WilsonDisc and SilverPlatter.

Current Contents. This is another ISI service. It is available in several formats: online as CURRENT CONTENTS SEARCH, on diskette as CURRENT CONTENTS ON DISKETTE, on CD-ROM, and in paper form. Updating is weekly in all formats. CC covers all subjects areas in seven subsets: Life Sciences; Clinical Medicine; Physical, Chemical and Earth Sciences; Engineering, Technology and Applied Sciences; Agriculture, Biology and Environmental Sciences; Social and Behavioral Sciences, and Arts and Humanities. The diskette version is available for all except the Arts and Humanities. The CD-ROM version, launched in early 1994, is available for four of the subject areas: Life Sciences; Clinical Medicine; Physical, Chemical and Earth Sciences; and Agriculture, Biology and Environmental Sciences. In these four areas abstracts are available on both the diskette and CD-ROM versions. CURRENT CONTENTS SEARCH is available online via OVID Technologies, DataStar, DIALOG and DIMDI.

DATABASES COVERING SPECIFIC ASPECTS OF ENGINEERING

Inspec. INSPEC is produced by the Institution of Electrical Engineers (UK) and corresponds to the three Science Abstracts print publications: *Physics Abstracts*, *Electrical and Electronics Abstracts*, and *Computer and Control Abstracts*. INSPEC is available on CD-ROM from UMI and online via OVID Technologies, DataStar, DIALOG, ESA/IRS, Questel.Orbit, STN and OCLC. As a companion to the CD-ROM product, UMI also offers a full-text service for the documents indexed in INSPEC that are produced by the IEE and IEEE (Institute of Electrical and Electronics Engineers (US)). INSPEC PERIODICALS ON DISC delivers facsimile images of the printed documents on CD-ROM. The discs can be loaded into large capacity jukeboxes and networked for more convenient access. UMI will also provide an Internet-based full-text delivery service from their headquarters in Michigan in the near future.

ISMEC: Mechanical Engineering Abstracts. ISMEC stands for Information Service in Mechanical Engineering and is produced by Cambridge Scientific Abstracts. The database indexes approximately 750 journals from throughout the world. Coverage is from 1973 to date. A selection of books, conference proceedings and reports are also indexed. ISMEC is available online via DIALOG and STN and on CD-ROM via SilverPlatter.

Metadex. METADEX is produced jointly by the Institute of Materials (UK) and ASM International (US). It is the principal metallurgy database with coverage of 3500 new documents per month since 1966. These include journals, books, conference papers and technical reports. It is equivalent to *Review of Metal Literature* and *Metals Abstracts*.

METADEX is available on CD-ROM from KR Information OnDisc using the name METADEX COLLECTION, which includes *Engineered Materials Abstracts* and covers polymers and ceramics. The online version is available via DataStar, DIALOG, Questel.Orbit and STN.

Fluidex (Fluid Engineering Abstracts). FLUIDEX indexes world literature on all aspects of fluid engineering. It is produced by the BHR Group in the UK. The coverage is from 1973 to date and includes more than 500 journals and many books, conference proceedings, British patents, standards and research reports. FLUIDEX is available online via Dialog.

Energy Science and Technology. This is the database of the US Department of Energy (formerly known as DOE Energy). All aspects of energy are covered from 1974 to the present. The database includes *Energy Research Abstracts* and *Coal Abstracts*. Document coverage includes journals, books, reports, conference papers, patents, dissertations and translations. The database is available online and on CD-ROM via DIALOG.

Aerospace Database. The AEROSPACE DATABASE is produced by the American Institute of Aeronautics and Astronautics/Technical Information Service (New York). It corresponds to two printed indexes: *Scientific and Technical Aerospace Reports* (from NASA) and *International Aerospace Abstracts* (from the AIAA under contract to NASA). The database covers the period from 1962 to the present. Documents indexed include technical reports, journal articles, books and conference papers. Though coverage is international access is limited to the United States and Canada. Access from other countries can be arranged through the AIAA. The AEROSPACE DATABASE is available online and on CD-ROM via DIALOG.

Chemical Engineering and Biotechnology Abstracts. This database is produced by the Royal Society of Chemistry (Cambridge, UK) and covers the period from 1971 to the present. It is available online via DataStar, DIALOG, Questel.Orbit and STN.

RAPRA Abstracts. RAPRA is devoted to rubber, plastics, adhesives and polymeric composites and is produced by RAPRA Technology Ltd. (Shropshire, UK). The database corresponds to the printed *RAPRA Abstracts* and *Adhesives Abstracts*. The coverage is from 1972 to the present. 'The database comprises a large collection of carefully produced, extensively indexed summaries covering a wide scope of subjects encompassing technical, academic, commercial and marketing aspects of the rubber and plastics industries.' (DIALOG Database Catalog 1994, p.124). RAPRA is available online via DataStar, DIALOG, ESA/IRS, Questel.Orbit and STN.

Water Resources Abstracts. WATER RESOURCES ABSTRACTS is produced by the US Department of the Interior, Geological Survey (Reston, Virginia) and covers from 1968 to the present. The database is equivalent to the printed *Selected Water Resources Abstracts*. It is available online via DIALOG and on CD-ROM from SilverPlatter and National Information Services Corporation (Baltimore, MD).

ICONDA. This database is produced in Germany by the Fraunhofer Society and provides international coverage of building construction, engineering geology, structural engineering, etc. The coverage is from 1976 to the present. ICONDA is available online via Questel.Orbit and STN, and on CD-ROM from SilverPlatter.

CITIS. CITIS incorporates two databases: INTERNATIONAL CIVIL ENGINEERING ABSTRACTS, which corresponds to the printed index of the same name, and SOFTWARE ABSTRACTS FOR ENGINEERS (SAFE). The former indexes about 500 journals from 1972 to date while the latter describes more than 4000 software programs from 1984 onwards. CITIS is available on CD-ROM from CITIS Ltd (Dublin, Ireland).

MathSci. The MATHSCI database is produced by the American Mathematical Society and covers from 1940 to the present. It is divided into seven subfiles: *Mathematical Reviews* and *Current Mathematical Publications* published by the American Mathematical Society; *ACM Guide to Computing Literature and Computing Reviews* published by the Association for Computing Machinery; *Technical Reports in Computer Science* compiled by Stanford University; *Current Index to Statistics* from the American Statistical Association and Institute of Mathematical Statistics; and *Index to Statistics and Probability* from Tukey and Ross. MATHSCI is available online via DIALOG and ESA/IRS and on CD-ROM from SilverPlatter.

Standards and Specifications Database. This database provides bibliographic access to all US government and industry standards since 1950. It also covers some international standards. It is produced by the National Standards Association (Gaithersburg, MD) and is available online from DIALOG and ESA/IRS.

Standards Infodisk. The STANDARDS INFODISK from ILI indexes standards from the US, UK, Germany, Japan, France, Sweden, Norway and Australia. International standards bodies such as ISO, ITU and CEN/CENELEC are also included. The database is available on CD-ROM from ILI (London and New Jersey).

Engineering information on the internet

The Internet is the name given to the global system of interconnected computer networks used for academic and research related purposes, and increasingly by companies, organizations and individuals throughout the world. It has developed over the years into a comprehensive and complex information access and dissemination system used by upwards of 25 million people. While it is, or can be, infuriating in its lack of organization, unreliability and overall performance, it has nonetheless transformed many aspects of the scholarly communication process and opened up new opportunities for communication and collaboration. The purpose of this section is to indicate sources of engineering information on the Internet.

The scholarly communication system, in general terms, involves both formal and informal processes. The former includes the production of books and journal articles while the latter refers to the direct contacts between researchers through correspondence or conference attendance. The Internet has affected both areas.

Formal communication

Many claims have been made for the Internet to transform the world of scholarly publishing. The notion of publishing journals electronically and delivering them directly to subscribers via their elecronic mail boxes is attractive in principle and technically feasible but in organizational and sociological terms it is more problematic.

Debates about the potential of network publishing often begin with the apparently absurd notion of the worldwide scholarly community toiling away to produce books and journal articles. These are then given free of charge to commercial publishers who then sell them back to the scholarly community through libraries. Stated thus, it is easy to see why many people propose electronic alternatives (Swinnerton-Dyer, 1992), but the early experience has been that authors are reluctant to write for electronic journals, preferring instead to see their name in 'print'. Doubt is also expressed about the ability or inclination of universities or other representative scholarly bodies to take on the organizational work of copyright and publishing and all that it entails. However, in these early days of network publishing there is still much to report.

The number of scholarly, refereed journals published on the Internet is currently small. There are seventy-four titles listed in a recent directory (Association of Research Libraries, 1994). The reasons for this relate partly to the factors mentioned above but they relate also to technical factors such as the ability of the network to transmit a series of complex images (pages), perhaps containing photographs and diagrams, at an acceptable speed. The techniques for achieving this are improving rapidly with the broadening bandwidth of the networks and with better data compression techniques. In contrast to the refereed journal situation the number of newsletters published on the Internet is large and growing with 366 titles listed in the ARL directory. One simple explanation of this is that newsletters, as their name implies, exist to communicate information to subscribers in a timely fashion. Thus for newsletters, the immediacy of Internet publishing with electronic mail as the delivery mechanism is very attractive.

REFEREED ELECTRONIC JOURNALS

Electronics Letters Online (ISSN 0013–5194). Published by the Institution of Electrical Engineers. This is an electronic equivalent of the printed *Electronic Letters*. Contact the editor, Jim Ashling, at: inspec@dm.rs.ch

Journal of Fluids Engineering (ISSN 0098–2202). This is a free journal published by the American Society of Mechanical Engineers and the Department of Engineering Science and Mechanics at Virginia Polytechnic Institute and State University (Blacksburg, Virginia). Contact the editor, Gail McMillan, at: gailmac@vt.edu

New York Journal of Mathematics. A free journal published by the Mathematics Department of the State University of New York (Albany, NY). Contact the editor, Mark Steinberger, at: mark@.sarah.albany.edu

ELECTRONIC NEWSLETTERS

ACM-NS Info Flash. Published by the Association for Computing Machinery. For information connect to: http://info.acm.org. (Free) Irregular.

ChE (Chemical Engineering) Electronic Newsletter. Published by Curtin University of Technology (Perth, Western Australia). For information contact the editor, Martyn S. Ray, at: trayms@cc.curtin.edu.au. (Free) Monthly.

E-letter on Systems, Control and Signal Processing. Published by Anton A. Stoorvogel, Department of Mathematics and Computing Science, Eindhoven University of Technology, Netherlands. (Tel: +31–10–472378) (Free) Monthly.

Another form of formal scholarly communication is preprints and technical reports. In those areas where preprint exchange was already popular, such as physics, the Internet has presented an attractive method of enhancing this process. The most notable example of an Internet preprint exchange service is the Los Alamos National Laboratory Physics Service. It is accessed by Gopher at mentor.lanl.gov. The area covered is nuclear and high energy physics. Other preprint services include the American Mathematical Society gopher service (at e-math.ams.org) and the CERN Preprint Server on the World Wide Web (at: http://darssrv1.cern.ch). A good example of a technical report service is the NASA Technical Report Service, also on the World Wide Web (http://techreports.larc.nasa.gov/cgi-bin/ntrs).

Informal communication

The expression 'invisible college' is often used to refer to the practice whereby scholars exchange information about their research. This may involve sending preprints of articles or simply brief descriptions of their conclusions. The process was invisible to all but the favoured few recipients on the researcher's mailing list. With the advent of electronic mail and other Internet services the concept of the invisible college is revisited with a vengeance.

Electronic mail provides the means for scholars to communicate on a direct one-to-one basis with colleagues throughout the world and this in itself represents a major advance. However, this simple concept has been developed further to open up the communication process to include large groups of scholars wherever they may be. This is known as an online discussion list or a computer mediated conference. It is a means of generating informal discussion among all those who have an interest in a particular topic and who are connected to the Internet. An online discussion list is simply a mailing list. A subscriber adds his or her name to a list by sending a one-line e-mail message to the list organizer saying 'subscribe list-name Joe Smith' where list-name is the name of the discussion list in question and Joe Smith is the name of the person wishing to subscribe.

Once included on the list Joe Smith will receive all messages sent to the list. Any subscriber can send a message. Typically they would be questions, requests for help with a particular problem, answers to questions, responses to the help messages, opinions, or announcements such as those of conferences. A subscriber has the option of simply observing the traffic so as to keep in touch with current concerns or occasionally participating directly with questions and answers of their own. Some lists are moderated, the informal equivalent to the refereeing process, to maintain a level of quality and relevance in the network traffic. Others, the majority, are open and may include much frivolous and irrelevant material. The above mentioned ARL directory listed nearly 1800 academic discussion lists. Among these are 39 general engineering lists and many more in related areas. The reader is referred to the appropriate sections in this directory for details.

The great advantage of this form of communication is its immediacy and simplicity. However, there is a downside to it. If you happen to subscribe to a 'busy' list which generates much debate and a large number of messages you could be inundated with hundreds of items of e-mail per day. The time taken to filter these for the few gems that might be hidden among them can be considerable. As a consequence, many e-mail software packages now include a filtering facility. This allows the user to establish recognition routines for incoming messages, for example, by source or subject as indicated in the header, and then to discard those that are unwanted. Another alternative is to avoid subscribing to the lists and use a news reader program instead. This will permit reading of list messages without receiving them directly by e-mail.

Information services

A third Internet service area to discuss is the information dissemination service. As well as a means of formal communication through electronic journals and informal communication via discussion lists, there is currently a rapid growth in general information services. Typically, an organization, such as NASA or the Electric Power Research Institute, would produce a collection of information about its activities, publications, conferences and meetings or courses, along with, perhaps, statistical and other factual information relating to the subjects of interest. The NASA information service is very comprehensive, covering all aspects of NASA activity and including photographs and animations of, for example, the recent Shoemaker-Levy comet collision with Jupiter and the repair of the Hubble Space Telescope. Universities, research groups, learned and professional societies and research institutes too are grasping the opportunity to 'publicize' their existence and activities through Internet information services.

The means by which these organizations disseminate their information or construct their services, are by use of the standard Internet navigation tools: Gopher, Wide Area Information Server (WAIS) and the World Wide Web. These are client-server software tools that allow an information provider to construct an information 'server' and for that server to be seen (discovered and used) by anybody connected to the Internet and using the appropriate 'client' software. The latter is simply a software package used at the user end of the process that allows viewing of a compatible 'server'. For example, if you or your organization wished to set up an information dissemination service either for local or Internet-wide use you could use, say, the Gopher software as your server. This is simply a piece of menu-making software. You set up your menu(s) and each item on the menu can point to a local collection of information or to other Internet services anywhere in the world. Anybody who wishes to look at your information server would then need the client version of the Gopher software to do so.

The means by which the connection is made from a Gopher menu to the information source in question is the Telnet command. This logs on to, and allows use of, the remote computer. Telnet is part of the general TCP/IP (Transmission Control Protocol/Internet Protocol) communications protocol for inter-networking different computers systems. TCP/IP is the language of the Internet; it is a pre-requisite for any computer wishing to connect to the Internet system.

The Gopher software has been credited with accounting for the rapid growth of the Internet in recent years and it is still very popular. It is also being improved all the time and recently a Gopher Plus package was introduced offering, through a series of viewers, the ability to observe graphic images as well as simple text. These enhancements are intended to help Gopher remain competitive with the new Internet superstar, the World Wide Web. The World Wide Web uses the same basic client server concept but allows colour graphic images, sound and video to be incorporated into documents and for hypertext links to be established between documents wherever they are located on the Internet. Thus a document on a Web server will do the same job as a Gopher server menu in that it points to other items (servers) elsewhere, but instead of doing so via a simple menu it can show a link by highlighted text anywhere in a document. The user simply clicks with the mouse to activate the link.

The World Wide Web, like Gopher, is a system for organizing information on the Internet. The Web system was developed at CERN (European Centre for Nuclear Research) in Geneva. It differs from Gopher by allowing links between Internet computers to be activated from anywhere within a document using a highlighted hypertext link. Web services can also incorprate graphics, colour, sound and video. Web services are created in HyperText Markup Language (HTML) and they are viewed by the user using a piece of software known as a Web browser (or client). The best known Web browsers are Mosaic and Netscape. Mosaic was created at the National Centre for Supercomputer Applications (NCSA) at the University of Illinois and released in late 1993. This event led to an explosion of interest in the WWW/Internet. Many other Web browsers are now available but it is Netscape Navigator, from Netscape Communications Corporation (http://www.netscape.com), that has assumed the dominant position in the market.

ENGINEERING INFORMATION VIA THE WORLD WIDE WEB

There is no official subject catalogue of information resources on the Internet. There can be no such product, given its present informality. There are, however, many unofficial listings which attempt to give a

Electronic sources of engineering information 119

reasonably accurate picture of the changing situation. Among the lists available are:

- Yahoo (http://www.yahoo.com)
- The Whole Internet Catalog (http://www.gnn.com)
- The World Wide Web Virtual Library (http://www.w3.org)
- The McKinley Internet Directory (http://www.mckinley.com)
- Excite (http://www/excite.com)

Of these, the two that are most comprehensive are Yahoo and the Web Virtual Library. Yahoo, produced at Stanford University, appears to have become the most popular, as reflected by its prominent inclusion in the Netscape browser's Internet Directory section. The Web Virtual Library, from the birthplace of the Web, CERN, Geneva, has been relegated to invisibility as far as Netscape is concerned but the list is still maintained and is very useful. As an indication of the breadth of coverage the Web Virtual Library divides its Engineering section into the following domain specific virtual libraries:

- Acoustics and Vibration
- Aerospace Engineering
- Amateur Radio
- Chemical Engineering
- Civil Engineering
- Control Engineering
- Electrical Engineering
- Engineering and Technology Management
- Environmental Engineering
- Industrial Engineering
- Materials Engineering
- Mechanical Engineering
- Naval Architecture and Ocean Engineering
- Nuclear Engineering
- Optical Science and Engineering
- Power Engineering
- Rapid Prototyping
- Software Engineering
- Technical Ceramics
- Wastewater Engineering
- Welding Engineering
- Information Resources Applicable Across Engineering Domains
- Standards
- Products and Services
- Academic and Research Institutions

This listing, and the others noted above, are updated on a continuous basis so that when the reader takes a look he or she will see a longer and more up to date description of engineering resources on the Internet.

Finally, there are two new engineering sites on the Internet worthy of mention.

Ei Village. This is a comprehensive, Internet-based, information service for engineers from Engineering Information Inc., the producers of COMPENDEX. As a subscriber, or 'village resident', you will have access to services such as:

- Ei Spotlights
- Ei Tech Alert
- Ei Connexion
- EiDDS
- Articles on Call
- Editor's choice
- Technical Librarian
- Ask Your Peers
- Senior Village Engineer
- Current trends in Engineering
- News and recent technical articles
- Access to COMPENDEX and 150 other databases
- Articles by Email service
- Express featured articles delivery service
- Site evaluations
- Library reference service
- Discussion groups and forums
- Help from a senior colleague

The Ei Village can be found at http://www.ei.org

EEVL (Edinburgh Engineering Virtual Library) Commencing in August 1995 this is a project to build a gateway for the higher education and research community to facilitate access to high quality information resources in Engineering. It is funded by the Electronic Libraries (or Elib) programme which in turn emanates from the *Follett Report* on university libraries (Joint Funding Councils' Libraries Review Group, 1993). EEVL will, when fully operational, allow users to browse through, or search for entries in the EEVL database of engineering resources, and dynamically connect to resources of interest. The majority of effort will be directed to collecting UK resources.

EEVL is based at Heriot-Watt University in Edinburgh and can be accessed at http://www.hw.ac.uk/libwww/eevl/eevlhome.html

References

Association of Research Libraries (1994) *Directory of electronic journals, newsletters and academic discussion lists* 4th ed. Washington, DC: Association of Research Libraries.
Joint Funding Councils' Libraries Review Group (1993). *Follett Report*. Bristol: Higher Education Funding Council for England.
Swinnerton-Dyer, P. (1992) A system of electronic journals for the UK. *Serials: the Journal of the United Kingdom Serials Group*, **5**(3), 33–35.

CHAPTER NINE

Standard reference sources

D. R. G. BRIMAGE

Reference works described in this chapter are chosen for their relevance to a wide range of engineering disciplines, and include encyclopaedias, dictionaries, handbooks and directories. Works relevant to more specialized areas are also covered in the appropriate chapters.

Guides to the literature

The two major guides to reference material are Mullay and Schlicke (1993) and Sheehy (1987). Volume one of Mullay and Schlicke (1993) is concerned with science and technology. This work, although international in scope, is biased towards British material and the entries are arranged in UDC (Universal Decimal Classification) order. Sheehy (1987) is published in the United States and is arranged in broad sections but is probably more biased towards the humanities and social sciences. A supplement has been published, Sheehy and Balay (1992), covering the years 1985–1990.

An introduction to literature work and reference sources in science and technology is given by Parker and Turley (1986). For reports and other forms of 'grey literature' Auger (1994) is a useful guide (see also chapter 2). The major information sources in energy technology are given in Anthony (1988). Information sources in materials engineering may be found in Reynard (1992) and the *Guide to Materials Engineering Data and Information* (1986).

Patents are an important source of engineering information and an introduction to patents and a guide to their literature may be found in Auger (1992). For those wishing to find trademark information Newton (1991) is a good starting point. Other guides to the literature may be found in the series Guides to Information Sources published by Bowker-Saur (see also chapter 4).

Encyclopaedias

Multi-volume encyclopaedias are often a useful starting point when approaching an unfamiliar subject. The general encyclopaedias often contain technical information. The *McGraw-Hill encyclopedia of science and technology* (1992) is probably the best known multi-volume encyclopaedia devoted to science and engineering. Published in twenty volumes it is updated with annual yearbooks. It is also available in CD-ROM format. Subsets of this work are published separately as single volumes as, for example, Parker (1993a). *The encyclopedia of physical science and technology* (Meyers, 1992) in eighteen volumes is a more technical work than *McGraw-Hill*, with longer articles at a higher level. Considine and Considine (1994), now in its eighth edition, is a well known single volume scientific encyclopedia. Wasserman (1995) is a recent wide ranging encyclopedia of physical science and engineering information.

Encyclopedias covering more specialist areas appear in the subject chapters but some examples are mentioned here. An encyclopaedic guide to engineering materials and processes in eight main volumes and three supplementary volumes, Bever (1986) is a very comprehensive work. In six volumes, Lee (1989) is a useful source of information on composite materials. Fifty articles on different techniques of materials characterization in Brundle et al. (1992) is useful to anyone involved in identifying materials. Mahajan and Kimerling (1992) contains articles, reproduced and updated, from the *Encyclopedia of materials science and engineering*. A useful single volume guide to materials, now in its thirteenth edition, is Brady and Clauser (1991), which is an encyclopedia for managers and technical professionals. For civil engineers the *Encyclopaedia of hydraulics, soil and foundation engineering* (Vollmer, 1991), is a useful work. Auger (1975) is an important alphabetical guide to named terms, machines and principles in mechanical engineering.

Dictionaries

Dictionaries are important in engineering as sources of definitions of terms used. Many dictionaries are available and some are grouped here under three main subheadings.

Defining dictionaries

Dictionaries covering the whole of science and engineering are usually very selective, but the best coverage can probably be found in Morris (1992) and Parker (1993b). On a slightly smaller scale Walker (1990) is still useful. *Industrial engineering terminology* (1991) is an ANSI standard (Z94.1) which defines terms in the field of industrial engineering. Booth (1993) is an ANSI standard and is a very comprehensive collection of electrical and electronic terms including those from other ANSI standards. A well established dictionary of electronics is Gibilisco (1994) which includes illustrations and many cross references. Markus and Sclater (1994) gives easy to understand up-to-date terms used in the electronics industry. It can serve as either a desktop or library volume. Nayler (1985) gives definitions in a wide range of mechanical engineering subjects and includes diagrams. Terms in the area of the science and technology of environmental protection and resource managemant are covered in Porteous (1992). A good source for definitions in materials engineering is Davis (1992).

A glossary of approximately 9000 building and civil engineering terms produced from all fifty-five parts of BS 6100 is given in British Standards Institution (1993). A more compact dictionary of civil engineering is Scott (1991). The terminology in the human factors field is well covered by Stramler (1993) which aims to eliminate any confusion about the current usage of terms. For definitions in the field of space technology, Williamson (1990) is thorough.

A useful guide to production management terminology is given in Bessant and Lamming (1991). Materials science and technology is served by Walker (1993).

Translating dictionaries

There are numerous translating dictionaries available, some of which are bilingual and some multilingual. These range from those covering technology and engineeering in general to those covering a very specific area. It would not be profitable to mention all of these here. Elsevier publish an excellent range of multilingual dictionaries on a wide range of subjects.

The main dictionary for German has been De Vries and Herrmann (1972) but *Routledge German technical dictionary* (1995), which is in two volumes, is much more up-to-date. In the same series *Routledge French technical dictionary* (1994) is probably the best recent work. For Russian Alford and Alford (1970) and Stoliarov and Kuzmin (1991) give good coverage of the language. A multilingual technical dictionary covering the main European languages is *Five language technology dictionary* (1993).

Dictionaries of symbols and abbreviations

There are many dictionaries of abbreviations and some of the more useful general ones are *The Oxford dictionary of abbreviations* (1992), Towell and Sheppard (1994), De Sola (1994), and Pugh (1987).

For engineering acronyms and abbreviations Keller and Erb (1989) is a good source. The two volume dictionary of abbreviations and acronyms of electronics, electrical engineering, computer technology and information processing, Wennrich (1992), is one of the most comprehensive works of its type. It contains mainly English language definitions but also includes other languages. Gordon and Singleton (1986) and Merkow (1990) contain acronyms in information technology and computer science.

Handbooks, tables and data sources

Bolz and Tuve (1973) is a valuable first source access to numerical data in applied engineering science. James and Lord (1992) is a composite collection of data in many disciplines. Sharpe (1995) is the standard work in its field and an essential reference text. A new edition of this work is published annually. Tapley and Poston (1990) has a broad range of data required to support the breadth of current engineering practice. Calculation routines to solve everyday problems in all branches of engineering are given in Hicks (1994).

Oberg *et al.* (1992) is a standard reference work in design and manufacturing. Kutz (1986) contains seventy-six chapters and is a very comprehensive work covering most aspects of mechanical engineering. A handbook for mechanical engineers more suitable for personal use but still of interest to libraries is Carvill (1993). Smith (1994) meets the needs of technologists in mechanical, production and chemical engineering. Baumeister (1986) contains lengthy articles and tables of data for mechanical engineers. Working designers will find the many formulas, tables, charts and graphs in Lingaiah (1994) valuable. Also useful and completely updated in the second edition, Rothbart (1985) is a guide for mechanical designers. A compilation of practical data detailing the specification and use of modern manufacturing equipment, Bakerjian (1993) is a thorough guide for manufacturing engineers.

A comprehensive three volume guide which aims to provide a system and the necessary information for the selection and specification of engineering materials and related component manufacturing processes is Waterman and Ashby (1991). A very good guide to engineered materials in four volumes is the *Engineered materials handbook* (1991). Designers, engineers and others concerned with the selection and use of materials will find the *Handbook of industrial materials* (1992) valuable. An important source for information on metals is the multi-volume *ASM handbook* (1991–), which is the continuation under the new title of the tenth edition of the *ASM metals handbook* series.

Aimed at the practising engineer in the telecommunications industry, Freeman (1994) is broken down into twenty-six subject areas to give a very comprehensive coverage. At an advanced level and also aimed at practising engineers in the telecommunications industry is Mazda (1993). A comprehensive guide to shock and vibration is Harris (1987).

Now in its fifteenth edition, and with thirty-six lengthy chapters each with bibliographies, Jones *et al.* (1993) is an important reference source for electrical engineers. Fink and Beaty (1993) is another well established handbook on electrical engineering and covers the generation, transmission, distribution, control, conservation and application of electrical power. To help workers in digital design, Giacomo (1990) is a guide to understanding digital bus networks. Completely revised since the first edition to reflect advances made since then, Skolnik (1990) is a comprehensive guide to radar technology. Process instruments and controls are covered in Considine (1993) and Lipták (1995). A comprehensive sourcebook for metallic materials specifications for all concerned with engineering materials is Ross (1992). Also giving specification of metallic materials Brandes and Brook (1992) is now in its seventh edition.

For those who make decisions affecting the selection of engineering materials and construction methods, Merritt (1995) gives the best civil engineering practices.

A useful guide to robotics is given by Dorf and Nof (1990). Blake (1989) is an important collection of state of the art reports on design and construction practice in the UK and overseas. It also contains many references and bibliographies to enable a reader to study a subject in greater depth.

Directories and yearbooks

A number of listings of directories have been published, including *Current British directories* (1993), Gilbert (1992) and *Dawson top 3000 directories and annuals* (1994).

Details of professional bodies, research and trade organizations in the United Kingdom are given in Henderson and Henderson (1992) and Millard (1994). On a Europe-wide scale Adams (1991) covers similar types of organizations. For worldwide coverage, Sachs (1990) is a comprehensive guide covering over 17 000 associations. An international guide covering mainly engineering organizations is Davis (1993).

Current research is often difficult to trace but research activities in engineering in universities and colleges in the UK may be found in volume 1 of the annual publication *Current research in Britain* (1994). Information on industrial research is given in *Industrial research in the United Kingdom* (1993). For information on research in Europe, European Research Centres (1993) is a valuable source. *European sources of scientific and technical information* (1994) is a guide to European organizations which may be sources of technical information. Data on organizations worldwide may be found in the work by the Union of International Organizations which is published annually. The *Europa world year book* (1994) gives details of international organizations. The standard guide to academic and research organizations worldwide is the *World of learning* (1995). For academic institutions in the Commonwealth there is the annual publication *Commonwealth universities yearbook*.

Organizations and research establishments in engineering are given in *Engineering Research Centres* (1993), which is an expanded new edition of Electronics Research Centres.

Official publications

A guide to government publications in Britain is given in Butcher (1991), and official statistics are covered in Dennis (1990). Sources of information on the European Communities are listed in Zolynski (1991) and *Directory of EEC information sources* (1991). The documentation of the European Communities is detailed in Thomson (1989). Information about all aspects of the European Community organization and policy is given in Roney (1993).

Biographical information

Biographical data of engineers and scientists can be found in the general annual series of Who's Who. There are however biographical works devoted specifically to engineers and scientists. The largest work is the multi-volume publication by Gillispie (1970–1980). In two volumes with over 25 000 entries is *Who's who in science in Europe* (1993). *Who's who of British engineers* (1982) has information on British engineers but is now less useful because of its age. For American engineers Davis (1995) is the best source. Association of Consulting Engineers (1995) is a useful annual publication, especially for civil and structural engineering. *Who's who in science and engineering, 1994–1995* (1994) is a more recent publication. Abbott (1985) contains volumes devoted to engineers and inventors, physicists and mathematicians.

References

Abbott, D. (ed.) (1985) *The biographical dictionary of scientists.* London: Blond Educational.

Adams, R. W. (ed.) (1991) *Directory of European industrial and trade associations* 5th ed. Beckenham: CBD Research.

Alford, M. H. T. and Alford, V. L. (1970) *Russian-English scientific and technical dictionary* 2 vols. Oxford: Pergamon.

Anthony, L. J. (ed.) (1988) *Information sources in energy technology.* London: Butterworths.

ASM handbook (1991–) Materials Park, OH: ASM International.

Association of Consulting Engineers (1995) *Consulting engineers who's who and year book.* London: Association of Consulting Engineers.

Auger, C. P. (1975) *Engineering eponyms* 2nd ed. London: Library Association.

Auger, C. P. (1994) *Information sources in grey literature* 3rd ed. London: Bowker-Saur.

Auger, C. P. (ed.) (1992) *Information sources in patents.* London: Bowker-Saur.

Bakerjian, R. (1993) *Tool and manufacturing engineers handbook* 4th ed., 7 vols. Dearborn, MI: Society of Manufacturing Engineers.

Baumeister, T. (1986) *Mark's standard handbook for mechanical engineers* 9th ed. New York: McGraw-Hill.

Bessant, J. and Lamming, R. (1991) *Macmillan dictionary of production management and technology.* London: Macmillan.

Bever, M. B. (ed.) (1986) *Encyclopedia of materials science and engineering* 8 vols (plus 3 supplementary vols). Oxford: Pergamon.

Blake, L. S. (ed.) (1989) *Civil engineers reference book* 4th ed. London: Butterworths.

Bolz, R. E. and Tuve, G. L. (eds) (1973) *CRC handbook of tables for applied engineering science* 2nd ed. Boca Raton, FL: CRC Press.

Booth, J. (ed.) (1993) *IEEE standard dictionary of electrical and electronic terms* 5th ed. (ANSI std. 100 – 1988). New York: IEEE.

Brady, G. S. and Clauser, H. R. (1991) *Materials handbook* 13th ed. New York: McGraw-Hill.

Brandes, E. A. and Brook, G. B. (eds) (1992) *Smithells metals reference book* 7th ed. Oxford: Butterworth-Heinemann.

British Standards Institution (1993) *Glossary of building and civil engineering terms.* Oxford: Blackwell Scientific.

Brundle, C. R., Evans, C. A. and Wilson, S. (1992) *Encyclopedia of materials characterization.* Boston: Butterworth-Heinemann.
Butcher, D. (1991) *Official publications in Britain* 2nd ed. London: Bingley.
Carvill, J. (1993) *Mechanical engineer's data handbook.* Oxford: Butterworth-Heinemann.
Commonwealth universities yearbook (annual) London: Gale.
Considine, D. M. and Considine, G. D. (eds) (1994) *Van Nostrand's scientific encyclopedia* 8th ed. New York: Van Nostrand-Reinhold.
Considine, D. M. (ed.) (1993) *Process/industrial instruments and control handbook* 4th ed. New York: McGraw-Hill.
Current British directories (1993) 12th ed. Beckenham: CBD Research.
Current research in Britain (1994) 9th ed. Harlow: Longman.
Davis, G. (1993) *International directory of engineering societies and related organizations* 14th ed. New York: American Association of Engineering Societies.
Davis, G. (ed.) (1995) *Who's who in engineering* 9th ed. Washington, DC: American Association of Engineering Societies.
Davis, J. R. (ed.) (1992) *ASM materials engineering dictionary.* Materials Park, OH: ASM International.
Dawson top 3000 directories and annuals (1994) 11th ed. Wellingborough: Dawson.
Dennis, G. H. (ed.) (1990) *Guide to official statistics* rev. ed. London: HMSO (Central Statistical Office).
De Sola, R. (1994) *Abbreviations dictionary* 9th ed. Boca Raton, FL: CRC Press.
De Vries, L. and Herrmann, T. M. (1972) *German-English and English-German technical and engineering dictionary.* New York: McGraw-Hill.
Directory of EEC Information Sources (1991) Genval: Euroconfidentie.
Dorf, R. C. and Nof, S. Y. (eds) (1990) *Concise international encyclopedia of robotics.* New York: Wiley.
Engineered materials handbook (1991) 4 vols. Materials Park, OH: ASM International.
Engineering Research Centres (1993) *A world directory of organizations and programmes* 3rd ed. Harlow: Longman.
Europa world year book (1994) 35th ed. London: Europa Publications.
European research centres (1993) 9th ed. Harlow: Longman.
European sources of scientific and technical information (1994) 11th ed. Harlow: Longman.
Fink, D. G. and Beaty, H. W. (eds) (1993) *Standard handbook for electrical engineers* 13th ed. New York: McGraw-Hill.
Five language technology dictionary (1993) London: Gale Research International.

Freeman, R. L. (1994) *Reference manual for telecommunications engineering* 2nd ed. New York: Wiley-Interscience.
Giacomo, J. D. (ed.) (1990) *Digital bus handbook.* New York: McGraw-Hill.
Gibilisco, S. (1994) *The illustrated dictionary of electronics* 6th ed. New York: McGraw-Hill.
Gilbert, J. (comp.) (1992) *Guide to directories at the Science Reference and Information Service* 3rd ed. London: British Library, SRIS.
Gillispie, C. C. (ed.) (1970–1980) *Dictionary of scientific biography.* New York: Scribner.
Gordon, M. and Singleton, C. (1986) *Dictionary of new information technology acronyms* 2nd ed. London: Kogan Page.
Guide to materials engineering data and information (1986) Materials Park, OH: ASM International.
Handbook of industrial materials (1992) 2nd ed. Oxford: Elsevier Advanced Technology.
Harris, C. M. (ed.) (1987) *Shock and vibration handbook* 3rd ed. New York: McGraw-Hill.
Henderson, G. P. and Henderson, S. P. A. (eds) (1994) *Directory of British associations* 12th ed. Beckenham: CBD Research.
Hicks, T. G. (1994) *Standard handbook of engineering calculations* 3rd ed. New York: McGraw-Hill.
Industrial Engineering Terminology (1991) rev. ed. ANSI standard Z94.0. Norcross, GA: Industrial Engineering and Management Press.
Industrial research in the United Kingdom (1993) 15th ed. Harlow: Longman.
James, A. M. and Lord, M. P. (1992) *Macmillan's chemical and physical data.* London: Macmillan.
Jones, G. R., Laughton, M. A. and Say, M. G. (eds) (1993) *Electrical engineer's reference book* 15th ed. Oxford: Butterworth-Heinemann.
Keller, H. and Erb, U. (1989) *Dictionary of engineering acronyms and abbreviations.* London: Adamantine.
Kettridge, J. O. (1990) *French-English and English-French dictionary of technical terms and phrases* 2nd ed. London: Routledge.
Kutz, M. (ed.) (1986) *Mechanical engineers handbook.* New York: Wiley-Interscience.
Lee, S. M. (ed.) (1989) *International encyclopedia of composites* 6 vols. New York: VCH.
Lingaiah, K. (1994) *Machine design data handbook.* New York: McGraw-Hill.
Lipták, B. G. (ed.) (1995) *Instrument engineers' handbook* 2 vols 2nd ed. Oxford: Butterworth-Heinemann.
Mahajan, S. and Kimerling, L. C. (eds) (1992) *Concise encyclopedia*

of semiconducting materials and related technologies. Oxford: Pergamon.
Mazda, F. (ed.) (1993) *Telecommunications engineer's reference book*. Oxford: Butterworth-Heinemann.
Markus, J. and Sclater, N. (1994) *McGraw-Hill electronics dictionary* 5th ed. New York: McGraw-Hill.
McGraw-Hill encyclopedia of science and technology (1992) 7th ed. New York: McGraw-Hill.
Merkow, M. S. (1990) *Breaking through technical jargon: a dictionary of computer and automation acronyms*. London: Van Nostrand Reinhold.
Merritt, F. S. (ed.) (1995) *Standard handbook for civil engineers* 4th ed. New York: McGraw-Hill.
Meyers, R. A. (ed.) (1992) *Encyclopedia of physical science and technology*. 2nd ed. San Diego, CA: Academic Press.
Millard, P. M. (ed.) (1994) *Trade associations and professional bodies of the United Kingdom* 12th ed. London: Gale Research International.
Morris, C. (ed.) (1992) *Academic Press dictionary of science and technology*. San Diego, CA: Academic Press.
Mullay, M. and Schlicke, P. (eds) (1993) *Walford's guide to reference material*, vol. 1. Science and technology, 6th ed. London: Library Association.
Nayler, G. H. F. (1985) *Dictionary of mechanical engineering* 3rd ed. London: Butterworths.
Newton, D. (1991) *Trade marks – an introductory guide and bibliography* 2nd ed. London: British Library, Science Reference and Information Service.
Oberg, E., et al. (1992) *Machinery's handbook* 24th ed. New York: Industrial Press.
Oxford dictionary of abbreviations (1992) Oxford: Clarendon Press.
Parker, C. C. and Turley, R. V. (1986) *Information sources in science and technology* 2nd ed. London: Butterworths.
Parker, S. P. (1993a) (ed.) *McGraw-Hill encyclopedia of engineering* 2nd ed. New York: McGraw-Hill.
Parker, S. P. (1993b) (ed.) *McGraw-Hill dictionary of scientific and technical terms* 5th ed. New York: McGraw-Hill.
Porteous, A. (1992) *Dictionary of environmental science and technology* rev. ed. Chichester: Wiley.
Pugh, E. (1987) *Pugh's dictionary of acronyms and abbreviations: abbreviations in management, technology and information science* 5th ed. London: Library Association.
Reynard, K. W. (1992) *UK materials information sources* 2nd ed. London: Design Council.

Roney, A. (1993) *The European Community fact book* 3rd ed. London: Kogan Page.
Ross, R. B. (1992) *Metallic materials specification handbook* 4th ed. London: Chapman and Hall.
Rothbart, H. A. (ed.) (1985) *Mechanical design and systems handbook* 2nd ed. New York: McGraw-Hill.
Routledge German technical dictionary (1995) 2 vols. London: Routledge.
Routledge French technical dictionary (1994) 2 vols. London: Routledge.
Sachs, M. (ed.) (1990) *World guide to scientific associations and learned societies* 5th ed. Munich: K. G. Saur.
Scott, J. S. (1991) *Penguin dictionary of civil engineering* 4th ed. London: Penguin.
Sharpe, C. (ed.) (1995) *Kempe's engineers year book* 100th ed. Tonbridge: M-G Information Services.
Sheehy, E. P. (ed.) (1987) *Guide to reference books* 10th ed. Chicago, IL: American Library Association.
Sheehy, E. P. and Balay, R. (eds) (1992) *Guide to reference books: covering materials from 1985–90*, A supplement to the 10th edition. Chicago, IL: American Library Association.
Skolnik, M. I. (ed.) (1990) *Radar handbook* 2nd ed. New York: McGraw-Hill.
Smith, E. H. (ed.) (1994) *Mechanical engineer's reference book* 12th ed. Oxford: Butterworth-Heinemann.
Stoliarov, D. E. and Kuzmin, I. A. (1991) *Comprehensive English-Russian scientific and technical dictionary*, 2 vols. New York: French and European Publications.
Stramler, J. H. (1993) *The dictionary for human factors/ergonomics*. Boca Raton, FL: CRC Press.
Tapley, B. D. and Poston, T. R. (eds) (1990) *Eshbach's handbook of engineering fundamentals* 4th ed. New York: Wiley.
Thomson, I. (1989) *The documentation of the European Communities: a guide*. London: Mansell.
Towell, J. E. and Sheppard, H. E. (1994) (eds) *International acronyms, initialisms and abbreviations dictionary* 3 vols, 2nd ed. Detroit, MI: Gale Research.
Union of International Organizations (annual) *Yearbook of international organizations* 3 vols. Munich: K. G. Saur.
Vollmer, E. (1991) *Encyclopaedia of hydraulics, soil and foundation engineering*. Amsterdam: Elsevier.
Walker, P. M. B. (ed.) (1993) *Chambers materials science and technology dictionary*. Edinburgh: Chambers.
Walker, P. M. B. (ed.) (1990) *Cambridge dictionary of science and technology*. Cambridge: Cambridge University Press.

Wasserman, S. R. (1995) *Encyclopedia of physical science and engineering information* 2nd ed. London: Gale.

Waterman, N. A. and Ashby, M. F. (eds) (1991) *Elsevier materials selector* 3 vols. London: Elsevier Applied Science.

Wennrich, P. (1992) *International dictionary of abbreviations and acronyms of electronics, electrical engineering, computer technology and information processing*, 2 vols. Munich: K. G. Saur.

Who's who in science in Europe (1993) 8th rev. ed. Harlow: Longman.

Who's who of British engineers (1982) 6th ed. London: Simon Books.

Who's who in science and engineering, 1994–1995 (1994) 2nd ed. New Providence, NJ: Marquis.

Williamson, M. (1990) *Dictionary of space technology*. Bristol: Hilger.

World of learning (annual). London: Europa.

Zolynski, B. (1991) *Basic sources on European Communities information*. Manchester: European Information Association.

CHAPTER TEN

Stress analysis

P. STANLEY AND J. SMART

Introduction

The 'strength' of a body or structure subjected to a load (e.g. an applied force) will depend on the magnitude of the stresses induced by the load and the strength of the material. (Stresses are defined as 'force per unit area'; when the force is normal to the area element the derived stresses are referred to as direct or normal stresses and when the force is in the plane of the area element the derived stresses are shear stresses.)

Techniques for the determination of these stresses and their appraisal comprise the subject area of Stress Analysis. This is a central topic in a wide range of engineering studies, especially design; it also has an important role in many of the physical sciences and in biomechanics (e.g. in the design of prostheses). Galileo recognized the importance of the subject, but essentially it is rooted in the classical work of a number of 19th century engineers and mathematicians, admirably reviewed by Timoshenko (1983). There is a very considerable overlap between Stress Analysis, Strength of Materials and Solid Mechanics, and increasingly nowadays the experimental part of Stress Analysis is referred to as Experimental Mechanics. (In 1985 the American Society for Experimental Stress Analysis changed its name to the Society for Experimental Mechanics, and in 1990 the European Permanent Committee for Stress Analysis became the European Permanent Committee for Experimental Mechanics.) This reflects the widely held view that Stress Analysis has a much broader scope than might be inferred from this title; it is also partly attributable to the inordinate growth and development in techniques, materials and applications which have occurred over the last two decades or so. Particularly prominent 'growth areas' have been Numerical Stress Analysis, Composite Materials and Fracture Mechanics. In a sense, the nomenclature change from Stress Analysis to Experimental Mechanics might be seen as unfortunate in that it separates off one form of stress analysis (i.e. experimental) and by implication precludes others. The emergence of numerical stress analysis techniques (Finite Element Analysis and Boundary Element Analysis) has completely transformed the 'tools' of stress analysis, and these techniques must figure prominently in any Stress Analysis review.

In this chapter the essentials of the different alternative stress analysis techniques are outlined, the relevant organizational/professional structure is described and details of a range of generally useful information sources are given. Information pertaining specifically to each technique in turn is then provided.

Stress analysis techniques

In approaching the choice of a stress analysis technique for a particular application, the material type, the way in which it responds to stress and the nature of the stress itself are important considerations. A material may be isotropic or anisotropic, homogeneous or composite and its response to an applied load may be elastic (linear or non-linear), plastic or viscoelastic. The loading may be such that the stresses are static, dynamic or possibly transient. The loading may be direct mechanical loading associated with applied forces and moments, or thermal loading associated with non-uniform temperatures or thermal mismatch. There may be stresses in an unloaded body (i.e. 'residual' stresses) associated, for example, with a manufacturing process, with bulk effects during solidification of a casting or with partial non-elastic behaviour during a prior loading. A further point of detail is that in many cases the measured quantity is strain (i.e. the relative deformation caused by a stress) and not stress *per se*; the 'stress' nomenclature persists nevertheless. Analytical techniques normally produce stress results.

Most stress analysis work relates to elastic behaviour due to mechanical loading in an isotropic solid, and any of the techniques mentioned may, in principle, be used for this purpose. Under other conditions, however, techniques may be highly specific and specialist knowledge may be required in their implementation. The principal features of the most widely used techniques are summarized below, firstly for experimental techniques which involve actual components or physical models, then for numerical techniques which nowadays are almost exclusively computer-based.

Experimental

PHOTOELASTIC STRESS ANALYSIS

In photoelastic work the optical interference effects in polarised light in a transparent model of the body in question, appropriately loaded, are studied to provide quantitative stress data. Techniques are available for two-dimensional and three-dimensional elastic stress problems. With appropriate model materials elastic-plastic stress studies are possible. Surface strain distributions (elastic and non-elastic) in metallic components and thermal stresses can also be studied.

ELECTRICAL RESISTANCE STRAIN GAUGES

These small devices are bonded directly to the surface of the component or structure under investigation, so that the surface strain is 'copied' into the gauge. They consist essentially of a metallic wire or foil, the electrical resistance of which varies with strain. By monitoring the change in resistance, or the effects of this change, in a balanced bridge circuit, the strain at the position of the gauges on the surface of the component or structure due to the applied load can be determined. These gauges are available in a wide variety of size and form. High-temperature and high-strain versions are available, and they can be used for the determination of residual stresses.

MOIRÉ FRINGE ANALYSIS

Moiré fringes are observed when a deformed grid of lines or grating (the specimen grating) is viewed through a nominally identical grid (the reference grating) which has not been deformed. (In practice, the specimen grating is applied to the surface of the component or specimen under investigation before the load is applied.) The fringes are formed as a result of geometric (or 'mechanical') interference and they relate directly to the surface displacements. Several versions of the technique are available; a high-sensitivity development (moiré interferometry) is noteworthy.

HOLOGRAPHIC INTERFEROMETRY AND ESPI

The physical basis of the holographic interferometry approach is the formation of displacement contours in the form of interference fringes produced as a result of optical interference between wavefronts from a strained body and the superimposed holographic reconstruction of the unstrained body. Techniques are available for static and dynamic studies.

A closely related technique is electronic speckle pattern interferometry (ESPI) or TV holography. This entails the study of the optical interference effects associated with the surface 'speckle' patterns observed when an optically rough surface is illuminated with coherent light. Video imaging and real-time signal processing are used.

THERMOELASTIC STRESS ANALYSIS

Under suitable conditions, a change of stress in an elastic solid causes a small change in temperature (this is referred to as the 'thermoelastic effect'). In exploiting this effect for stress analysis, the temperature changes associated with the cyclic loading of a specimen or component, in the elastic range, are monitored, without direct contact with the specimen, by means of a sensitive infra-red detector which 'scans' the specimen surface over a pre-selected 'frame'. The stress state within the scanned area of the specimen surface is then inferred from these measurements. The technique can be used with isotropic or anisotropic solids and with sinusoidal or random loading.

OTHER EXPERIMENTAL TECHNIQUES

Other experimental techniques are available but they may be less versatile or more specialized than those mentioned above.

The *brittle lacquer* approach depends on the observation of the position and orientation of cracks which develop in a brittle lacquer applied to the surface of a body as the applied load is gradually increased. This is not usually regarded as a quantitative technique in itself but is a valuable preliminary to an extensive strain gauge investigation.

X-ray diffraction and *neutron diffraction* techniques allow direct measurement of lattice dimensions (i.e. the structural lattice of the constituent crystallites or grains of a material) and thence strains and stresses. The latter in particular is very specialized but both methods are particularly relevant for the study of residual stresses.

The shadow optical method of *caustics* is based on the effects of thickness changes and refractive index changes in a stressed plate. The method has been used in the study of crack-tip stresses.

Numerical

Numerical stress analysis is used to analyse complicated bodies by making a mathematical model which is then solved by a computer. There are two important techniques in this area.

FINITE ELEMENT METHOD

In this method a body is treated as an assembly of small (finite) elements for which approximate equations can be written to relate the forces applied to the element with the resultant displacements. The elements can be one-dimensional e.g. bars and beams, two-dimensional membrane elements, three-dimensional e.g. bricks and tetrahedra, plates, shells, etc. The technique can be used for both linear and non-linear stress problems, and for static and dynamic problems. The original development of the method was for structural analysis but it has now been generalized as a technique for solving partial differential equations. Consequently, it can also be used for determining temperature fields in a body which may then be used to determine thermal stresses.

The method is extensively used in industry with the initial model being built in a CAD system and the results of the analysis then being obtained from post-processing programs. Programs are available which will automatically optimize a structure, although this is restricted generally to simply shaped bodies.

BOUNDARY ELEMENT ANALYSIS

In this method the surface of a body is divided into small elements and so the representation of a body is reduced by one dimension. For example, for a three-dimensional body only a two-dimensional surface needs to be represented.

This method is not as widely used as the finite element method for various reasons. These include (1) the mathematics are more complex and not as readily understood by engineers and (2) the method is most suited to linear problems of 'chunky bodies'; it is not very suitable for structures.

Because the major component of the cost of an analysis is the analyst's time, it is usually more efficient to use only one numerical technique. As the finite element method is more widely applicable, this is the technique which is normally used.

Professional organizations (UK)

Professional institutions and societies

The following professional organizations have some form of involvement in stress analysis activities in the UK:

British Society for Strain Measurement (BSSM), c/o Prof. I. M. Allison, The Surrey Technology Centre, The Surrey Research Park, Guildford, Surrey GU2 5YG, UK.

Engineering Integrity Society (EIS), 5 Wentworth Avenue, Sheffield, South Yorks S11 9QX, UK.

Institute of Marine Engineers, 76 Mark Lane, London EC3R 7JN, UK.

Institute of Materials, 1 Carlton House Terrace, London SW1Y 5DB, UK.

Institute of Physics, 47 Belgrave Square, London SW1X 8QX, UK.

Institution of Civil Engineers, Great George Street, London SW1F 3AA, UK.

Institution of Electrical Engineers, Savoy Place, London WC2R 0BL, UK.

Institution of Mechanical Engineers, 1 Birdcage Walk, London SW1H 9JJ, UK.

Institution of Structural Engineers, 11 Upper Belgrave Street, London SW1X 8BH, UK.

Royal Aeronautical Society, 4 Hamilton Place, London W1V 0BQ, UK.

Royal Institution of Naval Architects, 10 Upper Belgrave Street, London SW1X 7BQ, UK

The Welding Institute, Abington Hall, Abington, Cambridge CB1 6AL, UK.

The ways in which these organizations further the stress analysis interests of their members differ from one organization to another. Conferences, seminars, meetings and exhibitions are organized from time to time at both local and national levels. The British Society for Strain Measurement (BSSM) publishes the journal *Strain* which contains technical papers, information on new products, conference announcements and a particularly useful diary of forthcoming events. The Materials and Mechanics of Solids Group of the Institution of Mechanical Engineers and the Stress Analysis Group of the Institute of Physics are effective in promoting a range of activities in relevant areas.

Joint British Committee for Stress Analysis (JBCSA)

An important common feature of the organizations listed above is that each has a representative on the Joint British Committee for Stress Analysis (JBCSA). This Committee was established in 1960 with the prime objectives of liaising with and providing representation on the Permanent European Committee for Stress Analysis, coordinating stress analysis activities and interests in the UK, and sponsoring appropriate meetings and conferences. The Committee has fulfilled its role very effectively over the years, and continues to do so. A major initiative of the Committee was the founding of the *Journal of Strain Analysis* in 1965. The Committee meets regularly, usually at the Institution of Mechanical Engineers; its membership is listed on the inside back cover of the *Journal of Strain Analysis for Engineering Design*.

NAFEMS

NAFEMS was started at the National Engineering Laboratories in East Kilbride in the early 1980s. The acronym originally stood for National Agency for Finite Element Methods and Standards but now the organization is international and is simply known as NAFEMS. When it was formed its first aim was to formulate benchmarks which would test the suitability of commercial finite element programs. However, as quality assurance was taken more seriously in the 1980s, NAFEMS developed into training, education and quality assurance, and it has now formed a separate organization to perform quality assurance assessment of companies' finite element procedures.

NAFEMS still has its offices in East Kilbride and is supported in its work by a series of committees which cover the main areas of finite element usage. It publishes a quarterly magazine, *Benchmark* (with a diary of events and much useful information), and also other publications about finite elements. Of particular interest to beginners is a series of booklets in a 'How To' series, e.g. *How to get started with FEA*. Membership of NAFEMS is through companies although individuals may purchase its publications. The address is:

NAFEMS, Birniehill, East Kilbride, Glasgow G75 0QU, UK.

Other professional organizations

European Permanent Committee for Experimental Mechanics (EPCEM)

Previously known as the European Permanent Committee for Stress Analysis (EPCSA), this comprises representatives from some fifteen European countries. Its prime *raison d'être* is the organization of a major international conference on experimental mechanics every four years. Recent venues include Amsterdam (1986), Copenhagen (1990) and Lisbon (1994). The published conference proceedings provide an invaluable data source in the experimental mechanics field. A meeting of the Committee is held at each conference, where the principal task is to agree the venue for the next conference. The appropriate professional bodies in the chosen host country take on the organization of the next conference and provide the chairman of the EPCEM for the four-year period. This phoenix-like rejuvenation has proved extremely effective in ensuring the continuing success of the EPCEM activities since the Committee's inception in 1966.

European Structural Integrity Society (ESIS)

The Society was originally called the European Group on Fracture (EGF). It organizes the biennial European Conference on Fracture which provides a major European forum for technical interaction and exchanges in the fields of fracture mechanics and structural integrity. There are ten technical committees which cover the various fields of fracture, e.g. fatigue and environmentally assisted cracking, and also the various types of materials, e.g. ceramics and concrete. The technical committees organize meetings in their own fields. The Society publishes a very useful newsletter. The address of the secretariat is:

Prof. A. Bakker, (ESIS Secretariat), Materials Laboratory, PO Box 5025, 260 GA Delft, The Netherlands.

International Measurement Confederation (IMEKO)

IMEKO is a federation of national societies concerned with measurement science, technology and instrumentation, with headquarters and secretariat in Budapest. Its aims include the promotion of information exchange by organizing congresses and symposia and by publishing proceedings. It operates through a series of technical committees, the most relevant of which, in the present context, is TC15 on Experimental Mechanics. The address of the IMEKO secretariat is:

IMEKO Secretariat, PO Box 457, 1371 Budapest, Hungary.

International Union of Theoretical and Applied Mechanics (IUTAM)

IUTAM is an international organization which sponsors symposia in countries throughout the world. Attendance is usually by invitation. Symposia proceedings are published and often become valuable reference texts. The UK input to IUTAM activities is made through:

IUTAM Panel, The Royal Society, 6 Carlton House Terrace, London, SW1Y 5AG, UK.

National bodies

Many countries have a national organization covering professional interests in stress analysis and experimental mechanics. Prominent amongst these organizations is the American Society for Experimental

Mechanics (SEM). The Society was founded in 1943 and has headquarters at:

7 School Street, Bethel, CT 06801, USA.

It is organized on a divisional basis and is active over a very broad field. It publishes *Experimental Mechanics* (quarterly) and *Experimental Techniques* (bi-monthly), and its annual Spring Conference attracts papers and delegates from the stress analysis community worldwide.

Other active organizations (with a contact address) include:

Associazione Italiana per l'Analisi delle Sollecitazioni (ASIS), Fulvio di Marino (President), Dipartimento di Energetica, Universita di Trieste, Via A. Valerio, 10-I 34100 Trieste, Italy.

Associaçáo Portuguesa de Análise Experimental de Tensoes (APAET), Av. do Brasil 101, 1799 Lisboa, Portugal.

Groupement pour l'Avancement des Méthodes d'Analyse des Constraintes (GAMAC), Dr J. F. Julieu (President of GAMAC), Insa Lyon, DT Genie Civil, 20 Avenue Albert Einstein, 96921 Villuerbanne Cedex, France.

Gemeinschaft für Experimentelle Spannungsanalyse (GESA), Professor K-H Laermann, Bergische Universität – Gesamthochschule Wuppertal, Gauss-Strasse 20, Postfach 100127, 5600 Wuppertal 1, Germany.

Handbooks and data sources

Well established reference texts covering the stress analysis field are:

Benham, P. P. and Crawford, R. J. (1987) *Mechanics of engineering materials* (Longman).
Kobayashi, A. S. (ed.) (1993) *Handbook on experimental mechanics*, 2nd edn. (VCH).
Pople, J. (1979) *BSSM strain measurement reference book* (British Society of Strain Measurement).
Rohrbach, C. (ed.) (1989) *Handbuch für experimentelle spannungsanalyse* (VDI-Verlag).

A valuable series of more advanced tests has been produced by S. Timoshenko and co-authors, published by McGraw-Hill.

Textbooks and compilations of relevant data in the form of stress solutions and stress intensity factors include:

Ewalds, H. L. and Wanhill, R. J. H. (1984) *Fracture mechanics* (Edward Arnold).

Halpin, J. C. (1992) *Primer on composite materials: analysis*, 2nd ed. (Technomic Publishing Company).
Knott, J. F. (1973) *Fundamentals of fracture mechanics* (Butterworths).
Pendleton, R. L. and Tuttle, M. E. (1989) Manual on experimental methods for mechanical testing of composites (Society for Experimental Mechanics).
Peterson, R. E. (1973) *Stress concentration factors* (Wiley).
Rooke, D. P. and Cartwright, D. J. (1976) *Compendium of stress intensity factors* (HMSO).
Young, W. C. (1989) *Roark's formulas for stress and strain*, 6th Edn (McGraw-Hill).

A comprehensive index of publications is available from ESDU (Engineering Sciences Data Unit) giving, amongst other data, solutions to stress problems: Validated Engineering Data Index, ESDU International plc, 27 Corsham Street, London N1 6UA, UK.

The British Library Document Supply Centre (Boston Spa, Wetherby, West Yorkshire, LS23 7BQ) offers a valuable service by providing (1) Focus Bulletins. These are new monthly issues detailing recent reports, etc. covering a specified subject area. (2) British Thesis Service. Over 120 000 British doctoral theses are available. (3) Subject Search. Available reports, conferences and theses within a specified subject area are listed for a small charge.

Journals

Journals concerned entirely with Stress Analysis/Experimental Mechanics include:

Journal of Strain Analysis for Engineering Design, Mechanical Engineering Publications, Northgate Avenue, Bury St Edmunds, Suffolk IP32 6BW, UK.

Strain (Journal of the British Society for Strain Measurement), Prof. I. M. Allison (Hon Sec), Department of Civil Engineering, University of Surrey, Guildford GU2 5XH, UK.

Experimental Mechanics, Society for Experimental Mechanics, 7 School Street, Bethel, CT 06801, USA.

Experimental Techniques, Society for Experimental Mechanics, 7 School Street, Bethel, CT 06801, USA.

Journals with a broader remit but a strong Stress Analysis/Experimental Mechanics interest include:

International Journal of Mechanical Sciences, Elsevier Science, The Boulevard, Langford Lane, Kidlington, Oxford OX5 1GB, UK.

Österreichische Ingenieur – und Architekten – Zeitschrift, Redaktion der ÖIAZ, Eschenbachgasse 9, A-1010 Wien, Austria.

International Journal of Fatigue, Butterworth-Heinemann. PO Box 63, Westbury House, Bury Street, Guildford, Surrey GU2 5BH, UK.

Engineering Fracture Mechanics, Elsevier Science (see above).

International Journal of Fracture, Kluwer Academic Publishers, Postbus 17, 3300 AA, Dordrecht, The Netherlands.

Fatigue and Fracture of Engineering Materials and Structures, SIRIUS, Faculty of Engineering, University of Sheffield, Mappin Street, Sheffield S1 3JD, UK.

Communications in Numerical Methods in Engineering, John Wiley & Sons, Baffins Lane, Chichester, Sussex PO19 1UD, UK.

Finite Elements in Analysis and Design, Elsevier Science (see above).

International Journal for Numerical Methods in Engineering, John Wiley & Sons (see above).

International Journal of Pressure Vessels and Piping, Elsevier Science (see above).

Thin Walled Structures, Elsevier Science (see above).

Composites, Butterworth-Heinemann (see above).

Journal of Composite Materials, Technomic Publishing Co, 851 New Holland Ave, Box 3535, Lancaster, PA 17604, USA.

Materials and Structures, Secretariat General RILEM, 61 Av. du President Wilson, 94235 Cachan Cedex, France.

Measurement Science and Technology, Institute of Physics Publishing, Techno House, Redcliffe Way, Bristol BS1 6NX, UK.

The Aeronautical Journal, Royal Aeronautical Society, 4 Hamilton Place, London W1V 0BQ, UK.

Proceedings of the Institution of Mechanical Engineers, in particular:

Journal of Mechanical Engineering Science (Part C), *Journal of Automobile Engineering* (Part D), *Journal of Aerospace Engineering* (Part G), Mechanical Engineering Publications Ltd, Northgate Avenue, Bury St Edmunds, Suffolk IP32 6BW, UK.

Transactions of the ASME (American Society of Mechanical Engineers), in particular:

Journal of Applied Mechanics, Journal of Engineering Materials and Technology, Journal of Pressure Vessel Technology, American Society of Mechanical Engineers, 345 East 47th St, New York, NY 10017–2392, USA.

The American Society of Civil Engineers publishes:

Journal of Aerospace Engineering, Journal of Engineering Mechanics, Journal of Structural Engineering, American Society of Civil Engineers, 345 East 47th St, New York, NY 10017–2398, USA

Conferences

Well established annual conferences devoted to stress analysis and experimental mechanics are:

The Annual Conference of the BSSM. This is typically a two-day event, with a series of technical sessions for the presentation of papers. The papers are brought together in a bound volume of conference proceedings. The theme of the 1994 conference, held in Edinburgh and organized jointly with the SEM, was 'Advances in Engineering Measurement'. The 1995 conference was held in Sheffield, with the theme 'Automated Strain Measurement and Analysis'.

The Spring Conference of the SEM. This conference, held each year in early June, usually attracts several hundred delegates; the venue moves around the USA. The Proceedings of the 1995 Spring Conference held in Grand Rapids, Michigan, contains 145 papers arranged under 44 sessions headings.

The Danubia-Adria Symposium. The annual symposium is devoted to Experimental Methods in Solid Mechanics and is organized jointly by professional organizations in Austria, Croatia, the Czech Republic, Hungary and Italy. It is usually a three-day event. Extended summaries of the papers presented are distributed to all participants and selected papers may be published after the event. Further information is available from any of the organizing bodies (e.g. ASIS – see above) or from:

Dr Rudolf Beer, Technische Universität, Institut für Festigkeitslehre, A-1030 Vienna, Adolf-Blamauer-Gasse 1–3, Austria.

A noteworthy European event, held at four-yearly intervals, is the International Conference on Experimental Mechanics organized by a host country nominated by the European Permanent Committee for Experimental Mechanics (see above). The papers presented are published in full in the conference proceedings. The venue for the 1998 Conference will be Oxford.

In addition to the major events cited above there are many regular and *ad hoc* conferences, seminars and symposia organized by national bodies and independent organizations such as the universities and research establishments. It is not practicable to attempt to list such events here but many of them are publicized in the journals mentioned above. A 'Diary' published in each issue of the BSSM journal *Strain* is particularly useful in this respect.

Reference texts and suppliers

The key reference texts and equipment sources for the stress analysis techniques above are detailed below.

Photoelasticity

GENERAL

Durelli, A. J. and Riley, W. F. (1965) *Introduction to photomechanics* (Prentice Hall).
Frocht, M. M. (1941, 1948) *Photoelasticity*, volumes 1 and 2 (Wiley).
Kuske, A. and Robertson, G. (1974) *Photoelastic stress analysis* (Wiley).

SPECIALIST TEXTS

Aben, H. (1979) *Integrated photoelasticity* (McGraw-Hill).
Theocaris, P. S. and Gdoutos, E. E. (1979) *Matrix theory of photoelasticity* (Springer-Verlag).

EQUIPMENT

Polariscopes, optical accessories, photoelastic materials.

Sharples Photomechanics Ltd, Europa Works, Wesley St, Bamber Bridge, Preston PR5 4PB, UK.

146 *Stress analysis*

Measurements Group UK Ltd, Stroudley Road, Basingstoke, Hants RG24 8FW, UK.

Strain gauges

Window, A. L. (1978) *An Introduction to strain gauges*, 5th ed. (Welwyn Strain Measurement).

EQUIPMENT

Gauges, signal conditioners, amplifiers, loggers.

Fylde Electronic Laboratories Ltd, 49/51 Fylde Road, Preston PR1 2XQ, UK.

Graham & White Instruments Ltd, 135 Hatfield Road, St Albans, Herts AL1 4LZ, UK.

Measurements Group UK Ltd, Stroudley Road, Basingstoke, Hants RG24 8FW, UK.

Nobel Systems Limited, Murdock Road, Bedford, Bedfordshire MK41 7PQ, UK.

RDP Electronics Ltd, Grove St, Heath Town, Wolverhampton WV10 0PY, UK.

Strainstall Engineering Services Ltd, Denmark Road, Cowes, Isle of Wight PO31 7TB, UK.

Techni Measure, Alexandra Buildings, 59 Alcester Road, Studley, Warwickshire B80 7NJ, UK.

Holographic interferometry and ESPI

Jones, R. and Wykes, C. (1989) *Holographic and speckle interferometry* 2nd ed. (Cambridge University Press).
Vest, C. M. (1979) *Holographic interferometry* (Wiley).

EQUIPMENT

Optical instruments and accessories, positioning systems, lasers, fibre-optics, electro-optics.

Comar Instruments, 70 Hartington Grove, Cambridge CB1 4UH, UK.

Ealing Electro-Optics plc, Greycaine Road, Watford WD2 4PW, UK.

Laser Lines Ltd, Beaumont Close, Banbury, Oxon OX16 7TQ, UK.

Melles Griot Photon Control Ltd, Brookmount Court, Kirkwood Road, Cambridge CB4 2QH, UK.

Newport Micro-Controle, 4320 First Avenue, Newbury Business Park, London Road, Newbury, Berkshire RG13 2PZ, UK.

Spindler & Hoyer UK Ltd, 2 Drakes Mews, Crownhill, Milton Keynes, Buckinghamshire MK8 0ER, UK.

Moiré fringe analysis

Durelli, A. J. and Parks, V. J. (1970) *Moiré analysis of strain* (Prentice-Hall).
Theocaris, P. S. (1969) *Moiré fringes in strain analysis* (Pergamon).

EQUIPMENT

Most of the essential optical equipment is available from the suppliers listed above. High quality gratings are available from:

National Physical Laboratory, Department of Mechanical and Optical Metrology, Queens Road, Teddington, Middlesex TW11 0LW, UK.

Thermoelastic stress analysis

Harwood, N. and Cummings, W. M. (eds) (1991) *Thermoelastic stress analysis* (IOP Publishing).

EQUIPMENT

SPATE equipment, accessories.

Ometron Ltd, Kelvin House, Worsley Bridge Road, London SE26 5BX, UK.

SOFTWARE

Stress Photonics, 565 Science Drive, Madison WI 53711, USA.

Other experimental techniques

Introductory chapters on the brittle lacquer technique and the method of caustics are given in the Kobayashi Handbook (see above). The X-ray and neutron diffraction techniques are described in:

Allen, A. J., Hutchings, M. T. and Windsor, C. G. (1985) Neutron diffraction methods for the study of residual stress fields. *Advances in Physics*, **34**, 445–473.

Maeder, G., Lebrun, J. L. and Sprauel, J. M. (1981) Present possibilities for the X-ray diffraction method of stress measurement. *NDT International* **14**, 235–247.

Noyan, I. C. and Cohen, J. B. (1987) *Residual stress: measurement by diffraction and interpretation* (Springer-Verlag).

Smith, G. M. and Jolly, C. B. (1992) Residual stresses: their measurement and significance. In: *Experimental mechanics (technology transfer between high tech engineering and biomechanics)*, ed. E. G. Little (Elsevier).

EQUIPMENT

Brittle lacquer:

Stresscoat, International Sales, Electrix Industries Inc, PO Box F, Grayslake, IL 60030, USA.

X-ray diffraction:

Xray Associates, PO Box 25, Newbury, Berkshire RG16 8RY, UK.

Finite element method

INTRODUCTORY

NAFEMS (1986) *A finite element primer* (National Engineering Laboratory).

REFERENCE

Zienkiewicz, O. L. and Taylor, R. L. (1989, 1991) *The finite element method*, volumes 1 and 2, 4th ed. (McGraw-Hill).

PACKAGES

There are many commercially available packages for finite element analysis. Typical are:

ABAQUS, Hibbit, Karlsson & Sorenson Inc, 1080 Main St, Pawtucket, RI 02680, USA.

ANSYS, Swanson Analysis Systems Inc, Johnson Road, Houston, PA 15342–0065, USA.

LUSAS, FEA Ltd, Forge House, 66 High St, Kingston upon Thames KT1 1HN, UK.

PAFEC, PAFEC Ltd, Strelley Hall, Nottingham NG8 6PE, UK.

A program designed for optimization is:

RASNA, RASNA Corporation, 2590 North 1st St, Suite 200, San Jose, CA 95131, USA.

Boundary element analysis

Becker, A. A. (1992) *The boundary element method in engineering: a complete course* (McGraw-Hill).

PACKAGE

BEASY, Computational Mechanics Beasy, Ashurst Lodge, Ashurst, Southampton SO4 2AA, UK.

Reference

Timoshenko, S.P. (1983) *History of strength of materials*. New York: Dover.

CHAPTER ELEVEN

Kinematics, dynamics and machines

G. DRUCE

Introduction

The engineer involved with machinery will be concerned with aspects of the design, operation, reliability and safety of making and using the product. Appreciation of the operation requires knowledge of kinematics, 'the branch of theoretical mechanics dealing with the geometry of motion irrespective of the causes that produce the motion', and of dynamics, 'the branch of theoretical mechanics dealing with the motion and equilibrium of bodies and mechanical systems under action of forces'. [These definitions are reprinted from *Mechanism and Machine Theory*, **26**, 5, Leinonen, T. E. *et al.* International Federation for the Theory of Machines and Mechanisms Commission A, Terminology for the Theory of Machines and Mechanisms. (1991) with kind permission from Elsevier Science, Bampfylde Street, Exeter, EX1 2AH, UK]. Ability to analyse provides the foundation for synthesizing a new design and for specifying a procured item. Awareness of recommendations contained in British Standards Institution publications and Codes of Practice issued by the Health & Safety Executive, Lloyds Register, or other relevant body is also vital.

Therefore the study of machinery involves books, standards, journals and conference papers plus an awareness of relevant design data and software packages available from specialist firms and beginning to be supplied by manufacturers as supplements to their catalogues. Use of databases or abstracting sources (see Chapters 7 and 8) is strongly recommended to minimize the time occupied in literature searches and the danger of omitting relevant works. Recent publications reflect the fact that authors can now assume readers to be computer-literate and to have access to microcomputers. Hence recently published books place

152 Kinematics, dynamics and machines

greater emphasis on numerical analysis and the provision of flowcharts and listings, some are accompanied by floppy disks. The increasing complexity of modern machinery involving multi-disciplinary knowledge is reflected by the significant number of new books published on design methodology and on reliability, both also relevant to the increasingly stringent safety regulations being imposed in the UK under the aegis of the Health & Safety at Work Act, 1974.

Books and papers on kinematics, dynamics and machines are held by the Institution of Mechanical Engineers Library, London or can be obtained through local libraries from the British Library Document Supply Centre, Boston Spa. Currently, October 1995, there is open access reference to the large collection of books and journals at the British Library Science Reference Information Service (SRIS). Arrangements can usually be made to use University Libraries. The Health & Safety Executive (HSE) has public enquiry services at Sheffield, London and Bootle. General assistance in locating publications, software, university research and suitably qualified consultants can be obtained through the Machinery Design Centre, Aston Science Park. Longman Catermill publish a register of university research publications, patents and work in progress, also available as the BEST database.

The principles of standardization and the organization of the British Standards Institution (BSI) are outlined in BS 0: 1991. Preferred sizes, codes of practice, etc. published by BSI are identified by consulting the *BSI standards catalogue*, which also lists the many libraries holding complete, up-to-date sets of British Standards. Details of new and revised standards are announced in *BSI Update* and *BSI News*; the CD-ROM PERINORM service from BSI contains a complete set of BS, Deutsches Industrie-Norm (DIN) and Association Française de Normalisation (AFNOR) standards.

Besides the learned society activities of such organizations as the Institution of Mechanical Engineers (IMechE), the Institution of Engineering Designers (IED) and the American Society of Mechanical Engineers (ASME), conferences and seminars are organized by the International Federation for the Theory of Machines and Mechanisms (IFToMM). Commissions of IFToMM are responsible for compiling multi-lingual glossaries of terminology and similar work (see Crossley, 1970).

Organizations

American Society of Mechanical Engineers, 345 East 47th Street, New York, NY 10017, USA. Tel. +1 212 705 7722.

British Library Document Supply Centre, Boston Spa, Wetherby, West Yorks LS23 7BQ, UK. Tel: +44 1937 546080.

British Library Science Reference Information Service (SRIS), 25 Southampton Buildings, London WC2A 1AW, UK. Tel: +44 171 323 7494.

British Standards Institution Sales and Accounts, Linford Wood, Milton Keynes MK14 6LE, UK. Tel: +44 908 221166.

Institution of Engineering Designers, Courtleigh, Westbury Leigh, Westbury, Wilts BA13 3TA, UK. +44 1373 822801.

Institution of Mechanical Engineers, 1 Birdcage Walk, London SW1H 9JJ, UK. Tel: +44 71 222 7899; Library +44 71 973 1274.

Institution of Mechanical Incorporated Engineers, 3 Birdcage Walk, London SW1H 9JJ, UK. Tel: +44 171 799 1808.

International Federation for the Theory of Machines and Mechanisms, see *Mechanism and Machine Theory*. Exeter: Elsevier Science.

Longman Catermill Ltd., Technology Centre, St. Andrews, Fyfe KY16 9EA, UK. Tel: +44 334 77660.

Machinery Design Centre, Business and Innovation Centre, Aston Science Park, Aston Triangle, Birmingham B7 4BJ, UK. Tel: +44 121 359 0891.

Abstracts and databases

Papers in journals and conference proceedings are more difficult to find than books. Details of conference and symposia proceedings are found conveniently in *InterDok*. See Chapters 7 and 8 for details of hardcopy abstracts and databases; those directly applicable to machinery include:

Applied Mechanics Reviews (ASME).
Current Technology Index/CTI PLUS (Bowker-Saur).
Engineering Index/EI COMPENDEX PLUS (Engineering Information Inc.).
InterDok (InterDok Corporation).
Mechanical Engineering Abstracts (Formerly ISMEC: Mechanical Engineering Abstracts)/ISMEC (Engineering Information Inc. and Cambridge Scientific Abstracts).

Using a modem these can be accessed through DIALOG and ESA/IRS and are available, for example, at the SRIS and the professional engineering institutions, if not accessible at the reader's library or information unit.

Glossary

See Chapter 9 for dictionaries of general engineering terminology. One covering machinery and associated topics well is Nayler (1985) which has clear diagrams to amplify complex definitions. Two very helpful specialist sources facilitate both definition and accurate translation of terms encountered in kinematics, machinery and dynamics. One is Eurotrans (1982–1985) prepared in collaboration with the British Gear Manufacturers' Association (BGMA). It contains illustrated definitions of gear terminology including manufacture, assembly and testing written in English, German, Spanish, French, Italian, Dutch, Swedish and Finnish. The second is an authoritative and comprehensive collection of definitions covering the structure of machines and mechanisms, kinematics, dynamics, machine control and measurements and robotics compiled by Commission A of IFToMM. In English, French, German and Russian the text is arranged in four columns with numbered divisions to BS 5848: 1980; facilitating cross-reference (see Leinonen et al., 1991). BS glossaries of specific subject areas such as quality and reliability are found by consulting the *BSI standards catalogue*. The English language terminology of tribology is explained thoroughly by Kajdas, Harvey and Wilusz (1990).

Historical background

From the dawn of history human intelligence has devised means of reducing physical labour. Appreciation of the principles of leverage, illustrated in Egyptian tomb paintings c.1500BC, is as important to machinery as the invention of the wheel and led to machines such as those used for irrigation and for lifting large stones when building the Acropolis and medieval cathedrals. Crude gear trains were used by the Romans. The genius of Leonardo da Vinci (1452–1519) forecast many modern products, such as rolling bearings and screw-cutting machines. Drawings from the books by Agricola (1556) illustrating and describing the machinery used in the silver mines of Central Europe are well-known. The subsequent acceleration of development is recorded in the eight volume work begun by Singer et al. (1954–8) continued by Williams (1978); and Raper (1984) and also, more concisely, by Derry and Williams (1960). Books specifically on the development of machinery are those by Keller (1964) and Strandh (1979). The extensive introduction to his history of tribology, Dowson (1979) includes a substantial bibliography.

The subsequent development from the realization that mechanisms could transform motion through the development of both practical machinery for the generation and application of power and the extension of theoretical understanding of kinematics by Euler and subsequent work at the Ecole Polytechnique is surveyed briefly by Hartenberg and

Denavit (1964) and Meyer zur Capellen (1966). The fundamental classification of mechanism components by Reuleaux (1875) forms the basis of much subsequent work. For many years design continued to rely upon experience and intuition. The problem of relating several independent variables to meet specific requirements is exemplified by Hrones and Nelson (1951) whose atlas of coupler curves for the four-bar crank-rocker mechanism contains over 7000 displacement paths. Pioneer work, such as that by Kloomok and Muffley (1955), in the application of computers produced hardcopy tabular and graphical design aids to minimize reliance upon iterative procedures although voluminous data still result if more than three independent variables are involved. Beginning from the supply of design data for the aircraft industry, ESDU International now supplies hardcopy data and software for a wide range of aeronautical, mechanical and chemical engineering purposes. Contemporary research (Medland, 1992) involves the development of computer programs for synthesizing the dimensions of simple mechanisms to fit the specified input and output motions accurately. Commercial software packages are available for such purposes as gear design and the kinematic and dynamic analysis of complex spatial mechanisms (see Potter, 1992) and Tables 1 and 2. Notable research leading to advances in the knowledge of kinematics during the past forty years is reported by Erdman (1993) including a substantial number of citations.

The design process

Textbooks on machinery design tend to specialize on specific aspects of the design process, emphasizing the methodology or concentrating on either stress analysis or the kinematics and dynamics of machine parts. These in turn are the subject of specialist publications ranging from applied mathematics to explanations of engineering practice applied to particular components. Design guides, hardcopy data and software to aid component design are becoming increasingly available.

Conceptual design

The design process consists of two distinct phases, conceptual and detail. Starting from the customer's specification the conceptual phase involves:

(a) preparation of the detailed engineering design specification defining what is required, quoting quantities when possible;
(b) identification of alternative solutions;
(c) the criteria for selecting the preferred solution and hence converging upon the design chosen for manufacture.

At this stage it is inevitable that decisions have to be made with inadequate information which, together with the increasing complexity of modern machinery, provided the incentive for the development of design methodologies to assist these tasks. Methodologies aim to concentrate attention upon aspects of the construction, operation and maintenance of the proposal in a logical sequence. They can be particularly helpful for identifying the vital and desirable requirements of the product, the precautions to be taken and as an aid for devising the strategy and identifying the resources needed to solve a problem. Terry (1968) and Pitts (1973) explain the PABLA system developed by the Atomic Weapons Research Establishment but applicable to a wide range of products. Pahl and Beitz (1988) is a specialized text concentrating on methodologies whilst Ullmann (1992), Ertas and Jones (1993) and Cross (1994) all cover this topic well. Mittendorf (1990) gives an electrical engineer's approach.

Investigations into creative design thinking applied to engineering are reported in the proceedings of a workshop at Delft by Cross, Dorst and Roozenburg (1991) and of case-study related research by Lordan and Thompson (1994).

Guidance, with examples, on the preparation of Design Specifications is given by Cross (1994) Dieter (1991) and Pugh (1991). Methodologies intended for the generic design process prompt lateral thinking to identify new applications of proven concepts, but as work progresses it becomes increasingly difficult to follow a generalized approach. In particular, less guidance is available on the specialized procedure for identifying a variety of alternative solutions to a unique problem. For this purpose compilations of commercially-available components and of proven solutions to similar problems can be helpful. The four-volume series edited by Jones (1930, 1936 and 1951) and Newell and Horton (1967) cover a wide range of mechanisms grouped by generic type, e.g. cam mechanisms, or by purpose, e.g. differential mechanisms, which are illustrated and described, but not analysed. The index is comprehensive. Parmley (1985) surveys a wide range of commercially available mechanical components with examples of suitable applications, their advantages, precautions to be taken and theoretical analyses. He concludes with a helpful chapter on innovative design. More specialized collections of mechanisms are provided by Torafson (1986) and collections of devices by generic application have been published by Chironis (1991) who includes non-mechanical controls, and by Jensen (1991). The last provides fundamental analyses, worked examples and graphical design data to optimize solutions readily by relating dimensionless parameters. The index is supplemented by an extensive bibliography; many of the works cited are not in the English language.

The 'ti' Index (*Engineering Design and Manufacturing Index*), revised at six-monthly intervals, is a comprehensive collection of manufacturers' catalogues on microfilm by generic type. The accompanying index concisely identifies suppliers for specialized products as do the buyers' guides published under the names of the periodicals *Engineer* and *Machinery*, also Vol. 1 of *Kompass* published in collaboration with the Confederation of British Industry. *Kompass* is also available as a CD-ROM; other volumes provide corresponding information for other countries including all the European Union.

Awareness of new products, forthcoming exhibitions and conferences can be obtained from such journals as *Design Engineering*; *Design Products & Applications; Engineering; Eureka; OEM Design; Professional Engineering*; and *What's New in Design* and in *Technoshop '92*.

Periodicals

BSI News (British Standards Institution) monthly.
Design Engineering (Morgan Grampian) monthly.
Design Products and Applications (IML Group) monthly.
Engineering (Gillard Welch Associates) monthly.
Engineering Designer (Institution of Engineering Designers) monthly.
Eureka (Impress Ltd) monthly.
Journal of Engineering Design (Carfax Publishing) quarterly.
Journal of Mechanical Design. Transactions of the ASME (American Society of Mechanical Engineers) quarterly.
Machine Design International (Penton Publishing) fortnightly.
Mechanical Engineering (American Society of Mechanical Engineers) monthly.
Mechanical Incorporated Engineer (Institution of Mechanical Incorporated Engineers) bi-monthly.
OEM Design (OEM Design) monthly.
Professional Engineering (Mechanical Engineering Publications) 22 issues per annum.
What's New in Design (Morgan Grampian) monthly.

Buyers' guides

Kompass (1994/95) (Reed Information Services).
The Engineer buyers guide (1994) *(Benn Business Information Services)*.
Machinery buyers guide. (2 vols) (1994) (MachPress Ltd).
Engineering Design and Manufacturing Index (Technical Indexes Ltd) 2 issues per annum.

Analysis

Later stages of the feasibility study and the whole of the detail design stage when the complete instructions for manufacture and purchase of procured components are prepared involve a rigorous analysis of the motion and force transmission to define the material, strength and size

158 *Kinematics, dynamics and machines*

of every part. These considerations interact with stress analysis. Chow (1978), Marshek (1987) and Dimarogonas (1988) give helpful guidance for achieving high strength/mass and stiffness/mass ratios of parts subject to large accelerations and/or loads. Dimarogonas (1988) provides program listings.

Kinematics

The boundary between the applied mathematics of kinematics and dynamics and the analysis of machine parts is nebulous, consequently there is considerable difference of emphasis between current texts. In contrast to contemporary development of techniques to synthesize a solution most traditional texts on 'Theory of Machines' concentrate on analysis, including some machines now obsolete. Nevertheless the thorough treatment of machine theory in Green (1962) contains useful knowledge difficult to find elsewhere.

The application of kinematics to prove the constraint and optimize dimensions to refine the motion of a design whilst assuming rigid, weightless members usually precedes dynamic analysis. Drabble (1990) and Dyke and Whitworth (1992) are two recent introductions to the subject at first-year undergraduate level. The latest edition of Mabie and Reinholtz (1987) includes computing techniques. Beer and Johnson (1988) has useful commentaries on the worked examples and is accompanied by a floppy disk having 12 interactive tutorials. Both Grosjean (1991) and Kimbrell (1991) demonstrate the continuing value of velocity and acceleration diagrams and of models during the introductory learning process to establish the basis for appreciating synthesis techniques for linkages. Both include program listings for 4- and 6-bar linkages. Nikravesh (1988) developed numerical methods from the fundamental theories to develop general-purpose programs for planar kinematic and dynamic analysis giving particular attention to constraint whilst Gans (1991) adopts a complex variable approach to suit computing methods explained with pseudo-code listings. The clear presentation of Riley and Sturges (1993) emphasizes fundamental principles including generic methods of computational problem solving.

Of the more advanced texts, the translation of Hain (1961) (by Adams *et al.*, 1967); records the work done in Germany up to c.1960, including a vast bibliography. Dijksman (1976) and Hunt (1978) develop kinematics from fundamental geometric theorems, Bottema and Roth (1979) include the authors' research results, Sayer and Bones (1990) make extensive use of vector algebra and differentiation and of matrices. The vital abilities to construct mathematical models and appreciate their limitations are fostered by Meriam and Kraige (1993). Most of the above, including Nikravesh (1988) and Sayer and Bones

(1990) continue with dynamics and provide an introduction to spatial motion. Easthope (1964), Phillips (1984, 1990) and Chiang (1988) have written in-depth studies of spatial motion. The 'screw theory' of spatial mechanics, the subject of Phillips' second volume, is also explained by Hunt (1978).

Dynamics

As implied by their titles, many books on kinematics cited above also cover dynamics. Others are concerned with the kinematics and dynamics of machine components, some covering the range of bearings, power transmissions, and mechanisms, others making an in-depth study of one type (surveyed below by generic type). As with the treatment of kinematics many recent publications prefer analytical methods which lead to flowcharts or program listings. These also facilitate use of mathematical packages such as Mathcad, Mathematica or TK solver (reviewed by Ash, 1993).

Three books analysing the general theory of rigid multi-body planar and spatial dynamics are Roberson and Schwertassek (1988) who derive model equations from Newton's and Euler's laws, Shabana (1989) who provides algorithms and flowcharts and Huston (1990) who also includes a particularly extensive bibliography.

Authors who consider the application of kinematics and dynamics to a range of machinery include the handbook edited by Shigley and Mischke (1986) (also published in sections as Mechanical designers' workbooks), and Orthwein (1990) which contains helpful flowcharts and tabulated comparisons throughout the text to assist the design process and component selection.

In a book concentrating upon computer-oriented problem solving Doughty (1988) includes a thorough guide to the procedure and recommends precautions to be taken. He also quotes many listings and includes substantial appendices to aid program writing. A floppy disk is available separately to accompany Norton (1992) for kinematic and some dynamic analysis of linkages, cams and the engine mechanism. Wilson and Sadler (1993) include flowcharts and give worked examples illustrating the use of spreadsheets for velocity and acceleration analyses of linkages and of an epicyclic gear train.

Journals

Mechanism and Machine Theory (Elsevier Science) bi-monthly.
Proceedings of the Institution of Mechanical Engineers. Part C, Journal of Mechanical Engineering Science (Mechanical Engineering Publications) bimonthly.

Transactions of the ASME. Journal of Mechanical Design (American Society of Mechanical Engineers) quarterly.

Mathematics software packages

Mathcad (Adept Scientific Micro Systems Ltd).
Mathematica (Wolfram Research (UK) Ltd).
TK Solver (Universal Technical Software (UK) Ltd).

General machinery design

Comprehensive coverage of the analyses needed for general machinery design is provided by Jensen (1991) and Shigley and Uicker (1995), both of which provide graphical design data. The updated first volume of the valuable work by Erdman and Sandor (1991) has the merit of including comprehensive guidance on the conceptual and feasibility stages of mechanism design, including advice on trouble-shooting. An associated floppy disk has programs to aid the velocity and acceleration analysis of linkages, cam design and the kinematic synthesis of linkages.

Design Guides on specific aspects where university lecturers have identified gaps in existing information sources are in progress of publication by SEED (Sharing Experience in Engineering Design) and obtainable from SEED Ltd (Publications), P.O. Box 59, Loughborough, Leics. LE11 0FS. Primarily intended for undergraduate use, they merit a wider readership; some are cited below.

Handbooks

Engineers' handbooks containing summaries of standards, other tabulated data and general information on a wide variety of topics include Sharpe (1995), Smith (1994) and Avallone and Baumeister (1987). The primary content of Oberg *et al.* (1992) is production data, quoting both US and British Standards. Material and component data copied with acknowledgement from manufacturers' catalogues and British Standards is published in Mucci (1994). See Table 11.1.

Design data

In machine design it is frequently necessary to satisfy limiting conditions, each dependent upon several variables. Consequently the use of basic analytical methods involves repeated iteration often complicated by uncertainty of the relative significance of these parameters. This difficulty was not overcome until computer-generated graphical or

Table 11.1. Data, Guidance and/or Software available from specialist sources

Source:	British Standards	ESDU Int	Mucci	Neale	SEED	Shigley Mischke
Balancing		G S				
Bearings	D	D G S	D	D G	G	D G
Belts	D		D	D G	G	D G
Brakes				D G		D G
Chains	D		D	D G		D G
Cams		D G S		D G		D G S
Couplings	D			D G	G	D G
Clutches				D G		D G
Gears	D G	G S	D	D G	G	D G
Linkages		G S				G
Seals	D	D G	D	D G	G	D G
Springs	G	D G	D		G	D G
Type Selection					G	G
Notation:	D – Data					
	G – Guidance					
	S – Software or listing					

tabular data became available to determine critical dimensions directly. ESDU International is a specialist publisher of such data including a rapidly increasing provision of software to widen the scope to complex design and development problems. Realtime use of these programs facilitates rigorous comparison of alternative solutions using the 'what if?' principle and sensitivity analysis when optimizing dimensions. Hardcopy data provides guidance, examples and, within its limitations, graphical data for checking results. Readers' attention is drawn to other sources of specialist data in the appropriate section of this chapter.

Conferences

Besides those on specific topics conferences relevant to kinematics, dynamics and machinery include the major international conferences organized by the International Federation for the Theory of Machines & Mechanisms.

  1965 Varna, Bulgaria: *Collection of papers to the international conference of machines and mechanisms* (1965) 4 vols. (Mechanical and Electrotechnical Institute).
  1969 Zakopane, Poland: Oledzki, A. (ed.) (1969) *Proceedings, II international congress on the theory of machines and mechanisms.* 3 vols. (Zakladzie Graficznym Politechniki Warszawskiej).
  1971 Kupari, Yugoslavia: *Proceedings of the third world congress on the*

theory of machines and mechanisms (1971) 4 vols. plus 3 vols. 'communications'. (Jugoslovenski komitet za leoriju masina i mehanizarna).

1975 Newcastle, UK: Institution of Mechanical Engineers (1975) *Fourth world congress on the theory of machines and mechanisms.* (Mechanical Engineering Publications).

1979 Montreal, Canada: American Society of Mechanical Engineers (1979) *Fifth world congress on the theory of machines and mechanisms* (ASME).

1983 New Delhi, India: Rao, J. S. and Gupta, K. N. (eds) (1984) *The theory of machines and mechanisms.* 2 vols. (Wiley Eastern).

1987 Seville, Spain: Bautista, E., Garcia Lomas, J. and Navarro, A. (eds) (1987) *The theory of machines and mechanisms.* 3 vols. (Pergamon).

1991 Prague, Czech Republic: Okrouhlik, M. and Pust, L. (eds) (1991) *Proceedings of the eighth world congress on the theory of machines and mechanisms.* 6 vols. (Society of Czech Mathematicians and Physicists).

The sequence of International Conferences on Engineering Design (ICED) includes papers on machinery, the proceedings being published by the host organization(s).

1981 Rome, Italy: Hubka, V. and Eder, W.E. (eds) (1981) *Ergebnisse von ICED 81 (Rome).* [*Proceedings of ICED 81*]. (Heurista).

1983 Copenhagen, Denmark: Hubka, V. and Andreasen, M.M. (eds) (1983) *CAD, design methods, Konstruktionsmethoden: Proceedings ICED 83.* 2 vols. (Heurista).

1985 Hamburg, Germany: Hubka, V. (ed.) (1985) *Theory and practice of engineering design in international comparison.* 2 vols. (Heurista).

1987 Boston MA, USA: Eder, W.E. (ed.) (1987) *Proceedings of the 1987 International Conference on Engineering Design.* 2 vols. Boston, Mass., 17–20 August, 1987. (American Society of Mechanical Engineers).

1988 Budapest, Hungary: Hubka, V., Baratossy, J. and Pighini, U. (1988) *Proceedings of ICED 88 Budapest.* 3 vols. (Gepipari Tudomanyos Egyesulet(GTE)-Heurista).

1989 Harrogate, UK: *Proceedings of the Institution of Mechanical Engineers, Engineering Design ICED 89*, 2 vols. (Mechanical Engineering Publications).

1990 Dubrovnik, Yugoslavia: Hubka, V. and Kostelic, A. (eds) (1990) *ICED 90; International Conference on Engineering Design. Papers.* 4 vols. (Heurista; JUDEKO – Yugoslav Society of Machine Elements and Design).

1991 Zurich, Switzerland: Hubka, V. (ed.) (1991) *Proceedings of ICED 91; International Conference on Engineering Design.* (Heurista).

1993 The Hague, The Netherlands: Roozenberg, N.F.M. (ed.) (1993) *Proceedings of ICED 93. Ninth International Conference on Engineering Design.* The Hague. 17–19 August, 1993. (Heurista).

Proceedings of these conferences can be obtained from WDK Heurista, Postfach 102, CH–8029 Zurich, Switzerland.

Other conferences are organized by such organizations as the Institution of Mechanical Engineers, the Institution of Engineering Designers and the American Society of Mechanical Engineers. Thus Machine Systems contains papers from the IMechE 'EuroTech Direct '91' Conference and the two-volume collection edited by Pisano; McCarthy and Derby (1990); those from the ASME 21st bi-annual Mechanisms Conference at Chicago. Details of conference publications are to be found in the monthly publication *InterDok*. Papers presented to the 23rd ASME Mechanisms Conference have have been edited by Pennock (1994a and 1994b).

Machine components

Machine components are responsible for transmitting the power from its source to the end effector where useful work is done. This involves controls, connections, velocity ratios and transformation of the motion. Bearings transmit force through joints and maintain the correct location or path of moving parts. General coverage of the construction, application and theoretical analysis of machine components is to be found in Parmley (1985), Shigley and Mischke (1986), Jensen (1991) and Juvinall and Marshek (1991). Dimarogonas (1988) has a useful collection of computer program listings. Two books by Johnson (1978, 1980) describe the application of linear programming techniques to the optimization of detail part design. Journals covering these aspects well include *Design Engineering*, *Engineering*, *Engineering Designer*, *Machine Design International*, *Transactions of the ASME. Journal of Mechanical Design*, *Mechanism and Machine Theory* and the *Proceedings of the Institution of Mechanical Engineers. Part C Journal of Mechanical Engineering Science*.

Component design and analysis

In addition specialist books, monographs, design data and software concentrate upon specific machine components. Monographs, data and software such as the series in progress from SEED, and ESDU International are particularly relevant. Their scope is indicated in Table 11.1.

Shaft connections

The convenience of transmitting power through rotating shafts facilitates provision of velocity ratios and involves such units as clutches and couplings to connect separate items. Ready assembly relies upon adherence to BS preferred sizes. Techniques for coupling and aligning shafts including the precautions involved with foundations, the choice of flexible couplings, alignment and measurement, thermal movement and tolerances are given by Piotrowski (1986), Hamilton (1988b) and Neale, Needham and Horrell (1991). The design, selection, installation and maintenance of couplings and joints is covered by Mancuso (1986). Guidance on the specification of procured couplings is given in

BS 3170: 1972(1991); and BS 6613: 1985(1991). Recommendations on the design, construction and maintenance of the principal types of friction brakes and clutches together with theoretical analyses are contained in Orthwein (1986) and Baker (1992), the latter based on the well established Ferodo design manual.

Belts and chains

Although most of these items will be obtained from specialist manufacturers, appreciation of theory and good practice is vital for trouble-free operation. Considerations of power capacity, design, material selection, lubrication, examples of applications and comparisons of alternative types of belts are given by Erickson (1987) and by Hamilton (1988a), and of chains by the American Chain Association (1982). In addition, the catalogues and technical information services of many manufacturers are reliable sources of design guidance and up-to-date design data. The development of an expert system for chain drive design is reported by Wang, Zhou and Yu (1988).

Gears

The importance of gears in machinery is reflected by the number of authoritative works on this subject published since 1988 inspired by revisions of national standards. Recommendations for preferred dimensions and for calculating the strength and wear ratings are published by BSI, DIN, the American Gear Manufacturers' Association (AGMA), etc. Books covering the whole scope of gear analysis, type and material selection, manufacture and lubrication include Drago (1988), who uses AGMA standards and provides a helpful commentary through examples; the authoritative manual, also to USA standards, published by the Society of Automotive Engineers (1990) and the comprehensive undergraduate-level teaching material sponsored by the British Gear Association, Hofmann, Kohler and Munro (1991). A further revision of Dudley's *Handbook of practical gear design* by Townsend (1991) includes recent improvements in gear manufacture and transmission error measurements, together with the 1991 versions of the strength and wear ratings to the AGMA, DIN and ISO standards. A British publication, Stokes (1992), is similarly comprehensive and covers the BS gear ratings. Bartz (1993), translated from the German, is a specialist text on gear lubrication, considering both design and operational aspects. There are many references to DIN standards. Current British Standards cover the geometry and strength and wear rating of spur, helical, bevel and worm gears.

Guidance on the analysis and synthesis of epicyclic gear trains including design hints is given by Muller (1982), the kinematic design of epicyclic trains by ESDU International (1983) and on the design and manufacture of hypoid gears by Shtipelman (1978). The special considerations involved in the use of non-metallic materials is covered by Adams (1986).

Of the operational problems liable to be encountered, Smith (1983) surveys means of reducing vibration and noise, together with causes of gear tooth failure. The last is the sole topic of a well-illustrated book by Alban (1985), whilst Tallian (1991) has compiled a comprehensive and systematic atlas intended to assist the identification of causes of Hertz contact failures. See also Bartz (1993).

Software packages for gears

FIRST GEAR (Universal technical Software (UK) Ltd) MS-DOS v3.0 or higher and TK solver release 2.0 or higher. Menu-driven for preliminary geometric design and load rating of spur and helical gear sets.

Organization

British Gear Association, St. James House, Frederick Road, Birmingham B15 1JJ, UK. Tel: +44 121 456 3161.

Steplessly variable-ratio drives

An alternative to a number of fixed velocity ratios obtainable through a gearbox is to use a steplessly variable system. A number working on different principles and speed ranges are commercially available. Spitzer (1990) provides a general survey of types available in the USA whilst Heilich and Stube (1983) contains engineering and economic analyses supported with case studies of those drives which rely upon friction to transmit force. Some UK suppliers are identified in the 'ti' Index and in buyers' guides.

Periodicals

Drives and Controls (Kamtech Publishing) monthly.

Buyers' Guides

Engineering Design and Manufacturing Index (Technical Indexes Ltd) 2 per annum.

Transformation of motion

The operation of many machines involves the transformation of the power derived from the driving source into the required path and action of the end effector, the output doing useful work. Desirable features of the chosen mechanism include small transmission angles to ensure free movement, minimal force magnification and bearing loads, and good dynamic balance. The members need high strength/mass and stiffness/mass ratios. The generic type of mechanism, linkage, cam, etc., is decided during the conceptual design phase, informed decision making can be assisted by commercial software packages which facilitate rapid quantitative comparison of alternative designs and, by editing the input file to optimize the preferred solution. The control of mechanisms through transducer feedback to micro-processor or computer-controlled drives leads to the multi-disciplinary subject known as 'Mechatronics' including 'hybrid' machines powered by separate electronically controlled actuators. These are developing a literature of their own.

Methodologies of mechanism design

Mechanism design progresses through the stages of type, number and dimensional synthesis. Type synthesis, deciding the generic type of mechanism to be used in an application, and number synthesis, its construction and configuration, should be decided during the conceptual design phase of the feasibility study. Dimensional synthesis, finding the actual dimensions to obtain the required movements, strength and stiffness is appropriate at the detailing phase. The complexity of preparing guides for type synthesis has led to classifications based on the required input and output motions or comparisons of alternative generic types. An English translation of Hain (1967) includes a thorough classification of all variants of some basic mechanisms and extensive treatment of mechanism synthesis. Subsequent specialist linkage design methodologies are by Crossley (1980a and 1980b) and Olson, Erdman and Riley (1985). A special issue of *Mechanism and Machine Theory* edited by Marshek and Kanapan (1990) is dedicated to papers on this subject. In addition Erdman and Sandor (1991) provide a helpful guide including trouble-shooting techniques. The library, editing, animation and analysis facilities of software packages provide effective means of developing linkages with confidence (see Table 11.2).

Linkages

Development of methodologies for linkage design can be traced from the classification by Reuleaux (1875). The problem of generating the required output path accurately, even by coupler curves derived from a 4-bar crank-rocker linkage, is demonstrated by the 7000 solutions contained in the atlas by Hrones and Nelson (1951). Prior to the computer era the principles of kinematic synthesis were well explained by Hartenberg and Denavit (1964).

Sandor and Erdman (1984) are responsible for an advanced work on kinematic synthesis of linkages; other helpful sources include Nikravesh (1988), Gans (1991) and Grosjean (1991), and all are computer-oriented. Grosjean (1991) has a section on the effect of dimensional tolerances on the output motion including a program listing. These texts also cover the problems associated with force transmission and balancing of linkages. Darlow (1989) and Goodwin (1989) both specialize on these aspects, which is also covered in detail by the continuing series of ESDU hardcopy items and associated software packages. A second edition of Molian (1968) is in press.

Expert systems for mechanism synthesis have been reported by Soni, Dado and Weng (1986) and by Hoeltzel and Chieng (1990). The background to a commercial package, RuleCAD, which synthesizes the dimensions of a linkage to suit a specified output path accurately is reported by Medland (1992). See Table 11.2. for other software packages to aid mechanism design and analysis.

Cams

In contrast to linkages the cam mechanism provides a means of continuously and accurately controlling the displacement, velocity and acceleration of the output member but with the penalty of accurately manufacturing the complex curve forming the cam profile. Knowing the timing and output displacement, the motion is defined by selecting a suitable 'cam law' or a spline function relating the output displacement to the cam rotation. The cam size must suit the kinematic constraints imposed by both the maximum pressure angle and profile curvature. It is also vital to ensure that the limiting surface stress of the cam and roller materials and the load-speed capacity of the roller are not exceeded. In every case these parameters are functions of six or more independent variables, hence design data and/or software packages are invaluable for optimizing the specification. In simple cases, graphical data enable the least cam size to be found immediately from the kinematic constraints (but may need to be enlarged to suit the forces and/or stresses).

The specialized design procedure for cam mechanisms is outlined by Molian (1968) with a flow chart cross-referring to data sources by Druce and Watson (1991) and a sensitivity analysis by Mills, Notash and Fenton (1993). Many of the general texts on mechanisms, including Erdman and Sandor (1991), Kimbrell (1991) and Wilson and Sadler (1993) introduce the principles of selecting the cam law, analysing the output motion and calculating the profile shape with the support of program listings and/or disks. The special features of spline function

function profiles are covered by Angeles and Lopez-Cajun (1991) and Caulfield-Browne, MacCarthy and Syan (1991). The detailed analysis of the kinematic constraints is found in the classic work by Rothbart (1956) and more recently by Chen (1982) and Jensen (1987), the last providing program listings and a disk. Koloc and Vaclavik (1993) is an English translation of a mathematically-oriented text reporting east European work. It contains useful data and case studies.

Substantial graphical design data originated from a series of papers in *Product Engineering* by Kloomok and Muffley (1955 onwards). The basic principles of cam design and analyses accompanied by extensive graphical data have been published by ESDU International from 1979 onwards as part of their Mechanisms subseries. These are now being supplemented by software packages, known as ESDUpacs, for complex analyses, such as finding the cam dimensions for motions having intermediate precision points and the analysis of both contact stress in the cam disc and the load-speed capacity of roller followers. Reliability studies are supported by an atlas, Tallian (1991), which illustrates contact failures due to Hertz stress and suggests probable causes. Tribology, including the lubrication of cams, is covered well by Taylor (1993) and Zhu (1993). See also Proceedings of the Leeds-Lyon Symposia on Tribology (listed later).

Other commercial software for analysing and synthesizing plane cam mechanisms are Camlinks, Motion and OSCAM (see also Table 11.2). Typically the mechanism can be constructed by selecting and editing members from a library file; the output motion, pressure angle, profile curvature, driving torque and contact force computed; and the cam size optimized to suit a specified maximum pressure angle. Import and export facilities can include DXF and/or ASCII format.

Periodicals

Transactions of the ASME. Journal of Mechanical Design (American Society of Mechanical Engineers) quarterly.

Proceedings of the Institution of Mechanical Engineers. Part C, Journal of Mechanical Engineering Science (Mechanical Engineering Publications) bimonthly.

Mechanism and Machine Theory (Elsevier) bimonthly.

Springs

Springs are used in machinery to apply or balance forces. Besides the initial decision of the type to be used, primary design considerations include the stiffness, pre-tension, natural frequency, volume, type of end, limiting stress and fatigue strength or life. The well-known book by Wahl (1964) covers all these aspects, as does the more recent manual from the Society of Automotive Engineers (1989) which contains much useful information of general application. Helpful guides to coil spring design are contained in the three parts of BS 1726: 1987–1988; and the ESDU International series (1982, continuing). A sequence of books by Carlson (1978, 1980 and 1982) deals respectively with design, failures and manufacture. See also the monograph by Hurst (1990).

Whilst the calculations do not involve elaborate analysis the number of independent variables required to specify a spring justify software for optimizing the solution. Sources of programs suitable for microcomputers include Dietrich (1985) the suite produced by the Spring Research and Manufacturers' Association, and Associated Spring Ltd., both of which run on DOS, and Lee Spring Ltd. (Windows format).

The use of elastomers as springs is introduced in the principal texts cited above. The properties of these materials are covered by Allen, Lindley and Payne (1967) and Hepburn and Reynolds (1979). The special case of rubber springs with many examples of their application is to be found in the translation of Gobel (1974).

The catalogues and technical information services of manufacturers are another source of reliable guidance.

Software

Associated Spring SPEC Ltd.
Dietrich, A. (1985) *Spring design with an IBM PC*. New York: Marcel Dekker.
Lee Spring Ltd.
Spring Research and Manufacturers' Association.

Software Packages for Mechanism Analysis and Synthesis (see Table 11.2)

A range of software packages for mechanism design of varying complexity and cost are commercially available. Table 11.2 summarizes the principal features and cites references referring to or reviewing that package. Reviews of commercial software can be found in such journals as *CADCAM*, *Computer-Aided Design*, *Engineering*, and *What's New in Design*. ESDU International has introduced a user-friendly interface, ESDUview, for use on IBM-compatible personal computers with most of their software cited below.

Mechatronics and hybrid machines

The literature about the multi-disciplinary application of electronic controls to the control of machines already includes a specialist journal, *Mechatronics*, and conference papers. It was the subject of the 77th Thomas Hawksley Lecture to the Institution of Mechanical Engineers by Professor French (1992). A thorough introduction, including modelling, case studies, software, an extensive bibliography and a glossary

170 Kinematics, dynamics and machines

Table 11.2. Mechanisms Software Packages.

ADAMS (Automatic Dynamic Analysis of Mechanical Systems), Tedas, Barclays Venture Centre, University of Warwick, Science Park, Sir William Lyons Way, Coventry CV4 7EZ. (01203) 692252.	All UNIX-based workstations. 3-dimensional; multi-degree-of-freedom systems; component library; overconstraint check; kinematic, static, dynamic & modal analysis; modelling; animation; interfaces for most turnkey CAD/CAM systems and commercial analysis packages.
See: Sandor and Erdman, 1984; Wilson and Sadler, 1991; Potter, 1992; Sacks and Joskowicz, 1993.	
Kinepak and Dynapak, Algor Europe Ltd., Northumberland House, Drake Avenue, Staines, Middx TW18 2AP. (01784) 442246.	DOS or Sun SPARCstation. 2-dimensional, derived from ADAMS; kinematic/dynamic analysis of mechanisms and cams; animation; Interfaces include IGES; DXF; STL; and commercial CAD packages.
Applied Motion, Rasna UK, Ltd., Plessey Business Park, Technology Drive, Beeston, Nottingham NG9 2ND. (01159) 229005.	DEC; Hewlett-Packard; IBM; Silicon Graphics; SUN; PC-DOS Windows NT. 3-dimensional. User extendable-libraries of material properties, mass primitives, joints, forces, and motions. Add-on packages:-Applied Cams; Applied Loads. geometric modelling; kinematic and dynamic analysis; animation; DXF; IGES; VDA; MNF; Pro/Engineer; I-DEAS; CADDS5; Catia; interfaces.
See: Potter, 1992; Sacks and Joskowicz, 1993.	
CAMLINKS, Limaçon, Meadow Farm, Horton, Malpas Ches. SY14 7EU. (01829) 250278.	Use with MS-DOS v3.1 or later, Windows environment. 2-dimensional linkages comprised of cranks, dyads, sliders, dampers, springs, linear or rotary drives; cams; kinematic and dynamic analysis; animation; DXF import and export.
MOTION, Limaçon, Meadow Farm, Horton, Malpas Ches. SY14 7EU. (01829) 250278.	Use with MS-DOS v3.1 or later, Windows environment. Motion design based on timing diagram defining displacement, velocity, acceleration; 2-dimensional cam-driven and servo-driven mechanisms; kinematic synthesis; Use in conjunction with CAMLINKS.
See: Hammar, 1995.	

Kinematics, dynamics and machines 171

Table 11.2. Mechanisms Software Packages (cont.).

DADS (Dynamic Analysis and Design System), LMS Ltd., Cheddar Industrial Park, Wedmore Road, Cheddar. Som. BS27 3EB. (01934) 744222.	Most workstations including 486PC: Large program; 3-dimensional; multi-degree-of-freedom systems; kinematic & dynamic analysis; modelling; animation; import/export to CATIA, PRO-ENGINEER, IDEAS.

See: Chase and Sheth, 1973; Wilson and Sadler, 1991; Potter, 1992; Sacks and Joskowicz, 1993.

DE/Mec, Desktop Engineering, Ltd., Evenlode Court, Main Road, Long Hanborough, Oxon OX8 8LA. (01993) 883555.	PC-based, windows environment; 2-dimensional linkages comprised of cranks, dyads and slides; kinematic & dynamic analysis; animation; DXF import and export.

See: Hammar, 1995.

ESDUpac*
Available from ESDU International, 27 Corsham St, London N1 6UA. (0171) 490 5151.
* – Normally ESDUpac's are only available as part of the subseries.

Number	Title Linkages
A9007	Design of rotating counterweights for balancing planar linkages.
A9022	Kinetostatic force analysis of four-bar planar linkages.
	Cams
A8626	Evaluation of the coefficients of a minimum-order polynomial cam law.
A9126	Analysis of roller cam followers.
A9205	Minimum size of disc cams with radial translating roller followers.
A9214	Blending profiles of disc cams with radial translating roller followers to reduce segment angle, reduce reference circle radius or increase lift.
A9301	Contact stress in disc cams with roller followers.
A9302	Blending profiles of disc cams with radial translating roller followers to satisfy one follower displacement precision point.
B9302	Blending profiles of disc cams with radial translating roller followers to satisfy one follower constant velocity component.
A9408	Lubricant film thickness between disc cams & followers.
A9411	Contact stress in disc cams with domed or flat-faced followers.
A9501	Kinematic analysis of disc cams.

See: Hammar, 1995.

Kinematics, dynamics and machines

Table 11.2. Mechanisms Software Packages (cont.).

OSMEC, Osborn Technical Software, 20 High Street, Emberton, Olney, Bucks. MK46 5DH. Bucks. MK46 5DH. (01234) 240241.	MS-DOS v5.0 or later. 2-dimensional linkages comprising cranks, dyads, sliders, and some gears; models linear actuators, helical & torsion springs; library of mechanisms; animation; DXF export.
See: Wood, 1993; Hammar, 1995; Molian, 1995.	
OSCAM, Osborn Technical Software, 20 High Street, Emberton, Olney, Bucks. MK46 5DH. (01234) 240241.	MS-DOS v2.1 or later. 2-dimensional disc cam mechanisms; library of configurations and cam laws; kinematic synthesis; kinematic and dynamic analysis; animation; DXF export.
See: Shortlist, 1992; Hammar, 1995.	
Rosi, Carter Bennan Co., Jeffreys Building, Cowley Road, Cambridge CB4 4WS. (01223) 561970.	Sun4/Sparc; VAX ; windows-orientated; based on efficient spatial algebra; 3-dimensional, extendible joint, actuation, & control libraries; kinematic and dynamic analysis; animation; DXF import and export.
See: Featherstone, 1987.	
RuleCAD, CIMIO Ltd, Brunel Science Park, Coopers Hill Lane, Englefield Green, Surrey TW20 0JZ. (01764) 438038.	SUN/Sparc; fully 3-dimensional; input – path definition; output – optimum linkage to generate path; full multi-variable optimization; kinematic analysis; mechanism synthesis; animation; CAD files import.
See: McGarva and Mullineux, 1992; and Medland, 1992.	

is provided by Bradley *et al.* (1990). Another book edited by MacConaill, Drews and Robrock (1991) contains a useful collection of case studies, and the specialist firm Quin Systems (n.d.) have prepared a clear, non-commercial explanatory monograph. The controlled multiple drive of a hybrid machine is explained by Tokuz and Jones (1991), whilst Sanders (1993) emphasizes the close similarities between mechatronics and robotics.

Journals

Mechatronics (Elsevier Science) bimonthly.
Machine Design International (Penton Publishing) fortnightly.

Conferences

1990, Cambridge: *Mechatronics – designing intelligent machines*. London: Mechanical Engineering Publications
1992, Dundee: *Mechatronics – the integration of engineering design*. London: Mechanical Engineering Publications

Tribology

The study of the interaction between moving surfaces is known as tribology. It involves friction, wear, material properties, material selection, lubricants and lubrication systems; hence tribology is relevant to all types of machinery. Books cover the subject generally and specific aspects in depth, there are also several journals of repute and conferences are held regularly on this subject. These sources are supported by design guides and data from specialist publishers and bearing manufacturers which are beginning to be supplemented by software packages. A readable history, Dowson (1979), includes 24 biographies of prominent workers in this field and 44 pages of bibliography. It is an admirable non-mathematical introduction to tribology. The classic experimental study of the physical and chemical processes associated with relative sliding between solids was published in two volumes by Bowden and Tabor (1950 and 1964). Helpful general introductions to tribology are those by Jones and Scott (1983) and Arnell *et al.* (1991). 'Items' extending the scope of the ESDU International Tribology sub-series of bearing design guides and data, including software packages, continue to be issued. Rolling bearing manufacturers are beginning to provide software for design and draughting purposes. The practical application of tribology to clutches, brakes, belt drives, sliding bearings, gears and cams, partially complementing the ESDU International Tribology sub-series, is to be found in Stolarski (1990). A detailed study of friction and wear, covering abrasive, erosive and fretting wear including non-metallic materials and dry bearings is contained in Sarkar (1980).

Conferences, symposia and journals

Conferences and symposia on tribology are organized by the Institution of Mechanical Engineers, the American Society of Mechanical Engineers, the Society of Tribologists and Lubrication Engineers (USA) and the Institution of Engineers, Australia. In particular the Leeds-Lyon Symposia organized jointly by the Institute of Tribology, Leeds University, UK, and L'Institut National des Sciences Appliquées, Lyon, France are to be noted (see below). Specialist journals publishing refereed papers on tribology include *Tribology International*, *Tribology*

Transactions and *Wear* as well as the *Transactions of the ASME. Journal of Tribology*, and the new *Proceedings of the Institution of Mechanical Engineers. Part J Journal of Engineering Tribology*.

Proceedings – Leeds–Lyon symposia on tribology

Dowson, D., Godet, M. and Taylor, C. M. (eds) (1975) *Cavitation and related phenomena in lubrication*. London: Mechanical Engineering Publications.

Dowson, D., Godet, M. and Taylor, C. M. (eds) (1977) *Superlaminar flow in bearings*. London: Mechanical Engineering Publications.

Dowson, D. *et al.* (eds) (1978) *Wear of non-metallic materials*. London: Mechanical Engineering Publications.

Dowson, D. *et al.* (eds) (1978) *Surface roughness effects in lubrication*. London: Mechanical Engineering Publications.

Dowson, D. *et al.* (eds) (1979) *Elastohydrodynamics and related topics*. London: Mechanical Engineering Publications.

Dowson, D. *et al.* (eds) (1980) *Thermal effects in tribology*. London: Mechanical Engineering Publications.

Dowson, D. *et al.* (eds) (1981) *Friction and traction*. Guildford: IPC Business Press.

Dowson, D. *et al.* (eds) (1982) *The running-in process in tribology*. Guildford: Butterworths.

Dowson, D. *et al.* (eds) (1983) *Tribology of reciprocating engines*. Guildford: Butterworths.

Dowson, D. *et al.* (eds) (1984) *Numerical and experimental methods in tribology*. London: Butterworths.

Dowson, D. *et al.* (eds) (1985) *Mixed lubrication and lubricated wear*. London: Butterworths.

Dowson, D. *et al.* (eds) (1986) *Mechanisms and surface distress*. London: Butterworths.

Dowson, D. *et al.* (eds) (1987) *Fluid film lubrication – Osborne Reynolds centenary*. Amsterdam: Elsevier Science.

Dowson, D. *et al.* (eds) (1988) *Interface dynamics*. Amsterdam: Elsevier Science.

Dowson, D. *et al.* (eds) (1989) *Tribological design of machine elements*. Amsterdam: Elsevier Science.

Dowson, D., Taylor, C. M. and Godet, M. (eds) (1990) *Mechanics of coatings*. Amsterdam: Elsevier Science.

Dowson, D., Taylor, C. M. and Godet, M. (eds) (1991) *Vehicle tribology*. Amsterdam: Elsevier Science.

Dowson, D. *et al.* (eds) (1992) *Wear particles from the cradle to the grave*. Amsterdam: Elsevier Science.

Dowson, D. *et al.* (eds) (1994) *Dissipative processes in tribology*. Amsterdam: Elsevier Science.

Dowson, D. *et al.* (eds) (1995) *Lubricants and lubrication*. Amsterdam: Elsevier Science.

Journals

Proceedings of the Institution of Mechanical Engineers. Part J Journal of Engineering Tribology (Mechanical Engineering Publications) quarterly.

Transactions of the ASME. Journal of Tribology (American Society of Mechanical Engineers) quarterly.
Tribology International (Butterworth-Heinemann) bimonthly.
Tribology Transactions. (American Society of Tribologists and Lubrication Engineers) quarterly.
Wear (Elsevier Sequoia SA) 20 per annum.

Lubrication and sliding bearings

Aspects of lubrication theory are covered in depth by Fuller (1984), Gohar (1988), Bassani and Piccigallo (1992) and Williams (1994). The application to sliding bearings is well explained by Barwell (1979); helpful guides to their selection and design, including calculation methods and data have been published by Rowe (1983), Welsh (1983) and Constantinescu *et al*. (1985). Research into computerized bearing selection is reported by Rowe, Cheng and Ives (1991). Extensive data comparing theoretical and extensive test results of journal bearing performance compiled by the Japanese Society of Mechanical Engineers with instructions on the use of these data are edited by Someya (1989). Practical recommendations for lubrication under a variety of adverse conditions is contained in Billett (1979) and by Mortier and Orszulik (1992). Fluid film lubrication is covered in Hamrock (1994). A second edition of Neale's comprehensive *Tribology Handbook* is appearing in parts, retaining the previous format with updated content. Those on bearings, Neale (1993a) and lubrication, Neale (1993b) are available.

Software

ESDUpac* (ESDU International)
*Normally ESDUpac's are only available as part of the subseries:

A8507 Calculation of individual stress components in concentrated elastic contacts under combined normal and tangental loading.
A9137 Film thicknesses in lubricated Hertzian line contacts.
A9138 Film thicknesses in lubricated Hertzian point contacts.
A9234 Calculation methods for steadily loaded fixed-inclined-pad thrust bearings.
A9235 Calculation methods for steadily loaded, off-set pivot, tilting-pad thrust bearings.
A9237 Calculation methods for externally pressurised journal bearings with capillary restrictor control.
A9304 Calculation methods for steadily loaded central circumferential groove hydrodynamic journal bearings.
A9305 Calculation methods for steadily loaded axial groove hydrodynamic journal bearings.

A9306 Calculation methods for steadily loaded axial groove hydrodynamic journal bearings. Low viscosity process fluid lubrication.

A9434 Dimensions, deflections and stresses for Hertzian contacts. under combined normal and tangental loading.

Rolling bearings

Most rolling bearings are purchased from specialist manufacturers whose catalogues and in-house publications contain extensive data for determining load and speed capacity together with guidance on mounting and lubrication. Nevertheless detailed analyses of the operation, performance and stressing of rolling bearings are also available. Palmgren's classic work on the fatigue of rolling bearings (1959) continues to be cited. Gupta (1984) contains mathematical models of performance including flowcharts and a program listing. The English translation of Wan (1991) considers the geometry, kinematics, load distribution, stressing and deformation in a mathematical approach to the subject. Elastohydrodynamic lubrication of ball bearings and similar elliptical contacts is explained by Hamrock and Dowson (1981).

The principles of rolling bearing and housing design including examples are given by Eschmann, Hasbargen and Weigant (1985) of FAG, Schweinfurt. A large, well-illustrated volume edited by Harris (1991) has ample references, and extends the principles of rolling bearing theory to gyroscopes and space vehicles. Design software for rolling bearings is available from both FAG (W.A.S.) and SKF (CADalog), reported in *Design Engineering* (1992). Both calculate the ratings from the load and speed characteristics entered to select the most suitable bearings for that application. The interactive program Select-A-Nalysis described by Strong and Fahrni (1985) developed by the Timken Company is capable of analysing multi-shaft assemblies.

Journals

Ball and Roller Bearing Engineering (FAG (UK)) semi-annual.
Ball Bearing Journal (SKF (UK)) quarterly.

Software

CADalog (SKF (UK)).
Select-A-Nalysis (British Timken).
W.A.S. (FAG (UK)).

Materials and surface treatments

Whilst the behaviour and selection of tribological materials and surface treatments are contained in many of the books cited above, Bhusan and Gupta (1991) and Glaeser (1992) specialize on these aspects. Likewise

Tallian (1992) has prepared a copiously-illustrated atlas containing enlarged reproductions of black and white photographs showing a wide range of bearing failures. These are methodically arranged to facilitate identification and cross-reference, the load, material(s), lubricant and failure mode being defined with a brief description of notable features leading to suspected cause(s) for the failure. More concisely, Neale (1995) illustrates and suggests explanations for failures of plain and rolling bearings, gears, seals, clutches, brakes and pistons and piston rings.

Other bearings

Fuller (1984) devotes a chapter to gas bearings, which are also the subject of books by Grassam and Powell (1964) and Wilcock (1972). The proceedings of a symposium on magnetic bearings was edited by Schweitzer (1988). ESDU 'Item' 67023, 1967, provides guidance and data for the design of crossed flexure-pivots.

Organizations

Ball and Roller Bearing Manufacturers' Association, 136 Hagley Road, Birmingham B16 9PN, UK. Tel: +44 121 454 4141.

Institute of Tribology, University of Leeds, Leeds LS2 9JT, UK. Tel: +44 1132 431751.

The National Centre of Tribology, AEA Technology, Thomson House, Risley, Warrington WA3 6AT, UK. Tel: +44 1925 253508.

Quality and reliability of machinery

Economic and marketing incentives for the improvement of the quality and reliability of machinery, often associated with the influence of G. Taguchi, are reflected in the contemporary rate of publication on aspects of these subjects, including new British Standards. The concept of reliability is explained at length in BS 4778: Part 3: Section 3.1.: 1991 clause 13.1. The *Quality management handbook* (1994), in two parts, is a revision of the authoritative British Standards Handbook No 22, which contained all the BS standards and drafts for development (DD series) issued by the date of publication which apply to quality and reliability, including a glossary and all parts of BS 5750: 1987 Quality Systems (continued as BS EN ISO 9000– : 1994–). These standards provide effective guidance for improving reliability at the design stage and throughout the life of a product. Some examples are included below:

BS 5760. *Reliability of systems, equipment and components.*
Part 6: 1991. *Guide to programmes for reliability growth.*
Part 9: 1992. *Guide to the block diagram technique.*
BS 6143: *Guide to the economics of quality.*
Part 1: 1992: *Process cost model.*
BS 6548. *Maintainabilty of equipment.*
Part 2: 1992. *Guide to maintainability studies during the design phase.*
Part 3: 1992. *Guide to maintainability, verification and the collection, analysis and presentation of maintainability data.*
BS 7229: *Guide to quality systems auditing.*
Part 1: 1991: *Auditing.*
Part 2: 1991: *Qualification criteria for auditors.*
Part 3: 1991: *Managing an audit programme.*
BS 7373: 1991: *Guide to the preparation of specifications.*

Dale and Oakland (1991) have published a comprehensive guide to the use of these standards, and Rothery (1993) on ISO 9000. The revised edition of the American handbook by Kececioglu (1993) is comparable with the BSI one. It includes the relevant theory plus a wealth of examples.

General introductions to reliability studies are given by Carter (1986), Smith (1993), who includes useful failure rate and failure mode data and O'Connor (1991), the last based on experience in the aircraft industry. The theoretical basis of reliability theory is explained with the aid of examples by Crowder *et al.* (1991), in depth by Dai and Wang (1992) and by Ramakumar (1993), the last including a substantial appendix of associated analyses. Klaassen and van Peppen (1989) concentrate upon the physics and probability of failure of systems. The interaction of reliability with whole life cost with the aim of improving availability through corrective maintenance procedures is found in Ascher and Feingold (1984), Moss (1985) and Moubray (1991) the last extends concepts developed in the aircraft industry to other applications. Henley and Kumamoto (1985) and Fitch (1992) give techniques for fault prevention and optimization of reliability. Heideklang (1991) includes hazard identification, human aspects and US legislation. It is accompanied by a disk for Macintosh microcomputers.

Condition monitoring

A collection of papers edited by Holmberg and Folkeson (1991) involves accelerated testing, improvements in data collection and analysis, the development of condition monitoring and emphasizes the value of good communication between designers and purchasers. Papers about the current state of the art in condition monitoring from various viewpoints are edited by Patton, Frank and Clark (1989). See also Smith (1993), McEwan (1990) and Rao (1993). The Department of

Kinematics, dynamics and machines 179

Trade and Industry has established a Condition Monitoring and Sound Assessment Club (contact Dr. C. Bishop, A.E.A. Industrial Technology, Harwell Laboratory, B521.1. Oxon OX11 0RA. (01235) 435024). A practical guide to the procedures and equipment for such applications as the condition monitoring of lubricating oils is given by Hunt (1993).

Subjective analyses

Guides to the subjective 'fault tree' and 'failure modes, effect and criticality' analyses which are helpful for both design and fault diagnosis are included in Henley and Kumamoto (1985), Priest (1988), BS 5760: Part 7: 1991 and BS 5760: Part 5: 1991, respectively, Dai and Wang (1992) and Ramakumar (1993). Procedures for reliability audits and design reviews are covered well with a case study and sample assessment forms by Bloch and Geitner (1990). The Health and Safety Executive (1989) has published a concise guide on quantified risk assessment which cites relevant papers on specific applications.

Taguchi methods

The obvious citations for the Taguchi methods of 'robust design' evaluating the significance of variance in the parameters affecting the performance of machinery are Taguchi (1987) (who acknowledges the classic work of Fisher (1966) on the problems associated with the multitude of independent variables encountered in agricultural experimentation) and Taguchi (1993). Applications of Taguchi methods in a variety of applications are reported ably by Phadke (1989) and with well-illustrated examples from the automotive industry by Grove and Davis (1992). An assessment of the procedure is made by Logothetis and Wynn (1989) and a survey of software packages available in the USA is included in the explanation by Lochner and Matar (1990) who also provide a helpful glossary. Research into the application of Taguchi methods to engineering design is communicated by Pitts and Lewis (1993).

Case studies and failure rate data

Case studies are included in Ross (1988), Patton, Frank and Clark (1989), Barker (1990), O'Connor (1991), Cullen and Hollingum (1987), Oakland and Porter (1994) and an Institution of Mechanical Engineers Seminar (1990). They are also to be found in the specialist journals and conference proceedings. A useful set of failure rate data is included in Smith (1993), whose data is also available on disk. He also cites other sources of these data. Procedures and precautions recommended in the compilation of reliability data banks are made by Cannon and Bendell

(1991) and Flamm and Luisi (1992). The latter are the edited proceedings of a EuReDatA conference, the European Reliability Data Banks Association. Conferences and symposia are also organised by the Institution of Quality Assurance, the National Centre of Systems Reliability, and the Safety and Reliability Society.

British standards

BS 4778 *Quality vocabulary*.
BS 4778: Part 3: *Availability, reliability and maintainability terms*.
BS 4778: Part 3: Section 3.1: 1991. *Guide to concepts and related definitions*.
BS 5760 *Reliability of systems, equipment and components*.
Part 0: 1986: *Introductory guide to reliability*.
Part 5: 1991: *Guide to failure modes, effects and criticality analysis (FMEA and FMECA)*.
Part 7: 1991: *Guide to fault tree analysis*.

Journals

Quality and Reliability Engineering International (Wiley) bi-monthly.
Quality Today (Nexus Business Communications) monthly.
Reliability Engineering and System Safety (Elsevier) monthly.
Safety and Reliability (The Safety and Reliability Society) quarterly.

Organizations

Institute of Quality Assurance, 54 Princes Gate, Exhibition Road, London SW7 2PG, UK. Tel: +44 171 584 9026.

Institution of Mechanical Engineers, 1 Birdcage Walk, London SW1H 9JJ, UK. Tel: +44 171 222 7899.

National Centre of Systems Reliability, AEA Technology, Thomson House, Risley, Warrington WA3 6AT, UK. Tel: +44 192 525 2000.

Safety and Reliability Society, 59 Piccadilly, Manchester M1 2AQ, UK. Tel: +44 161 228 7824.

Safety of machinery

Besides humanitarian and financial considerations, the safety of machinery is governed by statute and by the requirements of insurance companies. The principal British legislation is enshrined in the Health and Safety at Work Act, 1974. A continuing series of Regulations is being issued under the aegis of this Act, an increasing number originating as Directives of the European Union. Hence one must ensure that current information is obtained and be aware of proposals for new

Regulations and Directives. It must be remembered that whilst the principles of safety are international, legislation is national.

This citation of sources is no guarantee of approval by the British Health and Safety Executive (HSE) or any other body. In Britain every individual has the responsibility of ensuring that the requirements and Regulations of the Health & Safety at Work Act, 1974 and all other relevant legislation and codes of practice and of any insurer are followed at all times.

General introductions to engineering safety, emphasizing incident prevention, are given by Stranks (1991) of the Royal Society for the Prevention of Accidents (RoSPA), and Terry (1991). The latter, like Thompson (1987) includes risk assessment, linking safety with reliability. Hunter (1992) approaches safety from the design viewpoint and includes US law, codes and standards. Details of the literature on the Control of Substances Hazardous to Health Regulations (1988) (COSHH) and subsequent amendments are given by Pantry (1993). The COSHH Regulations have recently been revised (1994). Kletz (1993) provides thought-provoking means of learning from experience. A guide to the application of the Machinery Directive, now mandatory in the European Economic Area, has been published by the Institution of Mechanical Engineers (1995). An updating service is promised.

Thorough surveys of current British health and safety requirements and impending changes are contained in *Croner's health and safety at work* and *Tolley's health and safety at work handbook*, Dewis and Murdoch (1995). The former has loose-leaf format to facilitate insertion of the bi-monthly amendments. It is supplemented by a fortnightly newsletter *Croner Health and Safety Briefing*. Another source is the *Barbour Index Environmental Health, Fire and Safety* microfile, issued alternately as main and intermediate revisions nine times a year. It contains a complete set of the HSE publications and EC Directives. Several journals are devoted to safety matters, lectures and conferences are organized by the specialist societies, and some universities, colleges and commercial organizations. They are advertised in the periodicals on safety.

Noise

The Noise at Work Regulations, 1989, apply to all equipment emitting sound in excess of 85dB[A]. The physics of sound emission and transmission, the terminology, means of measurement and control are explained with case studies in books by Bies and Hansen (1988), Norton (1989) and the Sound Research Laboratories (1991). Design principles for reducing sound emission are described by Lyon (1987). Practical advice to the engineer on the selection of noise measuring equipment, specification of measurements and analysis of the results, supported by a large bibliography including the ISO and national standards for 21 countries are contained in Yang and Ellison (1985). The artificial supression of sound is explored by Tokhi and Leitch (1992). The comprehensive handbook, Barber (1992), has been completely revised.

Official publications

Codes of practice and guidance documents are published by both BSI and the HSE. Those immediately relevant to machinery are the British Standards:

BS EN 292: *Safety of machinery. Basic concepts, general principles for design.*
  Part 1: 1991: *Basic terminology, methodology.* Part 2: 1991: *Technical principles and specifications.*
BS 5304: 1988 *Code of practice for safety of machinery.*

and the HSE Publications obtainable from HSE Books, PO Box 1999, Sudbury, Suffolk, CO10 6FS. (01 787) 881165:

Memorandum of guidance on the Electricity at Work Regulations (1990).
Display screen equipment work: guidance on regulations (1992).
Noise at work: noise assessment, information and control (1990).
Quantified risk assessment: its input to decision making (1989).
Work equipment: guidance on regulations (1992).

The booklet on risk assessment cites references to specific industries. Further guidance and information on the safety of machinery is contained in the General (GS), Legislation (L), and Plant and Machinery series (PM) of Guidance Notes and the Health and Safety: Guidance (HS(G)) and Health and Safety: Regulations (HS(R)) booklets obtainable through HMSO stockists. Advisory literature, mostly available free of charge, can be obtained from HSE Enquiry Points. It includes:

Access to occupational health and safety information.
HSC2 Health and Safety at Work etc., Act. The Act outlined.
HSC3 Health and Safety at Work etc., Act. Advice to employers.
HSC4 Health and Safety at Work etc., Act. Advice to employees.

The principal information centres of the HSE are:

Health and Safety Executive Library and Information Services, Broad Lane, Sheffield S3 7HQ. (01142) 892844.
Baynards House, 1 Chepstow Place, Westbourne Grove, London W2 4TF. (0171) 221 0870.
St. Hugh's House, Stanley Precinct, Trinity Road, Bootle, Merseyside L20 3OY. (0151) 951 4381.

Addresses of HSE area offices are available at public enquiry points and are listed in *Croner's health and safety at work.* The World Wide Web address for details of HSE publications is http://www.open.gov.uk/hse/hsehome.htm

European Union Directives

European Union Directives relevant to machinery (obtainable through HMSO outlets):
89/392/EEC (1989) Directive on the approximation of the laws of the Member States relating to machinery, as amended (Machinery Directive).
89/654/EEC (1989) Directive concerning the minimum safety and health requirements for the workplace (Workplace Directive).
89/655/EEC (1989) Directive on the minimum safety and health requirements for the use of work equipment by workers at work (Use of Work Equipment Directive).

Periodicals

Croner's Health and Safety Briefing. (Croner Publishing) fortnightly.
Environmental health, fire and safety. (Barbour Index) (microfile) 9 per annum.
Health and Safety at Work (Tolley Publishing Co.) monthly.
Health and Safety Practitioner (Paramount Publishing for the Institution of Occupational Safety and Health) monthly.
Occupational Safety and Health (Royal Society for the Prevention of Accidents) monthly.
RoSPA Bulletin (RoSPA) monthly.

Organizations

British Safety Council, 70 Chancellor's Road, London W6 9RS, UK. Tel:+44 181 741 1231.

Institution of Occupational Safety and Health, 222 Uppingham Road, Leicester LE5 0QG, UK. Tel: +44 1162 768424.

Royal Society for the Prevention of Accidents (RoSPA), Cannon House, Priory Queensway, Birmingham B4 6BS, UK. Tel: +44 121 200 2461.

Safety and Reliability Society, Clayton House, 59 Piccadilly, Manchester M1 2AQ, UK. Tel: +44 161 228 7824.

Conclusions

All aspects of kinematics, dynamics and machinery are covered by a good selection of recently written books aligned with computing techniques. Likewise the quantity of journals on the subject demands selection of those meeting the reader's needs.

An increasing number of new books contain program listings or, better, are accompanied by a disk. Since many of these are undergraduate textbooks there is duplication and scope tends to be limited, programs having a wider scope and some for complex specialized calculations are becoming available from software houses and publishers of design data. In addition some manufacturers are beginning to supplement their catalogues with software, some of whose programs are capable of complex calculations to extend the application of their data significantly. This improvement may be expected to develop rapidly, however current examples rely upon an incompatible variety of support. It is most desirable that an industry standard be agreed. Likewise it is surprising that so few books make extensive cross-reference to published data or software or coordinate analyses and examples with a mathematics software package.

References

Adams, C. E. (1986) *Plastics gearing*. New York: Marcel Dekker.
Agricola, G. [Bauer, G.] (1556) *De re metallica*. Basel. Trans. H. C. Hoover and L. H. Hoover (1950) New York: Dover.
Alban, L. E. (1985) *Systematic analysis of gear failures*. Metals Park, OH: American Society for Metals.
Allen, P. W., Lindley, P. B. and Payne, A. R. (eds) (1967) *Use of rubber in engineering*. London: MacLaren for the Natural Rubber Producers' Association.
American Chain Association (1982) *Chains for power transmission and material handling*. New York: Marcel Dekker.
Andreasen, M. M. and Hein, L. (1987) *Integrated product development*. Bedford: IFS Publications.
Angeles, J. and Lopez-Cajun, C. S. (1991) *Optimization of cam mechanisms*. Dordrecht, The Netherlands: Kluwer Academic.
Arnell, R. D. et al. (1991) *Tribology: principles & design applications*. Basingstoke: Macmillan.
Ascher, H. and Feingold, H. (1984) *Repairable systems reliability*. New York: Marcel Dekker.
Ash, N. I. (1993) Maths at the touch of a button. *Professional Engineering*, 6(6), 17–18.
Avallone, E. E. and Baumeister, T. III (1987) *Mark's standard handbook for mechanical engineers* 9th ed. New York: McGraw-Hill.
Baker, A. K. (1992) *Industrial brake and clutch design*. London: Pentech.
Barber, A. (1992) *Handbook of noise and vibration control* 6th ed. Oxford: Elsevier Advanced Technology.
Barker, T. B. (1990) *Engineering quality by design: interpreting the Taguchi approach*. New York: Marcel Dekker.
Bartz, W. (1993) *Lubrication of gears*, trans. P. Chatterly, A. J. Moore

(ed.) Ehningen bei Boblingen, Germany: Expert Verlag and London: Mechanical Engineering Publications.
Barwell, F. T. (1979) *Bearing systems: principles & practice.* Oxford: Oxford University Press.
Bassani, R. and Piccigallo, B. (1992) *Hydrostatic lubrication.* Amsterdam: Elsevier Science.
Beer, F. B. and Johnson, E. R. (1988) *Vector mechanics for engineers: dynamics.* (S.I.) 5th ed. New York: McGraw-Hill.
Beyer, R. (1963) *The kinematic synthesis of mechanisms*, trans. H. Kuenzel. London: Chapman and Hall.
Bhusan, B. and Gupta, B. K. (1991) *Handbook of tribology, materials, coatings and surface treatments.* New York: McGraw-Hill.
Bies, D. A. and Hansen, C. H. (1988) *Engineering noise control.* London: Unwin Hyman.
Billett, M. (1979) *Industrial lubrication.* Oxford: Pergamon.
Bloch, H. P. and Geitner, F. K. (1990) *An introduction to machinery reliability assessment.* New York: Van Nostrand Reinhold.
Bottema, O. and Roth, B. (1979) *Theoretical kinematics.* Series in applied mathematics and mechanics, no. 24. Amsterdam: North–Holland.
Bowden F. P. and Tabor, D. (1950, part I; 1964, part II) *Friction and lubrication of solids.* Oxford: Clarendon Press.
Bradley, D. A., et al. (1990) *Mechatronics: electronics in products & processes.* London: Chapman and Hall.
BS 1726 *Coil Springs.* Part 1: 1987: Guide for the design of helical compression springs. Part 2: 1988: Guide for the design of helical extension springs. Part 3: 1988: Guide for the design of helical torsion springs. Milton Keynes: British Standards Institution.
BS 3170: 1972 (1991) *Specification for flexible couplings for power transmission.* Milton Keynes: British Standards Institution.
BS 6613: 1985 (1991) *Methods for specifying characteristics of resilient shaft couplings.* Milton Keynes: British Standards Institution.
Cannon, A. G. and Bendell, A. (eds) (1991) *Reliability data banks.* London: Elsevier Applied Science.
Carlson, H. (1978) *Spring designer's handbook.* New York: Marcel Dekker.
Carlson, H. (1980) *Springs: trouble-shooting and failure analysis.* New York: Marcel Dekker.
Carlson, H. (1982) *Spring manufacturing handbook.* New York: Marcel Dekker.
Carter, A. D. S. (1986) *Mechanical reliability* 2nd ed. Basingstoke: Macmillan.
Caulfield-Browne, M., MacCarthy, B. L. and Syan, C. S. (1991) Interactive motion specification using splines. In Bera, H. and Gill, R. (eds) *Proceedings of the 6th International Conference on Robot-*

ics, CAD/CAM, and factories of the future. pp. 275–283. London: South Bank Press.

Chase, M. A. and Sheth, P. N. (1973) *Adoption of computer techniques to the design of mechanical dynamic machinery.* American Society of Mechanical Engineers paper No 73-DET-58. New York: American Society of Mechanical Engineers.

Chen, F. Y. (1982) *Mechanics & design of cam mechanisms.* New York: Pergamon.

Chiang, C.H. (1988) Kinematics of spherical mechanisms. Cambridge: Cambridge University Press.

Chironis, N. P. (1991) *Mechanisms and mechanical devices sourcebook.* New York: McGraw-Hill.

Chow, W. W.-C. (1978) *Cost reduction in product design.* New York: Van Nostrand Reinhold.

Comer, P. (ed.) (1990) *Advances in reliability technology, 11th. Symposium.* London: Elsevier Applied Science.

Computers help to find your bearings. (1992) *Design Engineering*, July, 29.

Constantinescu, V. N. *et al.* (1985) *Sliding bearings.* New York: Allertton Press.

Cossalter, V. *et al.* (1992) A simple numerical approach for optimum synthesis of a class of planar mechanisms. *Mechanism & Machine Theory*, **27**, 357–366.

Cross, N. (1994) *Engineering design methods: strategies for product design* 2nd ed. Chichester: Wiley.

Cross, N., Dorst, K. and Roozenburg, N. (1991) *Research in design thinking.* Delft, The Netherlands: Delft University Press.

Crossley, E. (1980a) A systematic approach to creative design. *Machine Design*, **52**(5), 150–153.

Crossley, E. (1980b) First step in a successful design. *Machine Design*, **52**(12), 128–130.

Crossley, F. R. E. (1970) The International Federation for the Theory of Machines and Mechanisms. *Journal of Mechanisms*, **5**, 133–145.

Crowder, M. J. *et al.* (1991) *Statistical analysis of reliability data.* London: Chapman and Hall.

Cullen, J. and Hollingum, J. (1987) *Implementing total quality.* Kempston: IFS Publications.

Dai, S.-H. and Wang, M.-O. (1992) *Reliability analysis in engineering applications.* New York: Van Nostrand Reinhold.

Dale, B. G. and Oakland, J. S. (1991) *Quality improvement through standards.* Cheltenham: Stanley Thornes.

Darlow, M. S. (1989) *Balancing of high-speed machinery.* New York: Springer-Verlag.

Derry, T. K. and Williams, T. I. (1960) *A short history of technology*. Oxford: Clarendon Press.
Design handbook. (1987) *Engineering guide to spring design*. Wickhamford: Associated Spring SPEC Ltd.
Dewis, M., Murdoch, J. (eds.) (1995) *Tolley's health and safety at work handbook* 5th edn. Croydon: Tolley Publishing and RoSPA.
Dieter, G. E. (1991) *Engineering Design: a materials and processing approach* 2nd ed. New York: McGraw-Hill.
Dietrich, A. (1985) *Spring design with an IRM PC*. New York: Marcel Dekker.
Dijksman, E. A. (1976) *Motion geometry of mechanisms*. Cambridge: Cambridge University Press.
Dimarogonas, A. D. (1988) *Computer-aided machine design*. New York: Prentice Hall.
Doughty, S. (1988) *Mechanics of machines*. New York: Wiley.
Dowson, D. (1979) *History of tribology*. London: Longman.
Drabble, G. E. (1990) *Dynamics*. Basingstoke: Macmillan.
Drago, R. J. (1988) *Fundamentals of gear design*. Boston, MA: Butterworths.
Druce, G. and Watson, A. C. (1991) Systematic design of machinery cams. In *Eurotech Direct '91 Machine Systems*, pp. 193–201. London: Mechanical Engineering Publications.
Dyke, P. and Whitworth, R. (1992) *Guide to mechanics*. Basingstoke: Macmillan.
Easthope, C. E. (1964) *Three-dimensional dynamics: a vectorial treatment* 2nd ed. London: Butterworths.
Erdman, A. G. (ed.) (1993) *Modern kinematics: developments in the last forty years*. New York: Wiley.
Erdman, A. G. and Sandor, G. N. (1991) *Mechanism design: Vol.1 analysis and synthesis* 2nd ed. Englewood Cliffs, NJ: Prentice Hall.
Erickson, W. D. (ed.) (1987) *Belt selection and application for engineers*. New York: Marcel Dekker.
Ertas, J. and Jones, J. C. (1993) *The engineering design process*. New York: Wiley.
Eschmann, P., Hasbargen, L. and Weigant, K. (1985) *Ball & roller bearings: theory, design & application*, trans. L. Hasbargen and J. Brandlein. 2nd ed. Chichester: Wiley.
ESDU International (1965 continuing) *Tribology* subseries. Vol. 1, Bearing selection. Rolling bearings. Vol. 2, Journal bearing calculations. Vol. 3, Thrust bearing calculations. Flexible elements. Vol. 4, Temperatures in bearings. Vol. 5, Contact stress. Vol. 6, Lubrication. Vol. 7, Seal selection. Material properties. Vol. 8, Design and material selection. Vol. S/W1, Software. London: ESDU International.
ESDU International, (1967) 'Item' no. 67023 *The design of crossed*

flexure-pivots. In *Tribology subseries,* Vol. 3. London: ESDU International.

ESDU International. (1979 to date) *Mechanisms subseries,* Vol. 3, Cams. (4 parts) London: ESDU International.

ESDU International. (1982) *Rolling bearings.* In *Tribology subseries,* Vol. 1. London: ESDU International.

ESDU International (1982 continuing) *Stress and Strength subseries.* Vol. 6, Fatigue strength of steels, Vol. 10, Disc and strip springs, and Vol. 11, Helical springs. London: ESDU International.

ESDU International (1983–1988) *Mechanisms subseries.* Vol. 1. Gears. London: ESDU International.

ESDU International (1989 ongoing) *Mechanisms subseries* Vol. 4b Linkages: synthesis and analysis, and Vol. 4c, Linkages: balancing. London: ESDU International.

Eurotrans (1982–1985) *Glossary of transmission elements* 3 vols. Berlin: Springer-Verlag.

Featherstone, R. (1987) *Robot dynamics algorithms.* Dordrecht: Kluwer Academic.

Fisher, R. A. (1966) *Design of experiments* 8th ed. Edinburgh: Oliver and Boyd.

Fitch, E. C. (1992) *Proactive maintenance for mechanical systems.* London: Elsevier Advanced Technology.

Flamm, J. and Luisi, T. (eds) (1992) *Reliability data collection and analysis.* Dordrecht: Kluwer Academic.

French, M. J. (1985) *Conceptual design for engineers.* London: Design Council.

French, M. J. (1992) Mechatronics and the imitation of nature. *Proceedings of the Institution of Mechanical Engineers. Part B Journal of Engineering Manufacture,* **206** (B1), 1–8.

Fuller, D. D. (1984) *Theory and practice of lubrication for engineers* 2nd ed. Chichester: Wiley.

Gans, R. F. (1991) *Analytical kinematics; analysis and synthesis of planar mechanisms.* Boston, MA: Butterworth-Heinemann.

Glaeser, W. A. (1992) *Materials for tribology.* Tribology series 20. Amsterdam: Elsevier Science.

Gobel, E. F. (1974) *Rubber springs design.* trans. A. M. Brichta London: Newnes-Butterworths.

Gohar, R. (1988) *Elastohydrodynamics.* Chichester: Ellis Horwood.

Goodwin, M. J. (1989) *Dynamics of rotor-bearing systems.* Boston: Butterworth-Heinemann.

Grassam, N. S. and Powell, J. W. (1964) *Gas lubricated bearings.* London: Butterworths.

Green, W. G. (1962) *Theory of Machines* 2nd ed. London: Blackie.

Grosjean, J. (1991) *Kinematics and dynamics of mechanisms.* London: McGraw-Hill.

Grove, D. M. and Davis, T. P. (1992) *Engineering quality and experimental design*. Harlow: Longman Scientific and Technical.

Gupta, P. K. (1984) *Advanced dynamics of rolling elements*. New York: Springer-Verlag.

Hain, K. (1961) *Angewordte Getriebelehre*. Düsseldorf: VDI Verlag GmbH.

Hain, K. (1967) *Applied kinematics*, trans. D. P. Adams *et al.* 2nd edn. New York: McGraw-Hill.

Hamilton, P. H. (1988a) *Unit selection – belt drive*. Loughborough: SEED.

Hamilton, P. H. (1988b) *Unit selection – shaft coupling*. Loughborough: SEED.

Hammar, J. (1995) Cam and followers. *Engineering, Technical File Theories*, **1**.

Hamrock, B. J. (1994) *Fundamentals of fluid film lubrication*. New York: McGraw Hill.

Hamrock, B. J. and Dowson, D. (1981) *Ball bearing lubrication: the elastohydrodynamics of elliptical contacts*. New York: Wiley.

Harris, C. M. (ed.) (1991) *Handbook of acoustical measurements and sound control* 3rd ed. London: McGraw-Hill.

Harris, T. A. (1991) *Rolling bearing analysis* 3rd ed. New York: Wiley.

Hart, I. B. (1961) *The world of Leonardo da Vinci*. London: Macdonald.

Hartenberg, R. S. and Denavit, J. (1964) *Kinematic synthesis of linkages*. New York: McGraw-Hill.

Health and Safety Executive (1989) *Quantified risk assessment: its input to decision making*. London: HMSO.

Heideklang, H. R. (1991) *Safe product design in law, management and engineering*. New York: Marcel Dekker.

Heilich, F. III and Stube, E. E. (1983) *Traction drives, selection and application*. New York: Marcel Dekker.

Henley, E. J. and Kumamoto, H. (1985) *Designing for reliability & safety control*. Englewood Cliffs, NJ: Prentice Hall.

Hepburn, C. and Reynolds, R. J. W. (1979) *Elastomers: criteria for engineering design*. London: Applied Science.

Hoeltzel, D. A. and Chieng, W-H. (1990) Knowledge-based approaches for the creative design of mechanisms. *Computer-Aided Design*, **22**, 57–67.

Hofmann, D. A., Kohler, H. and Munro, R. G. (1991) *Teaching pack on gear technology*. Vol. 1. Module 1, Drive systems and Module 2, Gear system design. Vol. 2. Module 3, Gear geometry and Module 4, Design and stress analysis of spur and helical gears. Vol. 3. Module 5, Gearbox design and Module 6, Manufacture and metrology of spur and helical gears. Birmingham: British Gear Association.

Holmberg, K. and Folkeson, A. (eds) (1991) *Operational reliability and systematic maintenance.* London: Elsevier Applied Science.

Hrones, J. A. and Nelson, G. L. (1951) *Analysis of the four-bar linkage.* New York: Technology Press of MIT and Wiley.

Hunt, K. H. (1978) *Kinematic geometry of mechanisms.* Oxford: Clarendon Press.

Hunt, T. M. (1993) *Handbook of wear debris analysis and particle detection in liquids.* London: Elsevier Applied Science.

Hunter, T. A. (1992) *Engineering design for safety.* New York: McGraw-Hill.

Hurricks, P. L. (1994) *Handbook of electromechanical product design.* Harlow: Longman.

Hurst, K. S. (1990) *Component selection – spring.* Loughborough: SEED.

Huston, R. L. (1990) *Multibody dynamics.* London: Butterworth-Heinemann.

Institution of Mechanical Engineers (1990) *Machine condition monitoring.* London: Mechanical Engineering Publications.

Institution of Mechanical Engineers (1991) *Machine systems.* London: Mechanical Engineering Publications.

Institution of Mechanical Engineers (1995) *A practical guide to the machinery directive.* London: Mechanical Engineering Publications.

Jensen, P. W. (1987) *Cam design and manufacture* 2nd ed. New York: Marcel Dekker.

Jensen, P. W. (1991) *Classical and modern mechanisms for engineers and inventors.* New York: Marcel Dekker.

Johnson, R. C. (1978) *Mechanical design synthesis: creative design and optimization* 2nd ed. New York: Krieger.

Johnson, R. C. (1980) *Optimum design of mechanical elements* 2nd ed. New York: Wiley.

Jones, F. D. (ed.) (1930, 1936, 1951) *Ingenious mechanisms for designers and inventors* Vols I, II, III. New York: Industrial Press.

Jones, M. H., and Scott, D. (eds) (1983) *Industrial tribology. The practical aspects of friction, lubrication and wear.* Tribology series 8. Amsterdam: Elsevier Scientific.

Juvinall, R. C. and Marshek, K. M. (1991) *Fundamentals of machine component design* 2nd ed. New York: Wiley.

Kajdas, C., Harvey, S. S. K. and Wilusz, E. (1990) *Encyclopedia of tribology.* Amsterdam: Elsevier Science.

Kececioglu, D. (1993) *Reliability and life testing handbook.* 2 vols. Englewood Cliffs, NJ: Prentice Hall.

Keller, A. G. (1964) *A theatre of machines.* London: Chapman and Hall.

Kimbrell, J. T. (1991) *Kinematics: analysis and synthesis*. New York: McGraw-Hill.

Klaassen, K. B. and van Peppen, J. C. L. (1989) *System reliability: concepts and applications*. London: Arnold.

Kletz, T. (1993) *Lessons from disaster: how organizations have no memory and accidents recur*. Rugby: Institution of Chemical Engineers.

Kloomok, M. and Muffley, R. V. (1955) Plate cam design: pressure angle analysis. *Product Engineering*, **26**(5), 155–160.

Knezevic, J. (1993) *Reliability, maintainability and supportability*. London: McGraw-Hill.

Koloc, Z. and Vaclavik, M. (1993) *Cam mechanisms*. Studies in mechanical engineering 14. Amsterdam: Elsevier.

Koster, M. P. (1974) *Vibrations of cam mechanisms*. London: Macmillan.

Landels, J. G. (1978) *Engineering in the ancient world*. London: Chatto & Windus.

Leinonen, T. E., et al. (1991) IFToMM Commission A for Standardisation of Terminology. Terminology for the theory of machines and mechanisms. *Mechanism & Machine Theory*, **26**(5), 435–539 and i-xxxix.

Lochner, R. H. and Matar, J. E. (1990) *Designing for quality*. London: Chapman and Hall.

Logothetis, N. and Wynn, H. P. (1989) *Quality through design*. Oxford: Clarendon Press.

Lordan, M. and Thompson, G. (1994) Creativity for plant design: an exploratory study. *Creativity and Innovation Management*, **3**(3), 177–183.

Lyon, R. H. (1987) *Machinery noise and diagnostics*. Boston, MA: Butterworths.

Mabie, H. H. and Reinholtz, C. F. (1987) *Mechanisms and dynamics of machinery* 4th ed. New York: Wiley.

MacConaill, P. A., Drews, P. and Robrock, K.-H. (1991) *Mechatronics and Robotics*, I. Amsterdam: IOS Press.

McEwan, J. R. (ed.) (1990) *Condition monitoring: international conference proceedings*. Oxford: Elsevier Science.

McGarva, J. R. and Mullineux, G. (1992) A new methodology for rapid synthesis of function generators. *Proceedings Institution of Mechanical Engineers, Part C, Journal of Mechanical Engineering Science*, **206**, 391–398.

Mancuso, J. R. (1986) *Couplings and joints: design, selection and applications*. New York: Marcel Dekker.

Marshek, K. M. (1987) *Design of machine and structural parts*. New York; Wiley.

Marshek, K. M. and Kannapan, S. M. (eds) (1990) Design Theories:

application to mechanism and machine design. *Mechanism and Machine Theory*, **25**, 243–394.

Medland, A. J. (1992) *The computer-based design process* 2nd ed. London: Chapman and Hall.

Meriam, J. L. and Kraige, L. G. (1993) *Engineering mechanics* Vol. 2. Dynamics, 3rd ed. New York: Wiley.

Meyer zur Capellen, W. (1966) Kinematics – a survey in retrospect and prospect. *Journal of Mechanisms*, **1**, 211–228.

Mills, J. H., Notash, L. and Fenton, R. G. (1993) Optimal design and sensitivity analysis of flexible cam systems. *Mechanism & Machine Theory*, **28**, 563–581.

Misra, K. B. (1992) *Reliability analysis and prediction: a methodology-oriented treatment.* Amsterdam: Elsevier Applied Science.

Mittendorf, W. H. (1990) *Design of devices and systems* 2nd ed. New York: Marcel Dekker.

Molian, S. (1968) *The design of cam mechanisms and linkages.* London: Constable.

Molian, S. (in press) *Mechanism design* 2nd ed. (Oxford: Elsevier Science).

Molian, S. (1995) Mechanism simulation using sub-assemblies. *Proceedings of the ninth world congress of the International Federation for the Theory of Machines and Mechanisms*, Milan, **4**, 2594–2597.

Mortier, K. M. and Orszulik, S. T. (eds) (1992) *Chemistry and technology of lubricants.* Glasgow: Blackie.

Moss, M. A. (1985) *Designing for minimal maintenance expense.* New York: Marcel Dekker.

Moubray, J. (1991) *Reliability-centred maintenance.* Oxford: Butterworth-Heinemann.

Mucci, P. (1994) *Handbook for engineering design using standard materials and components* 4th ed. Milton Keynes: British Standards Institution.

Muller, H. W. (1982) *Epicyclic drive trains.* trans. W. G. Mannhardt. Detroit, MI: Wayne State University Press.

Nayler, G. H. F. (1985) *Dictionary of mechanical engineering* 3rd ed. London: Butterworths.

Neale, M. J. (ed.) (1993) *Drives and seals. A tribology handbook.* Oxford: Butterworth-Heinemann.

Neale, M. J. (ed.) (1993a) *Bearings – a tribology handbook.* Oxford: Butterworth-Heinemann.

Neale, M. J. (ed.) (1993b) *Lubrication – a tribology handbook.* Oxford: Butterworth-Heinemann.

Neale, M. J. (ed.) (1995) *Component failures: maintenance and repair. A tribology handbook.* Oxford: Butterworth-Heinemann.

Neale, M., Needham, P. and Horrell, R. (1991) *Couplings and shaft alignment*. London: Mechanical Engineering Publications.

Newell, J. A. and Horton, H. L. (eds) (1967) *Ingenious mechanisms for designers and inventors*. Vol. IV. New York: Industrial Press.

Nikravesh, P. E. (1988) *Computer-aided analysis of mechanical systems*. Englewood Cliffs, NJ: Prentice Hall.

Norton, M. P. (1989) *Fundamentals of noise and vibration analysis for engineers*. Cambridge: Cambridge University Press.

Norton, R. L. (1992) *Design of machinery*. New York: McGraw-Hill.

Oakland, J. S. and Porter, L. J. (1994) *Cases in total quality management*. Oxford: Butterworth-Heinemann.

Oberg, E. et al. (1992) *Machinery's handbook* 24th ed. New York: Industrial Press.

O'Connor, P. D. T. (1991) *Practical reliability engineering* 3rd ed. Chichester; Wiley.

Olson, D. G., Erdman, A. G. and Riley, D. R. (1985) A systematic procedure for type synthesis of mechanisms with literature review. *Mechanism & Machine Theory*, **20**, 285–295.

Orthwein, W. C. (1986) *Clutches and brakes: design and selection*. New York: Marcel Dekker.

Orthwein, W. C. (1990) *Machine component design*. St. Paul, MN: West Publishing Co.

Pahl, G. and Beitz, W. (1988) *Engineering design: a systematic approach* 2nd ed. Trans. S. Pomerans and K. Wallace. London: Design Council.

Palmgren, A. (1959) *Ball and roller bearing engineering* 3rd ed. Philadelphia: SKF Industries.

Pantry, S. (1993) Health and safety. In *Information sources in chemistry* 4th ed., R. T. Bottle and J. F.B. Rowland (eds), pp. 231–249. London: Bowker-Saur.

Parmley, R. O. (ed.) (1985) *Mechanical components handbook*. New York: McGraw-Hill.

Patton, R., Frank, P. and Clark, R. (1989) *Fault diagnosis in dynamic systems*. Englewood Cliffs, NJ: Prentice Hall.

Pennock, G. R. (1994a) *Machine synthesis and analysis*. New York: ASME.

Pennock, G. R. (1994b) *Machine elements and machine dynamics*. New York: ASME.

Phadke, M. S. (1989) *Quality engineering using robust design*. Englewood Cliffs, NJ: Prentice Hall.

Phillips, J. (1984) *Freedom in machinery*, Vol. 1. Introducing screw theory. Cambridge: Cambridge University Press.

Phillips, J. (1990) *Freedom in machinery*, Vol. 2. Screw theory exemplified. Cambridge: Cambridge University Press.

Piotrowski, J. (1986) *Shaft alignment handbook*. New York: Marcel Dekker.

Pisano, A., McCarthy, M. and Derby, S. (eds) (1990) *Cams, gears, robot and mechanism design*. New York: American Society of Mechanical Engineers.

Pitts, G. (1973) *Techniques in engineering design*. London: Butterworths.

Pitts, G. and Lewis, S. M. (1993) Design modelling the Taguchi way. *Professional Engineering*, 6(4), 32–33.

Potter, C. D. (1992) Mechanism analysis moves with the times. *Computer Graphics World*, 15(5), 30–38.

Prentis, J. M. (1980) *Dynamics of mechanical systems* 2nd ed. Chichester: Ellis Horwood.

Priest, J. W. (1988) *Engineering design for producibility and reliability*. New York: Marcel Dekker.

Pugh, S. (1991) *Total design*. Wokingham: Addison-Wesley.

Quality management handbook (1994) Part 1: Quality assurance; Part 2; Reliability and maintainability. Milton Keynes: British Standards Institution.

Quin Systems (n.d.) *Intelligent motor control for industrial plan*. Wokingham: Quin Systems Ltd.

Ramakumar, R. (1993) *Engineering reliability: fundamentals and applications*. Englewood Cliffs, NJ: Prentice Hall.

Rao, B. K. N. (ed.) (1993) *Profitable condition monitoring*. Dordrecht: Kluwer Academic.

Raper, R. (comp.) (1984) *A history of technology*. Vol. 8, Consolidated indexes. Oxford: Clarendon Press.

Reuleaux, F. (1875) *Theoretische Kinematik [Theoretical kinematics]*. Braunschweig: Vieweg & Sohn.

Reuleaux, F. (1963) *The kinematics of machinery*, trans. A. B. W. Kennedy, New York: Dover.

Riley, W. F. and Sturges, L. D. (1993) *Engineering mechanics: dynamics*. New York: Wiley.

Roberson, R. E. and Schwertassek, R. (1988) *Dynamics of multibody systems*. Berlin: Springer-Verlag.

Ross, P. J. (1988) *Taguchi techniques for quality engineering*. New York: McGraw-Hill.

Rothbart, H. A. (1956) *Cams: design, dynamics and accuracy*. New York: Wiley.

Rothery, B. (1991) *ISO 9000* Aldershot: Gower Publishing Company Ltd.

Rowe, W. B. (1983) *Hydrostatic and hybrid bearing design*. London: Butterworths.

Rowe, W. B., Cheng, K. and Ives, D. (1991) A knowledge-based

system for the selection of fluid-film journal bearings. *Tribology International*, **24**, 291–297.
Sacks, E. and Joskowicz, L. (1993) Automated modeling and kinematic simulation of mechanisms. *Computer-Aided Design*, **25**, 106–118.
Sanders, D. A. (1993) *Making complex machinery move*. Robotics and Mechatronics Series, no. 1. Taunton: Research Studies Press.
Sandor, G. N. and Erdman, A. G. (1984) *Advanced mechanism design: analysis and synthesis*. Vol. 2. Englewood Cliffs, NJ: Prentice Hall.
Sarkar, A. D. (1980) *Friction and wear*. London: Academic Press.
Sayer, F. P. and Bones, J. A. (1990) *Applied mechanics: a modern approach*. London: Chapman and Hall.
Schweitzer, G. (ed.) (1988) *Magnetic bearings. International symposium*. Selected papers. Berlin: Springer-Verlag.
Shabana, A. A. (1989) *Dynamics of multibody systems*. New York: Wiley.
Sharpe, C. (ed.) (1995) *Kempe's engineers year-book* 100th ed. Tonbridge: M-G Information Services Ltd.
Shigley, J. E. and Mischke, C. R. (eds) (1986) *Standard handbook of machine design*. New York: McGraw-Hill.
Shigley, J. E. and Uicker, J. J. (1995) *Theory of machines and mechanisms* 2nd ed. New York: McGraw Hill.
Shortlist. (1992) OsCam 1.02. *CADCAM*, **10**(11), 51.
Shtipelman, B. A. (1978) *Design and manufacture of hypoid gears*. New York: Wiley.
Singer, C. *et al*. (1954) *A history of technology*. Vol. 1, From early times to fall of ancient empires; (1956) Vol. 2, The Mediterranean civilizations and the middle ages c700BC to cAD1500; (1957) Vol. 3, From the Renaissance to the Industrial Revolution c1500 to c1750; (1958) Vol. 4, The Industrial Revolution c1750 to c1850; (1958) Vol. 5, The late nineteenth century c1850 to c1900. Oxford: Clarendon Press.
Smith, D. J. (1993) *Reliability and maintainability and risk*. 4th. edn. Oxford: Butterworth-Heinemann.
Smith, E. H. (1994) (ed.) *Mechanical engineer's reference book* 12th ed. Oxford: Butterworth-Heinemann.
Smith, J. D. (1983) *Gears and their vibration*. New York: Marcel Dekker.
Smith J. D. (1989) *Vibration measurement and analysis*. London: Butterworths.
Society of Automotive Engineers (1979) *Universal joint and drive shaft manual*. Advances in engineering series, no. 7. Warrendale, PA: Society of Automotive Engineers.
Society of Automotive Engineers (1989) *Spring design manual*. Warrendale, PA: Society of Automotive Engineers.

Society of Automotive Engineers (1990) *Gear design, manufacturing and inspection manual*. Warrendale, PA: Society of Automotive Engineers.
Someya, T. (ed.) (1989) *Journal bearing databook*. Berlin: Springer-Verlag.
Soni, A. H., Dado, M. H. and Weng, Y. (1986) *An automated procedure for intelligent mechanism selection and dimensional synthesis*. American Society of Mechanical Engineers paper 86-DET-14. New York: American Society of Mechanical Engineers.
Sound Research Laboratories (1991) *Noise control in industry* 3rd ed. London: Spon.
Spitzer, D. W. (1990) *The application of variable-speed drives* 2nd ed. Research Triangle Park, NC: Instrument Society of America.
Stolarski, T. A. (1990) *Tribology in machine design*. Oxford: Heinemann Newnes.
Stokes, A. (1992) *Gear handbook; design and calculations*. Oxford: Butterworth-Heinemann.
Strandh, S. (1979) *Machines, an illustrated history*. London: Artists House.
Stranks, J. W. (1991) *RoSPA Handbook of health and safety practice* 2nd ed. London: Pitman.
Strong, S. R. and Fahrni, G. R., Jr. (1985) *Selecting bearings without a catalogue*. Society of Automotive Engineers. Paper No. 851510. Warrendale, PA: Society of Automotive Engineers.
Sundararajan, C. (1991) *Guide to reliability engineering: data, analysis, application, management*. New York: Van Nostrand Reinhold.
Taguchi, G. (1987) *System of experimental design*. 2 vols. White Plains, NY: Kraus International.
Taguchi, G. (1993) *Taguchi on robust technology development: bringing quality engineering upstream*. New York: ASME Press.
Tallian, T. E. (1991) *Failure atlas for Hertz contact machine elements* 2nd ed. New York: ASME Press.
Taylor, C. M. (1993) Valve train – cam and follower. Background and lubrication analysis. In C. M. Taylor (ed.) *Engine tribology*. Tribology Series 26. pp. 159–181. Amsterdam: Elsevier.
Technology Exchange Ltd. (1992) *Technoshop '92*. Silsoe: Technology Exchange Ltd.
Terry, G. J. (1968) A chart system to help designers. *Chartered Mechanical Engineer*, **15**(2), 56–59.
Terry, G. J. (1991) *Engineering system safety*. London: Mechanical Engineering Publications.
Thompson, J. R. (1987) *Engineering safety assessment*. Harlow: Longman Scientific and Technical.

Tokhi, M. O. and Leitch, R. R. (1992) *Active noise control*. Oxford: Oxford University Press.
Tokuz, L. C. and Jones, J. R. (1991) Programmable modulation of motion using hybrid machines. In *Eurotech Direct '91 Machine systems*, pp. 85–91. London: Mechanical Engineering Publications.
Torafson, L. E. (1986) A thesaurus of mechanisms. In *Standard handbook of machine design*. J. E. Shigley and C. R. Mischke (eds) pp. 39.1–39.28. New York: McGraw-Hill.
Townsend, D. P. (ed.) (1991) *Dudley's gear handbook* 2nd ed. New York: McGraw-Hill.
Ullmann, D. G. (1992) *The mechanical design process: the Capstone experience*. New York: McGraw-Hill.
Vogwell, J. (1992) *Component selection: standard gears*. Engineering design procedural guide MPT 5.3. Loughborough: SEED.
Wahl, A. M. (1964) *Mechanical springs* 2nd ed. London: McGraw-Hill.
Walter, M. H. and Cox, R. F. (eds) (1990) *Safety and reliability in the 90s: will past experience or prediction meet our needs?* London: Elsevier Applied Science.
Wan Changsen (1991) *Analysis of rolling element bearings*, trans. Wan Changsen and Zhang Zhaoying. London: Mechanical Engineering Publications.
Wang, Q., Zhou, J. and Yu, J. (1988) A chain-drive expert system and CAD system. In D. T. Pham, (ed.) *Expert systems in engineering*. Kempston: IFS Publications, pp. 315–323.
Welsh, R. J. (1983) *Plain bearing design handbook*. London: Butterworths.
Wilcock, D. F. (ed.) (1972) *Gas bearing design manual*. Latham, NY: Mechanical Technology Inc.
Williams, J. A. (1994) *Engineering tribology*. Oxford: Oxford University Press.
Williams, T. I. (ed.) (1978) *A history of technology*. Vol. VI, The twentieth century, c1900 to c1950, Part 1; (1978) Vol. VII, The twentieth century, c1900 to c1950, Part 2. Oxford: Clarendon Press.
Wilson, C. E. and Sadler, J. P. (1993) *Kinematics and dynamics of machinery* 2nd ed. New York: HarperCollins.
Wood, D. (1993) Design mechanisms. *CADCAM*, **12**(3), 67, 69.
Yang, S. J. and Ellison, A. J. (1987) *Machinery noise measurement*. Oxford: Clarendon Press.
Zhu, G. (1993) Valve trains – design studies, wider aspects and future developments. In C. M. Taylor (ed.) *Engine tribology*. Tribology Series 26. pp. 183–211. Amsterdam: Elsevier.

CHAPTER TWELVE

Thermodynamics and thermal systems
P. W. O'CALLAGHAN

Introduction

Thermodynamics is concerned with the ways in which substances behave as they are heated, cooled, expanded or compressed. In particular, it is concerned with the relationship between heat, work and other energy forms. The theory is based upon four basic laws.

The Zeroth Law of Thermodynamics states that if two substances are each in equilibrium with a third substance, then the two substances are in equilibrium with each other. The equilibrium state of a substance is defined by its temperature, pressure and chemical properties. If two connected substances are at different temperatures, heat will flow between them by conduction, convection and radiation until thermal equilibrium is established. This defines the concept of temperature and forms the basis of heat transfer. If two connected substances are at different pressures, mass will flow between them until pressure equilibrium is established, forming the basis of mass transfer. If two connected substances have different chemical compositions, chemical reactions will occur until chemical equilibrium is established. This constitutes the basis of chemical thermodynamics. All substances seek a general equilibrium state. Hence the tendency of materials to degrade by dispersion, decay or oxidation to a state in equilibrium with the environmental datum. For example, the release of stored energy in the combustion of fossil fuels converts energy from a relatively stable form to the unstable form of thermal energy, producing waste and pollution in the process.

The First Law of Thermodynamics states that energy and materials are always conserved, they can be neither created nor destroyed, only converted from one state or form to another. Work and heat appear at the boundary of a system undergoing a thermodynamic process (i.e.

heating, cooling, expansion or compression). When a system is heated, its temperature and internal energy rises and some work is performed. Conversely, when work is done on a system, its temperature and internal energy rises and some heat is released to the environment.

The heat absorbed by a system is equal to the work done by the system plus its change in internal energy.

When a gas expands, its pressure falls, its volume increases and it performs work in displacing its surroundings. The amount of work done is given by the area under the pressure versus volume curve. The relationship between pressure and volume can follow one of five processes,

- a constant volume process
- a constant pressure process
- a constant temperature process
- an adiabatic process, where no heat enters or leaves the system
- a polytropic process (PV^n = constant)

All engines are cyclic devices. This means that they must pressurize a fluid, such as air or steam, originally at environmental temperature and pressure, then allow the fluid to expand, performing work, and then return the fluid to its initial state in equilibrium with the environment, ready to start the cycle again. This is achieved in internal combustion engines by mixing air with oil or gas, igniting the mixture to raise its temperature and confining the mixture so that it pressurizes. Following the resisted expansion process which produces work, the low pressure mixture is rejected to the environment where it disperses and cools to environmental conditions, completing the cycle. In the steam engine, water at environmental conditions is boiled via the combustion of fossil fuels, the high pressure steam is expanded in a turbine to perform work and the low pressure steam is then condensed and cooled to the original state, ready for the next cycle.

Because heat must be rejected to the environment to complete an engine cycle, not all of the heat energy produced by combustion can be converted to work, i.e. an engine cycle which converts heat to work cannot be 100 per cent efficient. This leads to the Second Law of Thermodynamics which states that it is impossible to construct a system which will operate in a cycle, extract heat from a source, and do an equivalent amount of work on the surroundings. In order to receive heat, the system must be in contact with a thermal reservoir at a temperature higher than that of the working fluid at some point during the cycle. For heat to be rejected from the system, the fluid must, at some point in the cycle, be in contact with a thermal sink at a lower temperature than that of the working fluid. Thus, if a system is to undergo a cycle and produce work, it must operate between at least two reservoirs at different temperatures.

Heat cannot pass from a colder body to a hotter body without any other external effect occurring (i.e. heat cannot flow against a negative temperature gradient).

For an expansion process to be reversible, the pressure/volume path during expansion must be exactly retraceable during compression. If heat is transferred between the expanding fluid and the environment during the expansion process, a reverse compression process cannot bring the fluid back to its original state. A process can only be completely reversible if it is fully-resisted, and is either adiabatic (i.e. no heat is transferred between the system and its surroundings); or isothermal (i.e. the system temperature remains invariant and so the heat supplied to the system is converted totally into work). On the return path, the work supplied is converted totally into heat. Conditions which prevent a process being fully-reversible are

- the presence of friction
- heat transfer to the environment
- unresisted expansion
- paddle work

The Carnot cycle is an imaginary ideal cycle constructed from reversible adiabatic (isentropic) and isothermal expansions and compressions.

Heat exergy is that part of the quantity of heat which would be converted to work (the available energy, or thermodynamic availability) in an ideal Carnot cycle operating between the temperature of the heat source and the temperature of the environment.

Entropy may be considered as an indication of the amount of unavailable energy within a given system. Thermodynamic availability cannot be recycled. Hence, as much work and heat must be extracted from a degrading energy chain before the energy in transit assumes the mean properties of the environment (O'Callaghan, 1993).

Entropy is a measure of the disorder in a system (Angrist and Hepler, 1973). When a practical (irreversible) change occurs in a system, its entropy always increases.

The Third Law of Thermodynamics states that the entropy of a substance approaches zero as its thermodynamic temperature approaches zero degrees Kelvin.

Dictionaries and primers

James (1976) and Fenn (1982) contain many useful definitions and descriptions of thermodynamic terms, principles and mechanisms. Goodger (1984) contains a most useful glossary of terms.

Fundamentals

The fundamentals of thermodynamics are covered in many texts, including Walshaw (1947), Mooney (1955), Skrotzki (1963), Zemansky and Van Ness (1966), Howell and Buickius (1992), Keenan (1970), Spalding and Cole (1973), Zemansky, Abbot and Van Ness (1975), Van Wylen and Sonntag (1985), Reynold and Perkins (1977), Haywood (1980), Parker (1981), Cravalho and Smith (1981), Nag (1981), Zemansky and Dittman (1981), Whalley (1992), Burghardt (1987), Sonntag and Van Wylen (1991), Goodger (1984), Schmidt, Henderson and Wolgemuth (1993), Martin (1986), Jones and Hawkins (1986), Eastop and McConkey (1993), Joel (1987), Huang (1988), Look and Sauer (1988), and Rogers and Mayhew (1992). These are all introductory texts in thermodynamics and generally deal with concepts and definitions, the pure substance, the relationships between heat and work, the laws of thermodynamics, entropy, and power cycles. Kelly (1973), Landsberg (1978), Sears and Salinger (1975), Van Wylen and Sonntag (1985), Riedi (1976) and Zemansky and Dittman (1981) also deal with statistical thermodynamics.

Herzfeld (1962) and Plumb (1973) deal with the measurement and control of temperature. Zemansky and Dittman (1981) include a chapter on temperature measurement, as does O'Callaghan (1993), which also deals with measurements of transport properties.

Boxer (1976 and 1979) treat the subject almost entirely on worked examples. Haywood (1986) contains various analyses of engineering cycles and worked problems covering power, refrigeration and gas liquefaction plant. Bodsworth and Appleton (1965) work though many useful problems in applied thermodynamics, whilst Liley (1989) contains 2500 solved problems in mechanical engineering thermodynamics.

Rogers and Mayhew (1992) contains chapters on heat transfer, conduction, convection, radiation and combined modes of heat transfer.

Properties

Keenan, Chao and Kaye (1980) contains gas tables for the thermodynamic properties of air, and the products of combustion and component gases. Compressible flow functions are also included.

Hickson and Taylor (1980) provide the enthalpy-entropy diagram for steam.

Rogers and Mayhew (1988) contains tables for the thermodynamic and transport properties (e.g. density, specific heats, thermal conductivities, specific entropies, enthalpies and dynamic viscosities of fluids – e.g. water and steam, air, ammonia, CFC refrigerants, and other gases and vapours). Van Wylen and Sonntag (1985) contains a vast appendix of these properties, as well as extensive information on nitrogen, methane, the atomic weights of the elements, and thermodynamic and physical constants. Temperature-entropy charts for steam, nitrogen and oxygen and pressure-enthalpy charts for ammonia and some of the freons are included.

Hottell (1946) contains charts of thermodynamic properties of fluids encountered in internal combustion engine cycles. Newman and Allison (1966) demonstrate the direct calculation of specific heats and related thermodynamic properties of arbitrary gas mixtures and provide tabulated results.

Hippensteele (1978) lists a useful computer program for obtaining thermodynamic and transport properties of air and products from the combustion of fuels and air. O'Callaghan (1993) provides correlated equations for the variations of the transport properties of air and water with temperature. Moeckel and Weston (1958) deal with the composition and thermodynamic properties of air in chemical equilibrium. Badr, O'Callaghan and Probert (1985a and 1985b) list general equations for estimating the thermodynamic and thermophysical properties of organic working fluids for Rankine-cycle engines, whilst Herridge, O'Callaghan and Probert (1988) provide equations and computer program listings for the thermodynamic properties of fluids commonly used in refrigeration systems.

More advanced level treatments of thermodynamics are to be found in Benson (1977) and Emanuel (1987).

Some useful computer software routines for the analyses of thermodynamic systems are to be found in Bacon (1983).

Availability and exergy

The availability of energy, or *exergy* is a measure of the maximum useful work that can be performed by a system interacting with the environment (O'Callaghan 1981). The ideal thermodynamic efficiency of a process is defined as the ratio of the useful work performed to the amount of energy supplied to the process.

The Carnot efficiency is most useful when comparing the abilities of energy potentials to produce work. Work can be used to produce a stable and storable low entropy energy form (i.e. electricity, or a synthetic fuel or chemical), the energy content of which may be released at a higher temperature than was attained during the original combustion process. Thus fuels and electricity may be considered as the most pure energy forms.

There is no corresponding definition for the efficiency of a process, the purpose of which is to produce heat (for space heating, cooking, manufacturing processes, desalination, etc.). Nevertheless, it is intuitive that the collection and use at a temperature of 300K of 1MJ of heat from a source of low grade heat at 350K is an entirely different situation from the release and utilization of 1MJ of heat via the combustion of fuel at 2000K. Whilst the quantity of heat transferred is the same in each case, there is no doubt that the low-grade energy process is more commendable than the wastage of work-producing potential implicit in the high-grade energy process merely to satisfy a low-grade heating purpose. A descriptive parameter which rates the *quality* of energy is desirable. The exergetic potential fulfils this function.

Second Law, or Exergy Analyses (Bruges, 1959; Gaggioli, 1986) may be applied to systems, such as combined heat and power plant, to truly rate the overall system efficiency (Moore, 1981; Kotas, 1985; Moran, 1982; El-Masri, 1987; Chin and El-Masri, 1987; O'Callaghan and Probert, 1981).

Moore (1981) treats this subject in detail, including chapters on energy conservation, reversibility, entropy, available energy and energy degrading.

Process integration (1987–) is widely used in the design of complex thermal and chemical plants, where a total energy approach should be adopted and all opportunities for efficient and economic energy cascading should be sought (*Application of process integration to utilities, combined heat and power and heat pumps*, 1989).

Linnhoff (1982) has developed suitable exergy-based analytical techniques and introduced the concept of the *pinch-point* for integrated process plant. (See also *Process integration* 1987–). Chin and El-Masri (1987) have described the exergy analysis of combined cycles.

Thermal systems

A highly specialized compartment of thermal systems is involved in the thermal design of spacecraft and their services. Many conferences have been held by the American Institute of Aeronautics and Astronautics (e.g. Fletcher, 1978a and 1978b), and other bodies (e.g. the European Space Agency's 1978 conference, Spacecraft Thermal and Environmental Control Systems). Technical papers written in aspects of this speciality include Kozlov and Nusinov (1973), Joy and Goliazewski (1986) and Hong (1989).

According to Stoecker (1989), however, the designation 'thermal' implies systems based upon the principles of thermodynamics, heat transfer and fluid mechanics. Thus, the mechanisms involved are those associated with the heating, cooling, boiling, condensing, expansion and compression of fluids and are found within such industries as power generation, electric and gas utilities, refrigeration, air conditioning, heating and ventilating systems, compressed air services, desalination plant, and in the food, water, chemical and process industries. The hardware encountered includes furnaces, boilers, autoclaves, fluid heaters, thermal energy storage systems, heat exchangers, driers, refrigeration systems, engines, expanders, turbines, fans, gas compressors, pumps and refrigeration compressors.

General texts describing thermal physics and thermal analysis are provided by Kittel and Kromer (1980), Sprackling (1991), Adkins (1987), and Brown (1988). Hodge (1990) deals with the analyses and describes designs for energy systems, whilst Stoecker (1989) is a text concentrating upon thermal system design optimization and the microcomputer control of thermal and mechanical systems.

Haywood (1986) contains analyses of various engineering cycles and worked problems in power, refrigerating and gas liquefaction plant. Dryden (1982) covers most thermal systems in *The efficient use of energy*, which was produced in collaboration with the UK Institute of Fuel (now Energy) acting on behalf of the UK Department of Energy. The *Energy manager's workbook* (1982), based on the papers presented to the Energy Managers' Workshops organized jointly by the British Institute of Management and the Department of Energy, deals most comprehensively with many thermal energy flow systems.

Dinter, Geyer and Tamme (1990) covers the economics of using thermal energy for commercial applications. West and Kreith (1988) also deals with the economic analyses of thermal energy systems, particularly solar heat technologies. Threlkeld (1970) covers a range of thermal systems from heating applications to refrigeration systems.

Heat transfer and fluid mechanics

The American Heat Transfer and Fluid Mechanics Institute has held a number of conferences covering advanced aspects over the years.

FLUIDEX (1973–) produces abstracts dealing with heat and fluid flow. Useful introductory texts in heat transfer are Kreith and Black (1980), Kreith (1986), Janna (1986) and Incropera and Dewitt (1990).

Evett and Liu (1989) work through 2500 solved problems in fluid mechanics and hydraulics. Whilst the *Annual Review of Fluid Mechanics* (Annual Reviews Inc., 1969–) deal with advanced topics, useful introductory texts are Kinsky (1982) and Roy (1988). Bacon (1983) and Sharp (1988) contain computer listings for a variety of problems and solutions in fluid mechanics and heat transfer. Current research needs are identified in Jones and Telionis (1992). Journals covering heat and fluid flow include the *International Journal of Heat and Fluid Flow* and the *Journal of Thermophysics and Heat Transfer*.

Jones and Telionis (1992) considered basic research needs in fluid mechanics in the recent American Society of Mechanical Engineers conference.

Heating, furnaces and boilers

The American Society of Mechanical Engineers produces standards for boilers and pressure vessel design and maintenance. BS 845: Parts 1 and 2: 1987 describes methods for assessing the thermal performances of boilers for steam, hot water and high temperature heat transfer fluids. NIFES (1989) has produced a *Boiler operator's handbook* and condensing boilers are covered by Building Research Establishment (1988). Among their series of Fuel Efficiency Booklets, the UK Department of Energy produced a handbook (1984) dealing with the economic use of gas-fired boiler plant.

Davies (1970) contains calculations in furnace technology, Cone (1980) covers the energy management of industrial furnaces, and Rhine and Tucker (1990) the modelling of gas-fired furnaces, boilers and other industrial heating processes. Whalley (1987) covers boiling, condensation and gas-liquid flow. Gunn and Horton (1981) offers a practical guide to steam boiler design and operation.

A conference dealing with Energy Management in Buildings (Sherratt, 1986) was convened by the British Chartered Institution of Building Services Engineers.

Heat exchangers, heat recovery and thermal storage systems

Kays (1984) demonstrates the design of compact heat exchangers and contains many tables and design data. Shah, McDonald and Howard (1980) deals with the history, technological advancement and mechanical design problems in compact heat exchangers, Shah *et al.* (1992) covers compact heat exchangers for power and process industries, and Chisholm (1980) considers developments in heat exchanger technology.

Reay (1979) is a directory of equipment and techniques of heat recovery, Reiter (1983) describes a number of industrial and commercial heat recovery systems and Boyen (1975) deals with the problems of practical heat recovery. Department of Energy (1978a) is a Fuel Efficiency Booklet dealing with the recovery of waste heat from industrial processes, Sengupta and Lee (1983) report the proceedings of a conference dealing with the utilization and management of waste heat and Kreider and McNeil (1977) have produced a guidebook for waste heat management.

The Process Engineering Group and the Thermodynamics and Fluid

Mechanics Group of the Institution of Mechanical Engineers and the Department of Energy in London have held a conference, with *Energy recovery in process plants* (1976).

Ehringer, Hoyaux and Pilavachi (1983) contains papers on combustion, heat recovery and Rankine cycle machines from the 1982 EC contractors' meetings held in Brussels.

Kovach (1977) reports on a NATO Science Committee Conference on thermal energy storage systems.

Cooling systems

Air cooling plant, food freezing and refrigeration systems are dealt with by Stoecker (1989) and the *ASHRAE handbook* (1986–) covers refrigeration systems and applications in depth. Gosling (1980) deals with the principles and design of cooling plant. The report of the National Engineering Laboratory (1971) covers air coolers, cooling towers and evaporative coolers, and there is also a practical guide to the selection of refrigeration condensers (*Heat transfer*, 1985).

Refrigeration and air conditioning is covered by the *ASHRAE Handbooks*, *Refrigeration systems and applications* (1986), *Fundamentals* (1985), and *Heating, ventilating and air-conditioning systems and equipment* (1992), Croome and Roberts (1975), Langley (1986), Arora (1981), Gosney (1982), Stoecker and Jones (1982), Prasad (1983), Jones (1980 and 1985), Stanford (1988), Chatenever (1988), Trott (1989) and Olivo (1990).

The Building Services Research and Information Association has produced a great number of monographs and publications in the field of building services (e.g. 1992).

Booth (1970) provides a dictionary of refrigeration and air conditioning.

Journals covering the subject include the *BRE Digest* and *Building Services* (formerly the *Journal of the Institution of Heating and Ventilating Engineers*, *Building Services Engineer* and the *Journal of the Chartered Institution of Building Services*).

The Chartered Institution of Building Services Engineers produces the *CIBSE guide* (e.g. 5th ed. 1986) which is the British equivalent of the ASHRAE handbooks in the area of building services.

Heat pumps

The use of heat pumps for the simultaneous production of heat and cold is a rapidly developing technology and various conferences have taken place to stimulate research in this area, including Camatini and

Kester (1975), Department of Energy (1977), Berghmans (1980), Fitt and Moses (1984), Sherratt (1984), Commission of the European Communities (1985), Zimmerman and Powell (1987), and the British Hydromechanics Research Association (1987).

Loyd (1990) has produced an annotated bibliography with a survey of suppliers of heat pumps, whilst Berridge (1975) lists key references in the literature.

Texts include Sporn, Ambrose and Baumeister (1947), Ambrose (1966), Sumner (1976), Reay and Macmichael (1988), Lord, Ouellette and Cheremisinoff (1980), McMullan and Morgan (1981), McGuigan (1981), Armor (1981), Sauer and Howell (1983), and Moser and Schnitzer (1985).

Offenhartz (1979) dealt with methanol-based heat pumps for the storage of solar thermal energy.

The Building Research Establishment (1981) have produced a digest on heat pumps for domestic use and Sutphin (1987) considers the installation and trouble-shooting in respect of residential heat pumps.

Expansion and compression, fans and pumps

Specialist aspects of engines, expanders, gas turbines and steam turbines are covered by the US Society of Automotive Engineers Handbook (1995), whereas McMahon (1971) and Oates (1989) deal with aircraft engines, gas turbines and propulsion. The American Society of Mechanical Engineers produces the *Transactions of the ASME, Journal of Engineering for Gas Turbines and Power*.

The UK Energy Efficiency Office produces the Fuel Efficiency booklet, *Compressed air and energy use* (1984). *Design and operation of industrial compressors* (1978) is the proceedings of a conference sponsored by the Fluid Machinery Group of the Institution of Mechanical Engineers and the British Compressed Air Society.

Wallis (1961 and 1983) deals with the design and practical use of axial flow fans and Anderson (1980) and Lobanoff and Ross (1992) the design and application of centrifugal pumps.

Electricity generation, combined heat and power, co-generation and district heating

The technologies (Williamson, 1988) and comparative economics of electricity generation are described and discussed in Johansson (1989), Marsh (1980) and Munasinghe (1990).

The use of waste heat from power generation is receiving much attention. The UK Department of Energy set up a District Heating

Working Party and produced *District heating combined with electricity generation in the United Kingdom*, a discussion document, in 1977, followed by a further document in 1979. Loyd (1990) has produced an annotated bibliography covering combined heat and power.

Rice (1987) conducts a thermodynamic evaluation of gas turbine cogeneration cycles, Part I, Heat balance method analysis and Part II, Complex cycle analysis.

Wilkinson and Barnes (1980) deal with the cogeneration of electricity and useful heat, Diamant and Kut (1981) consider district heating and cooling for energy conservation, Mackenzie-Kennedy (1979) is a practical guide to district heating, thermal generation and distribution, Meador (1981) covers cogeneration and district heating and Marecki (1988) deals with combined heat and power generating systems. Horlock (1987) covers the thermodynamics and economics of combined heat and power.

The (British) Combined Heat and Power Association (e.g. 1992) and the District Heating Association (e.g. 1981) hold a number of meetings and conferences in this area, as does the (American) International District Heating and Cooling Association (e.g. 1985). This association also produces the quarterly journal, *District Heating and Cooling*. The (American) International District Heating Association (1983) has produced a design guide in the interest of the district heating and cooling industries. The British Petroleum Company (1976) has produced a technical bulletin, *District heating and technological developments*.

Gas Turbine World and *Cogeneration* are journals dedicated to gas turbines and the cogeneration of electrical power and heat. Another important journal in this area is the *Transactions of the ASME. Journal of Engineering for Gas Turbines and Power*.

New technologies and alternative energy systems

Ehringer, Hoyaux and Pilavachi (1983) edited the proceedings of the research contractors' meetings of the Energy Research and Development programme of the European Community and this covers novel thermodynamic and thermal systems. The research journal *Applied Energy*, published by Elsevier Science, often includes papers dealing with novel concepts and new technologies.

Blackburn (1987) considers the renewable energy alternative and how the United States and the world might prosper without nuclear energy or coal. Valette and Focquet (1982) perform this analysis for Europe.

West *et al.* (1984) edited the proceedings of a conference on alternative energy systems and Hackleman (1980) describes how alternative energy forms may be harnessed in a domestic situation. The UK

Department of Energy's report, *The development of alternative sources of energy* (1978b) was the British Government's reply to the third and fourth reports from the UK Select Committee on Science and Technology on the use of alternative energy in the United Kingdom. This was followed by *Renewable energy sources in the United Kingdom* in 1981. The (American) Center for Renewable Resources have produced a text dealing with the use of renewable energy in cities (1984). Taylor (1983) considered the use of alternative energy sources for the centralized generation of electricity. Denno (1989) dealt with power system design and applications for alternative energy sources and then examined (1990) the economics of alternative energy sources. Merrill and Gage (1978) is an energy primer dealing with solar energy, water power, wind power and the use of biofuels. Introductory texts in the subject include Sorensen (1979), Dunn (1986), Twidell and Weir (1986) and Laughton (1990).

Solar Thermal Systems are covered by Duffie and Beckman (1991), Howell, Bannerot and Vliet (1982), Lunde (1980), Reddy (1987) and Sukhatme (1984), who also considers the problem of the storage of solar energy. The American Society of Mechanical Engineers convened a conference on Heat Transfer and Fluid Flow in Solar Thermal Systems which was edited by Min and Chiou (1985). Another conference covering alternative energy systems was edited by West *et al.* (1984).

Societies and institutions

Agence Européenne d'Informations, AEI
American Institute of Aeronautics and Astronautics, AIAA
American Society of Heating, Refrigerating and Air Conditioning Engineers, ASHRAE
American Society of Mechanical Engineers, ASME
British Hydromechanics Research Association, BHRA
British Institute of Management, BIM
British Petroleum Company, BP
British Standards Institution, BSI
Building Research Establishment, BRE
Building Services Research and Information Association, BSRIA
Chartered Institution of Building Services Engineers, CIBSE
Combined Heat and Power Association, CHPA
Commission of the European Communities, CEE
District Heating Association, DHA
Energy Efficency Office, EEO
European Space Agency, ESA
Heat Transfer and Fluid Mechanics Institute, HTFMI
Institution of Electrical Engineers, IEE
Institute of Energy, InstE
Institution of Mechanical Engineers, IMechE

International District Heating Association, IDHA
International District Heating and Cooling Association, IDHCA
National Advisory Committee for Aeronautics (later became NASA), NACA
National Aeronautics and Space Administration, NASA
North Atlantic Treaty Organization, NATO
National Bureau of Standards, NBS
National Engineering Laboratory, NEL
National Industrial Fuel Efficiency Service, NIFES
Society of Automotive Engineers, SAE

Editor's note

Readers are referred to the Chapter on Energy Technology for comprehensive lists and information on indexing and abstracting journals, computer-based services and primary journals, which have not been repeated in this chapter as many of them are applicable to the subjects covered in both chapters.

References

Adkins, C. J. (1987) *An introduction to thermal physics* rev. ed. Cambridge: Cambridge University Press.
Ambrose, E. R. (1966) *Heat pumps and electric heating, residential, commercial, industrial year-round air conditioning.* New York: Wiley.
American Society of Heating, Refrigerating and Air Conditioning Engineers (1985) *ASHRAE handbook, fundamentals.* Atlanta, GA: ASHRAE.
American Society of Heating, Refrigerating and Air Conditioning Engineers (1986–) *ASHRAE handbook, refrigeration systems and applications* Atlanta, GA: ASHRAE.
American Society of Heating, Refrigerating and Air Conditioning Engineers (1992) *ASHRAE handbook, heating, ventilating and air-conditioning systems and equipment.* Atlanta, GA: ASHRAE.
American Society of Mechanical Engineers, Boiler and Pressure Vessel Committee (1993) *ASME boiler and pressure vessel code.* New York: ASME.
Anderson, H. H. (1980) *Centrifugal pumps*, 3rd ed. Morden: Trade and Technical Press.
Angrist, S. W. and Hepler, L. G. (1973) *Order and chaos, laws of energy and entropy.* Harmondsworth: Penguin.
Application of process integration to utilities, combined heat and power and heat pumps (1989) London: ESDU.
Armor, M. (1981) *Heat pumps and houses.* Dorchester: Prism.

Arora, C. P. (1981) *Refrigeration and air conditioning*, SI units. New Delhi: Tata McGraw-Hill.
Bacon, D. H. (1983) *BASIC thermodynamics and heat transfer*. London: Butterworths.
Badr, O., O'Callaghan, P. W. and Probert, S. D. (1985a) Thermodynamic and thermophysical properties of organic working fluids for Rankine-cycle engines. *Applied Energy*, **19**(1), 1–40.
Badr, O., Probert S. D. and O'Callaghan, P. W. (1985b) Selecting a working fluid for a Rankine cycle engine. *Applied Energy*, **21**(1) 1–42.
Benson, R. S. (1977) *Advanced Engineering Thermodynamics* 2nd ed. Oxford: Pergamon.
Berghmans, J. (ed.) *Heat pump fundamentals*, proceedings of the NATO Advanced Studies Institute on Heat Pump Fundamentals, Espinho, Portugal, September 1–12, 1980.The Hague: Martinos Nijhoff.
Berridge, G. L. C. (1975) *Heat Pumps: key references in the literature*. Boston Spa: Brainchild Information Services.
Blackburn, J. O. (1987) *The renewable energy alternative: how the United States and the world can prosper without nuclear energy or coal*. Durham, NC: Duke University Press.
Bodsworth, C. and Appleton, A. S. (1965) *Problems in applied thermodynamics*. London, Longman.
Booth, K. M. (1970) *Dictionary of refrigeration and air conditioning*. London: Applied Science.
Boxer, G. (1976) *Engineering thermodynamics*. London: Macmillan.
Boxer, G. (1979) *Applications of engineering thermodynamics*. London: Macmillan.
Boyen, J. L. (1975) *Practical heat recovery*. New York: Wiley.
British Petroleum Company (1976) *District heating and technological developments*. London: BP.
Brown, M. E. (1988) *Introduction to thermal analysis, techniques and applications*. London: Chapman and Hall.
Bruges, E. A. (1959) *Available energy and the second law analysis*. London: Butterworths.
BS 845: Parts 1 and 2: 1987 *Methods for assessing thermal performance of boilers for steam, hot water and high temperature heat transfer fluids*. London: BSI.
Building Research Establishment (1981) *Heat pumps for domestic use*. Watford: Building Research Advisory Service.
Building Research Establishment (1988) *Condensing boilers*. Watford: Building Research Advisory Service.
Building Services Research and Information Association (1992) *Refrigeration and the environment*. Bracknell: BSRIA.

Burghardt, M. D. (1987) *Engineering thermodynamics with applications* 3rd ed. New York: Harper and Row.
Camatini, E. and Kester, T. (1976) (eds) *Heat pumps and their contribution to energy conservation*. Leyden: Noordhoff.
Center for Renewable Resources (1984) *Renewable energy in cities*. New York: Van Nostrand Reinhold.
Chatenever, R. (1988) *Air conditioning and refrigeration for the professional*. New York: Wiley.
Chin, W. W. and El-Masri, M.A. (1987) Exergy analysis of combined cycles, – part 2, analysis and optimization of two-pressure steam bottoming cycles. *(ASME/86 JPGC-GT-10)*. In Transactions of the ASME. *Journal of Engineering for Gas Turbines and Power*, **109**, 237–243.
Chisholm, D. (ed.) (1980) *Developments in heat exchanger technology*. Barking: Applied Science.
Combined Heat and Power Association (1992) *Combined heat and power, an agenda for action*. London: Combined Heat and Power Association.
Combustion in steam raising plant (1992) London: Mechanical Engineering Publications (IMechE Seminar Series 1992–8).
Commission of the European Communities (1985) *Absorption heat pumps congress* Luxembourg: CEC.
Cone, C. (1980) *Energy management for industrial furnaces*. New York: Wiley.
Cravalho, E. G. and Smith, J. L. (1981) *Engineering thermodynamics*. Boston: Pitman.
Croome, D. J. and Roberts, B. M. (1975) *Air conditioning and ventilation of buildings*. Oxford: Pergamon.
Davies, C. (1970) *Calculations in furnace technology*. Oxford: Pergamon.
Denno, K. (1989) *Power system design and applications for alternative energy sources*. Englewood Cliffs, NJ: Prentice Hall.
Denno, K. (1990) *Engineering economics of alternative energy sources*. Boca Raton, FL: CRC Press.
Department of Energy, District Heating Working Party (1977a) *District heating combined with electricity generation in the United Kingdom*. London: HMSO.
Department of Energy, Energy Technology Support Unit (1977b) *UK workshop on heat pumps*, 30 June – 2 July 1976. Harwell: ETSU.
Department of Energy (1978a) *The recovery of waste heat from industrial processes*. London: Energy Efficiency Office.
Department of Energy (1978b) *The development of alternative sources of energy*. London: HMSO.
Department of Energy, Combined Heat and Power Group (1979) *Com-

bined heat and electrical power generation in the United Kingdom. London: HMSO.
Department of Energy (1981) *Renewable energy sources in the United Kingdom.* London: DoE.
Department of Energy (1984) *Economic use of gas-fired boiler plant.* London: Energy Efficiency Office.
Design and operation of industrial compressors (1978) London: Mechanical Engineering Publications.
Diamant, R. M. E. and Kut, D. (1981) *District heating and cooling for energy conservation.* London: Architectural Press.
Dinter, F., Geyer, M. A. and Tamme, R. (1990) *Thermal energy for commercial applications: a feasibility study on economic systems.* Berlin: Springer-Verlag.
District Heating Association (1981) *Practical aspects of district heating and combined heat and power.* London: District Heating Association.
Dryden, I. G. C. (ed.) (1982) *The efficient use of energy* 2nd ed. Sevenoaks: Butterworths.
Duffie, J. A and Beckman, W. A. (1991) *Solar engineering of thermal processes* 2nd ed. New York: Wiley.
Dunn, P. D. (1986) *Renewable energies, sources, conversion, and application.* London: Peregrinus.
Eastop, T. D. and McConkey, A. (1993) *Applied thermodynamics for engineering technologists* 5th ed. Harlow: Longman.
Ehringer, H., Hoyaux, G. and Pilavachi, P. A. (eds) (1983) *Energy conservation in industry, combustion, heat recovery, and Rankine cycle machines,* proceedings of the contractors' meetings held in Brussels on 10 and 18 June and 29 October 1982. Dordrecht: Reidel.
El-Masri, M. A. (1987) Exergy analysis of combined cycles, – part 1, air-cooled Brayton-cycle gas turbines. *Transactions of the ASME. Journal of Engineering for Gas Turbines and Power*, **109**, 228–236.
Emanuel, G. (1987) *Advanced classical thermodynamics.* New York: American Institute of Aeronautics and Astronautics.
Energy Efficiency Office (1984) *Compressed air and energy use.* London: Energy Efficiency Office.
Energy manager's workbook (1982) Cambridge: Energy Publications.
Energy recovery in process plants (1976) London: Mechanical Engineering Publications.
European Space Agency (1978) *Spacecraft thermal and environmental control systems.* Neuilly-sur-Seine: European Space Agency.
Evett, J. B. and Liu, C. (1989) *2500 solved problems in fluid mechanics and hydraulics.* New York: McGraw-Hill.

Fenn, J. B. (1982) *Engines, energy, and entropy, a thermodynamics primer.* San Francisco: W. H. Freeman.
Fitt, P. W. and Moses, R. T. (eds) (1984) *Directly-fired heat pumps for use in domestic and commercial premises*, proceedings of the International Conference held at the University of Bristol, 19–21 September 1984. Bristol: University of Bristol and the Institute of Refrigeration.
Fletcher, L. S. (ed.) (1987a) *Aerodynamic heating and thermal protection systems.* New York: American Institute of Aeronautics and Astronautics.
Fletcher L.S. (ed.) (1978b) *Heat transfer and thermal control systems.* New York: American Institute of Aeronautics and Astronautics.
Gaggioli, R. A. (ed.) (1986) *Computer-aided engineering of energy systems*, 3 vols. New York: ASME.
Goodger, E. M. (1984) *Principles of engineering thermodynamics* 2nd ed. London: Macmillan.
Gosling, C. T. (1980) *Applied air conditioning and refrigeration.* Barking: Applied Science.
Gosney, W. B. (1982) *Principles of refrigeration.* Cambridge: Cambridge University Press.
Gunn, D. and Horton F. (1981) *Industrial boilers.* Harlow: Longman.
Hackleman, M. A. (1980) *At home with alternative energy, a comprehensive guide to creating your own systems.* Culver City, CA: Peace Press.
Haywood, R.W. (1980) *Equilibrium thermodynamics for engineers and scientists.* Chichester: Wiley.
Haywood, R. W. (1986) *Analysis of engineering cycles: worked problems: power, refrigerating, and gas liquefaction plant.* Oxford: Pergamon.
Heat transfer (1985) Engineering sciences data, 85/022, Vol. 6. London: ESDU International.
Herridge, S. J., O'Callaghan, P. W. and Probert S. D. (1988) Thermodynamic properties of fluids commonly used in refrigeration system cycles. *Applied Energy*, 31, 161–188.
Herzfeld, C. M. (1962) *Temperature – its measurement and control in science and industry*, Vol. 3 parts 1, 2 and 3. New York: Van Nostrand Reinhold.
Hickson, D. C. and Taylor, F. R. (1980) *Enthalpy-entropy diagram for steam* 2nd ed. Oxford: Blackwell.
Hippensteele, S. A. (1978) *Computer program for obtaining thermodynamic and transport properties of air and products of combustion of ASTM-A-1 fuel and air.* Washington, DC: NASA.
Hodge, B. K. (1990) *Analysis and design of energy systems* 2nd ed. Englewood Cliffs, NJ: Prentice Hall.

Hong, S. I. (1989) *Automation techniques for thermal analysis of spacecraft systems*, Washington, DC: AIAA.

Horlock, J. H. (1987) *Cogeneration: combined heat and power (CHP): thermodynamics and economics*. Oxford: Pergamon.

Hottell, H. C. (1946) *Charts of thermodynamic properties of fluids encountered in calculations of internal combustion engine cycles*. Washington, DC: NACA. (NACA TN1026 appears in NACA TN 1026 to 1030).

Howell, J. R., Bannerot, R. B. and Vliet, G. C. (1982) *Solar-thermal energy systems: analysis and design*. New York: McGraw-Hill.

Howell, J. R. and Buickius, R. O. (1992) *Fundamentals of engineering thermodynamics* 2nd ed. New York: McGraw-Hill.

Huang, F. F. (1988) *Engineering thermodynamics: fundamentals and applications* 2nd ed. New York: Macmillan.

Incropera, F. P. and Dewitt, D. P. (1990) *Fundamentals of heat and mass transfer* 3rd ed. New York: Wiley.

International District Heating and Cooling Association (1985) *District heating and cooling*, **70**(4) Washington, DC: International District Heating and Cooling Association.

International District Heating Association (1983) *District heating handbook: a design guide* 4th ed. Washington, DC: DHA.

James, A. M. (1976) *A dictionary of thermodynamics*. London: Macmillan.

Janna, W. S. (1986) *Engineering heat transfer*. Boston, MA: PWS Engineering.

Joel, R. (1987) *Basic engineering thermodynamics* 4th ed. Harlow: Longman.

Johansson, T. B. (ed.) (1989) *Electricity: efficient end-use and new generation technologies and their planning implications*, Bromley: Chartwell-Bratt Publishing.

Jones, J. B. and Hawkins, G. A. (1986) *Engineering thermodynamics: an introductory textbook* 2nd ed. New York: Wiley.

Jones, O. C. and Telionis, D. P. (eds) (1992) *Basic research needs in fluid mechanics*, papers presented at the Fluids Engineering conference, Los Angeles, CA, June 21–26, 1992. New York: American Society of Mechanical Engineers.

Jones, W. P. (1980) *Air conditioning applications and design*. London: Arnold.

Jones, W. P. (1985) *Air conditioning engineering* 3rd ed. London: Arnold.

Joy, P. F. and Goliazewski, L. (1986) *Advanced thermal and power systems for the Satcom-Ku satellites*, 11th AIAA Communication Satellite Systems Conference. New York: AIAA.

Kays, W. M. (1984) *Compact heat exchangers* 3rd ed. New York: McGraw-Hill.

Keenan, J. H., Chao, J. and Kaye, J. (1980) *Gas tables: thermodynamic properties of air, products of combustion and component gases, compressible flow functions* 2nd ed. New York: Wiley.

Keenan, J. H. (1970) *Thermodynamics.* MIT Press.

Kelly, D. C. (1973) *Thermodynamics and statistical physics.* New York: Academic Press.

Kinsky, R. (1982) *Applied fluid mechanics.* New York: McGraw-Hill.

Kittel, C. and Kromer, H. (1980) *Thermal physics* 2nd ed., San Francisco: W. H. Freeman.

Kotas, T. J. (1985) *The exergy method of thermal plant analysis.* London: Butterworths.

Kovach, E. G. (ed.) (1977) *Thermal energy storage,* the report of a NATO Science Committee Conference held at Turnberry, Scotland, 1–5 March, 1976. Oxford: Pergamon.

Kozlov, L. V. and Nusinov, M. D. (1977) *Simulation of thermal systems of a space vehicle and its environment.* Washington, DC: NASA (NASA TT F-752).

Kreider, K. G. and McNeil. M. B. (eds) (1977) *Waste heat management guidebook,* Washington, DC: USGPO (NBS handbook 12).

Kreith, F. and Black, W. Z. (1980) *Basic heat transfer.* New York: Harper and Row.

Kreith, F. (1986) *Principles of heat transfer* 4th ed. New York: Harper and Row.

Landsberg, P. T. (1978) *Thermodynamics and statistical mechanics.* Oxford: Oxford University Press.

Langley, B. C. (1986) *Refrigeration and air conditioning* 3rd ed. Englewood Cliffs, NJ: Prentice Hall.

Large scale applications of heat pumps (1987) proceedings of the 3rd international symposium on the large scale applications of heat pumps, Oxford, 25–27 March, 1987. Bedford: British Hydromechanics Research Association.

Laughton, M. A. (ed.) (1990) *Renewable energy sources.* London: Elsevier for the Watt Committee on Energy.

Liley, P. E. (1989) *2500 solved problems in mechanical engineering thermodynamics.* New York: McGraw-Hill.

Linnhoff, B. (1982) *User guide on process integration for the efficient use of energy.* Rugby: Institution of Chemical Engineers.

Lobanoff, V. S. and Ross, R. R. (1992) *Centrifugal pumps, design and application* 2nd ed. Houston, TX: Gulf Publishing Co.

Logan, J. G. (1956) *The calculation of the thermodynamic properties of air at high temperatures.* Buffalo, CA: Buffalo Cornell Aeronautical Laboratory (CAL/AD-1052-A-1).

Look, D. C. and Sauer, H. J. (1988) *Engineering thermodynamics* SI ed. Wokingham: Van Nostrand Reinhold.

Lord, N. W., Ouellette, R. P. and Cheremisinoff, P. M. (1980) *Heat pump technology*. Ann Arbor, MI: Ann Arbor Science.

Loyd, S. (1981) *The heat pump: an annotated bibliography with a survey of suppliers* 2nd ed. Bracknell: Building Research and Information Association (BSRIA/LB 103/81).

Loyd, S. (1990) *Combined heat and power: an annotated bibliography*. Bracknell: Building Services Research and Information Association (BSRIA/LB-112/90).

Lunde, P. J. (1980) *Solar thermal engineering, space heating and hot water systems*. New York: Wiley.

Mackenzie-Kennedy, C. (1979) *District heating, thermal generation and distribution, a practical guide to centralised generation and distribution of heat services*. Oxford: Pergamon.

Marecki, J. (1988) *Combined heat and power generating systems*. London: Peregrinus.

Marsh, W. D. (1980) *Economics of electric utility power generation*. Oxford: Clarendon Press.

Martin, M. C. (1986) *Elements of thermodynamics*. Englewood Cliffs, NJ: Prentice Hall.

McGuigan, D. (1981) *Heat pumps: an efficient heating and cooling alternative*. Charlotte, VT: Garden Way Pub.

McMahon, P. J. (1971) *Aircraft propulsion*. London: Pitman.

McMullan, J. T. and Morgan, R. (1981) *Heat pumps*, Bristol: A. Hilger.

Meador, R. (1981) *Cogeneration and district heating: an energy-efficiency partnership*, Ann Arbor, MI: Ann Arbor Science.

Merrill, R. and Gage, T. (ed.) (1978) *Energy primer: solar, water, wind, and biofuels*. New York: Dell Publishing Co.

Min, T. C. and Chiou, J P. (eds) (1985) *Heat transfer and fluid flow in solar thermal systems*, Winter Annual Meeting of the American Society of Mechanical, Engineers, Miami Beach, Florida, November 17–22, 1985. New York: ASME.

Moeckel, W. E. and Weston, K. C. (1958) *Composition and thermodynamic properties of air in chemical equilibrium*. Washington, DC: NACA (NACA TN 4265, appears in NACA TN 4261 to 4270).

Mooney, D. A. (1955) *Introduction to thermodynamics and heat transfer*. Englewood Cliffs, NJ: Prentice Hall.

Moore D. F. (1981) *Thermodynamic principles of energy degrading*. London: Macmillan.

Moran, M. J. (1982) *Availability analysis, a guide to efficient energy use*. Englewood Cliffs, NJ: Prentice Hall.

Moser, F. and Schnitzer, H. (1985) *Heat pumps in industry*. Amsterdam: Elsevier.

Munasinghe, M. (1990) *Electric power economics*. London: Butterworth Scientific.

Nag, P. K. (1981) *Engineering thermodynamics.* New Delhi: Tata McGraw-Hill.
National Engineering Laboratory (1972) *Air coolers, cooling towers, and evaporative coolers,* Report of a meeting at NEL, 24 November 1971. East Kilbride: National Engineering Laboratory.
Newman, P. A. and Allison, D. O. (1966) *Direct calculation of specific heats and related thermodynamic properties of arbitrary gas mixtures with tabulated results.* Washington, DC: NASA (NASA TN D-3540).
NIFES (1989) *Boiler operator's handbook.* Rev. ed. London: Graham and Trotman.
Oates, G. C. (ed.) (1989) *Aircraft propulsion systems technology and design.* Washington, DC: AIAA.
O'Callaghan, P. W. and Probert, S. D. (1981) Exergy and economics. *Applied Energy,* **8**, 227–243.
O'Callaghan, P. W. (1981) *Design and management for energy conservation.* Oxford: Pergamon.
O'Callaghan, P. W. (1983) *Energy management.* London: McGraw-Hill.
Offenhartz, P. O'D. (1979) *Methanol-based heat pumps for storage of solar thermal energy,* Phase 1. Albuquerque: Sandia Laboratories.
Olivo, C. T. (1990) *Principles of refrigeration* 3rd ed. New York: Delmar.
Parker, J. (1981) *Elementary thermodynamics.* Cambridge: Cambridge University Press.
Plumb, H. H. (1973) *Temperature – its measurement and control in science and industry* vol. 4, parts 1, 2 and 3. Pittsburg: Instrument Society of America.
Prasad, M. (1983) *Refrigeration and air conditioning.* New Delhi: Wiley Eastern.
Process integration (1987–) London: ESDU International.
Reay, D. A. (1979) *Heat recovery systems, a directory of equipment and techniques.* London: Spon.
Reay, D. A. and Macmichael, D. B. A. (1988) *Heat pumps* 2nd ed. Oxford: Pergamon.
Reddy, T. A. (1987) *The design and sizing of active solar thermal systems.* Oxford: Clarendon Press.
Reiter, S. (1983) *Industrial and commercial heat recovery systems.* New York: Van Nostrand Reinhold.
Reynolds, W. C. and Perkins, H.C. (1977) *Engineering thermodynamics* 2nd ed. New York: McGraw-Hill.
Rhine, J. M. and Tucker, R. J. (1990) *Modelling of gas-fired furnaces and boilers and other industrial heating processes.* Maidenhead: McGraw-Hill.
Rice, I. G. (1987a) Thermodynamic evaluation of gas turbine cogener-

ation cycles, part I, heat balance method analysis. *Transactions of the ASME, Journal of Engineering for Gas Turbines and Power*, **109**, 1–7.

Rice, I. G. (1987b) Thermodynamic evaluation of gas turbine cogeneration cycles, part II, complex cycle analysis *Transactions of the ASME, Journal of Engineering for Gas Turbines and Power*, **109**, 8–15.

Riedi, P. C. (1988) *Thermal physics, an introduction to thermodynamics, statistical mechanics and kinetic theory* 2nd ed. Oxford: Oxford University Press.

Rogers, G. F. C., and Mayhew, Y. R. (1988) *Thermodynamic and transport properties of fluids*, SI units 4th ed. Oxford: Blackwell.

Rogers, G. F. C. and Mayhew, Y. R. (1992) *Engineering thermodynamics, work and heat transfer* 4th ed. Harlow: Longman.

Roy, D. N. (1988) *Applied fluid mechanics*. Chichester: Ellis Horwood.

Sauer, H. J. and Howell, R. H. (1983) *Heat pump systems*. New York: Wiley.

Schmidt, F. W. and Willmott, J. A. (1981) *Thermal energy storage and regeneration*. Washington, DC: Hemisphere.

Schmidt, F. W., Henderson, R. E. and Wolgemuth, C. H. (1993) *Introduction to thermal sciences: thermodynamics, fluid dynamics, heat transfer* 2nd ed. New York: Wiley.

Sears, F. W. and Salinger G. L. (1975) *Thermodynamics, kinetic theory and statistical thermodynamics* 3rd ed. Reading, MA: Addison-Wesley.

Sengupta, S. and Lee, S. S. (eds) (1983) *Waste heat, utilization and management*, Washington, DC: Hemisphere.

Shah, R. K. et al. *Compact heat exchangers for power and process industries*, 28th National Heat Transfer Conference and Exhibition, San Diego, CA. New York: ASME.

Shah R. K., McDonald, C. F. and Howard, C. P. (eds) *Compact heat exchangers – history, technological advancement, and mechanical design problems*, presented at the winter annual meeting of the American Society of Mechanical Engineers, Chicago, Illinois, November 16–21, 1980, and the 24th Annual International Gas Turbine Conference and First ASME Solar Energy Conference, San Diego, California, March 11–15, 1979. New York: ASME.

Sharp, J. J. (1988) *BASIC fluid mechanics*. London: Butterworths.

Sherratt, A. F. C. (ed.) (1984) *Heat pumps for buildings*. London, Hutchinson.

Sherratt, A. F. C. (1986) *Energy management in buildings*. London: Hutchinson.

Skrotzki, B. G. A. (1963) *Basic thermodynamics: elements of energy systems*. New York: McGraw-Hill.

Society of Automotive Engineers, (1995) *SAE handbook*. 4 vols. Warrendale, PA: Society of Automotive Engineers.
Sonntag R. E. and Van Wylen, G. (1991) *Introduction to thermodynamics: classical and statistical* 3rd ed. New York: Wiley.
Sorensen, B. (1979) *Renewable energy*. London: Academic Press.
Spalding, D. B. and Cole, E. H. (1973) *Engineering thermodynamics* 3rd ed. London: Arnold.
Sporn, P., Ambrose, E. R. and Baumeister, T. (1947) *Heat pumps*. New York: Wiley.
Sprackling, M. T. (1991) *Thermal physics*. Basingstoke: Macmillan Education.
Stanford, H. W. (1988) *Analysis and design of heating, ventilating, and air-conditioning systems*. Englewood Cliffs, NJ: Prentice Hall.
Stoecker, W. F. and Jones, J. W. (1982) *Refrigeration and air conditioning* 2nd ed. New York: McGraw-Hill.
Stoecker, W. F. (1989) *Design of thermal systems* 3rd ed. New York: McGraw-Hill.
Stoecker, W. F. and Stoecker, P. A. (1989) *Microcomputer control of thermal and mechanical systems*. New York: Van Nostrand Reinhold.
Sukhatme, S. P. (1984) *Solar energy: principles of thermal collection and storage*. New Delhi: Tata McGraw-Hill.
Sumner, J. A. (1976) *Domestic heat pumps*. Dorchester: Prism Press.
Sutphin, S. E. (1987) *Residential heat pumps: installation and troubleshooting*. Englewood Cliffs, NJ: Prentice Hall.
Taylor, R. H. (1983) *Alternative energy sources: for the centralised generation of electricity*. Bristol: A. Hilger.
Threlkeld, J. L. (1970) *Thermal environmental engineering* 2nd ed. Englewood Cliffs, NJ: Prentice Hall.
Trott, A. R. (1989) *Refrigeration and air conditioning* 2nd ed. London: Butterworths.
Twidell, J. and Weir, A. D. (1986) *Renewable energy resources*. London: Spon.
Valette, L. and Focquet, J. P. (1982) *Europe and renewable energy sources in the year 2000*. Bruxelles: Agence Européenne d'Informations.
Van Wylen, G. J. and Sonntag, R. E. (1985) *Fundamentals of classical thermodynamics* 3rd ed., SI version. New York: Wiley
Wallis, R. A. (1961) *Axial flow fans: design and practice*. London: Newnes.
Wallis, R. A. (1961) *Axial flow fans and ducts*. New York: Wiley.
Walshaw, A. C. (1947) *Applied thermodynamics*. London: Blackie.
West, R. E. and Kreith, F. (eds) (1988) *Economic analysis of solar thermal energy systems*. Cambridge, MA: MIT Press.
West, M., *et al.* (eds) (1984) *Alternative energy systems: electrical inte-

gration and utilisation, proceedings of the conference held at the Coventry (Lanchester) Polytechnic, 10–12 September 1984. Oxford: Pergamon.

Whalley, P. B. (1987) *Boiling, condensation, and gas-liquid flow*. Oxford: Clarendon Press.

Whalley, P. B. (1992) *Basic engineering thermodynamics*. Oxford: Oxford University Press.

Wilkinson, B. W. and Barnes. R. W. (eds.) (1980) *Cogeneration of electricity and useful heat*. Boca Raton, FL: CRC Press.

Williamson, A. C. (1988) *Introduction to electrical energy systems*. Harlow: Longman.

Zemansky, M. W. and Van Ness, H. C. (1966) *Basic engineering thermodynamics*. New York: McGraw-Hill.

Zemansky, M. W., Abbott, M. M. and Van Ness, H. C. (1975) *Basic engineering thermodynamics* 2nd ed. New York: McGraw-Hill.

Zemansky, M. W. and Dittman, R. H. (1981) *Heat and thermodynamics* 6th ed. New York: McGraw-Hill.

Zimmerman, K. H. and Powell, R. H. (1987) *Heat pumps: prospects in heat pump technology and marketing*, proceedings of the 1987 International Energy Agency Heat Pump Conference, Orlando, Florida, April 28–30, 1987. Chelsea, MI: Lewis Publishers.

CHAPTER THIRTEEN

Energy technology

J. T. McMULLAN

Introduction

Before examining information sources in energy technology, it is important to define the boundary conditions of the discussion, as energy is one of those terms which has very different meanings depending on the background of the reader. According to its precise scientific definition, it is the capacity of a system to do work, while in general use it takes on a much wider range of interpretations loosely related to the scientific definition, but coloured by the nature of the application. For example, it is very common to refer to *energy* supplies, when what is actually meant is *fuel* supplies and it is normal to talk of energy *consumption* and the need for energy *conservation* despite the fact that every schoolchild knows that energy *is* conserved and is not consumed. What has happened is that the term energy has come to be used as a catch-all for anything relating to the provision of heat and power, with the consequence that terms such as energy conservation, energy economics, renewable energies, rational use of energy, security of energy supplies, energy policy and energy technology have entered the language and are widely used but with very loose definitions.

It would be tempting to define energy technology as relating entirely to fuel technology, but this would then rule out areas such as renewable energy supplies or energy economics and policy, which are of crucial importance. It also overlooks the fact that energy technology is highly interdisciplinary and overlaps strongly with other engineering disciplines, with economics and with the environmental sciences. Because of this, energy technology will here be given a fairly wide brief and will include aspects of economics and policy formulation as well as engineering and science.

The information sources presented will be subdivided by energy technology area where possible and listed under the general headings of primary sources such as books, journals, reviews, reports, etc. and secondary sources such as abstracting services, online databases, bibliographies and encyclopedias. This will be preceded by a general review of the organizations which are major providers of energy related information on a national and international scale. In view of the greatly improved availability of and ease of access to electronic online and CD-ROM reference works, these will be treated separately.

One aspect of the interdisciplinary nature of the energy technology field is that the associated literature is so large that this chapter can only give a sample of what is available. Those who wish to delve deeper are recommended to turn to the abstracting services, the online databases and the citation indexes. These will normally provide the quickest routes into an unknown area. It is worth noting from the outset that all-embracing and completely unselective reference lists are readily available through general references such as *Books in Print* and through the catalogues of national libraries and government publications.

International agencies

The existence of civilized society represents a reduction in the disorder or entropy of human affairs. As such, it requires the manipulation of materials and systems and so depends on the availability of adequate means to achieve this objective. The primary requirement is for secure supplies of fuel which therefore assumes a central role in the affairs of nations, with the result that a number of international organizations have been created to foster cooperation. Some of these have wide-ranging briefs while others are concerned primarily with particular fuels or market sectors. All hold regular meetings and organize specialist working groups which produce reports and other publications which are normally available for distribution. The two best known examples of the wide spectrum agencies are the World Energy Council (WEC) and the International Energy Agency (IEA).

WEC is the major international forum for energy producers and consumers. Each of the member countries is represented on the committee and the secretariat is based in London. WEC holds triennial congresses and has a number of permanent and *ad hoc* specialist committees. Its publications are available through the British Library and other national libraries.

IEA is an autonomous agency set up by OECD. It provides a forum in which the OECD member states try to improve the world's energy supply and demand structure, provide a permanent information system on the international oil market and approach energy developments in a global context through international cooperation. It publishes annual and quarterly statistical reports on oil and gas, coal and overall energy supply, consumption, prices and taxes. IEA statistics are available online via the energy databases which will be discussed later.

IEA undertakes specialist reviews (e.g. District Heating and Combined Heat and Power, Heat Pumps, Coal Liquefaction) and a number of specialist organizations and programmes have been established including IEA Coal Research, IEA Greenhouse Gas Programme and IEA Biomass Conversion Technical Information Service. IEA publications are available from OECD and from national outlets such as HMSO in the UK. Additionally, OECD itself publishes occasional monographs on specific topics.

The European Union (formerly the European Economic Community) produces regular statistics on European energy production, consumption, trade and balances. It also publishes the results of studies carried out both internally and within the context of its energy R&D programmes.

United Nations Educational, Scientific and Cultural Organization (Unesco) collects and publishes information across a range of energy topics, including, for example its ENERGY database which is intended to provide access to information from 172 countries.

Among the fuel-specific organizations are IEA Coal Research, Conférence Internationale des Grandes Réseaux Electriques (CIGRE), Union for the Coordination of the Production and Transport of Electric Power (UCPTE), International Union of Producers and Distributors of Electrical Energy (UNIPEDE), International Gas Union (IGU), World Petroleum Congress (WPC), Organization of Petroleum Exporting Countries (OPEC), Organization of Arab Petroleum Exporting Countries (OAPEC), International Atomic Energy Agency (IAEA), Nuclear Energy Agency (NEA), International Solar Energy Society (ISES), International Association for Hydrogen Energy (IAHE), Commonwealth Regional Renewable Energy Resources Information Service (CRRERIS), and Renewable Energy Resources Information Centre (RERIC).

National organizations

Most national governments have departments or ministries with responsibility for the formulation and management of national energy policy and for the conduct of the appropriate research and development programmes. There are also nationalized and privatized energy industries, universities and professional organizations. By way of illustration, we will concentrate here on UK and US sources.

In the United Kingdom, the Department of Trade and Industry (DTI) has the responsibility for developing energy policy. The energy supply industries have been privatized in recent years and are now subject to regulation by regulatory boards rather than by the Department, as had been the case in earlier years. All of the research associations which supported the previously nationalized industries are also being privatized or becoming independent agencies, though most will continue to carry out their functions for the Department under contractual agreements. Most coordination of industrial energy R&D is carried out by the Energy Technology Support Unit (ETSU) which is located at the Harwell laboratory of the United Kingdom Atomic Energy Authority (UKAEA).

ETSU provides the government with a research and development service in the areas of energy conservation, renewable energy technologies and strategic studies relating to energy policy. It also undertakes project management of contracted R&D and the operation of projects for appropriate government departments. ETSU publishes a range of reports on energy topics which are available from the British Library.

The former nationalized industries British Gas, National Power, PowerGen and British Coal also support technical information facilities which will be discussed later. The Coal Research Establishment (CRE) carries out extensive research in coal technologies and acts, for example, as the project manager in the IEA Greenhouse Gas Programme.

The Institute of Energy is the professional body for those concerned with energy production, though there is considerable overlap with the Institutions of Electrical Engineers, Mechanical Engineers and Chemical Engineers. All publish regular journals and organize conferences, seminars and workshops on energy related topics.

The British Library provides an excellent information service through its Science Reference and Information Service (SRIS) and through its online service, BLAISE-LINE.

In the United States, the appropriate government department is the Department of Energy (DoE). Its Energy Information Administration (EIA) collects and publishes data on energy reserves, national patterns of production and consumption and the status of the energy companies. The USDoE Technical Information Centre cooperates with European bodies to compile an extremely comprehensive energy database which is available both in printed form and online.

Other official national agencies include the Department of the Interior (DOI), the National Petroleum Council (NPC) and the Nuclear Regulatory Commission (NRC). The National Technical Information Service provides the NTIS database of reports.

Other US information services are provided largely through industry-based agencies such as: American Petroleum Institute (API), American

Gas Association (AGA), National Coal Association (NCA), American Mining Congress (AMC), Electric Power Research Institute (EPRI) and American Society of Heating, Refrigeration and Air Conditioning Engineers (ASHRAE). All of these produce extensive reports, regular journals and databases relating to their spheres of activity.

Abstracting and indexing services

Abstracting and indexing services are useful tools for accessing the journal and conference literature. They provide a very convenient way of keeping abreast of a wide range of developments although, because of the time required for the abstracting and listing process, there may be a delay of several months between the appearance of the original paper and that of the abstract. Most of the major abstracting journals are also available as online databases.

Indexing services usually provide only the bibliographical information regarding the original work and so frequently process the data more quickly than the abstracting services.

General guides to the abstracting and indexing services themselves are:

Stephens, A. (ed.) (1986) *Inventory of abstracting and indexing services produced in the United Kingdom* 4th rev. ed. (British Library).
Burton, R. (comp.) (1991) *Scientific abstracting and indexing periodicals in the British Library* 4th ed. (British Library).
Ulrich's International Periodicals Directory (R. R. Bowker) annual.

Abstracting and indexing services which are of primary interest to the energy community include:

Current Bibliography on Science and Technology: Energy (Japan Information Center of Science and Technology) monthly.
Energy Abstracts (Engineering Information Inc.) monthly.
Energy Research Abstracts (USDoE Technical Information Centre) monthly.
Euro Abstracts (European Union) monthly.
Fuel and Energy Abstracts (Institute of Energy/Butterworths) bimonthly.
Science Citation Index (SCI) (Institute for Scientific Information) bimonthly. This is a unique reference which provides an extremely powerful method of building up a bibliography from the basis of a small number of key references.

More general scientific and technical abstracting and indexing services with a large energy relevance include:

Bulletin Signalétique (to 1983) (Centre National de la Recherche Scientifique).
Bibliographie Internationale (from 1984) (Centre National de la Recherche Scientifique) monthly.
Current Technology Index (Bowker-Saur) monthly.

228 Energy technology

Engineering Index (Engineering Information Inc.) monthly.
Science Abstracts (Institution of Electrical Engineers) monthly.
USSR Report (Various sections) (National Technical Information Service) monthly.

Other abstracting and indexing services which cover either particular energy areas or more general areas with a strong energy interest include:

Applied Mechanics Reviews (American Society of Mechanical Engineers) monthly.
Bibliography and Index of Geology (American Geological Institute) monthly.
Ceramic Abstracts (American Ceramic Society) bimonthly.
Chemical Abstracts (Chemical Abstracts Service) weekly.
Coal Abstracts (IEA Coal Research) monthly.
Current Contents: Engineering, Technology and Applied Sciences (Institute for Scientific Information) weekly.
Current Contents: Physical, Chemical and Earth Sciences (Institute for Scientific Information) weekly.
Engineered Materials Abstracts (Institute of Materials (UK) and American Society for Metals) monthly.
Gas Abstracts (Institute of Gas Technology) monthly.
Hydrogen Energy Quarterly Literature Review (Global Resources) quarterly.
INIS Atomindex (International Atomic Energy Agency) bimonthly.
International Building Services Abstracts (BSRIA) bimonthly.
International Petroleum Abstracts (Institute of Petroleum/Wiley) quarterly.
Literature Abstracts (American Petroleum Institute) weekly.
MIRA Automobile Abstracts (Motor Industry Research Association) monthly.
Petroleum Abstracts (University of Tulsa, Information Services Division) weekly.
Pollution Abstracts (Cambridge Scientific Abstracts) bimonthly.
Process and Chemical Engineering (Royal Society of Chemistry) monthly.
Wind Energy Abstracts (Wind Books) bi-monthly.

Online databases

Online databases provide rapid computer-based electronic access to a wide range of information services. Since they can be accessed remotely using computer terminals on the user's desk, these online services offer greatly increased convenience and access to information. The information available includes most of the abstracting services listed above, but with slightly different streamlined names. Other services include lists of materials, properties, market reports, library catalogues, business directories, financial news, directories of international standards, company reports, patents and so forth.

The information for each of the databases is provided by independent information providers such as the abstracting services. Access to the data is provided through hosts which provide a common access route, printed output, current awareness or selective dissemination (SDI) services and fax and electronic mail delivery facilities.

The range of services provided by the hosts is very wide, but, unfortunately, the online industry is in a constant state of flux. Some of the hosts are joining together and offering gateways to each other's databases, and some of the telecommunications companies are now acting as hosts and offering their own service.

Four widely available host services will be presented here. Two of them are American and two European.

The oldest and largest of the online hosts is DIALOG provided by Dialog Information Services, Inc., Palo Alto, California. DIALOG provides access to over 450 databases, of which the ones most appropriate to the energy field are:

APIBIZ (API ENERGY BUSINESS NEWS INDEX). Political, social and economic information relating to the energy industry (American Petroleum Institute) weekly.
APILIT. (Index to API Literature Abstracts) Non-patent literature relating to petroleum and petrochemical industry (American Petroleum Institute) monthly.
APIPAT. (Patent Index) Patent citations relating to petroleum and petrochemical industry (American Petroleum Institute) monthly.
CHEMICAL ENGINEERING AND BIOTECHNOLOGY ABSTRACT (CEBA). Industrial practice and theoretical chemical engineering (Royal Society of Chemistry) monthly.
COMPENDEX PLUS. Machine readable version of the *Engineering Index* (Engineering Information Inc.) monthly.
EI PAGE ONE. Covers worldwide engineering journals and conferences (Engineering Information Inc.) monthly.
ELECTRIC POWER DATABASE. R&D projects of interest to electric power industry (Electric Power Research Institute) monthly.
ENERGY SCIENCE AND TECHNOLOGY. One of the world's largest sources of literature on all aspects of energy and related topics (US Department of Energy) fortnightly.
ENGINEERED MATERIALS ABSTRACTS. Online version of printed publication (ASM International and The Institute of Materials) monthly.
ENVIROLINE. All aspects of the environment (Bowker) monthly.
ENVIRONMENTAL BIBLIOGRAPHY (EPB online). Human ecology, atmospheric studies, energy, land and water resources, nutrition (Environmental Studies Institute) bi-monthly.
FLUIDEX (Fluid Engineering Abstracts). Covers all aspects of fluid engineering including wind energy, river management, etc. (Elsevier) monthly.
GEOARCHIVE. Comprehensive geoscience database giving good emphasis on economic geology, energy sources, etc. (Geosystems) monthly.
GEOBASE. Worldwide coverage of literature on geography, geology, ecology, etc. (Elsevier) monthly.
GEOREF. Worldwide technical literature on geology and geophysics (American Geological Institute) monthly.

230 Energy technology

IHS INTERNATIONAL STANDARDS AND SPECIFICATIONS. National and international standards (Information Handling Services) weekly.

INSPEC. Online version of *Physics Abstracts, Electrical and Electronics Abstracts* and *Computer and Control Abstracts* (Institution of Electrical Engineers) weekly.

ISMEC: MECHANICAL ENGINEERING ABSTRACTS. Mechanical and production engineering and engineering management (Cambridge Scientific Abstracts) bi-monthly.

METADEX. Metallurgy (ASM International and The Institute of Materials) monthly.

NTIS. Results of US and non-US, Japan, UK, German and French government-sponsored research (US Department of Commerce, National Technical Information Service) fortnightly.

NUCLEAR SCIENCE ABSTRACTS (NSA). International nuclear science and technology literature to 1976 (US Department of Energy) closed file.

POLLUTION ABSTRACTS. Environment-related literature on pollution, its sources and its control (Cambridge Scientific Abstracts) bi-monthly.

SCISEARCH. Online version of the *Science Citation Index*, including additional records from the *Current Contents* series of publications which are not included in the printed *SCI* (Institute for Scientific Information) weekly.

STANDARDS AND SPECIFICATIONS DATABASE. US government and industry standards (National Standards Association) monthly.

ORBIT (now Questel.Orbit) is another large online host. Its energy-related databases include some of those already listed under DIALOG: (API ENERGY BUSINESS NEWS INDEX (APIBIZ), APILIT, CEBA (CHEMICAL ENGINEERING AND BIOTECHNOLOGY ABSTRACTS), COMPENDEX PLUS, ENERGYLINE, GEOBASE, GEOREF and INSPEC).

In addition, Questel.Orbit offers:

CHEMICAL ABSTRACTS (CA Search). The online version of the most extensive scientific abstracting reference (Chemical Abstracts Service) fortnightly.

ERTH (ENVIRONMENTAL RESOURCES TECHNOLOGY). Environmental issues relating to petroleum exploration, production and transportation (Petroleum Abstracts) weekly.

IPABASE. Petroleum and allied fields (Institute of Petroleum) quarterly.

TULSA. Online version of *Petroleum Abstracts* (Petroleum Abstracts) weekly.

ESA-IRS is the information retrieval service offered by the European Space Agency. It provides many of the databases offered by DIALOG and Questel.Orbit, for example, CHEMICAL ABSTRACTS (CHEMABS), COMPENDEX, ENERGYLINE, ENGINEERED MATERIALS ABSTRACTS (EMA), INSPEC, METADEX, NTIS, and in addition:

BRIX/FLAIR. Building research (BRIX) and fire research (FLAIR) topics (Building Research Establishment and Fire Research Station) monthly.

EDF-DOC. Production and applications of electricity (Electricité de France) monthly.

IBSEDEX. Mechanical and electrical services in buildings (BSRIA) monthly.

Energy technology 231

PASCAL. The online version of *Bulletin Signalétique* and *Bibliographie Internationale* (Centre National de la Recherche Scientifique) monthly.

STN International is another European host offered by FIZ Karlsruhe. Once again, it offers a similar wide range of databases with the same common core as DIALOG, Questel.Orbit and ESA-IRS. For example, it includes APILIT, APIPAT, CHEMICAL ASBSTRACTS (CA FILE) (including CAOLD which covers 1957–66), CHEMICAL ENGINEERING AND BIOTECHNOLOGY ABSTRACTS, COMPENDEX PLUS, ENGINEERED MATERIALS ABSTRACTS, INIS, INSPEC, among many others. One data base which STN offers which has interesting additional capabilities is SIGLE which is a guide to European 'grey' literature – literature which does not find its way into the conventional catalogues.

Because of the widespread availability of computerized databases and developments in compact discs, many of the information providers have begun to distribute their databases on computer readable CD-ROM. This has advantages for large companies and educational establishments who can subscribe to a CD-ROM updating service and provide a pseudo-online service via their own local area network. Information on CD-ROM sources is available in Finlay, M. (ed.) (1994) *The CD-ROM directory, international*. 11th ed. London: TFPL Publishing.

Information sources by energy area

The rest of this chapter will be comprised of lists of references subdivided within a fairly broad range of classifications. The lists include the key information sources in each area, and the list is intended to present a balanced resource list wherever possible.

Within each classification, the entries will be separated into the general categories of reference works which will include tables of statistics, general materials properties, etc., books and monographs, and journals and periodicals, though other categories may be used from time to time as the need arises.

General energy issues and multi-disciplinary aspects

This first classification is slightly atypical as it is taken to include a range of topics such as energy policy, energy economics, energy statistics and energy resources in a general sense, where they are not specifically related to one fuel or readily identifiable resource. It will also present (particularly in the journals section) a range of publications which allow for multi-disciplinary discussions and which try to integrate the technical and non-technical aspects of the subject. The journals especially contain material which is relevant to all of the technical areas covered in the rest of the chapter and should be included in any literature search.

REFERENCE WORKS AND STATISTICS

Advances in the economics of energy and resources Monographic series (JAI Press).
Annual bulletin of general energy statistics for Europe (United Nations).
Annual energy review (US Department of Energy).
Annual review of energy (Annual Reviews Inc.).
Annual review of energy and the environment (Annual Reviews Inc.).
Bolz, R. E. and Tuve, G. L. (1973) *Handbook of tables for applied engineering science*, 2nd ed. (CRL Press).
BP statistical review of world energy (British Petroleum Company) annual.
Counihan, M. (1981) *A dictionary of energy* (Routledge & Kegan Paul).
Digest of United Kingdom energy statistics (HMSO) annual.
Dryden, I. G. C. (ed.) (1982) *The efficient use of energy*, 2nd ed. (Butterworths).
Dryden, I. G. C. and Griffith, M. (1982) *The science and technology of fuel and energy: 60 year index (1922–1981)* (Butterworths).
EIA data index (US Department of Energy, Energy Information Administration).
Energy balances of OECD countries (OECD/IEA) annual.
Energy data base, subject thesaurus, permuted listing (1984). US Department of Energy, National Technical Information Service).
Energy map of Europe 2nd ed. (1993) (Petroleum Economist).
Energy map of the Middle East 2nd ed. (1993) (Petroleum Economist in association with Coopers & Lybrand).
Energy statistics and balances of non-OECD countries (OECD/IEA) annual.
Energy statistics of OECD countries. (OECD/IEA) annual.
Energy statistics yearbook (United Nations) annual.
Ernst, R. (1985) *Comprehensive dictionary of engineering and technology* (Cambridge University Press).
Eurostat energy statistical yearbook (Statistical Office of the Commission of European Communities Office for Official Publications of the European Communities) annual.
Gilpin, A. and Williams, A. (1982) *Dictionary of energy technology* (Butterworth Scientific).
Hicks, T. G. (1994) *Standard handbook of engineering calculations* 3rd ed. (McGraw-Hill).
Jenkins, G. (1989) *Oil economists' handbook*, 2 vols, 5th ed. (Elsevier).
Kaye, G. W. C. and Laby, T. H. (1995) *Tables of physical and chemical constants*, 16th ed. (Longman).
Mataré, H. F. (1989) *Energy: facts and future* (CRC Press).
Monthly Energy Review (Energy Information Administration, US Department of Energy).
Quarterly Energy Balance (OECD).
Quarterly Energy Review: Far East and Australasia (Economist Intelligence Unit).
Quarterly Energy Review: Latin America and the Caribbean (Economist Intelligence Unit).

Energy technology 233

Quarterly Energy Review: Middle East (Economist Intelligence Unit).
Quarterly Energy Review: North America (Economist Intelligence Unit).
Quarterly Energy Review: Western Europe (Economist Intelligence Unit).
Slesser, M. (1988) *Macmillan dictionary of energy*, 2nd ed. (Macmillan).
Stephens, J. H. and Ryder, C. (1994) *Energy users' year book* (Rodmell Press).
Wong, H. Y. (1977). *Handbook of essential formulae and data on heat transfer for engineers* (Longman).
World energy and nuclear directory (1993) 2nd rev. ed. (Longman).
World Energy Conference (1986) *Energy terminology: a multilingual dictionary*, 2nd ed. (Pergamon).
World Energy Council (1993) *Energy for tomorrow's world* (Kogan Page).

BOOKS AND MONOGRAPHS

Anderson, V. (1993) *Energy efficiency policies* (Routledge).
Brown, G. C. and Skipsey, E. (1986). *Energy resources: geology, supply and demand* (Open University Press).
Carraro, C. and Siniscalco, D. (eds) (1993) *The European carbon tax: an economic assessment* (Kluwer Academic).
Challoner, J. (1993) *Energy* (Dorling Kindersley).
Dienes, L., Dobozi, I. and Radetzki, M. (1994) *Energy and economic reform in the former Soviet Union: implications for production, consumption and exports, and for the international energy markets* (St. Martin's Press).
Hayes, P. and Smith, K. (eds) (1993) *The global greenhouse regime: who pays? – science, economics and north-south politics in the climatic change convention* (Earthscan Publications).
IBA (1988) *Energy Law. Proc. Advanced Seminar on Petroleum, Minerals and Energy Law*, Sydney, March 1988. International Bar Association (Graham and Trotman).
Kraushaar, J. J. and Ristinen, R. A. (1993) *Energy and problems of a technical society*, 2nd ed. (Wiley).
McMullan, J. T., Morgan, R. and Murray, R. B. (1975) *Energy resources and supply* (Wiley).
McMullan, J. T., Morgan, R. and Murray, R. B. (1983) *Energy resources* 2nd ed. (Arnold).
Mashburn, W. H. (1992) *Managing energy resources in times of dynamic change*, 2nd ed. (Fairmont Press).
Munasinghe, M. and Meier, P. (1993) *Energy policy analysis and modeling* (Cambridge University Press).
Smith, H. B. (1993) *Energy sources, applications and alternatives* (Goodheart-Willcox).
Umana, Q. A. (1989) Greenhouse economics, global resources and the political economy of climate change. *Environmental Policy and Law*, **19**, 154–161.
Vellinga, P. and Grubb, M. (eds.) (1993) *Climate change policy in the European Community: report of a workshop held in October 1992* (Royal Institute of International Affairs).

JOURNALS AND PERIODICALS

Advances in Energy Systems and Technology (Academic Press) irregular.
Annual Technical Report: Energy Materials Coordinating Committee (United

States Department of Energy, National Technical Information Service) annual.
Applied Energy (Elsevier Science) monthly.
CADDET Newsletter (Centre for the Analysis and Dissemination of Demonstrated Energy Technologies) quarterly.
China Energy Report (American Chamber of Commerce Hong Kong) annual.
DRI Energy Bulletin (Data Resources Inc., McGraw-Hill) biannual.
Eastern Bloc Energy (Eastern Bloc Research Ltd.) monthly.
EC Energy Monthly (Financial Times Business Information) monthly.
Energy (Pergamon) monthly.
Energy Analysis. Monograph Series (American Gas Association) irregular.
Energy Conversion and Management (Pergamon) monthly.
Energy and Fuels (American Chemical Society) bimonthly.
Energy Daily, The (King Publishing Group) daily.
Energy Economics (Butterworth-Heinemann) quarterly.
Energy Economics and Climate Change (Cutter Information Corporation) monthly.
Energy Economist (Financial Times Business Information) monthly.
Energy Engineering, Journal of the Association of Energy Engineers (Fairmont Press) bi-monthly.
Energy Exploration and Exploitation (Multi Science Publishing Corporation) bi-monthly.
Energy in Japan (Institute of Energy Economics) bimonthly.
Energy Journal (International Association of Energy Economics) quarterly.
Energy Letters (International Energy Society) quarterly.
Energy Papers Monographic series (HMSO) irregular.
Energy Policy (Butterworth-Heinemann) monthly.
Energy Prices and Taxes (International Energy Agency) quarterly.
Energy Research (Elsevier) irregular.
Energy Sources (Taylor and Francis) quarterly.
Energy Systems and Policy (Taylor and Francis) quarterly.
Energy Today (Trends Publishing Inc.) monthly.
Energy Trends (UK Department of Trade and Industry) monthly.
Energy World (Institute of Energy) 10 per annum.
Entropie (Association Entropie) 7 per annum.
Europe Energy (Europe Information Service) fortnightly.
European Energy Report (Financial Times Business Information) fortnightly.
Geopolitics of Energy (Canadian Energy Research Institute) monthly.
International Journal of Energy, Environment, Economics (Nova Science Publishers Inc.) quarterly.
International Journal of Energy Research (Wiley) monthly.
International Journal of Global Energy Issues (InterScience Enterprises Ltd.) quarterly.
International Journal of Power and Energy Systems (International Association of Science and Technology for Development) 3 per annum.
Journal of Energy and Development (International Research Center for Energy and Economic Development) biannual.
Journal of the Institute of Energy (Institute of Energy) quarterly.
Perspectives in Energy (Pion) quarterly.

Revue de l'Energie (Editions Techniques et Economiques France) monthly.
TERA Analysis (Total Energy Resource Analysis) (American Gas Association) irregular.
United Kingdom Energy Statistics (HMSO) monthly.

Combustion, gas and steam turbines, steam properties, fluidized beds, etc.

This section includes references to material covering aspects generally relating to combustion and steam raising. Thus, fluidized bed combustion is included, as are the properties of steam. An attempt is made to avoid fuel-specific references, but this is obviously not possible in the case of fluidized bed combustion.

BOOKS AND REFERENCE MATERIALS

Bain, R. W. (1964). *Steam tables 1964: physical properties of water and steam, 0–800°C, 0–1000 bar* (HMSO).
Barnard, J. A. and Bradley, J. N. (1985) *Flame and combustion*, 2nd ed. (Chapman and Hall).
Bastress, E. K. (ed.) (1977) *Gas turbine combustion and fuels technology*, Winter Annual Meeting, Atlanta, GA, November 27–December 2, 1977 (American Society of Mechanical Engineers).
Basu, P. (ed.) (1984) *Fluidized bed boilers: design and application* (Pergamon).
Beer, J. M. and Chigier, N. A. (1972) *Combustion aerodynamics* (Applied Science).
Benson, R. S. (1982) *The thermodynamics and gas dynamics of internal combustion engines*, Vol. 1 (Clarendon Press).
Benson, R. S. and Whitehouse, N. D. (1979) *Internal combustion engines* (Pergamon).
British Standards Institution (1990) *Boilers and pressure vessels: an international survey of design and approval requirements*, 5th ed. (British Standards Institution).
Bryers, R. W. (ed.) (1978) *Ash deposits and corrosion due to impurities in combustion gases* (McGraw-Hill).
Cohen, H., Rogers, G. F. C. and Saravanamuttoo, H. I. H. (1987) *Gas turbine theory*, 3rd ed. (Longman).
Collier, J. G. (1981). *Convective boiling and condensation*, 3rd rev. ed. (revised by J. R. Thorne) (Oxford University Press).
Commission of the European Communities (1983) *Fluidised bed systems; proceedings of the Contractors' Meeting held in Brussels, 12–13 October, 1982* (Reidel).
De Renzo, D. J. (ed.) (1983). *Cogeneration technology and economics for the process industries* (Noyes Data Corporation).
Electrical Research Association (1967) *1967 steam tables: thermodynamic properties of water and steam; viscosity of water and steam; thermal conductivity of water and steam* (Edward Arnold).
Ewing, J. A. (1926) *The steam engine and other heat engines*, 4th ed. (Cambridge University Press).

Faulkner, E. A. (1987) *Guide to efficient burner operation: gas, oil and dual fuel*, 2nd ed. (Fairmont Press).
Ferguson, C. R. (1986) *Internal combustion engines: applied thermosciences*. (Wiley).
French, D. N. (1993) *Metallurgical failures in fossil-fired boilers*, 2nd ed. (Wiley Interscience).
Glassman, I. (1986) *Combustion*, 2nd ed. (Academic Press).
Goodall, P. M. (ed.) (1980) *The efficient use of steam* (IPC Science and Technology).
Goodger, E. M. (1977) *Combustion calculations: theory, worked examples and problems* (Macmillan).
Haar, L., Gallagher, J. S. and Kell, G. S. (1984) *NBS/NRC steam tables. Thermodynamic and transport properties and computer program for vapor and liquid states of water in SI units* (Hemisphere).
Haywood, R. W. (1991) *Analysis of engineering cycles: power, refrigeration and gas liquefaction plant*, 4th ed. (Pergamon).
Horlock, J. H. and Winterbone, D. E. (eds) (1986) *The thermodynamics and gas dynamics of internal combustion engines*, Vol. 2 (Clarendon Press).
Howard, J. R. (ed.) (1983) *Fluidised beds: combustion and applications* (Applied Science).
Irvine, T. F. and Liley, P. E. (1984) *Steam and gas tables with computer equations* (Academic Press).
Jaffe, R. I. (ed.) (1983) *Corrosion fatigue of steam turbine blade materials* (Pergamon).
Kanury, A. M. (1975) *Introduction to combustion phenomena for fire, incineration, pollution and energy applications* (Gordon and Breach).
Kearton, W. J. (1958) *Steam turbine theory and practice: a textbook for engineering students*, 7th ed. (Pitman).
Kearton, W. J. (1964) *Steam turbine operation: a textbook on the installation, running, maintenance and testing of steam turbines*, 7th ed. (Pitman).
Kowalewicz, A. (1984) *Combustion systems of high speed piston I.C. engines* (Elsevier).
Kuo, K. K. (1986) *Principles of combustion* (Wiley).
Lees, B. (1976) *High temperature corrosion in oil-fired plant* (British Petroleum Company).
Lefebvre, A. H. (ed.) (1980) *Gas turbine combustor design problems* (Hemisphere).
Lefebvre, A. H. (1983) *Gas turbine combustion* (Hemisphere).
Lewis, B. and von Elbe, G. (1987) *Combustion flames and explosion of gases*, 3rd. ed. (Academic Press).
Meyer, C. A., et al. (1993) *ASME steam tables: thermodynamic and transport properties of steam comprising tables and charts for steam and water*, 6th Edn (American Society of Mechanical Engineers).
Moore, M. J. and Sieverding, C. H. (eds) (1976) *Two-phase steam flow in turbines and separators: theory, instrumentation, engineering* (Hemisphere).
National Coal Board (1985). *Fluidised bed combustion of coal*, 2nd ed. (National Coal Board).
National Industrial Fuel Efficiency Service (1989) *Boiler operators handbook*, 2nd ed. (Graham and Trotman).

Singer, J. G. (ed.) (1981) *Combustion, fossil power systems: a reference book on fuel burning and steam generation*, 3rd ed. (Combustion Engineering Inc.).
Sitkei, G. (1974) *Heat transfer and thermal loading in internal combustion engines*. (Akademiai Kiado).
Skinner, D. G. (1971) *The fluidised combustion of coal* (Mills and Boon).
Stambuleanu, A. (1976) *Flame combustion processes in industry* (Abacus Press).
Strehlow, R. A. (1984) *Combustion fundamentals* (McGraw-Hill).
Stultz, S. C. and Kitto, J. B. (eds) (1992) *Steam: its generation and use*, 40th ed. (Babcock and Wilcox Inc.).
Taylor, C. F. (1985) *The internal combustion engine in theory and practice*, 2 vols, 2nd ed. (MIT Press).
Weinberg, F. J. (ed.) (1986) *Advanced combustion methods* (Academic Press).
Whalley, P. B. (1987) *Boiling, condensation and gas-liquid flow* (Clarendon Press).
Williams, F. A. (1985) *Combustion theory*, 2nd ed. (Benjamin-Cummings)
Yaverbaum, L. (1977) *Fluidised bed combustion of coal and waste materials* (Noyes Data Corp.)

JOURNALS

Combustion and Flame. (Combustion Institute, Elsevier Science) 16 per annum.
Combustion Science and Technology (Gordon and Breach Science) 48 per annum.
Progress in Energy and Combustion Science (Pergamon) bimonthly.

Coal, coal gasification and coal liquefaction

This section is concerned with various aspects of coal extraction, combustion and conversion. Some of the material relating to fluidized bed combustion is repeated and is linked with references to synthetic fuels, liquefaction and gasification, and to combined cycle systems.

REFERENCE MATERIALS

Annual bulletin of coal statistics for Europe (UN, Economic Commission for Europe).
Coal conversion systems technical data book (1978) (United States Department of Energy, Office of Energy Technology).
National Coal Association, *Coal data* (US Government Printing Office) annual.
Quarterly Coal Report (US Department of Energy) quarterly.

BOOKS

Anderson, L. L. and Tillman, D. A. (1979) *Synthetic fuels from coal: overview and assessment* (Wiley).
Attia, Y. A. (ed.) (1986) *Processing and utilisation of high sulphur coals* (Elsevier Applied Science).

Badin, E. J. (1984) *Coal combustion chemistry: correlation aspects* (Elsevier Applied Science).
Berkowitz, N. (1985) *The chemistry of coal* (Elsevier Applied Science).
Bouska, V. (1982) *Geochemistry of Coal* (Elsevier Applied Science).
Central Fuel Research Institute (1981–82) *Indian coals*, 8 vols (Central Fuel Research Institute).
Commission on Energy and the Environment (1981) *Coal and the environment* (HMSO).
Cooper, B. R. and Ellingson, W. A. (1984) *The science and technology of coal and coal utilisation* (Plenum Press).
Cusumano, J. A. et al. (1978) *Catalysis in coal conversion* (Academic Press).
Farmer, I. W. (1984) *Coal mine structures* (Chapman and Hall).
Fettweis, G. B. (1979) *World coal resources: methods of assessment and results* (Elsevier).
Gavalas, G. R. (1982) *Coal pyrolysis* (Elsevier).
Gibson, J. (1984) *Coal and the environment* (Science Reviews).
Grainger, L. and Gibson, J. (1981) *Coal utilisation: technology, economics and policy* (Graham and Trotman).
IEA (1982) *Coal liquefaction: a technology review* (OECD).
IEA (1984) *Coal transport infrastructure: a study prepared by the Coal Industry Advisory Board* (OECD).
IEA (1985) *Coal quality and ash characteristics: a study by the IEA Coal Industry Advisory Board* (OECD).
Meadowcroft, D. B. and Manning, M. L. (eds) (1983) *Corrosion resistant materials for coal conversion systems* (Applied Science).
National Coal Board (1985) *Fluidised bed combustion of coal*, 2nd ed. (National Coal Board).
Nowacki, P. (1981) *Coal gasification processes* (Noyes Data Corporation).
Peng, S. S. (1986) *Coal mine ground control*, 2nd ed. (Wiley).
Peng, S. S. and Chiang, H. S. (1984) *Longwall mining* (Wiley).
Petrakis, L. and Grandy, D. W. (1983) *Free radicals in coals and synthetic fuels* (Elsevier Applied Science).
Qader, S. A. (1985) *Natural gas substitutes from coal and oil* (Elsevier Applied Science).
R. M. Parsons Co. (1985) *Evaluation of the British Gas Corporation/Lurgi slagging gasifier in gasification-combined-cycle power generation* (EPRI).
Raask, E. (1985) *Mineral impurities in coal combustion: behavior problems and remedial measures* (Hemisphere).
Singer, S. (1984) *Pulverized coal combustion: recent developments* (Noyes Data Corporation).
Skinner, D. G. (1971) *The fluidised combustion of coal* (Mills and Boon).
Smoot, L. D. and Smith, P. J. (1985) *Coal combustion and gasification* (Plenum).
Trantolo, D. J. and Wise, D. L. (eds) (1989) *Energy recovery from lignin, peat and lower rank coals* (Elsevier Applied Science).
Volborth, A. (ed.) (1986) *Coal science and chemistry* (Elsevier Applied Science).
Vorres, K. S. (ed.) (1986) *Mineral matter and ash in coal* (American Chemical Society).

JOURNALS AND PERIODICALS

Braunkohle (Rheinische-Bergische Drueckerei und Verlagsgesellschaft) monthly.
Bulletin of the International Peat Society (International Peat Society) irregular.
Coal Calendar (IEA Coal Research) bi-monthly.
Coal Information (OECD) annual.
Coal Outlook (Pasha Publications) weekly.
Coal Preparation (Gordon and Breach) monthly.
Coal Science and Technology (Elsevier) irregular.
Coal Week (McGraw-Hill) weekly.
Coal Week International (McGraw-Hill) weekly.
Fuel (Butterworths-Heinemann) monthly.
Fuel Science and Technology International (Marcel Dekker) monthly.
International Coal Letter (International Coal Letter) fortnightly.
International Coal Report (Financial Times Business Information) fortnightly.
International Journal of Mineral Processing (Elsevier) monthly.
Journal of Coal Quality (Center for Coal Science, West Kentucky University) quarterly.

Oil and gas

In view of the importance of oil and gas to the world economy, coupled with the fact that much of the total reserve is situated in regions which are inhospitable either through reasons of climate or through potential political instability, economic factors form an important part of the literature. The references included here will provide a basic guide to the area, while other considerations are dealt with under combustion, synthetic fuels and power generation.

STATISTICS

Jenkins, G. (1989) *Oil economists' handbook*, 2 vols, 5th ed. (Elsevier Applied Science).
Oil and energy trends (Blackwell) annual.
OPEC Review (Pergamon) quarterly.
Quarterly Oil Statistics and Energy Balances (OECD) quarterly.

BOOKS

Al-Chalabi, F. J. (1989) *Opec at the crossroads* (Pergamon).
Anderson, R. O. (1984) *Fundamentals of the petroleum industry* (University of Oklahoma Press).
Archer, J. S. and Wall, C. G. (1986) *Petroleum engineering: principles and practice* (Graham and Trotman).
Bacha, J. D., Newman, J. W. and White, J. L. (eds) (1986) *Petroleum-derived carbons* (American Chemical Society).
Baughman, G. L. (1978) *Synthetic fuels data book: US oil shale, US coal, oil sands*, 2nd Edn (Cameron Engineers).

Din, F. (ed.) (1962) *Thermodynamic functions of gases* (Butterworths).
Dohr, G. (1981) *Applied geophysics: introduction to geophysical prospecting* (Wiley).
Ewing, W. M., Jardetzky, W. S. and Press, F. (1957) *Elastic waves in layered media* (McGraw-Hill).
Fitch, A. A. (ed.) (1985) *Developments in geophysical exploration methods* (Elsevier Applied Science).
Gallick, E. C. (1993) *Competition in the natural gas pipeline industry – economic policy analysis* (Praeger).
Graff, W. J. (1981) *Introduction to offshore structures: design, fabrication and installation* (Gulf Publishing).
Hobson, G. D. (1984) *Modern petroleum technology*, 2 vols, 5th ed. (Wiley).
IBA (1988) *Energy Law. Proc. Advanced Seminar on Petroleum, Minerals and Energy Law*, Sydney, March 1988 (International Bar Association and Graham and Trotman).
Katz, D. L. et al. (1959) *Handbook of natural gas engineering* (McGraw-Hill).
Lang, K. R. and Donohue, D. A. T. (1986) *A first course in petroleum technology* (International Human Resources Development Corp.).
Melvin, A. (1987) *Natural gas: basic science and technology* (Adam Hilger).
Merrick, D. (1984) *Coal combustion and conversion technology* (Macmillan).
Penner, S. S. et al. (1982) *New sources of oil and gas; gases from coal; liquid fuels from coal, shale, tar sands and heavy oil sources* (Pergamon).
PSTI. (1993) *European oil and gas demonstration project inventory – non-EC supported projects* (Petroleum Science and Technology Institute for the Commission of the European Communities).
Rees, J. and Odell, P. (eds) (1986) *The international oil industry – an interdisciplinary perspective* (Macmillan).

JOURNALS AND PERIODICALS

Chemical and Petroleum Engineering (Plenum) monthly.
Fuel (Butterworth-Heinemann) monthly.
Fuel Science and Technology International (Marcel Dekker) monthly.
Gas Energy Review (American Gas Association) monthly.
Gas Engineering and Management (Institution of Gas Engineers) 10 per annum.
Gas Separation and Purification (Butterworth-Heinemann) quarterly.
GWF: das Gas und Wasserfach. Gas, Erdgas (R. Oldenbourg Verlag) monthly.
Hydrocarbon Processing (Gulf Publishing) monthly.
International Chemical Engineering (American Institute of Chemical Engineers) quarterly.
Journal of Petroleum Geology (Scientific Press) quarterly.
Journal of Petroleum Science and Engineering (Elsevier) 8 per annum.
Journal of Petroleum Technology (Society of Petroleum Engineers of AIME) monthly.
Offshore Engineer (Thomas Telford) monthly.
Oil and Gas Journal (PennWell Publishing) weekly.
Petroleum Economist (Petroleum Economist) monthly.
Petroleum Review. (Institute of Petroleum) monthly.

Petrostrategies (Petrostrategies) 48 per annum.
World Oil (Gulf Publishing) monthly.

Hydrogen and synthetic fuels

Synthetic fuels covers the whole area of secondary fuels derived from other feedstocks. Hydrogen is fairly straightforward in this respect, but the area includes all other derived liquid and gaseous fuels irrespective of their source: coal, oil or biomass. Once again, the material presented here relates directly to the area, but associated material will be found under combustion, coal, oil and gas and renewable energy resources.

BOOKS

Anderson, L. L., and Tillman, D. A. (1979) *Synthetic fuels from coal: overview and assessment* (Wiley).
Baughman, G. L. (1978) *Synthetic fuels data book: US oil shale, US coal, oil sands* (Cameron Engineers).
Glynn, P. (1986) *Prospects for hydrogen from advanced water electrolysis* (Commission of the European Communities).
Hunt, V. D. (1982) *Synfuels handbook* (Industrial Press).
Imarisio, G. and Bemtgen, J. M. (eds) (1988) *Progress in synthetic fuels* (Graham and Trotman for the Commission of the European Communities).
Meyers, R. A. (1984) *Handbook of synfuels technology* (McGraw-Hill).
Penner, S. S. et al. (1982) *New sources of oil and gas; gases from coal; liquid fuels from coal, shale, tar sands and heavy oil sources* (Pergamon).
Qader, S. A. (1985) *Natural gas substitutes from coal and oil* (Elsevier Applied Science).
Veziroglu, T. N., Zhu Yajie and Bao Deyou (1986) *Hydrogen systems*, Proceedings of Beijing International Symposium on Hydrogen Systems, Beijing, China, 7–11 May 1985 (Pergamon).
Veziroglu, T. N. and Takahashi, P. (1990) *Hydrogen energy progress VIII*. Proceedings of the 8th World Hydrogen Energy Conference, Hawaii, 1990, 3 vols (Pergamon).

JOURNALS AND PERIODICALS

Hydrogen Energy Coordinating Committee Annual Report (United States Department of Energy, National Technical Information Service) annual.
International Journal of Hydrogen Energy (Pergamon) monthly.
Pace Synthetic Fuels Report (Pace Company) quarterly.

Renewable energy resources – general

The renewable energy area is extremely diverse and is now generally taken to include waste treatment, particularly municipal waste. There is a sizeable bibliography covering what could be classed as 'general' renewable energies, that is, covering a range of topics or including the associated environmental or social issues. These aspects will be covered in this section, leaving more detailed treatment of each of the recognized renewable areas to be treated separately.

BOOKS

Burnham, L. and Johansson, T. B. (eds) (1993) *Renewable energy: sources for fuels and electricity* (Island Press).

IEE (1993) *International Conference on Renewable Energy – Clean Power 2001*, 17–19 November 1993, London (Power and Science, Education and Technology Divisions of the Institution of Electrical Engineers in association with the British Wind Energy Association, IEE).

Karekezi, S. and Mackenzie, G. A. (eds) (1993) *Energy options for Africa: environmentally sustainable alternatives* (Zed Books in association with the African Energy Policy Research Network (AFREPREN), The Foundation for Woodstove Dissemination (FWD), UNEP Collaborating Centre on Energy and Environment and Riso National Laboratory).

Mustoe, J. E. H. (1984) *An atlas of renewable energy resources in the United Kingdom and North America* (Wiley).

Rosenberg, P. (1993) *The alternative energy handbook* (Fairmont Press).

Sayigh, A. A. M. (ed.) (1992) *Renewable energy – technology and the environment. Proceedings of the Second World Renewable Energy Congress*, Reading, 13–18 September 1992 (Pergamon).

Sodha, M. S., Mathur, S. S. and Malik, M. A. S. (1987) *Reviews of renewable energy resources*, Vol. 3 (Wiley).

Twidell, J. W. and Weir, A. D. (1986) *Renewable energy resources* (Spon).

UNECSO (1993) *International directory of new and renewable energy information sources and research centres*, 3rd rev. ed. (James and James).

Wrixon, G. T., Rooney, A. M. E. and Palz, W. (1993) *Renewable energy 2000* (Springer-Verlag).

Wilbur, L. C. (ed.) (1985). *Handbook of energy systems engineering: production and utilisation* (Wiley).

JOURNALS AND PERIODICALS

Alternative Energy (Technology Forecasts) monthly.
Alternative Energy Digests (International Academy at Santa Barbara, CA) 8 per annum.
Energy and Environment (MultiScience Publishing) quarterly.
International Journal of Ambient Energy (Ambient Press) quarterly.
Renewable Energy News Digest (c/o Sandra Oddo, 861 Central Parkway, Schenectady, NY) monthly.
Renewable Energy (Pergamon) 8 per annum.
Renewable Sources of Energy (Tycody International Publishing) quarterly.
Reviews of Renewable Energy Resources (Wiley) irregular.

Solar energy

The solar energy area is interesting both from a technological viewpoint and from a market perspective. Rapid technological advances are occurring, while the market is developing rapidly both through natural demand and through stimulation. A high level of public interest is playing an important part in maintaining this momentum. The literature includes market surveys, atlases and data books of insolation levels, while the technical treatment ranges from detailed scientific treatises on solar radiation and capture devices to manuals describing the construction and installation of solar panels.

REFERENCE MATERIALS

Palz, W. (ed.) (1984) *European solar radiation atlas*, Vol. 1, Global radiation on horizontal surfaces (Commission of the European Communities, Köln-Verlag TUV).
Derrick, A. *et al.* (eds) (1993) *Photovoltaics: a market overview* (James and James).
Meteorological Office (1980) *Solar radiation data for the United Kingdom (1951–75)* (Meteorological Office).

BOOKS

Alawi, H. *et al.* (eds) (1986) *Solar energy prospect in the Arab World: Proceedings of the 2nd Arab International Solar Energy Conference*, Bahrain, 15–21 February, 1986 (Pergamon).
Arden, M. E., Burley, S. M. A. and Coleman, M. (1992) *1991 Solar World Congress*, Proceedings of the International Solar Energy Society, Denver, CO, 19–23 August, 1991 (Pergamon).
Berliand, G. T. (1980) *Components of solar irradiance of horizontal and inclined surfaces* (International Energy Agency).
Carter, C. and De Villers, J. (1987) *Principles of passive solar building design* (Pergamon).
Chadwick, A. T. (1981) *Economics of solar water heating* (Energy Technology Support Unit).
Chandrasekhar, S. (1960) *Radiative transfer* (Dover).
De Vos, A. (1992) *Endoreversible thermodynamics of solar energy conversion* (Oxford University Press).
Dixon, A. E. and Leslie, J. D. (eds) (1979) *Solar energy conversion* (Pergamon).
Duffie, J. A. and Beckman, W. A. (1991) *Solar engineering of thermal processes*, 2nd ed. (Wiley).
Fleagle, R. C. and Businger, J. A. (1980) *An introduction to atmospheric physics*, 2nd ed. (Academic Press).
Goulding, J. R., *et al.* (1992) *Energy in architecture: the European passive solar handbook*, rev. ed. (Batsford for the Commission of the European Communities).
Granqvist, C. G. (ed.) (1991) *Materials science for solar energy conversion systems* (Pergamon).
Hord, R. M. (1984) *Handbook of space technology: status and projections* (CRC Press).
Howe, E. D. (1974) *Fundamentals of water desalination* (Marcel Dekker).

Howell, J. R., Bannerot, R. B. and Vliet, G. (1982) *Solar-thermal energy systems: analysis and design* (McGraw-Hill).
Kondratyev, K. Y. (1965) *Radiative heat exchange in the atmosphere* (Pergamon).
Kondratyev, K. Y. (1972) *Radiation processes in the atmosphere* (World Meteorological Organization).
Lane, G. A. (1986) *Solar heat storage: latent heat materials*, 2 vols (CRC Press).
Meinel, A. B. and Meinel, M. P. (1976) *Applied solar energy* (Addison-Wesley).
O'Leary, B. (ed.) (1982) *Space industrialization*, 2 vols (CRC Press).
Pulfrey, D. L. (1978) *Photovoltaic power generation* (Van Nostrand Reinhold).
Reddy, T. A. (1987) *Design and sizing of active solar thermal systems* (Oxford University Press).
Sayigh, A. A. M. and McVeigh, J. C. (eds) (1992) *Solar air conditioning and refrigeration* (Pergamon).
Sodha, M. S. et al. (1987) *Solar crop drying*, 2 vols (CRC Press).
Sodha, M. S., Mathur, S. S. and Malik, M. A. S. (eds) (1987) *Reviews of renewable energy resources*, Vol. 3 (Wiley).
Spiegler, K. S. and Laird, A. D. K. (eds) (1980) *Principles of desalination*, 2 vols, 2nd ed. (Academic Press).
Starr, M. R. and Palz, W. (1983) *Photovoltaic power for Europe: an assessment study* (Reidel for the Commission of the European Communities).
Steemers, T. C. (ed.) (1987) *Solar energy applications to buildings and solar radiation data. Proceedings of the European Contractors Meeting*, Brussels, 13–14 November, 1986 (Reidel for the Commission of the European Communities).
Tabb, P. (1984) *Solar energy planning* (McGraw-Hill).
Takahashi, K. and Konagai, M. (1986) *Amorphous silicon solar cells* (North Oxford Academic Publishers).
Tiller, J. and Creech, D. B. (1988) *Energy design and construction. A manual for energy efficient and passive solar homes* (Governor's Division of Energy, Agriculture and National Resources).
Williams, A. F. (1986) *The handbook of photovoltaic applications: building applications and system design considerations* (Fairmont Press).
Yuncu, H., Payko, E. and Yener, Y. (eds) (1987) *Solar energy utilisation*, NATO Advanced Science Institute Series E, No. 129 (Martinus Nijhoff).

JOURNALS AND PERIODICALS

Advances in Solar Energy (Plenum) annual.
International Journal of Ambient Energy (Ambient Press) quarterly.
International Journal of Solar Energy (Harwood Academic Publishers) 8 per annum.
Journal of Solar Energy Engineering (American Society of Mechanical Engineers) quarterly.
Journal of Solar Sciences (Van Nostrand-Reinhold) quarterly.
International Journal of Solar Energy (Harwood Academic Publishers) 8 per annum.

Journal of Solar Energy Engineering (American Society of Mechanical Engineers) quarterly.
Progress in Batteries and Solar Cells (JEC Press) annual.
PV News (Photovoltaic Energy Systems Inc.) monthly.
SERI Journal (Solar Energy Research Institute) quarterly.
SERI Materials Branch Semiannual Report (Solar Energy Research Institute) biannual.
Solar Energy Materials and Solar Cells (Elsevier) 16 per annum.
Solar Energy (Pergamon) monthly.
Solar Progress (Australian and New Zealand Solar Energy Society) quarterly.
Solar Thermal Energy Technology (National Technical Information Service) bi-monthly.
Sonnenenergie und Warmetechnik (Bielefelder Verlagsanstalt KG) bi-monthly.
Sunworld (International Solar Energy Society) quarterly.

Geothermal energy

Geothermal energy is exploited by sinking boreholes into the earth and extracting heat using a heat transfer fluid. This can be the ground water itself in the case of conventional geothermal exploitation or, with the experimental hot dry rocks approach, water is injected from the surface and circulated through a fracture zone. The references below give a good introduction to the subject and address the primary areas of concern.

BOOKS AND REFERENCE MATERIALS

Armstead, H. C. H. (1983) *Geothermal energy*, 2nd ed. (Spon).
Aureille, M. (1982) *Geothermal heating* (Commission of the European Communities).
Berman, E. R. (1975) *Geothermal energy* (Noyes Data Corporation).
Cheremisinoff, P. N. and Morresi, A. C. (1976) *Geothermal energy technology assessment* (Technomic).
De Bremaecker, J. C. (1985) *Geophysics: the earth's interior* (Wiley).
Economides, M. J. and Ungemach, P. O. (eds) (1987) *Applied Geothermics* (Wiley).
Ellis, A. J. and Mahon, W. A. J. (1977) *Chemistry and geothermal systems* (Academic Press).
Harrison, R., Mortimer, N. D. and Smarason, O. B. (1990) *Geothermal heating: handbook of engineering economics* (Pergamon).
Kruger, P. and Otte, C. (eds) (1973) *Geothermal energy: resources, production and stimulation* (Stanford University Press).
Milora, S. L. and Tester, J. W. (1977) *Geothermal energy as a source of electrical power: thermodynamic and economic design criteria* (MIT Press).
Parker, R. H. (ed.) (1989) *Hot dry rock geothermal energy*, Phase 2B final report of the Camborne School of Mines, 2 vols (Pergamon).
Shock, R. A. W. (1986) *An economic assessment of hot dry rocks as an energy source for the UK* (Energy Technology Support Unit).

JOURNALS AND PERIODICALS

Bulletin of the Geothermal Resources Council (Geothermal Resources Council) 11 per annum.
Geothermal Energy (United States Department of Energy, National Technical Information Service) bi-monthly.
Geothermics (Pergamon) bi-monthly.
Transactions – Geothermal Resources Council (Geothermal Resources Council) annual.

Wind power

The development of wind power raises a number of issues over and above the obvious ones of resource estimation, power extraction and management of the wind energy generator. These include aspects of systems integration, grid-management, performance measurement and safety. This diversity is reflected in the sources listed below.

BOOKS

AWEA (1985) *Standard performance testing of wind energy conversion systems* (American Wind Energy Association).
Barltop, N. D. P., Ward, I. P. and Daw, D. J. (1993) *A fatigue costing study of horizontal axis wind turbines* (Energy Technology Support Unit).
BWEA (1987) *Wind turbines – meteorology, siting and monitoring* (British Wind Energy Association).
Clayton, B. R. (ed.) (1992) *Wind energy conversion 1992. Proceedings of the 14th British Wind Energy Association Conference*, Nottingham, 25–27 March, 1992 (Mechanical Engineering Publications).
Dickson, E. M. and Loperena, G. A. (1984) *Wind power parks 1983 survey* (Electric Power Research Institute).
Eggleston, D. M. and Stoddard, F. S. (1987) *Wind turbine engineering design* (Van Nostrand Reinhold).
Foster, J. E. (ed.) (1993) *Wind energy penetration into weak electricity networks* (British Wind Energy Association).
Freris, L. L. (ed.) (1990) *Wind energy conversion systems* (Prentice Hall).
Garrad Hassan & Partners Ltd (1993) *Further development of a finite element code for use by the wind turbine industry* (Energy Technology Support Unit for the Department of Trade and Industry).
Hunter, R. and Elliot, G. (eds) (1994) *Wind-diesel systems* (Cambridge University Press).
Johnson, G. L. (1985) *Wind energy systems* (Prentice Hall).
Kovarik, T., Pipher, C. and Hurst, J. (1980) *Wind energy* (Prism Press).
Le Gouriérés, D. (1982) *Wind power plants, theory and design* (Pergamon).
Lowson, M. V. (1993) *Systematic comparison of predictions and experiment for wind turbine aerodynamic noise* (Department of Trade and Industry).
McGuigan, D. (1978) *Small scale wind power* (Prism Press).
Melaragno, M. (1982) *Wind in architectural and environmental design* (Van Nostrand-Reinhold).

Moreno, R. (1991) *Guidelines for assessing wind energy potential* (World Bank).
Nacfaire, H. (ed.) (1988) *Grid-connected wind turbines* (Elsevier Applied Science).
Nacfaire, H. (ed.) (1989) *Wind-diesel and wind-autonomous energy systems* (Elsevier Applied Science).
Roots, G. (1993) Tilting at windmills: development control and renewable energy, *Journal of Planning and Environment Law*, June, 515–520.
Schmid, J. and Klein, H. P. (1991) *Performance of European wind turbines: a statistical evaluation from the European wind turbine database Eurowin* (Elsevier Applied Science).
Taylor, G. J. (1990) *Wake measurements on the NIBE turbines in Denmark* (Energy Technology Support Unit).
Twidell, J. (ed.) (1987) *A guide to small wind energy conversion systems* (Cambridge University Press).
Van Hulle, F. J. L., Smulders, P. T. and Dragt, J. B. (eds) (1991) *Wind energy, technology and implementation* (Elsevier).
Warne, D. F. (1983) *Wind power equipment* (Spon).

JOURNALS AND PERIODICALS

American Wind Energy Association Wind Energy Weekly (American Wind Energy Association) weekly.
Wind Energy News (Wind Books Inc.) monthly.
Windkraft Journal (Verlag Natürliche Energie GmbH) quarterly.
Windpower Monthly (Windpower Monthly) monthly.

Water power: hydro, wave and tidal

Despite their differences, the different technologies for generating power from water have been grouped together. This is partly because of the degree of overlap between the underlying technologies and partly because of the difficulties encountered when trying to separate them in a systematic way. All involve aspects of civil and electrical engineering, and all have associated environmental concerns which are addressed both in this section and under the environmental aspects.

BOOKS

American Society of Civil Engineers (1989) *Civil engineering guidelines for planning and designing hydroelectric developments*, 5 vols (American Society of Civil Engineers).
American Society of Civil Engineers (1992) *Guidelines for rehabilitation of civil works of hydroelectric plants* (American Society of Civil Engineers).
Avery, W. H. and Wu Chi (1994) *Ocean thermal energy conversion* (Oxford University Press).
Baker, A. C. (1991) *Tidal power* (Peregrinus for Institution of Electrical Engineers).
Bermacsek, G. M. (1984) *Dam design and operation to optimise fish pro-*

duction in impounded river basins (Food and Agriculture Organization of the United Nations).

Brin, A. (1980) *Energy and the oceans* (Westbury House).

Burns, T. R. and Midttun, A. (1985) *Economic growth, environmentalism and social conflict: a case study on hydro-power planning in Norway* (International Institute for Environment and Safety).

Charlier, R. H. (1982) *Tidal energy* (Van Nostrand Reinhold).

Charlier, R. H. and Justus, J. R. (1993) *Ocean energies: environmental, economic and technological aspects of alternative power sources* (Elsevier).

Clare R. (ed.) (1992) *Proceedings of the Fourth Conference on Tidal Power. Tidal power: trends and developments*, London, 19–20 March, 1992 (Thomas Telford).

Crabb, J. A. (1983) *Assessment of wave power available at key United Kingdom sites: a description of work undertaken in the Department of Energy's wave energy programme* (Institute of Oceanographic Sciences).

Davis, G. H. (1985) *Water and energy: demand and effects* (Unesco).

Fritz, J. J. (ed.) (1984) *Small and mini hydropower systems, a resource assessment and project feasibility* (McGraw-Hill).

Garzon, C. E. (1984) *Water quality in hydroelectric projects: considerations for planning in tropical forest regions* (World Bank).

Goldin, A. (1980) *Oceans of energy: reservoir of power for the future* (Harcourt Brace Jovanovich).

Gulliver, J. S. and Arndt, R. E. A. (eds) (1991) *Hydropower engineering handbook* (McGraw-Hill).

Harvey, A. et al. (1993) *Micro-hydro design manual: a guide to small-scale water power schemes* (Intermediate Technology).

Institution of Mechanical Engineers (1991) *Wave energy.* (Mechanical Engineering Publications).

Jog, M. G. (1989) *Hydro-electric and pumped storage plants* (Wiley).

Leliavsky, S. (1982) *Hydro-electric engineering for civil engineers* (Chapman and Hall).

Lennard, D. E. (1987) *Prospects and potential for ocean thermal energy conversion* (World Energy Conference).

McCormick, M. E. (1981) *Ocean wave energy conversion* (Wiley).

McGuigan, D. (1979) *Small scale water power* (Prism Press).

Moniton, L., Le Nir, M. and Roux, J. (1984) *Microhydroelectric power stations* (Wiley).

Office of Ocean Minerals and Energy. (1985) *Final environmental impact statement for commercial ocean thermal energy conversion (OTEC) licensing* (US Department of Commerce).

Sassaman, J. F., Ouellette, R. P. and Cheremisinoff, P. N. (1983) *Low-head hydropower* (Technomic Publishing).

Select Committee on Science and Technology, House of Commons (1977) *The exploitation of tidal power in the Severn Estuary*, 4th report (HMSO).

Seymour, R. J. (ed.) (1992) *Ocean energy recovery: the state of the art* (American Society of Civil Engineers).

Shaw, R. (1982) *Wave energy, a design challenge* (Ellis Horwood).

Warnick, C. C. et al. (1984) *Hydropower engineering.* (Prentice Hall).

JOURNALS AND PERIODICALS

Ocean Engineering (Pergamon) bi-monthly.
Purpa Lines (Hydro Consultants Inc.) bi-weekly.

Biomass energy

Energy from biomass has been the subject of considerable development effort over the last twenty years and has consequently been in a state of considerable flux. In many cases, the earlier work has been overtaken by technology or events. Because of this, the literature cited in this section is all of fairly recent origin and represents the current state of the subject.

BOOKS

Bridgwater, A. V. (ed.) (1993) *Advances in thermochemical biomass conversion* (Elsevier Applied Science).
Bridgwater, A. V. and Grassi, G. (eds) (1991) *Biomass pyrolysis liquids upgrading and utilisation* (Elsevier Applied Science).
Ferrero, G. L. *et al.* (1989) *Pyrolysis and gasification* (Elsevier Applied Science).
Frankena, F. (1992) *Strategies of expertise in technical controversies: a study of wood energy development* (Lehigh University Press).
Grassi, G., Gosse, G. and dos Santos, G. (eds) (1990) *Biomass for energy and industry* (Elsevier Applied Science).
Hall, D. O. (1992) *Biomass* (World Bank).
Hogan, E. *et al.* (eds) (1992) *Biomass thermal processing* (CPL Press).
Mitchell, C. P. *et al.* (eds) (1990) *Forestry, forest biomass and biomass conversion: the IEA bioenergy agreement summary reports* (Elsevier Science Publishers).
Pasztor, J. and Kristoferson, L. A. (eds) (1990) *Bioenergy and the environment* (United Nations Environment Programme, Westview Press).
Richards, G. E. (ed.) (1992) *Wood, fuel for thought* (Harwell Laboratories).
Richard, G. E. (ed.) (1993) *Wood, energy and the environment* (Harwell Laboratories).
Saddler, J. N. (ed.) (1993) *Bioconversion of forest and agricultural plant residues* (CAB International).
Scott, C. D. (ed.) (1986) *Eighth Symposium on Biotechnology for Fuels and Chemicals, Proceedings*, Gatlinburg, Tennessee, May 13–16 (Wiley).
UNIDO (1990) *Workshop on Biomass Thermal Processing Projects 1990, London* (United Nations Industrial Development Organization and Institute of Energy).

JOURNALS AND PERIODICALS

Biomass and Bioenergy (Pergamon) monthly.
Biomass Bulletin (MultiScience) quarterly.
Biomass Energy Directory (Independent Energy) annual.
Bioresource Technology (Elsevier Applied Science) monthly.

Earth Energy (National Alcohol Fuel Producers Association) monthly.
Energy from Biomass and Wastes (Institute of Gas Technology) annual.
Wood Energy Monthly Update (G. V. Olsen Associates) monthly.

Nuclear power

Nuclear power has a vast bibliography covering fuel extraction and processing, fission and fusion reactor development, materials, reactor safety and environmental impacts, radiation and radioactive waste disposal. A representative selection is listed below.

BOOKS AND REFERENCE MATERIALS

Bennet, D. J. and Thomson, J. R. (1981) *The elements of nuclear power*, 3rd ed. (Longman).
Bittencourt, J. A. (1986) *Fundamentals of plasma physics* (Pergamon).
Blix, H. (1989) Nuclear power and the environment. *Environmental Policy and Law*, **19**, 29–30.
Brookes, L. G. and Motamen, H. (eds) (1984) *Economics of nuclear energy* (Chapman and Hall).
Frost, B. R. T. (1982). *Nuclear fuel elements: design, fabrication and performance* (Pergamon).
Glasstone, S. and Sesonske, A. (1994) *Nuclear reactor engineering*, 2 vols, 4th rev. ed. (Van Nostrand Reinhold).
Gross, R. A. (1984) *Fusion energy* (Wiley).
Harms, A. A. (1987) *Principles of nuclear science and engineering* (Research Studies Press).
Institution of Mechanical Engineers. (1983) *Decommissioning of radioactive facilities* (Mechanical Engineering Publications).
Institution of Mechanical Engineers (1987) *The management and disposal of intermediate and low level radioactive waste* (Mechanical Engineering Publications).
International Atomic Energy Agency (1980) *Guide to the safe handling of radioactive wastes at nuclear power plants* (IAEA).
International Atomic Energy Agency (1985) *IAEA safeguards: implementation at nuclear fuel cycle cacilities* (IAEA).
Lamarsh, J. R. (1983) *Introduction to nuclear engineering*, 2nd ed. (Addison-Wesley).
Lewins, J. D. and Gittus, J. H. (eds) (1993) *Research and development in the nuclear industry* (Research Studies).
Marshall, W. (ed.) (1983) *Nuclear power technology*, Vol. 1. Reactor technology, Vol 2. Fuel cycle, Vol. 3. Nuclear radiation (Clarendon Press).
OECD (1984) *Nuclear legislation. Analytical study for a regulatory institutional framework for nuclear activities* (Nuclear Energy Agency, OECD Publications and Information Centre).
Pershagen, B. and Bowen, M. (1989) *Light water reactor safety* (Pergamon).
Openshaw, S. (1986) *Nuclear power: siting and safety* (Routledge and Kegan Paul).
Ott, K. O. and Spinrad, B. I. (eds) (1985) *Nuclear energy: a sensible alternative* (Plenum Press).

Raeder, J. et al. (1986) *Controlled thermonuclear fusion: fundamentals of its utilisation for energy supply* (Wiley).
Stacey, W. M. (1984) *Fusion: an introduction to the physics and technology of magnetic confinement fusion* (Wiley).
Ursu, I. (1985) *Physics and technology of nuclear materials* (Pergamon).
Wolfson, R. (1993) *Nuclear choices: a citizen's guide to nuclear technology*, rev. ed. (MIT Press).

JOURNALS AND PERIODICALS

Annals of Nuclear Energy (Pergamon) monthly.
Atomic Energy Clearing House (Congressional Information Bureau) weekly.
Fusion Engineering and Design (Elsevier Sequoia SA) monthly.
Fusion Technology (American Nuclear Society) 8 per annum.
Fusion Power Program Quarterly Progress Report (United States Department of Energy, Argonne National Laboratory) quarterly.
Fusion Power Report (Business Publishers Inc.) monthly.
IEEE Transactions on Nuclear Science (IEEE) bimonthly.
Journal of Fusion Energy (Plenum) quarterly.
Journal of Nuclear Materials (Elsevier) 33 per annum.
Journal of Nuclear Materials Management (Institute of Nuclear Materials Management) quarterly.
Nuclear Energy (Thomas Telford) bimonthly.
Nuclear Energy Data (OECD Nuclear Energy Agency) annual.
Nuclear Engineering and Design (Elsevier Sequoia SA) 24 per annum.
Nuclear Fuel Cycle (United States Department of Energy, National Technical Information Service) monthly.
Nuclear Plant Journal (EQES Inc.) bimonthly.
Nuclear Science and Engineering (American Nuclear Society) monthly.
Nuclear Technology (American Nuclear Society) monthly.
Progress in Nuclear Energy (Pergamon) quarterly.
Revue Général Nucleaire (Revue Générale de l'Electricité SA) bi-monthly.
Transactions of the American Nuclear Society (American Nuclear Society) 3 per annum.

Electricity and magnetohydrodynamics (MHD)

Electrical power generation is covered in detail elsewhere in this book and is also closely linked to the combustion, fuels and nuclear power sections of this chapter. Because of this, only a limited treatment is given here, with some emphasis being placed on magnetohydrodynamics, an area like nuclear fusion which offers great potential but which has not yet been realized on a demonstration scale.

BOOKS AND REFERENCE MATERIALS

Bateman, G. (1978) *MHD instabilities* (MIT Press).
Blums, E., Mikhailov, Y. A. and Ozols, R. (1987) *Heat and mass transfer in MHD flows* (World Scientific).
Branover, H. (1978) *Magnetohydrodynamic flow in ducts* (Wiley).
British Electricity International (ed.) (1992–1993) *Modern power station prac-*

tice, 11 Volumes: Vol. A, Station planning and design; Vol. B, Boilers and ancillary plant; Vol. C, Turbines, generators and associated plant; Vol. D, Electrical systems and equipment; Vol. E, Chemistry and metallurgy; Vol. F, Control and instrumentation; Vol. G, Station operation and maintenance; Vol. H, Station commissioning; Vol. J, Nuclear power generation; Vol. K, EHV transmission; Vol. L, System operation; Vol. M, Index (Pergamon).

Coombe, R. A. (ed.) (1964) *Magnetohydrodynamic generation of electrical power* (Van Nostrand-Reinhold).

Helm, D. *et al.* (eds) (1993) *Generation in the 1990s: electricity capacity and new power projects* (Oxford Economic Research Associates).

Hicks, T. G. (ed.) (1987). *Power generation calculations reference guide* (McGraw-Hill).

Johanson, N. R. and Chapman, J. N. (1991) *Magnetohydrodynamics (MHD) power generation 1991* (American Society of Mechanical Engineers).

Kirillin, V. A. and Sheyndlin, A. E. (eds) (1986) *MHD energy conversion: physicotechnical problems* (American Institute of Aeronautics and Astronautics).

McGowan, F. (1993) *The struggle for power in Europe: competition and regulation in the EC electricity industry* (Royal Institute of International Affairs, Energy and Environmental Programme).

Manheimer, W. M. and Lashmore-Davies, C. N. (1984) *MHD instabilities in simple plasma configuration* (Naval Research Laboratory).

Manheimer, W. M. and Lashmore-Davies, C. N. (1989) *MHD and microinstabilities in confined plasma* (Adam Hilger).

Moreau, R. (1990) *Magnetohydrodynamics* (Kluwer Academic).

Polovin, R. V. and Demutskii, V. P. (1990) *Fundamentals of magnetohydrodynamics* (Consultants Bureau).

JOURNALS AND PERIODICALS

Annual Bulletin of Electric Energy Statistics for Europe (European Commission for Europe) annual.
Electric Machines and Power Systems (Taylor and Francis) monthly.
Electric Power Systems Research (Elsevier Sequoia SA) 9 per annum.
Electric Vehicle-Battery Technology (Business Communications Co.) monthly.
Electrical World (McGraw-Hill) monthly.
EPRI Journal (Electric Power Research Institute) 8 per annum.
European Transactions on Electrical Power Engineering (VDE-Verlag) bi-monthly.
IEE Proceedings B. Electric Power Applications (Institution of Electrical Engineers) bi-monthly.
IEE Proceedings C. Generation, Transmission and Distribution (Institution of Electrical Engineers) bi-monthly.
IEEE Power Engineering Review (IEEE) monthly.
IEEE Transactions on Energy Conversion (IEEE) quarterly.
IEEE Transactions on Power Electronics (IEEE) quarterly.
IEEE Transactions on Power Systems (IEEE) quarterly.
IEEE Transactions on Vehicular Technology (IEEE) quarterly.
International Power Generation (Argus Business Publications) bi-monthly.

Magnetohydrodynamics: Journal of the International Liaison Group on MHD Electrical Power Generation (Taylor and Francis) quarterly.
Modern Power Systems (Wilmington Publishing) monthly.
Power Engineering Journal (Institution of Electrical Engineers) bi-monthly.
Proceedings of the IEEE (IEEE) monthly.
Proceedings of the Institution of Electrical Engineers (Institution of Electrical Engineers) bi-monthly.
Revue Générale de l'Electricité (Revue Générale de l'Electricité) 11 per annum.

Energy in buildings, heat pumps, combined heat and power, district heating

This section and the next one deal essentially with the provision of energy supplies to the end-user application and with management of energy consumption. Because of this, they are both highly interactive and could logically be linked together. This would produce a very large and unwieldy grouping, however, and so the rather arbitrary division indicated in the captions has been used. This section deals essentially with the provision of energy supplies to buildings and with certain related areas. Essential reference material such as the properties of water and steam and psychrometric tables are included here because of their importance for heat load and comfort level calculations.

BOOKS

ASHRAE (1986) *Terminology of heating, ventilation, air-conditioning and refrigeration* (American Society of Heating, Refrigerating and Air-Conditioning Engineers).
Bansal, N. K. (1994) *Passive building design, a handbook of natural climatic control* (Elsevier).
Chadderton, D. V. (1991) *Building services engineering* (Chapman and Hall).
Chng-Kuo Ho, Jui-Tien Chi (eds) (1986) *Energy conservation in buildings in China and Sweden* (Swedish Council for Building Research).
Diamant, R. M. E and Kut, D. (1981) *District heating and cooling for energy conservation* (Architectural Press).
EEO (1993) *Energy efficiency in buildings. Energy appraisal of existing buildings: a handbook for surveyors* (Energy Efficiency Office, RICS).
Elmahdy, A. H. (ed.) (1985) *Energy-efficient office buildings. Report of Canadian case studies* (National Research Council of Canada).
Heap, R. D. (1983) *Heat pumps*, 2nd ed. (Spon).
Holdsworth, W. and Sealey, A. (1992) *Healthy buildings, a design primer for a living environment* (Longman).
Holland, F. A., Watson, F. A. and Devotta, S. (1982) *Thermodynamic design data for heat pump systems: a comprehensive data base and design manual* (Pergamon).
Horlock, J. H. (1987) *Cogeneration: combined heat and power, thermodynamics and economics* (Pergamon).

Howard, R., Winterkorn, E. and Cooper, I. (1993) *Building environmental and energy design survey* (Building Research Establishment).
Kut, D. (1968) *Heating and hot water services in buildings* (Pergamon).
Kut, D. (1970) *Warm air heating* (Pergamon).
Limaye, D. R. (1985) *Planning co-generation systems* (Fairmont Press).
Mackenzie-Kennedy, C. (1979) *District heating: thermal generation and distribution: a practical guide to centralised generation and distribution of heat services* (Pergamon).
McMullan, J. T. and Morgan, R. (1981) *Heat pumps* (Adam Hilger).
McQuiston, F. C. and Parker, J. D. (1994) *Heating, ventilating and air conditioning: analysis and design*, 4th ed. (Wiley).
Meador, R. (1981) *Cogeneration and district heating: an energy-efficiency partnership* (Ann Arbor Science).
Moore, F. (1993) *Environmental control systems: heating, cooling, lighting* (McGraw-Hill).
National Audubon Society (1994) *Audubon House, building the environmentally responsible energy-efficient house* (Wiley).
Panzhauser, E. and Hoglund, I. (eds) (1988) *Energy conservation by improving building envelopes and HVAC equipment in buildings* (Metrica).
Porges, F. (1982) *Handbook of heating, ventilating and air conditioning*, 8th ed. (Butterworths).
Reay, D. A. and Dunn, P. (1993) *Heat pipes*, 4th ed. (Pergamon).
Reay, D. A. and Macmichael, D. B. A. (1988) *Heat pumps*, 2nd ed. (Pergamon).
Rizzi, E. A. (1980) *Design and estimating for heating, ventilating and air conditioning* (Van Nostrand-Reinhold).
Saito, T. and Igarashi, Y. (eds) (1990) *Heat pumps: solving energy and environmental challenges. Proceedings of Third International Energy Agency Heat Pump Conference*, Tokyo, 12–15, March (Pergamon).
Sauer, H. J. and Howell, R. H. (1983) *Heat pump systems* (Wiley).
Schueman, D. (ed.) (1992) *The residential energy audit manual*, 2nd ed. prepared by the US Department of Energy, Office of the Assistant Secretary for Conservation and Solar, Office of Building and Community Systems, with contributions from Oak Ridge National Laboratory, University of Massachusetts Cooperative Extensive Service Energy Education Center and the Solar Energy Research Institute (Fairmont Press).
Tiller, J. and Creech, D. B. (1988) *Energy design and construction. A manual for energy efficient and passive solar homes* (Governor's Division of Energy, Agriculture and National Resources).
Von Cube, H. L. and Steimle, F. (1981) *Heat pump technology* (Butterworths).
Wexler, A., Hyland, R. and Stewart, R. (1983) *Thermodynamic properties of dry air, moist air and water and SI psychrometric charts* (American Society of Heating, Refrigerating and Air-Conditioning Engineers).

JOURNALS

Buildings Energy Technology (United States Department of Energy, Office of Scientific and Technical Information) monthly.
Energy Design Update (Cutter Information Corporation) monthly.

Heat Engineering (Foster Wheeler Corporation) quarterly.
Heat Recovery Systems and CHP (Pergamon) bi-monthly.
International Journal of Multiphase Flow (Pergamon) bi-monthly.
Journal of Energy Engineering (American Society of Civil Engineers) 2–3 per annum.
Journal of Energy, Heat and Mass Transfer (Indian Institute of Technology) quarterly.
Thermal Engineering (Interperiodica Publishing) monthly
Sonnenenergie und Warmepumpe (Bielefelder Verlagsanstalt KG) bi-monthly.

Energy conservation and storage, energy management, environmental issues

As has already been noted, this section and the previous one are closely linked, dealing essentially with the provision of energy supplies to the end-user application and with management of energy consumption. This section has been devoted to energy conservation and environment-related issues, but it should not be read in isolation; elements of all of the previous sections are appropriate – particularly in relation to environmental arguments and the rising profile of renewable energy sources.

BOOKS AND REFERENCE WORKS

Banister, D. (1981) *Transport policy and energy: perspectives, options and scope for conservation in the passenger transport sector* (University College London).
Bermacsek, G. M. (1984). *Dam design and operation to optimise fish production in impounded river basins* (Food and Agriculture Organization of the United Nations).
Bolin, B. et al. (eds) (1986) *The greenhouse effect, climate change and ecosystems* (Wiley).
Boos, B. (1986) *Energy conservation in towns; the Swedish concept* (Swedish Council for Building Research).
Burns, T. R. and Midttun, A. (1985) *Economic growth, environmentalism and social conflict: a case study on hydro-power planning in Norway* (International Institute for Environment and Society).
Capehart, B. L., Turner, W. C. and Kennedy, W. J. (1994) *Guide to energy management* (Fairmont Press).
Centre for Environmental Management and Planning. (1987) *Environmentally sound development in the energy and mining industries* (Centre for Environmental Management and Planning).
Charlier, R. H. and Justus, J. R. (1993) *Ocean energies: environmental, economic and technological aspects of alternative power sources* (Elsevier).
Eastop, T. D. and Croft, D. R. (1990) *Energy efficiency for engineers and technologists* (Longman).
Energy Information Centre. (1991) *British Gas directory of energy saving equipment* (Cambridge Information and Research Services).
Friends of the Earth (1993) *Energy and the environment* (Friends of the Earth).

Garzon, C. E. (1984) *Water quality in hydroelectric projects: considerations for planning in tropical forest regions* (World Bank).
Genta, G. (1985) *Kinetic energy storage: theory, and practice of advanced flywheel systems* (Butterworths).
Gibbons, J. H. and Chandler, W. U. (1981) *Energy, the conservation revolution* (Plenum).
Gopalakrishnan, C. (1994) *Economics of energy in agriculture* (Avebury).
Hedges, A. (1991) *Attitudes to energy conservation in the home: report on a qualitative study* (HMSO).
IEA. (1987) *Energy conservation in IEA countries* (OECD).
Jackson, M. (ed.) (1993) *Innovative energy and environmental applications* (Fairmont Press).
Jensen, J. (1982) *Energy storage* (Butterworth Scientific).
Katzev, R. D. and Johnson, T. R. (1987) *Promoting energy conservation: an analysis of behavioral research* (Westview Press).
Kelley, D. R. (1977) *The energy crisis and the environment: an international perspective* (Praeger).
Krause, F., Bach, W. and Koomey, J. (1990) *Energy policy in the greenhouse* (Earthscan).
Kut, D. (1982) *Dictionary of applied energy conservation* (Kogan Page).
Leggett, J. K. (1991) *Energy gap* (Heinemann).
Liu, P. I. (1993) *Introduction to energy and the environment* (Van Nostrand Reinhold).
Long, R. E. (1989) *Energy and conservation* (H. W. Wilson).
Moffat, D. W. (1993) *Plant engineer's portable problem solver* (Prentice Hall).
Nemetz, P. N. and Hankey, M. (1984) *Economic incentives for energy conservation* (Wiley).
O'Callaghan, P. W. (1993) *Energy management* (McGraw-Hill).
Office of Ocean Minerals and Energy (1985) *Final environmental impact statement for commercial ocean thermal energy conversion (OTEC) licensing* (US Department of Commerce).
Ottinger, R. L. *et al.* (1991) *Environmental costs of electricity* (Oceana).
Peltz, L. (1993) *Take the heat off the planet: how you can really help stop climate change* (Friends of the Earth).
Petrecca, G. (1993) *Industrial energy management* (Kluwer Academic).
Pilavachi, P. A. (ed.) (1993) *Energy efficiency in process technology* (Elsevier Applied Science).
Price, B. (1982) *Lead in petrol, an energy analysis* (Friends of the Earth).
Richards, G. E. (ed.) (1993) *Wood: energy and the environment* (Harwell Laboratories).
Schipper, L. *et al.* (1992) *Energy efficiency and human activity: past trends, future prospects* (Cambridge University Press).
Smil, V. (1987) *Energy, food, environment: realities, myths, options* (Clarendon Press).
Snow, D. A. (ed.) (1991) *Plant engineers' reference book* (Butterworth-Heinemann).
Thompson, P. (ed.) (1991) *Global warming, the debate* (Wiley for Strategy Europe Ltd.).
Thumann, A. (1983) *Plant engineers and managers guide to energy conser-

vation – the role of the energy manager, 2nd ed. (Van Nostrand Reinhold).
Thumann, A. (1991) *Plant engineers and managers guide to energy conservation 5th ed.* (Fairmont Press).
Thumann, A. (1992) *Handbook of energy audits*, 3rd ed. (Fairmont Press).
Turner, W. C. (1993) *Energy management handbook*, 2nd ed. (Fairmont Press).
Vine, E. and Crawley, D. (eds) (1991) *State of the art of energy efficiency – future directions* (American Council for an Energy Efficient Economy and University of California, Berkeley).
Vine, E., Crawley, D. and Centolella, P. (eds) (1991) *Energy efficiency and the environment: forging the link* (American Council for an Energy Efficient Economy and University of California, Berkeley).

JOURNALS AND PERIODICALS

E-Notes: Quarterly Newsletter of the International Institute for Energy Conservation (International Institute for Energy Conservation) quarterly.
Energy Conservation Digest (Energy Resources Inc.) 24 per annum.
Energy Conservation in IEA Countries (OECD) irregular.
Energy Conservation News (Business Communications Company) monthly.
Energy Conversion and Management (Pergamon) monthly.
Environment (Heldref Publications) monthly.
Environmental Impact Assessment Review (Elsevier) bimonthly.
Environmental Management (Springer-Verlag) bimonthly.
Environmetrics (Wiley) quarterly.
Strategic Planning for Energy and the Environment (Fairmont Press) quarterly.

CHAPTER FOURTEEN

Nuclear engineering

J. HARRIS

Introduction

Nuclear engineering is primarily concerned with achieving the safe, reliable and efficient conversion into electrical energy, of the thermal energy that can be released in a nuclear reactor, this being carried out in a nuclear power station. Just as in any 'conventional' coal, oil or gas-fired power station the electrical energy is produced by steam-driven turbo-alternators (although these are usually rather different in their detailed design, from those of conventional plant, in order optimally to exploit the particular steam conditions available). The heat required to generate the steam is not, however, derived by the burning of a fuel but by the highly exothermic neutron-induced fission of nuclei of atoms of uranium (and also of the plutonium which results, eventually, from the capture of neutrons by some of the uranium). It is the design, construction, operation and control of the nuclear reactor in which this occurs, the utilization of the resulting heat and the processing of the various special materials involved, which are the distinguishing activities of nuclear engineering.

The primary fuel for the fission process is uranium, which has two naturally occurring isotopic forms: the lighter U^{235}, which is readily fissile, and the heavier and more stable U^{238}, which is not. Only about 0.7 per cent of natural uranium is U^{235}. There is a high probability that if a neutron encounters and is absorbed by a U^{235} nucleus, the resulting formation will be so unstable that it will split apart into two (sometimes three) new and smaller nuclei (of barium and krypton, say) known as fission products. These two product nuclei will possess, as kinetic energy, most of the energy released by the reaction and this will eventually appear as heat in the surrounding material as they are slowed down by collision with surrounding nuclei, such heat being the main source of the heat derived from a reactor. (Many of these fission products are highly radioactive and are the major hazard in the handling, processing and long-term disposal of used nuclear fuel.).

Also released at the instant of fission are two or three new neutrons (known as fission neutrons). If the assembly, the reactor, is sufficiently large and not too thin in any direction (so that the neutrons will not too easily escape from the assembly), and if there is enough U^{235} and not too much neutron-absorber (such as boron or cadmium) present, then, on average, one or more of the neutrons released per fission will induce a further fission, i.e. a fission chain reaction will be generated which will be sustained at a constant level (a critical reaction, producing heat at a constant rate) or will be exponentially increasing (a supercritical reaction, producing heat at an increasing rate).

When first produced the fission neutrons have a very high energy and will not very readily induce further fissions of uranium nuclei. However, as they slow down by collision with surrounding nuclei their probability of doing so greatly increases. Neutrons are most efficiently slowed down by collision with the lightest nuclei – i.e. of the smallest atoms such as hydrogen, deuterium (heavy hydrogen) or carbon. Inclusion in the reactor of materials such as water, heavy water or graphite – which are rich in these elements and are known as neutron moderators – therefore greatly increases the likelihood of inducing a fission chain reaction and so eases the design requirements regarding U^{235} concentration. At the present time almost all of the world's power reactors employ such a moderator (and hence are termed thermal (fission) reactors because most of the fissions occurring in them are induced by neutrons which have slowed down to thermal equilibrium with the material of the reactor). The exceptions are a few prototype unmoderated systems (such as the French 'Phénix' reactor) which are termed fast (fission) reactors because most of their fissions are induced by neutrons which have not slowed down. To achieve criticality, fast reactors must have fuel which contains a very much higher proportion (20 per cent, say) of fissile atoms (U^{235} and/or plutonium) than does natural uranium (0.7 per cent). The fuel of reactors which are moderated by ordinary water also needs to contain additional fissile atoms (about 3 per cent) to compensate for the significant neutron absorptivity of the water. Producing such enriched fuel is itself a significant part of nuclear engineering activity.

A power-producing reactor also needs a coolant, to transfer the reactor's heat to the primary side of boilers (usually termed, in this context, steam generators) which, on their secondary side, convert feedwater into the steam which then drives the turbo-alternators. Coolants used may be gaseous (e.g. carbon dioxide, as in most UK reactors) or liquid (e.g. water, heavy water, liquid sodium, etc.), choice of coolant being determined not only by its thermal properties but also by such factors as its chemical compatibility, neutron absorptivity and, very importantly, the power density (Mw of heat per m^3) of the reactor core.

The wide range of options available to the designer when making his choice of the three basic reactor components: fuel (metal? oxide? U^{235} enrichment?), moderator (graphite? water? heavy water? none?), coolant (gas? water? liquid metal?) – is reflected in the surprisingly wide range of reactor types currently employed. The table below lists the main half dozen or so and some of their distinguishing characteristics. Worldwide, by far the most dominant type is undoubtedly the PWR, which originated in the USA as a power unit for naval submarines. (The Russian VVER is not listed because it is essentially a variant of the PWR).

The countries which have made the most significant contribution to the development of civil nuclear energy and/or have shown the greatest commitment to its exploitation are probably the USA, the UK, the former USSR, France, Canada, Germany and Japan. The organizations and information sources listed are therefore drawn mainly from these countries. Several other nations (e.g. Spain, Sweden, Bulgaria, Italy, South Korea) also have substantial nuclear power programmes; details of these may be found in various of the listed sources, e.g. in *Nuclear Power: Status and trends* or in *World nuclear industries handbook* (see 'Handbooks, etc.').

A further source of nuclear power is nuclear fusion, the joining together of the lightest nuclei – those of hydrogen, deuterium or tritium atoms – to form rather heavier ones. So far, however, only its military application, the triggering of an uncontrolled explosive reaction – as in the so-called 'Hydrogen Bomb' – has been achieved. The conditions for controlled nuclear fusion have yet to be attained and the engineering of a practical fusion power reactor remains an activity for the future.

Main reactor types

Reactor	Fuel	Moderator	Coolant	Coolant Pressure (MPa)	Coolant Max Temp(°C)	Power Density (MW/m^3)
PWR	UO_2	H_2O	H_2O	15.5	332	104
BWR	UO_2	H_2O	H_2O	7.2	286	56
Candu	UO_2	D_2O	D_2O	10.5	310	12
Magnox	U metal	graphite	CO_2	2.7	360	0.9
AGR	UO_2	graphite	CO_2	4.4	635	2.7
RBMK	UO_2	graphite	H_2O	7.8	284	4.2
LMFBR	$(U,Pu)O_2$	none	Na	0.25	545	285

Note on names and country of origin of reactor types

PWR = pressurized water reactor (USA)
BWR = boiling water reactor (USA)
Candu = Canadian deuterium-uranium reactor (Canada)
Magnox = Magnesium (non-oxidizing), the alloy used for the cans in which the uranium fuel rods are enclosed (UK)
AGR = advanced gas-cooled reactor (UK)
RBMK = (in Russian) high power pressure-tube reactor (USSR)
LMFBR = liquid-metal-cooled fast breeder reactor (UK, France)

Principal UK companies

AEA Technology, Harwell Laboratory, Didcot, Oxfordshire OX11 0RA, UK. Tel: +44 1235 821111. Fax: +44 1235 832591.

British Nuclear Fuels plc (BNFL), Risley, Warrington, Cheshire WA3 6AS, UK. Tel: +44 1925 832000. Fax: +44 1925 822711.

National Nuclear Corporation Ltd (NNC), Booths Hall, Chelford Road, Knutsford, Cheshire WA16 8QZ, UK. Tel: +44 1565 633800. Fax: +44 1565 633659.

Nuclear Electric plc, Barnett Way, Barnwood, Gloucestershire GL4 7RS, UK. Tel: +44 1452 652222. Fax: +44 1452 652776.

Scottish Nuclear Ltd, Redwood Crescent, Peel Park, East Kilbride, Glasgow G74 5PR, UK. Tel: +44 1355 262000. Fax: +44 1355 262626.

Representative international organizations

Forum Atomique Européene (FORATOM, European Nuclear Forum), 15 rue d'Egmont, B–1050 Brussels, Belgium. Tel: +32 2 502 4595. Fax: +32 2 502 3902.

International Atomic Energy Agency (IAEA), Vienna International Centre, Wagramerstrasse 5, PO Box 100, A–1400 Vienna, Austria. Tel: +43 1 2360. Fax: +43 1 234564.

International Commission on Radiological Protection (ICRP), PO Box 35, Didcot, Oxfordshire OX11 0RJ, UK. Tel: +44 1235 833929. Fax: +44 1235 832832.

International Nuclear Law Association (INLA), 29 Square de Meeus, B–1040 Brussels, Belgium. Tel: +32 2 513 68 45. Fax: +32 2 513 38 33.

OECD Nuclear Energy Agency (NEA), Le Seine-Saint Germain, 12 boulevard des Iles, F–92130 Issy les Moulineaux, France. Tel: +33 1 45 24 82 00. Fax: +33 1 45 24 11 10.

Organisation des Producteurs d'Energie Nucleaire (OPEN, Association of Nuclear Energy Producers), 20 rue de Lisbonne, F–75008 Paris, France. Tel: +33 1 42 93 26 47. Fax: +33 1 40 42 50 73.

World Association of Nuclear Operators (WANO), Co-ordinating Centre, King's Buildings, 16 Smith Square, London SW1P 3JG, UK. Tel: +44 171 828 2111. Fax: +44 171 828 6691.

Representative national organizations

Governmental

Atomic Energy Commission (AEC), 2–2–1 Kasumigaseki, Chiyoda-ku, Tokyo, Japan. Tel: +81 3 3581 2585. Fax: +81 3 3581 2487.

Atomic Energy of Canada Ltd (AECL), 334 Slater Street, Ottawa, Ontario K1A 0S4, Canada. Tel: +1 613 237 3270. Fax: +1 613 563 9499.

Commissariat a l'Energie Atomique (CEA), 31–33 rue de la Federation, 75752 Paris, Cedex 15, France. Tel: +33 1 40 56 10 00.

Ministry of Atomic Energy of the Russian Federation (MINATOM), B. Ordinka 26, 10100-Moscow, Russia. Tel: +70 095 239 4545. Fax: +70 095 230 2420.

Industrial

British Nuclear Forum (BNF), 22 Buckingham Gate, London SW1E 6LB, UK. Tel: +44 171 828 0116. Fax: +44 171 828 0110.

Canadian Nuclear Association (CNA), 144 Front Street West, Suite 725, Toronto, Ontario M5J 2L7, Canada. Tel: +1 416 977 6152. Fax: +1 416 979 8356.

Deutsches Atomforum eV (DAtF), Heussallee 10, D–53113 Bonn, Germany. Tel: +49 228 507 0. Fax: +49 228 507 219.

Nuclear engineering

Electrical Power Research Institute (EPRI), 3412 Hillview Avenue, PO Box 10412, Palo Alto, CA 94303, USA. Tel: +1 415 855 2000. Fax: +1 415 855 2954.

Forum Atomique Francaise (FAF), 48 rue de la Procession, F–75015 Paris, France. Tel: +33 1 44 49 60 00. Fax: +33 1 44 49 60 11.

Institute of Nuclear Power Operations (INPO), 700 Galleria Parkway, Atlanta, GA 30339–5957, USA. Tel: +1 404 644 8000. Fax: +1 404 644 8549.

Japan Atomic Industrial Forum (JAIF), 6th Floor, Toshin Building, 1-1-13 Shinbishi, Minato-ku, Tokyo 105, Japan. Tel: +81 3 3508 2411. Fax: +81 3 3508 2094.

Regulatory authorities

Canadian Atomic Energy Control Board (AECB), PO Box 1046 Station B, 280 Slater Street, Ottawa, Ontario K1P 5S9, Canada. Tel: +1 613 995 5894. Fax: +1 613 995 5086.

Direction de la Sûreté des Installations Nucleaires (DSIN), Ministere de l'Industrie, 99 rue de Grenelle, F–75753 Paris, France. Tel: +33 1 43 19 36 36. Fax: +33 1 43 19 48 69.

HM Nuclear Installations Inspectorate (Health and Safety Executive) (NII, HSE), Baynards House, 1 Chepstow Place, Westbourne Grove, London W2 4TF, UK. Tel: +44 171 243 6000. Fax: +44 171 727 4116.

(Japan) Nuclear Safety Commission (NSC), 2–2–1 Kasumigaseki, Chiyoda-ku, Tokyo, Japan. Tel: +81 3 3581 5271. Fax: +81 3 3581 2487.

National Radiological Protection Board (NRPB), Chilton, Didcot, Oxfordshire OX11 0RQ, UK. Tel: +44 1235 831600. Fax: +44 1235 833891.

Nuclear Regulatory Commission (NRC), Washington, DC 20555, USA. Tel: +1 301 492 1759. Fax: +1 301 492 0275.

The State Committee for the Supervision of Nuclear and Radiation Safety under the President of Russia (Gosatomnadzor), Taganskaya str. 34, 109147 Moscow, Russia. Tel: +7 095 272 47 10. Fax: +7 095 278 80 90.

Learned societies

American Nuclear Society (ANS), 555 North Kensington Avenue, La Grange Park, IL 60525, USA. Tel: +1 708 352 6611. Fax: +1 708 352 0499.

Atomic Energy Society of Japan (AESJ), 1-1-13 Shimbashi, Minato-ku, Tokyo 105, Japan. Tel: +81 3 3508 1261. Fax: +81 3 3581 6128.

British Nuclear Energy Society (BNES), 1–7 Great George Street, London SW1P 3AA, UK. Tel: +44 171 222 7722. Fax: +44 171 630 9177.

Canadian Nuclear Society (CNS), 144 Front Street West, Suite 725, Toronto, Ontario M5J 2L7, Canada. Tel: +1 416 977 7620. Fax: +1 416 979 8356.

European Nuçlear Society (ENS), PO Box 5032, CH–3001, Berne, Switzerland. Tel: +41 31 21 61 11. Fax: +41 31 22 92 03.

Ex-USSR Nuclear Society (EUNS), c/o IV Kurchatov Institute, Kurchatov Square, 123182 Moscow, Russia. Tel: +7 095 196 9900/7300. Fax: +7 095 196 2073.

Kerntechnische Gesellschaft e. V. (KTG), Heussallee 10, D–53113, Bonn 1, Germany. Tel: +49 228 50 72 59. Fax: +49 228 50 72 19.

Société Francaise d'Energie Nucleaire (SFEN), 48, rue de la Procession, F–75724 Paris, France. Tel: +33 1 44 49 60 00. Fax: +33 1 44 49 60 11.

The Institution of Nuclear Engineers (INucE), Allan House, 1 Penerley Road, London, SE6 2LQ, UK. Tel: +44 181 698 1500. Fax: +44 181 695 6409.

Authoritative and/or comprehensive textbooks.

Benedict, M., Pigford, T. H. and Levi, H. W. (1980) *Nuclear chemical engineering*, 2nd ed. (McGraw-Hill).
Bennet, D. J. and Thomson, J. R. (1989) *The elements of nuclear power*, 3rd ed. (Longman Scientific and Technical).
Duderstadt, J. J. and Hamilton, L. J. (1976) *Nuclear reactor analysis* (Wiley).
Farmer, F. R. (ed.) (1977) *Nuclear reactor safety*. (Academic Press).

Glasstone, S. and Sesonske, A. (1994) *Nuclear reactor engineering*, Vol. 1: Reactor design basics, Vol. 2: Reactor systems engineering 4th rev. ed. (Van Nostrand Rheinhold).
Lamarsh, J. R. (1983) *Introduction to nuclear engineering*, 2nd ed. (Addison-Wesley).
Marshall, W. (ed.) (1983) *Nuclear power technology*, Vol. 1: Reactor technology, Vol. 2: Fuel cycle, Vol. 3: Nuclear radiation (Oxford University Press).
Thompson, T. J. and Beckerley, J. G. (eds.) (1964, 1973) *The technology of nuclear reactor safety*, Vol. 1: Reactor physics and control, Vol. 2: Reactor materials and engineering (MIT Press).
Weinberg, A. M. and Wigner, E. P. (1958) *The physical theory of neutron chain reactors* (University of Chicago Press).

Public reports

Flowers, Sir Brian (1976) *Nuclear power and the environment: sixth report of the royal commission on environmental pollution*, Cmnd.6618 (HMSO).
Gittus, J. H. et al. (1988) *The Chernobyl accident and its consequences*, 2nd ed. UKAEA NOR 4200 (UKAEA, Available from HMSO).
Kemeny, J. G. (1979) *Report of the President's Commission on the accident at Three Mile Island. The need for change: the legacy of TMI*. (Pergamon).
Layfield, Sir Frank (1987) *Sizewell B public inquiry*, 8 vols. (HMSO).
US Nuclear Regulatory Commission (1975) *Reactor safety study: an assessment of accident risks in US commercial nuclear power plants*, main report including executive summary ('The Rasmussen report') WASH-1400 (NUREG 75/014) (US Department of Commerce, National Technical Information Service).

Learned journals

Advances in Nuclear Science and Technology (Plenum) irregular reviewing journal.
Nuclear Energy (Thomas Telford) bi-monthly, published by the British Nuclear Energy Society.
Nuclear Engineering and Design (Elsevier) fortnightly, affiliated with, and obtainable on subscription from, the European Nuclear Society.
Nuclear Science and Engineering (American Nuclear Society) monthly.
Nuclear Technology (European and American Nuclear Societies) monthly.
Progress in Nuclear Energy *(Pergamon)* quarterly.

Periodicals

Atom (AEA Technology) monthly.
Atoms in Japan (Japan Atomic Industrial Forum) monthly.
IAEA Bulletin (International Atomic Energy Agency) quarterly.
Nuclear Engineer (Institution of Nuclear Engineers) bi-monthly.

Nuclear Engineering International (Reed Business Publishing) monthly.
Nuclear News (American Nuclear Society) monthly.
Nuclear Safety (Office of Scientific and Technical Information) quarterly, published on behalf of the US Department of Energy and the US Nuclear Regulatory Commission.
Nucleonics Week (McGraw-Hill) weekly.

Abstracting and indexing journals

INIS (International Nuclear Information System) Atomindex (IAEA Vienna) semi-monthly. Available for computer search through STN International, DIALOG, ESA/IRS. Also available on CD-ROM (SilverPlatter).
Energy Research Abstracts (US Department of Energy, Office of Scientific and Technical Information) semi-monthly. Available for computer search on STN International.

Handbooks, databooks, directories

Sube, R. (comp.) (1985) *Dictionary of nuclear engineering* (English/French/German/Russian) (Elsevier).
Nuclear power: status and trends (IAEA) digest of data published annually.
World nuclear industry handbook (Nuclear Engineering International) listings of companies, representative and governmental organizations, nuclear power stations, performance statistics, etc. Published annually.
Meetings on Atomic Energy (IAEA) a quarterly worldwide list of conferences, exhibitions and training courses in atomic energy.

CHAPTER FIFTEEN

Chemical engineering

M. J. PITT

Introduction

The special feature of chemical engineering is that its practitioners design processes rather than equipment, so the term 'process engineering' is sometimes used. Likewise, chemical engineering skills are applied in many areas apart from chemicals, such as food, biotechnology, materials, water treatment. These are the so-called process industries.

In general a process involves the transfer and conversion of matter and energy. Chemical reactions are often not involved at all. The task of a chemical engineer is principally to analyse and optimize the mass and energy flows, then consider different portions, the so-called 'unit operations' and specify (rather than design) the equipment required. Actual equipment may be bought off the shelf or designed by specialists.

The chemical engineer therefore draws upon fundamental engineering which is common to other disciplines. Unfortunately books on these topics specifically intended for chemical engineers are quite likely to end up elsewhere in bookshops and libraries. Thus information on heat transfer may be shelved by the library in the section for mechanical engineering, fluid flow texts may be split between civil and mechanical engineering, whereas thermodynamic data may be in chemistry or physics!

However, chemical engineers commonly have to deal with other engineers in order to get their processes built, so they may be expected to have some familiarity with the other disciplines, and at least know where to look things up.

In practice, much useful information relating to processes tends to be defined by the end use. So far as the chemical engineer is concerned, pumping chocolate or pumping concrete is much the same but relevant data and know-how will be found in quite different sources. Similarly, a heat exchanger to boil custard and another to boil nitric acid may have similar throughputs and thermal duties, but will require information from separate sources about materials compatibility and practical details of operation before they can be specified.

The technology is progressing and it is important to keep up to date, so this chapter concentrates on recent publications, except where earlier items are still in print and have not been superseded. For a substantial list of earlier publications, consult Ray (1990).

Institutions

There are two main institutions which have members worldwide and produce the majority of professional texts. They have a reciprocal agreement to supply each other's publications.

In the UK is the Institution of Chemical Engineers (known as the IChemE), 165–171 Railway Terrace, Rugby CV21 3HQ, UK. Tel: +44 1788-578214. Fax: +44 1788-547262. It publishes:

Symposium Series: more than 130 collections of conference proceedings; Guides: short books on professional topics; many related to safety, but also practical operation of equipment, costing and management techniques; Safety training modules: text, slides and videos for self or group instruction; and a small range of other publications such as monographs on technical topics.

In the USA is the American Institute of Chemical Engineers (known as the AIChE), 345 East 47th St, New York, NY 10017, USA, which publishes Symposium Series: more than 100 collections of conference proceedings; Guidelines: substantial books mainly on safety topics; and a small range of other publications.

Other relevant institutional publishers are:

The American Petroleum Institute, 1271 Avenue of the Americas, NY 10020, USA (known as the API).

The Institute of Petroleum, 61 New Cavendish Street, London W1M 8AR, UK (known as the IP).

The Instrument Society of America, 67 Alexander Drive, P.O. Box 12277, Research Triangle Park, NC 27709, USA (known as the ISA).

They produce standards and codes of practice which are used in chemical engineering both in and outside the petroleum industry.

Basic references

The best-known general book is Perry and Green (1984) which is a compendium of data and short descriptions of equipment and chemical engineering practice. A useful recent book is Lydersen and Dahlo (1992) – which is more an explanation of terms than a dictionary, with over 1000 entries, with graphs and diagrams as appropriate, and also the equivalent terms in French, German and Spanish. A similar effort in English (or German) only is Noether and Noether (1993). There is a number of bilingual dictionaries of chemistry and industrial chemistry, sometimes specifically including chemical engineering. The widest range (English, French, Spanish, Italian, Dutch, German) is covered by Clason (1968).

The most elementary introductions are given by: Field (1988), Heaton (1991) and Felder and Rousseau (1986). A major source of data and process descriptions is *Kirk-Othmer encyclopaedia of chemical technology* (1991–) 4th ed. (Wiley). The 3rd edition is in 24 volumes, and is also available in computer-readable form as an online database or CD-ROM. The 4th edition will be in 27 volumes when completed in 1998. A shorter version is available, namely *Kirk-Othmer concise encyclopaedia of chemical technology* (1985) 3rd ed., 2 vols (Wiley).

The *Encyclopedia of chemical processing and design* (1976–) published by Dekker has reached Volume 46.

Another source is *Ullmann's encyclopedia of industrial chemistry*, 5th ed. 36 vols, a German production but available in English from VCH. By 1995 there will be 26 'A' volumes, and 8 'B' volumes, of which B1 to B4 are on basic chemical engineering thus:

B1: Fundamentals of chemical engineering (1990)
B2: Unit operations I (1988)
B3: Unit operations II (1988)
B4: Principles of chemical reaction engineering and plant design (1992).

The most widely used textbook set is: Coulson, J. M. and Richardson, J. F. (eds) *Chemical Engineering* (Pergamon), of which the individual volumes are:

Vol. 1: Backhurst, J. R. and Harker, J. H. (1990) Fluid flow, heat transfer and mass transfer, 4th ed.
Vol. 2: Backhurst, J. R. and Harker, J. H. (1991) Unit operations, 4th ed.
Vol. 3: Richardson, J. F. and Peacock, D. G. (1994) Chemical and biochemical reactors design and process control, 3rd ed.

Vols 4 and 5 combined: Backhurst, J. R. and Harker, J. H. (1993) *Solutions to the problems in Volume 1*, 2nd ed.
Vol. 6: Sinnott, R. K. (1993) *An introduction to chemical engineering design*, 2nd rev. edn.

Some books that cover a wide part of the field are:

Chopey, N. P. and Hicks, T. G. (eds) (1984) *Handbook of chemical engineering calculations* (McGraw-Hill).
Douglas, J. M. (1988) *Conceptual design of chemical processes* (McGraw-Hill).
Himmelblau, D. M. (1989) *Basic principles and calculations in chemical engineering*, 5th ed. (Prentice Hall).
Robinson, R. N. (1988) *Chemical engineering reference manual*, 4th ed. (Professional Publications), which has a companion volume *Robinson, R. N. (1990) Solutions manual for the chemical engineering reference manual* (Professional Publications).
Shaheen, E. I. (1983) *Basic practice of chemical engineering*, 2nd ed. (International Institute of Technology).
A serial publication by Academic Press is *Advances in chemical engineering*, of which Volume 20 (ed. M. Kwank) was published in 1994.

Data compilations

The most widely used sources of physical and chemical data must be Perry's *Handbook* referred to above, and the Chemical Rubber Co's *Handbook of chemistry and physics* (annual).

Some insight into the sources and method of compilation is given in a symposium proceedings edited by Jankowski and Selover (1986). There has, however, been considerable development since that time.

The largest single source [14 volumes, but a total of over 40 bound books] is the DECHEMA *Chemistry data series* (1978–), available from DECHEMA e.V. or Scholium International, New York. The text is in English.

Physical properties of pure substances for chemical engineering use are given in 9 volumes (but 23 books) of *Physical data, chemical engineering* by ESDU International plc.
Some compilations are:

Barin, I. (1992) *Thermochemical data of pure substances*, 2nd ed. (VCH).
Christensen, J. J., Rowley, R. L. and Izatt, R. M. (1988) *Handbook of heats of mixing* (Wiley).
Daubert, T. E. and Danner, R. P. (1989) *Physical and thermodynamic properties of pure chemicals* (Hemisphere).
Gess, M. A., Danner, R. P. and Nagvekar, M. (1991) *Thermodynamic analysis of vapor-liquid equilibria: recommended models and standard database* (American Institute of Chemical Engineers).

Pedley, J. B., Naylor, R. D. and Kirby, S. P. (1986) *Thermochemical data of organic compounds*, 2nd ed. (Chapman and Hall).
Reid, R. C., Prausnitz, J. M. and Poling, B. E. (1987) *The properties of gases and liquids*, 4th ed. (McGraw-Hill).
Stull, D. R., Westrum, E. F. and Sinke, G. C. (1969) *The chemical thermodynamics of organic compounds* (Wiley).

Specialist data for the oil and gas industries is given in:

American Petroleum Institute technical data book – petroleum refining (1992) 5th ed. (API).
Gallant, R. W. and Yaws, C. L. (1992–1993) *Physical properties of hydrocarbons*. Vol. 1, 2nd ed. (1992); Vol. 2, 3rd ed. (1993), Vol. 3 (1993) (Gulf Publishing).
Starling, K. E. (1973) *Fluid thermodynamic properties for light petroleum systems* (Gulf Publishing).
Yaws, C. L. (1992) *Thermodynamic and physical property data* (Gulf Publishing).

It is increasingly common for chemical engineers to access data by computer. Physical property data is often provided as part of the package or as an add-on for design programs such as simulators which require it. Separate data compilations can be obtained from the Institution of Chemical Engineers (Rugby, UK), DECHEMA e.V. (Frankfurt, Germany) and ESDU International (PO Box 1633 Manassas VA 22110, USA; or 27 Corsham Street London N1 6UA, UK).

Unit operations and general design

Chemical process plant is designed by reducing it to a set of subprocesses which can be individually dealt with. A mass and energy balance is carried out on each of these unit operations and is balanced for the process overall. An important group of unit operations involves material passing from one physical phase to another, which is known as mass transfer. Unit operations involving the transfer of energy and/or material are sometimes called transport processes.

Process analysis is concerned with understanding the overall requirements and the interactions of different parts. Process synthesis means putting portions together to make an effective overall design. The chemical engineering design of a unit operation will specify the physical conditions (temperature, pressure, etc.), the flows and general necessary dimensions, plus the number of stages or the amount of treatment (e.g. number of distillation stages, or length of catalyst bed) and materials requirement. It is then necessary to select standard items such as pumps or have a mechanical design procedure for equipment fabricated.

Chemical engineering

Bloch, H. P. (ed.) (1989) *Process plant machinery* (Butterworth-Heinemann).
Douglas, J. M. (1988) *Conceptual design of chemical processes* (McGraw-Hill).
Edgar, T. F. and Himmelblau, D. M. (1988) *Optimization of chemical processes* (McGraw-Hill).
Geankoplis, C. J. (1983) *Transport processes and unit operations*, 2nd ed. (Allyn and Bacon).
Giddings, J. C. (1991) *Unified separation science* (Wiley).
Govind, R. and Mocsny, D. (1992) *Modern process synthesis* (Prentice Hall).
Harnby, N., Edwards, M. F. and Nienow, A. W. (eds.) (1992) *Mixing in the process industries*, 2nd ed. (Butterworth-Heinemann).
Hewitt, G. F. (1982) *Liquid-gas systems: handbook of multiphase systems* (Hemisphere).
King, C. J. (1979) *Separation processes*, 2nd ed. (McGraw-Hill).
Kister, H. Z. (1989) *Distillation operation* (McGraw-Hill).
Kister, H. Z. (1992) *Distillation design* (McGraw-Hill).
Li, N. N. and Strathmann, H. (eds) (1988) *Separation technology* (American Institute of Chemical Engineers).
Liu, Y. A., McGee, A. and Epperly, W. R. (eds) (1987) *Recent developments in chemical process and plant design* (Wiley).
Luyben, W. L. and Wenzel, L. A. (1988) *Chemical process analysis: mass and energy balances* (Prentice Hall).
Masters, K. (1991) *Spray drying handbook*, 5th ed. (Longman).
McCabe, W. L., Smith, J. C. and Harriott, P. (1993) *Unit operations of chemical engineering*, 5th ed. (McGraw-Hill).
Mullin, J. W. (1992) *Crystallization*, 3rd ed. (Butterworth-Heinemann).
Naumann, E. B. (1990) *Introductory systems analysis for process engineers* (Butterworth-Heinemann).
Peters, M. S. and Timmerhaus, K. D. (1991) *Plant design and economics for chemical engineers*, 4th ed. (McGraw-Hill).
Prentice, G. (1991) *Electrochemical engineering principles* (Prentice Hall).
Rautenbach and Albrecht, R. (1989) *Membrane processes* (Wiley).
Rousseau, R. W. (1987) *Handbook of separation process technology* (Wiley).
Schweitzer, P. (ed.) (1988) *Handbook of separation techniques for chemical engineers*, 2nd ed. (McGraw-Hill).
Smith, R. (ed.) (1988) *Understanding process integration II* (Institution of Chemical Engineers).
Sohnel, O. and Garside, J. (1992) *Precipitation: basic principles and industrial Applications* (Butterworth-Heinemann).
Speight, J. G. (1993) *Gas processing: environmental aspects and methods* (Butterworth-Heinemann).
Strigle, R. F., Jr (1987) *Random packings and packed towers* (Gulf Publishing).
Svarovsky, L. (1990) *Solid-liquid separation*, 3rd ed. (Butterworth-Heinemann).
Treybal, R. E. (1981) *Mass transfer operations*, 3rd ed. (McGraw-Hill).
Walas, S. M. (1988) *Chemical process equipment: selection and design* (Butterworth-Heinemann).
Wesselingh, J. A. and Krishna, R. (1990) *Mass transfer* (Ellis Horwood).

Chemical reactors

The design of industrial equipment for carrying out chemical reactions involves many physical processes as well. The chemical engineer needs to balance chemical kinetics with diffusion, agitation, heat transfer and other factors not covered in purely chemical books. (Biochemical reactors are given later.)

Aris, R. (1990) *Elementary chemical reactor analysis* (Butterworth-Heinemann)
Cheremisinoff, N. P. (ed.) (1986) *Handbook of heat and mass transfer*, Vol. 2: Mass transfer and reactor design (Gulf Publishing).
Cheremisinoff, N. P. (ed.) (1989) *Handbook of heat and mass transfer*, Vol. 3: Catalysis, kinetics and reactor engineering (Gulf Publishing).
Cheremisinoff, N. P. (ed.) (1990) *Handbook of heat and mass transfer*, Vol. 4: Advances in reactor design and combustion Science (Gulf Publishing).
Deckwer, W. D. (1991) *Bubble column reactors* (Wiley).
Fogler, H. S. (1991) *Elements of chemical reaction engineering*, 2nd ed. (Prentice Hall).
Froment, G. F. and Bischoff, K. (1990) *Chemical reactor analysis and design*, 2nd ed. (Wiley).
Holland, C. D. and Anthony, R. G. (1989) *Fundamentals of chemical reaction engineering*, 2nd ed. (Prentice Hall).
Kastanek, F. et al (1992) *Chemical reactors for gas-liquid systems* (Ellis Horwood).
Rase, H. F. (1990) *Fixed-bed reactor design and diagnostics: gas-phase reactions* (Butterworth-Heinemann).
Scott Fogler, H. (1991) *Elements of chemical reaction engineering*, 2nd ed. (Prentice Hall).
Westerterp, K. R., Van Swaaij, W. P. M. and Beenackers, A. A. C. M. (1987) *Chemical reactor design and operation*, 2nd ed. (Wiley).

Physical chemistry

Many standard chemical textbooks, and texts on thermodynamics for mechanical and general engineering may be used, but the following are especially relevant.

Boudart, M. (1991) *Kinetics of chemical processes* (Prentice Hall).
Cardew, M. H. (1990) *Thermodynamics for chemists and chemical engineers* (Prentice Hall).
Kyle, B. G. (1992) *Chemical and process thermodynamics*, 2nd ed. (Prentice Hall).
Prausnitz, J. M. and Lichtenthaler, R. N. (1985) *Molecular thermodynamics of fluid-phase equilibria*, 2nd ed. (Prentice Hall).
Reid, C. E. (1990) *Chemical thermodynamics* (McGraw-Hill).

Smith, J. M. (1980) *Chemical engineering kinetics*, 3rd ed. (McGraw-Hill).
Smith, J. M. and Van Ness, H. C. (1987) *Introduction to chemical engineering thermodynamics* 4th ed. (McGraw-Hill).
Walas, S. M. (1990) *Reaction kinetics for chemical engineers* (Butterworth-Heinemann).

Fluid mechanics

The movement of fluids is fundamental to many processes. The basic theory is covered in books intended for general engineering. Some that are specially written for, or well suited to, chemical engineering or cover special applications are as follows.

Cheremisinoff, N. P. (ed.) (1986–1990) *Encyclopedia of fluid mechanics*, 10 vol. (Gulf Publishing).
Churchill, S. W. (1989) *Viscous flows: the practical use of theory* (Butterworth-Heinemann).
de Nevers, N. (1991) *Fluid mechanics for chemical engineers*, 2nd ed. (McGraw-Hill).
Denn, M. M. (1980) *Process fluid mechanics* (Prentice Hall).
Fan, L.-S. (1989) *Gas-liquid-solid fluidization engineering* (Butterworth-Heinemann).
Holland, F. A. (1973) *Fluid flow for chemical engineers* (Arnold).
Kim, S. and Karrila, S. J. (1991) *Microhydrodynamics: principles and selected applications* (Butterworth-Heinemann).
Kunii, D. and Levenspiel, O. (1991) *Fluidization engineering*, 2nd ed. (Butterworth-Heinemann).
Probstein, R. F. (1989) *Physicochemical hydrodynamics: an introduction* (Butterworth-Heinemann).
Stanek, V. (1992) *Fixed bed operations: flow distribution and efficiency* (Prentice Hall).
Upp Loy, E. and Daniel Industries Staff (1993) *Fluid flow measurement* (Gulf Publishing).

In addition, ESDU International has a set of regularly updated data books (11 volumes, 16 books) *Fluid mechanics, internal flow*, which is specifically for chemical engineering design usage.

Heat exchangers, heat pumps and refrigeration

Devices for heating and cooling are of major concern to chemical engineers. Many are used to conserve energy by transferring energy gained in one operation to material in the same or a different operation which requires heat input. These are called heat exchangers, but the

Chemical engineering 277

same general term and technology is used for simple heaters, coolers, boilers and condensers.

Heat pumps are machines for redistributing energy. This generally means to produce a higher temperature than the energy source, although in fact refrigerators are actually heat pumps producing lower temperatures.

The basic physics of heat transfer and heat pumping is common to other fields, notably mechanical engineering, physics and fuel and energy studies (see Chapters 12 and 13). However, texts particularly relevant to chemical engineering are as follows.

American Institute of Chemical Engineers (1993) *Heat transfer*, Vol. 89 (annual symposium, previous volumes available) (AIChE).
Bott, T. R. (1990) *Fouling notebook: a practical guide to minimising fouling in heat exchangers* (Institution of Chemical Engineers).
Brodowicz, K. and Dyakowski, T. (1993) *Heat pumps* (Butterworth-Heinemann).
Cheremisinoff, N. P. (ed.) (1986) *Handbook of heat and mass transfer*; Vol. 1: Heat transfer operations (Gulf Publishing).
Cheremisinoff, P. N. and Cheremisinoff, N. P. (1993) *Heat transfer equipment* (Prentice Hall).
Eastop, T. D. and Croft, D. R. (1990) *Energy efficiency for engineers and technologists* (Longman).
Foumeny, E. A. and Heggs, P. J. (eds) (1991) *Heat exchange engineering*, 2 vols: Vol. 1 – Design of heat exchangers; Vol. 2 – Compact heat exchangers: techniques of size reduction (Ellis Horwood).
Fraas, A. P. (1989) *Heat exchanger design*, 2nd ed. (Wiley).
Gunn, D. and Horton, R. (1989) *Industrial boilers* (Longman).
Heat transfer: 3rd UK National Conference incorporating 1st European Conference on Thermal Sciences, University of Birmingham, 16–18 September 1992 (Institution of Chemical Engineers).
Incropera, F. P. and Dewitt, D. P. (1990) *Fundamentals of heat and mass transfer*, 3rd Edn (Wiley).
Incropera, F. P. and Dewitt, D. P. (1990) *Introduction to heat transfer*, 2nd ed. (Wiley).
Institution of Chemical Engineers (1982) *User guide on process integration for the efficient use of energy* (IChemE).
Kakac, S. (ed.) (1991) *Boilers, evaporators and condensers* (Wiley).
Levenspiel, O. (1984) *Engineering flow and heat exchange* (Plenum).
Ozisik, M. N. (1993) *Heat conduction*, 2nd ed. (Wiley).
Ozisik, M. N. (1985) *Heat transfer: a basic approach* (McGraw-Hill).
Saunders, E. A. D. (1988) *Heat exchangers: selection design and construction* (Longman).
Welty, J. R., Wicks, C. E. and Wilson, R. E. (1984) *Fundamentals of momentum, heat and mass transfer*, 3rd ed. (Wiley).

A set of data books (10 volumes, 12 books) for heat transfer and design of heat exchangers is available from ESDU International.

Particle technology

There are comparatively few publications devoted to the handling of solid materials in a chemical engineering context. The following are substantially relevant.

British Materials Handling Board (1987) *Draft code of practice for the design of silos, bins, bunkers and hoppers* (BMHB).
Davidson, J. F., Clift, R. and Harrison, D. (eds) (1985) *Fluidization*, 2nd ed. (Academic).
Heiskanen, K. I. (1993) *Particle classification* (Chapman and Hall).
Kwauk, M. (1992) *Fluidization: idealised and bubbleless with applications* (Ellis Horwood).
Marcus, R. D. *et al.* (1990) *Pneumatic conveying of solids* (Chapman and Hall).
Mills, D. (1990) *Pneumatic conveying design guide* (Butterworth Scientific).
Molerus, O. (1993) *Principles of flow in disperse systems* (Chapman and Hall).
Rhodes, M. J. (ed.) (1990) *Principles of powder technology* (Wiley).
Rumpf, H. and Bull, F. A. (translator) (1990) *Particle technology* (Chapman and Hall).
Shamlou, P. A. (1988) *Handling of bulk solids: theory and practice* (Butterworths).
Stanek, V. (1993) *Fixed bed operations : flow distribution and efficiency* (Prentice Hall).
Woodcock, C. R. and Mason, J. S. (eds.) (1987) *Bulk solids handling: an introduction to the practice and technology* (Hill).

Biochemical engineering

Engineering involving living cells (or materials such as enzymes derived from them) has special requirements, both in the biochemical processes and in subsequent operations such as separation and purification.

Bailey, J. E. and Ollis, D. F. (1986) *Biochemical engineering fundamentals*, 2nd ed. (McGraw-Hill).
Belter, P. A., Cussler, E. L. and Hu, W.-S. (1988) *Bioseparations: downstream processing for biotechnology* (Wiley).
Bungay, H. R. and Belfort, G. (eds.) (1987) *Advanced biochemical engineering* (Wiley).
Dunn, I. J. (1992) *Biological reaction engineering : principles, applications and modelling with PC simulation* (VCH).
Jackson, A. T. (1990) *Process engineering in biotechnology* (Open University).
Kennedy, J. F. and Cabral, J. M. S. (eds) (1993) *Recovery processes for biological materials* (Wiley).
Lee, J. M. (1992) *Biochemical engineering* (Prentice Hall).
McMillan, G. K. (1987) *Biochemical measurement and control* (Instrument Society of America).

Mitchell, W. J. and Slaughter, J. C. (1989) *Biology and biochemistry for chemists and chemical engineers* (Ellis Horwood).
Schugerl, K. (1987) *Bioreaction engineering*, Vol. 1: Fundamentals, thermodynamics, formal kinetics, idealised reactor types and operation modes (Wiley).
Schugerl, K. (1991) *Bioreaction engineering*, Vol. 2: Characteristic features of bioreactors (Wiley).
Schuler, M. L. and Kargi, F. (1992) *Bioprocess engineering: basic concepts* (Prentice Hall).
Scragg, A. H. (ed.) (1991) *Bioreactors in biotechnology: a practical approach* (Ellis Horwood).
Stephanopoulos, G. (ed.) (1993) *Biotechnology*, Vol. 3: Bioprocessing, 2nd ed. (VCH).

Control

Process control can mean different things in different disciplines. Note that the Institute of Measurement and Control (London) organizes occasional symposia and the Instrument Society of America (ISA) publishes many guides. The following are relevant to chemical engineering.

Andrew, W. G., Williams, H. B. and Zoss, L. M. (1979–1993) *Applied instrumentation in the process industries*, Vol. 1, 2nd ed. (1979) A survey; Vol. 2 (1980) Practical guidelines; Vol. 3 (1993) Engineering data and resource material; Vol. 4 (1982) Control, systems, theory, trouble-shooting and design (Gulf Publishing).
Bishop, D. N. (1992) *Electrical systems for oil and gas production facilities*, 2nd ed. (Instrument Society of America).
Borer, J. (1985) *Instrumentation and control for the process industries* (Elsevier Applied Science).
Buckley, P. S., Luyben, W. L. and Shunta, J. P. (1985) *Design of distillation column control systems* (Instrument Society of America).
Center for Chemical Process Safety (1992) *The safe automation of chemical processes* (American Institute of Chemical Engineers).
Clevett, K. J. (1986) *Process analyzer technology* (Wiley).
Coughanowr, D. R. (1991) *Process systems analysis and control*, 2nd ed. (McGraw-Hill).
Deshpande, P. B. (1985) *Distillation dynamics and control* (Arnold).
Driskell, L. R. (1982) *Selection of control valves and other final control Devices* (Instrument Society of America).
Dukelow, S. G. (1991) *The control of boilers*, 2nd ed. (Instrument Society of America).
Fisher, T. G. (1990) *Batch control systems: design, application, and implementation* (Instrument Society of America).
Instrument Society of America (1991) *Standards and recommended practices for instrumentation and control*, 11th ed. (Instrument Society of America).

Luyben, W. L. (1990) *Process modeling, simulation and control for chemical engineers*, 2nd ed. (McGraw-Hill).
McGreavy, C. *et al.* (eds) (1990) *Computer integrated process engineering* (Institution of Chemical Engineers).
McMillan, G. K. (1985) *pH control* (Instrument Society of America).
Morari, M. and Zafiriou, E. (1989) *Robust process control* (Prentice Hall).
Murrill, P. W. (1991) *Fundamentals of process control theory*, 2nd ed. (Instrument Society of America).
Najim, K. (ed.) (1989) *Process modeling and control in chemical engineering* (Marcel Dekker).
Najim, K. (1988) *Control of liquid-liquid extraction columns* (Gordon and Breach).
Newell, R. B. and Lee, P. L. (1989) *Applied process control: a case study* (Prentice Hall).
Nichols, G. D. (1988) *On-line process analyzers* (Wiley).
Nisenfeld, A. E. (1985) *Industrial evaporators: principles of operation and control* (Instrument Society of America).
Pitt, M. J. and Preece, P. E. (eds) (1990) *Instrumentation and automation in process control* (Ellis Horwood).
Prett, D. M. and Garcia, C. E. (1988) *Fundamental process control* (Butterworth-Heinemann).
Prett, D. M. and Morari, M. (1988) *Shell process control workshop* (Butterworth-Heinemann).
Prett, D. M., Garcia, C. E. and Hamaker, B. L. (1989) *The second Shell process control workshop* (Butterworth-Heinemann).
Quantrille, T. E. and Liu, Y. A. (1992) *Artificial intelligence in chemical engineering* (Academic Press).
Roffel, B., Vermeer, P. J. and Chin, P. A. (1989) *Simulation and implementation of self-tuning controllers* (Prentice Hall).
Sawyer, P. (1993) *Computer controlled batch processing* (Institution of Chemical Engineers).
Seborg, D. E., Edgar, T. F. and Mellichamp, D. A. (1989) *Process dynamics and control* (Wiley).
Smith, C. A. and Corripio, A. B. (1985) *Principles and practice of automatic process control* (Wiley).
Stephanopoulos, G. (1984) *Chemical process control: an introduction to theory and practice* (Prentice Hall).
Tompkins, W. G. (1992) *Conceptual design analysis applied to offshore control systems* (Instrument Society of America).

The following journals are relevant:.

Control and Instrumentation (Morgan-Grampian) monthly; news and commercial articles
Flow Measurement and Instrumentation (Butterworth-Heinemann) quarterly; research papers and technical articles.
Journal of Process Control (Butterworth-Heinemann) quarterly; research papers and technical articles.

Safety

There are probably more publications on safety related to chemical engineering than any other branch of engineering. These are therefore covered in the chapter on safety engineering.

Software

More than 2000 programs and data sets are available commercially. Most are listed and briefly described in an annual supplement to the journal *Chemical Engineering Progress*, and may be obtained separately from the AIChE. It is increasingly common for software versions of books to be produced, and certain books have computer disks with them.

The Internet computer network has a newsgroup sci.engr.chem.

Journals

UK: published by the Institution of Chemical Engineers

The Chemical Engineer: topical news and technical articles; fortnightly.
Transactions: learned journal, refereed research papers
  Part A: Chemical Engineering Research and Design; bimonthly.
  Part B: Process Safety and Environmental Protection; quarterly.
  Part C: Food and Bioproducts Processing; quarterly.
Loss Prevention Bulletin: accident reports and articles on safety; bimonthly.
Environmental Protection Bulletin; bimonthy.
European Bulletin; quarterly.
What's On: lists of meetings and conferences relevant to the profession; quarterly.

UK: for the Society of Chemical Industry (Wiley)

Chemistry & Industry news, reviews and short papers; fortnightly.
Journal of Chemical Technology & Biotechnology technical papers; monthly.

USA: published by the American Institute of Chemical Engineers

AIChE Journal technical papers; monthly.
Biotechnology Progress technical papers; bimonthly.
Chemical Engineering Progress news and general articles; monthly.
International Chemical Engineering translations; quarterly.
Process Safety Progress (formerly Plant/Operations Progress) technical papers; quarterly.

Other journals

Publications of research findings are given in:

Canadian Journal of Chemical Engineering (Canadian Society of Chemical Engineering) monthly.
Chemical & Engineering Technology (VCH) bimonthly.
Chemical Engineering Science (Pergamon) fortnightly.
Combustion and Flame The Journal of the Combustion Institute (Elsevier) monthly.
Computers and Chemical Engineering (Pergamon) monthly.
Filtration and Separation (Elsevier for Filtration Society) 8 per annum.
Fuel (Butterworth-Heinemann) monthly.
Fuel Processing Technology (Elsevier) quarterly.
Gas Separation & Purification (Butterworth-Heinemann) quarterly.
Industrial & Engineering Chemistry Research (American Chemical Society) monthly.
International Journal of Heat and Mass Transfer (Pergamon) monthly.
International Journal of Multiphase Flow (Pergamon) bi-monthly.
Journal of Chemical Engineering Data (American Chemical Society) quarterly.
Journal of Chemical Engineering of Japan (Society of Chemical Engineering Japan) bi-monthly.
Journal of Process Control (Butterworth-Heinemann) quarterly.
Journal of the Chemical Society, Faraday Transactions (Royal Society of Chemistry) fortnightly.
Powder Technology (Elsevier) 15 per annum.
Process Biochemistry (Elsevier) 8 per annum.
Separations Technology (Butterworth-Heinemann) quarterly.

News items and technical review articles are given in:

Chemical & Engineering News (American Chemical Society) weekly.
Chemical Engineering (McGraw-Hill) monthly.
Chemical Engineering Education (American Society for Engineering Education) quarterly.
Chemical Plants + Processing (Konradin Verlag Robert Kohlhammer GmbH) 3 per annum.
Chemie-Ingenieur-Technik (VCH Germany) (German with English abstracts) monthly.
Hydrocarbon Processing (Gulf Publishing) monthly.
International Journal of Heat and Fluid Flow (Butterworth-Heinemann) quarterly.
Institute of Petroleum Quarterly Journal of Technical Papers (IP) quarterly.
Journal of Chemical Thermodynamics (Academic) monthly.
Journal of Loss Prevention in the Process Industries (Butterworth-Heinemann) 5 per annum.
Oil & Gas Journal (Pennwell Publishing) weekly.
Processing (IML) monthly.

Chemical engineering 283

Abstracts

An important literature database is: CHEMICAL ENGINEERING AND BIO-TECHNOLOGY ABSTRACTS CEBA, a joint collaboration of the IChemE (UK), RSC (UK), DECHEMA (Germany) and FIZ CHEMIE (Germany). These are available in printed form as sections as follows:

Biotechnology Apparatus, Plant and Equipment
(formerly Biotechnologie-Verfahren, Anlagen, Apparate) monthly.
Current Biotechnology
(formerly Current Biotechnology Abstracts) monthly.
Process and Chemical Engineering
(formerly Chemical Engineering Abstracts) monthly.
Theoretical Chemical Engineering
(formerly Theoretical Chemical Engineering Abstracts) monthly.

In addition, the following are relevant:

Chemical Abstracts (American Chemical Society) weekly.
Chemical Hazards in Industry (Royal Society of Chemistry) monthly.
Fuel and Energy Abstracts (Butterworth-Heinemann for the Institute of Energy) bi-monthly.

There are some 15 literature and patent indexes published by American Petroleum Institute, 275 Seventh Avenue, New York NY 10001–6708.

References

Clason, W. E. (1968) *Elsevier's dictionary of chemical engineering.* Amsterdam: Elsevier.
Felder, R. M. and Rousseau, R. W. (1986) *Elementary principles of chemical processes* 2nd ed. New York: Wiley.
Field, R. W. (1988) *Chemical engineering: introductory aspects.* Basingstoke: Macmillan Education.
Heaton, C. A. (1991) *An introduction to industrial chemistry* 2nd ed. Glasgow: Blackie Academic.
Jankowski, D. A. and Selover, T. B. Jr (1986) *Chemical engineering data sources.* New York: AIChE.
Lydersen, A. L. and Dahlo, I. (1992) *Dictionary of chemical engineering.* Chichester: Wiley.
Noether, D. and Noether, H.(1993) *Encyclopedic dictionary of chemical technology.* New York: VCH.
Perry, R. H. and Green, D. W. (eds) (1984) *Perry's chemical engineers' handbook* 6th ed. New York: McGraw-Hill.
Ray, M. S. (1990) *Chemical engineering bibliography (1967–1988).* Park Ridge, NJ: Noyes Publications.

CHAPTER SIXTEEN

Fluid mechanics and fluid power systems

J. WATTON

Introduction

Fluid power covers a wide spectrum of applications and embraces both pneumatic and hydraulic techniques, although the term 'fluid power' is often perceived as oil hydraulics. Indeed, it is this area that produces the majority of research and development resulting in published papers, books, and conferences. The main reason for this is that pneumatics tends to be applied to low-power sequencing applications common to automated manufacturing and assembly. Hence, although pneumatics has a vast market to supply and the technology is continually being developed, particularly with respect to microprocessor control, such as programmable logic controller interfacing, there has not been a requirement for detailed performance studies. There are, however, signs that this is changing since pneumatic closed-loop position control systems are now emerging with the potential to compete with the more usual electrohydraulic type. Given the superior stiffness of oil hydraulic systems compared with pneumatic systems, this pneumatic development is to be welcomed bearing in mind the lower cost of equivalent-power pneumatic components and the simplicity of discharging 'used' air to atmosphere. General design information on pneumatics may be found in the following publications:

Barber, A. (1989) *Pneumatic handbook*, 7th ed. (Trade and Technical Press).
Valves, piping and pipelines handbook (1986) 2nd ed. (Trade and Technical Press).
Fluid power engineers data book (no date) (British Fluid Power Association).

Electrohydraulic fluid power systems are used in a wide variety of applications ranging from precision control systems, such as robotics and aerospace, to heavy industry systems, such as forging presses and steel rolling mills, and mobile systems, such as agricultural machines and civil engineering plant. The relatively high power-to-weight ratio of electrohydraulics together with the advantages of electronic signal processing produces a flexible and efficient means of power transfer and control. This marriage of hydraulics and electronics is now being complemented by the developments in low-cost microcomputer control. Fluid power control has also entered the entertainment business with real-time computer control and monitoring being used in sophisticated fairground rides. It therefore perhaps comes as no surprise to discover that life-size sharks and prehistoric animals created for the latest cinema films are activated by electrohydraulic robotic-type systems. This not only produces spectacular realistic movement, but spectacular profits for the film business entrepreneurs.

The remainder of this chapter will therefore concentrate on fluid mechanics aspects applicable to components and systems using either mineral oil or its equivalent substitution. It is therefore worthwhile to review the basic concepts of electrohydraulics systems theory and establish the terminology used by control system engineers. A fluid power control system usually consists of a standard arrangement of interconnected elements as shown in Figure 16.1.

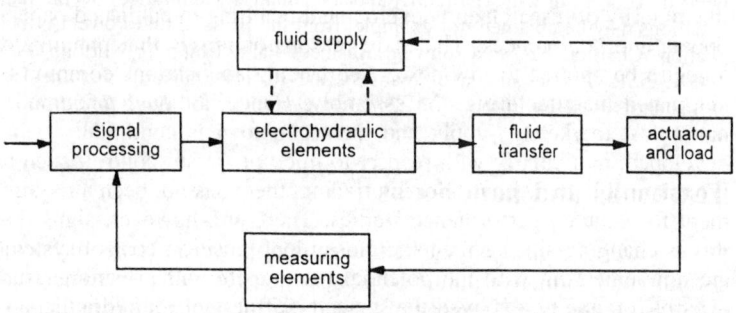

Figure 16.1 Interconnection of System Elements

The electrohydraulic elements may be variable displacement pumps, relief valves with proportional control, servovalves and actuators. In a multivariable control system there may be several feedback parameters required for optimum performance and/or control, and these are usually represented as shown in a single feedback loop. For many systems applications it will be sufficient to have analog control elements, in which case the signal processing shown may be a summing amplifier plus power stage. In more advanced systems this processing stage may be a microprocessor capable of comparing signals and computing control algorithms and/or adjusting electronic elements to always ensure optimum performance under varying operating conditions.

From a fundamental performance point of view there are many challenging issues confronting the fluid power engineer. Component static and dynamic performance improvement via fundamental fluid mechanics studies using computational fluid dynamics packages, component miniaturization, design for higher power levels, integrated systems, environmental issues of noise and fluids, improved materials for life and cost, reliability, ease of maintenance, greater productivity, are just some of the areas now receiving attention.

Fluid power is an international business with most of the research and development work being undertaken in-house by companies often having significant activities in many countries. It is therefore desirable to have a consistent approach to component terminology/standards, and a number of sources are available such as:

Graphic symbols for fluid power diagrams (1967) ASME Y32.10 (American Society of Mechanical Engineers).

Graphic symbols and circuit diagrams for fluid power systems and components – specification for graphic symbols. BS 2917: Part 1: 1993 (British Standards Institution).

Fluid power systems and components – Vocabulary, ISO 5598: 1985 (International Standards Organization).

CETOP (European Oil Hydraulic and Pneumatics Committee) recommendations may be obtained, for example, from the British Fluid Power Association, Cheriton House, Cromwell Business Park, Chipping Norton, Oxon OX7 5SR. A comparison of many of the British Standards and the ISO Standards is given in the *BFPA Fluid Power Engineers Data Book*.

Textbooks and handbooks

Fluid power is the poor relation to electrical power in terms of books, several good formative publications now being out of print. There are not many books currently available, and the following short list should prove useful:

Pippenger, J. J. (1989) *Zero downtime hydraulics* (Amalgam Publishing Company. Also available from Sun Hydraulics Corporation).
Prokes, J. (1977) *Hydraulic mechanisms in automation* (Elsevier Scientific).
Thoma, J. U. (1990) *Simulation by bondgraphs* (Springer-Verlag).
Pinches, M. J. and Ashby, J. G. (1988) *Power hydraulics* (Prentice Hall).
Pippenger, J. and Hicks, T. G. (1979) *Industrial hydraulics* (McGraw-Hill).
Banks, D. D. and Banks, D. S. (1988) *Industrial hydraulic systems* (Prentice Hall).
Reed, E. W. and Larman, I. S. (1985) *Fluid power with microprocessor control* (Prentice Hall).

288 Fluid mechanics and fluid power systems

The hydraulic trainer: instruction and information on oil hydraulics (1978) (G. L. Rexroth GmbH).

Rohner, P. (1988) *Industrial hydraulic control*, 3rd ed. (Australasian Educational Press).

Barber, M. J.(comp.) (1986) *Handbook of power cylinders, valves and controls* (Trade and Technical Press).

Watton, J. (1992) *Condition monitoring and fault diagnosis in fluid power systems* (Ellis Horwood).

Watton, J. (1989) *Fluid power systems – modeling, simulation, analog and microcomputer control* (Prentice Hall).

The reader is advised to check the current catalogues of the major publishing houses since there is evidence that some books currently out of print may be revived.

Fluid properties

In many industrial applications and mobile applications mineral oil is the preferred fluid due to the inherent lubricating characteristic. However, in applications such as steel processing and mining, the risk of fire must be minimized and an appropriate alternative fluid must be used. Fire resistant fluids fall into four classes defined briefly as follows:

HFA – Oil-in-water emulsion, typically 95 per cent water and 5 per cent oil by volume
HFB – Water-in-oil emulsion, typically 40 per cent water and 60 per cent oil
HFC – Water-Glycol solution, typically 40 per cent water and 60 per cent Glycol
HFD – Synthetic fluid containing no water

A vast range of mineral oils and fire-resistant fluids are available to satisfy the particular application. Performance features of a good mineral oil must embrace thermal stability, oxidation resistance, anti-wear, filterability, low friction, good water separation, acceptable air release and anti-foam properties. Continually changing Health and Safety Executive legislation is now forcing more hydraulics companies to consider the operation of standard components using water-based fluids, and the use of other materials such as ceramics is offering some potential.

When considering water-based fluids, the user should pay particular attention to manufacturers' data and the precautions needed to be observed when designing hydraulic circuits. Some of the issues are discussed in Watton (1992).

When considering the actual performance, particularly dynamics of hydraulic circuits, then fluid compressibility is of crucial importance. Bulk modulus is a measure of the compressibility of a fluid and is inevitably required to calculate hydraulic undamped natural frequencies in a system. It is perhaps the one fluid parameter that causes most concern in its numerical evaluation due to other effects which modify it. The basic definition of fluid bulk modulus arises by considering the compression of a fluid initially at atmospheric pressure and volume to a new volume and pressure. It may be defined in a number of ways, but the one most commonly used is the isentropic tangent bulk modulus, where the isentropic bulk modulus is defined at the thermodynamic condition of constant entropy. A finite difference form is often used in practice and does allow calculations to be made on volume changes during compression/decompression of large industrial fluid power systems. Bulk modulus may be determined from manufacturers' data, but care must be taken in interpreting the data in practice. If the bulk modulus quoted is measured under static conditions then the result may well be different under dynamic conditions and in practical circuits where the fluid will contain dissolved air and possibly entrained air. The effect of dissolved air is usually considered to be of secondary importance at realistic fluid power pressures. However, at low pressures, for example, where dynamic effects cause the pressure to fall towards atmospheric conditions, the fluid bulk modulus may well be reduced. The introduction of entrained air bubbles is more serious and suggests a deterioration in bulk modulus, particularly at low pressures. However, there is again evidence that the deterioration may not be too drastic at realistic fluid power pressures. The main problems are concerned with assessing the amount of dissolved and/or entrained air together with a correct description of the thermodynamic process for the air equation of state. At rapid rates of change of pressure typical in fluid power systems the literature points towards an adiabatic process, and at constant temperature various expressions for bulk modulus have been developed. Empirical constants in the appropriate equations must be determined by experiment. For a fluid in a pipe or container having flexibility then it may be shown that individual bulk moduli may be added in parallel to give an effective bulk modulus.

Experimentally, bulk modulus of a fluid and container can be determined by a variety of methods such as direct impedance measurement, determination of line delay time via signal cross-correlation, or by the use of transmission line theory using inherent pump ripple. Further information may be obtained from the following papers:

Smith, L. H., Peder, R. L. and Bend, L. H. (1960) *Hydraulic bulk modulus – its effect on system performance and techniques for physical measurement.* 16th National Conference on Industrial Hydraulics, 179–197 (Illinois Institute of Technology).

Hayward, A. T. J. (1961) *Aeration in hydraulic systems – its assessment and control.* Proceedings of the Institution of Mechanical Engineers. Conference on Oil Hydraulic Power Transmission and Control, 216–224.

Margolis, D. L. and Brown, F. T. (1976) Measurement of the propagation of long-wavelength disturbances through turbulent flow in tubes. *Transactions of the ASME. Journal of Fluids Engineering*, **98**, 70–78.

Johnston, D. N. and Edge, K. A. (1991) In-situ measurement of the wavespeed and bulk modulus in hydraulic lines. *Proceedings of the Institution of Mechanical Engineers. Part I Journal of Systems and Control Engineering*, **205**, 191–197.

Yu, Jinghong. (1991) Measurement of oil effective bulk modulus in hydraulic systems. *Chinese Journal of Fluid Power Engineering*, **3**, 46–48.

Watton, J. and Xue, Y. (1994) A new direct-measurement method for determining fluid bulk modulus in oil hydraulic systems. Proceedings FLUCOME'94, 4th Triennial International Symposium on Fluid Control, Measurement and Visualization, 543–545.

Generalized continuity equations for lumped components

Considering mass flow rate into and out of an arbitrary control volume and including the equation of state for a fluid leads to the flow continuity equation that may be applied to lumped elements. Following the previous section it may be necessary to include a pressure-dependent bulk modulus characteristic. It is common experience that the analysis of fluid power systems performance will inevitably embrace the general flow continuity equation which cannot be solved without the inclusion of force equations and additional flow continuity equations. Motors and cylinders controlled, for example, by servovalves will require two flow continuity equations and one load torque or force equation. It is not possible to state the appropriate torque of force equations to be used since internal characteristics of cylinders and particularly pumps and motors vary with the component type. Considering the hydraulic force generated in a cylinder will lead to an equation having reasonably well known terms. Since the piston moves inside the cylinder, the radial clearance may well generate a resisting viscous force proportional to velocity. Friction is a complex quantity to determine and is inevitably determined by measurement. It may contain both stiction and Coulomb-friction type components created by the piston seals and the rod seal. The equation for a motor will contain similar terms to those defined for a linear actuator. Continuity equations have been applied in some detail, but before references are presented it is worth considering the flow continuity equation applied to a motor. The inlet and outlet flow equations will contain contributions from cross-port leakage and leakage from the body of the machine via a number of leakage paths such as piston clearance, port plate side leakage, piston slipper lubrication hole, etc. These are complex flow passages which will be influenced by pressures and the machine speed. Manufacturers do spend a great deal of research and development time and money on improving performance, which includes both efficiency improvement and noise reduction, and performance characteristics are made aware to the user. The reader should therefore check manufacturers' data for the component being used. These data often represent the overall performance that may be expected, but not specific details that may be required for further computer analysis or CAD of circuits operating in a dynamic mode. Many research institutions and universities have over the years studied in detail the performance of components, and published literature from this source is particularly valuable.

When considering pressure and flow control valves, the main issues with respect to understanding the dynamic behaviour are still concerned with the precise mechanisms of viscous friction and flow reaction forces. In two-stage proportional pressure control valves, for example, the force balance across the main poppet may well be dominated by the flow reaction force. Measurements by the author have shown this force could be a factor of two greater than that calculated using ideal momentum change theory. Care has also to be taken when considering the direction of flow forces, the contribution of static and dynamic elements, and even whether they actually exist in some cases. The pressure drop/flow equation across valve ports and similar restrictions is reasonably well understood, and the Bernoulli form of equation is used with an appropriate flow coefficient selected. It is impossible to cover the vast range of papers published in the general area of fluid mechanics applied to fluid power components, but the following list embraces several hundred relevant articles:

Watton, J. (1989) *Fluid power systems – modeling, simulation, analog and microcomputer control* (Prentice Hall).

Watton, J. (1992) *Condition monitoring and fault diagnosis in fluid power systems* (Ellis Horwood).

BHRA Fluid Power Conference series. This is an internationally supported Conference series currently standing at No. 10. The Conference titles have changed over the years but information on all the Conference Proceedings may be obtained from BHR Group Ltd., Cranfield, Bedford, MK43 0AJ, UK.

JHPS International Symposium on Fluid Power, Tokyo, Japan. The first Symposium was held in March 1989 and the second in September 1993. Details obtainable from the Japan Hydraulic and Pneumatics Society, Kikaishinko Kaikan No. 301–3, 5–8.

Proceedings of the Conference on Fluids in Control and Automation, Toronto, Canada, 1976. Details from BHR Group Ltd., Cranfield, Bedford, MK43 0AJ, UK.

FLUCOME International Conference series. This is a Conference series on fluid control, measurement, and visualization. No. 1 was held in Japan 1985, No. 2 in Sheffield, U. K., 1988, No. 3 in San Francisco, USA, 1991, and No. 4 in Toulouse, France, 1994.

Journal of Fluid Control (Delbridge Publishing Co) quarterly.

Proceedings of the Institution of Mechanical Engineers, particularly *Part I, Journal of Systems and Control Engineering* (Mechanical Engineering Publications) quarterly.

Transactions of the ASME, Journal of Dynamic Systems, Measurement and Control (American Society of Mechanical Engineers) quarterly.

JSME International Journal (Japan Society of Mechanical Engineers) quarterly.

Mechatronics (Pergamon) 8 per annum.

Aachen Colloquium on Fluid Power Technology. This is a Conference series currently standing at No. 10 1992. Details from Rheinisch-Westfälische Technische Hochschule Aachen, Steinbachstrasse 53, D-5100 Aachen, Germany.

In particular the JHPS Conference series, the *Journal of Fluid Control*, the *Proceedings of the Institution of Mechanical Engineers*, and the FLUCOME Conference series contain a wealth of detailed fluid mechanics research on a variety of components related to fluid power.

Transmission line dynamics

In industrial applications where control valves and actuators are connected by long lines, the dynamic characteristics of the lines become important, and could dominate the overall system response. In mining applications it is unavoidable that the power pack is separated from the hydraulic roof supports and cutting equipment by long distances resulting in typically 1000m of hose in the supply and return lines. Large pressure surges have been reported in such systems, often causing repeated oscillations. Electrohydraulic control of machine tools requires a knowledge of system dynamics to reduce the vibration level due to separation of components, and it has been shown that the introduction of correct pipe lengths can improve the vibration level. In modern aircraft systems the desire to use thin flexible wings from aerodynamic considerations has resulted in the need to separate the servovalve from control surface actuator by large distances. In high power applications such as forging presses and bridge structure automatic welding systems, the high noise levels of the hydraulic components, plus their physically large size, means that again separation of actuators and control components is inevitable. Power supplies, together with some control valves, are often placed below ground working level in effectively a sound-isolated room. It is also desirable in some applications to separate servovalve and actuator for maintenance reasons, and it is known that in some cases, e.g. spring rolling machines, there is a deterioration in system performance.

A transmission line element is uniquely defined by two components referred to as the series impedance per unit length and the shunt admittance per unit length. There are a variety of techniques available to solve these equations in both the frequency domain and the time domain, and using any of the 3 models available for series impedance and shunt admittance. A large amount of research has been directed towards air and water systems, and the references that follow adequately cover the important contributions in terms of systems dynamics.

Franke, M. E. and Drzewiecki, T. M. (eds) (1981) *Fluid transmission line dynamics*. Presented at the Winter Annual Meeting of the ASME, November 15–20, 1981, Washington, DC (American Society of Mechanical Engineers).

Franke, M. E. and Drzewiecki, T. M. (eds) (1983) *Fluid transmission line dynamics*. Presented at the Winter Annual Meeting of the ASME, November 13–18, 1983, Boston, MA (American Society of Mechanical Engineers).

With respect to oil hydraulics systems dynamics, work has been directed towards conventional system analysis together with the use of transmission line theory for investigating noise and high-frequency impedance characteristics of components. Some of the latest publications are as follows:

Watton, J. (1986) The stability of electrohydraulic servomotor systems with transmission lines and non linear motor friction effects. Part A system modelling. *Journal of Fluid Control*, **16**(2), 118–136.

Watton, J. (1986) The stability of electrohydraulic servomotor systems with transmission lines and non linear motor friction effects. Part B system stability. *Journal of Fluid Control*, **16**(3), 137–151.

Watton, J. and Tadmori, M. J. (1988) A comparison of techniques for the analysis of transmission line dynamics in electrohydraulic control systems. *Applied Mathematical Modelling*, **12**, 457–466.

Watton, J. (1988) Modelling of electrohydraulic systems with transmission lines using modal approximations. *Proceedings of the Institution of Mechanical Engineers*, **202B**, 153–163.

Edge, K. A. and Johnston, D. N. (1990) The 'secondary source' method for the measurement of pump pressure ripple characteristics. Part 1 description of method. *Proceedings of the Institution of Mechanical Engineers*, **204A**, 33–40.

Edge, K. A. and Johnston, D. N. (1990) The 'secondary source' method for the measurement of pump pressure ripple characteristics. Part 2 experimental results. *Proceedings of the Institution of Mechanical Engineers*, **204A**, 41–46.

Johnston, D. N. and Edge, K. A. (1991) The impedance characteristics of fluid power components: restrictor and flow control valves. *Proceedings of the Institution of Mechanical Engineers*, **201I**, 3–10

Edge, K. A. and Johnston, D. N. (1991) The impedance characteristics of fluid power components: relief valves and accumulators. *Proceedings of the Institution of Mechanical Engineers*, **2011**, 11–22

Computational fluid dynamics (CFD) modelling

There has been a great detail of experimental work extending over many years in connection with fluid mechanics applied to typical flow paths in hydraulic-type components. This will be evident from references quoted in this chapter. Even the most simple of flow geometries offers a computationally complex problem, particularly if jet reattachment occurs, or the flow is highly turbulent. There is a small number of commercially-available software packages available, FLUENT being one of the more advanced versions. This has been extensively used by the writer and colleagues in a wide variety of applications embracing 2D/3D geometry, steady-state and transient flow, incompressible and compressible flow, laminar and turbulent flow. The user must, however, be most careful in interpreting simulation results, and it is wise to validate experimental data/simulation data via logical steps before attempting to solve a complex geometrical or flow problem. It is also important to select the appropriate solution techniques for the problem of concern, and the software company can be extremely helpful in this respect. Further details on FLUENT may be obtained from:

Fluent Incorporated, Centerra Resource Park, 10 Cavendish Court, Lebanon, NH 03766, USA.

or

Fluent Europe Ltd, Hutton's Buildings, 146 West Street, Sheffield, S1 4ES, UK.

Computer simulation (CAD)

The use of fluid mechanics to analyse component and system performance is of little value if it does not lead to a better understanding of the behaviour of the component or system. Experimental data will inevitably be required when considering the modelling of fluid power components, particularly with the more specialized and hence more complex pressure and flow control valves. A great deal of progress has been made in this area over many years via frequency domain techniques and the development of finite difference techniques in the time domain. The parallel development of numerical techniques now means that time domain analysis of performance presents few computational problems, and there are many software packages that are suitable for the analysis of fluid power circuits. For simple systems the designer may develop a numerical solution technique that may be adequate. Complex systems may be analyzed in this way although the program can also be complex when non-linearities are introduced. However, the 'feel' of the hydraulic system has been lost since the original circuit has been transformed to a set of mathematical statements. This may now be overcome by the next-generation CAD approach using software which may be run on conventional low-cost workstations. The workstation runs continuous simulations from a hierarchy of block diagrams drawn on its colour graphics screen. A mouse is used to position simulation diagram symbols, and pull down menus provide overall control. Realtime in-the-loop operation as well as purely computational simulations are supported. The model is entered onto the workstation graphically by building simulation block diagrams from icons on the screen. Thus the program of the model and the block diagram of the model are one and the same. The system provides a hierarchical structure for the implementation of large simulations in a rational structured manner, and simulation block diagrams can contain icons representing submodels. These submodel icons can in turn themselves contain submodels. Such a hierarchy can continue to depths sufficient for even the most complex models. A library of standard submodels can be built up and shared between various members of a department. Interaction is primarily through its tactile sensing mouse. The mouse is used to access the menus for initiating an editing session. The user then draws simulation block diagrams by dragging icons from the border of the screen onto the screen and connecting them as required. Practically all of the functions of continuous simulation languages are provided through the icons. In addition, an appropriate programming language may be entered into a user defined algebraic icon, thus creating special functions. The component equations are therefore set up using circuits that appear like conventional control systems, using block diagram algebra. Further information may be obtained from the following condensed list of references:

MATRIXx – Available from Integrated Systems Inc., Santa Clara, California 95054, USA.
XANALOG Continuous Simulation Workstation – Available from Scientific Computers Ltd., Burgess Hill, West Sussex, UK, or Xanalog Corporation, 300 Wild Wood Street, Woburn, MA 01801, USA.
Watton, J. (1989) *Fluid power systems – modeling, simulation, analog and microcomputer control* (Prentice Hall).
Haiping, Z. (1988) HYCAD – a general hydraulic simulation package. *Journal of Fluid Control*, **18**(2), 36–44.
Kinoglu, F. et al. (1982) Streamlining hydraulic circuit designs with computer aid. *Computers in Mechanical Engineering*, **1**(2), 21–26.
Watton, J. (1989) Fluid power design by CAD. *Professional Engineering*, **2**, 33–34.

Watton, J., Salters, D. G. and Nelson, R. J. (1990) *New software techniques for the CAD of fluid power systems*. Proceedings 9th International Fluid Power Symposium, 1990, 323–330 (Scientific and Technical Information).
Richard, C. W. et al. (1990) *A second generation package for fluid power systems*. ibid., 315–322.
Watton, J. and Salters, D. G. (1990) *Transient response involvement of a pressure relief valve via spindle modification using a CAD package*. Proceedings 1990 American Control Conference, 3, 2622–2625 (Institute of Electrical and Electronics Engineers).
Vilenius, M. J., Luomaratna, M. K. and Rinkimen, J. A. (1986) *CATSIM – A new kind of CAD program for hydraulic circuit design*. Proceedings 7th International Fluid Power Symposium, Bath, September 16–18, 1986, 65–72 (British Hydromechanics Research Association).
Backe, W. and Hoffmann, W. (1981) *DSH program for digital simulation of hydraulic systems*. Proceedings 6th International Fluid Power Symposium, Cambridge, U. K., 1981, Paper C1, chapter 29, 95–114 (British Hydromechanics Research Asssociation).
Leaney, P. G. (1990) *HYCAAF – A computer-aided analysis facility for hydraulic servo systems*. Proceedings 9th International Symposium on Fluid Power 1990, 293–314 (Scientific and Technical Information).
Handroos, H. M. (1990) *A library of component models in fluid power circuit performance simulation program – aspects and methods for improving*. ibid., 331–342.

Further reading

There are just a few new conferences emerging plus technical publications and trade shows which combine commercial and technical aspects of fluid power. The following information therefore provides an excellent source of material:

COMADEM. Condition Monitoring and Diagnostic Engineering Management Conference Series, currently No. 5, 1993. Details from University of West England, Bristol, UK.
The Scandinavian International Conference on Fluid Power series, currently No. 3, 1993. Details from Department of Mechanical Engineering Linköping University S-58183, Sweden.
Journal of the Japan Hydraulics and Pneumatics Society (JHPS), Kikaishinko Kaikan No. **301**(3), 5–8, Japan.
Mechanical Engineering (American Society of Mechanical Engineers) monthly.
Fluid Power (Intech Promotions Ltd) 3 per annum.
International Off-Highway and Power Plant Congress and Exposition, held usually in September, Milwaukee, USA. Further details on this plus a range of international exhibitions can be obtained from The British Fluid Power Association (BFPA), 235–237 Vauxhall Road, London SW1V 1EJ, UK.

Fluid Power Transmission and Control Conference series, China, currently No. 3, 1993. Organized by the Institute for Fluid Power, Zhejiang University, Hangzhou, China.

Fluid Abstracts – Process Engineering. (Elsevier) An abstracting and indexing service published monthly.

Öleodinamica-Pneumatica-Lubrificazione. (Techiche Nuove srl) Technical publication, in Italian, published monthly.

Olhydraulik und Pneumatik (Vereinigte Fachverlage) Technical publication, in German, published monthly.

Bath International Fluid Power Workshop series, currently No. 8, 1995. These volumes represent different aspects of fluid power, individual themes being set for each workshop. Published by Research Studies Press, Taunton, Somerset, UK.

Fundamental fluid mechanics studies

Many of the conferences, journals and textbooks quoted earlier include detailed research findings concerned with the fundamentals of fluid mechanics applied to fluid power control. A measure of the detail of such work may be found in the following references which also contain further references indicating other valuable background sources.

Banieghbal, M. R. (1985) Experimental and theoretical analysis of flow characteristics of servovalve orifices, under steady and unsteady flow. PhD Thesis, Liverpool University.

Baylet, V., O'Doherty, T. and Watton, J. (1993) *Reversed flow characteristic of a cone seated poppet valve.* Proceedings of the 5th International Symposium on Refined Flow Modelling and Turbulence Measurements, Paris, September, 1993, 867–874 (Presses Ponts et Chaussées).

Brown, F. T. (1962) Transient response of fluid lines. *Transactions of the ASME. Journal of Basic Engineering*, **84**, 547–553.

Kokoshima, Y., Shimizu, A. and Murao, T. (1991) *Numerical analysis of annular turbulent jet impinging on a flat plate.* Proceedings of FLUCOME '91, 3rd Triennial International Symposium on Fluid Control, Measurement and Visualization, 205–210 (American Society of Mechanical Engineers).

Lichtarowicz, A. (1973) *Flow and force characteristics of flapper valves.* Proceedings 3rd International Fluid Power Symposium, Turin, Italy, 1973, B1-1–B1-24 (British Hydromechanics Research Association).

Lichtarowicz, A., Duggins, R. K. and Markland, E. (1965) Discharge coefficients for incompressible non-cavitating flow through long orifices. *Journal of Mechanical Engineering Science*, **7**, 210–219.

Lugowski, J. (1993) *Experimental investigations of the origin of flow forces in hydraulic piston valves.* Proceedings 10th International Conference on Fluid Power, Brugge, Belgium, 1993, 233–244 (Mechanical Engineering Publications).

Nakano, N. (1992) Experimental study for the compensation of axial flow force in a spool valve. *Journal of Fluid Control* **21** (2–3), 7–26.

Neno, H., Okazima, A. and Muromiya, Y. (1993) *Visualisation of cavitating flow and numerical simulation of flow in a poppet valve*. Proceedings 2nd JHPS Symposium on Fluid Power, 1993, 385–390 (Japan Hydraulics and Pneumatics Society/Spon).

Ohmi, M. et al. (1980) Flow pattern and frictional losses in pulsating pipe flow – Part 1 effect of pulsating frequency on the turbulent flow pattern. *Bulletin of the JSME*, **23**, 2013–2020.

Ohmi, M. and Iguchi, M. (1980) Flow pattern and frictional losses in pulsating pipe flow – Part 2 effect of pulsating frequency on the turbulent frictional losses. *Bulletin of the JSME*, **23**, 2021–2028.

Oki, M. et al. (1991) *Flow around a circular cylinder with grooves: Comparison with circular cylinder without grooves*. Proceedings FLUCOME '91, 3rd Triennial International Symposium on Fluid Control, Measurement and Visualisation, 211–215 (American Society of Mechanical Engineers).

Oshima, S. and Tsuneo, I. (1983) Measurement of the unsteady flow force of a poppet valve. *Transactions. Japan Society of Mechanical Engineers (B)*, **49**, 2473–2477.

Shimizu, S. and Nakosumi, M. (1993) *Discrete vortex simulation of two-dimensional and axisymmetric flow in a disk valve*. Proceedings 2nd JHPS Symposium on Fluid Power, 1993, 605–670 (Japan Hydraulics and Pneumatics Society/Spon).

Suzuki, K., Matsui, K. and Machimaru, Y. (1989) *Bubble elimination by swirl flow*. Proceedings 1st JHPS Symposium on Fluid Power, 1989, 513–516. (Japan Hydraulics and Pneumatics Society).

Taft, C. K. and Twill, J. P. (1978) An analysis of the three-way underlapped hydraulic spool servovalve. *Transactions of the ASME. Journal of Dynamic Systems, Measurement and Control*, **100**, 117–123.

Takahashi, K., Ishigawa, S. and Wang, Z. (1989) *Viscous flow between parallel discs with a time varying gap width and central fluid source*. Proceedings 1st JHPS Symposium on Fluid Power, 1989, 407–414 (Japan Hydraulics and Pneumatics Society/Mita Press).

Tsukiji, T., Soshino, M. and Yonezawa, Y. (1993) *Numerical and experimental flow visualisation of unsteady flow in a spool valve*. Proceedings 2nd JHPS Symposium on Fluid Power, 1993, 379–384 (Japan Hydraulics and Pneumatics Society).

Watton, J. (1983) *Steady-state and dynamic flow characteristics of positive displacement pumps using laser Doppler anemometry*. Conference on Optical Techniques in Process Control, The Hague, 1983, 165–178 (British Hydromechanics Research Association).

Watton, J. (1987) The effect of drain orifice damping on the performance characteristics of a servovalve flapper/nozzle stage. *Transactions of the ASME. Journal of Dynamic Systems, Measurement and Control*, **109**, 19–23.

Weston, W. and Lalor, M. J. (1979) Determination of the dynamic flow characteristics of an electrohydraulic servovalve using a laser Doppler velocimeter. *Journal of Physics D*, **12**, 203–219.

References

Watton, J. (1992) *Condition monitoring and fault diagnosis in fluid power systems.* Chichester: Ellis Horwood.

CHAPTER SEVENTEEN

Materials information sources for engineers

P. SARGENT

Introduction

Engineers need different kinds of information and different kinds of information sources depending on the stage they are at in the product design process.

At the initial, conceptual, stage, when an idea is just forming, only gross characteristics are required, e.g. 'do plastics float?', 'does silicon carbide conduct heat?', 'can I make a bicycle chain from polythene?', etc. As a design progresses, the quantity of information to be dealt with expands enormously and many decisions are committed for parts of the product which render the design of other parts problematic, so questions arise such as: 'can I use iron screws to attach the copper bottom to the boat or would that cause any unfortunate electrochemical corrosion?'. Further on, the materials selection issues are committed and manufacturing engineers (who would ideally already have given their approval) now have to face detailed questions such as 'what is the yield stress for this specific alloy at 350°C so that I can adjust the rolling mill accordingly?'. Finally questions of use and disposal must be addressed: 'is it legal to use cadmium in batteries to be sold in the European Union?', 'can I make all the internal automotive plastics components from polypropylene to aid recycling?', 'is organic mercury poisonous?'

All these questions have different time scales and costs (in terms of designers' time) associated with them, so different sources of information are appropriate. Thus in early conceptual design the engineer will use overview publications (favourite volumes, often years out of date) because accuracy is not necessary but ease of access is.

A further important class of users is those who are learning to become practising engineers: students and engineers in training. These people need not just access to data, but access to guidelines and methods to help them interpret the information usefully. Even professional engineers increasingly have to work with materials which are unfamiliar to them, so that they too need to learn new methods for accessing and using the information.

Most important sources

There are two types of data source which stand out significantly above all the others in terms of their degree of use by engineers: current company data sheets describing materials which can be purchased immediately (with prices and telephone numbers), and national or international standards. An example of such a standard is A741:1990, 'Specification for zinc-coated steel wire rope and fittings for highway guardrail', one of the forty specifications set down for coated steel wire alone (among about 120 specification) in volume 01.06 of the annual ASTM book of standards. Note the degree of specificity: A741:1940 is for highway guardrails, different specifications are relevant for railway guardrails or sports-stadium guardrails.

A complete central technical library will hold all sixty-eight of the ASTM volumes covering all materials, a dozen volumes of the ASM Handbooks on metals, and equivalent volumes on 'engineered materials' such as composites, ceramics and polymers. Practising engineers often possess their own copies of summary handbooks, e.g. the desk edition of the *ASM metals handbook* (ASM, 1985) or

Brandes, E. A. and Brock, G.B. (eds) (1992) *Smithells metals reference book* 7th ed. (Butterworth-Heinemann).

Standards

Most countries define materials-related standards by the industrial community of users rather than the suppliers, so a standard composition of copper pipe for hot-water central heating systems will be documented (perhaps as a footnote) in a standard on hot water systems. Thus, even keeping track of all the standards that relate to one type of material in one country presents problems. All large industrialized countries have their own independent materials-defining standards bodies. The compositions and thermo-mechanical treatments of alloys for identical uses are thus different in different countries.

'Equivalent alloys' volumes are thus essential for organizations designing equipment to be exported. These volumes, such as *Stahl-*

Schlüssel which lists German and Swedish standards, lists steel equivalents and near-equivalents. Sometimes two different alloy specifications are equivalent for some uses but not for others, e.g. equivalent for use in fresh-water swimming pools but not in salt-water pools.

Wegst, C. W. (1995). *Stahlschlüssel: key to steel* 17th ed. (Verlag Stahlschlüssel Wegst).

Relevant organizations

Every major type of mature engineering material has a trade organization, usually international in scope with local branches in each industrialized country. There are often several for each material specializing in different aspects: technical information, refining, promotion, etc. Examples are the International Tin Association, the International Copper Research Association, the Copper Development Association, the Aluminium Association of America, the Rubber and Plastics Research Association, the Association of British Plywood and Veneer Manufacturers, the Ductile Iron Pipe Association, etc.

More recent engineering materials such as polymers, composites and fine ceramics have fewer trade associations and there are several organizations, usually professional institutions, which cover broader ranges of materials, such as the Welding Institute (which researches joining methods for all materials) and the UK Institute of Materials (a merger of institutes of metals, composites, glass and ceramics).

Comprehensive lists of such organizations are given in the *Aslib directory of information sources in the UK* (Aslib, 1994) and in *Information sources in metallic materials* (Bowker-Saur, 1989). Also in the following two books for materials information sources:

Houldcroft, P. T. (ed.) (1987) *Materials data sources* (Mechanical Engineering Publications for the Institution of Mechanical Engineers).
Reynard, K. (1993) *UK materials information sources* 2nd ed. (The Design Council).

Databases

The 83 papers in the volumes of proceedings from the three international conferences organized by the ASTM provide an unparalleled resource for the study of materials databases. The years since the mid-1980s have seen a very great many 'grey publications' on materials databases from *ad hoc* meetings of industry groups, in-house reports of international working groups and the like.

The three ASTM conference proceedings, together with the following:

Sargent, P. M. (1991) *Materials information for CAD/CAM* (Butterworth-Heinemann).
Newton, C. H. (ed.) (1993) *Manual on the building of materials databases*, ASTM MNL 19 (American Society for Testing and Materials).

provide the only consistently available textbooks on the development of materials databases.

Ashby, M. F. (1992) *Materials selection in mechanical design* (Pergamon).

also provides useful insight on how to assess the validity and usefulness of results retrieved from a database and which databases to use for which types of data. The other significant body of literature consists of reports issued over the last twenty years by NIST in Gaithersberg, USA, by various European Community research laboratories and by the Materials Databases Task Group of the VAMAS organization (the Versailles Agreement on Advanced Materials and Standards).

Between 1984 and 1989 the Commission of the European Communities (CEC) funded a research and development programme which supported eleven such online materials databases. One of the most startling conclusions of this project was that although materials property data was very expensive to produce, companies were extremely reluctant to access anything other than a free database. The reason is apparently because it is hard to justify the cost of accessing the data at the time, the value only becomes apparent much later in the design and manufacturing process. In addition, all engineers are used to having essentially free access to data books and handbooks. The problem of finding ways for otherwise willing users to pay sufficient money to support the (expensive) development of materials databases continues to this day.

Design aids and sales aids

The most successful materials databases are those which are funded as part of the sales and marketing operations of materials producing companies. The amounts of money spent in sales is sufficiently high for the cost of producing expensive databases to be readily covered, whereas requiring end-user engineers to pay for data on an as-need basis was never adequate to pay for their development. This is because while design engineers may implicitly commit large expenditures in the course of the design decisions they make, companies generally give them very little liquid cash for their expenses.

The most successful initiative in producing materials databases of any kind is the CAMPUS project, originated by Polydata (a commercial company based in Dublin and Aachen). This project gained agreement (eventually) from 26 polymer companies to produce test data measured using exactly the same standards for comparative purposes. The sales

Materials information sources for engineers 305

divisions of these companies then used the same software package to contain and display the data, which was distributed to potential clients along with the data on the same floppy disk. Participating companies include Akzo, BASF, Ciba Geigy, Dow, Du Pont, EMS Chemie, Enichem, General Electric, Hoechst, Hüls, Monsanto, Rhône Poulenc and Solvay. (ICI use Polydata's software but did not agree to support the CAMPUS standards.) Each company distributes its own data freely, but the entire collection is only available (for sale) from Polydata.

Online networks

In the early 1980s it was tacitly assumed that the appropriate distribution mechanism for numerical materials data was via computer networks and communications links, in the same way that bibliographic databases had been successfully deployed. A significant conclusion drawn from the extensive CEC research exercise was that online systems were not appropriate for most users. The reason is that engineers are infrequent users of materials databases and therefore very unlikely to have easy access to a computer system connected to the international analog-based packed-switch system for data transfer (based on X.25 standard protocols and modems).

Online databases for materials information managed to continue to exist, but usually only when cross-subsidized by a bibliographic online database or provided as a service by a professional organization or government institution. The case of France is an exception to the general rule because the near-universal Minitel service provides a universal access point for dial-up databases which is easy to link into and to use, even for occasional enquiries. The Internet is now changing this situation again (see below).

The UK Institute of Materials and the American Society for Metals operate a joint online data business 'Materials Information'. This publishes a number of abstracting journals and books (see further, below) and also makes the data available on several hosts as an online service.

Free databases and the Internet

The sales-aid databases are publicly available, as are a number of free databases produced by publicly-funded laboratories over the years. These latter collections have historically been publicized only by word of mouth and distributed by floppy disks via airmail (assuming an IBM compatible personal computer for their use).

A number of data collections is now available via the Internet; specifically via the World Wide Web which is easier to use than traditional Usenet conference groups eg. sci.materials and sci.engr.metallurgy, and ftp (file transfer protocol). One of the earliest gopher servers containing materials property data is gopher@eng.cam.ac.uk where there is an example database taken from undergraduate textbooks in one of the indexed reports available.

Floppy disks and CD-ROM

During the mid-1980s, personal computers based on the IBM model became widespread in all industrialized countries, even those with very rudimentary telecommunications systems. Also, by the later years of the decade, the capacity of a single standard floppy disk (1.4 MB) was adequate to hold a significant and useful quantity of numerical data. Since then we have seen the price of pressing a CD-ROM (600 MB) drop below the price of a floppy disk (in sufficient quantities).

The other significant advantage of using floppy disk or CD-ROMs to distribute data is that it is the same method that is used to distribute sales information and catalogues. Thus it is administratively easy to manage and updates can naturally be distributed at the same time that new materials become available or new prices are set.

Some consider that online networks are inherently superior because the data are always absolutely up-to-date on a day to day basis. However materials property data do not change rapidly: there is a significant lag in the system of producing, evaluating and condensing experimental measurements into publishable information. Also, even on an online system, updates are batched for administrative ease and performed every few months.

Many databases previously only available online are now available by annual subscription on CD-ROM. One example is METADEX, a cumulative database of all the abstracts which appear in the journal *Metals Abstracts*. Similar bibliographic databases produced by the same organization (Materials Information, see above) are EMA, cumulative abstracts of 'engineered' materials (polymers, composites, ceramics), and MATERIALS BUSINESS FILE, cumulative articles from a series of 'business alert' publications. Property data are also available from this same archive: METALS DATAFILE is a database of numerical data culled from, and referenced to, published papers and articles. It is an unusual property database in that it makes no attempt at consistency or completeness, it merely records all measurements which have been published in the open literature with no later editing or refinement. For this reason it has some unique uses but is rarely the first resource which an engineer would wish to consult.

Databases for analysis

One area where there is a great need for materials data in numerical form is in stress and deformation analysis by numerical methods, typically finite element analysis. One supplier of analysis software (McNeal-Schwendler Corp./PDA) Engineering sells a software package specifically designed for storing, manipulating and refining materials information for this purpose. It also sells several databooks of numerical data in machine-readable format suitable for use with this system: MSC/MVISION. Examples of such handbooks are the American military handbooks MIL-HDBK-5F (metal alloys), MIL-HDBK-17A (composites) and PMC-90 (carbon and SiC composites). This commercial system currently has no competitors in its chosen market and is priced accordingly. MSC/MVISION has an early version of a STEP data transfer interface (pending publication of ISO 10303 Part 45, the materials schema).

Handbooks and databooks

There is a mature 'industry' of publishers and professional engineering bodies supplying metals references handbooks in great profusion at all levels of detail. However the polymers, rubbers, composites, glasses and ceramics materials are far less-well covered. Partly this is because the materials are new so there is little stability in the data. Every few months new grades of polymers and resins for composites production are brought to market. Very slight changes in the operating conditions of fine, engineering ceramics production can make great differences to the properties of the final product – for identical chemical composition.

Without reproducibility of material for an extended length of time it is difficult to accumulate sufficient data to assemble a handbook since data from slightly different materials cannot be pooled, and having to describe the precise differences between every material grade and test would make a handbook far too voluminous. Nevertheless people have tried.

For polymers, the role of the professional bodies in providing databooks has been taken on by the materials suppliers: the chemical companies. Their handbooks are intended to support their sales and so may only use their own products as examples, but the technical content is perceived to be unbiased. Simple introductory textbooks, e.g.

Birley, A. W., Heath, R. J. and Scott, M. J. (1988) *Plastics materials: properties and applications*, 2nd ed. (Blackie).

can give very adequate overviews of the major classes of polymers and their usual uses.

Textbooks

Students and others learning to design and learning to select and specify materials have a great quantity of textbooks to aid them. Nearly all introductory texts in materials for engineers cover materials selection in some detail, and nearly all do it in the same way: chapters on the major classes of materials (steels, concrete, etc.) followed by a series of case studies for specific components (gas turbine blades, pressure vessels, electric kettles, bicycle clips, etc.). A minority of textbooks devote space to the formal method of selection itself, e.g.

Crane, F. A. A. and Charles, J. A. (1989) *Selection and use of engineering materials*, 2nd ed. (Butterworth-Heinemann).
Dieter, G. E. (1991). *Engineering design: a materials and processing approach*, 2nd ed. (McGraw-Hill).

These present simple weighting algorithm for ranking materials selected against multiple competing goals, e.g. strong and light, stiff and tough. Unfortunately many of these types of methods have deep flaws (reviewed in Sargent, 1991) and an entirely different type of selection is presented by Ashby (1993): the subject of his entire book and also used as the major introductory method for the *Elsevier materials selector* 3 vols (Elsevier, 1991) databook. All textbooks since Ashby have followed his approach so earlier editions should be avoided.

At least two databases have been produced largely for use by students learning about materials selection. The CAMBRIDGE MATERIALS SELECTOR contains highly selected, comparable data from 300 or so materials covering the full range of metals, ceramics, glasses, foams, composites, timber, bone, etc. It uses Ashby's selection methodology which is built into the way the data are displayed and the method of restricting selection to the most useful materials for a particular use. A more conventional database, MATERIALS DATABASE FOR STUDENTS, became available from the UK Institute of Materials in 1994.

Conferences and journals

Conferences are not always useful sources of materials information for design engineers except when the data requirement has become highly determined and specialized. Even then, it is very likely to be the case that only a single number out of a whole volume is relevant.

Academic journals similarly are not always useful for practising design engineers, even though the data that engineers may eventually use first appear there. The difficulty is one of overload and indexing, which is why engineers usually only access academic journals and conferences via abstracting services or bibliographic databases (see below).

Useful journals include *Metallurgical Transactions, Journal of*

Materials information sources for engineers

Materials Science, Transactions of the ASME. Journal of Engineering Materials and Technology, Proceedings of the Institution of Mechanical Engineers, Eureka, Materials World (Institute of Materials members' magazine), *Materials and Design, Computer-Aided Design, Computer-Aided Engineering Journal.*

Trade journals are of great use to design engineers because they consist largely of advertisments of current and recent developments, together with articles looking a few years ahead as to what may be expected in the near and medium term.

Conferences are much more appropriate for extraction and manufacturing engineers who are always working with the same materials, particularly where there is a complex process, such as electro-refining of aluminium, casting bronze, rolling steel, autoclaving carbon-fibre composites, pultruding glass-fibre, etc.

Abstracting and indexing journals

The most significant abstracting journal is *Metals Abstracts*, published by the 'Materials Information' business unit of ASM and the UK Institute of Materials. Each monthly issue contains more than 3000 abstracts which are collated into a cumulative database (see above). Materials Information also produces a similar journal for ceramics, polymers and composites and a set of Business Alerts taken from hundreds of trade magazines and journals. A number of related publications, such as thesauri of technical terms and lists of technical journals related to materials, are also available from Materials Information on paper only (i.e. not via online databases or on CD-ROM).

Data transfer methods

The MSC/MVISION database mentioned above is the only one which purports to support any kind of neutral data format for transferring numerical data from a database to any other database or software package. Many freely distributed databases do provide an 'export' or 'report' production function (including the excellent CAMPUS packages) and with some manual care in 'cutting and pasting' this can avoid the need to retype tables of numbers. However the most dangerous aspect of transferring data is the possibility of using the data in a way which violates the assumptions or conditions under which they were initially derived. Here the STEP standard makes explicit provision which no other transfer mechanism has attempted.

STEP

Whereas past transfer standards have purely addressed the need to exchange computerized engineering drawings, the new international Standard for the Exchange of Product Data (STEP) is focused on exchanging entire product descriptions ('product models') which contain sufficient information to be used directly by CAD/CAM systems.

The international standard Standard for the Exchange of Product Data (STEP, ISO 10303) aims for a complete specification of engineering products, including civil engineering, ship-building, power electrics, electronics, machines and process engineering. It covers project management, version control and records of the engineering design process. Part 45, formally approved in 1994, covers materials specifications and the recording of conditions under which properties are tested and measured.

Geometry and topology descriptions form the core of STEP. In addition to geometry, it will support tolerances, material properties, design 'features' and surface finish specifications. This will enable the standard to be used to communicate a description of an engineering component part or assembly of parts for the purposes of design, analysis, manufacture, quality control, test, inspection, maintenance and final disposal.

In principle, very few people would actually be aware of the inner workings of STEP, they would just interact with STEP-compatible software packages which would ensure that any data transfer that was necessary would be performed invisibly and painlessly.

Advanced procurement and logistics systems

Computerizing the internal affairs of a single organization, even if some management techniques extend to supplier chains, is not the whole story by any means. Software-aided management of the entire procurement and logistics support of multi-organization projects is now a reality, led by the US Department of Defense in the CALS initiative.

These new ways of working require formal methods for specifying engineering products and this is where the new STEP standard (ISO 10303: 1994) is applicable. These formal, computerized descriptions also have to be integrated with less formal, human readable documents; a combination of databases, calculations and desktop publishing.

CALS

Computer Aided Acquisition and Logistic Support (CALS) is a US Department of Defense (DoD) and industry programme to aid the integration of technical information for weapon system design, manufacturing, purchasing and maintenance. In 1988 the DoD gave notice that CALS compliance was required for all weapon systems entering production from 1990. This is currently the case. The CALS programme aims to completely computerize all technical data relating to the acquisition of systems. The method is to adopt sub-sets of existing standards for data transfer and archiving and to change these adopted sub-sets as new standards are developed. The CALS standards themselves are unclassified and publicly available worldwide. CALS also generally adopts international rather than purely domestic standards. Although the purpose of the programme is for weapon systems, the CALS standards are also directly applicable to non-weapons commercial engineering products where 'response to the market' substitutes for military preparedness. In this guise they are known as Advanced Procurement and Logistics Systems (APLS). Non-military industry has enthusiastically welcomed the CALS initiative because it permits greater control of engineering information, reduces waste and should permit more rapid response to market forces.

CALS adopted standards

CALS uses MIL-STD-1840A for numerical data (very rudimentary and almost useless for materials information) and Group III facsimile for images. CALS will migrate to using STEP as it becomes available. Many of the CALS standards are international ISO standards but CALS restricts the user to using a particular subset of each. The list will be extended to include STEP and possibly EDIFACT (see below).

EDI: Electronic data interchange

EDI is usually used to mean trade data interchange: invoices, kitting lists, shipment notices, bills of lading, etc. Some materials-related information is present in these data, e.g. test measurements performed to validate the quality of a shipment of aluminium alloy, but there is no way at present for the receiving engineer to access this information and to use it profitably. The dominant standard is EDIFACT, ISO 9735: 1987.

Expert systems

During the 1980s, materials selection attracted a great deal of attention from those looking for applications of the new technology of expert systems. A number of proofs-of-concept were devised but almost all projects failed to 'scale' adequately as the size of the knowledge base had to be increased to engineeringly-useful proportions. In one classic case, perhaps a million dollars was spent developing a technically successful knowledge-based system to advise on the generation of new alloys (Sargent, 1991). However, the alloy research group for which the software was aimed could not afford to employ the two full-time knowledge-refiners the system required to maintain the currency and accuracy of its data. This is a further example of the difficulty of matching the resources required for developing materials information support systems for engineers with the funding that such engineers usually have at their disposal.

Most expert systems which have enjoyed a degree of success have specialized on one of the areas of corrosion where there is sufficient theory to make a simple database inadequate by comparison, but where the science is sufficiently poorly understood or measured to make it describable by a few simple formulas.

The most used expert system is possibly a decision-tree system developed by Permabond Ltd as a sales aid (see above, for other examples of the success of this funding method) to advise on the selection of adhesives.

Reference information

Internet

Meltsner, K. J. *Understanding the Internet: a guide far materials scientists and engineers*, Journal of Metals, **47** (4)(1995), 9.

Many materials suppliers now provide sales and materials information via their World-Wide Web Home pages. Also try using an automated Web search engine such as www.yahoo.com or www.webcrawler.com.

Data and design in product development

In addition to those mentioned earlier, the following are useful sources:

Carter, D. E. and Baker, B. S. (1991) *Concurrent engineering: the product development environment for the 1990s* (Addison-Wesley).
Cornish, E. H. (1987) *Materials and the designer* (Cambridge University Press).
Pahl, G. and Beitz, W. (1988). *Engineering design*, rev. edn. (The Design Council).
Ullman, D. G. (1992). *The mechanical design process* (McGraw-Hill).

Databooks and design methods

Bringas, E. A. (ed) (1992) *The metals black book*, Vol. 1: Ferrous edition (CASTI Publishing).
Edwards, L. and Endean, M. (eds) (1990) *Manufacturing with materials* (Open University).
Engineered materials handbook, Vol. 4: Ceramics and glasses. (1991) (ASM International).

Evetts, J. (ed.) (1992) *Concise encyclopedia of magnetic and superconducting materials* (Pergamon).
Gall, T. (ed.) (1993) *Metals handbook: desk edition* (American Society for Metals).
Gibson, L. J. and Ashby, M. F. (1988) *Cellular solids: structure and properties* (Pergamon).
Handbook of industrial materials, 2nd ed. (1992) (Elsevier).
Hufnagel, W. (1991) *Key to aluminium alloys: aluminium schlüssel*, 4th ed. (Aluminium-Verlag).
Kelly, A. (ed.) (1989). *Concise encyclopedia of composite materials*, 2nd ed. (Pergamon).
Murphy, J. (ed.) (1991) *New horizons in plastics* (Institute of Materials).
Robb, C. (1987) *Metals databook* (Institute of Metals).
Tsai, S. W. (1988) *Composites design*, 4th. ed. (Thick Composites).
Waterman, N. A. and Ashby, M. F. (eds) (1991) *Elsevier materials selector*, 3 vols (CRC Press).

Database literature

Barry, T. I. and Reynard, K. W. (eds) (1992) *Computerization and networking of materials databases*, Vol. 3 (American Society for Testing and Materials, ASTM STP 1140).
Glazman, J. S. & Rumble, J. (eds) (1989) *Computerization and networking of materials databases* (American Society for Testing and Materials, ASTM STP 1017).
Kaufman, J. G. and Drago, V. J. (1992) Direct access to material properties for modeling and simulation. *Modelling and Simulation in Materials Science and Engineering*, **1**, 335–347.
Kaufman, J. G. and Glazman, J. S. (eds) (1991) *Computerization and networking of materials databases*, Vol. 2. (American Society for Testing and Materials, ASTM STP 1106).

Database suppliers

Bayer AG, Plastics Business group, Marketing (CAMPUS), Building B207, D–5090 Leverkusen, Germany.

Cambridge Materials Selector, EDC, Engineering Dept, Trumpington Street, Cambridge CB2 1PZ, UK.

DataPLAS, Modern Plastics, 43rd Floor, 1221 Avenue of the Americas, New York, NY 10020, USA.

EPOS: ICI Engineering, Sales Office, PO Box 90, Wilton, Middlesbrough TS6 8JE, UK.

High Temperature Materials Database, CEC Joint Research Centre, Petten, Netherlands.

Hüls AG D-W-4370 Marl, PO. Box 1320, Germany.

Mat.DB, ASM International, Metals Park, OH 44073, USA.

Materials Information, 1 Carlton House Terrace, London SW1Y 5DB, UK.

MATUS: Engineering Information Company Ltd., 15/17 Ingate Place, London SW8 3NS, UK.

MSC/MVISION. MSC/PDA Engineering, Redhill Road, Costa Mesa, CA 92626, USA.

Permabond Adhesives Locator (PAL), Permabond Ltd, Woodside Rd, Eastleigh SO5 4EX, UK.

PLASCAMS : RAPRA Ltd. (Rubber and Plastics Research Association) Shawbury, Shrewsbury, Shropshire SY4 4NR, UK.

Polydata Ltd, (CAMPUS coordinator), Unit 16, IDA Enterprise Centre, 111 Pearse Street, Dublin 2, Eire.

Progetim (CETIM) Paris Nord II, Micro Park, 33 rue des Chardonnerets, BP 60028, F–95971 Roissy CDG Cedex, France.

THERM: c/o Bob Bailey, Lawrence Livermore Laboratory, PO Box 808, Livermore, CA 94550, USA.

Standards and CALS

Annual book of ASTM standards (1995). Section 01.06: coated steel products (American Society for Testing and Metals).

Computer-aided acquisition and logistic support (CALS) program implementation guide, MIL-HDBK-59 (1988) (US Department of Defense).

International Organization for Standardization (1993) *Standard for the Exchange of Product Data*, (STEP) ISO 10303 (International Organization for Standardization).

Smith, J. M. (1990) *An introduction to CALS: the strategy and the standards* (Technology Appraisals).

CHAPTER EIGHTEEN

Safety engineering

M. J. PITT

Introduction

Much safety is achieved by sound basic engineering – things which are well designed and made are less likely to fail. This is an important reason for national standards and codes of practice, which encapsulate much experience and understanding.

These must be considered some of the most basic safety sources. However, they deal with prescribed devices and situations, but the professional engineer has to take responsibility for new devices, new materials and complex situations with many interactions. It is therefore increasingly important that engineers have some formal understanding of risk assessment techniques and active methods for achieving safety. This has been expressed in a 10-point code by the (UK) Engineering Council (1993) *Engineers and risk issues: code of professional practice*. The same organization publishes (1993) *Risk guidelines for engineers*, an expansion and background to the code, with a bibliography.

The most general text is by Brauer (1990) which provides a wide background. Burditt (1993) includes legal aspects as well as technical ones, but is more concerned with products rather than processes or large structures. Hunter (1992) focuses on mechanical and structural aspects of products. The book edited by Blockley (1992) includes some general theory but is most suitable for civil engineers as the applications are mainly concerned with large structures and nuclear power stations. Gloss and Wardle (1984) is a slightly older introduction.

The accident prevention manual for industrial operations Vol. 2: Engineering and technology, 10th ed. (1992) produced by the

National Safety Council particularly deals with industrial handling equipment.

A substantial source with an updating service is *Best's loss control engineering manual* (1993) (Best) which deals with hazards and their control in various industrial and commercial categories. For buildings and industrial installations, information is available in looseleaf form (currently 11 volumes) with an updating service from an insurance company: *Loss prevention data by factory mutual engineering research* (Factory Mutual Engineering Corp). The coverage varies greatly among disciplines, being greatest in chemical engineering. At the other extreme, in a recent survey of transport engineering books Pant (1993) noted the lack of any text on highway safety, though Cebon and Mitchell (1992) have remedied this.

It should be noted that the content is not always obvious from the title, due in part to specialists having a narrow interpretation of certain words, and publishers looking for title phrases which will sell.

Terminology

Formal definitions of terms are given in Abercrombie (1988) and Jones (1992).

Generally speaking, an unwelcome occurrence such as fire or structural failure is a hazard, the chance of it happening is the risk, and the overall combination of risk and hazard constitutes the danger. (Some groups call this the risk, being the product of the probability and the hazard). Total safety is impracticable, but a safe system implies an acceptably low level of danger. Often safety is understood to mean the safety of people. The term loss prevention is used to indicate measures to reduce damage to equipment and goods (i.e. financial loss).

Most texts on *industrial safety* are concerned with human activities (office, factory, construction work) and the prevention of accidents. Publications on *occupational health* are concerned with the exposure of people to harmful agents such as vapours, dust and disease. As there are a great many publications of this nature (often of a non-technical or low-level technical content) they will not be reviewed here, unless especially relevant to engineering issues.

Risk assessment

Formal techniques for recognizing hazards and evaluating risks are given below. There have been significant developments in a number of these, so earlier documents are not listed where they have been superseded.

Safety engineering

American Institute of Chemical Engineers (1987) *Fire and explosion index hazard classification guide*, 6th ed. (American Institute of Chemical Engineers).
Browning, R. L. (1980) *The loss rate concept in safety engineering* (Dekker)
Cebon, D., Mitchell, C. G. B. (eds) (1992) *Heavy vehicles and roads: technology, safety and policy* London: (Thomas Telford).
Center for Chemical Process Safety (1989) *Guidelines for chemical process quantitative risk analysis* (American Institute of Chemical Engineers).
Center for Chemical Process Safety (1992) *Guidelines for hazard evaluation procedures*, 2nd ed. (American Institute of Chemical Engineers).
Doran, P. (1993) *The Mond index: how to identify, assess and minimise potential hazards on chemical plant units for new and existing processes*, 2nd ed. (Imperial Chemical Industries).
Elia, F. A., Jr, (ed.) (1991) *Engineering applications of risk analysis III: presented at the winter annual meeting of the American Society of Mechanical Engineers*, Atlanta, GA, December 1–6, 1991 (American Society of Mechanical Engineers).
Engineering Employers' Federation (1993) *Practical risk assessment* (EEF)
Greenberg, H. R., and Cramer, J. J. (eds) (1991) *Risk assessment and risk management for the chemical process industry* (Van Nostrand Reinhold).
Harns-Ringdahl, L. (1992) *Safety analysis: principles and practice in occupational safety* (Elsevier).
Hochreiter, L. E. and Shiralkar, B. S. (eds) (1992) *Best estimate safety analysis, presented at the 28th National Heat Transfer Conference and Exhibition*: San Diego, CA, August 9–12, 1992 (American Society of Mechanical Engineers).
International Study Group on Risk Analysis (1985) *Risk analysis in the process industries* (Institution of Chemical Engineers).
Kayes, P. J. (ed.) (1985) *Manual of industrial hazard assessment techniques* (Technica).
Kletz, T. A. (1992) *Hazop and Hazan: identifying and assessing process industry hazards*, 3rd ed. (Institution of Chemical Engineers).
Roland, H. E. and Moriarty, B. (1990) *System safety engineering and management*, 2nd ed. (Wiley).
Technica Ltd. (1986) *Techniques for assessing industrial hazards: a manual* (World Bank).
US Nuclear Regulatory Commission (1981) *Fault tree handbook NUREG-0492* (NRC).

Case studies

Much can be learned from proper studies of things that have gone wrong. A key author is T. A. Kletz, who writes in the main about the chemical industry but includes many other examples from other industries plus civil and natural disasters in very readable and helpful books. The American Society for Metals has published a number of books of case histories of material failures.

Kharbanda, O. P. and Stallworthy, E. A. (1988) *Safety in the chemical industry: lessons from major disasters* (Butterworth-Heinemann).

Kletz, T. A. (1988) *What went wrong: case histories of process plant disasters*, 2nd ed. (Gulf Publishing).
Kletz, T. A. (1989) *Learning from accidents in industry* (Butterworth-Heinemann).
Kletz, T. A. (1990) *An engineer's view of human error*, 2nd ed. (Institution of Chemical Engineers).
Kletz, T. A. (1990) *Critical aspects of safety and loss prevention* (Butterworth-Heinemann).
Kletz, T. A. (1992) *Lessons from disaster: how organisations have no memory and accidents recur* (Institution of Chemical Engineers).
Lees, F. P. (1981) *Loss prevention in the process industries: hazard identification, assessment and control.* (Butterworth-Heinemann).
Lees, F. P. and Ang, M. L. (eds) (1989) *Safety cases within the control of industrial major accident hazards* (CIMAH) (Butterworth-Heinemann).
Marshall, V. C. (1987) *Major chemical hazards* (Ellis Horwood).
Mosey, D. (1990) *Reactor accidents: nuclear safety and the role of institutional failure* (Butterworth-Heinemann).
Nishida, S.-I. (1992) *Failure analysis in engineering applications* (Butterworth-Heinemann).
Perkins, J. L. and Rose, V. E. (eds) (1987) *Case studies in industrial hygiene* (Wiley).
Petroski, H. (1994) *Design paradigms: case histories of error and judgment in engineering* (Cambridge University Press).
Piesold, D. D. A. (1990) *Civil engineering practice: engineering success by analysis of failure* (McGraw-Hill).
Sanders, R. E. (1993) *Management of change in chemical plants: learning from case histories* (Butterworth-Heinemann).
Toft, B. and Reynolds, S. (1994) *Learning from disasters* (Butterworth-Heinemann).
van der Schaaf, T. W., Lucas, D. A. and Hale, A. R., (eds) (1991) *Near miss reporting as a safety tool* (Butterworth-Heinemann).

Reliability

In certain areas there is actual statistical data on the breakdown of equipment which may be used for estimates of the probability of failure, while in all areas efforts are being made to increase reliability. It should be noted that there is a difference between improving the performance and improving the reliability of engineering. This is a subject of great complexity and subtlety. Some sources are thus:

Bamford, W. H. *et al.* (eds) (1991) *Fatigue, fracture and risk 1991* (American Society of Mechanical Engineers).
Bamford, W. H. *et al.* (eds) (1992) *Fatigue, fracture and risk 1992* (American Society of Mechanical Engineers).
Bass, L. (1986) *Products liability: design and manufacturing defects* (Shepard's/McGraw-Hill).
Center for Chemical Process Safety (1989) *Guidelines for process equipment reliability data with data tables* (American Institute of Chemical Engineers).

Safety engineering 319

Cruse, T. A. (1992) *Reliability technology, 1992: presented at the Winter Annual Meeting of the American Society of Mechanical Engineers*, Anaheim, CA, November 8–13, 1992 (American Society of Mechanical Engineers).
Davidson, J. (ed.) (1988) *The reliability of mechanical systems* (Institution of Mechanical Engineers).
Gertman, D. I. and Blackman, H. S. (1994) *Human reliability and safety analysis data handbook* (Wiley).
Goble, W. M. (1992) *Evaluating control systems reliability – techniques & applications* (Instrument Society of America).
Henley, E. J. and Kumamoto, H. (1981) *Reliability engineering and risk assessment* (Prentice Hall).
IEEE Standard 500 (1984) *Reliability handbook* (Institute of Electrical and Electronics Engineers).
Klaassen, K. B. and van Peppen, J. C. (1989) *System reliability, concepts and applications* (Arnold).
Misra, K. B. (1992) *Reliability analysis and prediction* (Elsevier).
Nuclear Regulatory Commission (1982) *Handbook for probabilistic risk assessment for nuclear power plants NUREG/CR-2300* (NRC).
O'Connor, P. D. T. (1990) *Practical reliability engineering*, 3rd ed. (Wiley).
Pangborn, R. N. et al. (eds) (1990) *Damage assessment, reliability, and life prediction of power plant components: presented at the 1990 Pressure Vessels and Piping Conference, Nashville, TN, June 17–21, 1991* (American Society of Mechanical Engineers).
Service, T. H. (ed.) (1991) *Reliability, stress analysis, and failure prevention 1991: presented at the 1991 ASME design technical conference*, Miami, FL, September 21–25 (American Society of Mechanical Engineers).
Smith, D. J. (1993) *Reliability, maintainability and risk: practical methods for engineers*, 4th ed. (Butterworth-Heinemann).
Villemeur, A. (1992) *Reliability, availability, maintainability and safety assessment*, 2 vols (Wiley).
Yao, J. T. P. (1985) *Safety and reliability of existing structures* (Pitman).

The possible failure of electronics or software (e.g. for process control or emergency action) is considered in the following:

Engineering Equipment and Materials Users Association (1989) *Safety-related instrument systems for the process industries* (EEMUA).
Fisher, T. G. (ed.) (1991) *Control systems safety* (Instrument Society of America).
Health and Safety Executive (1987) *Programmable electronic systems in safety related applications* (HMSO).
Institution of Electrical Engineers and British Computer Society (1989) *Software in safety-related systems* (IEE).
Institution of Electrical Engineers (1991) *Formal methods in safety-critical systems* (IEE).
Institution of Electrical Engineers (1992) *Safety-related systems: a professional brief* (IEE).
Safety-Related Computers. EWICS (European Workshop on Industrial Computer Systems), (1985) *TC7: Systems Reliability safety and Security* (Verlag TuV Rheinland).

Electrical Safety.

It is important to distinguish between hazards due to electric current and those due to static electricity (which can cause faults in electronic devices and also be a source of ignition: see section on fire hazards below). There are also significant differences between direct and alternating current (ac and dc), high voltage, mains voltage and low voltage work. Electric current used for power purposes is generally around mains voltage 100–500V, and this is the subject of much legislation and many regulations. In particular, many UK texts refer to the *IEE wiring regulations*, 16th ed. This is a national code which has been adopted as a British Standard, namely BS 7671: 1992 *Requirements for electrical installations*, effective from 1993.

US regulations include a National Electrical Code and OSHA requirements. These are covered in a text suitable for engineers by Winburn (1988).

In other areas, much safety engineering is concerned with keeping the levels of electric current below those which would be harmful to people or would ignite flammable substances. In addition the failure of electric or electronic devices (including computers) can cause other dangers, physical, chemical, etc.

The term *intrinsically safe* is used for specially constructed low power devices and circuits which will not produce a spark or hot surface capable of ignition even when accidentally exposed to full mains voltage or when the device itself fails. They are used, for example, in coal mines and in the petroleum industry to make measurements and carry signals in potentially flammable atmospheres.

Bass, H. G. (1984) *Intrinsic safety: instrumentation for flammable atmospheres* (Quartermaine House).
Boylston, R. (1993) *Electrical safety-related work practices* (Lewis Publishers) (book and video).
Buschart, R. J. (1991) *Electrical and instrumentation safety for chemical processes* (Van Nostrand-Reinhold).
Electricity Council (1981) *Power system protection*, 2nd ed. 3 vols (Peregrinus).
Fordham Cooper, W. (1986) *Electrical safety engineering*, 2nd ed. (Butterworth-Heinemann).
Garside, R. (1990) *Electrical apparatus and hazardous areas* (Hexagon Technology).
Garside, R. (1988) *Intrinsically safe instrumentation – a guide to the use of electrical apparatus*, 2nd ed. (Dowty Safety Technology).
Greenwald, E. K. (ed.) (1991) *Electrical hazards and accidents: their cause and prevention* (Van Nostrand Reinhold).
Hall, J. R. (1985) *Intrinsic safety in British coal mines* (Marylebone Press).
Hasse, P. (1992) *Overvoltage protection of low-voltage systems* (Peregrinus).
Horvath, T. (1991) *Computation of lighting protection* (Wiley).

Safety engineering 321

Institution of Electrical Engineers (1992) *Guidance notes on the 16th edition of the wiring regulations*, 6 vols (IEE).
Institute of Petroleum (1990) *Model code of safe practice in the petroleum industry: part 15, area classification code for petroleum installations* (Wiley).
Institute of Petroleum (1991) *Model code of safe practice in the petroleum industry: part 1, electrical safety code* (Wiley).
Jones, T. B. and Thomas, B. (1991) *Powder handling and electrostatics – understanding and preventing hazards* (Lewis Publishers).
Luttgens, G. and Glor, M. (1989) *Understanding and controlling static electricity* (Expert Verlag GmbH).
Magison, E. C. (1984) *Intrinsic safety* (Instrument Society of America).
Magison, E. C. and Calder, W. (1983) *Electrical safety in hazardous locations* (Instrument Society of America)
National Fire Protection Association (1993) *National electrical code 1990 handbook* (NFPA).
Underwriters Laboratories (1989) *Standard for safety for power supplies*, 4th ed. (UL).

Fire hazards

Fire is in practice the greatest danger, and there is an abundance of books, journals and individual articles.

In the UK, the Loss Prevention Council (140 Aldersgate Street, London EC1A 4HY) publishes hundreds of documents on behalf of itself and other organizations such as the Fire Protection Association. In particular a regularly updated collection of approximately 300 data sheets is sold as the FPA *Compendium of fire safety*.

In the USA, the National Fire Protection Association (NFPA) is a major source of standards and other documents on fire prevention and fire fighting. The Underwriters Laboratories (UL) produces standards for fire protection and electrical safety as an insurance industry body.

Publications on fire engineering tend to deal with three general topics: (1) prevention of fires (2) passive protection, e.g. fire-resisting walls, and (3) active systems such as sprinklers.

Abbott, J. A. (1990) *Prevention of fires and explosions in dryers*, 2nd ed. (Institution of Chemical Engineers).
Baker, W. E. and Tang, M. J. (1991) *Gas, dust and hybrid explosions* (Elsevier).
Bartknecht, W. (1989) *Dust explosions: cause, prevention, protection* (Springer-Verlag).
Bodurtha, F. T. (1980) *Industrial explosion prevention and protection* (McGraw-Hill).
Bond, J. (1991) *Sources of ignition* (Butterworth-Heinemann).
Cashdollar, K. L. and Hertzberg, M. (eds) (1987) *Industrial dust explosions. ASTM special technical publication 958* (American Society for Testing and Materials).

Cassidy, K. A. (1992) *Fire safety and loss prevention* (Butterworth-Heinemann).
Cote, A. E. and Linville, J. L. (eds) (1991) *Fire protection handbook*, 17th ed. (National Fire Protection Association).
Dean, A. E. and Tower, K. (1991) *Fire protection guide on hazardous materials*, 10th ed. (National Fire Protection Association).
Dow Chemical Company (1987) *Fire and explosion index hazard classification guide*, 6th ed. (American Institute of Chemical Engineers).
Eckhoff, R. K. (1991) *Dust explosions in the process industries* (Butterworth-Heinemann).
Glor, M. (1988) *Electrostatic hazards in powder handling* (Research Studies Press).
Institute of Petroleum (1994) *Model code of safe practice in the petroleum industry part 19: fire precautions at petroleum refineries and bulk storage installations* (IP).
International Conference on Materials and Design against Fire (1992) *Materials and design against fire. International Conference* 27–28 October 1992, Institution of Mechanical Engineers, London (Mechanical Engineering Publications).
James, D. (1986) *Fire prevention handbook* (Butterworth).
Jones, T. B. and Thomas, B. (1991) *Powder handling and electrostatics* (Lewis Publishers).
Medard, L A. (1989) *Accidental explosions*: Vol. 1: Physical and Chemical Properties; Vol. 2: Types of Explosive Substances (Ellis Horwood).
National Fire Protection Association (1991) *Fire hazard properties of flammable liquids, gases and volatile solids* (NFPA).
National Fire Protection Association (1990) *Identification of the fire hazards of materials, 704–90* (NFPA).
Pratt, T. H. (1991) *Electrostatic ignition hazards* (Burgoynes Inc)
Schofield, C. and Abbott, J. A. (1988) *Guide to dust explosion prevention and protection – part 2: ignition prevention, containment, inerting, suppression and isolation* (Institution of Chemical Engineers).
Schultz, N. (1985) *Fire and flammability handbook* (Van Nostrand Reinhold).
Tuhtar (1989) *Fire and explosion protection: a systems approach* (Ellis Horwood).
Walls, W. L. (ed.) (1986) *Liquefied petroleum gases handbook* (National Fire Protection Association).

Chemical hazards

Chemical substances can be dangerous by their direct effects (e.g. being toxic, explosive) or by reaction with other materials (including water, air and people). The hazard is not always obvious. For example a mild reaction can cause excessive pressure leading to rupture of a closed container. Furthermore, many reactions are very different on a large scale (particularly if heat cannot be dissipated readily) and some may be unexpectedly accelerated by contaminants. Care should therefore be

Safety engineering 323

taken when consulting chemical texts which commonly refer to small scale and pure materials.

Sources of information on single chemicals include:

Carson, P. A. and Mumford, C. J. (1992) *Newnes hazardous chemicals pocket book* (Butterworth-Heinemann).
Carson, P. A. and Mumford, C. J. (1993) *Hazardous chemicals handbook* (Butterworth-Heinemann).
Environmental Protection Agency (USA) (1987) *Extremely hazardous substances list 40 CFR part 355* (National Technical Information Service).
International Technical Information Institute (1984) *Toxic and hazardous industrial chemicals: safety manual for handling and disposal with toxicity and hazard data* (ITII, Tokyo).
NIOSH pocket guide to chemical hazards (1990) (National Institute for Occupational Safety and Health).
Richardson, M. I. (ed.) (1993) *Directory of substances and their effects* (Royal Society of Chemistry).
Royal Society of Chemistry (1989–) *Chemical safety data sheets*: Vol. 1 – Solvents (1989); Vol. 2 – Main group metals and their compounds (1989); Vol. 3 – Corrosives and irritants (1990); Vols 4a and 4b – Toxic chemicals (1991); Vol. 5 – Flammable chemicals (1993) (RSC).
Sax, N. I. and Lewis, R. J., Sr (1992) *Dangerous properties of industrial materials*, 8th ed. (Van Nostrand Reinhold).
Sax, N. I. and Lewis, R. J., Sr (1993) *Hazardous chemicals desk reference*, 3rd ed. (Van Nostrand Reinhold).
Sax, N. I. and Lewis, R. J., Sr (1994) *Rapid guide to hazardous chemicals in the workplace*, 3rd ed. (Van Nostrand Reinhold).
US Coast Guard (1990) *CHRIS hazardous chemicals data manual* (Lab Safety Supply).
Weiss, G. (ed.) (1986) *Hazardous chemicals data book*, 2nd ed. (Noyes Data Corporation).

The pre-eminent source on dangerous reactions is Bretherick (1990) which is also available on computer disk and CD-ROM. Other sources are:

Medard, L. A. (1989) *Accidental explosions*, 2 vols (Ellis Horwood).
National Fire Protection Association (1975) *Manual of hazardous chemical reactions*, 5th ed. (NFPA).
Yoshida, T. (1980) *Handbook of hazardous reactions with chemicals* (Tokyo Fire Dept).

For industrial scale use of chemicals and emergency treatment of chemical releases, the following have specific information:

American Petroleum Institute (1990) *Management of process hazards API recommended practice 750* (American Petroleum Institute).
Association of American Railroads (1987) *Emergency handling of hazardous materials in surface transportation* (AAR).
Barton, J. and Rogers, R. (eds) (1993) *Chemical reaction hazards* (Institution of Chemical Engineers).

Bennett, G. F., Feates, F. S. and Wilder, I. (eds) (1982) *Hazardous materials spills handbook* (McGraw-Hill).
Benuzzi, A. and Zaldivar, J. M. (eds) (1991) *Safety of chemical batch reactors and storage tanks* (Kluwer Academic).
Buzzi, R. A. (1992) *Chemical hazards at water and wastewater treatment plants* (CRC Press).
Carson, P. A. and Mumford, C. J. (1988) *The safe handling of chemicals in industry*. Vol. 1, 2 and 3 (1995) (Longman).
Center for Chemical Process Safety (1992) *Chemical reactivity evaluation and application to process design* (American Institute of Chemical Engineers).
Center for Chemical Process Safety (1988) *Guidelines for safe storage and handling of high toxic hazard materials* (American Institute of Chemical Engineers).
Conlon, P. C. and Mason, A. M. (eds) (1984) *Emergency action guides* (Association of American Railroads).
Fawcett, H. H. and Wood, W. S. (1982) *Safety and accident prevention in chemical operations*, 2nd ed. (Wiley).
Foden, C. R. and Weddell, J. L. (1992) *Hazardous materials: emergency action data* (CRC Press).
Henry, M. F. (ed.) (1992) *Hazardous materials response handbook* (National Fire Protection Association).
Hoffman, J. M. and Maser, D. C. (eds) (1985) *Chemical process hazard review: ACS Symposium Series 274* (American Chemical Society).
Lipton, S. and Lynch, J. (1987) *Health hazard control in the chemical process industry* (Wiley).
Meyer, E. (1989) *Chemistry of hazardous materials*, 2nd ed. (Prentice Hall).
Neely, W. B. (1992) *Emergency response to chemical spills* (CRC Press) Software and manual.
Sacarello, H. L. A. (1993) *The comprehensive handbook of hazardous materials: regulations, handling, monitoring and safety* (CRC Press).
Seaton, W. H., Freedman, E. and Treweek, D. N. (1974) *CHETAH: the ASTM Chemical Thermodynamic and Energy Release Potential Evaluation Program DS51* (American Society for Testing and Materials).
Weaver, L. A. (1991) *Techniques for hazardous chemical and waste spill control* (L. A. Weaver Co).
Yoshida, T. (1987) *Safety of reactive chemicals* (Elsevier).

Nuclear and radioactive safety

Safety is of course an overwhelming consideration in nuclear engineering and is included in many publications. Some particular references are:

Ebert, K. and von Ammon, R. (eds) (1989) *Safety of the nuclear fuel cycle* (VCH).
Institution of Mechanical Engineers (1993) *Nuclear power plant safety standards: towards international harmonization. Proceedings of an international conference*, 26–28 October, 1993 (Mechanical Engineering Publications).
International Atomic Energy Agency (1986) *Manual on maintenance of systems & components important to safety* (IAEA).

International Atomic Energy Agency (1989) *Fire protection and fire fighting in nuclear installations* (IAEA).
International Atomic Energy Agency (1991) *Emergency power systems at nuclear power plants: a safety review* (IAEA).
Orn, M. K. (1992) *Handbook of engineering control methods for occupational radiation protection* (Prentice Hall).
See also under 'Reliability': Bamford *et al.* (1991, 1992) and Nuclear Regulatory Commission (1982).

Design, management and operation

Actual techniques for achieving safety or for dealing with emergencies in engineering are covered in the following:

Algar, P. (1993) *Managing industrial emergencies – a planning and communications guide* (Financial Times Business Information).
American Petroleum Institute (1990) *Management of process hazards API recommended practice 750* (American Petroleum Institute).
American Society of Mechanical Engineers (1984) *An instructional aid for occupational safety and health in mechanical engineering design* (ASME).
British Plastics Federation Thermosetting Material Group (1979) *Guidelines for the safe production of phenolic resins* (British Plastics Federation).
Center for Chemical Process Safety (1993) *Guidelines for auditing process safety management systems* (American Institute of Chemical Engineers).
Center for Chemical Process Safety (1993) *Guidelines for engineering design for process safety* (American Institute of Chemical Engineers).
Center for Chemical Process Safety (1992) *Guidelines for investigating chemical process incidents* (American Institute of Chemical Engineers).
Center for Chemical Process Safety (1992) *Guidelines for safe automation of chemical processes* (American Institute of Chemical Engineers).
Center for Chemical Process Safety (1989) *Guidelines for technical management of chemical process safety* (American Institute of Chemical Engineers).
Center for Chemical Process Safety (1991) *Plant guidelines for technical management of chemical process safety* (American Institute of Chemical Engineers).
Chemical Manufacturers' Association (1985) *Process safety management: control of acute hazards* (CMA).
Compressed Gas Association (1990) *Handbook of compressed gases*, 3rd ed. (Van Nostrand Reinhold).
Cox, S. J. and Tait, N. R. S. (1991) *Reliability, safety and risk management: an integrated approach* (Butterworth-Heinemann).
Croner's Manual Handling Operations (1993) (updated annually) (Croner Publications).
Crowl, D. A. and Louvar, J. F. (1990) *Chemical process safety: fundamentals with applications* (Prentice Hall).
Cumo, M. and Naviglio, A. (eds) (1989) *Safety design criteria for industrial plants* (CRC Press).

Fawcett, H. H., & Wood, W. S., (1982) *Safety and accident prevention in chemical operations*, 2nd ed. (Wiley).

Hammer, W. (1989) *Occupational safety management and engineering*, 4th ed. (Prentice Hall).

Institute of Petroleum (1981–1993) *Model code of safe practice in the petroleum industry* (Wiley).

Institution of Mechanical Engineers (1993) *PROCESS TECH 93: The design and operation of safe and profitable process plant*, a conference (Mechanical Engineering Publications).

Institution of Mechanical Engineers (1993) *The successful management for safety:* papers presented at a meeting organized by the Engineering Manufacturing Industries Division of the Institution of Mechanical Engineers held at the Institution of Mechanical Engineers, 12–13 October, 1993 (Mechanical Engineering Publications).

International Labour Office (1981) *Civil engineering work: a compendium of occupational safety practice* (ILO).

Kenney, W. F. (1993) *Process risk management systems* (VCH).

King, R. W. (1990) *Safety in the process industries* (Butterworth-Heinemann).

King, R. W. and Hudson, R. (eds) (1985) *Construction hazard and safety handbook* (Butterworth-Heinemann).

Mill, R. C. (ed.) (1992) *Human factors in process operation* (Institution of Chemical Engineers).

Petersen, D. (1988) *Safety management: a human approach*, 2nd ed. (Aloray).

Petersen, D. (1989) *Techniques of safety management: a systems approach*, 3rd ed. (Aloray).

Ridley, J. R., (ed.) (1990) *Safety at work*, 3rd ed. (Butterworth-Heinemann).

Roland, H. E., and Moriarty, B. (1990) *System safety engineering and management*, 2nd ed. (Wiley).

Scott, D. and Crawley, F. (1992) *Process plant design and operation* (Institution of Chemical Engineers).

Shell, R. L. and Simmons, R. J. (1990) *An engineering approach to occupational safety and health in business and industry* (Institute of Industrial Engineers).

Theodore, L., Reynolds, J. P. and Taylor, F. B. (1989) *Accident and emergency management* (Wiley).

Thomen, J. R. (1991) *Leadership in safety management* (Wiley).

Townsend, A. (1992) *Maintenance of process plant: a guide to safe practice*, 2nd ed. (Institution of Chemical Engineers).

US Department of Transport (1990) *Hazardous materials emergency response guidebook: DOT-P-5800.5* (US Department of Transport).

Wadden, R. A. and Scheff, P. A. (1987) *Engineering design for the control of workplace hazards* (McGraw-Hill).

Equipment and structures

Bennett, P. (ed.) (1992) *Safety aspects of computer control* (Butterworth-Heinemann).

Blockley, D. I. (1980) *The nature of structural design and safety* (Ellis Horwood).

Safety engineering

British Cryogenics Council (1991) *Cryogenics safety manual*, 3rd ed. (Butterworth-Heinemann).
Coburn, A. W. and Spence, R. J. S. (1992) *Earthquake protection* (Wiley).
Cubbon, R. C. P. (1991) *Health and safety in ceramics: a guide for educational workshops and studios*, 3rd ed. (Institute of Ceramics).
Dowrick, D. J. (1987) *Earthquake resistant design: for engineers and architects*, 2nd ed. (Wiley).
French, D. N. (1993) *Metallurgical failures in fossil-fired boilers* 2nd ed. (Wiley).
Hart, G. C. (1982) *Uncertainty analysis, loads, and safety in structural engineering* (Prentice Hall).
Krinitzsky, E. L., Gould, J. P. and Edinger, P. H. (1993) *Fundamentals of earthquake resistant construction* (Wiley).
Lindley, J. (1987) *User guide for the safe operation of centrifuges*, 2nd rev. ed. (Institution of Chemical Engineers).
Madsen, H. O., Krenk, S. and Lind, N. C. (1986) *Methods of structural safety* (Prentice Hall).
National Safety Council (1981) *Guards illustrated: ideas for mechanical safety*, 4th ed. (NSC).
Parry, C. F. (1991) *Relief systems handbook* (Institution of Chemical Engineers).

Institutions

Some relevant organizations are:

American Society of Safety Engineers, 1800 East Oakton Street, Des Plaines, IL 60018, USA.
Fire Protection Association, 140 Aldersgate Street, London EC1A 4HX, UK.
National Fire Protection Association, 1 Batterymarch Park, Quincy, MA 02269, USA.
National Safety Council, 1121 Spring Lake Drive, Itasca, IL 60143, USA.
Safety Equipment Institute, 1901 N. Moore Street, Suite 808, Arlington, VA 22209, USA.
Safety and Loss Prevention Subject Group, Institution of Chemical Engineers, 165–171 Railway Terrace, Rugby CV21 3HQ, UK (open to non-members of the IChemE).

Journals

Some journals largely devoted to safety in industry are:

Accident Analysis and Prevention (Pergamon) bimonthly.
Chemical Hazards in Industry (Royal Society of Chemistry) monthly.
Diagnostic Engineering (Institution of Diagnostic Engineers) bimonthly.
Engineering Failure Analysis (Pergamon) quarterly.
Fire and Materials (Wiley) bimonthly.

Fire Safety Engineering (formerly Fire Surveyor) (Paramount) bimonthly.
Health and Safety Newsline (Engineering Employers' Federation) 8 per annum.
Journal of Loss Prevention in the Process Industries (Butterworth-Heinemann) 5 per annum.
Loss Prevention Bulletin (Institution of Chemical Engineers) bimonthly.
National Safety Council Industrial Section Newsletters: (National Safety Council) there are 26 industry-specific newsletters on occupational safety and health, ranging from Aerospace to Textiles, each bimonthly.
Process Safety Progress (American Institute of Chemical Engineers) quarterly.
Professional Safety (American Society of Safety Engineers) monthly.
Safety Management (British Safety Council) 11 per annum.

Software

There is an increasing number of safety-related software items, ranging from databases to expert systems. The American Society of Safety Engineers produces a *Directory of safety related computer resources* (ed. R. Brauer, 1993, updated annually).

For hazards due to software failures, see 'Reliability' above.

References

Abercrombie, S. A. (1988) *Dictionary of terms used in the safety profession*, 3rd ed. Des Plaines, IL: American Society of Safety Engineers.
Blockley, D. I. (ed.) (1992) *Engineering safety*. London: McGraw-Hill.
Brauer, R. L. (1990) *Safety and health for engineers*. New York: Van Nostrand Reinhold.
Bretherick, L. (1995) *Bretherick's handbook of reactive chemical hazards* 5th ed. Oxford: (Butterworth-Heinemann).
Burditt, M. F. (ed.) (1993) *Product safety management and engineering*, 2nd ed. Des Plaines, Ill.: American Society of Safety Engineers.
Gloss, D. S. and Wardle, M. G. (1984) *Introduction to safety engineering*. New York: Wiley.
Hunter, T. A. (1992) *Engineering design for safety*. New York: McGraw-Hill.
Jones, D. A. (ed.) (1992) *Nomenclature for hazard and risk assessment in the process industries*, 2nd ed. Rugby: Institution of Chemical Engineers.
Pant, P. D. (1993) Transportation engineering textbook needs, *ITE Journal*, **63**(4), 33–39.
Winburn, D. C. (1988) *Practical electrical safety*. New York: Marcel Dekker.

CHAPTER NINETEEN

Design and ergonomics

P. J. HICKS

Introduction

The strategic role of design in industry is well recognized. Two particular reasons are that the design function is an integrating one, also that, often, a major proportion of total project cost is committed at the design stage. Because design is central to engineering, the majority of engineering graduates are likely to interface with it during their subsequent careers, whilst others will become professional designers. Consequently, and because of its importance to the economy, a well-founded appreciation of design is essential.

The strategic importance of design has been addressed at national level over the years in many reports including those by Feilden (1963), Moulton et al. (1976), Corfield (1979), Finniston (1980), Lickley et al. (1983) and CNAA (1984). These have all discussed the linkages between academia and industry.

It is the aim of this chapter to refer to the spectrum of publications that may be of use in understanding 'the discipline of design' and the many relevant associated interfaces. Design is clearly a discipline in its own right, as well as being an integrating and management facility. Also, ergonomics is an important factor in many design considerations.

Publications concerned with engineering analysis are not included since they are covered in the specialist chapters within this text. However, this chapter clearly sets the bedrock from which specialized and detailed analytical work will be effective because an appreciation of the discipline of design, the techniques and methods, together with the realities of engineering judgement present a fundamental basis for setting up pragmatic analytical models.

Where reference to a publication is made, it is not implied that the reference is the only one on such a topic, or that it contains only information on that topic.

What is design

Definitions of design are many and varied because of the scope and range of both technical and non-technical factors. It is that total activity necessary to provide an artefact or system, often in the context of cost effectively meeting a market need. It is this 'total activity', sometimes defined as 'coming to grips with complexity', that is found to be so challenging and rewarding to many.

Design is seen to have the following characteristics: central to engineering, transdisciplinary, highly complex, highly iterative, highly interactive. It demands of practitioners the following attributes: ability to communicate, creative as well as analytical skills, ability to integrate, use of judgement, management ability.

The underlying principles, concepts and general methods which can be employed for whole classes of problems should be capable of application in a disciplined and natural manner. The usual design problem-solving situation is one in which there is a need to progress from an initial situation of little understanding (the design brief) to a final situation of a successful solution, fully specified, achieved through the time and effort expended. It is convenient to use a product-oriented model of the activity with various related phases identified. One such model is that due to SEED (Sharing Experience in Engineering Design) (1985) and is presented in Figure 19.1.

While following the project development vertically from top to bottom, the model also identifies the iterative nature of design. Useful information sources about the nature of the identified phases of the design activity are available.

Appropriate techniques abound; such as creativity, evaluation and decision-making, quality and reliability considerations, communication and management, to identify but a few.

Direct reference material is also prolific; typically codes and standards, ergonomic data, patents, trade journals, technical reports, technical journals, conference papers, manufacturing and materials data.

It is an understanding of the disciplined approach, together with the appropriate techniques, that forms the basis for good design across the branches of engineering and, indeed, out into the sciences, and commerce.

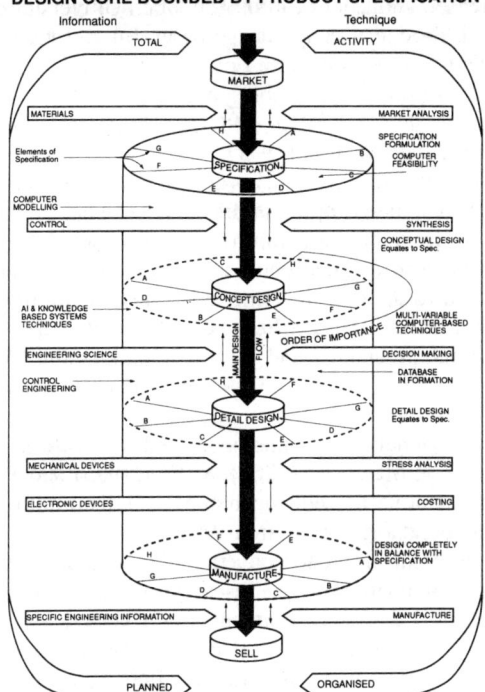

Figure 19.1 Design Activity Model

Design activity

This section encompasses such terminology as design methods, the design process, product design, total design. It considers the design activity; either through product-based or process-based models.

A readable and broadbased coverage is given in Jones (1981) and Ertas and Jones (1993). Cross (1994) includes a detailed comparison of various design methods. A light-hearted but penetrating soft systems approach is presented by Koberg and Bagnall (1981). An overview of design methods, liberally laced with practical examples, is given by Hawkes and Abinett (1984) and Suh (1990). Simon (1975), Middendorf (1990) and Pugh (1991) link design methods to the realities of industrial requirements at a broad level, whilst Pahl and Beitz (1995) go into greater detail with respect to procedural and machine element considerations. Mechanical engineering oriented approaches are covered in Ullman (1992) and an electrical engineering viewpoint is presented in Wilcox et al. (1990).

Systematic design is considered by Meredith et al. (1985) and

Hubka, Andreasen and Eder (1988). The Open University course T392 on Engineering Product Design has some useful and varied case study material. Broadbased texts include Love (1986), Starkey (1988), Ulrich and Eppinger (1994) and Roozenburg and Eekels (1994). Of more general interest reading are Glegg (1969), Newell and Simon (1971), Broadbent (1973) and Walton (1991).

A comprehensive interdisciplinary design approach which deals with a wide variety of related topics is presented in Dieter (1991). French (1985) adopts a more creative approach to conceptual design methods with some linkage to detailed engineering analysis. French (1992 and 1994) moves on to address the deep and subtle relationships between design and engineering science. Burgess (1984) puts a design assurance viewpoint through procedural and control considerations. A broadbased text linking design methods, analytical and management techniques is Shigley (1990).

The formal sequencing of elements of the design activity is discussed in Asimow (1962), Matousek (1963), SEED (1985) and Polak (1991). The direct application of a sequenced approach, incorporating elements of Matchett's Fundamental Design Method into PABLA (Problem Analysis By Logical Approach), is found in Latham (1965). Archer (1971) presents the continuous development of models through iteration and Morgan (1972) considers element interdependencies.

Examples of texts relating to decision theory are Lindgren (1971), Siddall (1972), Davis (1983) and Starkey (1992).

The use of design methods and techniques for the management of design and projects is covered later in this chapter.

Creativity

In the efforts to encourage innovation and creativity in engineering design, sources look at factors inducing psychological set, and at techniques for increasing the solution space searched. The psychological aspects are covered in Gruber (1967) and Vernon (1970). Broadening techniques are presented in Gordon (1961), Osborn (1963) and de Bono (1970). Thring and Laithwaite (1977) present how to invent.

Alger and Hays (1964), Edel (1967), Walker, Dagger and Roy (1991) and Lumsdaine and Lumsdaine (1995) look at creative problem solving. Bailey (1978) considers disciplined creativity. Broader considerations are by Gregory (1972).

Gardiner and Rothwell (1985) and Holt (1987) consider innovation whilst Piatier (1984) considers barriers to innovation. Design and innovation linked to the marketplace is presented by Walsh *et al.* (1992).

Design management

This section encompasses such terminology as; corporate strategy, product development, product planning, design reviews, quality and reliability, dealing with risk, concurrent engineering, optimization and sensitivity analysis.

The strategic role of design in industry is without question. One of the reasons for its importance is that in many cases up to 80 per cent of the total project cost is committed at the design stage. Also, many companies design then sub-contract much of the manufacture, hence it is crucial that the design is 'right first time' and can be costed.

Linked directly to the design activity model already identified, SEED (1985) is a design management facility. It is an aid to multi-disciplinary group teamworking, concurrent engineering (see separate chapter), audits, quality, project control, and in line with legislative implications for the designer.

In overviewing the design activity model presented above, it is useful to identify the key stages. In simple terms, the designer usually produces several possible solutions to the problem posed, then decides which is the 'best' solution measured against particular criteria. Now what of these criteria? Where are they to be found and appropriately quantified? This datum for all decision making is the PRODUCT DESIGN SPECIFICATION (PDS) which is the developed result of the needs analysis phase and feedback from other phases.

The PDS sets out the desirable outcomes and general restraints for a design which has yet to be accomplished. In particular, it is useful to separate the PDS into mandatory (must have) and optional (would like) sections. In assuming that an open-minded and creative approach has resulted in the proposal of several viable alternative *ideas* then *evaluation* can proceed. Here, the ideas for total systems, or sub-systems of the whole, are measured against the parameters specified in the PDS.

Guidance on the preparation of the PDS. is in SEED's *Specification phase* (1986). The use of the PDS as a datum for good design cannot be overemphasized. Other related and linked publications from SEED are *Quality and reliability* (1989c), *Manufacturing phase* (1989b), and *Detail design phase* (1993b).

British Standards publications; BS 7000: Part 1: 1989, BS EN ISO 9000: 1994, BS 5760: 1986, BS 7373: 1991 have relevant procedural and information guidelines.

Corporate strategies and management in industry should interface with design. Broad considerations are covered in Lorenz (1990), Heskett (1989) and Olins (1989). Texts specifically addressing the management of design include those by Leech (1972), Topalian (1980), Oakley (1984), Roy and Wield (1989), Ettlie and Stoll (1990), Kicherer (1990) and Hales (1993). A more general text is by Siddall (1972).

Designers usually operate in a situation of endeavouring to balance competing requirements (variables) in a cost and time effective way. It is important that the significance of these variables is assessed in an organized manner and priority given to those that have most influence on the 'total' situation being considered. This involves sensitivity analysis, Pareto principles, logic judgement and justification. The mathematical modelling of multi-variable systems for optimization and sensitivity considerations ranges from linear programming techniques, simultaneous and sequential searches, through to simplex search, gradient search and quadratic convergence techniques.

Optimization techniques are dealt with in Johnson (1980), Siddall (1982) and Arora (1989).

Examples of less esoteric techniques are CPA (Critical Path Analysis) and PERT (Programme Evaluation by Review Technique) which consider the design and optimization of sequenced dependent events. Taguchi (1986) links design variables to cost considerations with a view to realizing 'robust designs' of enhanced quality at reduced cost. This is developed by Barker (1990).

Quality in design is covered as an Open University Course (PT619) by Bruce (1986). A thorough reference text that covers quality and reliability in the context of design influences and interfaces is by SEED (1989c). In-depth sources linking design and quality include Juran (1988), DeVor, Chang and Sutherland (1992) and Cortada and Woods (1995). Broader texts are by Raheja (1991), Logothetis (1993) and Oakland (1993).

Risk and safety considerations are dealt with in Lowrance (1976), Blakstad (1979), Griffiths (1981), Health and Safety Executive (1989) and Hunter (1992).

Legislative implications on the designer are significant. The Health and Safety at Work Act (1974) requires designers, amongst others, to show that they have considered all relevant factors, at that state of the art, in their decision making. The Health and Safety Executive issue advisory literature. Particularly useful general references are Croner's *Health and Safety at Work* encyclopaedia and the *Barbour Microfile: Health and Safety*. Similarly, product liability legislation makes it important that a record of all work carried out and decisions taken be recorded formally in a project working file.

Abbott (1980 and 1987) and Wright (1989) specifically address product liability and the management of design risks. Broader texts are by Miller (1988) and Clark (1989). European considerations are presented in Kelly and Attree (1992).

Computer-aided design

The use of computers for modelling, analysis, manufacturing and related alpha-numeric information handling is well established. The integration of these activities in a manner which allows iteration and development across the total engineering operation has become well developed. However, the transfer of data between software packages through such facilities as IGES (International Graphics Exchange System) can still be problematical. This 'integrated' approach has helped concurrent engineering concepts and at the hub of this is the setting up of the engineering design through computer modelling. More recent developments include the place and use of artificial intelligence (AI) techniques.

As an engineering designer, one should never forget that computer-aided activities are but yet another tool to assist. The statement 'rubbish in–rubbish out!' clearly identifies the need for the human to ensure that input data are valid. Given that it is, then 'what-if' scenarios, sensitivity analysis, simulation and cross-discipline integration are excellent facilities available through the use of computers.

Computer-aided design and manufacture (CAD/CAM) is specifically addressed in Knox (1983) and Preston, Crawford and Coticchia (1984). Groover and Zimmers (1984), Haigh (1985), Medland and Mullineux (1988), Besant and Lui (1991), Rooney and Stedman (1991) and Medland (1992) have a broad coverage of computer-aided design and the interface to manufacture. Jones (1992) and Tizzard (1994) give detailed consideration to CAD/CAM and its implementation within industry. More detailed computer graphics considerations are presented in Rodriguez (1992), Earle (1993) and Ryan (1994).

Hsu and Sinha (1992) consider the integration of analytical techniques through computers. The interactions between design and analytical techniques are dealt with by Berk (1988).

The specific use of Fortran, other languages and applications packages for the solution of engineering and mathematical problems is covered in Jewell (1991).

Artificial intelligence interfaces to design are presented in Simon (1981), Rich and Knight (1991), Winston (1993) and Dym (1994). Ergonomics considerations feature in some of these texts.

Ergonomics

The designer should never forget the place of the human as maker, user, maintainer and indeed mis-user of equipment and systems. One of the objectives of good design should be to endeavour to reduce human fatigue. The manifestation of this ranges from increased productivity, reduced tiredness and stress, to the amelioration of RSI (repetitive strain injury) in the human. Less tangible are the psychological and sociological impacts.

The marketplace and commercial success of products and systems is very often linked to a thorough consideration of ergonomic factors through design. Ergonomics can be viewed as comprising the three areas of anthropometrics, cybernetics and environmental considerations. The three are interlinked in many ergonomic situations with which designers are involved.

Broad and thorough coverage of ergonomics is found in Galer (1987), Pheasant (1987), Sanders and McCormick (1993), SEED (1993a), Oborne (1994) and Bridger (1995). Most of these have large bibliography and reference sections. A useful introduction is by Garner (1991). Industrially related publications are from Alexander (1986), Wykes (1988) and Costanzo (1992). Shigley and Mischke (1986) consider hazards of human origin and safety aspects. The need for products that customers find easy to use and the resulting psychological impact is presented by Norman (1988).

Anthropometrics and related data are presented in Croney (1980) Easterby (1982), Knight (1984), Diffrient *et al.* (1987) and Pheasant (1984 and 1986).

Considerations of human-computer interaction appear in Downton (1991) and Johnson (1992), and also in some of the computer-aided texts referenced previously.

Industrial design

This subject is sometimes referred to as aesthetics in design, three-dimensional design, engineering product design, product development.

It relates to the more aesthetic aspects of design, as compared to the functional and performance aspects. Since many products and systems sell on the basis of their attractive looks as well as, or indeed in preference to, their performance, it is an important subject area for consideration.

A review of industrial design education was carried out by The Design Council (1977). Broad texts include Ashford (1970), Maldonado (1972), Loewy (1979), Heskett (1980), Pye (1982), Flurscheim (1983), Papanek (1984) and Dormer (1990).

The relevance of the whole spectrum of aesthetics considerations throughout the design activity is addressed in SEED (1992). More directed material is by Manzini (1989), Russell (1990) and Bayley (1991).

Industrial interfaces

The strategic importance of design in industry has already been identified and the legislative and management considerations for industry have been dealt with. This section particularly identifies costs, profitability, optimization and sensitivity analysis in the context of design, and also the interfaces between design, manufacture and materials. Broadbased texts are by Cox and Webb (1978), Ray (1985) and SEED (1989a). Design and cost interactions are addressed in Heller (1971), Radke (1972), Chow (1978), Malstrom (1981) and Humphreys and Wellman (1986). Comprehensive cost estimating is presented in Stewart, Wyskida and Johannes (1995). Value analysis and value engineering are considered in Miles (1972). Design for profit concepts are presented by Leech and Turner (1985), and Service, Hart and Baker (1989).

In industry, the links between design, manufacture and materials need to be strong for competitive edge. BSI (1981) produced a guide addressing design for economic production and another very useful text is by Cooke et al. (1984).

A thorough coverage of design for manufacture with pertinent case studies is presented in Corbett et al. (1991). Bralla (1986) provides many 'design for manufacture guidelines' as do Ludema, Caddell and Atkins (1987). Boothroyd and Dewhurst (1987), Andreason et al. (1988) and Redford and Chal (1993) cover design for assembly. Black (1995) presents detailed considerations of design, manufacture and their integration for effective product development.

Useful papers are by Stoll (1988) on American design for manufacturing, and by Whitney (1988) on design and manufacture links.

Materials considerations in design are addressed by Ashby and Jones (1987), Cornish (1987), Charles and Crane (1989), Ashby (1992) and Derby, Hills and Ruiz (1992). A broadbased text in integrating materials, manufacture and design is by Farag (1989). Design for recyclability is presented in Henstock (1988).

Detail design

It is so true of many aspects of engineering that 'the devil is in the detail'. Good detail design can be the difference between company profitability and failure and it often links to the careful assessment of related materials and manufacturing considerations. Typically, effective dimensioning and tolerancing of components and the correct selection and use of 'standard parts' enhance the quality, reliability and cost-effectiveness of product and system realization.

Parker (1991) (in collaboration with the British Standards Institution) is an excellent reference on these matters. So too is the coverage of how to ensure good detail design in SEED (1993b).

Simmons and Maguire (1995) present engineering drawing, tolerancing and standard components. BS 308: Parts 1, 2 and 3: 1985–1993 covers detail drawing and tolerancing conventions. Mucci (1994) is a very useful quick reference source for the application of a broad spectrum of standard components and materials.

It is, of course, important that the details of a design are well communicated both verbally and in writing. Advice on effective communication is presented in Whiteman (1982), Eyre (1983), Scott (1984) and Turk and Kirkman (1989).

An excellent publication covering reading, listening, graphics, interpersonal, oral and technical writing skills is SEED (1988).

Journals

The nature, breadth and depth of design is such that coverage of the topic through journals is vast. The listing below is a representative sample of journals and periodicals.

Engineering design

Design Engineering (Morgan Grampian) monthly.
Design Engineering (Maclean Hunter) 10 per annum.
Design Management Journal (Design Management Institute) quarterly.
Design News (Cahners) bimonthly.
Design Products and Applications (IML Group) monthly.
Design Studies (Butterworth-Heinemann) quarterly.
Engineer (Morgan Grampian) 32 per annum.
Engineering (Gillard Welch) monthly.
Engineering Design Graphics Journal (Ohio State University) 3 per annum.
Engineering Designer (Institution of Engineering Designers) bimonthly.
Eureka: Engineering Materials and Design (Innopress) monthly.
Health and Safety at Work (Tolley Publishing) monthly.
International Journal of Technology and Design Education (Kluwer Academic Publishers) 3 per annum.
International Journal of Vehicle Design (Interscience Enterprises) bimonthly.
Journal of Design and Manufacturing (Chapman and Hall) quarterly.
Journal of Engineering Design (Carfax Publishing) quarterly.
Journal of Mechanical Design. Transactions of the ASME (American Society of Mechanical Engineers) quarterly.
Journal of Strain Analysis for Engineering Design (Mechanical Engineering Publications) quarterly.
Machine Design. (Penton Publishing) 23 per annum.
Materials and Design (Butterworth-Heinemann) bimonthly.
Mechanical Engineering (American Society of Mechanical Engineers) monthly.
OEM Design (Wilmington Publishing) 11 per annum.
Proceedings of the Institution of Electrical Engineers (Institution of Electrical Engineers) various parts, bimonthly.
Proceedings of the Institution of Mechanical Engineers (Mechanical Engineering Publications) various parts and frequency of publication.

Professional Engineering (Mechanical Engineering Publications) 22 per annum.
Quality and Reliability Engineering International (Wiley) bi-monthly.
Research in Engineering Design (Springer-Verlag) quarterly.
Transactions of the ASME (American Society of Mechanical Engineers) various parts, quarterly.
What's New in Design (Morgan Grampian) monthly.
What's New in Industry (Morgan Grampian) monthly.

Ergonomics

Advances in Human Factors Ergonomics (Elsevier) irregular.
Applied Ergonomics (Butterworth-Heinemann) bimonthly.
Contemporary Ergonomics: Proceedings of the Ergonomics Society Annual Conference (Taylor and Francis) annual.
Ergonomics: The Official Publication of the Ergonomics Research Society (Taylor and Francis) monthly.
Ergonomics in Design (Human Factors and Ergonomics Society) quarterly.
Human-Computer Interaction (Lawrence Erlbaum) quarterly.
Human Factors: The Journal of the Human Factors Society (Human Factors and Ergonomics Society) quarterly.
Human Movement Science (Elsevier) bimonthly.
Human Systems Management (IOS Press) quarterly.
International Journal of Human-Computer Interaction (Ablex Publishing) quarterly.
International Journal of Human-Computer Studies (Academic Press) monthly.
International Journal of Human Factors in Manufacturing (Wiley) quarterly.
International Journal of Industrial Ergonomics (Elsevier) monthly.
Proceedings of the Human Factors and Ergonomics Society Annual Meeting (Human Factors and Ergonomics Society) annual.

Computer-aided activities

Artificial Intelligence for Engineering Design, Analysis and Manufacturing: AIEDAM (Cambridge University Press) 5 per annum.
CAD/CAM (EMAP Readerlink) monthly.
Computer-Aided Design (Butterworth-Heinemann) monthly.
Computer Aided Design Report (CAD/CAM Publishing) monthly.
Computer-Aided Engineering: CAE (Penton Publishing) monthly.
Computer Aided Geometric Design (Elsevier) 8 per annum.
Computer Graphics (ACM Association for Computing Machinery) quarterly.
Computer Graphics Forum: A Journal of the European Association for Computer Graphics (Blackwell) 5 per annum.
Computers and Graphics (Pergamon) bimonthly.

Standards and guides

Guidance on the searching for and use of information is covered by Wall (1986) and SEED (1987). Specific consideration of information access and storage for design has been given by Court, Culley and McMahon (1993).

National and international standards cover many procedural and analytical aspects of design and related activities. Handbooks listing all the standards are usually produced annually; typically the *BSI standards catalogue*, and the *Annual book of ASTM standards*. CD-ROM services are also available.

Product and component information is available from many sources, including Technical Indexes (Bracknell) and RS Catalogues (Corby, Northants)

General reference books and buyers' guides include:

ASM handbook, 10th ed. (1990–1) (American Society for Metals).
Bralla, J. D. (ed.) (1986) *Handbook of product design for manufacturing* (McGraw-Hill).
Engineer buyers guide (annual) (Morgan Grampian).
Green, R. E. (ed.) (1992) *Machinery's handbook*, 24th ed. (Industrial Press).
Institution of Engineering Designers – Official reference book and buyers' guide (annual) (IML Group).
Kempe's engineers yearbook (annual) (Morgan Grampian).
Kompass: the authority on British industry (annual) (Kompass).
Jones, G. R., Laughton, M. A. and Say, M. G. (1993) *Electrical engineer's reference book*, 15th ed. (Butterworth-Heinemann).
Lingaiah, K. (1994) *Machine design data handbook* (McGraw-Hill).
Machinery buyers guide (annual) (Machpress).
Parmley, R. O. (ed.) (1985) *Mechanical components handbook* (McGraw-Hill).
Parmley, R. O. (ed.) (1989) *Standard handbook of fastening and joining* 2nd ed. (McGraw-Hill).
Rothbart, H. A. (ed.) (1985) *Mechanical design and systems handbook* 2nd ed. (McGraw-Hill).
Smith, E. H. (ed.) (1994) *Mechanical engineer's reference book* 12th ed. (Butterworth-Heinemann).
Thomas register of American manufacturers. Thomas Publishing (annual).

Directories available include:

Adams, R. W. (ed.) (1991) *Directory of European Industrial and trade Associations*, 15th ed. (CBD Research).
Henderson, S. P. A. and Henderson, A. J. W. (eds) (1994) *Directory of British Associations*, 12th ed. (CBD research).
Murphy, S. and Henderson, C. A. P. (eds) (1993) *Current British Directories*, 12th ed. (CBD Research).
O'Connor, T. (ed.) (1994) *Current European Directories*, 3rd ed. (CBD Research).

Procedural guides include:

Engineering Sciences Data Unit (ESDU) guides on many selection and detail design activities.

Design and ergonomics 341

Roark's formulae for use in stress and strain (Roark and Young, 1989).

SEED Design procedural guides covering: Operational Amplifiers, Planar Mechanisms, Springs, Rotary Power Transmission, Electric Motor, Shaft Coupling, Gearbox, Clutches, Shaft/Hub Connections, Belt Drive, Seals, Rolling Element Bearing, Standard Gears, Shaft for Strength and Rigidity, Shaft with Fluctuating Loads.

European documentation centres

The European Union Directives are increasingly important for design engineers. European Documentation Centres (EDCs) receive one copy of every published work of the European Union institutions. There are 44 EDCs in the UK located in academic institutions.

Patents

Full details are contained in a separate chapter in this book. The British Library Science Reference and Information Service (SRIS) and thirteen major public libraries throughout the UK have major holdings of patent information. These cover UK, European, American and International patent information via CD-ROM, fiche and online facilities.

Computer sources/databases/abstracts

This subject is covered in other chapters of this book and so brief details only are presented here.

Important on line and CD-ROM computer databases for information on engineering include COMPENDEX, INSPEC and ISMEC. More specific databases are METADEX and RAPRA for materials, and WELDASEARCH for production processes.

Abstracting and indexing journals include:

Catchword and Trade Name Index
Computer and Control Abstracts
Current Technology Index (formerly *British Technology Index*)
Electrical and Electronics Abstracts
Engineered Materials Abstracts
Engineering Index
Government Reports Announcements and Index
Mechanical Engineering Abstracts
Metals Abstracts

Networked computer information exchange has been revolutionized by e-mail, Internet and World Wide Web facilities. In particular, an electronic 'bulletin-board' is available on MAILBASE and subscribers can select subject areas from which they will receive all circulated e-mail. For example, engineering-design@mailbase.ac.uk is a subject area.

Conferences and exhibitions

Major international and national conferences include:
ICED (International Conference on Engineering Design) held biannually, the ASME Design Engineering Division 'Design Technical Conference' held annually in the USA, the National Science Foundation 'Design and Manufacturing Systems Conference' held annually in the USA, SEED (Sharing Experience in Engineering Design) held annually in the UK.

Examples of exhibitions include:
ICAT – Computer-aided design and manufacturing hardware and software. National Exhibition Centre, Birmingham (annual).
MULTIMEDIA – Video conferencing, business presentations and authoring. Earls Court, London (annual).
NETWORKS – Computer networking systems. National Exhibition Centre, Birmingham (annual).

Learned societies/organizations

American Society of Mechanical Engineers (ASME), New York
British Standards Institution (BSI), Milton Keynes
Confederation of British Industry (CBI), London
Design Council (DC), London
Department of Trade and Industry (DTI), London
Engineering Council (EC), London
Institution of Electrical Engineers (IEE), London
Institute of Electrical and Electronics Engineers (IEEE), New York
Institution of Engineering Designers (IED), Westbury
Institution of Mechanical Engineers (IMechE), London
Japan Society of Mechanical Engineers (JSME), Tokyo
Production Engineering Research Association (PERA), Melton Mowbray, Leicestershire
Royal Academy of Engineering (RAE), London
Sharing Experience in Engineering Design (SEED), Loughborough
VDI – Gesellshaft Produktionstechnik (VDI–GP), Düsseldorf
Workshop Design – Konstruktion (WDK) – Switzerland

References

Abbott, H. (1980) *Safe enough to sell*. London: Design Council.
Abbott, H. (1987) *Safer by design*. London: Design Council.

Alexander, D. C. (1986) *The practice and management of industrial ergonomics*. Englewood Cliffs, NJ: Prentice Hall.
Alger, J. R. M. and Hays, C. V. (1964) *Creative synthesis in design*. Englewood Cliffs, NJ: Prentice Hall.
Andreasen, M. M. et al. (1988) *Design for assembly*, 2nd ed. Bedford: IFS.
Archer, L. B. (1971) *Technological innovation – a methodology*. Frimley, Surrey: Inforlink.
Arora, J. S. (1989) *Introduction to optimum design*. New York: McGraw-Hill.
Ashby, M. F. and Jones, D. R. H. (1980) *Engineering materials: an introduction to their properties and applications*. Oxford: Pergamon.
Ashby, M. F. (1992) *Materials selection in mechanical design*. Oxford: Pergamon.
Ashford, F. C. (1970) *The aesthetics of engineering design*. London: Business Books.
Asimow, M. (1962) *Introduction to design*. Englewood Cliffs, NJ: Prentice Hall.
Bailey, R. L. (1978) *Disciplined creativity for engineers*. Ann Arbor, MI: Ann Arbor Science.
Barker, T. B. (1990) *Engineering quality by design: interpreting the Taguchi approach*. New York: Marcel Dekker.
Bayley, S. (1991) *Taste: the secret meaning of things*. New York: Pantheon Books.
Berk, A. A. (1988) *Computer aided design and analysis for engineers*. London: BSP Professional.
Besant, C. B. and Lui, C. W. K. (1991) *Computer aided design and manufacture* rev. ed. Chichester: Ellis Horwood.
Black, R. M. (1995) *Design and manufacture: an integrated approach*. Basingstoke: Macmillan.
Blakstad, M. (1979) *The risk business*. London: Design Council.
Boothroyd, G. and Dewhurst, P. (1987) *Product design for assembly*. Wakefield, RI: Boothroyd Dewhurst Inc.
Bralla, J.D. (ed.) (1986) *Handbook of product design for manufacturing*. New York: McGraw-Hill.
Bridger, R.S. (1995) *Introduction to ergonomics*. New York: McGraw-Hill.
Broadbent, G. (1973) *Design in architecture*. London: Wiley.
Bruce, M. (1986) *Designing for quality* (Open University Course PT619). Milton Keynes: Open University.
BSI (1981) *Management of design for economic production* (PD6470). Milton Keynes: British Standards Institution.
BS 308: Parts 1, 2 and 3: 1985–1993. *Engineering drawing practice*. Milton Keynes: British Standards Institution.

BS 5760: 1986 *Reliability of systems, equipment and components*. Milton Keynes: British Standards Institution.
BS 7000: Part 1: 1989 *Guide to managing product design*. Milton Keynes: British Standards Institution.
BS 7373: 1991 *Guide to the preparation of specifications*. Milton Keynes: British Standards Institution.
BS EN ISO 9000 (1994) *Quality management and quality assurance standards*. Milton Keynes: British Standards Institution.
Burgess, J. A. (1984) *Design assurance for engineers and managers*. New York: Marcel Dekker.
Charles, J. A. and Crane, F. A. A. (1989) *Selection and use of engineering materials* 2nd ed. Oxford: Butterworth-Heinemann.
Chow, W. W.-C. (1978) *Cost reduction in product design*. New York: Van Nostrand Reinhold.
Clark, A. M. (1989) *Product liability*. London: Sweet and Maxwell.
CNAA (1984) *Managing design: an initiative in management education*. London: Council for National Academic Awards.
Cooke, P. et al. (1984) *A guide to design for production*. London: Institution of Production Engineers.
Corbett, J. et al. (eds) (1991) *Design for manufacture: strategies, principles and techniques*. Wokingham: Addison Wesley.
Corfield, K. G. (1979) *Product design*. London: National Economic Development Office.
Cornish, E. H. (1987) *Materials and the designer*. Cambridge: Cambridge University Press.
Cortada, J. W. and Woods, J. A. (eds) (1995) *The quality yearbook*. New York: McGraw-Hill.
Costanzo, L. (1992) *Taking the strain, vehicle engineering and design*. London: Design Council.
Court, A. W., Culley, S. J. and McMahon, C. A. (1993) *A survey of information access and storage amongst engineering designers* internal report number 008/1993. Bath: Design Group, University of Bath.
Cox, B. and Webb, R. (1978) *An engineer's guide to costing*. London: Institution of Production Engineers.
Croney, J. (1980) *Anthropometry for designers* 2nd ed. London: Batsford.
Cross, N. (1994) *Engineering design methods* 2nd ed. Chichester: Wiley.
Davis, M. D. (1983) *Game theory: a non-technical introduction* rev. ed. New York: Basic Books.
de Bono, E. (1970) *Lateral thinking: a textbook of creativity*. London: Ward Lock Educational.
Derby, B., Hills, D. A. and Ruiz, C. (1992) *Materials for engineering: a fundamental design approach*. Harlow: Longman.

Design Council (1977) *Industrial design education in the United Kingdom*. London: Design Council.
DeVor, R. E., Chang, T. and Sutherland, J. W. (1992) *Statistical quality design and control*. Toronto: Maxwell Macmillan.
Dieter, G. E. (1991) *Engineering design: a materials and processing approach* 2nd ed. New York: McGraw-Hill.
Diffrient, N. *et al.* (1987) *Humanscale*. Cambridge, MA: MIT Press.
Dormer, P. (1990) *The meanings of modern design*. London: Thames and Hudson.
Downton, A. (ed.) (1991) *Engineering the human-computer interface*. London: McGraw-Hill.
Dym, C. L. (1994) *Engineering design: a synthesis of views*. Cambridge: Cambridge University Press.
Earle, J. H. (1993) *Engineering design graphics* 8th ed. Reading, MA: Addison Wesley.
Easterby, R. *et al.* (eds) (1982) *Anthropometry and biomechanics: theory and applications*. New York: Plenum.
Edel, D. H. (1967) *Introduction to creative design*. Englewood Cliffs, NJ: Prentice Hall.
Ertas, A. and Jones, J. C. (1993) *The engineering design process*. New York: Wiley.
Ettlie, J. E. and Stoll, H. W. (1990) *Managing the design–manufacturing process*. New York: McGraw-Hill.
Eyre, E. C. (1983) *Effective communication made simple*. London: Heinemann.
Farag, M. M. (1989) *Selection of materials and manufacturing processes for engineering design*. London: Prentice Hall.
Feilden, G. B. R. (1963) (Chairman) *Engineering design*. Report of a committee appointed by the Council for Scientific and Industrial Research. London: HMSO.
Finniston, M. (1980) (Chairman) *Engineering our future*: report of the Committee of Inquiry into the Engineering Profession. Cmnd 7794. London: HMSO.
Flurscheim, C. H. (ed.) (1983) *Industrial design in engineering*. London: Design Council.
French, M. J. (1985) *Conceptual design for engineers* 2nd ed. London: Design Council.
French, M. J. (1992) *Form, structure and mechanism*. Basingstoke: Macmillan.
French, M. J. (1994) *Invention and evolution: design in nature and engineering* 2nd ed. Cambridge: Cambridge University Press.
Galer, I. A. R. (ed.) (1987) *Applied ergonomics handbook* 2nd ed. Oxford: Butterworth-Heineman.

Gardiner, P. and Rothwell, R. (1985) *Innovation*. London: Design Council.
Garner, S. (1991) *Human-factors*. Oxford: Oxford University Press.
Glegg, G. L. (1969) *The design of design*. Cambridge: Cambridge University Press.
Gordon, W. J. J. (1961) *Synectics: the development of creative capacity*. New York: Harper and Row.
Gregory, S. A. (ed.) (1972) *Creativity and innovation in engineering*. London: Butterworths.
Griffiths, R. F. (1981) *Dealing with risk*. Manchester: Manchester University Press.
Groover, M. P. and Zimmers, E. W. (1984) *CAD/CAM*. Englewood Cliffs, NJ: Prentice Hall.
Gruber, H. E. (1967) *Imagination and thinking*. New York: Atherton Press.
Haigh, M. J. (1985) *An introduction to computer aided design and manufacture*. Oxford: Blackwell Scientific.
Hales, C. (1993) *Managing engineering design*. Harlow: Longman.
Hawkes, B. and Abinett, R. (1984) *The engineering design process*. London: Pitman.
Health and Safety Executive (1989) *Quantified risk assessment*. London: HMSO.
Heller, E. D. (1971) *Value management, value engineering and cost reduction*. Reading, MA: Addison-Wesley.
Henstock, M. E. (1988) *Design for recyclability*. London: Institute of Metals.
Heskett, J. (1980) *Industrial design*. London: Thames and Hudson.
Heskett, J. (1989) *Philips: a study of the corporate management of design*. London: Trefoil.
Holt, K. (1987) *Innovation: a challenge to the engineer*. Amsterdam: Elsevier.
Hsu, T. R. and Sinha, D. K. (1992) *Computer aided design: an integrated approach*. St Paul, MN: West Publishing Company.
Hubka, V., Andreasen, M. M. and Eder, W. E. (1988) *Practical studies in systematic design*. London: Butterworths.
Humphreys, K. K. and Wellman, P. (1986) *Basic cost engineering* 2nd ed. New York: Marcel Dekker.
Hunter, T. A. (1992) *Engineering design for safety*. New York: McGraw-Hill.
Jewell, T. K. (1991) *Computer applications for engineers*. Chichester: Wiley.
Johnson, P. (1992) *Human-computer interaction*. Maidenhead: McGraw-Hill.
Johnson, R. C. (1980) *Optimum design of mechanical elements* 2nd ed. New York: Wiley.

Jones, J. C. (1981) *Design methods*. Chichester: Wiley.
Jones, P. F. (1992) *CAD/CAM features, applications and management*. London: Macmillan.
Juran, J. M. (ed.) (1988) *Quality control handbook* 4th ed. London: McGraw-Hill.
Kelly, P. and Attree, R. (1992) *European product liability*. London: Butterworths.
Kicherer, S. (1990) *Olivetti, a study of the corporate management of design*. London: Trefoil.
Knight, I. (1984) *The heights and weights of adults*. London: HMSO.
Knox, C. S (1983) *CAD/CAM systems planning and implementation*. New York: Marcel Dekker.
Koberg, D. and Bagnall, J. (1981) *The universal traveler* rev. ed. Los Altos, CA: Kaufmann.
Latham, R. W. (1965) *Problem analysis by logical approach*. Harwell: Atomic Weapons Research Establishment, United Kingdom Atomic Energy Authority.
Leech, D. J. (1972) *Management of engineering design*. Chichester: Wiley.
Leech, D. J. and Turner, B. T. (1985) *Engineering design for profit*. Chichester: Ellis Horwood.
Lickley, R. L. et al. (1983) *Report of the Engineering Design Working Party*. London: Science and Engineering Research Council.
Lindgren, B. W. (1971) *Elements of decision theory*. New York: Macmillan.
Loewy, R (1979) *Industrial design*. London: Faber and Faber.
Logothetis, N. (1993) *Managing for total quality, from Deming to Taguchi and SPC*. Englewood Cliffs, NJ: Prentice Hall.
Lorenz, C. (1990) *The design dimension: product strategy and the challenge of global marketing* rev. ed. Oxford: Blackwell.
Lowrance, W. W. (1976) *Of acceptable risk*. Los Altos, CA: Kaufmann.
Love, S. F. (1986) *Planning and creating successful engineered designs*. Los Angeles, CA: Advanced Professional Development.
Ludema, K. C., Caddell, R. M. and Atkins, A. G. (1987) *Manufacturing engineering: economics and processes*. Englewood Cliffs, NJ: Prentice Hall.
Lumsdaine, E. and Lumsdaine, M. (1995) *Creative problem solving*. New York: McGraw-Hill.
Maldonado, T. (1972) *Design, nature and revolution*. New York: Harper and Row.
Malstrom, E. M. (1981) *What every engineer should know about manufacturing cost estimating*. New York: Marcel Dekker.
Manzini, E. (1989). *The material of invention: materials and design*. London: Design Council.

Matousek, R. (1963) *Engineering design*. London: Blackie.
Medland, A. J. (1992) *The computer-based design process* 2nd ed. London: Chapman and Hall.
Medland, A. J. and Mullineux, G. (1988) *Principles of CAD*. London: Kogan Page.
Meredith, D. D. et al. (1985) *Design and planning of engineering systems* 2nd ed. Englewood Cliffs, NJ: Prentice Hall.
Middendorf, W. H. (1990) *Design of devices and systems* 2nd ed. New York: Marcel Dekker.
Miles, L. D. (1972) *Techniques of value analysis in engineering* 2nd ed. New York: McGraw-Hill.
Miller, C. J. (1988) *Product liability and safety encyclopaedia*. London: Butterworths.
Morgan, J. R. (1972) *AIDA : a technique for the management of design*. Coventry: Institute for Operational Research.
Moulton, A. E. et al. (1976) *Engineering design education*. London: Design Council.
Mucci, P. (1994) *Handbook for engineering design using standard materials and components* 4th ed. Milton Keynes: British Standards Institution.
Newell, A. and Simon, H. A. (1971) *Human problem solving*. Englewood Cliffs, NJ: Prentice Hall.
Norman, D. A. (1988) *The psychology of everyday things*. New York: Basic Books.
Oakland, J. S. (1993) *Total quality management* 2nd ed. Oxford: Butterworth-Heinemann.
Oakley, M. (1984) *Managing product design*. London: Weidenfeld and Nicolson.
Oborne, D. J. (1994) *Ergonomics at work: human factors in design and development* 3rd ed. Chichester: Wiley.
Olins, W. (1989) *Corporate identity: making business strategy visible through design*. London: Thames and Hudson.
Osborn, A. F. (1963) *Applied imagination* 3rd ed. New York: Scribner's.
Pahl, G. and Beitz, W. (1995) *Engineering design: systematic approach* 2nd rev. ed. Berlin: Springer-Verlag.
Papanek, V. (1984) *Design for the real world* 2nd ed. London: Thames and Hudson.
Parker, M. (ed.) (1991) *Manual of British Standards in engineering drawing and design* 2nd ed. London: British Standards Institution and Stanley Thornes.
Pheasant, S. T. (1984) *Anthropometrics: an introduction for schools and colleges* (PP7310). London: British Standards Institution.
Pheasant, S. T. (1986) *Bodyspace: anthropometry, ergonomics and design*. London: Taylor and Francis.

Pheasant, S. T. (1987) *Ergonomics: standards and guidelines for designers* (PP7317). Milton Keynes: British Standards Institution.
Piatier, A. (1984) *Barriers to innovation.* London: Frances Pinter.
Polak, P. (1991) *Engineering design elements.* London: McGraw-Hill.
Preston, E. J., Crawford, G. W. and Coticchia, M. E. (1984) *CAD/CAM systems: justification, implementation, productivity measurement.* New York: Marcel Dekker.
Pugh, S. (1991) *Total design.* Wokingham: Addison-Wesley.
Pye, D. (1982) *The nature and aesthetics of design.* New York: Van Nostrand Reinhold.
Radke, M. (1972) *Manual of cost reduction techniques.* New York: McGraw-Hill.
Raheja, D. (1991) *Assurance technologies: principles and practice.* New York: McGraw-Hill.
Ray, M. S (1985) *Elements of engineering design.* Englewood Cliffs, NJ: Prentice Hall.
Redford, A. H. and Chal, J. (1993) *Design for assembly.* London: McGraw-Hill.
Rich, E. and Knight, K. (1991) *Artificial Intelligence.* New York: McGraw-Hill.
Roark, R. J. and Young, W. C. (1989) *Roark's formulas for stress and strain* 6 rev. ed. New York: McGraw-Hill.
Rodriguez, W. (1992) *Modelling of design ideas.* New York: McGraw-Hill.
Rooney, J. and Steadman, P. (eds) (1991) *Principles of computer-aided design.* London: Pitman.
Roozenburg, N. F. M. and Eekels, J. (1994) *Product design, structure and methods.* Chichester: Wiley.
Roy, R. and Wield, D. (eds) (1989) *Product design and technological innovation* rev. ed. Milton Keynes: Open University.
Russell, D. (1990) *Colour in industrial design.* London: Design Council.
Ryan, D. L. (1994) *Computer aided graphics and design* 3rd ed. New York: Marcel Dekker.
Sanders, M. S. and McCormick, E. J. (1993) *Human factors in engineering and design* 7th ed. New York: McGraw-Hill.
Scott, W. (1984) *Communication for professional engineers.* London: Thomas Telford.
SEED (1985) *Curriculum for design: proceedings of working party.* Loughborough: SEED publications.
SEED (1986) *Specification phase: preparation material for design teaching.* Loughborough: SEED Publications.
SEED (1987) *Information retrieval: preparation material for design teaching.* Loughborough: SEED Publications.

SEED (1988) *Communication in design: preparation material for design teaching*. Loughborough: SEED Publications.

SEED (1989a) *Costing in design: preparation material for design teaching*. Loughborough: SEED Publications.

SEED (1989b) *Manufacturing phase: preparation material for design teaching*. Loughborough: SEED Publications.

SEED (1989c) *Quality and reliability: preparation material for design teaching*. Loughborough: SEED Publications.

SEED (1992) *Aesthetics in engineering design: preparation material for design teaching*. Loughborough: SEED Publications.

SEED (1993a) *Ergonomics in engineering design: preparation material for design teaching*. Loughborough: SEED Publications.

SEED (1993b) *Detail design phase: preparation material for design teaching*. Loughborough: SEED Publications.

SEED (various) *Design procedural guides series*. Loughborough: SEED Publications.

Service, L. M., Hart, S. J. and Baker, M. J. (1989) *Profit by design*. London: Design Council.

Shigley, J. E. (1990) *Mechanical engineering design* 5th ed. New York: McGraw-Hill.

Shigley, J. E. and Mischke, C. R. (eds) (1986) *Standard handbook of machine design*. New York: McGraw-Hill.

Siddall, J. N. (1972) *Analytical decision making in engineering design*. Englewood Cliffs, NJ: Prentice Hall.

Siddall, J. N. (1982) *Optimal engineering design: principles and applications*. New York: Marcel Dekker.

Simmons, C. H. and Maguire, D. E. (1995) *A manual of engineering drawing practice* rev. ed. London: Edward Arnold.

Simon, H. A. (1975) *A student's introduction to engineering design*. Cambridge, MA: Pergamon.

Simon, H. A. (1981) *The sciences of the artificial* 2nd ed. London: MIT Press.

Starkey, C. V. (1988) *Basic engineering design*. London: Edward Arnold.

Starkey, C. V. (1992) *Engineering design decisions*. London: Edward Arnold.

Stewart, R. D., Wyskida, R. M. and Johannes, J. D. (eds) (1995) *Cost estimators reference manual* 2nd ed. New York: Wiley.

Stoll, H. W. (1988) Design for manufacture. *Manufacturing Engineering*, **100**(1), 67–73.

Suh, N. P. (1990) *The principles of design*. Oxford: Oxford University Press.

Taguchi, G. (1986) *Introduction to quality engineering*. Tokyo: Asian Productivity Organization.

Thring, M. W. and Laithwaite, E. R. (1977) *How to invent*. London: Macmillan.
Tizzard, A. (1994) *An introduction to computer-aided engineering*. London: McGraw-Hill.
Topalian, A. (1980) *The management of design projects*. London: Associated Business Press.
Turk, C. and Kirkman, J. (1989) *Effective writing* 2nd ed. London: Spon.
Ullman, D. G. (1992) *The mechanical design process*. New York: McGraw-Hill.
Ulrich, K. T. and Eppinger, S. D. (1994) *Product design and development*. New York: McGraw-Hill.
Vernon, P. E. (ed.) (1970) *Creativity: selected readings*. Harmondsworth: Penguin.
Walker, D. J., Dagger, B. K. J. and Roy, R. (1991) *Creative techniques in product and engineering design*. Cambridge: Woodhead.
Wall, R. A. (ed.) (1986) *Finding and using product information*. Aldershot: Gower.
Walsh, V. *et al.* (1992) *Winning by design*. Oxford: Blackwell Business.
Walton, J. W. (1991) *Engineering design: from art to practice*. St Paul, MN: West Publishing.
Whiteman, M. F. (ed.) (1982) *Writing: the nature, development and teaching of written communications* 2 vols. Hillsdale, NJ: Erlbaum.
Whitney, D. E. (1988) Manufacturing by design. *Harvard Business Review*, **66**, July–August, 83–91.
Wilcox, A. D. *et al.* (1990) *Engineering design for electrical engineers*. Englewood Cliffs, NJ: Prentice Hall.
Winston, P. H. (1993) *Artificial Intelligence* 3rd ed. Reading, MA: Addison-Wesley.
Wright, C. J. (1989) *Product liability: the law and its implications for risk management*. London: Blackstone.
Wykes, K. (1988) *Ergonomist* (includes video). London: Design Council.

CHAPTER TWENTY

Manufacturing techniques

J. CORBETT
(Additional research by S. A. Jones)

Introduction

The manufacturing activity is inextricably linked with virtually all aspects of engineering and will therefore permeate throughout the other chapters in this book. A modern definition for manufacturing from the CAM-1 Organization (Arlington,TX) states that it is 'a series of interrelated activities and operations involving design, materials selection, planning, production, quality assurance, management and marketing of discrete consumer goods'. This definition shows clearly the close interaction of several disciplines.

Modern manufacturing techniques often call for the incorporation and integration of numerous technologies. This is seen to be essential in order to compete in world markets which are developing and introducing products in shorter and shorter time scales. There is also the need to reduce overall manufacturing costs including those for tooling, the machining of new and novel materials, and to reduce the steps required to produce a component/product. The introduction of new and advanced materials (e.g. ceramics and inter-metallics) and new products (e.g. micro- and opto-electronics) can be exploited fully only by creating new manufacturing techniques or by adapting existing ones. Therefore the next generation of manufacturing techniques will have a major impact on these issues. However, while recognizing recent developments, it is extremely unlikely that these new techniques will reduce the dominance of metal cutting tools in meeting the overall needs of manufacturing companies in the foreseeable future.

Two distinct technological thrusts in advanced manufacturing techniques over the last forty years have been (a) the automation of manufacturing, which is aimed at reducing costs, and is covered in Chapter

354 Manufacturing techniques

21, and (b) manufacturing with higher precision. The latter is probably less obvious but is becoming recognized as a strategic issue in many industries as it is the dimensioned geometry of components which determines the nature and quality in the way products function. Tighter tolerances can have a major effect on the manufacturing technique selected as well as the quality control function, including measurement and inspection, which forms an integral and fundamental role in manufacturing. The scope of manufacturing, however, and of this chapter, also includes joining, assembly, planning, tooling, manufacturing systems and manufacturing management.

The importance of information in such a dynamic, extensive and interactive activity as manufacturing cannot be overstated. As more relevant information becomes available to an organization, more alternatives can be explored with an increased chance of arriving at an optimum solution.

Abstracting and indexing

With the ever increasing volume of printed manufacturing information the use of a high quality abstracting service is invaluable. These are available in published form, although the growing accessibility of computerized online database systems, at relatively low cost, enables rapid access to huge amounts of data.

Technical journals and conference proceedings are two of the most valuable sources of manufacturing information and the major databases include abstracts from these, together with selected reports and books.

Two of the most useful online databases and services are COMPENDEX (Engineering Information Inc.) and BIDS (University of Bath). COMPENDEX provides information dating from 1970 and includes worldwide coverage of engineering and technology literature from approximately 4500 journals, plus selected government reports and books. Specific manufacturing and process engineering information, derived from COMPENDEX, is available on CD-ROM through DIALOG EI MANUFACTURING. Coverage dates back to 1984. BIDS includes abstracts from databases covering over 7000 journals and over 4000 scientific and technical conferences per year. Each year has its own index so that the user is connected to a particular year at any one time.

The major online databases cover wide fields of engineering and manufacturing. Information is either specifically indexed or the user may have to search deeper on other related areas. Fortunately, almost all major databases have a comprehensive index and include a thesaurus which provides related terms for keywords. INSPEC (Institution of Electrical Engineers) contains abstracts from approximately 4200 journals, conference proceedings, books, reports and dissertations from all over the world. It encompasses physics, electrical engineering, electronics, computing, information technology and computers in manufacturing. The manufacturing information relates mainly to the more sophisticated areas of computer control and computer applications. In addition the physics abstract section covers nanotechnology and nano-fabrication.

Information on materials is often relevant to manufacturing and two particularly useful online databases, both provided by 'Materials Information', a joint service of ASM International (USA) and the Institute of Materials (UK) are:

METADEX covering the science and practice of metallurgy, including machining and forming.

ENGINEERED MATERIALS ABSTRACTS (EMA) covering polymers, ceramics and composites, including material properties, products and forms of these materials.

Smaller but nevertheless useful online databases include:

MECHANICAL ENGINEERING ABSTRACTS (1973–) (Cambridge Scientific Abstracts and Engineering Information Inc.), includes all aspects of production engineering with good coverage of production processes, transport and handling.

NTIS (1964–) (US National Technical Information Service) provides access to information from US government-sponsored research and also results of government-sponsored research outside the USA, including the Japanese Ministry of International Trade and Industry (MITI); the UK laboratories administered by the Department of Trade and Industry (DTI); the German Federal Ministry of Research and Technology (BMFT); the French National Centre for Scientific Research (CNRS); and others.

RAPRA (1972–) (Rubber and Plastics Research Association) is dedicated exclusively to rubber, plastics, adhesives and polymeric composites. Abstracts include properties and testing, environmental effects, machinery and test equipment, processing technology, etc.

WELDASEARCH (1967–) (Welding Institute) is the world's largest online database on welding, allied processes and the science of joining.

For those without access to online facilities there are numerous published abstracting and indexing journals. Two of the major ones are *Engineering Index*, which is the printed equivalent of COMPENDEX, and INSPEC's, *Science Abstracts*. Other useful ones include:

Applied Mechanics Reviews (American Society of Mechanical Engineers) includes manufacturing, assembly, materials processing, etc.

International Abstracts in Operations Research (International Federation of Operational Research Societies and Macmillan) includes production, scheduling, Just-in-Time, Group Technology, Flexible Manufacturing Systems, etc.

Metals Abstracts (ASM International and the Institute of Materials) the published form of METADEX.

RAMB (Nottingham Consultants Ltd, UK) a specialist manufacturing database, which is also available as a personal computer database.

Journals

Two valuable quarterly publications for research engineers and manufacturing practitioners looking for papers on the latest manufacturing concepts, including technological and scientific aspects of good practice, are the *Proceedings of the Institution of Mechanical Engineers. Part B, Journal of Engineering Manufacture*; and the *Transactions of the American Society of Mechanical Engineers. Journal of Engineering for Industry*. Two other research journals published more frequently are the *International Journal of Production Research* and the *International Journal of Machine Tools and Manufacture*. Both of these give good coverage of the design of manufacturing systems and management issues. The *International Journal of Advanced Manufacturing Technology* covers the middle ground between the pure research and the more practical journals, with worldwide papers covering all aspects of manufacturing. *Production Planning and Control* also claims this middle ground but has more technical articles and case studies, so will be of particular interest to manufacturing engineers and managers as well as to manufacturing researchers. The professional engineering institutions provide good coverage of news and recent developments. These journals include the Institution of Electrical Engineers' (IEE) *Manufacturing Engineer*, which has superseded the *Production Engineer*, and the Society of Mechanical Engineers' *Manufacturing Engineering*. *Professional Engineering*, which has replaced the *Chartered Mechanical Engineer*, from the Institution of Mechanical Engineers also demonstrates a regular interest in manufacturing techniques. For manufacturing engineers who wish to keep in touch with the latest news, including topical special reports, trends and coming events, *Machinery and Production Engineering* and *Metalworking Production*, both UK publications, provide excellent coverage. Two similar US journals are *American Machinist* and *Production*.

There is a multitude of journals which are of a more specialist nature and, in the area of Manufacturing Systems, these include *Computer Integrated Manufacturing* and *Artificial Intelligence*. The *Journal of Manufacturing Systems* from the Society of Mechanical Engineers (USA) provides an insight into the theories and applications of recent integrated manufacturing systems research as well as indicating future directions of manufacturing systems research. *Computer Integrated Manufacturing Systems (incorporating Advanced Manufacturing Engineering)* and the *International Journal of Computer Integrated Manufacturing* both include original research papers as well as practical applications for specific manufacturing situations. A less theoretical coverage, more appropriate for design engineers and manufacturing managers, is given in *Integrated Manufacturing Systems*, which has information on computer integrated manufacturing, just-in-time and group technology methods, and *Computers and Industrial Engineering*, which provides computer solutions for solving practical manufacturing problems. *CAD/CAM International* provides news, latest product information and articles covering applications and strategies. Journals with a wide coverage of intelligent manufacturing, including recent methods, case studies and surveys are *Journal of Intelligent Manufacturing*, *Engineering Applications of Artificial Intelligence* and *Artificial Intelligence in Engineering*.

Narrowing the scope of manufacturing techniques still further *Assembly Automation* charts international developments in both fully automated and flexible assembly. *Welding International*, which includes the English translation of articles selected from leading CIS, Japanese and European journals, and the *International Journal of Adhesion and Adhesives* cover recent research developments into the joining of metals and other materials.

With the introduction of new materials, with particular machining and processing problems, two important journals which include original research papers describing new and adapted traditional manufacturing techniques are *Processing of Advanced Materials* and the *Journal of Materials Processing and Manufacturing Science*. *Advanced Materials & Processes* concentrates more on the latest high performance materials but also has valuable information on techniques for their processing. The niche areas of composite processing and surface treatments are also well covered by *Composites Manufacturing* and *Metal Finishing* respectively. A publication from the former Soviet Union, in English translation, which gives access to some of the country's most important research is *Russian Castings Technology* which includes information on casting alloys and special casting techniques.

The development and processing of new materials has been particularly important in the area of ultra precision manufacturing. *Precision Engineering*, the journal of the American Society of Precision Engineering and the *Japan Society for Precision Engineering International Journal* are both valuable sources of the latest research developments in this area, including diamond turning and grinding, elastic emission machining, electro-discharge machining, electro-chemical machining, and energy beam machining. *Precision Engineering* which is published quarterly has a particularly valuable citation and abstract section.

Research papers in the area of quality can be found in *Measurement* which covers all aspects of measurement related to planing and processes, and *NDT & E International* (formerly *NDT International*) which includes information on fundamental research in non-destructive testing

358 Manufacturing techniques

as well as industrial applications. Other useful journals in these areas include the *European Journal of Non Destructive Testing* and the *British Journal of Non-Destructive Testing*. The latter incorporates condition monitoring and diagnostic technology. More general news, issues and applications can be found in the monthly journals *Quality Progress* from the American Society for Quality Control and *Quality News*, from the UK's Institute of Quality Assurance. *Quality Forum* is also a valuable journal, published quarterly by the Institute of Quality Assurance, which includes original articles covering issues of product liability, maintenance and statistical process control. More general news, equipment information and announcements can be found in *Quality Today*.

Journals which address management aspects of manufacturing as opposed to the more technical aspects are the *Journal of the Operational Research Society*, which is published monthly and includes theoretical papers and case studies, the *International Journal of Production Economics* (now incorporating the *Journal of Manufacturing and Operations Management*), which covers issues from the complete product life cycle from research through the material flow cycle to distribution and disposal. Others include the *International Journal of Operations and Production Management* and *Work Study* a more specialized publication covering important planning and decision-making processes.

Conference proceedings

Conferences provide an excellent way of communicating and keeping in touch with the latest manufacturing technology and techniques. There are regular annual and biennial conferences and also many arranged on a less regular basis but which are directed at more specialist and/or topical matters. For details of the latter regular attention should be paid to the events and diary sections in publications from the major professional institutions and other major journals.

The *CIRP annals* from the International Institution for Production Engineering Research provide a considerable number of papers, divided into specialist areas such as assembly, cutting, design, electrophysical and chemical processes, forming, abrasive processes, machines, optimization of manufacturing systems, precision engineering and metrology, and surfaces. The *CIRP annals* appear in a series of bound volumes and include contributions from the world's leading academic proponents. The proceedings from the annual International Machine Tool Design and Research Conference (MATADOR), organized principally by the University of Manchester Institute of Science and Technology (UMIST), also provide a considerable number of papers on machine

Manufacturing techniques 359

tool design, metal cutting and metal forming processes. Other regular major conferences include the International Conference on Metal Forming; the International Machine Tool Engineers Conference (IMEC) with translations in Japanese and English; the National Science Foundation (USA) Design and Manufacturing Systems Conference published by the Society of Manufacturing Engineers (SME); Manufacturing Research organized by the SME, North American Manufacturing Research Institution; and Advanced Manufacturing Technology organized by the Irish Manufacturing Committee (IMC).

There are also regular conferences in more specialist areas. The American Society of Mechanical Engineers (ASME) runs a regular joint Japan/USA symposium, Flexible Automation. ASME's Design Engineering Division also runs a conference on Flexible Assembly Systems. Computer integrated manufacturing is well served by conferences which include the International Conference on CAD/CAM (MICAD) which has good coverage of computer graphics and computer-aided technologies with French and English translation; CAD/CAM, Robotics and Factories of the Future organized by the International Society for Productivity Enhancement; Realising CIM's Industrial Potential and Computer Integrated Manufacturing are both annual conferences organized by the Commission of the European Communities; University Programs in Computer-Aided Engineering, Design and Manufacture (UPCAEDM) is an annual US conference.

The growing area of precision manufacturing is catered for by the bi-annual International Precision Engineering Seminar (IPES) organized by Cranfield Unit for Precision Engineering and Butterworth-Heinemann; the annual American Society for Precision Engineering (ASPE) Conference; and the tri-annual International Conference on Ultra-Precision in Manufacturing Engineering (UME) jointly organized in Germany by Forschungsgemeinschaft Ultrapräzisionstechnie e.V (UPT) and Physikalisch-Technische Bundesanstalt (PTB). A more specialist conference which reports on the machining of brittle materials is Progress in Fine Grinding Technology organized by the Society of Powder Technology, Japan, with abstracts in Japanese and English.

With the growing use of ceramics in component manufacture, a conference useful to those wishing to gain information on ceramic processing science and technology is the International CERAMITEC Symposium, with English and German translations, organized by Deutschen Keramischen Gesellschaft. Other appropriate processing conferences are Materials Processing organized jointly by Nanyang Technological University and the National University of Singapore, and the International Conference on Surface Finishing (ISFEC).

Leading conferences relevant to manufacturing quality include the Annual Quality Congress organized by the American Society for Quality Control, and the Annual Instrumentation and Measurement Technology Conference (IMTC) organized by the IEEE Instrumentation and Measurement Society.

Two international conferences organized by the Welding Institute (UK) are Advances in Welding Processes and Computer Technology in Welding, which incorporates EUROWELD.

Books

General introduction and descriptive

There is a large number of books which give a sound introduction to manufacturing techniques and materials processing. Some very good ones, and there will be many more, include the following.

A particularly thorough introduction is given by Edwards and Endean (1990) who encourage the reader to learn by doing through the provision of exercises and self-assessment questions. The book focuses on how materials are processed into different shapes and discusses process selection. Excellent diagrams and figures are included and a novel feature is a set of data cards provided with each book. These include detailed process information which enable several processes to be considered without losing one's place in the text.

Kalpakjian (1991) is widely used as an undergraduate textbook in the USA. The emphasis is on the analysis of manufacturing processes rather than the more comprehensive approach supplied in books by Amstead, Ostwald and Begeman (1987) and Schey (1987).

An excellent technical handbook of machine tools by Weck (1984) contains four volumes derived from lectures given at the Technical University for the Rhineland and Westphalia in Aachen, Germany. The volumes include (i) types of machine tools; (ii) their construction; (iii) automation and controls; and (iv) metrology and performance testing.

Another popular book by Lissaman and Martin (1982) treats technical aspects of manufacturing with respect to metal machining and press forming. It starts with a consideration of specification and standardization and goes on to deal analytically with the main aspects of the manufacturing processes giving due attention to the crucial matters of quality and cost. Machine tool design aspects, cutting, forming, measurement, tool design, time and cost estimating are all included.

DeGarmo, Black and Kohser (1988) also provides a solid introduction to the fundamentals of manufacturing. This revised edition includes thirty per cent new text and includes polymers and the manufacture of electronic components. The quality section has been expanded to include non-destructive testing and inspection, statistical process control and new quality assurance concepts.

Gillespie (1988) provides a guide designed to be used in troubleshooting manufacturing problems including machining and forming, assembly, cleaning, surface engineering, materials (including composites) and quality systems.

The importance of competitive cost-effective manufacturing is demonstrated by Weck et al. (1991). The book focuses on the need to use advanced manufacturing techniques more profoundly than in the past, not least with respect to their environmental effect. Intelligent machines, systems, strategies for the factory of the future, environmental responsibility and quality assurance are all included.

In order to retain long-term competitiveness manufacturing research is essential. Tanner (1991) includes, in a revised edition of an introductory text, practical information on manufacturing research which emphasizes the functions, facilities required and goods and objectives of this important aspect of manufacturing engineering.

Other excellent fundamental textbooks are provided by Alexander, Brewer and Rowe (1987) and Lindberg (1988).

Assembly and joining

A first-class book on automatic assembly has been produced by Boothroyd (1991) who has recently co-authored another book Boothroyd, Dewhurst and Knight (1994) which emphasizes product design for manufacture and assembly. Two other very good books which concentrate on design for assembly come from Andreasen et al. (1988), which concentrates mainly on mechanical and automatic assembly, and Redford and Chal (1993). The last provides a guide to design for assembly linking the technologies of manual, dedicated and flexible assembly to the rules which have been formulated for them.

Electronics assembly is a rapidly growing and extremely competitive field. Riley (1987) has brought together many of the best articles, originally published in the monthly journal *Electronic Packaging and Production*, covering assembly, joining and testing of electronic components. A valuable feature of the book is the opinions expressed by many practitioners on the various aspects of assembly techniques.

A strategic study of the European contract electronics manufacturer is provided by Fletcher (1994). This book examines the emerging contract electronic manufacturing industry in Europe (which provides the service of assembling electronic components onto printed circuit boards), from the point of view of both the original equipment manufacturer (OEM) and the contract electronics manufacturer. Further books which will be of interest to those involved in electronics assembly include texts from Hughes (1992) which looks at experimentation methods, and Herrick (1992) which concentrates on soft soldering and wire wrapping.

362 Manufacturing techniques

An excellent book which deals with the subject of machine soldering in a clear definitive way is one by Woodgate (1988), who provides enough theory to give the readers a good understanding of the subject. A broad coverage of joining methods is contained in a book edited by Buckley and Stein (1986) which includes welding, brazing, soldering, mechanical fastening, explosive welding, solid state bonding and adhesive bonding.

The increased use of plastics over the last decade has led to a wide number of joining techniques for plastics. Watson (1988) provides a solid introduction for those who need to understand the joining of plastics and composites in his book which is based on presentations given at a series of seminars.

A comprehensive study on welding is contained in two volumes from Davies (1992). The first includes basic material on the physics and chemistry of welding and also has an excellent section on non-destructive and destructive testing. The second volume provides an extensive survey of welding methods including cutting processes and the welding of plastics and duplex stainless steels. Other books which provide a good insight into the fundamentals of welding are provided by Lucas (1990) who touches on most aspects of TIG and plasma welding, Cary (1993) in a comprehensive presentation on present welding technology, and Norrish (1992) who examines advanced welding processes.

CAD/CAM/CIM/FMS

Computers are now used extensively in the manufacturing industry in the areas of both design and manufacture. Besant and Lui (1986) introduce the subject of computer-aided design and manufacture (CAD/CAM) by taking the reader through from the basics of computers to their application in real engineering design and manufacture. Further sound treatment of this important topic is provided by Shahinpoor (1990), Zeid (1991) and a two-volume series from Dwivedi et al. (1991). Jones (1992) gives a good introduction to the latest techniques for managers of CAD systems and those called upon to procure a new system. A wide range of topics are covered including data security.

CAD/CAM is an integral part of computer integrated manufacture (CIM) and Mitchell (1991) provides a sound introduction to the wider role of CIM covering a range of topics including CAD's role in manufacture, flexible manufacturing systems (FMS), communication networks, process planning and analysis tools in simulation. Practical examples and experiences of CIM users in Germany and the USA are presented in an excellent book by Scheer (1991) who also discusses the components of CIM plus information management aspects and potential future developments in this rapidly growing technology. A

Manufacturing techniques 363

number of case studies on leading Japanese manufacturers, including Hitachi, Mazda, Kobe Steel and Toyota, is included in a text by Asai and Takashima (1994). It provides an overview of advanced manufacturing technology in Japan, describing the modern manufacturing concepts and highlighting current applications, technologies and systems in Japanese manufacturing industry. Looking towards the future, the book reports on Japanese industry's progress towards computerized intelligent manufacturing management systems (IMMS). Further insights into the principles of CIM can be obtained from books by Waldner (1992) and Weatherall (1992), while Bertain (1991) provides a comprehensive guide to the implementation of CIM.

As part of their overall CIM strategy, companies are looking increasingly at flexible manufacturing systems (FMS). A good book for non-specialists is provided by Parrish (1990) who, in only 147 pages, supplies a broad overview and explains the technology, the applications and advantages. Talavage and Hannam (1988) give a wider introduction, dealing with the design, operation and simulation of FMSs. They have aimed their book at practising manufacturing engineers and university students. Finally, for those interested in the management of this technology as well as the technology itself, Maleki (1991) explains both aspects in a clear concise manner.

Machining and forming

Two comprehensive texts covering basic machining principles and machine tool parameters are provided in texts from Boothroyd and Knight (1989) and Baril (1987). The latter also presents an exciting look into the future of machining technology including descriptions of laser cutting, artificial machine vision and automated guided vehicles (AGVs). In another book Schwartz (1989) has selected some of the best articles from the technical literature and provides sections on traditional and non-traditional machining methods as well as new developments, including high speed machining, adaptive control and advances in tooling. Walker (1993) and Oxley (1989) also provide sound texts dealing with the fundamentals and mechanics of machining.

A comprehensive coverage on research and production developments in high rate metal removal is provided by King (1985), who presents a wealth of basic theory and production data. The book covers the principal cutting techniques including milling, turning, grinding and drilling as well as the less widespread laser technology.

In a book devoted mainly to the presentation of non-conventional methods of material removal, McGeough (1988) includes energy beam, electrochemical, electrodischarge, plasma arc, ultrasonic water jet and other non-specialized machining methods. An excellent description of energy beam processes is provided by Taniguchi et al. (1989). This book provides an introduction to advanced manufacturing processes using new energy sources. It starts with a discussion on the basic concept of energy beam processing and moves on to discuss in detail photon-beam processing, electron-beam processing and ion-beam processing. The use of laser machining is increasing in manufacturing industry, and because of the laser's unique characteristics the education and training of staff at different levels is required. Crafer and Oakley (1993) have addressed this problem in their book which covers carbon dioxide and solid state lasers as well as the more recently introduced excimer lasers. The theory and practice of laser machining are covered in books by Chryssolouris (1991) and Gaillard and Quenzer (1989). Other electron beam processes are discussed in books by Bly (1990) and Schiller, Heisig and Panzer (1982).

There are many metal forming techniques, all of them having the primary objective to produce a desired shape change. Avitzur (1983) provides a practical insight to many of these processes in a book aimed to give the reader a thorough understanding of the behaviour of shaping by plastic deformation processes. Forging, extrusion, wire tube and can making are comprehensively covered, as well as a complete review of friction, pressure induced ductility, soft tooling and high energy rate forming. The metallurgical aspects, as well as a review of the necessary mechanics, are included in a book by Hosford and Caddell (1993), which was based upon courses developed for senior graduate students. Other books supplying a solid introduction to metal forming are supplied by Altan et al. (1983) and Lange (1985).

Finite element methods (FEM) offer significant potential in mathematically modelling and analysing forming processes, and Kobayashi, Oh and Altan (1989) present the fundamentals and applications of FEM in metal forming analysis and technology. This book, although written mainly for graduate students and researchers, will be useful for practising engineers if they have an understanding of FEM.

In a more specialized book on the pressing of sheet metal, Pearce (1991) discusses a wide range of subjects appropriate for obtaining an understanding of this technology. The text is divided into three sections, (a) processes, (b) materials and (c) tests. Marciniak and Duncan (1992) also provide an excellent text on the mechanics of sheet metal forming.

An extremely thorough book on casting technology is presented by Campbell (1991). The book examines the events that occur from the 'melt' to finished product. It aims to provide the knowledge necessary to avoid common mishaps and to encourage a more professional approach to the design and manufacture of castings. Sections are included on metal reactions, fluid dynamics, mould dynamics, solidification dynamics, casting shrinkage, casting accuracy and structure, defects and properties of the finished casting.

Close-to-finish form including precision casting is becoming an important competitive technology, resulting in increasing demands for improved dimensional accuracy and consistency, improved surface finish and improved metallurgical integrity being placed on the foundry industry. Clegg (1991) reviews the factors which influence the ability to meet these demands in a book which also describes the principal casting production processes available to the foundry industry. Although the book places a special emphasis upon the casting process it also includes a look at casting design, and quality control procedures in the production of precision castings.

For those interested in injection moulding, two books by Whelan (1984) and Whelan and Goft (1986) will provide a sound knowledge of the technology and machines used in this process. Powder injection moulding is generating considerable interest in industry and a book by German (1990) provides a useful starting point. It includes coverage of the principles associated with the process including powder fabrication and selection, bonder formulation, mixing, rheological considerations, moulding and sintering.

Management

Effective manufacturing depends upon sound management techniques. Resources such as information, people, power, materials, machines and finance must be organized successfully in any profit making and competitive company, and fortunately, there are many excellent texts on manufacturing management. For example, Amrine, Ritchey and Moodie (1993) in the 6th edition of their book, as in previous editions, familiarize the reader with the principles, practices and functions of manufacturing management. The text is presented in a simple, easy to follow style and deals mainly with the manufacturing areas but also includes sections on finance, marketing and personnel. In a very concise text of only 135 pages Macbeth (1989) offers a choice of different routes to manufacturing excellence. This is done through taking account of modern manufacturing philosophies, such as reducing inventory sizes and lead times, and the elimination of waste, particularly that waste associated with poor quality. Vollman, Berry and Whybark (1988) also reflect on the way manufacturing has changed in their revised edition, with a detailed treatment of just-in-time, new sections on production planning, the scheduling of flexible manufacturing systems, a description of optimized production technology (OPT) techniques and some new mathematical models for aggregate planning.

Although principally about managing design, the text by Ettlie and Stoll (1990) indicates clearly the critical relationship between the design and manufacturing areas. It includes seven chapters which describe the case histories of some companies which have achieved success in organizing these two areas to reflect this important connection. These companies include Black and Decker, General Motors, General Electric, IBM, A.B. Chance and Northern Telecom. The case histories are followed by a forecast of what future-oriented manufacturing companies can do.

Performance measurement is essential for the achievement of world class manufacturing, and a timely text by Maskell (1991) defines and presents in detail the new performance measures currently being adopted for achieving this. The book supplies practical assistance for the development of dependable and relevant measures of performance for those working towards world class competitiveness.

Finally, for those involved in research and teaching in the management of advanced manufacturing a book by Gerwin and Kolodny (1992) will be of value. An emphasis is placed on the management of new types of computerized technologies and the text suggests future research directions and also includes some successful, and some unsuccessful case studies of actual industrial systems. These are closely analyzed to highlight reasons for success as well as sources of failure.

Other useful books recently published include those from Hill (1993) on strategy, East (1994) on systems integration, and Vanderspek (1992) on factory automation.

Quality

This area covers overall quality principles, procedures and management as well as measurement and inspection techniques. In a particularly well written book Feigenbaum (1988) provides a basis for an understanding of the fundamental principles and practices of 'total quality' systems which have been incorporated in many companies over a substantial period of time. For those interested in understanding and applying quality systems in accordance with BS 5750: 1987, or ISO 9000: 1987, Hall (1988) has produced a guide which includes a set of ready made forms which can be directly incorporated into a particular company's systems.

Tannock (1992) presents a framework for automating the quality function in a way which is appropriate to the concepts of total quality management. This book will be of interest to quality and manufacturing engineers and managers, as well as students. The emphasis is on flexible automation and integration of the quality function and the text includes sections on the design and improvement of quality control systems; an examination of quality data analysis and management, including statistical process control (SPC) software and shop floor data collection.

In a book written mainly for industrial practitioners responsible for the creation and management of quality systems, Menon (1992) provides an overall perspective of total quality management, based on the experiences of the author in the automotive industry. Information is included on acquiring and stablizing new machinery, gauging, data collection and processing equipment, design of experiments, SPC, failure modes and effects analysis (FMEA). The approach taken by the author is very much user-oriented and aims at 'hands-on' applications.

Two books with an emphasis on the fundamentals of engineering measurement come from Adams (1975) and Galyer and Shotbolt (1990). Adams covers a wide range of basic measurement topics appropriate for aeronautics, mechanical and production engineers, and assumes that the reader has knowledge of basic mechanics, electrics and heat theories. Galyer and Shotbolt's book is a mainly mechanical text which is comprehensively illustrated and includes sections on mechanical measurements (size, form and surface texture), machine tool metrology and SPC. In a more recent book Dally, Riley and McConnell (1993) look at the many changes that have occurred over recent years in engineering measurement. These include the rapid advancements of digital instruments, the widespread availability of personal computers for data acquisition and improved methods for the analysis of vibrations.

Other recent books giving a good coverage to quality aspects are provided by Logothetis (1993), who covers techniques from Deming to Taguchi; Christou (1994), who looks at microelectronic manufacturing; Barkman (1989), on in-process quality control and Harvey (1993) who considers the implications of integrating just-in-time methods, materials requirement planning (MRP) and total quality management.

Standards

For any person involved in manufacturing it will almost certainly be necessary, at some time, to consult relevant British or International standards. The British Standards Institution (BSI) produces an annual catalogue which will assist in identifying appropriate manufacturing standards, many of which are now incorporated into international ISO standards. Some of those which are likely to be referred to most frequently include the following:

BS 3800: 1990–1991 General Tests for Machine Tools. Part 1 defines a code of practice for treating geometric accuracy of machines operating under no load or finishing conditions. Part 2 describes statistical methods for the determination of accuracy and repeatability of machine tools. Part 3 defines methods for testing the performance of machines operating under loaded conditions in respect of thermal distortion.

BS 4656: 1971–1993 Accuracy of Machine Tools and Methods of

368 Manufacturing techniques

Test. This extensive standard (37 parts) specifies the accuracy and methods of test for a wide range of standard machine tools including numerically controlled machines.

BS 4813: 1972 Method of Measuring Noise from Machine Tools – excluding testing in anechoic chambers.

BS 5304: 1988 Code of Practice for Safety of Machinery. This standard identifies hazards arising from the use of machinery and describes methods to eliminate or reduce these hazards. It includes machine design, guard design, safety and interlocking devices.

BS 5750 (ISO 9000): 1987 Quality Systems. This standard provides detailed guidance to both supplier and purchaser of the requirements of implementing and maintaining quality standards.

BS 5760: 1991–1995 Reliability of Systems, Equipment and Components. The essential features of a comprehensive reliability and maintainability programme are identified. Examples of how some of the principles are applied in industry are included.

BS 6101: 1981–1992 Machine Tool Ball Screws. The methods for calculating load and life ratings are included together with specifications for dimensions and accuracy.

BS 6143: 1990–1992 Guide to the Economics of Quality. This standard defines an appropriate process cost model.

BS 7850: 1992–1994 Total Quality Management. Part 1 provides a guide to management principles and Part 2 a guide to quality improvement methods.

From time to time the BSI arrange for useful handbooks to be published which are a compilation of standards on a particular area. For example Handbook 22 – Quality Assurance (1992), contains fourteen BSI publications on reliability and maintainability.

A manual of British standards in engineering metrology edited by K. Brooker, BSI (1984), published by Hutchinson and Co brings together material from over 40 British standards and other related publications which were current at the end of April 1983.

A useful database STANDARDS AND SPECIFICATIONS provided by the National Standards Association in the USA is available via the DIALOG (File 113) electronic information system. This provides bibliographic access to all US government and industry standards, specifications and related documents since 1950. These specify terminology, performance testing, safety, materials, products, or other requirements and characteristics of interest to a particularly technology or industry.

Reference books

An excellent compilation of practical information which details the specification and use of modern manufacturing techniques is presented in the 6 volumes of the *Tool and manufacturing engineers handbook*,

4th ed., published in the USA by the Society of Manufacturing Engineers. The volumes are:

Volume 1 (1983) Machining: Gives a broad coverage of important aspects of modern machining technology.

Volume 2 (1984) Forming: Provides an in-depth description of more than 50 different forming techniques (sheet metal, forging, casting, powder metallurgy, etc.).

Volume 3 (1985) Materials, finishing and coating: Discusses fundamental and advanced techniques on the application of traditional and non-traditional heat treatment, finishing and coating processes and properties of many engineering materials.

Volume 4 (1987) Quality control and assembly: The quality control section includes planning for quality, SPC, dimensional metrology, non-destructive testing. The assembly section covers part feeding methods, mechanical fastenings, joining methods and automated assembly.

Volume 5 (1988) Manufacturing management: A wide range of management techniques are discussed including capital equipment justification, process planning, CIM, automation, cost estimating and make or buy decisions.

Volume 6 (1992) Design for manufacturability: This provides extensive discussion on ways to reduce manufacturing costs at the design stage.

Other reference books which should be available to every manufacturing engineer include the following works.

Salvendy (1992) presents modern approaches for increasing productivity and quality and emphasizes the effective management of resources. Over 4000 references are cited and 1500 figures and tables included.

The machining data handbook from Metcut Research Associates Inc. (1980) provides important machining parameters for many different materials and processes. Initial machining recommendations (speeds, feeds, tool material, etc.) for conventional and non-traditional processes are included.

A standard international reference source is *Machinery's handbook*. Its 24th edition was published in 1992 by Industrial Press, and it provides a substantial amount of general information and mechanical data, representing modern design and manufacturing practice.

Juran and Gryna (1988) is an excellent book for all those involved in the quality of products and services. It follows the product through its life cycle examining ways of managing quality in various national cultures, and includes applications of concepts, methods and tools appropriate for various industries.

Other excellent reference texts are provided by Lotter (1990) on assembly, Matisoff (1986) on electronics manufacture and Tijunelis and McKee (1987) on high technology.

370 Manufacturing techniques

Directories

Probably the most widely used directory is the *Machinery buyers guide* published annually by Machpress Ltd. It is arguably the most up-to-date directory of engineering products and services. An international directory *Engineering research centres*, 3rd ed. (1993) (Longman) details around 8000 research centres in more than 70 countries which undertake, promote or fund research in engineering, including manufacturing. Information on the size and nature of manufacturing organizations is included in the *Directory of European industrial and trade associations*, 5th ed. (1991) (CBD Research), and *Trade associations and professional bodies of the United Kingdom*, 12th ed. (1994). The latter text includes in many cases the member profile and objectives of the organization and where appropriate, their publications. Finally the *Engineering design and manufacturing index* provided by Technical Indexes Ltd proves details of suppliers of materials, components and semi-finished products of interest to designers and manufacturers. The product data book is used together with a microfilm system which provides information direct from suppliers catalogues.

Dictionaries

Students and young engineers often require definitions for lessons associated with manufacturing. In addition, more senior practitioners will also occasionally require definitions for new terms associated with new or fast moving technologies. A few useful dictionaries in various areas of manufacturing include the following:

Ernst R. (1980) *Dictionary of engineering and technology*, Vol. 1: German–English, 4th rev. ed. (Brandstetter).
Ernst, R. (1985) *Comprehensive dictionary of engineering and technology*, Vol. 1: French–English (Cambridge University Press).
John V. (ed.) (1990) *Dictionary of materials and manufacturing* (Nichols Publishing Company).
Junge, H. D. (1992) *Dictionary of machine tools and mechanical engineering*, 2 vols: English/German, German/English (VCH).
Preston, E. J., Crawford, G. W. and Cotticchia, M. E. (1985) *CAD/CAM dictionary* (Marcel Dekker).
Veilleux, R. F. (ed.) (1987) *Dictionary of manufacturing terms* (Society of Manufacturing Engineers).

Representative organizations

CIRP (International Institution for Production Engineering Research), whose headquarters are in Paris, is an international organization composed of leading manufacturing research and development experts from all over the world. CIRP's purpose is to improve manufacturing productivity through cooperation and exchange of information among the members, and it publishes its *Annals* paper annually.

Both the Institution of Mechanical Engineers, and Institution of Electrical Engineers have strong manufacturing groups. They each act as professional qualifying bodies and as learned societies involved in the professional development of engineers and as international centres for technology transfer. Extensive library and information services are available and they offer an online search service to members and non-members.

The Advanced Manufacturing Technology Research Institute (AMTRI), engages in research, development and consultancy work with regard to machine tools, general purpose machines and systems, and advanced manufacturing technology applications for industry.

The Institute of Measurement and Control is a qualifying body whose objective is to advance measurement and control science and its applications.

The British Quality Foundation promotes total quality management via a prestigious quality award scheme, courses, conferences and publications.

The Machine Tool Technologies Association (MTTA) undertakes the research, preparation and distribution of economic and statistical information and also offers assistance on matters concerning the environment and the promotion of standardization and safety.

The International Institute of Welding provides for exchange of scientific and technical information relating to welding research and education.

References

Adams, L. F. (1975) *Engineering measurements and instrumentation.* London: English Universities Press.
Alexander, J. M., Brewer, R. C. and Rowe, G. W. (1987) *Manufacturing technology* 2 vols. Chichester: Ellis Horwood.
Altan, T. *et al. (1983) Metal forming: fundamentals and applications.* Metals Park, OH: American Society for Metals.
Amrine, H. T., Ritchey, J. A. and Moodie, C. L. (1993) *Manufacturing organization and management* 6th ed. Englewood Cliffs, NJ: Prentice Hall.
Amstead, B. H., Ostwald, P. F. and Begeman, M. L. (1987) *Manufacturing processes* 8th ed. New York: Wiley.
Andreasen, M. M. *et al.* (1988) *Design for assembly* 2nd ed. Berlin Springer-Verlag.

372 Manufacturing techniques

Asai, K. and Takashima, S. (eds) (1994) *Manufacturing, automation systems and CIM factories.* London: Chapman and Hall.

Avitzur, B. (1983) *Handbook of metal forming processes.* New York: Wiley.

Baril, R. (1987) *Modern machining technology.* New York: Delmar.

Barkman, W. E. (1989) *In-process quality control for manufacturing.* New York: Marcel Dekker.

Bertain, L. (ed.) (1991) *CIM implementation guide* 3rd ed. Dearborn, MI: Society of Mechanical Engineers.

Besant, C. B. and Lui, C. W. K. (1986) *Computer-aided design and manufacture* 3rd ed. Chichester: Ellis Horwood.

Bly, J. H. (1990) *Electron beam processing.* Yardley, PA: International Information Association.

Boothroyd, G. and Knight, W. A. (1989) *Fundamentals of machining and machine tools* 2nd ed. New York: Marcel Dekker.

Boothroyd, G. (1991) *Assembly automation and product design.* New York: Marcel Dekker.

Boothroyd, G., Dewhurst, P. and Knight, W. A. (1994) *Product design for assembly and manufacture.* New York: Marcel Dekker.

Buckley, J. D. and Stein, B. A. (1986) *Joining technologies for the 1990's.* Park Ridge, NJ: Noyes Data Corporation.

Campbell, J. (1991) *Castings.* Oxford: Butterworth-Heinemann.

Cary, H. B. (1993) *Modern welding technology* 3rd ed. Englewood Cliffs, NJ: Prentice Hall.

Christou, A. (1994) *Integrating reliability into microelectronics manufacturing.* Chichester: Wiley.

Chryssolouris, G. (1991) *Laser machining: theory and practice.* Berlin: Springer-Verlag.

Clegg, A. J. (1991) *Precision casting processes.* Oxford: Pergamon.

Crafer, R. C. and Oakley, P. J. (eds) (1993) *Laser processing in manufacturing.* London: Chapman and Hall.

Dally, J. W., Riley, W. F. and McConnell, K. G. (1993) *Instrumentation for engineering measurements* 2nd ed. New York: Wiley.

Davies, A. C. (1992) *The science and practice of welding* 10th ed. 2 vols. Cambridge: Cambridge University Press.

DeGarmo, E. P., Black, J. T. and Kohser, R. A. (1988) *Materials and processes in manufacturing* 7th ed. New York: Macmillan.

Dwivedi, S. N. *et al.* (1991) *CAD/CAM, robotics and factories of the future '90.* Berlin: Springer-Verlag.

East, S. (1994) *Systems integration: a management guide for manufacturing engineers.* London: McGraw-Hill.

Edwards, L. and Endean, M. (eds) (1990) *Manufacturing with materials.* London: Butterworths.

Ettlie, J. E. and Stoll, H. W. (eds) (1990) *Managing the design/manufacturing process.* New York: McGraw-Hill.

Feigenbaum, A. V. (1991) *Total quality control* 3rd rev. ed. New York: McGraw-Hill.
Fletcher, A. (1994) *The European contract electronics assembly industry, 1993-1997.* Oxford: Elsevier Advanced Technology.
Gaillard, M. L. and Quenzer, A. (1989) *High power lasers and laser machining technology.* Bellingham, WA: Society of Photo-Optical Instrumentation Engineers.
Galyer, J. F. W. and Shotbolt, C. R. (1990) *Metrology for engineers* 5th ed. London: Cassell.
German, R. M. (1990) *Powder injection moulding.* Princeton, NJ: Metal Powder Industries Federation.
Gerwin, D. and Kolodny, H. (1992) *Management of advanced manufacturing technology: strategy, organization and innovation.* New York: Wiley.
Gillespie, L. K. (ed.) (1988) *Troubleshooting manufacturing processes* 4th ed. Dearborn, MI: Society of Manufacturing Engineers.
Hall, R. W. (1988) *Attaining manufacturing excellence.* Homewood, IL: Dow Jones-Irwin.
Harvey, B. (1993) *High velocity manufacturing: integrating JIT, MRP II and total quality management.* Dearborn, MI: Oliver Wight.
Herrick, G. (1992) *Electronic assembly.* Englewood Cliffs, NJ: Prentice Hall.
Hill, T. (1993) *Manufacturing strategy: the strategic management of the manufacturing function* 2nd ed. Basingstoke: Macmillan.
Hosford, W. F. and Caddell, R. M. (1993) *Metal forming: mechanics and metallurgy* 2nd ed. Englewood Cliffs, NJ: Prentice Hall.
Hughes, F. W. (1992) *Electronic assembly.* Englewood Cliffs, NJ: Prentice Hall.
Jones, P. F. (1992) *CAD/CAM: features, applications and management.* Basingstoke: Macmillan.
Juran, J. M. and Gryna, F. M. (1988) *Juran's quality control handbook* 4th ed. New York: McGraw-Hill.
Kalpakjian, S. (1991) *Manufacturing processes for engineering materials* 2nd ed. Reading, MA: Addison-Wesley.
King, R. I. (1985) *Handbook of high-speed machining technology.* New York: Chapman and Hall.
Kobayashi, S., Oh, S. I. and Altan, T. (1989) *Metal forming and the finite element method.* New York: Oxford University Press.
Lange, K. (ed.) (1985) *Handbook of metal forming.* New York: McGraw-Hill.
Lindberg, A. (1988) *Materials and manufacturing technology.* Englewood Cliffs, NJ: Prentice-Hall.
Lissaman, A. J. and Martin, S. J. (1982) *Principles of engineering production* 2nd ed. London: Hodder and Stoughton.

Logothetis, N. (1993) *Managing for total quality: from Deming to Taguchi and SPC*. London: Prentice Hall.
Lotter, B. (1989) *Manufacturing assembly handbook*. London: Butterworths.
Lucas, W. (1990) *TIG and plasma welding*. Cambridge: Abington Publishing.
Macbeth, D. K. (1989) *Advanced manufacturing: strategy and management*. Bedford: IFS.
Maleki, R. A. (1991) *Flexible manufacturing systems: the technology and management*. Englewood Cliffs, NJ: Prentice Hall.
Marciniak, Z. and Duncan, J. L. (1992) *Mechanics of sheet metal forming*. London: Edward Arnold.
Maskell, B. H. (1991) *Performance measurement for world class manufacturing*. Cambridge, MA: Productivity Press.
Matisoff, B. S. (1986) *Handbook of electronics manufacturing engineering* 2nd ed. New York: Van Nostrand Reinhold.
McGeough, J. A. (1988) *Advanced methods of machining*. London: Chapman and Hall.
Menon, H. G. (1992) *TQM in new product manufacturing*. New York: McGraw-Hill.
Mitchell, F. H. (1991) *CIM systems: an introduction to computer integrated manufacturing*. Englewood Cliffs, NJ: Prentice Hall.
Norrish, J. (1992) *Advanced welding processes*. Bristol: Institute of Physics.
Oxley, P. L.B. (1989) *The mechanics of machining: an analytical approach to assessing machinability*. Chichester: Ellis Horwood.
Parrish, D. J. (1990) Flexible manufacturing. London: Butterworth-Heinemann.
Pearce, R. (1991) *Sheet metal forming*. Bristol: Adam Hilger.
Redford, A. H. and Chal, J. (1993) *Design for assembly* New York: McGraw-Hill.
Riley, F. J. (ed.) (1987) *Electronic assembly*. Berlin: Springer-Verlag.
Salvendy, G. (ed.) (1992) *Handbook of industrial engineering* 2nd ed. New York: Wiley.
Scheer, A. W. (1991) *CIM: computer integrated manufacturing: towards the factory of the future* 2nd rev. ed. Berlin: Springer-Verlag.
Schey, J. A. (1987) *Introduction to manufacturing processes*. New York: McGraw-Hill.
Schiller, S., Heisig, U. and Panzer, S. (1982) *Electron beam technology*. New York: Wiley.
Shahinpoor, M. (1990) *CAD/CAM systems*. New York: Wiley.
Schwartz, M. M. (ed.) (1989) *Machining source book: a collection of outstanding articles from the technical literature*. Metals Park, OH: ASM International.

Talavage, J. and Hannam, R. G. (1988) *Flexible manufacturing systems in practice*. New York: Marcel Dekker.
Taniguchi, N. *et al.* (1989) *Energy-beam processing of materials*. Oxford: Clarendon Press.
Tanner, J. P. (1991) *Manufacturing engineering: an introduction to the basic processes* 2nd ed. New York: Marcel Dekker.
Tannock, J. D.T. (1992) *Automating quality systems*. London: Chapman and Hall.
Tijunelis, D. and McKee, K. E. (1987) *Manufacturing high technology handbook*. New York: Marcel Dekker.
Vanderspek, P. G. (1992) *Planning for factory automation: a management guide to world-class manufacturing*. New York: McGraw-Hill.
Vollmann, T. E., Berry, W. L. and Whybark, D. C. (1988) *Manufacturing planning and control systems* 3rd ed. Homewood IL: Irwin.
Waldner, J. B., Duffin, W. J. (trans.) (1992) *CIM: principles of computer integrated manufacturing*. Chichester: Wiley.
Walker, J. R. (1993) *Machining fundamentals: from basic to advanced techniques*. South Holland, IL: Goodheart Willcox.
Watson, M. N. (ed.) (1988) *Joining plastics in production*. Cambridge: Welding Institute.
Weatherall, A. (1992) *Computer integrated manufacturing: a total company competitive strategy*. 2nd ed. Oxford: Butterworth-Heinemann.
Weck, M. (1984) *Handbook of machine tools*. Chichester: Wiley.
Weck, M. *et al.* (1991) *Production engineering the competitive edge*. Oxford: Butterworth-Heinemann.
Whelan, A. (1984) *Injection moulding machines*. London: Elsevier Applied Science.
Whelan, A. and Goff, J. P. (eds) (1986) *Developments in injection moulding*. London: Elsevier Applied Science.
Woodgate, R. W. (1988) *The handbook of machine soldering: a guide for the soldering of electronic printed wiring assemblies* 2nd ed. New York: Wiley.
Zeid, I. (1991) *CAD/CAM theory and practice*. New York: McGraw-Hill.

CHAPTER TWENTY-ONE

Robotics and automated manufacture

R. S. McMASTER AND B. G. BOYLE

Introduction

There are as many definitions of the word 'robot' as there are robots in existence. Today's robots are the science fiction of a time past. Joseph Engelberger, the so-called 'Father of Robotics', along with the Robots Institute of America, came up with the definition that:

> A robot is a reprogrammable multifunctional manipulator designed to move material, parts, tools, or specialized devices through variable programmed motions for the performance of a variety of tasks.

That was many years ago, and since then there have been some changes and refinements to the 'definition' of a robot, but the idea remains the same. Robots are now used in almost every industry in the world, performing a multitude of tasks in a host of environments, and have come a long way towards the fulfilment of their literary beginnings in the stories of Capek and Asimov.

The modern robot was born in the minds of Engelberger and a man named George C. Devol, responsible for the combination of the application of digital control from the newly developed 'Logic Machines' with the multi-axis 'manipulator' technology that had developed from the nuclear industry in the 1950s. By 1961 'Unimates' were being manufactured by hand and sold to the Ford Motor Company for use in the die-casting parts of production. Robots continued to develop, gaining first senses, like vision, and then intelligence with the first use of computer control in a robot by the mid-1970s. At this time however, labour was plentiful and cheap, whereas robots were generally expensive and deemed unreliable, and hence only in Japan did the robot explosion begin. The robot proceeded to develop and become stronger, more dextrous, accurate and diverse, gaining the senses of touch and force. The controlling computers also improved immensely, and robots are now reliable and in many cases irreplaceable. From the automobile industry, robot automation spread to all types of manufacture. They are used for their pinpoint accuracy, their brute strength, their ability to go where humans cannot, or to repeat a task endlessly without rest or complaint.

More recently the robot has developed in different ways, especially through the means by which it is programmed. Offline programming and simulation are prevalent. The use of the robot has diversified into many unautomated fields, including the boom industry associated with off shore oil and other underwater work, miniaturization for surgery, military uses such as bomb disposal, construction and more. Combined with the advances in artificial intelligence spawned by the growth in processing power, today's robots are starting to make the intelligent android of 1940s' science fiction a possibility of the foreseeable future.

Journals

There is a plethora of publications, ranging from the weekly to the annual, in this field. Their range of style and content is also diverse, and for these reasons, given below are the main journals in publication and a short description of their content, market and style.

IEEE Transactions on Robotics and Automation is standard for all types of research into the field including dynamics and control theory and practice. *IEEE Transactions on Systems, Man and Cybernetics* may be of interest to the less industrial robot user. Received free by members of the appropriate IEEE Society, both are bi-monthly. Every quarter the International Federation of Robotics prints its *IFR Robotics Newsletter*, and the Australian Robot Association also publishes a regular newsletter. *Advanced Robotics*, bi-monthly from the Robotics Society of Japan, gives coverage of robotics in science and engineering. Another Japanese bi-monthly publication, *JIRA Robot News* concentrates more on industrial robots and application systems. The Association Française de Robotique Industrielle publishes the quarterly *AFRI Liaison* which gives international coverage of the industry and includes market news, developments, automation and various features. *Robotics Today* is a quarterly publication from the Society of Manufacturing Engineers (SME) and offers comprehensive information on the use of robots in automated manufacture. The SME Computer and Automated Systems Association also offers the *Journal of Manufacturing Systems* every two months. The *International Journal of Robotics and Auto-*

mation is produced by the International Association of Science and Technology for Development and published quarterly by Acta Press. It covers all aspects of its title, including modelling, simulation, design and social implications and applications.

The *International Journal of Advanced Manufacturing Technology* (Springer-Verlag) bi-monthly. Aims to bridge the gap between pure research journals and more practical publications. Covering applications-based research, topics in manufacturing include robotics, AI, grippers, auto test equipment, flexible manufacturing systems simulation and much more.

Robotics and Computer-Integrated Manufacturing (Elsevier) quarterly. Prints papers, theoretical, experimental and applied, in the field of its title with an emphasis on flexible manufacturing systems and includes editorials, reviews, etc., but also a novel 'New patents' section listing the abstracts of related new entries from patent search databases.

Robotica (Cambridge University Press) bi-monthly. Claims to be the international journal of information, education and research in robotics and AI. It covers mainly software, kinematics and dynamics in robot design, modelling, problems and task planning.

The *International Journal of Robotics Research* (MIT Press) bi-monthly. Deals with the theoretical and experimental developments in the robotics field. Not as directed towards robots for manufacture, its field is wide and of general interest.

Robotics and Autonomous Systems (Elsevier) 8 issues per annum. Carries articles describing 'fundamental developments' in the field, encompassing both symbolic and sensory-based control and learning in the context of autonomous systems.

Assembly Automation (MCB University Press) quarterly. Journal of assembly technology and management. Industry-based with editorial and guest viewpoint, company news and contributed features, both in depth and abbreviated.

Industrial Robot (MCB University Press) bi-monthly. Deals with the industry from a robotics viewpoint.

Controls & Systems (Penton Publishing) monthly, *Robotics World* (Argus Business) quarterly and even *Manufacturing Systems* (Chilton Publishing) monthly, are all a mix of industry news, views, developments and discussions interspersed with an array of advertisements. They serve the business user by keeping abreast of development and competition, and allow academics to see final products and new requirements.

Manufacturing Automation is a compact monthly newsletter published by Vital Information Publications and consists of three or four interview style articles dealing with the people working with new developments.

Abstracting and indexing

There is a host of these services available, in either print, CD-ROM format or indeed online, for a much more up-to-date coverage. A selection of the most relevant to the field of robotics and automated manufacture are given here.

COMPENDEX*PLUS on CD-ROM or online through DIALOG, covers 1970 to the present, and is updated monthly by Engineering Information Inc. It is the machine-readable version of their publication *Engineering Index*, the main provider of abstracts from the world's significant literature in engineering and technology. The British equivalent is INSPEC, also on CD-ROM or online through DIALOG. It is updated weekly, by IEE in London, and is made up from their *Science Abstracts* publication, which is divided into *Physics Abstracts, Electrical & Electronics Abstracts* and *Computer and Control Abstracts* publications. MECHANICAL ENGINEERING ABSTRACTS from Cambridge Scientific Abstracts and compiled by Engineering Information Inc. is the mechanical engineering equivalent to INSPEC. ROBOTICS is an online database from Bowker A&I Publishing covering robots and their applications, and is available through ESA-IRS. A newcomer on the British scene is RAM (Recent Advances in Manufacturing) from Nottingham Trent University. It covers journal articles and new books on a range of manufacturing subjects. DERWENT WORLD PATENTS INDEX, available through DIALOG and ORBIT (now Questel.Orbit) although not strictly an abstracting service, does provide abstracts and files on over 12 million patents and patent applications from over 30 countries, updated weekly. SCISEARCH is the electronic version of the *Science Citation Index*, a monthly publication by the Institute for Scientific Information, but is more science than engineering-based.

Current Technology Index is a purely British indexing source, published by Bowker-Saur, which combined with the mainly American *Applied Science and Technology Index* provides a cover of the main English language journals. *The Applied Science and Technology Index* is supplied monthly by H. W. Wilson and covers a wide range, most of which is, however, covered already in the more comprehensive *Engineering Index* and *Current Technology Index*. *Current Contents: Engineering, Technology and Applied Sciences* is more of a weekly awareness service by the publishers of the *Science Citation Index*.

Organizations

There are as many organizations in the robotics and automation field as there are spheres of interest. Many of the interests of these learned society overlap, or in fact are wholly encompassed by other more general groups. Most of the societies have the proceedings of their conferences, symposiums and meetings listed in the conferences section of this chapter.

In Britain the contributors to the field are the British Robot Association (BRA), the Institution of Electrical Engineers (IEE) (mainly concentrated in the Computing and Control Division – professional group C15), the Institution of Mechanical Engineers (IMechE) and the Machine Tool Technologies Association (MTTA).

The international groups include the American Institute of Electrical and Electronics Engineers (Robotics and Automation Society), Society of Manufacturing Engineers (Computers and Automated Systems Association), and the American Society of Mechanical Engineers. Also very important is the International Federation of Robotics (IFR). The national associations of the major countries, such as the Japanese Industrial Robots Association (JIRA) and the Robotics Society of Japan, or the Association Française de Robotique Industrielle are also very active. A list of the addresses of these British and international organizations in given at the end of this chapter.

Conference proceedings

There are numerous conferences dedicated to, or involving, the field of robotics and automated manufacture. One of the major conferences in the robotics field is the International Symposium on Industrial Robots (ISIR) held annually by the International Federation of Robotics and published by IFS publications. It presents numerous papers from scores of countries worldwide. Of the other international institutions, the IEEE International Conference on Robotics and Automation, as run by their Robotics and Automation Society, is also a major publication. It is published in volumes by the IEEE Computer Society Press. Another American organization running an annual conference in the field is SME Robotics International, the papers of which they publish and distribute through North-Holland outside the USA. Robotics and Manufacturing is the annual symposium from the ASME: they also publish the papers from their winter conference in Anaheim, CA for that year under the title of *Advances in robotics*. In Britain the IEE Computing and Control Division (C15) holds conferences on various related topics, and the British Robot Association (BRA) has an annual conference, the proceedings of both of which are available. The European Conference on Automated Manufacture is held and published biennially, and triennially for the IFAC Symposium on Robot Control (SYROCO).

Of other conferences, here is a list of the more recent publications.

Autofact 92: Conference of the Society of Manufacturing Engineers, Computer & Automated Systems Association, Detroit, MI, 1992.

Automation and Robotics in Construction: 10th International Symposium, ISARC, Houston, TX, 1993.

Automation and Robotics in Construction: 9th International Symposium, JIRA, Tokyo, 1992.

Advances in Robotics and Automation International Symposium of the International Association of Science & Technology for Development, San Francisco, CA, 1984.

Manufacturing and Robotics 14th International Symposium, IASTED, Lugano, Switzerland, 1991.

Manufacturing Automation International Conference held at the University of Hong Kong, Department of Mechanical Engineering, Hong Kong, 1992.

ICARCV'92 2nd International Conference on Automation Robotics and Computer Vision, Institution of Engineers, Singapore, 1992.

Robots & Vision Automation International Conference of the Robotic Industries Association, Automated Imaging Association, National Service Robot Ass., Global Automation Information Network, Detroit, MI, 1993.

Applications in Manufacturing & Robotics 5th International Symposium on AI–AI: The Transfer. Conference held at Instituto Technologico y de Estudios Superiores de Monterrey, Cancun, Mexico, 1992.

Reference works

There are annual publications in the form of a statistical analysis in the field of industrial robotics and manufacture. The IFR each year produces the *World industrial robot statistics*. It is based on the information supplied to it by the national robotic societies of its member states, and, to some extent, the data bank of the United Nations Economic Commission for Europe. It serves to show the relative growth and distribution of robots in worldwide industry, where the robots came from, and to what uses they are put. Although comprehensive, this publication is very slow to be produced, not usually appearing until two years after the period with which it deals, and is prohibitively priced to non-members of the IFR. The UK's national society, the BRA, also publishes an annual document under the title of *Robot facts*, containing graphically represented statistics and analysis of the industry in the UK and its place in the European and world market. It includes population, growth, applications, cost analysis and geographic breakdown of installations. The BRA *Robot facts* publication tends to come out the year following that of its subject, therefore a year ahead of the IFR, and is free to all members of the BRA, placing it in a more attainable price bracket.

Standards

There are specific standards relating to robots in industry. BS 7228: 1991 Parts 1–6 cover industrial robots, as do the international standards ISO 9283: 1990, ISO /TR 8373: 1988, ISO 9787: 1990 and ISO 9946: 1990.

Textbooks

There are many textbooks on the subject of robotics and automated manufacture, far too numerous to be included here. In order to rationalize this, only modern texts or books of classical importance have been included. In this selection the books have been categorized into broad application groups and related issues.

General texts

A well balanced and readable book suitable for the professional engineer or postgraduate student is Asfahl (1992). This applications-oriented text surveys the wide spectrum of automated systems available to improve manufacturing productivity, including robots, numerical control machines, programmable controllers and microprocessor-based automated systems. It features industrial case studies, product design for manufacture, CAD, karnaugh maps, CIM and a problems section.

A book which encompasses the entire range of robotics and provides a comprehensive yet detailed lead in to the design and application of industrial robots is McCloy and Harris (1986). This is a classic text for engineering students and manufacturing engineers. It provides an authoritative and integrated coverage of the many disciplines involved; mechanical, electrical, electronic and software aspects of robots are all covered, together with the closely related technologies of pick and place devices, walking machines, teleoperated systems and prosthetics. Paul (1981) is a commonly referenced book in many robotics research works. It covers much of the theory of robotics control, and has particular significance when dealing with control mathematics, giving a standard means of positional calculation through forward and inverse kinematics and the standard formulae for the matrices implementation.

In Ross (1992) the emphasis is on the practical application of automation for maintaining a competitive edge in the industrial market. Aimed mainly at the professional engineer and management, it examines the implementation of flexible automation, the skill requirements and the problems that may arise. Included are such topics as; factory automation as a business concept, automation proposal identification, creation and analysis, the requirements specification and cost justification techniques. This practical theme is carried right the way through to cover detail design, building, debugging, testing and commissioning. Robot specifications and reference material are also included. Aimed at engineering students, Sandler (1991) emphasizes and answers such questions as: how do these devices work?; what are the means for solving specific technical problems in this domain?; what are the difficulties and obstacles in creating a successfully functioning automatic machine?; and what must be done to design a highly productive industrial robot? While the main accent of the book is on industrial applications of these machines, the levels of intelligence of such machinery are analysed and concepts relevant to robotics are addressed throughout. Step by step, the reader is guided through the proper sequence for constructing automatic mechanisms. Such issues as the relationship between level of robot intelligence and the type of product produced are discussed. Types, levels and structure of robots are detailed and the dynamics of manipulators are presented.

There are few books available on the ergonomics of automation but this is fully addressed in Rahimi and Karwowski (1992). As robotic systems become widespread in the manufacturing and service industries, this text addresses the key issue of how they interact with people. The book is organized around three central themes; the strength and weakness of human operators in interactive robotic environments; safety issues, including robot accidents, operator intrusions and system reliability; and the design of robots for different applications. The topics covered include questions of occupational safety, productivity, process design, scheduling, operation and repair; everywhere that the differing abilities and characteristics of human and robotic systems are seen. Highlighting the important aspects of human-robot interaction, this book is designed to provide easily retrievable information and is applicable to a wide range of engineers and scientists who have an involvement in robotics. The text is also suitable for specialists in human factors and students interested in robotic systems.

A review of recent research advances and state of the art industrial applications is given in MacConaill, Drews and Robrock (1991). Special consideration is given to subjects like intelligent robot systems, mobile robots, new control strategies, sensors, planning and programming plus various industrial applications. The book is primarily aimed at engineers and scientists who want to update their knowledge but may also be of interest to postgraduate students specializing in this subject.

Manufacturing applications

A classic text for students studying industrial robotics as part of a course in mechanical, electrical or manufacturing engineering is

Groover et al. (1987). This was one of the first such volumes on robotics to be designed specifically as a textbook. It provides a comprehensive survey of the applications, technology and computer science of robotics, adopting an interdisciplinary approach and including separate chapters on machine vision and artificial intelligence.

A guide to how and when automation technology should be applied is given in Williams (1994). Manufacturing industries need to respond to ever-increasing demands for flexibility, quality, speed and economy in order to remain competitive. This text shows how the hardware and software of automation technology can be harnessed to build successful manufacturing systems. The opening discussion demonstrates how programmable automation bears on manufacturing problems and the early chapters describe the building blocks of the system. The technology of these elements is analysed to lead to their integration into sub-systems and factory automation cells. This sets the stage for explaining how a complete programmable manufacturing facility is organized with full treatment of the control and computing functions involved. The book provides a clear account of steps to automation for engineering students and practising engineers. An extension of the principles given here is provided in Williams and Rogers (1991) and is a suitable text for research and postgraduate work.

An overview of the techniques and information needed by manufacturing executives for modernization of their manufacturing enterprises is given in Vanderspek (1992). This book combines automation and planning in a single integrated approach. Coverage of the main technologies, methodologies and software systems are presented and specific steps given in order to prepare an update programme. Definitions of automation equipment and processes are pursued in detail. This includes: CNC machines tools, robots, automated guided vehicles, conveyors, storage and retrieval systems, bar coding, machine vision and radio frequency tagging systems. Clear planning guidelines are presented and advice on how to avoid potential pitfalls is given.

In Goetsch (1991) a wide range of introductory topics is covered, aimed at students in the early stages of a manufacturing technology degree course or practising engineers wishing to gain a basic understanding of advanced manufacturing technology. Special features include: how automated assembly and automated guided vehicles are used in the modern manufacturing environment; the use of CAD/CAM and computers in manufacturing; programmable logic controllers, their history, configuration, operation and application, and an in-depth chapter on CIM which identifies problems and benefits of this philosophy.

Aimed mainly at managers and technical personnel, Maleki (1990) explains the component technologies used and explores their relationship with each other, and flexible manufacturing systems as a whole. Principal technologies and systems discussed include: robotics, programmable logic controllers and automated guided vehicles. Also covered are such issues as technology management of FMS, planning, control and economic justification.

A useful insight into advanced manufacturing technology in Japan is presented in Asai and Takashima (1993). The book describes the prevalent manufacturing engineering concepts and highlights the current applications, technologies and systems in Japanese manufacturing industry. In Japan, the progress of numerically controlled automation systems has resulted in practical manufacturing systems characterized by a high degree of flexibility and productivity. This in turn has resulted in the establishment of reliable, high quality factory automation systems. Numerous case studies from leading Japanese manufacturers, including Toyota, Hitachi, Mazda and Kode Steel, are featured. Looking towards the future, the book reports on Japanese industry's progress towards computerized intelligent manufacturing management systems. This work will appeal mainly to practising engineers, postgraduate students and researchers.

Weston (1989) discusses the relationship between robots and arc welding from a number of angles, including the economics, basic principles, safety and future trends, and gives a series of case histories from a range of industries. The book is divided into sections dealing with the use of robotics, industrial installations, vision-based sensor systems, flexible manufacture and CIM, and finally the problems of design for robot welding. The book would be of interest to readers both inside and outside the field, from the manager seeking insight into the field, to the student looking for an overview of the place of robotics in manufacture.

An overview of common production welding processes is presented in Norrish (1992). Welding has traditionally been regarded as a craft rather than a technological manufacturing process. This reputation has not been helped by the dependence of conventional joining techniques on highly skilled manual operators and the relatively high cost of certain welding processes. Developments in recent welding technology have been devoted to an improved understanding of the basic welding processes. This volume covers all of these issues and discusses recent developments in electronic power regulation, computer control, automation, robotics and process monitoring. The text shows how advances in welding technology may be used to improve cost-effectiveness and quality of joining techniques. The book will be of use to manufacturing engineers and roboticists who need detailed process information when involved with robotic and automatic welding systems.

References

Asai, K. and Takashima, S. (eds) (1993) *Manufacturing automation systems and CIM factories* London: Chapman and Hall.

Asfahl, C. R. (1992) *Robots and manufacturing automation* 2nd ed. New York: Wiley.
Goetsch, D. L. (1990) *Advanced manufacturing technology*. Albany, NY: Delmar.
Groover, M. P. et al. (1986) *Industrial robotics: technology, programming and applications*. New York: McGraw-Hill.
MacConaill, P. A., Drews, P. and Robrock, K. H. (1991) *Mechatronics and robotics* Amsterdam: IOS Press.
Maleki, R. A. (1990) *Flexible manufacturing systems*. Englewood Cliffs, NJ: Prentice Hall.
McCloy, D. and Harris, D. M. J. (1986) *Robotics an introduction*. Milton Keynes: Open University Press.
Norrish, J. (1992) *Advanced welding processes*. Bristol: Adam Hilger.
Paul, R. P. (1981) *Robot manipulators: mathematics, programming and control: the computer control of robot manipulators*. Cambridge, MA: MIT Press.
Rahimi, M. and Karwowski, W. (eds) (1992) *Human robot interaction*. London: Taylor and Francis.
Ross, A. (1992) *Putting robots to work*. New York: McGraw-Hill.
Sandler, B. Z. (1991) *Robotics: designing the mechanisms for automated machinery*. Englewood Cliffs, NJ: Prentice Hall.
Vanderspek, P. G. (1992) *Planning for factory automation*. New York: McGraw-Hill.
Weston, J. (ed.) (1989) *Exploiting robots in arc welded fabrication*. Cambridge: Welding Institute.
Williams, D. J. (1994) *Manufacturing systems: an introduction to the technologies* 2nd ed. London: Chapman and Hall.
Williams, D. J. and Rogers, P. (eds) (1991) *Manufacturing cells: control, programming and integration*. Oxford: Butterworth-Heinemann.

Contact addresses of relevant organizations

International Federation of Robotics (IFR), Sveriges Verkstadsindustrier, Storgatan 19, Box 5506, S–114 85, Stockholm, Sweden.

Institute of Electrical and Electronics Engineers (IEEE), 445 Hoes Lane, PO Box 1331, Piscataway NJ 08855–1331, USA.

Robotics International of SME, 1 SME Drive, PO Box 930, Dearborn, MI 48121, USA.

International Association for Science and Technology for Development (IASTED), PO Box 25, Station G, Calgary, AB T3A 2G1, Canada.

Association Française de Robotique Industrielle (AFRI), 61 Avenue Du President Wilson – 94230 Cachan, France.

Japanese Industrial Robot Association (JIRA), Kikai-Shinko Bldg. 3–5–8 Shiba-Koen, Minato-ku, Tokyo, Japan.

British Robot Association (BRA), Aston Science Pk, Love Lane, Birmingham B7 4BJ, UK.

Institution of Electrical Engineers (IEE), Savoy Place, London WC2R 0BL, UK.

Institution of Mechanical Engineers (IMechE), 1 Birdcage Walk, London SW1H 9JJ, UK.

Machine Tool Technologies Association (MTTA), 62 Bayswater Rd, London W2 3PS, UK.

CHAPTER TWENTY-TWO

Control engineering

T. P. SATTAR

Outline of developments in control engineering

The systematic design and use of control systems has become an important part of engineering since World War II. An extensive theory of control systems has been developed and applied for many purposes including the regulation of manufacturing processes, motion control of machines, and the operation of complex systems such as economic and management systems. Control is used to improve system performance with regard to the positioning of plant variables at desired levels, tracking of desired motion trajectories, and reduction of the effect of disturbances on plant variables. The control task is usually made difficult by the complex nature of the dynamics of real systems. The complexity is due to the high dimensionality of differential equations used to model the dynamics, non-linearities in the dynamics, interaction of system variables, and possible time variations in the dynamics. As a result of this complexity the designer usually has incomplete knowledge of the dynamics which adds uncertainty into the design process. Feedback control provides the means to deal with uncertainty about real systems by using the error between desired and actual system variables to make adjustments to inputs into the system.

The earliest control methods did not attempt to model the system but simply provided a few tuning knobs with which the operator could tune the system response heuristically to what he considered to be the best one. These methods were successful essentially because they were simple, robust and quick to implement. For example, proportional, integral and derivative (PID) controllers tune only three parameters. They have been around for nearly fifty years and still continue to be widely used in the process industries despite the numerous developments which have taken place during this period.

The next phase of development in control system theory was the model-based approach to control system design. Mathematical modelling of continuous-time systems using differential equations and of discrete-time and sampled data systems by difference equations opened up a Pandora's box of analysis and design methods. Transformation methods such as the Laplace transform enabled algebraic manipulations and frequency response designs to be performed with system transfer functions. Development of state space and matrix methods of system representation were particularly suitable and powerful tools for computer analysis and design. The analytic approach to control design led inevitably to the search for optimal control laws. Optimality was usually defined as the minimization of some performance cost functional. Earlier deterministic models were extended to include the effect of stochastic noise and disturbances on process plants. This led to the development of stochastic control theory which attempted to minimize the variance of plant variables. The best improvements in control performance were made when the physical system was fully understood using a combination of physical system modelling, simulation and experimental verification. System identification and estimation techniques were developed which used histories of plant input-output data to refine models obtained initially from balance equations. These identification and estimation techniques were combined with control laws to build online self-tuning and adaptive control algorithms.

However, the success of all these model-based design methods still rested on adequate modelling of the underlying physical system. After a few decades of the model-based approach to control design it was eventually accepted that uncertainties about the dynamics of real physical systems would always remain. The best approach would be to seek control laws which remained within some defined performance bounds despite a mismatch between the physical system and its model. The past decade has seen an effort to develop robust controllers which can maintain suitable performance for a range of models between which the real system may lie.

The H2 and H-infinity design techniques have become dominant in industrial control applications and adaptive prediction control techniques have offered benefits in a range of real applications. The model-based design has increasingly begun to be augmented by online tuning knobs which can eliminate or reduce the effect of model mismatch. Automatic tuning of these knobs and monitoring of control performance is now being attempted by using artificial intelligence and self-learning control methods. Artificial neural networks have come to be seen as very good tools for pattern recognition with which classification of system dynamics or disturbances can be obtained. Neural networks have also provided a generic modelling tool to allow rapid and cost effective development of complex and non-linear chemical/biochemical processes.

Fuzzy logic control has been found to be particularly valuable in applications where formal methods are needed to interpret the manual operators procedures. There is strong interest in techniques associated with linear matrix equalities and techniques which reduce these equations into a set of matrix inequalities which can be easily solved by numerically efficient computer algorithms. These design processes are being built into advanced computer-aided design control packages. Developments in computer technology also means that computers have become cheap enough and reliable enough to replace single loop analog controllers. They can be embedded into control loops in order to implement control laws of ever increasing numerical sophistication. Computers also make possible the implementation of supervisory or optimizing control. The supervisory computer can be placed in a secondary loop to the primary plant control loop and used to trim plant operation and provide setpoints for an optimizing control.

Some information sources to trace these developments in control engineering are given below. References to textbooks which have been repeatedly cited in the research literature are given as a first source of information on a new topic. This is usually the quickest introduction. This can be followed up by looking at the first or seminal papers on the subject. Periodicals can then be searched for recent or current developments.

System modelling

Physical modelling of dynamic systems is covered in almost all control text books with some useful ones being MacFarlane (1970), Shearer, Murphy and Richardson (1971) and Wellstead (1979). The text by Ogata (1992) offers good coverage of the analysis and design of mechanical, electrical, pneumatic, hydraulic and thermal systems. The background theory of stochastic systems is excellently and concisely described in the text by Davis and Vinter (1985), and Caines (1988). An excellent text on stochastic control theory is Astrom (1970). Linear systems and state space methods are comprehensively covered in Kailath (1980).

System identification

Mathematical models obtained from balance equations can be refined further by estimating parameters of the system by fitting input-output data obtained experimentally from the real system. Methods for estimating system parameters are described in the texts Norton (1986), Ljung (1987), Soderstrom and Stoica (1989). The Kalman filter

(Kalman, 1960a) has become a fundamental tool in modern control to estimate the states of a system from noisy measurements. Kalman filtering is described in the texts by Anderson and Moore (1979), Bozic (1979) and Aoki (1990). An easy to read text is Brown and Hwang (1992) which gives the introductory theory of random signals and applied Kalman filtering. A good collection of applications papers can be found in Sorenson (1985). The computational burden of the Kalman filter is heavy, especially for broadband realtime applications, so that parallel computation is seen as a means of easing the computational bottleneck. Gaston and Irwin (1990) gives an overview of standard and square root systolic Kalman filtering for state estimation.

System simulation

Simulation is an important modelling technique for analysing and solving complex problems (see, for example, the survey by Christy and Watson, 1983). The system under study is modelled and its behaviour observed via simulation. It is allowed to run for a set time to validate its behaviour against that of the real system. If there is close correspondence between the variables of the model and the corresponding variables of the real system then the model can be used to provide controlled experiments. The parameters of interest can be systematically varied and the simulation rerun. The book by Neelamkavil (1986) conveniently collects together the many aspects of computer simulation and modelling and leads the reader through continuous system simulation techniques. In addition to numerous references an extensive bibliography is also provided. A comprehensive survey of existing simulation tools for the modelling of continuous systems is given in Matko, Karba and Zupanic (1992). Computer-aided simulation modelling and artificial-intelligence techniques are increasingly being used to automate parts of the process of simulation modelling (see, for example, Balmer and Paul, 1986, and Paul and Doukidis, 1987).

Linear control systems

CONTINUOUS-TIME CONTROL SYSTEMS

The seminal results of Ziegler and Nichols (1942) are used in the tuning of proportional and integral (PI) controllers for single-input/single-output plants. Several automatic methods for determining PID controller parameters have been developed. These employ information simply obtained from open-loop step responses, for example, the Coon-Cohen reaction curve method and from frequency responses of the plant, for example, the Ziegler-Nichols method (see Kinney, 1983 for these methods). The Astrom-Hagglund phase margin method (Astrom, 1984), and the refined Ziegler-Nichols method (Hang, Astrom and Ho 1991), have been recent proposals to improve PID tuning.

There are numerous good textbooks covering the basic theory of model-based feedback control for linear time-invariant systems. Some very good ones are by Franklin, Powell and Workman (1994) and Hostetter, Savant and Stefani (1989). Warwick (1988) is an excellent text for reference. and acquiring fundamental knowledge of a broad range of control methods. For a broad overview of the place occupied by control engineering in the larger concerns of industrial engineering see the handbook edited by Salvendy (1992), in particular the section covering the technology of information systems, decision support systems, knowledge-based systems, control models and manufacturing engineering.

An exposition of distributed-parameter control systems can be found in Tzafestas (1982) and the techniques and applications of multidimensional systems in Tzafestas (1986).

DIGITAL AND COMPUTER CONTROL SYSTEMS

Digital control theory is covered in numerous texts. A comprehensive treatment which covers system identification of parametric and nonparametric models plus the design of digital multivariable, optimal and non-linear control systems is given in Franklin, Powell and Workman (1990). Design techniques which convert continuous-time controllers into equivalent digital ones and state space design methods are well covered. A good treatment is given of quantization effects and the application of digital control. For more information on discrete-time systems consult related material from the areas of digital signal processing. A classic book is by Oppenheim and Schafer (1975). An excellent treatment of the practical problems which are encountered in the application of digital control theory to computer-based process control can be found in Astrom and Wittenmark (1990). The analysis, design and implementation of deterministic as well as stochastic control systems is described. The book gives a wealth of design methods ranging from digital PID controllers and their tuning, model-based pole placement and minimum variance regulators, self-tuning and adaptive controllers, state space linear-quadratic (Gaussian) regulators, parameter and state identification.

An advanced text by Kucera (1991) shows the complementary nature of state-space and transfer function approaches to the design of discrete linear control systems. Leigh (1992) is a very good applied digital control text dealing with adaptive and robust control implementations on personal computers and VME-based systems.

Computer control of realtime processes is well covered in Bennett and Virk (1990). Application of microprocessors to control tasks is given in Sinha (1986). Application of control computers in hierarchical structures for control of all plant functions is covered in Williams (1985).

Non-linear control systems

Every system must be regarded as non-linear either due to its design or physical nature. The construction of a unified theory of non-linear analysis and design has been prevented by the complexity of the mathematical models generated for systems with significant non-linearity plus the wide range of non-linear characteristics encountered in practice. Few truly non-linear control design problems have been solved so far, with most progress being made in the control of rigid link robot manipulators (see Craig, 1986, Spong and Vidyasagar, 1989). An excellent text providing analytical foundations of non-linear systems analysis and latest research findings is Vidyasagar (1992). Slotine and Li (1990) gives a good perspective on non-linear control design and analysis methods.

Multivariable control systems

The simplest approach to designing feedback controllers for multivariable systems is to close the loops sequentially one by one without explicitly considering the cross-couplings present in the system. Each time a controller is designed for the loop which is being closed the effects of those loops which have been already closed are taken into account. This method can work with proper assignment of inputs to outputs and the order in which the loops are closed. It can be improved by allowing static cross-coupling before closing the individual loops in a method proposed by Mayne (1979).

These methods are described in the book by Maciejowski (1989). Application of multivariable system techniques is in Whalley (1990). The development of state-space techniques for control analysis by Kalman (1960a, 1960b), Gilbert (1963) and the frequency-domain work of Rosenbrock (1970) made possible a rigorous analysis of multivariable linear processes. A text describing multivarible system theory and design from this period is by Patel and Munro (1982). The state-space approach furthered the theory of linear quadratic optimal control which now provides complete design and synthesis methods for multivariable systems. The linear quadratic (LQ) and linear quadratic Gaussian (LQG) design utilizes a quadratic performance cost index to obtain optimality. Excellent books describing early work in this area are by Kwaakernak and Sivan (1972) and Anderson and Moore (1971). The review by Johnson and Grimble (1987) covers subsequent theoretical advances using frequency-domain analyses to more fully understand the robustness and integrity properties of LQG regulators and the selection of performance-index weighting matrices. The text by Lewis (1992) covers optimal and robust design methods for continuous-time and digital multivariable control systems. Jamshidi, Tarokh and Shafai (1992) introduces computer-aided design (CAD) of linear multivariable and large-scale control systems with a review of twenty-two CAD environments from around the world.

Self-tuning and adaptive control systems

Development of model-based approaches to control system design led to the subsequent realization that mathematical modelling of dynamical systems from balance equations inevitably leads to approximate models because of uncertainty about system parameters such as friction coefficients, etc. The model can be improved by logging input-output data and performing an offline estimate of the system model. However, linear models obtained from offline system identification of essentially non-linear systems using data obtained at a particular operating point can lead to an inadequate model if the control system designed from this model operates at some other point. The self-tuning approach to control system design was developed in the 1970s and 1980s to deal with this problem. In this approach an online recursive estimation of system parameters is performed and the control law synthesized from the estimates. The advantage of the method is that the estimation and control are done at the same operating point. In addition, if the system is time varying, either due to changes in the operating point or due to parameter drifts, then the estimation algorithm can update the system model and hence keep the control performance invariant. Such a system is termed an adaptive control system.

The earliest self-tuning control system was proposed by Kalman (1958) but it was a paper by Peterka (1970) which started the modern interest in self-tuning. Earlier work which is now of historical interest is described in Mishkin and Braun (1961). For a theoretical overview of modern self-tuning control, excellent books are by Landau (1979), Goodwin and Sin (1984), and Astrom and Wittenmark (1988). A lighter read and a good overview is Harris and Billings (1985). The book by Wellstead and Zarrop (1991) provides access to the main algorithms with which self-tuning systems are constructed and gives a more applications-oriented treatment of the subject. Useful survey articles on self-tuning and adaptive control systems are by Wellstead and Zanker (1982), Astrom (1983), and Seborg, Edgar and Shah (1986). For more information in the areas of adaptive filtering see Widrow and Stearns (1985) with useful implementation information for commercial signal processing chips in Treichler, Johnson and Larimore (1987). The technical problems relating to the stability and convergence of adaptive algorithms are described in the book by Ljung and Soderstrom (1983) which focuses on the stochastic convergence problem using the ordinary differential equation approach. More recent convergence results based on averaging methods can be found in Anderson et al. (1986) and the issues of stability, convergence and robustness are described in Sastry and Bodson (1989).

An excellent review paper on the adaptive control of robot manipulators is given in Ortega and Spong (1988).

Fuzzy logic and self-organizing controllers

An attractive approach to controller design for complex systems is to try to encapsulate an experienced operator's control strategy in a linguistic model. Fuzzy logic is employed to build the linguistic model. For a review of fuzzy systems in control engineering see Tong (1977). Success depends on a good linguistic model and all significant process changes being in the operators experience. Real-world problems usually have domains of incomplete understanding or uncertain knowledge so that even experts may not be able to build adequate models. A self-organizing fuzzy logic controller, first proposed by Procyk and Mamdani (1979), incorporates a learning algorithm which modifies control rules based on an evaluation of system performance. A survey of techniques to handle uncertainties and conflicts in real-world problems is given in Pang, Bigham and Mamdani (1987). Learning control involves finding computational methods for acquiring new knowledge, new skills and new ways to organize existing knowledge (see Michalski, Carbonell and Mitchell 1983). Neural networks approaches of mapping stimuli vector into response vector, applying rules for learning, error correcting and unsupervised learning are under investigation (see Matheus and Hohensee, 1987).

Artificial intelligence for control

The analytic and numerical approach to modelling of engineering systems seldom allows incorporation of the experience and intuitions of the design expert. As a result a dichotomy has existed between control theoreticians who have tended to use numerical methods and practising engineers who have favoured an empirical approach. Recently the techniques of artificial intelligence have started to overcome the gap between the analytic and empirical approaches. Two distinct artificial intelligence approaches have begun to emerge.

The first approach constructs an expert system and attempts to automate the empirical knowledge of the design expert through condition/action or condition/event pairs typical of production systems. The expertise is put on top of existing control system designs and corrects the system when it goes wrong. For a review of knowledge-representation tools and techniques see Jackson (1987) and for expert control see Astrom, Anton and Arzen (1986). A guide to expert systems is given in Waterman (1986).

The second approach attempts to extend the analytic approach to modelling by building a qualitative model of the physical world such that the purpose for which the model is to be used is explicitly included. Some fundamental techniques which are used in online qualitative modelling are surveyed in Milne (1987) and Leitch (1987). Applications to process regulation and control, PI tuning, fault detection and project management control are reported in the special issue on artificial intelligence (IEE, 1987).

Basic AI techniques are in Nilsson (1982), Winston (1984), Eisenstadt and O'Shea (1984) and Aleksander (1985). An introduction to expert systems is in the text by Jackson (1990). A handbook of intelligent control is White and Sofge (1992).

Robust control systems

The robust control of linear systems with parameter perturbations leads to the problem of eigenstructure assignment with minimum eigenvalue sensitivity and has recently attracted a lot of attention. A general parametric approach to eigenstructure assignment in multivariable systems via output feedback has been proposed by Duan, Wu and Huang (1991) and Duan (1992). The problem of obtaining robust performance in the face of system uncertainty is elegantly presented in Doyle, Francis and Tannenbaum (1992) and contains useful references to work in this area.

Computer-aided design

So much control theory is available for control system design that even an experienced designer has difficulty choosing appropriate strategies at each stage of the design process and remembering all the detail required for applying a particular method. An expert system can be a valuable design aid in those areas where good sets of design rules exist. An expert system approach to multivariable design is described in the text by Pang and MacFarlane (1987). Boyle and Maciejowski (1992) describe a knowledge-based support system for designing multivariable control systems.

Computationally complex design techniques become acceptable when the computing power to support them is available. Before the widespread use of computers the design techniques had to be simple in order to be of any use. For example, one of the first design techniques was the Nyquist Stability criterion which was suitable for simple, repetitive hand computations. The sequential computer made possible the design process in optimization and the LQG (linear model, quadratic integral criterion and Gaussian noise) approach to optimal control (see Anderson and Moore, 1971, 1989). More recently computing power has made it possible to implement the numerically complex H infinity design process which minimizes the H infinity norm of a transfer function (from a disturbance to the output) over the set of all stabilizing controllers. It extends easily to multivariable systems and has the ability to cope with some model uncertainty. A good exposition of this design method is given by Francis (1987) and a state-of-the-art survey by Stoorvogel (1992). The design process can be applied to a wide range of practical control problems. Two applications of the method can be found in Postelthwaite, O'Young and Gu (1987). Parallel computation could have the same affect on H infinity design that sequential computation had on LQG design.

Process control and instrumentation

An encyclopedia of instrumentation and control (Considine, 1971) and a handbook of process instruments and control by the same author (Considine, 1974), are early sources of information on the subject. Important control signal sensing and conversion techniques are covered in Considine and Considine (1985). A reference book on instrumentation is Noltingk (1988).

Control of distillation columns, because of its importance and complexity, has received considerable attention in the literature of process engineering and control (see Shinskey, 1984).

Parallel processing for real-time control

The appearance of parallel processors on the market has made possible control of faster systems through increased computational speed which is the primary benefit of parallel processing. It has also afforded implementation possibilities for complex control algorithms for optimal control, the adaptive control of variable systems and the control of non-defined or partly defined systems. With increased speed existing analysis techniques are beginning to emulate real control systems so that they can be used in real-time controller implementation for analysis and correction. More importantly, fault tolerance can be realized in a parallel processing system by distributing the computational operations so that a failure results in performance degradation rather than complete breakdown. The IEE (1989) and the IEEE (1987) have identified concurrency as an issue of strategic importance for control. Parallel processing in control is the subject of the text by Fleming (1988). A very readable general text describing the hardware and software requirements for concurrent systems including realtime distributed process control is Bacon (1993).

Sources of information in control engineering

A comprehensive source of information about systems and control is the eight volume encyclopedia edited by Singh (1987). The encyclopedia contains articles on various aspects of the theory, technology and applications of control written by leading experts in the field. Each article is provided with an extensive bibliography.

Many guides and sourcebooks are published around the world to provide information about engineering literature, world meetings, manufacturers, products, academic and research institutions, etc. In addition, information is provided by encyclopedias, handbooks, dictionaries, glossaries, abstracts, review sources, technical journals, international book series, conference proceedings, standards, specifications, patents, engineering tables and education courses. Numerous disciplines which require or contribute to control theory have their own information sources which can be used as sources of information about control engineering. For a broad treatment of information sources in systems and control see Tzafestas (1987). The article lists abstracts, review sources, technical journals, international book series, conference proceedings, directories of scientists and their expertise, sources of standards, patents and engineering tables from the areas of mechanical, electrical, electronic and information engineering.

Indexing and abstracting journals, and databases

Articles from primary journals can be traced via printed indexing and abstracting publications, CD-ROMs or online databases. The majority of searches in control engineering could be satisfied by using one of two databases. These are COMPENDEX and INSPEC.

The INSPEC database has a wider coverage of electrical, electronic and control engineering than COMPENDEX. INSPEC is produced by the Institution of Electrical Engineers (IEE) in the UK and has some input from the Institute of Electrical and Electronics Engineers (IEEE) in the USA. It is the electronic version of the three hardcopy abstracting services produced by the IEE:

(i) *Computer and Control Abstracts* (monthly)
(ii) *Electrical and Electronics Abstracts* (monthly)
(iii) *Physics Abstracts* (bimonthly)

INSPEC publications:

(1) *Computer and Control Abstracts*: subjects covered are computers and computing, control engineering and instrumentation.
(2) *Current Papers in Computing and Control*: this monthly current awareness bulletin provides the titles of articles and full bibliographic details of the source papers.
(3) *Key Abstracts*: a series of 22 monthly journals, each providing about 250 summaries of the more important journal and conference

papers. The journals of interest for control engineering are: *Robotics and Control, Artificial Intelligence, Business Automation, Factory Automation* and *Human-Computer Interaction*.

COMPENDEX (the computerized version of *Engineering Index*) is produced in the USA by Engineering Information Inc. The database record consists of citations plus abstracts, primarily to journal articles. Another version of COMPENDEX called COMPENDEX*PLUS includes all of COMPENDEX and also records on citations plus abstracts to papers given at a number of international conferences, symposia and meetings and includes a review of each conference.

Other services which cover journals relevant to control engineering include:

CURRENT TECHNOLOGY INDEX
APPLIED SCIENCE AND TECHNOLOGY INDEX
CHEMICAL ENGINEERING AND BIOTECHNOLOGY ABSTRACTS
ACM GUIDE TO COMPUTING LITERATURE
JAPIO
JICST
NTIS
ARTIFICIAL INTELLIGENCE
FLUIDEX (Includes Fluid Abstracts: Process Engineering)
SCISEARCH
ZDE (Zentralstelle Dokumentation Elektrotechnik Database).

Periodicals, serials and symposia

Information on the abstracting and indexing journals and databases which cover the following periodicals, serials and symposia can be found in various sources including the *Serials directory (on CD-ROM)* (EBSCO), *Ulrich's periodicals directory* (Bowker) and in the *Directory of periodicals online: indexed, abstracted and full text, science & technology*, 2nd ed. (1990) (Info Globe, John Deyell Co.).

Examples of these secondary sources are given in brackets after each title.

Automation, instrumentation and control

Automatic Control and Computer Sciences (Allerton Press) bimonthly (INSPEC).
Automatica (Pergamon) bimonthly (INSPEC; COMPENDEX).
Automation and Control (Matrix Publishing) 11 per annum (INSPEC).
Automation and Remote Control (Plenum) bimonthly (INSPEC; COMPENDEX).
Computing and Control Engineering Journal (Institution of Electrical Engineers) bimonthly (INSPEC; COMPENDEX).
Control Engineering (Cahners) monthly (APPLIED SCIENCE AND TECHNOLOGY INDEX; INSPEC; SCISEARCH).

Control and Dynamic Systems (Academic Press) irregular (INSPEC).
Controls and Systems (Penton Publishing) monthly (INSPEC).
Cybernetics and Systems (Taylor and Francis) bi-monthly (ACM GUIDE TO COMPUTING LITERATURE; MECHANICAL ENGINEERING ABSTRACTS; INSPEC).
Design Automation Conference, Proceedings (Institute of Electrical and Electronics Engineers) irregular (INSPEC).
Engineering and Automation (Siemens) quarterly (COMPENDEX; INSPEC).
Foundations of Computing and Decision Science (ARS Polona) quarterly (MECHANICAL ENGINEERING ABSTRACTS; INSPEC).
IEE Control Engineering Series (Peter Peregrinus) irregular (INSPEC).
IEE Proceedings D: Control Theory and Applications (Institution of Electrical Engineers) bimonthly (APPLIED SCIENCE AND TECHNOLOGY INDEX; COMPENDEX; CURRENT TECHNOLOGY INDEX; INSPEC; SCISEARCH).
IEEE Control Systems Magazine (Institute of Electrical and Electronic Engineers) 7 per annum (INSPEC; MECHANICAL ENGINEERING ABSTRACTS).
IEEE Transactions on Automatic Control (Institute of Electrical and Electronics Engineers) monthly (COMPENDEX; INSPEC; MECHANICAL ENGINEERING ABSTRACTS; SCISEARCH).
International Journal of Adaptive Control and Signal Processing (Wiley) quarterly (COMPENDEX; INSPEC).
International Journal of Control (Taylor and Francis) monthly (CURRENT TECHNOLOGY INDEX; INSPEC; SCISEARCH).
Institute of Measurement and Control, Transactions (Institute of Measurement and Control) quarterly (CURRENT TECHNOLOGY INDEX; INSPEC).
Optimal Control Applications and Methods (Wiley) quarterly (ACM GUIDE TO COMPUTING LITERATURE; INSPEC; SCISEARCH).
SIAM Journal on Control and Optimization (Society for Industrial and Applied Mathematics) bimonthly (INSPEC; SCISEARCH).
Studies in Automation and Control (Elsevier) irregular (INSPEC).
Systems and Control Letters (North-Holland) monthly (ACM GUIDE TO COMPUTING LITERATURE; INSPEC; SCISEARCH).
Transactions of the ASME. Journal of Dynamic Systems, Measurement & Control (American Society of Mechanical Engineers) quarterly (APPLIED SCIENCE AND TECHNOLOGY INDEX; MECHANICAL ENGINEERING ABSTRACTS; INSPEC; SCISEARCH).

Control of robot manipulators

International Journal of Robotics Research (MIT Press) bimonthly (ACM GUIDE TO COMPUTING LITERATURE; INSPEC).
Journal of Robotic Systems (Wiley) bimonthly (MECHANICAL ENGINEERING ABSTRACTS; INSPEC; SCISEARCH).
IEEE Transactions on Robotics and Automation (Institute of Electrical and Electronics Engineers) quarterly (COMPENDEX; INSPEC; SCISEARCH).

Process control

I & CS (Instrumentation and Control Systems) (Chilton) monthly (APPLIED SCIENCE AND TECHNOLOGY INDEX; INSPEC; SCISEARCH).

402 Control engineering

Process Engineering (Morgan Grampian) monthly (APPLIED SCIENCE AND TECHNOLOGY INDEX; CURRENT TECHNOLOGY INDEX).

Instrumentation and measurement

Control and Instrumentation (Morgan Grampian) monthly (CURRENT TECHNOLOGY INDEX; INSPEC; SCISEARCH).
Keisoku to Seigyo/Society of Instrument and Control Engineers. Journal (Society of Instrument and Control Engineers) monthly (INSPEC).
IEEE Transactions on Instrumentation and Measurement (Institute of Electrical and Electronics Engineers) bimonthly (COMPENDEX; INSPEC; MECHANICAL ENGINEERING ABSTRACTS; SCISEARCH).
Sensor Review (MCB University Press) quarterly (INSPEC; MECHANICAL ENGINEERING ABSTRACTS).
Sensor Technology (Technical Insights Inc.) monthly (PTS NEWLETTER DATABASE).
Sensors and Actuators; A Physical, B Chemical (Elsevier Sequoia) 18 per annum/21 per annum (MECHANICAL ENGINEERING ABSTRACTS; INSPEC; SCISEARCH).

Self-learning, expert and knowledge-based control

IEEE Expert: Intelligent Systems and their Applications (Institute of Electrical and Electronics Engineers) quarterly (INSPEC; MECHANICAL ENGINEERING ABSTRACTS; SCISEARCH).
IEEE Transactions on Knowledge and Data Engineering (Institute of Electrical and Electronics Engineers) quarterly (INSPEC).
International Journal of Expert Systems, Research and Applications (JAI Press Inc.) quarterly (INSPEC).
Expert Systems: the International Journal of Knowledge Engineering and Neural Networks (Learned Information) quarterly (SCISEARCH; INSPEC; COMPENDEX).
International Journal of Neural Systems (World Scientific Publishing Co.) quarterly (INSPEC).
Knowledge Engineering Review (Cambridge University Press) quarterly (INSPEC).
Knowledge-Based Systems (Butterworth-Heinemann) bimonthly (INSPEC; SCISEARCH, COMPUMATH CITATION INDEX).
Neural Networks (Pergamon) 9 per annum (INSPEC; SCISEARCH; ENGINEERING INDEX; COMPUMATH CITATION INDEX).

Other journals, magazines and conferences covering this topic include:

Artificial Intelligence (Elsevier) 16 per annum (INSPEC; COMPENDEX; COMPUMATH CITATION INDEX; APPLIED SCIENCE AND TECHNOLOGY INDEX).
International Journal of Human-Computer Studies (Academic Press) monthly (INSPEC; COMPENDEX).
Byte (McGraw-Hill) monthly (INSPEC; COMPUMATH; COMPENDEX).
AI Magazine (American Association of Artificial Intelligence) quarterly (INSPEC; COMPENDEX).

Proceedings of the National Conference on Artificial Intelligence (MIT Press) annual (INSPEC).

Fuzzy Sets & Systems (Elsevier) 24 per annum (INSPEC; MATHEMATICAL REVIEWS; COMPUMATH CITATION INDEX; COMPUTING REVIEWS).

IASTED International Conference Proceedings (International Association of Science and Technology for Development) irregular (INSPEC).

Proceedings of IEEE Conference Decision & Control (IEEE) annual (INSPEC).

Computer (IEEE) monthly (INSPEC).

Related and specialist journals and symposia

Simulation Symposium (IEEE Computer Society Press) annual, (INSPEC).

Real-Time Systems (Kluwer Academic) bimonthly (INSPEC).

Simulation (Society for Computer Simulation) monthly (INSPEC).

Associations in the UK

Association of Control & Automation Manufacturers, London.
United Kingdom Automation Council, London.
Heating, Ventilating and Air Conditioning Manufacturers Association, Bourne End, Bucks.
Building Energy Management Systems Centre, Bracknell, Berks.
Energy Systems Trade Association, Stroud, Gloucestershire.

References

Aleksander, I. *et al.* (eds) (1985) *Advanced digital information systems.* Englewood Cliffs, NJ: Prentice Hall.
Anderson, B. D. O. and Moore, J. B. (1971) *Linear optimal control.* Englewood Cliffs, NJ: Prentice Hall.
Anderson, B. D. O. and Moore, J. B. (1979) *Optimal filtering.* Englewood Cliffs, NJ: Prentice Hall.
Anderson, B. D. O. and Moore, J. B. (1989) *Optimal control: linear quadratic methods.* Englewood Cliffs, NJ: Prentice Hall.
Anderson, B. D. O. *et al.* (1986) *Stability of adaptive systems – passivity and averaging analysis.* Cambridge, MA: MIT Press.
Aoki, M. (1990) *State space modelling of time series* 2nd ed. Berlin: Springer-Verlag.
Astrom, K. J. (1970) *Introduction to stochastic control theory.* New York: Academic Press.
Astrom, K. J. (1983) Theory and applications of adaptive control – a survey. *Automatica,* **19,** 471–86.
Astrom, K. J. (1984) Automatic tuning of simple regulators with specifications on phase and amplitude margins. *Automatica,* **20,** 645–651.
Astrom, K. J., Anton, J. J. and Arzen, K. E. (1986) Expert control. *Automatica,* **22,** 277–286.

Astrom, K. J. and Wittenmark, B. (1988) *Adaptive control*. Reading, MA: Addison-Wesley.
Astrom, K. J. and Wittenmark, B. (1990) *Computer controlled systems: theory and design* 2nd ed. Englewood Cliffs, NJ: Prentice Hall.
Bacon, J. (1993) *Concurrent systems: an integrated approach to operating systems, database and distributed systems.* Wokingham: Addison-Wesley.
Balmer, D. W. and Paul, R. J. (1986) CASM – The right environment for simulation. *J. Operational Research Society*, **37**, 443–452.
Bennett, S. and Virk, G. S. (eds) (1990) *Computer control of real-time processes*. London: Peregrinus.
Bose, B. K. (1986) *Power electronics and AC drives*. Englewood Cliffs, NJ: Prentice Hall.
Boyle, J. M. and Maciejowski, J. M. (1992) Expert aided sequential design of multivariable systems. *IEE Proceedings D*, **139**, 471–80.
Bozic, S. M. (1979) *Digital and Kalman filtering*. London: Edward Arnold.
Brown, R. G. and Hwang, P. Y. C. (1992) *Introduction to random signals and applied Kalman filtering* 2nd ed. New York: Wiley.
Caines, P. E. (1988) *Linear stochastic systems*. New York: Wiley.
Christy, D. P. and Watson, H. J. (1983) The application of simulation: a survey of industry practice. *Interfaces*, **13**, 47–52.
Considine, D. M. (ed.) (1971) *Encyclopedia of instrumentation and control*. New York: McGraw-Hill.
Considine, D. M. (ed.) (1974) *Process instruments and controls handbook*. New York: McGraw-Hill.
Considine, D. M. and Considine, G. D. (eds) (1985) *Process instruments and controls handbook* 3rd ed. New York: McGraw-Hill.
Craig, J. J. (1986) *Introduction to robotics: mechanics and control* 2nd ed. Reading, MA: Addison-Wesley.
Davis, M. H. A. and Vinter, R. B. (1985) *Stochastic modelling and control*. London: Chapman and Hall.
Doyle, J. C., Francis, B. A. and Tannenbaum, A. R. (1992) *Feedback control theory*. New York: Macmillan.
Duan, G. R., Wu, G. Y. and Huang, W. H. (1991) Eigenstructure assignment for time-varying linear systems. *Scientia Sinica/Science China* (Series A, English edition), **34**, 246–256.
Duan, G. R. (1992) Simple algorithm for robust pole assignment in linear output feedback. *IEE Proceedings D*, **139**, 465–469.
Eisenstadt, M. and O'Shea, T. (1984) *Artificial intelligence: tools, techniques and applications*. New York: Harper and Row.
Fleming, P. J. (1988) (ed.) *Parallel processing in control: the transputer and other architectures*. London: Peregrinus.

Francis, B. A. (1987) *A course in H-infinity control theory*. Berlin: Springer-Verlag.
Franklin, G. F., Powell, J. D. and Emami-Naeini, A. (1994) *Feedback control of dynamic systems* 3rd ed. Reading, MA: Addison-Wesley.
Franklin, G. F., Powell, J. D. and Workman, M. L. (1990) *Digital control of dynamic systems* 2nd ed. Reading, MA: Addison-Wesley.
Gaston, F. M. F. and Irwin, G. W. (1990) Systolic Kalman filtering: an overview. *IEE Proceedings D*, **137**, 235–244.
Gilbert, E. G. (1963) Controllability and observability in multivariable control systems. *Society for Industrial and Applied Mathematics. Journal on Control. Series A*, **1**, 128–151.
Goodwin, G. C. and Sin, K. S. (1984) *Adaptive filtering, prediction and control*. Englewood Cliffs, NJ: Prentice Hall.
Hang, C. C., Astrom, K. J. and Ho, W. K. (1991) Refinements of the Ziegler-Nichols tuning formula. *IEE Proceedings D*, **138**, 111–118.
Harris, C. J. and Billings, S. A. (1985) *Self-tuning and adaptive control: theory and applications* 2nd ed. London: Peregrinus.
Hostetter, G. H., Savant, C. J. and Stefani, R. T. (1989) *Design of feedback control systems* 2nd ed. New York: Saunders College Publishing.
IEE (1987) Artificial intelligence: the emergence of a formal basis for engineering. (Special Issue). *IEE Proceedings D*, **134**, 217–300.
IEE (1989) Strategic directions for control. *IEE Colloquium Digest* 1989/106.
IEEE (1987) Challenges to control: a collective view. *IEEE Transactions on Automatic Control*, **AC-32**, 275–285.
Jackson, P. (1990) *Introduction to expert systems* 2nd ed. Wokingham: Addison-Wesley.
Jackson, P. (1987) Review of knowledge-representation tools and techniques. *IEE Proceedings D*, **134**, 224–230.
Jamshidi, M., Tarokh, M. and Shafai, B. (1992) *Computer-aided analysis and design of Linear control systems*. London: Prentice Hall.
Johnson, M. A. and Grimble, M. J. (1987) Recent trends in linear optimal quadratic multivariable control system design. *IEE Proceedings D*, **134**, 53–71.
Kailath, T. (1980) *Linear systems*. Englewood Cliffs, NJ: Prentice Hall.
Kalman, R. E. (1958) Design of a self-optimizing control system. *Transactions of the ASME. Journal of Basic Engineering*, **80D**, 468–478.
Kalman, R. E. (1960a) A new approach to linear filtering and prediction problems. *Transactions of the ASME, Journal of Basic Engineering*, **82D**, 35–45.
Kalman, R. E. (1960b) On the general theory of control systems. *Auto-*

matic and Remote Control. Proceedings 1st International Congress of IFAC, Moscow, 1960. Vol. 1, 481–492. London: Butterworths.

Kinney, T. B. (1983) Tuning process controllers. Chemical Engineering, **90**, 67–74.

Kucera, V. (1991) Analysis and design of discrete linear control systems. London: Prentice Hall.

Kwakernaak, H. and Sivan, R. (1972) Linear optimal control systems. New York: Wiley-Interscience.

Landau, Y. D. (1979) Adaptive control: the model reference approach. New York: Marcel Dekker.

Leigh, J. R. (1992) Applied digital control 2nd ed. London: Prentice Hall.

Leitch, R. R. (1987) Modelling of complex dynamic systems. IEE Proceedings D, **134**, 245–250.

Leonhard, W. (1985) Control of electrical drives. Berlin: Springer-Verlag.

Levi, S. T. and Agrawala, A. K. (1990) Real-time system design. New York: McGraw-Hill.

Lewis. F. L. (1992) Applied optimal control and estimation: digital design and implementation. Englewood Cliffs, NJ: Prentice Hall.

Ljung, L. and Soderstrom, T. (1983) Theory and practice of recursive identification. Cambridge, MA: MIT Press.

Ljung, L. (1987) System identification: theory for the user. Englewood Cliffs, NJ: Prentice Hall.

Macfarlane, A. G. J. (1970) Dynamical system models. London: Harrap.

Maciejowski, J. M. (1989) Multivariable feedback design. Wokingham: Addison-Wesley.

Matheus, C. J. and Hohensee, W. E. (1987) Learning in artificial neural systems. Computational Intelligence, **3**, 283–294.

Matko, D., Karba, R. and Zupanic, B. (1992) Simulation and modelling of continuous systems. New York: Prentice Hall.

Mayne, D. Q. (1979) Sequential design of linear multivariable systems. IEE Proceedings, **126**, 568–572.

Michalski, R. S., Carbonell, J. G. and Mitchell, T. M. (eds) (1983) Machine learning: an AI approach. Palo Alto, CA: Tioga Publishing.

Milne, R. (1987) Artificial intelligence for online diagnosis. IEE Proceedings D, **134**, 238–244.

Mishkin, E. and Braun, L. (1961) Adaptive control systems. New York: McGraw-Hill.

Neelamkavil, F. (1986) Computer simulation and modelling. Chichester: Wiley.

Nilsson, N. J. (1982) *Principles of artificial intelligence.* New York: Springer-Verlag.
Noltingk, B. E. (ed.) (1988) *Instrumentation reference book.* London: Butterworths.
Norton, J. P. (1986) *An introduction to identification.* London: Academic Press.
Oppenheim, A. V. and Schafer, R. W. (1975) *Digital signal processing.* Englewood Cliffs, NJ: Prentice Hall.
Ogata, K. (1992) *System dynamics* 2nd ed. Englewood Cliffs, NJ: Prentice Hall.
Ortega, R. and Spong, M. (1988) Adaptive motion control of rigid robots: a tutorial. *Proceedings of the 27th IEEE Conference on Decision and Control.* Austin, TX., 7–9 December, 1988, **2**, 1575–1584.
Pang, D., Bigham, J. and Mamdani, E. H. (1987) Reasoning with uncertain information. *IEE Proceedings D,* **134**, 231–237.
Pang, G. K. H. and MacFarlane, A. G. J. (1987) *An expert systems approach to computer-aided design of multivariable systems.* Berlin: Springer-Verlag.
Patel, R. V. and Munro, N. (1982) *Multivariable system theory and design.* Oxford: Pergamon.
Paul, R. P. (1981) *Robot manipulators: mathematics, programming and control.* Cambridge, MA: MIT Press.
Paul, R. J. and Doukidis, G. I. (1987) Artificial intelligence aids in discrete-event digital simulation modelling. *IEE Proceedings D,* **134**, 278–286.
Peterka, V. (1970) Adaptive digital regulation of noisy systems. *Second IFAC Symposium on Identification and Process Parameter Estimation,* paper 6.2. Prague: Academia.
Postelthwaite, I., O'Young, S. D. and Gu, D. W. (1987) *Stable-H user's guide.* Report OUEL 1687/87. Oxford: Department of Engineering Science, Oxford University.
Procyk, T. J. and Mamdani, E. H. (1979) A liguistic self-organizing process controller. *Automatica,* **15**, 15–30.
Rosenbrock, H. H. (1970) *State-space and multivariable theory.* London: Nelson.
Salvendy, G. (ed.) (1992) *Handbook of industrial engineering* 2nd ed. New York: Wiley.
Sastry, S. and Bodson, M. (1989) *Adaptive control – stability, convergence and robustness.* Englewood Cliffs, NJ: Prentice Hall.
Seborg, D. E., Edgar, T. F. and Shah, S. L. (1986) Adaptive control strategies for process control: a survey. *AIChE Journal,* **32**, 881–913.
Shearer, J. L., Murphy, A. T. and Richardson, H. H. (1971) *Introduction to system dynamics.* Reading, MA: Addison-Wesley.

Shinskey, F. G. (1984) *Distillation control* 2nd ed. Westport, CT: Greenwood Press.
Singh, M. G. (ed.) (1987) *Systems & control encyclopedia: theory, technology, applications.* 8 vols. Oxford: Pergamon.
Sinha, N. K. (ed.) (1986) *Microprocessor-based control systems.* Dordrecht: Reidel.
Slotine, J. E. and Li, W. (1990) *Applied nonlinear control.* London: Prentice Hall.
Soderstrom, T. and Stoica, P. (1989) *System identification.* New York: Prentice Hall.
Sorenson, H. W. (ed.) (1985) *Kalman filtering: theory and application.* New York: IEEE Press.
Spong, M. W. and Vidyasagar, M. (1989) *Robot dynamics and control.* New York: Wiley.
Stoorvogel, A. A. (1992) *The H infinity control problem.* New York: Prentice Hall.
Tong, R. M. (1977) A control engineering review of fuzzy systems. *Automatica,* **13**, 559–569.
Treichler, J. R., Johnson, C. R. and Larimore, M. G. (1987) *Theory and design of adaptive filters.* New York: Wiley.
Tzafestas S. G. (ed.) (1982) *Distributed-parameter control systems: theory and application.* Oxford: Pergamon.
Tzafestas S. G. (ed.) (1986) *Multidimensional systems: techniques and applications.* New York: Marcel Dekker.
Tzafestas, S. G. (1987) Information sources in systems and control. In M. G. Singh *(ed.) Systems & control encyclopedia: theory, technology, applications.* Oxford: Pergamon, **8**, 5575–5585.
Vidyasagar, M. (1992) *Nonlinear systems analysis* 2nd ed. Englewood Cliffs, NJ: Prentice Hall.
Warwick, K. (1988) *Control systems: an introduction.* New York: Prentice Hall.
Waterman, D. A. (1986) *A guide to expert systems.* Reading, MA: Addison-Wesley.
Wellstead, P. E. (1979) *Introduction to physical system modelling.* London: Academic Press.
Wellstead, P. E. and Zanker, P. (1982) Techniques of self-tuning. *Optimal Control Applications and Methods,* **3**, 305–22.
Wellstead, P. E. and Zarrop, M. B. (1991) *Self-tuning systems: control and signal processsing.* Chichester: Wiley.
Whalley, R. (ed.) (1990) *Application of multivariable system techniques (AMST 90).* London: Elsevier.
White, D. A. and Sofge, D. A. (eds) (1992) *Handbook of intelligent control: fuzzy, neural and adaptive approaches.* New York: Van Nostrand Reinhold.

Widrow, B. and Stearns, S. D. (1985) *Adaptive signal processing*. Englewood Cliffs, NJ: Prentice Hall.
Williams, T. J. (ed.) (1985) *Analysis and design of hierarchical control systems*. Amsterdam: Elsevier.
Winston, P. H. (1984) *Artificial intelligence* 2nd ed. Reading, MA: Addison-Wesley.
Ziegler, J. G. and Nichols, N. B. (1942) Optimum settings for automatic controllers. *Transactions of the ASME*, **64**, 759–768.

CHAPTER TWENTY-THREE

Electronics

J. DELL

Introduction

Electronics has a history which can be traced back to the turn of the twentieth century, starting with the discovery by Edison that negative electric charges, later known as electrons, are emitted by a heated filament and that they are attracted to a positively charged metal plate when both are enclosed in a partial vacuum. A few years later Flemming put this phenomenon to practical application in a primitive radio receiver. A range of electronic devices known as 'thermionic valves' or 'electron tubes' was developed over the succeeding years and continued in wide use until the 1950s. They enabled significant developments in communications and radio and later the introduction of television. Even the early computers used these elements to provide their logic functions but it was soon found that the reliability of such systems left much to be desired. The major milestones in the subsequent development of electronics are summarized below.

The invention of the transistor in 1947 by Bardeen, Brattain and Shockley brought about a fundamental change in the technology utilized for the fabrication of electronic devices. These devices depend on the semi-conducting properties exhibited by germanium and silicon when certain impurities are present in very small quantities. These impurities give rise to a small surplus or deficiency of electrons. It was found that a three-terminal, current-controlled device could be fabricated when these materials were formed into a minute three-layer structure. Their basic characteristics had many parallels with the earlier thermionic valves without the need for a heated filament and high voltage supplies. As a result, vastly superior reliability was achieved as well as an enormous reduction in physical size. The use of transistors became prevalent in the 1960s when almost all electronic applications made use of them.

It was soon discovered that several transistors, as well as other components, could be fabricated on the same small slice of high purity silicon. This technique, which became known as the integrated circuit, also improved reliability because fewer external connections were required and the circuit was able to operate at lower power levels. A steady development in the number of transistors which could be fabricated on the same 'chip' took place during the 1970s and 1980s and still continues today. This led to improvements in the functionality and ease of application of these components. Today many thousands of transistors are fabricated within a few square millimetres of silicon in the most advanced analogue and digital designs.

The parallel development of digital computers during this period, starting from the original ENIAC machine which was built in the 1940s and used only thermionic valves, depended heavily on the development of integrated circuits to provide reliable logic, memory and arithmetic functions, each of which utilized many thousands of electronic logic gates. These developments continued with the level of integration that could be practically achieved, eventually leading to the advanced microprocessors and powerful computing machines which are an established feature of so many systems today.

Communications and telecommunications was another branch of electronics to benefit greatly from the advanced components which became available during this period. Most telecommunications channels now exploit integrated circuits and digital techniques to enable many 'conversations' to share the same physical circuits, which in turn helped to accommodate the explosion in demand for these services. The introduction of the mobile telephone network was made possible by the availability of advanced analogue and digital integrated circuit functions. Another development which has been of particular benefit to communications can be traced back to the realization of a semiconductor laser in 1961 using fabrication techniques similar to those employed for transistor and integrated circuit manufacture. The subsequent development of a low-loss optical fibre for guiding the laser beam achieved practical performance over several kilometres in the early 1970s and enabled the implementation of communication systems which can operate at hundreds of million bits per second. These systems now form the main links between telephone exchanges throughout the UK and in other countries with a well developed communications industry. Together with satellite communication channels, high speed data links are readily available to many parts of the world.

The consumer electronics industry has also seen rapid developments with the availability of advanced components enabling applications such as video recorders, calculators, watches and electronic games. The recent developments in consumer electronics which have taken place, such as the introduction of the compact disc (CD) and the development of high-definition television systems (HDTV), have also been dependent on the availability of advanced electronic components to provide the high speed digital circuits on which these systems are based.

Electronics as a subject has developed at an ever increasing rate since its early history and the volume of information from professional and other sources has mirrored or exceeded this rate of increase. Electronics has an impact on almost every individual in the developed world and constitutes a major world industry. In the UK alone the turnover of electronics companies runs into many millions of pounds. What follows is a short review of professional organizations as well as primary and secondary information sources that can be consulted to gain an insight into various aspects of the subject.

Professional organizations as sources of information

Learned societies

Two major learned societies, both of which have long histories, now occupy central positions in the structure of the electronics industry as well as its education, literature and standards.

The Institution of Electrical Engineers (IEE) was founded in the UK in 1871, although its present title dates back to 1888. The IEE was awarded a Royal Charter in 1921 and now has a membership of some 134 000 men and women throughout the world.

The remit of the IEE is to 'promote the advancement of electrical, electronic and manufacturing science and engineering and to facilitate the exchange of information and ideas'. To this end, together with its associated publishing company Peter Peregrinus and its Information Services for the Physics and Engineering Communities (INSPEC) (see below, under 'Abstracting and indexing services'), it publishes a wide range of books and periodicals on electrical and electronic engineering, physics, computing, control and software engineering as well as related subjects in information technology. The IEE library in London, which incorporates the British Computer Society Library and that of the Institution of Manufacturing Engineers, offers an extensive collection of books, journals and conference proceedings from which photocopies can be obtained. Its Technical Information Unit specialists can answer enquiries on almost any aspect of the subject and the unit has online access to over 1000 databases from which customized searches can be made.

The Institute of Electrical and Electronics Engineers (IEEE), based in the USA, is the world's largest technical professional society.

Founded in 1884 it is now comprized of over 320 000 members and its objectives focus on the advancement of theory and practice, as do those of the IEE. To realize its objectives it sponsors technical conferences, symposia, workshops and local meetings throughout the world, publishing nearly 25 per cent of the world's technical papers in this and closely related subjects. It has 35 technical societies which publish more than 50 *Transactions* as well as various magazines and journals, each having up to twelve issues per annum.

Other small learned societies such as the Institution of British Telecommunications Engineers publish their own specialist journals, in this case *British Telecommunications Engineering*. The Institution of Electronics and Electrical Incorporated Engineers (IEEIE) publishes a monthly news sheet and sponsors a series of public lectures for its members.

Trade associations

A number of trade associations represents the interests of the electronics equipment and component manufacturing industry. The Federation of the Electronics Industry (FEI), London, is the main UK trade organization, recently formed from the merger of the Electronic Components Industry Federation (ECIF) and the Electronic Engineering Association (EEA), 'to create a forum for developing policies for those within the electronics industry who provide enabling technology, and those whose products, in turn, enable many industries to operate effectively and flourish'.

The interests of American electronics manufacturers, as well as those engaged in research and development, are served by the American Electronics Association (AEA), Palo Alto, California, and the Electronic Industries Association (EIA), Washington, who publish a useful analysis of the electronics market. They both have a significant role in training and sponsoring exhibitions as well as the definition of standards and certification of components for high-integrity systems.

Research associations

ERA Technology, whose origins go back to a Research Committee which was set up by the IEE in 1912, is located at Leatherhead in Surrey and provides technical support and consultancy services for the electrical and electronic engineering industry. It publishes various reports on its current work, such as the recent *Buyers' guide to uninterruptible power supplies* (UPS), and organizes conferences and symposia on specialist subjects. An annual review provides an effective summary of their work.

Electronic measurement, measuring instruments and control systems are an essential part of many diverse industries and the SIRA Institute, formerly the British Scientific Instrument Research Association, provides information on measurement, instruments and controlling devices. The SIRA Institute is located at Chislehurst in Kent and, as well as its consultancy services, it publishes regular technical reviews, organizes meetings and provides training courses for professional engineers.

Official organizations

The Department of Trade and Industry (DTI) sponsors various programmes in electronics and appoints various working parties with the particular objective of transferring the results of electronics research to industry, improving the competitive edge in the process. The Engineering and Physical Sciences Research Council (EPSRC) provides support for academic research which includes many projects in electronics. Initiatives involving industrial partners are strongly supported, a recent example being the Joint Framework for Information Technology (JFIT) which funded programmes covering the five broad areas of Advanced Devices and Materials, Systems Architecture, Communications and Distributed Systems, Systems Engineering and Very Large Scale Integrated circuit (VLSI) Technology. In this way the EPSRC supports basic research, technology transfer and standards as well as education and training throughout the UK.

Primary sources of information

Journals

LEARNED JOURNALS

The publishing of electronics learned journals is dominated by the IEE and IEEE. The *IEE Proceedings* is historically the most important, its origins dating back to 1872, and is divided into eleven bi-monthly parts. The most important for practising electronic engineers are *Computers and Digital Techniques, Circuits, Devices and Systems, Communications Speech and Vision* and *Optoelectronics*. The fortnightly learned journal *Electronic Letters* provides rapid publication of short contributions describing ongoing research in electronic engineering, science, telecommunications and optoelectronics, reflecting current research across the industry.

The IEE publishes a series of six engineering journals, bi-monthly, which provide an in-depth coverage of new work which will be both informative and accessible to a wide spectrum of engineers. The most important of these are *Electronics and Communication Engineering Journal* and *Computing and Control Engineering Journal*. The IEE also publishes two journals jointly with the British Computer Society as well as two more general journals; *IEE Review* (11 issues per annum), dealing with advances in electronic engineering, and *IEE News* (monthly), dealing mostly with the electronic engineering profession.

Similarly, the IEEE publishes an extensive series of journals which includes the excellent, award-winning magazine *Spectrum* (monthly), where advances in technology and their significance in society are discussed from a professional viewpoint. The *IEEE Proceedings* (monthly) contains papers of broad significance to electronic engineers as well as papers describing original work, reviews and tutorials. Over fifty *Transactions* (usually monthly) are produced by the thirty-five IEEE specialized technical societies, and these cover an extremely wide area of interest. A number of other specialist journals, usually bimonthly, such as *IEEE Journal of Solid-State Circuits* and the *IEEE Micro*, dealing with 'chips, systems, software and applications', cater for more specific areas of interest.

Several commercial publishing houses produce high-quality learned journals containing articles on theoretical electronics and research, the monthly *International Journal of Electronics* (Taylor and Francis), the monthly *Digital Signal Processing* (Academic Press) and the monthly *Microprocessors and Microsystems* (Butterworth-Heinemann) are good examples.

COMMERCIAL JOURNALS

Commercial industry-oriented journals are produced by a number of publishers. These fall between learned journals proper and advertising papers. They are often described as controlled-circulation journals.

The monthly journal *Electronics* (McGraw-Hill), first published in 1930 and one of the most widely known, includes technical articles, news and reviews of new equipment. *Electronic Engineering* (Morgan Grampian) and *Electronic Design* (Penton Publishing), both published monthly, provide technical articles including application ideas, product reviews and news of the electronic industry. *Components in Electronics* (TAS Publishing) provides a monthly update on component and industry news.

The weekly journals *Electronics Weekly* (Reed Business Publishing) and *Electronic Times* (Morgan Grampian) cover news of products, people, jobs and the industry.

POPULAR JOURNALS

There are many popular journals aimed at the home electronics enthusiast which contain construction projects and technical feature articles.

The monthly *Electronics World and Wireless World* (Reed Business Publishing) has a long history and commands a good reputation, and its contributors provide high-quality articles. Other examples such as *Electronics Today International* (Argus Technical Publishing) and *Everyday with Practical Electronics* (Wimbourne Publishing) have a wide circulation and usually contain a number of electronics projects.

HOUSE JOURNALS

Many large electronics companies publish house journals to bring their technical developments to a worldwide audience and promote internal communications. Examples include the bi-monthly *AT & T Technical Journal* (formerly *Bell System Technical Journal*) (American Telephone and Telegraph Company), the quarterly *IBM Systems Journal* (IBM), the monthly *Philips Journal of Research* (Philips), the bi-monthly *Siemens Review* (Siemens) and the quarterly *BT Technology Journal* (British Telecom).

Conferences

The majority of conferences, symposia and workshops are organized by the major professional bodies who also publish their proceedings. All papers presented at such events are reviewed by a panel of professionals working in related subject areas, thus ensuring the highest possible technical quality.

The IEE is again of central importance and typically organizes about twenty major conferences each year, the present series started in 1961, as well as more than one hundred and fifty colloquia. Comprehensive conference proceedings are published, recent examples in 1993 include the Seventh International Conference on Mobile Radio and Personal Communications (no. 387) and the Fifth European Conference on Power Electronics and Applications (no. 377).

The IEEE has an extensive programme of some three hundred conferences annually, many of which are held outside the USA, and comprehensive conference proceedings are published for each. Recent examples in 1993 include the International Solid-State Circuits Conference, USA, the International Conference on Solid-State Devices and Materials, Japan, the Workshop on VLSI Signal Processing, the Netherlands, and the International Symposium on Semi-conductor Manufacturing (USA).

International conferences are frequently organized jointly by the IEE and IEEE as well as other bodies. The annual International Broadcasting Convention (IEE, IEEE, EBU) held in Europe is a good example.

Electronic product data

Accurate and comprehensive product data are essential to any practising engineer and manufacturers go to great lengths to provide adequate information to facilitate design with their products. In the UK the two most important publications are the annual *Electrical and electronics trades directory* (The Blue Book) (Peregrinus), which provides a classified listing of products materials and services, and the *Electronics and instruments directory*, 22nd edition (1988) (Morgan Grampian) which contains an extensive classified listing of manufacturers and their agents, electronic equipment and other supporting hardware.

Directories of electronic products, components and sub-assemblies are also available for companies outside the UK. For European sources the *European electronics directory 1993* (Elsevier) provides information on suppliers of components and sub-assemblies and the *International electronics directory '90* (Elsevier) provides a guide to European manufacturers, their agents and applications. Information about manufacturers in the USA can be found in the *1992 US electronics industry directory* (Harris Publishing) and for Japan the *Japan electronics buyers' guide 1990* (Dempa Publications). See also *Diskette of World Electronics Data 1993* (Elsevier) for information on America, Japan and Asia/Pacific regions.

Component data

The detailed information required for original designs is best obtained from the component manufacturers themselves. For example, Philips and Texas Instruments publish comprehensive sets of data books running into many volumes. In general each volume is dedicated to a particular type of component or application function. A good example is the *TTL data book* (Texas Instruments), an indispensable reference for any designer of digital logic systems, which will be found in most electronics laboratories. The major companies now provide component data on computer disk to facilitate search and selection based on parameters of direct relevance to the application in question. Most components are supplied through distributors, except when very large volumes are required, and such companies can frequently provide a fast, efficient and comprehensive data service on their range of products.

It is worth noting that full details of an extensive range of components which have been approved for use in government and defence industries can be obtained from the CODUS computer database (Fretwell-Downing Data Systems).

Cross-reference information listing equivalent types and a short summary of the main specification parameters, including connection details, is provided by several companies D.A.T.A. International Inc. is the foremost publisher in this field and produces a series of eleven *Data digests*. These cover the main component classifications from manufacturers in all parts of the world and include discrete devices, integrated circuits and special application functions. The listings, which group components by generic family, include both current and discontinued types, outline drawings and manufacturers' addresses. With more than 1000 pages in each *Digest* the volume of data, which is frequently updated each year, can be appreciated. Other publications such as *I.C. master* (Hirst Business Communications) and the *Semicon Indexes* (International semiconductor data summaries) have a more restricted content.

Application circuits

Component manufacturers often produce useful applications information on their products but it is frequently more convenient in the first instance to consult a compendium of applications such as Hughes (1986) or Lancaster (1988). With the more advanced and highly integrated components applications information and suggested circuits are essential. For devices such as the Texas Instruments family of digital signal processors, applications information is contained in a complete book which covers the hardware, software and interface requirements.

Electronic materials

Electronics would not have achieved the importance it now holds without a plentiful supply of the raw materials having both appropriate and well-defined properties. The realization of the transistor in 1948 was only possible as a result of nearly a decade of research and development centred on the materials needed for its construction. Most electronic materials can be classified under one of the following types: conductor, semi-conductor, insulator, dielectric or magnetic. In many applications of dielectric or magnetic materials a synthetic material is favoured over a naturally occurring material on account of the superior properties which can be obtained. A good example is found in magnetic materials; naturally occurring iron has very poor properties and an alloy is needed even to form an effective permanent magnet. In order to meet the stringent demands of digital read/write heads found in disk storage systems various synthetic ceramic materials are now universally employed in this application to provide the required performance in the very small physical dimensions.

The electrical and magnetic properties of natural materials and the most common synthetic materials can be found in standard texts such as Kaye and Laby (1995). A detailed explanation of electronic materials, the physical theory and their applications is given in Warnes (1990). This also includes a very useful collection of information on the properties of semi-conductor materials as well as many references to other information sources.

For information on the highly developed magnetic materials which can be used in the construction of inductors and other electro-magnetic devices it is best to consult the data books provided by the manufacturers. For example the *Philips (Mullard) Data Book 3* contains information on a wide range of electronic components, materials and assemblies. In particular parts 2, 3 and 4 of this volume focus on magnetic materials and components as well as various inductor cores.

Research into new electronic materials continues in many institutions and recent advances enabled the realization of optical fibre communication systems, mentioned previously, and high-temperature superconductors. Current research papers can be found in the monthly *Journal of Electronic Materials* published by the IEEE as well as subject-related journals such as the bi-monthly *IEEE Transactions on Magnetics*.

Standards and specifications

Standards and specifications are extremely important in the electronics industry, for example the European Directive (89/336/EEC) on Electro-Magnetic Compatibility (EMC) affects every manufacturer of electronic equipment even when their product is embedded within other products like a car or washing machine.

European standards are set by CENELEC (European Committee for Electrotechnical Standardization), which was founded in 1960 and covers the EEC and EFTA countries. In order to harmonize standards across Europe CENELEC produces a series of standards under the CECC (CENELEC Electronic Components Committee) which supersede national standards in the member countries.

The former national standards in the UK were produced by the British Standards Institution and a number of these were pertinent to electronics, notably BS 9000: 1989–1991 General Requirements for a System for Electronic Components of Assessed Quality. During an interim period BS standards will be used when a CECC standard has not been issued.

The IEEE has an important role in establishing standards and has numerous technical committees working in this area. A recent example of this work is the series of standards under the IEEE 802 banner which cover local area computer network implementation technologies. Another example is the IEEE 488: 1978 standard which applies to the interconnection of programmable measuring instruments. This ensures that an automated system can be built up with equipment from a diversity of manufacturers. The IEE has a similar role in developing standards and providing training for engineers, such as the recent series of courses on the European EMC Directive, its implications on equipment design and testing procedures. This course is available as a series of videos for distance learning.

Many other organizations in Europe and the USA are involved in creating standards which affect the electronics industry, the most influ-

Electronics 421

ential being the European Telecommunications Standards Institute (ETSI), the UK Department of Trade and Industry (DTI), the Ministry of Defence (MOD), the Civil Aviation Authority (CAA), and Verein Deutscher Ingenieure (VDI) in Germany. In the USA the Federal Communications Commission (FCC), the Radio Technical Commission for Aeronautics (RTCA) and the Radio Technical Commission for Maritime Services (RTCM) are of major importance.

Secondary sources of information

Handbooks and reference books

Gibilisco and Sclater (1990) covers nearly 7000 topics in a nonacademic format and is a thoroughly indexed and cross-referenced encyclopedia.

Useful handbooks are available from several publishers. Fink and Christiansen (1989) and Mazda (1989) cover the full spectrum of components, assemblies and systems used in linear and digital systems. Kaufman and Seidman (1984) provides useful reference material on electronic laboratory measuring instruments.

Abstracting and indexing services

The IEE provides one of the world's major abstracting publications through its INSPEC database service, which was established in 1898 as *Science Abstracts* and currently contains records for more than 4.2 million scientific and technical papers. The contents of over 4200 journals and some 1000 published conference proceedings as well as books and reports are regularly scanned, and records are added at a rate of 250 000 per year. The database is divided into sections, Electrical and Electronics Engineering (Section B) covers electronic components and technology, telecommunications, power engineering and instrumentation. The powerful classification is devised by INSPEC (INSPEC *Classification*) and includes terms from the INSPEC *Thesaurus*.

Electrical and Electronics Abstracts can be accessed in several different ways. The abstracts journal is published monthly and includes a series of indexes. A cumulative index is provided every six months and these are collected into a four-yearly index. *Key Abstracts* are provided in a series of twenty-two journals highlighting the key articles from periodicals and conference proceedings in subject areas of particular relevance such as electronic circuits, semiconductor devices and electronic instrumentation. A current awareness service can be provided where particular subjects which have been carefully selected from dynamic areas of the database to appeal to a wide audience are updated every two weeks. A similar service known as SDI allows a customer to specify a narrower field of interest and thus receive information according to a personalized profile.

The complete INSPEC database can be accessed online via a number of major host systems throughout the world allowing simple, fast and efficient interactive searches to be made. These include OVID Technologies (formerly BRS and CDP Online) (USA), CAN/OLE (Canada), CEDOCAR (France), Data Star (Switzerland), DIALOG (USA), ESA-IRS (Italy), ORBIT (now Questel.Orbit) (USA), STN (Germany) and STIC (Taiwan).

The INSPEC database from 1989 onwards is available on CD-ROM and can be searched using special software running on an IBM-PC compatible. Each year of the database is on one disc and the current year is updated quarterly. INSPEC is also available on tape so that organizations with suitable information services can run their own profiles of retrospective searches.

The Technical Information Unit of the IEE has access to over 1000 online databases covering technical, business, marketing, company and news data. It compiles its own database on sales and market forecasts in the electrical, electronic, computing and information technology sectors. Examples of some other useful databases are given below:

COMPENDEX*PLUS includes coverage of electrical, electronic and control engineering as well as the full range of engineering subjects. It has been operating since 1970 and contains over 2.9 million records.
COMPUTER DATABASE covers computers, telecommunications and electronics and contains the full text of over 70 magazines from which product evaluations, comparisons and buyers' guides can be obtained.
REUTER TEXTLINE contains facts, figures and comment from a selection of news publications around the world which include specialist sources for computing and electronics.

Textbooks and monographs

The enormous breadth of electronics as a subject means that no textbook can possibly provide an adequate overview of the field. Even when the subject of interest is identified precisely, a choice between introductory and advanced texts will have to be made. Some illustrative examples are given below for the particular subject areas that are most frequently encountered. It should be noted that electronics textbooks do not reflect the current state of development for very long because rapid advances are taking place in nearly all aspects of the subject. However, some books are now established as standard texts and are widely recommended in academic circles.

Electronics textbooks are available from very many publishers, of particular note McGraw-Hill have published an extensive series on electronics and many new titles are added each year as the field advances. Other major publishers such as Prentice Hall, John Wiley, Addison-Wesley, Chapman and Hall and Macmillan produce texts on both introductory and advanced topics which are widely recommended. The significant contribution of other publishers should not be overlooked.

The professional organizations, the IEE and the IEEE, are also involved in the publication of electronics textbooks. In general these are aimed at the more specialized subject areas which have a smaller readership. An extensive list of new titles are produced each year as major developments and advances are made.

A general introduction to modern electronic components and their application in system design can be found in Horowitz and Hill (1989) and Millman and Grabel (1988).

At a fundamental level electronics is divided between two distinct types of system and the textbooks follow this division. Analogue systems form one branch and at a simple level the op-amp is an almost universal building block. Digital systems form the other branch and make extensive use of the logic components. A wide range of textbooks on these and more specialized subjects is available, a few examples of these are given below. For digital systems, Stonham (1987) and Wakerly (1990) are recommended. For analogue systems, Franco (1988) and Wait, Huelsman and Karn (1992) are recommended.

Some sample textbooks which deal with more specific subject areas such as communications, instrumentation, semi-conductor devices and opto-electronics will now be given. For an introduction to communications techniques, O'Reilly (1989) is recommended and a more advanced reference is Lathi (1989). For instrumentation systems, Bentley (1988) is an excellent reference. In the field of semiconductor devices, works such as Streetman (1990) and Sze (1986) are good references. A reference for opto-electronics is Gower (1993).

References

Bardeen, J. and Brattain, W. E. (1948) The transistor: a semiconducting triode. *Physical Review*, **74**, 230–231.
Bentley, J. P. (1988) *Principles of measurement systems* 2nd ed. Harlow: Longman.
Fink, D. G. and Christiansen, D. (eds) (1989) *The electronics engineers' handbook* 3rd ed. New York: McGraw-Hill.
Franco, S. (1988) *Design with operational amplifiers and analogue integrated circuits.* New York: McGraw-Hill.
Gibilisco, S. and Sclater, N.J. (eds) (1990) *Encyclopedia of electronics* 2nd ed. Blue Ridge Summit, PA: TAB Books.

Horowitz, P. and Hill, W. (1989) *The art of electronics* 2nd ed. Cambridge: Cambridge University Press.
Hughes, F. W. (1986) *Op-amp handbook* 2nd ed. Englewood Cliffs, NJ: Prentice Hall.
Kaufman, M. and Seidman, A. H. (eds) (1984) *The handbook for electronics engineering technicians* 2nd ed. New York: McGraw-Hill.
Kaye, G. W. C. and Laby, T. H. (1995) *Table of physical and chemical constants* 16th ed. London: Longman.
Lancaster, D. (1988) *CMOS Cookbook* 2nd ed., rev. H. M Berlip. Indianapolis, IN: H. W. Sams.
Lathi, B. P. (1989) *Modern digital and analogue communication systems* 2nd ed. Philadelphia, PA: Holt, Rinehart and Winston.
Mazda, F. (ed.) (1989) *Electronics engineer's reference book* 6th ed. London: Butterworths.
Millman, J. and Grabel, A. (1988) *Microelectronics* 2nd ed. London: McGraw-Hill.
O'Reilly, J. J. (1989) *Telecommunication principles* 2nd ed. New York: Van Nostrand-Reinhold.
Stonham, T. J. (1987) *Digital logic techniques* 2nd ed. New York: Van Nostrand-Reinhold.
Streetman, B. G. (1990) *Solid state electronic devices* 3rd ed. Englewood Cliffs, NJ: Prentice Hall.
Sze, S. M. (ed.) (1986) *VLSI technology* 2nd ed. New York: McGraw-Hill.
Wait, J. V., Huelsman, L. P. and Korn G. A. (1992) *Introduction to operational amplifier theory and applications* 2nd ed. New York: McGraw-Hill.
Wakerly, J. F. (1990) *Digital design: principles and practices*. Englewood Cliffs, NJ: Prentice Hall.
Warnes, L. A. A. (1990) *Electronic materials*. Basingstoke: Macmillan.

CHAPTER TWENTY-FOUR

Electrical power systems and machines

A. M. PARKER

Introduction

The large-scale generation of electrical energy, its bulk transmission and distribution to consumers and its utilization in the form of mechanical work represents an engineering enterprise on a massive scale, demanding huge amounts of physical, financial and human resources. The energy delivered by electrical power systems is vital to the operation and sustenance of modern life and in fact the usage of this energy per capita is often taken as a major determinant of the degree of development of a society.

The fact that these systems require such large investment – often a significant proportion of gross national product – means that their planning, operation and control are of great importance. Marginal changes in technical performance can have profound effects on resource inventories and on cash flows, placing great responsibilities on the engineers entrusted with their design and operation.

Virtually all electricity generated in commercial quantities results from energy conversion on a massive scale utilizing rotating electrical machines. In developed countries, well over half of the energy supplied by distributed electrical power systems to consumers eventually is converted into mechanical form, again by rotating machines.

We shall be considering the power generation, transmission and distribution process and the electrical machines involved in its utilization. Traditionally this combination has been grouped under the heading of 'Power Engineering', and some of the sources we shall discuss do in fact cover this whole field. In keeping with all other areas of technology, however, there is an inexorable movement towards ever greater specialization, with specialist journals and reference works appearing at regular intervals.

In this chapter we will be dealing first with the general power field, and then following the path of the electrical energy as it is generated, transmitted and distributed to consumers and finally used in rotating machines. We shall also consider the power-conditioning equipment which forms an increasingly important part of most electrical drive systems.

A central problem in recommending a literature source in any area of technology is that potential users of the information cover a very wide range from research specialists to those seeking a general introduction to the area and who may have a commercial or other non-engineering background. I shall attempt to cater for this diverse readership with a variety of sources at each stage.

Electrical power engineering

Literature searches in all fields are now immeasurably helped by international databases compiled by commercial organizations. To an increasing extent these are made more accessible by being computer readable, either online or using CD-ROM technology. In the area of power engineering, the most universal of thse databases is INSPEC, which originates from the Institution of Electrical Engineers (IEE) in the UK. The subset of its holdings containing electrical engineering entries is Section B, which is obtainable online using a number of host systems throughout the world, most notably DIALOG. In print form, these entries can be scanned in the monthly *Electrical and Electronics Abstracts*. In more specialized areas, as we shall see later, there are monthly *Key Abstracts* which provide a more direct path to important new information. In CD-ROM form, the INSPEC – ELECTRONICS AND COMPUTING ONDISC is available, compatible with ProQuest software from University Microfilms International which provides a powerful search facility.

Of the remainder of the readily accessible databases, perhaps the most useful to power engineers is ELECTRIC POWER DATABASE (EPD) produced by the Electric Power Research Institute in the USA. The coverage here is all information sources in USA, Canada, Japan and Mexico. This is again accessible online through DIALOG and on CD-ROM.

There are two pre-eminent English-language sources for the promulgation of research results in electrical and electronic engineering: the *IEE Proceedings* in the UK and the *Transactions of the Institute of Electrical and Electronics Engineers* (IEEE) based in the USA. These together form the primary information base for researchers, and should be the starting point for those seeking specialized information. These publications in both cases are subdivided into specialist categories, and we shall encounter these again at the appropriate stages in the chapter. These institutions combine to produce a CD-ROM of IEEE/IEE PUBLICATIONS ONDISC (IPO) which contains scanned images of all journal and conference papers since 1988 and which may be searched using ProQuest.

As well as these specialist papers, both of these bodies produce magazine-style journals designed for a general readership, and the monthly *IEE Review* and *Spectrum* (IEEE) are well worth perusing on a regular basis to keep in touch with current developments, often concerned with power engineering. In recent years, the IEE has sought to bridge the gap between the often highly-mathematical *Proceedings* and popular magazines by publishing a series of journals designed to review areas of technology and aimed at a 'technical generalist' rather than a specialist.

The *Power Engineering Journal* was one of the first of this series, and is a highly readable publication appearing bi-monthly. Since its appearance in 1987 it has covered virtually all aspects of the field, and it is recommended as a starting point for anyone with a general technical background seeking an introduction to topics in the power area. Tutorial articles from time to time provide the theoretical background necessary to understand key recent developments.

A good general survey of the power field for the non-specialist is also given in Wildi (1990), which includes enough analysis to be able to predict the performance of power equipment and systems, without obscuring the salient practical features of their operation.

A number of publications cater for those seeking up-to-date commercial and technical news on recent developments in the power industry. These are naturally of most immediate relevance in their country of origin, and include, for the UK *Electrical Times* and *Electrical Review*, both published by Reed Business Publishing, *Electrical World* (McGraw-Hill) in the USA and *Electrical Engineer* (Thomson Publishing) in Australia.

Power systems

In this section we will be considering the processes which span the generation of electrical power in central power stations, its transmission in bulk across geographical regions and its distribution to consumers. There is a number of books whose scope covers this whole area and which form a good starting point for those entering the field.

Weedy (1987) has proved a reliable source for many years in its several editions, and provides a balanced and comprehensive overview of the field. Although published in the UK, mention is frequently made of practices and conditions in other parts of the world. A useful further reference from a US perspective is Elgerd (1982) which covers largely the same field.

Both of the above texts concentrate on the technical features which constrain the design and operation of power systems, and place less emphasis on the economic context in which long-term planning and day-to-day management of these large systems have to exist. This area is covered very well in Wood and Wollenberg (1984), which forms a good starting point for understanding the concepts and models used. A key issue in economic planning is that of the reliability of plant and systems and in this the work of R. Billinton and his collaborators is pivotal. Billinton and Allen (1988) is a recommended overview of the techniques used in this type of study.

Research papers in the field of power systems are most conveniently monitored by the *INSPEC Key Abstracts – Power Systems and Applications* which appears monthly. The two major institutions have specialist sections of their publications devoted to this area. In the case of IEE, the *IEE Proceedings. Generation, Transmission and Distribution* is directly relevant, while the *IEEE Transactions on Power Systems* and the *IEEE Transactions on Power Delivery* should be consulted, with the latter placing more emphasis on the physical plant involved in transmission and distribution networks.

Electrical power generation

All electrical power generation is essentially a process of converting energy from one form to another. By far the greatest proportion of electrical energy generated throughout the world originates in chemical form within fossil fuels. The traditional generation process then involves releasing a large proportion of this chemical energy by combustion with oxygen, transferring the heat to a fluid medium (invariably water) and deriving mechanical work from expanding the resultant high-temperature, high-pressure steam through a series of turbines. The work done on the turbines then serves as the input to the rotating electrical generator which is the last link in the conversion chain.

Each of these energy conversion processes is constrained by the practical limitations of materials and economics, and less than half of the energy potential of the fuel is realized in electrical form in conventional power stations. Nevertheless, great advances have been made in the technology employed in power station design and construction, and improvements in energy efficiency over the years have resulted in a far more effective usage of finite fuel resources.

The literature on power generation has been immeasurably strengthened in recent years by a significant contribution from the former Central Electricity Generating Board (CEGB) in the United Kingdom. As one of its final corporate acts before the privatization of the UK electricity supply industry dispersed its various elements, the CEGB decided to publish its in-house training and instructional material on power station (and transmission system) design, operation and control in the form of a twelve-volume set *Modern power station practice* (British Electricity International). Although previous editions with this title have appeared over the years, the third edition of 1991 represents an undertaking on a much larger scale than its predecessors and is a unique source in the field.

For each topic, basic theory is followed by a wealth of practical detail which has not before been accessible to a general readership within a single published work. The value of this contribution is such that the individual volume titles are worth recording for those seeking specialist sources:

Volume A: Station planning and design
Volume B: Boilers and ancillary plant
Volume C: Turbines, generators and associated plant
Volume D: Electrical systems and equipment
Volume E: Chemistry and metallurgy
Volume F: Control and instrumentation
Volume G: Station operation and maintenance
Volume H: Station commissioning
Volume J: Nuclear power generation
Volume K: EHV transmission
Volume L: System operation
Volume M: Index.

The type of power stations dealt with in this series has existed essentially in this form since the turn of the century. There are however growing pressures from a number of directions which are causing the technology of power generation to change more rapidly than ever before. The processes which lead to the formation of fossil fuels (or nuclear fuel isotopes) take place over geological time scales, and so these fuel sources must be treated as non-renewable for planning purposes. The finite proved reserves of these fuels must therefore be husbanded more carefully than before and the changing economics of the situation are producing new optimal solutions in the generation process.

Thus we are seeing increased penetration of technologies such as combined-cycle combustion where the high temperature products of combustion are first passed through a gas turbine to extract some mechanical and hence electrical energy, before using the exhaust from this process in relatively conventional steam-raising plant. This system is capable of utilizing far more of original chemical potential, and is particularly suited to the use of natural gas fuel.

A further development which is now moving from the research laboratory to commercial-scale plant is that of fluidized-bed combustion, in which fuel and air are bubbled through a granular matrix in which highly efficient mixing and combustion take place.

An unavoidable result of the combustion of fossil fuels is the production of huge amounts of solid and gaseous emissions, a process which is causing increasing concern on a worldwide scale and which gives rise to such undesirable consequences as acid rain and the greenhouse effect. The statutory limitations which are being imposed on power generation authorities throughout the world to reduce or mitigate the effects of these by-products are forcing measures such as flue gas desulphurization to be introduced and are changing the economics of the industry. It will indeed be surprising if further radical changes in technology are not introduced in the coming years to meet the new industry conditions.

As well as forcing design changes on conventional methods of power generation, the changing circumstances are also causing a re-appraisal of alternative approaches using renewable energy resources. As the name implies, these approaches do not involve the depletion of long-term fuel reserves and in the main are less susceptible to environmental damage.

The first such method to find wide commercial acceptance was hydroelectric generation. This of course needs a geographical area with a suitable rainfall regime and a topography which will enable either impoundment of large quantities of water in a high-level reservoir or the exploitation of fast-flowing, high-volume rivers. The immense civil works involved in constructing these schemes means that they have a higher capital cost per unit output than thermal plants, balanced of course by minimal operating costs. Although there are no conventional pollutants, there are environmental costs associated with the alienation of the land required for water storage and civil works and with the diversion of natural water courses. Useful reviews of the technology involved may be found in Jog (1989) or Gulliver and Arndt (1991).

An alternative candidate which is enjoying increasing penetration in most areas of the world is the use of wind power for electricity generation. Following almost twenty years of commercial experience, largely in the USA, in which unit sizes have reached 5 megawatts at times and the market has been distorted by unusual tax and regulatory conditions, the technology may now be regarded as mature. The optimum size of the wind turbine generation unit has now stabilized at a few hundred kilowatts, and reliable designs at this size are now being installed in wind farms in all parts of the world where suitable wind regimes exist and in which the economics (dominated largely by savings in the fossil fuels replaced) are favourable. A good coverage of a wide range of issues in the area is given by Freris (1990) and the high rate of take-up and of developments in the technology can be tracked in the bi-monthly UK periodical *Wind Engineering* (Multi-Science Publishing). A reader seeking up-to-date information on the commmercial aspects of power generation is recommended to read *International Power Generation* (International Trade Publications), a UK-based publication appearing bi-monthly which contains useful survey articles as well as market and contract news from around the world.

Transmission and distribution

The transmission and distribution networks in operation today are the result of almost a century of continuous development. Although the functions of the basic components of overhead lines, underground cables and transformers have remained unchanged in that time, there have been significant advances in the materials and control technology applied to them, and the revolutionary changes in analytical and computational tools within this time have given rise to new classes of controllable devices to improve the performance of the system for both steady-state and transient purposes.

Useful reviews of these areas are included in the general power systems books mentioned earlier, but there are more specialist works available. A particularly useful source of information about current practices and problems is the Conference International des Grands Reseaux Electriques (CIGRE). This organization comprises representatives from most major organizations responsible for high-voltage transmission networks throughout the world. It publishes biennially the proceedings of its General Assembly and forms several Study Committees to investigate specialist problems of mutual concern. Its periodical publication *Electra*, bi-lingual in English and French, provides a valuable summary of its deliberations and reflects current practices and trends.

The plant involved in transmission and distribution is the subject of a range of publications, some produced by the manufacturers of the plants themselves. For many years, practising engineers in the UK have found the *J. and P. transformer book*, 11th ed. (1988) (Butterworths) and *J. and P. switchgear book*, 8th ed. (1987) (Butterworths) invaluable examples of the latter, while their US counterparts have relied upon *Westinghouse transmission and distribution reference book*, 4th ed. (1982) (Westinghouse Electric Corporation).

Protection systems play a vital role in maintaining the security of power systems, and their continuing development is an important feature in improving system reliability. An extremely valuable summary of current practice is to be found in Horowitz (1992), which is a compilation of strategically chosen review papers from the *IEEE Trans-*

actions. This is a welcome modern reference to place alongside the classic work by Warrington (1977). An alternative treatment, written from the point of view of a major equipment supplier, is by GEC Measurements (1987).

Since the late 1950s High-Voltage Direct Current (HVDC) transmission has become recognized as enhancing the capabilities of the traditional alternating current networks in certain circumstances. The interconnection of two large ac systems without stability problems, the transmission of power over very long distances or the negotiating of long underwater crossings are specific instances in which HVDC transmission might well be the most economic or even the only technically feasible solution.

Arrillaga (1983) gives an effective background to the subject, with practical details as well as theoretical analysis. In addition, the IEE has held a series of international conferences on all aspects of the integration of ac and dc systems since 1964 and their Conference Publication series is worth perusing for contributions of topical interest.

Power system analysis

The size and complexity of modern power systems have required specialized techniques to be developed in dealing with the analysis of complete networks or sub-networks under operational conditions. In general, the problems encountered in the analysis fall into three main categories: (1) the solution of the network equations under given loading and generation conditions to give the complete loadflow pattern; (2) a similar analysis under specified fault conditions, both types of analysis being carried out in the frequency domain under steady-state conditions; (3) analysis using the dynamic representation of the network, to obtain the time response to a specified disturbance or the stability margins associated with a particular set of operating conditions. A good survey of the analytical requirements and some solution strategies are provided in Gross (1986).

There have been three distinct stages in the development of analytical methods: the postulation of simplified models and sub-systems amenable to hand or machine-assisted calculation (largely in the 1920s to 1940s), the exploitation of large mainframe computers in batch-processing mode (1950s to 1970s), and the use of personal computers and workstations in an interactive environment (1980s onwards). In current practice, virtually all power system analysis is conducted using commercially-available computer software, and the range of increasingly user-friendly packages is continually growing.

Particularly useful surveys on the types of analysis available are to be found in Arrillaga and Arnold (1990), while the earlier text by Stagg and El-Abiad (1968) is still a worthwhile reference for algorithms and general approaches. Virtually all of the methods used in modern computer software can be traced to pioneering work from the earlier eras mentioned above, and it is often instructive to return to some of the classic works to obtain the theoretical basis for the approach or some physical insight into the models used.

In the field of steady-state circuit analysis which underpins loadflow and fault analysis software, significant early contributions are by Clarke (1943), in which symmetrical component theory and per-unit calculations are thoroughly investigated, and the work of Gabriel Kron. Although some of Kron's output is not easy to assimilate, Kron (1959) is recommended as an example of his contributions. A journal paper from the 1970s, Stott (1974), gives a comprehensive overview on the theoretical basis for all current methods.

For the dynamic behaviour of generators which dominates the transient response of power systems and which is at the heart of stability software, Kimbark (1968) and Concordia (1951) provide insights into the physical phenomena occurring within the machines and Adkins and Harley (1975) introduces the formulation of the dynamic equations and the simplifying assumptions normally incorporated into the computer models.

One problem with power system software is the difficulty of verifying the models or the algorithms used. Occasionally the results of full-scale system tests are made available for benchmarking purposes, and in the case of system dynamics selected results have been published over the years by the CEGB in the UK. Results from these tests can be found in Busemann and Casson (1958), Shackshaft (1963) and Shackshaft and Neilson (1972). It is salutary to read Chorlton and Shackshaft (1972) in which the last quoted system test is used to benchmark commonly used software packages of the time, with very sobering results!

In the USA the IEEE has assisted in the verification process by publishing its *IEEE Transactions* data sets which have become standards for comparing different computer models. For loadflow calculations, the IEEE standard circuits have been used by a large number of researchers over very many years, and the data have been published many times. One of the more accessible references for these data is Freris and Sasson (1968). For system reliability the benchmark is IEEE Reliability Test System (1979) which has been updated and extended in Allan, Billinton and Abdel-Gawad (1986).

Electrical machine drives

Although the provision of mechanical drives from electrical machinery was historically one of the first applications of electricity in industry, and the basic forms of the electric motor were almost all invented well before the end of the nineteenth century, it is in this branch of power engineering that the most fundamental changes have occurred and in which the present rate of development is at its most rapid.

The advent of widespread and cheap computing power, together with advances in the capabilities and competitiveness of solid-state switching devices have caused a fundamental shift in philosophy in mechanical drives. The traditional means of providing a given torque/speed characteristic was to design a specific machine for the purpose. The varying demands of different industries therefore created a market for a wide range of machine types, and labour-intensive manufacturing plants supplied the needs of this industry, often with quite small production runs.

It is now a virtually universal rule that the 'smart' part of the drive design is in the computer-controlled power electronics which condition the electrical power supplied to a motor which may well be a basic, cheap and robust design. In this process only a few types of machine are required, almost invariably those which do not require elaborate manufacturing stages.

Some reference sources cover the whole of the motor/drive field. The principal research organ to address the whole field is the *IEE Proceedings. Electric Power Applications*. In addition, there is a biennial International Conference on Electrical Machines (ICEM) held in even-dated years in either Europe or North America whose proceedings form a fruitful source of review. The *IEEE Transactions* are subdivided into specialist areas which will be covered at the appropriate time.

A well written book which serves to introduce the whole of this field is Sen (1989).

Electric motors

The basic principles by which electric motors work – the interchange of electrical and mechanical energy in the presence of a magnetic field – is the subject of a number of books, of which two notable examples are Fitzgerald, Kingsley and Umans (1992) and Slemon (1992). These give a good coverage of comparative machine types and their essential characteristics, and are an ideal starting point for exploring the field. At a rather more detailed level, the companion volumes Say and Taylor (1986), and Say (1984) provide a very effective insight into machine design with a wealth of practical detail.

Researchers interested in rotating machines should consult the *IEEE Transactions on Energy Conversion*.

Since its discovery by Tesla, the induction motor has become the workhorse of industry, and its cheap and robust design means that its share of the drive market is growing as increasingly sophisticated controllers are able to match its output to more and more esoteric applications. A classic work on induction motor design and performance is Alger (1965), while Laithwaite (1966) introduces some alternative geometries for the basic motor as well as an interesting treatise on the concept of the goodness of a machine.

For smaller drive sizes there is a number of competitors for the traditional induction, synchronous or dc types. The emergence of permanent-magnet materials such as neodymium-iron-boron with good energy products and very high coercive force makes the use of permanent-magnet synchronous machines (or brushless dc machines) in conjunction with variable-frequency inverters an attractive proposition. This subject is well treated in Kenjo and Nagamori (1985).

The ability of modern controllers to provide fast switching and current control facilities has also enabled the switched reluctance drive to become competitive in a range of applications. This uses a very cheap and robust construction without current-carrying windings on the rotor, the basic principles and characteristics being found in the review article by Miller (1987).

Power electronics for drives

The steady evolution of the technology of switching devices over a period of some fifty years has been the underlying cause for the revolution in machine drives. Once the early and inefficient methods of rheostatic control of traction motors had been replaced by controlled mercury-arc rectifiers, the way was cleared for a progressive adoption of complete power conditioner/motor packages to satisfy the controlled-speed drive market.

The successive generations of solid-state switching devices from thyristors, bipolar transistors, power MOSFETS, gate turn off thyristors, insulated gate transistors and so on have enabled these controlled drives to reach hitherto inaccessible regions as switching performance and device voltage and current ratings have increased. The rate of advance of technology is now so rapid that 'state of the art' reviews in the literature quickly become out of date, and the reader is commended to current periodicals to keep abreast of the latest device capabilities. A topical review of this nature is contained in the IEE *Power Engineering Journal* (February 1994), an issue devoted to power electronics topics. There is, however, a landmark compilation of significant review and research papers in Thollot (1993), which will serve as a valuable introduction to the area for some time to come.

The *IEEE Transactions* include topics relating to power electronics and motor drives under several headings, and readers seeking the latest references in this general area are advised to monitor the issues on

Energy Conversion, Industry Applications, Industrial Electronics and *Power Electronics*. The basic topology of power electronic circuits is subject to less rapid change than that of the devices they contain, and good overall coverage of most commonly used drive circuits may be found in books such as Rashid (1993) and Williams (1992).

The first controlled-speed drive application was a phase-controlled rectifier drive for a dc motor. This uses the fact that the speed of a dc motor is directly proportional to its armature voltage when the field current is held constant. As device capabilities improved, this variable voltage was often obtained from a diode rectifier followed by a variable mark-space-ratio chopper because of its better power factor and efficiency at low speed levels.

For a considerable period of time speed control was synonymous with dc drives, but the market has now largely passed to ac types, dominated by the induction motor. The few remaining dc areas of traction, crane drives and servo motors are coming under increasing threat from ac alternatives.

The advantages of supplying a cheap induction motor with a variable-frequency supply for speed control purposes have long been apparent. In conjunction with control over frequency, a simultaneous control over the rms stator voltage must be maintained to preserve the motor's air-gap flux at its design level. With early switching devices, the variable output frequency was obtained by the controlled switching of an intermediate dc stage, derived from the controlled rectification of the ac mains. Forced commutation using auxiliary circuits had to be used to switch off the thyristors of the day, resulting in physically cumbersome circuits and switching frequencies initially limited to one conduction period per half-cycle and a 'quasi-square' waveform. As device performance improved, multiple switchings within each cycle became feasible, with pulse-width-modulation (PWM) strategies being employed to reduce the harmonics imposed on the motor.

It is possible to convert a fixed-frequency ac supply directly into a variable low frequency without using an intermediate dc stage by the use of cycloconvertor circuits, the subject being comprehensively treated in Pelly (1971).

A number of reference books cover the whole ac drive area, with two notable examples being Bose (1986) and Murphy and Turnbull (1988).

The most recent area of controlled drives to fall to the ubiquitous induction motor is that traditionally satisfied by the dc servo motor, in which a quick response to rotor position demand is required. The feature of a dc motor which makes it a natural choice for this type of application is the fact that the axis of the flux-producing current (the field current) is orthogonal to that of the torque-producing current (the armature current), and so the two functions are separable and torque control is a straightforward matter. With an induction motor, not only is there no section of the machine which is associated with either torque or flux uniquely, but the flux vector rotates in space at synchronous speed. However, the combination of fast online processing power in microprocessors and fast switching action in semiconductor devices makes it possible to design a stator switching strategy in which the component of current orthogonal to the flux (and hence proportional to torque) may be isolated and independently controlled, once the instantaneous air-gap flux position is either monitored or inferred. This type of control is known alternatively as vector control or field-oriented control, for which a good reference is Vas (1990).

As mentioned above, the power electronics revolution has enabled a number of new machine types to be exploited, especially in the smaller drive sizes. The drive strategies used in these applications can be found in Miller (1989).

References

Adkins, B. and Harley, R. G. (1975) *The general theory of alternating current machines.* London: Chapman and Hall.
Alger, P. L. (1965) *The nature of induction machines.* New York: Gordon and Breach.
Allan, R. N., Billinton, R. and Abdel-Gawad, N. M. K. (1986) The IEEE reliability test system – extensions to and evaluation of the generating system. *IEEE Transactions on Power Systems,* **1**, 1–8.
Arrillaga, J. (1983) *High voltage direct current transmission.* London: Peregrinus.
Arrillaga, J. and Arnold, C. P. (1990) *Computer analysis of power systems.* Chichester: Wiley.
Billinton, R. and Allen, R. N. (1988) *Reliability assessment of large electric power systems.* Dordrecht: Kluwer Academic.
Bose, B. K. (1986) *Power electronics and A.C. drives.* Englewood Cliffs, NJ: Prentice Hall.
Busemann, F. and Casson, W. (1958) Results of full-scale stability tests on the British 132 kV grid system. *Proceedings of the Institution of Electrical Engineers,* **105A**, 347–362.
Chorlton, A. and Shackshaft, G. (1972) Comparison of accuracy of methods for studying stability: Northfleet exercise. *Electra,* **23**, 9–49.
Clarke, E. (1943) *Circuit analysis of A.C. power systems.* New York: Wiley.

Concordia, C. (1951) *Synchronous machines – theory and performance*. New York: Wiley.
Elgerd, O. I. (1982) *Electric energy systems theory: an introduction* 2nd ed. New York: McGraw-Hill.
Fitzgerald, A. E., Kingsley, C. and Umans, S. D. (1992) *Electric machinery* 5th ed. (SI units) London: McGraw-Hill.
Freris, L. L. (ed.) (1990) *Principles of wind energy conversion systems*. New York: Prentice Hall.
Freris, L. L. and Sasson, A. M. (1968) Investigation of the load flow problem. *Proceedings IEE*, **115**, 1459–1470.
GEC Measurements (1987) *Protective relays: application guide* 3rd ed. Stafford: GEC Measurements.
Gross, C. A. (1986) *Power system analysis* 2nd ed. New York: Wiley.
Gulliver, J. S. and Arndt, R. E. A. (1991) *Hydropower engineering handbook*. New York: McGraw-Hill.
Horowitz, S. H. (ed.) (1992) *Protective relaying for power systems* II. New York: IEEE Press.
IEEE Reliability Test System (1979) *IEEE Transactions on Power Applications and Systems*, **98**, 2047–2054.
Jog, M. G. (1989) *Hydro-electric and pumped storage plants*. New York: Wiley.
Kenjo, T. and Nagamori, S. (1985) *Permanent-magnet and brushless D.C. motors*. Oxford: Clarendon Press.
Kimbark, E. W. (1968) *Power system stability: synchronous machines*. New York: Dover.
Kron, G. (1959) *Tensors for circuits* 2nd ed. New York: Dover.
Laithwaite, E. R. (1966) *Induction machines for special purposes*. London: Newnes.
Miller, T. J. E. (1987) Brushless reluctance motor drives. *IEE Power Engineering Journal*, **1**, 325–331.
Miller, T. J. E. (1989) *Brushless permanent-magnet and reluctance motor drives*. Oxford: Clarendon Press.
Murphy, J. M. D. and Turnbull, F. G. (1988) *Power electronic control of A.C. motors*. Oxford: Pergamon.
Pelly, B. R. (1971) *Thyristor phase-controlled convertors and cycloconvertors: operation, control, and performance*. New York: Wiley.
Rashid, M. H. (1993) *Power electronics; circuits, devices and applications* 2nd ed. Englewood Cliffs, NJ: Prentice Hall.
Say, M. G. (1984) *Alternating current machines* 5th ed. London: Pitman.
Say, M. G. and Taylor, E. O. (1986) *Direct current machines* 2nd ed. London: Pitman.
Sen, P. C. (1989) *Principles of electric machines and power electronics*. New York: Wiley.

Shackshaft, G. (1963) General-purpose turbo-alternator model. *Proceedings IEE*, **110**, 703–713.
Shackshaft, G. and Neilson, R. (1972) Results of stability tests on an underexcited 120MW generator. *Proceedings IEE*, **119**, 175–188.
Slemon, G. R. (1992) *Electric machines and drives*. Reading, MA: Addison-Wesley.
Stagg, G. W. and El-Abiad, A. H. (1968) *Computer methods in power system analysis*. New York: McGraw-Hill.
Stott, B. (1974) Review of load flow calculation methods. *Proceedings IEEE*, **62**, 916–929.
Thollot, P. A. (ed.) (1993) *Power electronics technology and applications 1993* IEEE Technology Update Series. New York: IEEE.
Vas, P. (1990) *Vector control of A.C. machines*. Oxford: Clarendon Press.
Warrington, A. R. van C. (1977) *Protective relays: their theory and practice* 3rd ed. London: Chapman and Hall.
Weedy, B. M. (1987) *Electric power systems* 3rd ed. Chichester: Wiley.
Wildi, T. (1990) *Electrical machines, drives and power systems* 2nd ed. Englewood Cliffs, NJ: Prentice Hall.
Williams, B. W. (1992) *Power electronics: devices, drivers, applications and passive components* 2nd ed. Basingstoke: Macmillan.
Wood, A. J. and Wollenberg, B. F. (1984) *Power generation operation and control*. New York: Wiley.

CHAPTER TWENTY-FIVE

Communications engineering

R. L. BREWSTER

Communication is the process of exchanging information and telecommunications is the science of communicating over distances where the basic modes of communication, such as speech and vision, are no longer feasible. The most widely used method of communicating over such distances is by the use of electrical signals, either over cables or through free space using radio waves.

The art of telecommunications is passing through an exciting stage of development and around the turn of the century we are likely to see many new developments which will significantly affect our whole way of life. There have been a number of recent developments in telecommunications technology which have brought considerable change to the existing communications network, with many of the traditional modes of communication now fading into obsolescence. Recently, the use of light signals carried over optical fibres has become a practical alternative to wire cables, with considerable implications for the future of telecommunications.

The best known, and by far the most widely used method of person-to-person telecommunications, is the telephone. The telephone converts speech pressure waves into electrical signals, which are transmitted over the telephone network and then converted back into sound waves in the telephone earpiece. The electrical signals are complex waveforms which can take on any value. Such signals are known as continuous, or analogue, signals. The disadvantage of transmitting analogue signals over a network is that they become impaired by noise during the transmission process. However, this can be largely overcome by converting the continuously variable analogue signal into discrete signal levels representing the amplitude of samples of the analogue waveform and transmitting these signal levels in the form of a sequence of digital (usually binary) characters. The process of converting the analogue speech signals into digital form is known as pulse code modulation (PCM).

Over the last two decades, with the rapid growth of data communications, the telecommunications network has moved from being almost exclusively used for telephony to becoming a fully integrated digital network. The integrated services digital network (ISDN) represents the current position in about a hundred years of evolutionary growth of the worldwide telecommunications infrastructure. This evolution is by no means complete and in the next year or two we will see the emergence of the 'broadband' ISDN (B-ISDN) as the next stage of the evolutionary development.

A further recent development in telecommunications is the move towards mobility of the user equipment. This has given rise to cordless and cellular mobile radio telephony and an increasing interest in the wider issues of personal communications networks (PCN).

The increased flexibility demanded by the increase in the range and type of service available in telecommunications and the growing use of computing technology within the telecommunications network has stimulated an interest in intelligent networks (IN), where inherent network intelligence can be used to adapt the network criteria to suit the user and traffic demands in an efficient and user-friendly way. The main use of network intelligence is to enable the network to manage its resources to meet the demands of the network user by the most economical and technically efficient means possible. There is therefore a growing widespread interest in the topic of network management (NM).

The rapid changes that are taking place in the world of telecommunications is accompanied, not surprisingly, by a rapid increase in the amount of published literature in the field. In this chapter, we review the available literature under a number of topic headings. Because of the rapid rate of publication and the rapid rate with which published material becomes out of date, it is impossible to include all possible references under each heading. The selection is based mainly on those that I have personally found useful, omission of a particular title should not be taken to imply criticism, simply that I have not personally had an opportunity to make use of that particular title. Some titles appear under more than one heading.

General communications theory

There is a plethora of books on the basic concepts of communications theory. The following is a small selection of the most helpful works:

Communications engineering 443

Brewster, R. L. (1986) *Telecommunications technology* (Ellis Horwood).
Stremler, F. G. (1990) *Introduction to communication systems*, 3rd ed. (Addison-Wesley).
Tomasi, W. (1992) *Advanced electronic communication systems*, 2nd ed, (Prentice Hall).
Dunlop, J. and Smith, D. G. (1989) *Telecommunications engineering* 2nd ed. (Van Nostrand Reinhold).
O'Reilly, J. J. (1989) *Telecommunication principles*, 2nd ed. (Van Nostrand Reinhold).
Taub, H. and Schilling, D. L. (1986) *Principles of communication systems*, 2nd ed. (McGraw-Hill).
Carlson, A. B. (1986) *Communication systems* (McGraw-Hill).
Schwartz, M. (1990) *Information transmission, modulation and noise*, 4th ed. (McGraw-Hill).

IEE telecommunications series

This is the most comprehensive collection of books covering the complete range of telecommunications topics in a single series:

Volume 1: Flood, J. E. (ed.) (1975) *Telecommunication networks*.
Volume 2: Bear, D. (1988) *Principles of telecommunication traffic engineering*, 3rd ed.
Volume 3: Hills, M. T. and Kano, S. (1976) *Programming electronic switching systems*.
Volume 4: Bylanski, P. and Ingram, D. G. W. (1980) *Digital transmission systems*, rev. ed.
Volume 5: Roberts, J. H. (1977) *Angle modulation: the theory of system assessment*.
Volume 6: Welch, S. (1981) *Signalling in telecommunication networks*, rev. ed.
Volume 7: Littlechild, S. C. (1979) *Elements of telecommunication economics*.
Volume 8: Takamura, S., Kawashima, H. and Nakajima, N. (1979) *Software design for electronic switching systems*.
Volume 9: Robins, W. P. (1982) *Phase noise in signal sources*.
Volume 10: Griffiths, J. M. (ed.) (1983) *Local telecommunications*.
Volume 11: Ralphs, J. D. (1985) *Principles and practice of multi-frequency telegraphy*.
Volume 12: Skaug, R. and Hjelmstad, J. F. (1985) *Spread spectrum in communication*.
Volume 13: Creasey, D. J. (ed.) (1985) *Advanced signal processing*.
Volume 14: Holbeche, R. J. (ed.) (1985) *Land mobile radio systems*.
Volume 15: Gosling, W. (ed.) (1986) *Radio receivers*.
Volume 16: Brewster, R. L. (ed.) (1986) *Data communications and networks*.
Volume 17: Griffiths, J. M. (ed.) (1988) *Local telecommunications 2 – into the digital era*.
Volume 18: Evans, B. G. (ed.) (1991) *Satellite communication systems*.
Volume 19: Farr, R. E. (1988) *Telecommunications traffic, tariffs and costs*.

444 Communications engineering

Volume 20: Dalgleish, D. I. (1989) *An introduction to satellite communications*.
Volume 21: Redmill, F. J. and Valdar, A. R. (1990) *SPC digital telephone exchanges*.
Volume 22: Brewster, R. L. (ed.) (1989) *Data communications and networks II*.
Volume 23: Withers, D. J. (1991) *Radio spectrum management*.
Volume 24: Evans, B. G. (ed.) (1991) *Satellite communication systems II*.
Volume 25: Macario, R. C. V. (ed.) (1991) *Personal and mobile radio systems*.
Volume 26: Manterfield, R. J. (1991) *Common-channel signalling*.
Volume 27: Flood, J. E. and Cochrane, P. (eds) (1991) *Transmission systems*.
Volume 28: Everett, J. L. (ed.) (1992) *VSATs: very small aperture terminals*.
Volume 29: Cuthbert, L. G. and Sapanel, J.-C. (1993) *ATM: the broadband telecommunications solution*.

Telecommunications networks

The backbone of the national and international telecommunications infrastructure is the telephone network. Even where there is no demand for 'Potential Advanced Network Services' (PANS), there is still a requirement for 'Plain Old Telephone Service' (POTS). The following books give a broad coverage of the POTS and PANS of telecommunications, their main emphasis being on the provision of telephony services and extended services which derive from the technology of telephony:

Flood, J. E. (ed.) (1975) *Telecommunication networks* (Peregrinus).
Flood, J. E. and Cochrane, P. (eds) (1991) *Transmission systems* (Peregrinus).
Redmill, F. J. and Valdar, A. R. (1990) *SPC digital telephone exchanges* (Peregrinus).

Signalling

With a network of the size of the international telecommunications network, it is essential that efficient network signalling is provided to keep the network under control. The definitive book on signalling is:

Welch, S. (1981) *Signalling in telecommunications networks*, rev. ed. (Peregrinus).

Packet switching

Although the topic of packet switching is generally covered in some detail in the books listed in the section 'Data networks', the following book is a useful handbook for those with a special interest in the detail of the X25 protocol for packet-switched networks:

Deasington, R. J. (1985) *X25 explained* (Ellis Horwood).

ISDN and broadband ISDN

The latest stage in the evolution of the telecommunications network is towards an integrated services digital network (ISDN). Current developments in optical fibre technology mean that much higher transmission rates are becoming a practical reality. This has led to the development of proposals for a broadband ISDN (B-ISDN). Each of the following books gives a comprehensive overview of ISDN and, in most cases, an indication of the way B-ISDN is currently emerging and of its application for general use in the immediate future:

Brewster, R. L. (1993) *ISDN technology* (Chapman and Hall).
Griffiths, J. M. et al. (1992) *ISDN explained*, 2nd ed. (Wiley).
Helgert, H. J. (1991) *Integrated services digital networks* (Addison-Wesley).
Ronayne, J. (1987) *The integrated services digital network; from conception to application* (Pitman).
Stallings, W. (1992) *ISDN and broadband ISDN* (Macmillan).
van Duuren, J., Schoute, F. C. and Kastelein, P. *Telecommunication networks and services.* (Addison-Wesley).

Information theory and coding

No review of literature regarding telecommunications would be complete without a mention of information theory and coding. There are numerous books available covering various aspects of these topics, many of them very specialized. The two books given below are those found to be most useful as general reviews with a practical approach:

Hamming, R. W. (1986) *Coding and information theory*, 2nd ed. (Prentice Hall).
Wade, J. G. (1987) *Signal coding and processing* (Ellis Horwood).

Data networks

With the rapid growth in the use of computers and the resultant need for data interchange, there have been considerable developments in data network technology. A wide variety of technologies, strategies and protocols exist for a broad range of applications. The following list contains books covering most topics associated with a wide range of data networks:

Beauchamp, K. G. (1990) *Computer communications*, 2nd ed. (Chapman and Hall).
Bleazard, G. B. (1982) *Handbook of data communications* (NCC Publications).
Brewster, R. L. (1989) *Communication systems and computer networks* (Ellis Horwood).
Brewster, R. L. (ed.) (1989) *Data communications and networks 2* (Peregrinus).
Bylanski, P. and Ingram, D. G. W. (1976) *Digital transmission systems* (Peregrinus).
Cheong, V. E. (1983) *Local area networks* (Wiley).
Clark, A. P. (1976) *Principles of digital data transmission* (Pentech).
Davies, D. W. and Barber, D. L. A. (1973) *Communication networks for computers* (Wiley).
Deasington, R. J. (1984) *A practical guide to computer communications and networking*, 2nd ed. (Ellis Horwood).
Halsall, F. (1992) *Data communications, computer networks and open systems*, 3rd ed. (Addison-Wesley).
Haykin, S. S. (1988) *Digital communications* (Wiley).

A book on performance modelling:

Woodward, M. E. (1993) *Communication and computer networks* (Wiley).

Some older books that are still worth consulting, although probably out of print if you wish to purchase:

Jesty, P. H. (1985) *Networking with microcomputers* (Blackwell).
Marshall, G. J. (1980) *Principles of digital communications* (McGraw-Hill).
Purser, M. (1987) *Computers and telecommunications networks* (Blackwell).
Sherman, K. (1990) *Data communication, user guide*, 3rd ed. (Prentice Hall).
Stallings, W. (1984) *Local networks* (Macmillan).

Mobile communications

The last decade has seen both the birth and the growth to maturity of mobile cellular radio technology for telephony and a demand for data over cellular is now emerging. A number of books on the topic are just beginning to appear on the market. The following is a selection of useful books:

Lee, W. C. Y. (1989) *Mobile cellular telecommunications systems* (McGraw-Hill).
Steele, R. (ed.) (1992) *Mobile radio communications* (Wiley) A comprehensive handbook of the latest 'mobile' technology. A reference work rather than a student textbook.
Lee, W. C. Y. (1993) *Mobile communications design fundamentals*, 2nd ed. (Wiley) A useful designer's handbook.
Jagoda, A. and de Villepin, M. (1993) *Mobile communications* (Wiley) A review of the services and commercial aspects, rather than a technical treatise.

Satellite systems

The following two books are those found most useful on the topic of satellite systems:

Gordon, G. D. and Morgan, W.L. (1993) *Principles of communications satellites* (Wiley).
Maral, G. and Bousquet, M. (1992) *Satellite communications systems*, 2nd ed. (Wiley).

Radio systems

Communication systems frequently employ radio technology to reach locations where cable access would not be viable, and sometimes completely impossible. The increasing demand for terminal mobility and personal communications is also placing demands on radio technology The following is a list of books on various aspects of radio systems which will be helpful. Again, the list is not exhaustive:

Dunlop, J. and Smith, D. G. (1989) *Telecommunications engineering*, 2nd ed. (Van Nostrand Reinhold).
Olver, A. D. (1992) *Microwave and optical transmission* (Wiley).
Ramo, S., Whinnery, J. R. and Van Duzer, T. (1984) *Fields and waves in communication electronics*, 2nd ed. (Wiley).
Connor, F. R. (1989) *Antennas*, 2nd ed. (Edward Arnold).
Collin, R. E. (1985) *Antennas and radiowave propagation* (McGraw-Hill).
Tomasi, W. (1993) *Advanced electronic communications systems*, 2nd ed. (Prentice Hall).
Maral, G. and Bousquet, M. (1992) *Satellite communications systems*, 2nd ed. (Wiley).
Dalgleish, D. I. (1989) *An introduction to satellite communications* (Peregrinus).
Gordon, G. D. and Morgan, W. L. (1993) *Principles of communications satellites* (Wiley).
Steele, R. (ed.) (1992) *Mobile radio communications* (Wiley).

Optical communications

Optical fibre technology is revolutionizing the transmission of information, providing high capacity links of high reliability. The following suggestions provide a comprehensive coverage of the topic:

Senior, J. M. (1992) *Optical fiber communications*, 2nd ed. (Prentice Hall).
Keiser, G. (1991) *Optical fiber communications*, 2nd ed. (McGraw-Hill).
Gowar, J. (1993) *Optical communication systems*, 2nd ed. (Prentice Hall).

Supplementary reading:

Jones, W. B. (1988) *Introduction to optical fiber communication systems* (Holt Rinehart and Winston).

Recommended background textbook on optics:

Guenther, R. (1990) *Modern optics* (Wiley).

Two thorough theoretical treatments of components and fibre propagation:

Yariv, A. (1991) *Optical electronics*, 4th ed. (Saunders College).
Ghatak, A. K. and Thyagarajan, K. (1989) *Optical electronics* (Cambridge University Press).

A wide-ranging, systems-oriented book with lots of data in summary form:

Hoss, R. J. (1990) *Fiber optic communications design handbook* (Prentice Hall).

Two volumes that provide an overview of all aspects of fibre-optic communications from an advanced but practical perspective:

Miller, S. E. and Chynoweth, A. G. (eds) (1979) *Optical fiber telecommunications* (Academic Press).
Miller, S. E. and Kaminow, I. P. (eds) (1988) *Optical fiber telecommunications II* (Academic Press).

Advanced – definitive in their respective areas:

Snyder, A. W. and Love, J. D. (1983) *Optical waveguide theory* (Chapman and Hall).
Casey, H. and Panish, M. B. (1978) *Heterostructure lasers* (Academic Press).

A particularly good historical perspective:

Midwinter, J. E. and Guo, Y. L. (1992) *Optoelectronics and lightwave technology* (Wiley).

Journals

There is a wide range of technical and commercial regular publications aimed at the telecommunications industry. A selection of these is as follows:

IEE Proceedings. Computers and Digital Techniques
IEE Proceedings. Microwaves, Antennas and Propagation
IEE Proceedings. Communications
IEE Proceedings. Optoelectronics

Electronics Letters
Electronics and Communication Engineering Journal
IEEE Journal on Selected Areas in Communications
IEEE Transactions on Communications
IEEE Transactions on Signal Processing
Signal Processing

House journals

BT Technology Journal (British Telecom)
Electrical Communication (Alcatel)

Free circulation

Communications Networks
Communications News
Datacom (the networking magazine)
Network (the data communications magazine)
LAN Magazine.
Lightwave (fibre optics technology and applications worldwide)
Microwaves & RF
Mobile and Cellular magazine
Telecommunications

Standards

In the UK, as with the rest of Europe and much of the rest of the world, telecommunications equipment is generally produced in accordance with CCITT (International Telegraph and Telephone Consultative Committee) recommendations. In fact the CCITT has recently been merged with the CCIR (International Radiocommunications Consultative Committee) and renamed ITU-T (International Telecommunication Union – Telecommunication Standards Sector). The CCITT recommendations are published every four years in various 'coloured' books. The latest issue is the 'blue' book, based on the 1988 plenary sessions. The next issue, based on the 1992 plenary sessions, is now due for publication. A European standards body has recently been formed (1988) with a similar remit to CCITT but covering only countries in the European Community. This body is known as ETSI (European Telecommunications Standards Institute). ETSI standards are known as ETS or sometimes, especially among French speakers, as NETs.

The principal standards organization in the USA is ANSI (American National Standards Institute). Rather than make its own standards, it accredits those of other standards bodies. Many of the American standards originate with the IEEE (Institute of Electrical and Electronics

Engineers), the American professional body for electrical and electronic engineers.

International standards are overseen by the ISO (International Standards Organization). In the case of LAN (local area network) standards, the ISO specification numbers are derived from the IEEE numbers by the addition of a preceeding 8. Thus IEEE 802.3 is identical with ISO 8802.3, and so on.

The standards business in telecommunications is exceedingly complex and equipment develops at such a rate that *de facto* standards are frequently introduced before international agreement has been reached. Thus, the ISO is anxious that a policy of open systems interconnection (OSI) is established to enable widespread interconnection of equipment produced by a variety of manufacturers from various countries throughout the world.

Conferences

The most important conferences in the area of communications are those organized by the Professional Groups of the Electronics Division of the Institution of Electrical Engineers (IEE) and those organized by the Institute of Electrical and Electronics Engineers (IEEE). Both of these institutions organize international conferences which take place all over the world. The highlight of the IEE conference programme is the International Telecommunications Conference held bi-annually and usually hosted by one of the Universities in the UK. Full details of IEE conferences and colloquia are regularly published in the *IEE News*.

Databases

By far the most important and useful database covering the subject of communications is the IEE Information Services for the Physics and Engineering Communities (INSPEC). INSPEC scans over 4000 scientific and technical journals and some 1400 conference publications each year. The INSPEC database can be accessed online or is available for lease on magnetic tape or on CD-ROM.

CHAPTER TWENTY-SIX

Mathematics and statistics for engineers

N. DEAN

Introduction: the role of mathematics and statistics in engineering

Mathematics and statistics are not in themselves branches of engineering, but yet are essential to all areas of engineering. Most engineers use but a small fraction of mathematics and statistics, yet there is a vast wealth of knowledge available for application. With such a paradoxical situation, it is difficult to decide just what should be included in a chapter on 'Mathematics and statistics for engineers'.

Modern engineering depends heavily on the concept of the mathematical model. A model can be used to help solve a real world problem; often the model is in the form of a computer model. This model may be deterministic (no chance element involved) or stochastic (in which case probability will be used); it may well contain parameters whose values will have to be determined from measured data using statistical estimation techniques. Furthermore, a stochastic model may need to be confirmed as appropriate using statistical tests (hypothesis tests). The model will present a mathematical problem; this can be solved using conventional mathematical techniques (calculus, algebra, trigonometry, transforms, etc.) or by numerical methods. In some modern engineering methods such as finite element analysis or boundary element analysis, the numerical method is implicit in the form of the model. Discrete mathematics, based on set theory and logic, is also playing an increasing role in engineering; in particular graph theory and combinatorics.

Evidently there is a wide variety of mathematical and statistical topics which an engineer might need to use. But what sort of knowledge is required? A common traditional view is that engineers need only know and be proficient in mathematical techniques. This point of view, however, can no longer be regarded as valid, if indeed it ever could. There are two main reasons for this. Firstly, in order to build and use models effectively, it is necessary to have a sound understanding of the relevant mathematical concepts. Secondly, the advent of a wide range of powerful computer software has meant that the knowledge and expertise in mathematical techniques is much less important to the engineer now than in the past; and in fact to use the software effectively requires a much better understanding of the underlying mathematics.

Computer software

The engineer will of necessity use computer software to build models and solve problems. It is important to stress that a major 'source' of mathematical and statistical information is essentially contained in such software.

Any attempt to give a list of software will necessarily be selective and will rapidly become outdated; but in statistics there are Minitab (Minitab Inc.), Statgraphics (Statistical Graphics Corporation) which seem to be well established and quite useful for engineers.

In mathematics, formulae are stored and can be manipulated by computer algebra systems – these can also be used to produce graphs. A basic system is Derive (Soft Warehouse, Inc.) which is also available on pocket calculators. More extensive systems include Maple (Waterloo Maple Software) and Mathematica (Wolfram Research).

There are also more 'traditional' numerical packages such as NAG (Numerical Algorithms Group Ltd.), Numerical Recipes (Cambridge University Press), Matlab (The MathWorks, Inc.) and MathCAD (MathSoft, Inc.). The current trend is now to incorporate these with computer algebra systems. For example, MathCAD v4, a modelling package, includes a small subset of Maple so that mathematical information can be called up from the computer rather than needing to refer to a book in the course of building a mathematical model.

The use of books for formulae (including integral tables), graphs and statistical tables may be regarded by some workers as obsolescent, since much of this information can now be called up from software in computers and even pocket calculators. In fact this argument has long been applicable to statistics tables, yet such tables are still published. It would seem likely that there will always be a place for such information in printed form.

Organizations

Professional engineering bodies usually have sections devoted to the more mathematical and statistical aspects, and mathematical societies usually have sections concerned with the application of mathematics in engineering and industry. A major international mathematical association is the American Mathematical Society (AMS), based in Providence, Rhode Island, USA.

There are also some organizations which are more specifically oriented towards application of mathematics. In the UK, the Institute for Mathematics and its Applications (IMA), encourages the industrial applications of mathematics by way of publications, meetings and conferences. It is based at Southend-on-Sea, Essex. In the USA, the Society for Industrial and Applied Mathematics (SIAM), has a similar role but on a rather larger scale; it is based in Philadelphia, Pennsylvania. Several publications of the IMA and SIAM are listed below.

Professional bodies in statistical science have a more general concern than just engineering; the techniques required for engineering have much in common with those for other applications in biological and social sciences. In the UK the main professional body is the Royal Statistical Society, or (RSS), based in London. In the USA there is the American Statistical Association in Alexandria, Virginia and the American Society for Quality Control in Milwaukee, Wisconsin.

Books

Bibliographies and quick reference books

There are surprisingly few bibliographies; one of the most useful, though now sadly dated is by Dorling (1977). Perhaps part of the reason for the lack of bibliographies is the wealth of reference books, many of which contain substantial bibliographies. The nine volume encyclopaedia edited by Hazewinkel (1988–1993) contains a mixture of short facts, long articles and extensive lists of references to a wide range of mathematical topics. Other general mathematical reference books include the compact handbook by Rade and Westergren (1990), the single volume dictionaries by Sneddon (1976), James and James (1992), and the four volume dictionary by Ito (1987). The major reference for statistics is the nine volume encyclopedia (Kotz and Johnson, 1982–1989), which also contains excellent bibliographies. A useful handy reference is that by Marriott (1990). There are, in addition, several handbooks specifically on mathematics relevant to engineers including those by Tuma (1989) and Kurtz (1991). A rather unusual book is that by Atherton and Borne (1992), which concentrates on modelling and simulation. Another useful specialist handbook on Fourier theorems is that by Champeney (1987).

There are many books of mathematical tables and lists of standard formulae. Advanced reference works include those by Gradshteyn and Ryzhik (1980), Beyer (1987) and Beyer (1991a). The handbook by Abramowitz and Segun (1972) is notable for its comprehensive coverage of tables, formulae, graphs and properties of functions. Seggern (1993) has compiled a useful compendium of curves and surfaces along with the corresponding equations.

A useful compendium of statistical and probability tables is that by Beyer (1991b).

Foreign language dictionaries

The standard polyglot dictionary of mathematical terms (English, German, French and Russian) is the two volume work by Eisenreich and Sube (1982). The statistical dictionary published by the Hungarian Central Statistical Office (1962) covers seven languages (including Russian, Hungarian, English, French and German).

A Chinese–English glossary of the mathematical sciences is that by DeFrancis (1964). There is also the Russian–English/English–Russian glossary of statistical terms by Kotz (1971).

Textbooks and monographs

Encyclopedic series

A mathematical encyclopedia may be organized as a series of conventional monographs or textbooks rather than as an encyclopedic dictionary. The *Handbook of applicable mathematics* (Ledermann, 1991) consists of eight volumes together with an index volume, and covers both mathematics and statistics.

A similar philosophy underlies the series edited by Rota (1976–1993), which comprises a continuing series of more than forty six monographs on areas of mathematics which the editor believes should be made more accessible to the users of mathematics. The series is very useful for the engineer who wishes to become acquainted with any branch of mathematics which happens to be covered. There is no overall index, however.

General books

There are many well loved textbooks on engineering mathematics including Kovach (1982), Cakmak, Botha and Gray (1987), Englefield (1987), Jeffrey (1990) and Croft, Davison and Hargreaves (1992). The book by Bajpai, Mustoe and Walker (1980) covers very advanced topics, being the third book in a series by the same authors.

INDUSTRIAL APPLICATIONS AND THE USE OF COMPUTER ALGEBRA

Although numerous books cover applications, two in particular deserve special comment; both consider a range of industrial applications and both give useful insights to mathematical modelling. Friedman (1988–1992) concentrates more on the mathematics and modelling, while Cohen (1993) is specifically about the use of computer algebra software in handling mathematical models. Atherton and Borne (1992) also cover mathematical modelling.

CURVES AND SURFACES

Differential geometry applied to curve and surface design by Nutbourne and Martin (1988) covers its subject matter from the point of view of the practitioner using computer-aided design software.

The mathematical description of shape and form by Lord and Wilson (1984) covers curve and surface design, graph theory and curve/surface fitting.

NUMERICAL METHODS

Conventional numerical techniques are well covered in the book by Faires and Burden (1993), which comes with software in both Pascal and Fortran. A companion book (Burden and Faires, 1993) gives a readable and thorough account of the mathematical theory of these methods. Numerical recipes, in C, Pascal or Fortran, by Press et al. (1986) concentrates on the algorithms and the program code.

Finite element analysis is a popular approach to solving systems of partial differential equations numerically. Among many books on the subject are those by Bathe (1982), Irons and Shrive (1983) and Rao (1989). Wait and Mitchell (1985) give an account of the theory. An advanced monograph covering theory and applications is by Schwarz (1988). Bettess (1992) considers the related area of infinite elements and includes a disk of software with the monograph.

A more recent alternative to finite element analysis is boundary element analysis; it is based on the mathematics of integral equations and Green's functions. Useful and comprehensive introductions are given by Becker (1992) and Banerjee and Butterfield (1993). Examples of applications are given by Aliabadi and Brebbia (1993), Banerjee and Butterfield (1979), Banerjee and Butterfield (1982), Banerjee and Mukherjee (1984), Brebbia (1985), Banerjee and Watson (1986), Banerjee and Wilson (1989), Banerjee and Marino (1990) and Brebbia and Chaudouet-Miranda (1990).

VECTOR AND TENSOR ANALYSIS

Engineering field theory with applications by Setian (1992) covers vector analysis and gives an introduction to finite differences and finite element analysis. Other books on vector analysis for engineering include those by Reddy and Rasmussen (1982) and Schilling and Lee (1988). Farrashkhalvat and Miles (1990) describe tensor methods.

FUNCTIONAL ANALYSIS

The textbook by Oden (1979) requires a fair degree of mathematical sophistication.

DIFFERENTIAL EQUATIONS AND VARIATIONAL METHODS

The underlying concepts of differential equations are covered by standard mathematical textbooks, while the practical solution of systems of equations is usually a matter for appropriate selection of software. Many books on differential equations generally seem to be of greater interest to the mathematician than the engineer and will not be considered here.

Quinney (1987) gives a good overview of the numerical methods for solving differential equations. Tyn (1987) gives a thorough account, relevant to engineering, of partial differential equations. Variational methods (and an introduction to finite element analysis) is given by Reddy (1984).

INTEGRAL EQUATIONS

Corduneanu (1991) describes both theory and applications.

COMPLEX VARIABLES

An introductory textbook by Jeffrey (1992) includes many examples of engineering applications. A major reference work on complex analysis are the two books by Gonzalez (1992a, 1992b), though there are no applications given.

LINEAR ALGEBRA; LINEAR SYSTEMS AND CONTROL

Goult (1978) covers both the mathematical theory of linear algebra and applications, including Fourier series and eigenvalues. The standard work on matrix theory (in two volumes) is by Gantmacher (1977). Barnett and Cameron (1985) specialize in the application to control theory.

Mathematics and statistics for engineers 457

TRANSFORMS AND GENERALIZED FUNCTIONS

Guest (1991) gives the mathematical theory of Laplace transforms, and Lighthill (1958) does likewise for Fourier analysis; both authors describe generalized functions (sometimes known as distributions) such as the Dirac delta. Solymar (1988) has written specifically about Fourier series. Other (advanced) books on Fourier methods are by Cartwright (1990) and Folland (1992), the latter including Green's functions (upon which boundary element methods are based).

WAVELETS

Wavelets are an exciting recent development in applicable mathematics (invented in the early 1980s) and find particular application in data and image compression. Several books are now available such as the series edited by Chui (1992a, 1992b).

CHAOS AND NON-LINEAR SYSTEMS

Chaos is another modern development in mathematical theory; chaos theory is needed to handle non-linear dynamical systems. Although there are many books on chaos (and fractals), several of these are fairly general books without any substantial mathematics; the book by Peitgen, Jurgens and Saupe (1992) is a good example of this type of work. An introductory (mathematical) textbook is by Devaney (1992). More advanced books (with applications), are by Marek and Schreiber (1991), Arrowsmith and Place (1992, 1990) and Kim and Stringer (1992).

CATASTROPHE THEORY

Catastrophe theory arises in the study of non-linear systems; Gilmore (1981) expounds the theory for engineers and scientists.

DISCRETE MATHEMATICS, MODERN ALGEBRA AND LOGIC

Stanat and McAllister (1977) give a thorough grounding in the foundations of discrete mathematics, including set theory, logic and algebraic structures. Many books, e.g. Biggs (1989), give a wide-ranging overview of discrete mathematics and its many specialized areas. Books on these specific areas are by Galton (1990) for logic; Foulds (1992) for graph theory; Hill (1986), Cover and Thomas (1991) and Roman (1992) for coding and information theory; Bezdek (1987) and Bothe (1992) for fuzzy logic and fuzzy systems.

PROBABILITY, STOCHASTIC MODELLING AND STATISTICS

In addition to many general books on probability and statistics, there are several which are directed more towards engineering; these include those by Chatfield (1983), Smith (1986), Ross (1989), Lapin (1990) and Scheaffer and McClave (1990). Stochastic models and processes, such as queuing theory and reliability theory, are discussed by Soong (1981), Helstrom (1990) and Ross (1989); Helstrom (1990) also includes useful bibliographies. Dupraz (1986) has written a monograph dealing specifically with signals and noise. Arnold (1974) deals with stochastic differential equations – theory and application.

Congresses and symposia proceedings

Numerical Methods

Recent symposia devoted to numerical methods in engineering include: Gruber, Periaux and Shaw (1989) *International symposium on numerical methods in engineering* (Computational Mechanics); Pande and Middleton (1990) *Numerical methods in engineering: theory and applications* (Elsevier); C. A. Brebbia (ed. 1990) *International seminar on recent advances in boundary element methods* (Butterworths); C. A. Brebbia (ed., 1978, 1980–1982) *International conferences on boundary element methods in engineering* (Springer Elsevier); C. A. Brebbia (ed., 1985–1987, 1989–1994) *Conferences on boundary element technology* (Computational Mechanics).

Also note Tanaka, Brebbia and Shaw (1990) *Advances in BEM in Japan and USA*.

Periodicals

GENERAL

SIAM Review contains expository and survey papers, and book reviews.

Bulletins of various mathematical societies often contain survey papers on engineering and industrial applications of mathematics, for example *IMA Bulletin*. Sometimes whole issues may be devoted to a particular topic, for example, the November 1992 issue of *Nieuw Archief voor Wiskunde* was devoted to image processing.

LINEAR ALGEBRA, VECTOR ANALYSIS AND TENSOR ANALYSIS

SIAM Journal on Matrix Analysis and Applications is very theoretical.

Tensor contains research papers and advanced study papers on vector and tensor analysis and their application.

NUMERICAL METHODS

International Journal for Numerical Methods in Engineering and the corresponding *Communications in Applied Numerical Methods* concen-

Mathematics and statistics for engineers 459

trate on numerical methods rather than new techniques. They encompass finite element analysis and boundary integral techniques.

Mathematics of Computation is largely concerned with the theoretical aspects of numerical methods and finite element techniques. There are two periodicals published by SIAM with papers on numerical methods: *SIAM Journal on Numerical Analysis*; *SIAM Journal on Scientific and Statistical Computing*.

MODELLING/APPLIED MATHEMATICS

Journal of Engineering Mathematics comprises papers which describe mathematical models, but with the emphasis on those features which are applicable to various branches of engineering. The editorial policy refers to 'the intrinsic unity, through mathematics, of the fundamental problems of applied and engineering science.'

Mathematical Methods in the Applied Sciences seems to have a similar editorial policy, though the areas of application are wider than just engineering.

Quarterly of Applied Mathematics, although primarily concerned with fluid dynamics and mechanics, often contains some more generally applicable mathematics.

Zeitschrift für angewandte Mathematik und Mechanik comprises research papers, reviews and survey papers, mostly in English. It is an excellent source for papers on mathematical modelling.

Other periodicals on mathematics applied to mechanics and material properties include: *Acta Mechanica*; *Archive for Rational Mechanics and Analysis*; *IMA Journal of Applied Mathematics*; *The Quarterly Journal of Mechanics and Applied Mathematics*; *Studies in Applied Mathematics*; *Zeitschrift für angewandte Mathematik und Physik*; *Applications of Mathematics*.

The *SIAM Journal on Applied Mathematics* not only covers mechanics, but other areas as well e.g. semiconductor lasers.

Mathematics and modelling is, of course, to be found in many periodicals devoted to specific aspects of engineering, such as control theory or signal processing; as a general rule these periodicals are not listed in this chapter. There are, however, some journals which are more biased towards mathematics. In control theory, these include: *SIAM Journal on Control and Optimization*; *IMA Journal of Mathematical Control and Information*; *Mathematics of Control Signals and Systems*.

Networks – an international journal is devoted to the mathematics and probability theory of networks.

MATHEMATICAL ANALYSIS AND DIFFERENTIAL EQUATIONS

There are many journals on analysis and differential equations. Among those relevant to the engineer are: *SIAM Journal on Mathematical*

Analysis; *Journal of Differential Equations; Journal of Mathematical Analysis and Applications.*

Non-linearity covers the mathematics of solving non-linear problems and includes applications to non-linear systems.

DISCRETE MATHEMATICS

SIAM Journal on Discrete Mathematics is largely concerned with graph theory and its application in software engineering.

Discrete Applied Mathematics is biased toward optimization and networks.

Periodicals on fuzzy systems include: *IEEE Transactions on Fuzzy Systems; Fuzzy Sets and Systems.*

Naval Research Logistics covers logistics and probabilistic models.

PROBABILITY

Periodicals on probability tend to be biased towards applications in the biological and social sciences. *Advances in Applied Probability* does however also contain some papers on engineering applications, such as switching networks and image analysis.

Probability in the Engineering and Informational Sciences tends to be oriented towards software engineering.

Stochastics and Stochastic Reports is concerned with probability theory applied to systems and control.

STATISTICS

As with probability, most journals are concerned with biological and social science applications. Among those which have some papers relating to engineering are: *Statistical Science – a Review Journal; Annals of the Institute of Statistical Mathematics; Applied Statistics – Journal of the Royal Statistical Society, Series C. Technometrics* is specifically concerned with the application of statistics in the physical sciences and engineering.

Abstracts and indexes

The two main sources of mathematical abstracts are *Mathematical Reviews* (also available on CD-ROM) and *Mathematics Abstracts*. Both publications consist of specially written reviews of papers, articles, books and conference proceedings.

Current Mathematical Publications contains facsimiles of tables of contents, and an index of current titles.

Current Index to Statistics – Applications, Methods and Theory is an annual publication listing articles, reviews, books and conference proceedings. An effective index allows for easy finding of publications relevant to engineering.

CompuMath Citation Index is arguably the major source for finding publications in mathematics and statistics. It is published semi-annually by the Institute for Scientific Information Inc. As well as the publications list, citation index, source index and corporate index, it has an extremely useful 'research front speciality index'. This speciality index lists the current specialist areas of research in the mathematical sciences together with the core literature in each area; recent publications referring to the core literature in any area are also listed.

News and book reviews

International Mathematical News is a German-language publication giving notification of meetings, conferences and courses, and reports on these. It also includes extensive lists of new books and book reviews.

References

Abramowitz, M. and Segun I. E. (eds) (1972) *Handbook of mathematical functions with formulas, graphs and mathematical tables*. New York: Wiley.
Aliabadi, M. H. and Brebbia, C. A. (eds) (1993) *Advanced formulations in boundary element methods*. Southampton: Computational Mechanics; Barking: Elsevier.
Arnold, L. (1974) *Stochastic differential equations – theory and applications*. New York: Wiley.
Arrowsmith, D. K. and Place, C. M. (1992) *Dynamical systems*. London: Chapman and Hall.
Arrowsmith, D. K. and Place, C. M. (1990) *An introduction to dynamical systems*. Cambridge: Cambridge University Press.
Atherton, D. P. and Borne, P. (eds) (1992) *Concise encyclopedia of modelling and simulation*. Oxford: Pergamon.
Bajpai, A. C., Mustoe, L. R. and Walker, D. (1980) *Specialist techniques in engineering mathematics*. Chichester: Wiley.
Banerjee, P. K. and Butterfield, R. (eds) (1979) *Developments in boundary element methods* Vol. 1. London: Elsevier.
Banerjee, P. K. and Butterfield, R. (eds) (1982) *Developments in boundary element methods* Vol. 2. London: Elsevier.
Banerjee, P. K. and Mukherjee, S. (eds) (1984) *Developments in boundary element methods* Vol. 3. London: Elsevier.
Banerjee, P. K. and Watson, J. O. (eds) (1986) *Developments in boundary element methods* Vol. 4. London: Elsevier.
Banerjee, P. K. and Wilson R. B. (eds) (1989) *Developments in bound-*

ary element methods Vol. 5. Industrial applications of boundary element methods. London: Elsevier.

Banerjee, P. K. and Marino, L. (eds) (1990) *Developments in boundary element methods* Vol. 6. Boundary element methods in nonlinear fluid dynamics. London: Elsevier.

Banerjee, P. K. and Butterfield, R. (1993) *Boundary element methods in engineering science*. London: McGraw-Hill.

Barnett, S. and Cameron, R. G. (1985) *Introduction to mathematical control theory* 2nd ed. Oxford: Clarendon Press.

Bathe, K.-J. (1982) *Finite element procedures in engineering analysis*. London: Prentice Hall.

Becker, A. A. (1992) *The boundary element method in engineering – a complete course*. London: McGraw-Hill.

Bettess, P. (1992) *Infinite elements*. Sunderland: Penshaw.

Beyer, W. H. (1987) *CRC Handbook of mathematical sciences* 6th ed. Boca Raton, FL: CRC Press.

Beyer, W. H. (1991a) *CRC Standard mathematical tables and formulae* 29th ed. Boca Raton, FL: CRC Press.

Beyer, W. H. (1991b) *CRC Standard probability and statistical tables and formulae*. Boca Raton, FL: CRC Press.

Bezdek, J. C. (ed.) (1987) *Analysis of fuzzy information* 3 vols. Boca Raton, FL: CRC Press.

Biggs, N. L. (1989) *Discrete mathematics* 2nd rev. ed. Oxford: Clarendon Press.

Bothe, H.-H. (1992) *Fuzzy logik – einführung in theorie und anwendungen*. Berlin: Springer-Verlag.

Brebbia, C. A. (1985) *Boundary element research*. Southampton: Computational Mechanics.

Brebbia, C. A. and Chaudouet-Miranda, A. (eds) (1990) *Boundary elements in mechanical and electrical engineering*. Southampton: Computational Mechanics.

Burden, R. L. and Faires, J. D. (1993) *Numerical analysis* 5th ed. Boston, MA: PWS-Kent.

Cakmak, A. S., Botha, J. F. and Gray, W. G. (1987) *Computational and applied mathematics for engineering analysis*. Southampton: Computational Mechanics.

Cartwright, M. (1990) *Fourier methods for mathematicians, scientists and engineers*. Chichester: Ellis Horwood.

Champeney, D. C. (1987) *A handbook of Fourier theorems*. Cambridge: Cambridge University Press.

Chatfield, C. (1983) *Statistics for technology* 3rd ed. London: Chapman and Hall.

Chui, C. K. (ed.) (1992a) *An introduction to wavelets*. London: Academic Press.

Chui, C. K. (ed.) (1992b) *Wavelets; a tutorial in theory and applications*. London: Academic Press.
Cohen, A. M. (ed.) (1993) *Computer algebra in industry*. Chichester: Wiley.
Corduneanu, C. (1991) *Integral equations and applications*. Cambridge: Cambridge University Press.
Cover, T. M. and Thomas, J. A. (1991) *Elements of information theory*. New York: Wiley.
Croft, A., Davison, R. and Hargreaves, M. (1992) *Engineering mathematics – a modern foundation for electronic, electrical and control engineers*. Wokingham: Addison-Wesley.
DeFrancis, J. (1964) *Chinese–English glossary of the mathematical sciences*. Providence, RI: American Mathematical Society.
Devaney, R. L. (1992) *A first course in chaotic dynamical systems*. Redwood City, CA: Addison-Wesley.
Dorling, A. R. (ed.) (1977) *Use of mathematical literature*. London: Butterworths.
Dupraz, J. (1986) *Probability, signals and noise*, trans. A. Howie. London: North Oxford Academic.
Englefield, M. J. (1987) *Mathematical methods for engineering and science students*. Sevenoaks: Edward Arnold.
Eisenreich, G. and Sube, R. (1982) *Dictionary of mathematics*. Amsterdam: Elsevier.
Faires, J. D. and Burden, R. L. (1993) *Numerical methods*. Boston, MA: PWS-Kent.
Farrashkhalvat, M. and Miles, J. P. (1990) *Tensor methods for engineers*. New York: Ellis Horwood.
Folland, G. B. (1992) *Fourier analysis and its applications*. Pacific Grove, CA: Brooks/Cole.
Foulds, L. R. (1992) *Graph theory applications*. Berlin: Springer-Verlag.
Friedman, A. (1988–1992) *Mathematics in industrial problems* Vols 1–5. London: Springer-Verlag with Institute of Mathematics and Its Applications.
Galton, A. (1990) *Logic for information technology*. Chichester: Wiley.
Gantmacher, F. R. (1977) *Matrix theory*. New York: Chelsea.
Gilmore, R. (1981) *Catastrophe theory for scientists and engineers*. New York: Wiley.
Gonzalez, M. O. (1992a) *Classical complex analysis*. New York: Marcel Dekker.
Gonzalez, M. O. (1992b) *Complex analysis: selected topics*. New York: Marcel Dekker.
Goult, R. J. (1978) *Applied linear algebra*. Chichester: Ellis Horwood.
Gradshteyn, I. S. and Ryzhik, I. M. (1980) *Table of integrals, series,*

and products edited, corrected and enlarged by A. Jeffrey, rev. ed. New York: Academic Press.

Gruber, R., Periaux, J. and Shaw, R. P. (eds) (1989) *Proceedings of the fifth International Symposium on Numerical Methods in Engineering*. Southampton: Computational Mechanics.

Guest, P. B. (1991) *Laplace transforms and an introduction to distributions*. Chichester: Ellis Horwood.

Hazewinkel, M. (ed.) (1988–1993) *Encyclopaedia of mathematics* 9 vols. Dordrecht: Kluwer Academic.

Helstrom, C. W. (1990) *Probability and stochastic processes for engineers* 2nd ed. New York: Macmillan.

Hill, R. (1986) *A first course in coding theory*. Oxford: Oxford University Press.

Hungarian Central Statistical Office (1962) *Statistical dictionary*. Budapest: Hungarian Central Statistical Office.

Irons, B. and Shrive, N. (1983) *Finite element primer*. Chichester: Ellis Horwood.

Ito, K. (ed.) (1987) *Encyclopedic dictionary of mathematics* 2nd ed. Cambridge, MA: MIT Press.

James, G. and James, R. C. (eds) (1992) *Mathematics dictionary* 5th ed. New York: Van Nostrand Reinhold.

Jeffrey, A. (1990) *Linear algebra and ordinary differential equations*. Oxford: Blackwell.

Jeffrey, A. (1992) *Complex analysis and applications*. Boca Raton, FL: CRC Press.

Kim, J. H. and Stringer, J. (eds) (1992) *Applied chaos*. New York: Wiley.

Kotz, S. (1971) *Russian–English/English–Russian glossary of statistical terms*. Edinburgh: Oliver and Boyd.

Kotz, S. and Johnson, N. L. (eds) (1982–1989) *Encyclopedia of statistical sciences* 9 vols. New York: Wiley.

Kovach, L. D. (1982) *Advanced engineering mathematics*. Reading, MA: Addison-Wesley.

Kurtz, M. (1991) *Handbook of applied mathematics for engineers and scientists*. New York: McGraw-Hill.

Lapin, L. L. (1990) *Probability and statistics for modern engineering* 2nd ed. Boston, MA: PWS-Kent.

Ledermann, W. (ed.) (1991) *Handbook of applicable mathematics* 8 vols. Chichester: Wiley.

Lighthill, M. J. (1958) *Introduction to Fourier analysis and generalised functions*. Cambridge: Cambridge University Press.

Lord, E. A. and Wilson, C. B. (1984) *The mathematical description of shape and form*. Chichester: Ellis Horwood.

Marek, M. and Schreiber, I. (1991) *Chaotic behaviour of deterministic dissipative systems*. Cambridge: Cambridge University Press.

Marriott, F. H. C. (ed.) (1990) *A dictionary of statistical terms* 5th ed. Harlow: Longman.
Nutbourne, A. N. and Martin, R. R. (1988) *Differential geometry applied to curve and surface design* Vol. 1. Foundations. Chichester: Ellis Horwood.
Oden, J. T. (1979) *Applied functional analysis.* Englewood Cliffs, NJ: Prentice Hall.
Pande, G. N. and Middleton, J. (eds) (1990) *NUMETA90 – numerical methods in engineering: theory and applications.* Barking: Elsevier.
Peitgen, H. O., Jurgens, H. and Saupe, D. (1992) *Chaos and fractals.* New York: Springer-Verlag.
Press, W. H. et al. (1986) *Numerical recipes.* Cambridge: Cambridge University Press.
Quinney, D. (1987) *An introduction to the numerical solution of differential equations* rev. ed. Letchworth: Research Studies Press.
Rade, L. and Westergren, B. (1990) *Beta mathematics handbook* 2nd ed. Boca Raton, FL: CRC Press.
Rao, S. S. (1989) *The finite element method in engineering* 2nd ed. Oxford: Pergamon.
Reddy, J. N. and Rasmussen, M. L. (1982) *Advanced engineering analysis.* Chichester: Wiley.
Reddy, J. N. (1984) *Energy and variational methods in applied mechanics.* Chichester: Wiley.
Roman, S. (1992) *Coding and information theory.* Berlin: Springer-Verlag.
Ross, S. M. (1989) *Introduction to probability models* 4th ed. Boston, MA: Academic Press.
Rota, G.-C. (ed.) (1976–1993) *Encyclopedia of mathematics and applications (1976–1993).* (Vols 1–23) Reading, MA: Addison-Wesley (Vols 24–) Cambridge: Cambridge University Press.
Scheaffer, R. L. and McClave, J. T. (1990) *Probability and statistics for engineers* 3rd ed. Boston, MA: PWS-Kent.
Schilling, R. J. and Lee, H. (1988) *Engineering analysis: a vector space approach.* Chichester: Wiley.
Schwarz, H. R. (1988) *Finite element methods,* trans. C. M. Whiteman. London: Academic Press.
Seggern, David von (1993) *CRC standard curves and surfaces.* Boca Raton, FL: CRC Press.
Setian, L. (1992) *Engineering field theory with applications.* Cambridge: Cambridge University Press.
Smith, G. N. (1986) *Probability and statistics in civil engineering.* London: Collins Professional and Technical.
Sneddon, I. N. (ed.) (1976) *Encyclopaedic dictionary of mathematics for engineers and applied scientists.* Oxford: Pergamon.

Solymar, L. (1988) *Lectures on Fourier series.* Oxford: Oxford University Press.
Soong, T. T. (1981) *Probabilistic modeling and analysis in science and engineering.* Chichester: Wiley.
Stanat, D. F. and McAllister, D. F. (1977) *Discrete mathematics in computer science.* Englewood Cliffs, NJ: Prentice Hall.
Tanaka, M., Brebbia, C. A. and Shaw, R. (eds) (1990) *Advances in BEM in Japan and USA.* Southampton: Computational Mechanics.
Tuma, J. J. (1989) *Handbook of numerical calculations in engineering.* London: McGraw-Hill.
Tyn, Myint-U. and Debnath, L. (1987) *Partial differential equations for scientists and engineers* 3rd ed. New York: North-Holland.
Wait, R. and Mitchell, A. R. (1985) *Finite element analysis and applications.* Chichester: Wiley.

CHAPTER TWENTY-SEVEN

Marine technology

C. KUO

What is MARINE technology?

For many years 'marine technology' was the term used to refer to engineering activities relating to marine sciences, for example, instrumentation required to measure various properties of the ocean. The more traditional marine activities came under the following headings:

Naval architecture: This term denotes the science and engineering of a floating ship. Typical subjects covered range from ship design, hydrostatics, resistance, propulsion, strength and motions, to general arrangement, contracting and commissioning. The word 'architecture' was adopted to convey the idea of tasks involving both aesthetic and technical features.

Marine engineering: This term is employed to refer to technical items relating to the machinery of a ship. Typical topics include the design of main and auxiliary power plants, transmission of the power to the propeller, vibration of the system and methods of controlling the various functions of the machinery. Planned maintenance is also an important subject.

Nautical studies: This term is used to deal with subjects such as ship operation, navigation and associated activities.

Clearly, the main thrust of these disciplines is towards the technical aspects of the ship. In the present context this is recognized as a floating body with a sharp leading end (the bow) for reducing resistance as it goes forward and a 'rounded' form (stern) in the trailing end where the propulsive device is situated.

As exploration for and production of oil and gas moved from land to offshore locations, and particularly as a result of the hydrocarbon activities in the North Sea in the late 1960s and 1970s, it was

468 Marine technology

recognized that the above terms were too restrictive to represent all scientific and technological activities in the widest possible sense. This led to the following terms being introduced:

Ocean engineering: The term is used to represent all engineering activities associated with oceans, involving in particular the exploitation of living and non-living resources.

Off shore engineering: This term is suggested as an alternative to 'ocean engineering', which is regarded as too broad, and to place emphasis on those activities nearer to the coastline.

Ocean technology: This is another variation of the terms ocean engineering and off shore engineering, but places the emphasis on technologies from both the scientific and the engineering points of view.

In the early 1970s when the Science and Engineering Research Council decided to designate hydrocarbon exploration and exploitation in the North Sea as a priority area, and a suitable name had to be selected to encapsulate the vast range of activities involved, 'marine technology' was the term chosen, and over the subsequent two decades it has taken on a new role that embraces all the traditional disciplines and the development of others which are needed for the responsible exploitation of off shore resources.

What, therefore, is 'marine technology'? A suitable definition is given here, as follows:

All scientific and engineering aspects associated with the responsible exploitation and management of ocean resources, involving tasks ranging from performing work in the ocean, transporting goods and people and providing for leisure activities, to cost-benefit decision making, conservation and environmental pollution control.

Two points are worth noting. Firstly, this is a very broad subject which interacts with many other disciplines. Secondly, although it is mainly concerned with scientific and engineering aspects, marine technology also has to take into account the implications of economic, social and ecological factors.

Approach and topic range

In view of the wide scope of marine technology it was felt necessary to devise an approach that would provide ready assistance in gaining access to information on the subject. After exploration of various possibilities, it was decided to adopt a strategy comprising the following steps:

Step 1: Definition of objective. Start by defining the objective for which the information is required, e.g. to gain an appreciation of the subject, to obtain access to research advances, etc.

Marine technology 469

Table 27.1. General topic guide.

Ref. No	Main topic heading	Related subject areas
1	CONSTRUCTION	Assembling; Burning; Cutting; Quality Control; Welding
2	DESIGN	Concept; Detail; Feasibility; Decision; Information
3	LEGISLATIVE MATERIAL	Insurance; Law; Liability; Regulations
4	MARINE CONTROL SYSTEMS	Lifting Surfaces; Manoeuvring; Steering
5	MARINE ENGINEERING	Auxiliary Engines; Power Plants; Transmission
6	MARINE HYDRODYNAMICS	Flow; Propulsion; Resistance; Streamlined; Velocity Potential
7	MARITIME BUSINESS	Economics; Finance; Marketing; Management
8	MATERIALS FOR MARINE USE	Composites; Corrosion; Properties; Steel
9	MOTIONS AND DYNAMICS	Heave; Roll; Pitch; Sway; Surge; Yaw
10	OPERATIONS	Communications; Navigation; Planned Maintenance; Training
11	SAFETY	Dangerous Cargo; Human Factors; Methodology; Stability
12	STRUCTURES	Analysis; Finite Element Methods; Strain; Stress
13	SURFACES AND FORMS	Fairing; Graphics; Hull Shape; Surface Fitting
14	TRANSPORTATION	Cargo Handling; Freight Rates; Ports
15	VIBRATION	Acceleration; Amplitude; Excitation; Natural Frequency

Step 2: Subject of interest. Select the subject of interest with the aid of the Topic Guide contained in Tables 27.1 and 27.2. Table 27.1 covers general topics such as design and hydrodynamics while Table 27.2 deals with more specialized aspects such as ocean resources. Each main topic area is supplemented by a list of major related subject areas.

Step 3: Examine the classified sources. Sources are classified in such a way that the usefulness of each category can be examined to see how it would meet the stated objective.

Step 4: Concentrate on the principal source. Once the principal source is identified the reader's search should be concentrated on this source before moving to examine others.

This chapter will concentrate on Step 3.

470 *Marine technology*

Table 27.2. Special topic guide

Ref. No	Main topic heading	Related subject areas
16	ENVIRONMENTAL PROTECTION	Control; Detection; Pollution; Protection Techniques
17	MAINTENANCE AND REPAIR	Manned; Planned Techniques; Unmanned
18	MARINE COMMISSIONING	Sea Trial; Techniques; Topside Facilities; Tow Out
19	MARINE DECOMMISSIONING	Clean-up; Environmental Effects; Equipment; Techniques
20	OCEANOGRAPHY	Current; Seabed Properties; Soil Mechanics; Waves; Wind
21	OCEAN RESOURCES	Chemical; Living; Mineral; Non-living; Plant
22	OFFSHORE INSTALLATIONS	Fixed; Floating; Semi-flexible; Semi-submersible
23	OILFIELD DEVELOPMENT	Exploratory Drilling; Extraction Rate; Prospecting; Reserves; Secondary Recovery
24	PIPELINE TECHNOLOGY	Cleaning; External Protection; Installation; Maintenance
25	POLAR TECHNOLOGY	Ice Breakers; Ice Properties; Low Temperature Work; Supply
26	PROCESSING FACILITIES	Failure Detection; Operational Control; Plant Design
27	PRODUCTION AND PLANNING	Critical Path Methods; Handling; Procurement; Scheduling
28	SHIPS AND MARINE CRAFT	Function; Specification; High Speed; Ship Types
29	SUBSEA WORK & INSTALLATIONS	Inspection; Installation; Power Supply; Repair
30	UNMANNED VEHICLES & ROBOTICS	Free-swimming; Positioning; Tethered Vehicles

Classification of publication sources

There are many ways of classifying publication sources, and after careful consideration it was decided to group these into four types, for the following reasons:

Publications in book form

Books are classed into the following categories:

TEXT BOOKS

These books provide 'proven' information to the readers. However, because it takes time to write books, the authors tend to concentrate

Marine technology 471

on basics and on material that will last for some time. These serve as a starting point for the subject of interest.

BOOKS FROM CONFERENCES

The main output of most conferences consists of one or more volumes of papers reporting on research advances and results, usually referred to as conference proceedings. These can be very useful to people working on similar or other related research projects. Other types of paper given include: keynote addresses on specific major topics, reviews providing state-of-the-art analyses, fresh methodologies for tackling problems, etc.

REFERENCE BOOKS

These books are usually found in libraries and other central points, providing a wide range of basic factual information as required.

Publications by professional institutions

Professional institutions publish in the following main categories;

MAJOR WORKS

The materials included in this category provide the members of the institutions concerned with technical advances and proven information. Their role is similar to that of texts and books from conferences.

JOURNALS

Most research advances are published in professional journals and periodicals. The orientation of some journals can be very theoretical, but others focus more on practical topics.

OTHER FORMS OF PUBLICATION

These usually contain material stemming from the work of specialized groups or committees and are published as notes, codes of practice or reports for the membership of a particular institution.

Publications from specialized maritime organizations

These organizations specialize in maritime matters. The main groups include:

GOVERNMENTS AND RELATED BODIES

Materials from these sources relate to legislative matters, the findings of public inquiries on disasters, etc. These provide useful background information for many types of practical usage.

CLASSIFICATION SOCIETIES

These organizations publish detailed guidance standards to be met in the construction of ships and other marine vehicles. They also produce technical reports on particular topics, and these can often be obtained on request to the society.

RESEARCH INSTITUTIONS

The reports of these bodies are often a valuable source of information for researchers and practitioners. Some reports, however, are classified as 'confidential' and are thus available only to the sponsors of the project in question.

ACADEMIC INSTITUTIONS

The reports here fall into two main groups, usually called research reports and contract reports. The former may be obtained from the body concerned, but more often their research results are published in technical journals, the transactions of professional institutions, conference proceedings, etc. The latter are generally confidential to the sponsors.

OTHER ORGANIZATIONS

These organizations issue specialized reports on particular subjects, which can generally be purchased direct from the organization's offices. However, they can be quite expensive as the number of copies of a given report is usually very limited.

General publications

In this group consideration is given to publications which do not readily fit into any of the other three. The sub-headings used include the following:

ABSTRACTS

A number of organizations provide abstracts of publications in order to condense the volume of information that needs to be read. These may be drawn from both national and international sources, providing researchers and senior research managers with precise information and offering them a quick and easy way of keeping up-to-date with peripheral fields.

NEWSPAPERS AND NEWSLETTERS

These sources provide a considerable amount of topical information, and can give a lead to other sources as well.

TRADE JOURNALS

Useful and up-to-date design and operational information and commercial data on equipment can be obtained from these periodicals.

SPECIALIZED SUBJECT JOURNALS

Researchers will find these journals a useful source of advanced research results.

COMMERCIAL BROCHURES

These brochures are produced for the guidance of possible purchasers of equipment, but they provide valuable data for other interested readers and researchers. However, as these are of an ephemeral nature and of only short-term interest, non-current items may be difficult to obtain.

OTHER FORMS OF PUBLICATION

Useful general, marine-related information can be obtained from this source, often written for the non-technical audience and highly readable as a consequence.

Each of the four main sections will now be considered separately.

Publications in book form

Textbooks

Compared with many other main-stream types of engineering, marine technology has only a limited range of textbooks. One of the main reasons for this is the size of the market and the effort needed to write a book. However, some useful books have been written for teaching purposes and others dealing with research. Typical examples include the following:

Rawson and Tupper (1983) is a general textbook for use in undergraduate courses, which has been revised several times since it was first published.

Barltrop and Adams (1991) includes many case examples in support of a comprehensive study.

Baxter (1992) is a reprint of a book first published in 1959, which is very useful for anyone wishing to learn the basics of the subject.

Faltinsen (1990) is a book for both undergraduate and postgraduate use. Some of the material is highly advanced, and would be helpful to researchers as well.

Kuo (1992) (which contains many marine-related examples) is a general text for engineers.

474 *Marine technology*

Books from conferences

The modern trend is to print the proceedings of conferences, particularly those of international conferences, in the form of a book. In many cases these are published by companies specializing in conference material related to marine technology. The main special contributions of this group are *special subject series*. The better known series include:

Proceedings of the International Ship Structures Congress (ISSC). The conferences have been held on a three-yearly basis since 1961. The 1991 conference was held in China and the most recent *Proceedings* are edited by Hsu and Wu (1991) and 1994 in St John's Newfoundland, Canada.

Proceedings of the International Conference on Computer Aided Shipbuilding (ICCAS). The series began in 1973. It comes under the aegis of the International Federation for Information Processing (IFIP) and is published by North-Holland. The most recent conference in this three-yearly series was held in Brazil in 1991 and the *Proceedings* were edited by Vieira *et al.* (1992).

Proceedings of the Conferences on the Stability of Ships and Ocean Vehicles. This series began in 1975 but has not adopted a fixed three- or four-year cycle. The fourth conference was held in Naples, Italy, in 1990 and the fifth in Florida, USA in 1994. The most recent *Proceedings* are edited by Cassella *et al.* (1990).

Proceedings of the Conference Series on the Practical Design of Ships (PRADS). The most recent was held at Newcastle upon Tyne, UK, in 1992 and the Proceedings were edited by Caldwell and Ward (1992).

In the offshore topic areas there are various annual or biennial conferences which constitute a major source of information. The key ones are as follows:

The Offshore Technology Conference (OTC). This annual conference has been held in Houston, Texas, during the first week of May since 1970, and attracts attendances of up to 25 000 for the presentations and exhibitions. The annual publication consists of four volumes of technical reports covering up to sixty subject sessions.

Offshore Europe. A European equivalent to OTC is the Offshore Europe series which takes place every two years in Aberdeen, Scotland. The output of this conference is a two-volume set of reports which concentrates on specific areas, e.g., subsea processing, safety, exploration, inspection, maintenance and repair.

Reference books

Books published in this category serve as reference for a number of purposes. Typical examples include the following, all of which are published annually:

Books about ships and offshore structures, e.g.: Jane's books entitled: *Jane's fighting ships, Jane's ocean technology.*

Marine technology 475

Directories of marine facilities, e.g.: *Lloyd's ports of the world*; *The offshore service vessel register*.
Directories of marine equipment and supplies, e.g.: *Worldwide pipelines and constructors' directory*, *Offshore fire safety*.
Vessel information, e.g.: *Standby vessels of the world*; *ROV review*.

Publications of professional institutions

Basic activities

Professional institutions such as the Royal Institution of Naval Architects (RINA), the Institute of Marine Engineers (IMarE) and the Society of Naval Architects and Marine Engineers (SNAME) are all involved in a wide variety of activities. These range from setting standards and serving the needs of members to responding to requests for information from external sources, and representing their membership on other related bodies. However, one of the high-profile activities is the publication of books, journals, conference papers and reports. For this reason professional institutions are a very good source of information. Although the range of publications varies enormously from one organization to another, it is always worthwhile to examine what is available on a topic of interest in the areas of major works, journals, and so on. Table 27.3 gives a selected international list of institutions that actively publish technical information.

TABLE 27.3.

Country	Names of institutions
France	Association Technique Maritime et Aeronautique
Germany	Schiffbautechnische Gesellschaft
Italy	Associazione Italiana di Tecnica Navale
Japan	The Society of Naval Architects of Japan
Korea	The Society of Naval Architects of Korea
The Netherlands	Koninklijke Instituut van Ingenieurs Afdeling Maritieme Technologie
United Kingdom	Institution of Engineers and Shipbuilders in Scotland (IESS)
	The Nautical Institute (NI)
	The Royal Institution of Naval Architects (RINA)
	The Society for Underwater Technology (SUT)
United States	American Society of Mechanical Engineers (ASME)(Ocean Engineering Section)
	International Society of Ocean and Polar Engineering (ISOPE)
	The Marine Technology Society (MTS)
	The Society of Naval Architects and Marine Engineers (SNAME)
	The Society of Naval Engineers (SNE)

Major works

Under this heading two types of publication are considered: transactions and books. Transactions are a collection of papers presented at a meeting of an Institution or prepared for written discussion, usually by some of its members. They appear annually, and typical examples would be:

Transactions of the Royal Institution of Naval Architects
Transactions of the Society of Naval Architects And Marine Engineers

Some institutions also publish books by their members, which are sold to both members and non-members, with the former getting reduced rates. Typical examples of SNAME'S publications include: Lewis (1988), Hughes (1988), Allmendinger (1990), Benford (1991) and Harrington (1992).

In addition, SNAME is the only institution which publishes a complete catalogue of all the papers published by it up to a given date.

Journals

Many institutions also publish journals on a regular basis of four to six times a year. The better-known journals include the following:
General areas of interest:

Journal of SNAJ (SNAJ) biannual.
Marine Technology (SNAME) quarterly.
Naval Architect (RINA) 10 per annum.
Marine Engineers' Review (IMarE) monthly.
Seaways (NI) quarterly.
Offshore Technology (IMarE) quarterly.
Journal of MTS (MTS) quarterly.

Special areas of interest:

Journal of Ship Research (SNAME) quarterly.
Journal of Ship Production (SNAME) quarterly.
Underwater Technology (SUT) quarterly.
Ship & Boat International (RINA) quarterly.
Offshore Engineer (ICE) monthly.

Other forms of publication

There are many other forms of publication by the professional institutions, and some of the main ones are highlighted here.
Special conference papers:

Fourteenth STAR symposium: 21st century – ship and offshore vessel design, production and operation (1989) (SNAME).
Warship 93 – new developments in naval submarines (1993) (RINA).

Research reports:

Notes on ship slamming: Panel HS-2 report (1993) (SNAME).
Ship design bulletins: SWATH ships (1993) (SNAME).

Publications of specialized maritime organizations

Governments and related bodies

This heading would embrace all governments with a significant maritime interest, e.g. having ships sailing under its flag or with oil-drilling in its national waters. It would also include the International Maritime Organization, which comprises representatives of all governments with maritime interests. Examples of their publications would include the following.

Books and booklets. A typical example is: *Successful health and safety management* (1991) (HMSO).

Codes of Practice. A typical example is: *A guide to the offshore installations (safety case) regulations* (1992) (HMSO).

Statutory regulations. A typical example is: *Statutory instruments: (dangerous goods) regulations* (1990) (HMSO).

Commission of inquiry reports. The best known recent example is: *The public inquiry into the Piper Alpha disaster* Cullen Report, Cm.1310, 2 vols (1990) (HMSO).

Classification societies

These societies classify ships and offshore installations so that given standards can be met and, in general, every major maritime nation has its own classification society. Publications include the following:

RULES FOR CONSTRUCTION

These are published in a form that will enable organizations to build ships and offshore structures according to the rules of the chosen society, e.g.:

Rules for building and classing of steel ships (American Bureau of Shipping) annual.
Rules for classification: high speed and light craft (Det norske Veritas) annual.

TECHNICAL REPORTS

Reports on research projects carried out by societies are often published at their conferences, but they also have technical reports prepared for internal use, e.g.:

Ferguson, J. M., Cheng, Y. F. and Purdey, B. (1992–1993) *On Bulkhead Carrier Safety* Paper No.5 (Lloyd's Register).

ANNUAL REPORTS

Each society usually publishes an annual report providing factual information on its activities, finance and organization, e.g.:

Annual Report of Nippon Kaiji Kyokai (1991) (NK – Register of Shipping) This provides useful data on the location of their offices around the world, and the different committees in operation. It also contains a number of technical articles.

MARITIME GUIDES

Typical examples are:

Maritime guide (Lloyd's Register of Shipping). Gives data on shipbuilders, marine insurers, drydocks, etc., around the world and is published annually.

Register of ships (Lloyd's Register of Shipping). Published annually and contains full data about every ship registered with the organization – from previous owners, to hull data and types of machinery installed.

Research organizations

In most countries there are research organizations with a prime interest in technical research, whose principal output consists of technical reports together with charts and tables prepared on the basis of experimental studies. Research organizations with significant publications in the marine/maritime sector include the following:

MARIN (The Marine Research Institute of the Netherlands)
MARINTECH (Marine Technology Centre, Trondheim, Norway)
British Maritime Technology, UK and its associated companies
The David W. Taylor Naval Ship Research and Development Center, Washington, DC, USA
The Danish Ship Research Institute, Lyngby, Denmark
The Krylov Shipbuilding Research Institute, St Petersburg, Russia
The Technical Institute of Research, Tokyo, Japan
The National Research Council of Canada, St John's Newfoundland, Canada
Hamburg Ship Model Basin, Hamburg, Germany.

Academic institutions

From the information point of view, the principal output of academic institutions usually consists of technical reports and lecture notes. The majority of the former eventually become papers published at conferences or in journals. Generally academic institutions that incorporate research institutes of various kinds will have an extensive range of technical reports available to other interested parties, e.g.:

International Shipbuilding Progress published quarterly by Delft University Press, Delft, The Netherlands.

Examples of other institutions active in the marine/maritime field include:

Kyushu University, Fukuoka, Kyushu, Japan
Tokyo University and its Industrial Institute, Tokyo, Japan
The Norwegian Technical University, Trondheim, Norway
The University of Michigan, Ann Arbor, MI, USA.

Another form of publication arises from a group of universities offering courses on specialized subjects, using their combined expertise. One such active group in Europe is Western European Graduate Education in Marine Technology (WEGEMT). Nearly twenty schools have been held under its auspices since 1978. One of the more recent major sets of lecture notes is entitled: *Underwater technology* (1991) MTD Ltd, Helsinki, Finland.

Other organizations

A number of private companies specialize in maritime activities, and the various types include the following:

MARKET RESEARCH ORGANIZATIONS

Their reports provide forecasts on the demand for ships or trades and other trading information. These reports are usually very expensive to purchase because they are highly specialized and their 'shelf-life' is relatively short, e.g.: *Shipping Statistics and Economics* (Drewry Shipping Consultants) monthly.

USER ASSOCIATIONS

Associations are formed by particular groups of people to look after their interests, and these organizations bring out publications of various kinds. One example is the UK Oil Operators' Association (UKOOA), which is a 'trade' organization for oil companies. It publishes reports by its committee and makes press releases of opinions on specific issues.

General publications

Abstracting publications

A number of organizations specialize in providing abstracts of reports, technical papers and articles. These are gathered together as publications to be read by whoever requires to keep up-to-date with advances. The better known of these include:

480 Marine technology

BMT Abstracts formerly called the *Journal of Abstracts of the British Ship Research Association* (British Maritime Technology).
Cargo Handling Abstracts (International Cargo Handling Coordination Association).
International Petroleum Abstracts incorporating *Offshore Abstracts* (Wiley).
Maritime Abstracts (National Maritime Research Center).

Newspapers and newsletters

A regular update on marine affairs can be obtained from newspapers specializing in general maritime affairs, e.g.:

Lloyd's List International (Lloyd's of London Press) The most respected daily newspaper in this category. It provides the total range of information including movements of ships, accidents and topics of current interest.
Maritime Monitor (Rigas Publishing) A weekly international shipping and trade paper published in Piraeus, Greece.
Fairplay International Shipping Weekly (Fairplay Publications) Offers shipping news and statistics and is published weekly.

The second form of publication is the newsletter. These fall into two main groups. Those in the first group are usually published as a supplementary feature of a conventional newspaper in order to concentrate on a special topic, e.g.:

North Sea Letter (Financial Times Business Information) Its contents include the status of various oilfields and other topical news of field developments. It is published on a biweekly basis.

The newsletters in the second group are published by an individual company and circulated to its members and clients. These provide news of interest from the company and selected technical material, e.g:

Newsletter (MTD Ltd).
Newsletter (Metocean, Ltd).
Oil and Gas Technology (Petroleum Science and Technology Institute for the European Commission).

Trade journals

The publications under this heading are journals which contain a mixture of short articles, information about equipment, regional news items, production data, reviews of new technology and extensive advertising material. Typical examples of monthly journals would be:

Offshore incorporating *The Oilman* (Pennwell Publishing).
Marine Log (Simmons-Boardman).
Safety at Sea International (International Trade Publications).

Special subject journals

Typical examples of this group include:

Ocean Engineering (Pergamon) bimonthly.

Marine Structures, Design Construction and Safety (Elsevier) bimonthly.
Schiffstecknik (Schiffahrts-Verlag Hansa) bimonthly.

Commercial brochures

Valuable information can often be gathered from a study of promotional brochures produced by organizations in the area of interest. These vary greatly in size, shape, design and quality. There are, however, a number of common features, and the key ones are: insight into the organization's policies and objectives; presentation of its major activities, often in some detail; a list of the main clients and users, or sectors, being targeted; ready availability at exhibitions or by writing to the organization concerned. The following is a description of a typical example:

NSW Offshore is a division of Nordeutsche Seekabelwerke AG, and its brochure carries an attractive sketch of the products supplied to the offshore industry – submarine cables, electromagnetic cables, umbilicals for diving and underwater vehicles. Further details of their products are given in Factsheets that include information about materials used, specifications, and drawings of cross-sections of the various types of cable.

Other forms of publication

Under this can be found a variety of items, but three are of particular significance, and these are:

GRAPHICAL DISPLAYS

These include maps, charts and sketches. A typical example would be:

Oil Map produced by the Highlands and Island Development Board, Scotland. This map shows the various fields in the UK sector of the North Sea on a block-by-block basis, and also contains some publicity material.

COMPANY MAGAZINES AND ANNUAL REPORTS

Most large companies, particularly the multinationals, publish magazines containing well-written articles about the organization's business and some of its activities, e.g.:

Horizon (BP Exploration) biannual.
Marine News (World Ship Society) monthly.

The annual report of a company is prepared for the shareholders, and contains a review of the company's activities over the preceding year together with a copy of its financial accounts. It often provides a valuable insight into the company, e.g.:

Annual Report of P&O Shipping.
Annual Report of Total Marine.

Acknowledgements

I wish to thank Dr H. Cargill Thompson and Miss C. Hutcheon for their assistance in the preparation of this text.

Bibliography

Allmendinger, E. E. (1990) *Submersible vehicle systems design.* Jersey City, NJ: Society of Naval Architects and Marine Engineers.
Barltrop, N. D. P. and Adams, A. J. (1991) *Dynamics of fixed marine structures* 3rd ed. Oxford: Butterworth-Heinemann.
Baxter, B. (1992) *Teach yourself naval architecture.* Warsash, Southampton: Warsash Nautical Bookshop.
Benford, H (1991) *Naval architecture for non-naval architects.* Jersey City, NJ: Society of Naval Architects and Marine Engineers.
Caldwell, J. B. and Ward, G. (eds) (1992) *Practical design of ships and mobile units.* 5th International Symposium on Practical Design of Ships and Mobile Units University of Newcastle upon Tyne, 2 vols. Barking: Elsevier.
Cassella, P. et al. (eds) (1990) *Proceedings of STAB 90 – The Fourth International Conference on the Stability of Ships and Ocean Vehicles,* Department of Naval Engineering of University 'Federico II' of Naples.
Churchill, R. and Ulfstein, G. (1992) *Marine management in dispute areas: the case of the Barents Sea.* London: Routledge.
Earney, F. C. F. (1990) *Marine mineral resources.* London: Routledge.
Faltinsen, O. M. (1990) *Sea loads on ships and offshore structures.* Cambridge: Cambridge University Press.
Harrington, R. L. (ed.) (1992) *Marine engineering* rev. ed. Jersey City, NJ: Society of Naval Architects and Marine Engineers.
Howard-Williams, J. (1988) *Sails* 6th ed. London: Adlard Coles.
Hsu, P. H. and Wu, Y. S. (eds) (1991) *Proceedings of the 11th International Ship Structures Congress.* Barking: Elsevier.
Hughes, O. F. (1988) *Ship structural design: a rationally-based computer aided optimization approach.* Jersey City, NJ: Society of Naval Architects and Marine Engineers.
Kuo, C. (1992) *Business fundamentals for engineers.* London: McGraw-Hill.
Lewis, E. V. (ed.) (1988) *Principles of naval architecture* 3 vols. Jersey City, NJ: Society of Naval Architects and Marine Engineers.
Lloyd, A. R. J. M. (1989) *Seakeeping: ship behaviour in rough weather.* Hemel Hempstead: Ellis Horwood.
Marcus, H. S. (1986) *Maritime transportation management.* London: Croom Helm.

Morgan, N. (ed.) (1990) *Marine technology reference book*. Sevenoaks: Butterworths.
Rawson, K. J. and Tupper, E. (1983) *Basic ship theory* 2 vols, 3rd ed. Harlow: Longman.
Stopford, M. (1992) *Maritime economics*. London: Routledge.
Storch, R. L., Hammon, C. P. and Bunch, H. M. (1988), *Ship production*. Centreville, MD: Cornell Maritime Press.
Vieira, C., Martins, P. and Kuo, C. (1992) *ICCAS 91 – computer applications in the automation of shipyard operation and ship design*. Amsterdam: North-Holland.

CHAPTER TWENTY-EIGHT

Environmental control engineering

G. K. ANDERSON AND D. J. ELLIOTT

The general professional title 'environmental engineer' is so wide-ranging in its meaning that it is not possible to study or practice in depth all of those aspects to which the title refers. Environmental engineering literature is also produced at such a rate as to make it impossible for any one person to keep pace with it, let alone appreciate all of the implications of the books and technical papers published in the field.

It has been found necessary here to be selective in noting specific examples of sources of information and also to divide the subject into groupings within which the environmental engineers will normally specialize. It is thereby hoped that by doing this, anyone wishing to carry out a literature search in their specialized area will be able to begin the investigation using only a limited number of specific but valuable sources of information. Only texts published in English are quoted, but this is not intended to imply that there are not many sources available in other languages. It would also be remiss not to make reference to the increasing use of computerized literature search systems which are now available in most technical libraries and which have been designed to provide a comprehensive and wide-ranging database of papers based upon the use of keywords. Of particular interest are the following CD-ROM packages :

AQUALINE: which covers many aspects of water-related engineering and is particularly useful for water and wastewater treatment, and environmental engineering in general. It currently goes back to 1960 and each record provides an abstract of the relevant paper.

COMPENDEX: which is the computerized version of *Engineering Index* providing bibliographic references and abstracts of a wide range of journal articles, conference papers, books, etc. and covering most aspects of engineering.

NATIONAL TECHNICAL INFORMATION SERVICE: (NTIS) database contains information, including abstracts, on US government reports from 1980 onwards. These have been sponsored by organizations such as the National Aeronautical and Space Administration (NASA), the Environmental Protection Agency and over 300 others. The database has a bias towards engineering and technology.

CURRENT TECHNOLOGY INDEX: contains details of articles published in British periodicals on engineering and technology going back to 1980 but does not give abstracts.

BIOLOGICAL ABSTRACTS: covers biological and biomedical journals.

Environmental engineering has its origins rooted mainly in the nineteenth century among those scientists and laymen who were able to see the need for cleaning up the air, water and land around them, but who at that time did not have either the technology or legislative power available to see that their demands were carried out. One outcome of this was that the work fell between that of the chemical engineer and the civil engineer and as a result has often been an art rather than a science. It is only over the past 20 years or so that environmental engineers have organized themselves into a recognized profession in their own right, and with this has come a vast improvement in the quality of literature specifically produced for and often by the environmental engineer.

Examination and standards

Since much of the work is associated in some way or other with the protection of the environment, it is essential that the quality of waters, wastewaters, air, etc. be measured by recognized, reproducible standard techniques. A number of textbooks are available which enable an analyst to carry out standard examination procedures which will be identical with those performed by other analysts in a variety of widespread laboratories. In water and wastewater analysis the most widely used books are Greenberg, Clesceri and Eaton (1992) and the many booklets in the continuing series *Methods for the examination of waters and associated materials (1976–)*. For similar analysis in saline water, Parsons et al. (1984) has proved invaluable. Further analytical techniques and their application may be found in the following textbooks:

Sawyer, C. N., McCarty, P. L. and Parkin, G. F. (1994) *Chemistry for environmental engineering*, 4th ed. (McGraw-Hill).

Sterritt, R. M. and Lester, J. N. (1988) *Microbiology for environmental and public health engineers* (Spon).

With respect to the setting, designing for and implementation of standards, it may be necessary to study Acts of Parliament, Bills of Congress and EU Directives, etc. in order to determine those standards applicable, but a number of references may be used as guidelines. For example, the Control of Pollution Act (1974) and the Environmental Protection Act (1990) are invaluable in water pollution control but two books in particular provide invaluable sources of information:

M. J. Suess (ed.) *Examination of water for pollution control: a reference handbook*, 3 vols (Pergamon).

Harrison, R. M. (ed.) (1990) *Pollution: causes, effects and control*, 2nd ed. (Royal Society of Chemistry).

Water pollution

Textbooks

Although it is often difficult to decide exactly into which category a particular textbook comes, there are certain which may be considered to deal basically with water pollution and its many facets. In the past, the most widely used of the many books available were the three volumes by Klein (1959, 1962, 1966). Recently, a more mathematical and scientific approach has been made to the study of water pollution, notably by James (1992), and Mason (1991).

Other, more specific books relating to water pollution include:

Abel, P. D. (1989) *Water pollution biology* (Ellis Horwood).
Alabaster, J. S. and Lloyd, R. (1980) *Water quality criteria for freshwater fish* (Butterworths).
Boon, P. J., Calow, P. and Petts, G. E. (eds) (1992) *River conservation and management* (Wiley).
Moss, B. (1988) *Ecology of fresh waters*, 2nd ed. (Blackwell).
Wilson, J. G. (1988) *The biology of estuarine management* (Croom Helm).

Journals

No satisfactory journal exists which deals exclusively with water pollution as such, but there are many which touch on the subject either in detail or briefly, including:

Environmental Pollution
Environmental Science and Technology
European Environmental Law Review
Integrated Environmental Management
Journal of Environmental Engineering (ASCE)
Journal of Environmental Management
Marine Pollution Bulletin
Water and Environmental Management
Water Environment Research
Water International
Water Quality International
Water Research
Water Science and Technology

Wastewater treatment

Possibly nowhere in environmental engineering is a particular subject covered by so many books as in wastewater treatment. Since each situation is different, depending upon location and wastewater characteristics, the ideal design book does not exist although a number provides both theoretical and practical approaches to the problem. Overall, Metcalfe and Eddy Inc. Staff and Tchobanglous (1990) is the most comprehensive and most widely used text. Hammer (1986) and Tebbutt (1992) provide useful information while Reynolds (1982) covers the subject area from a unit process design point of view.

Of a more specific nature the following books should be consulted :

Vesilind, P. A. (1980) *Treatment and disposal of wastewater sludges*, 2nd ed. (University of Michigan Press).
Horan, N. J. (1990) *Biological wastewater treatment systems* (Wiley).
Cooper, P. F. and Findlater, B. C. (eds) (1990) *Constructed wetlands in water pollution control* (Pergamon).
Read, G. F. and Vickridge, I. (eds) (1992) *Sewers: rehabilitation and new construction* (Edward Arnold).

A set of books which covers all aspects of wastewater treatment plants has been published in the USA with the general title *Operation of municipal wastewater treatment plants* (1990).

A number of books has also been written dealing with wastewater engineering problems in developing countries, notably:

Mara, D. D. (1976) *Sewage treatment in hot climates* (Wiley).
Cairncross, S. and Feachem, R. G. (1983) *Environmental health engineering in the Tropics* (Wiley).
Ludwig, H. F. et al. (1991) *Manual of environmental technology in developing countries* (South Asia Publications).
Wastewater stabilization ponds: principles of planning and practice (1987) (World Health Organization).

Journals

The journal *Water Environment Research* is devoted almost exclusively to wastewater treatment and effluent disposal and as such has a wide range of technical papers at a very high level. Other journals which cover the subject are:

Industrial Waste Management
Journal of Environmental Engineering (ASCE)
Process Biochemistry International
Water and Environmental Management
Water and Waste Treatment
Water and Wastewater International
Water Quality International
Water Research

Water Science and Technology
World Water and Environmental Engineer

Water treatment

Textbooks

The lack of high quality available water is now one of the major problems of both the developed and developing world. This, together with more stringent water quality standards is forcing the water engineer to provide water of sufficient quantity to greater populations which is wholesome and safe from health hazards. Many disciplines are concerned in this area, including the civil engineer, chemical engineer, biologist, chemist and hydrologist. Probably the most widely used textbook is that by Twort *et al.* (1993). Also of value to the design engineer would be a book produced by the American Water Works Association (1990). Other texts which are of general value include:

Hammer, M. J. (1986) *Water and wastewater technology*, 2nd ed. (Wiley).
McGhee, T. J. (1991) *Water supply and sewerage*, 6th ed. (McGraw-Hill).
Water treatment handbook, 2 vols, 6th ed., (1991) (Intercept).
Lorch, W. (1987) *Handbook of water purification*, 2nd ed. (Ellis Horwood).
Smethurst, G. (1988) *Basic water treatment: for application worldwide*, 2nd ed. (Thomas Telford).

Journals

Specialized coverage of the subject is provided in the following journals:

Journal (American Water Works Association)
Water & Environmental Management
Water and Waste Treatment
Water and Wastewater International
Water Bulletin
Water Engineering and Management
Water Quality International
Water Record
Water Science and Technology
World Water and Environmental Engineer

Solid and hazardous wastes

Probably more than in any other field of environmental engineering, the collection, treatment and disposal of solid waste has been treated as an art rather than a science, and the recent public concern over the environmental issues with regard to the safe disposal of hazardous wastes has made this area the subject of considerable academic and professional interest. A wide range of disciplines is associated with the problem but relatively few personnel receive specialized training in the subject nor have access to the relevant literature.

A number of books has been produced on the subject but it is recommended that the series of publications produced by the Department of Environment, *Waste Management Papers 1–28* should be studied since they cover all aspects of waste management from specific industrial wastes to landfill as well as volumes on Landfill Gas and Waste Re-cycling. The books are in a series entitled Waste Management and published by HMSO.

A comprehensive text on urban solid waste management has been produced by Pescod, Anderson and Elliott (1991).

A number of other books are available, including:

Suess, M. J. (ed.) (1985) *Solid waste management* (World Health Organization).
Henstock, M. E. (ed.) (1983) *Disposal and recovery of municipal solid waste* (Butterworths).
Holmes, J. R. (ed.) (1984) *Managing solid wastes in developing countries* (Wiley).
Porteous, A. (ed.) (1985) *Hazardous waste management handbook* (Butterworths).
Suess, M. J. and Huismans, J. W. (1983) *Management of hazardous waste: policy guidelines and code of practice* (World Health Organization).
Hazardous waste treatment processes (1990) (Water Pollution Control Federation).

Journals

A number of journal publications now deal with solid and hazardous waste management in a proficient way, of which the following are important reference material:

Haznews
Industrial Waste Management
Material, Recovery Weekly
Recycling World
Surveyor
Waste Management
The Waste Manager

It may also be useful to search such journals as *New Civil Engineer*, *Process Biochemistry* and other related journals, for information concerning specific solid and hazardous waste disposal problems.

Air pollution control

Air pollution is currently perceived as a growing problem requiring control of a wide variety of mobile and static emission sources. Among many books dealing with air pollution a compendium of 8 volumes by Stern (1968–1986) provides a comprehensive treatise covering all aspects of air pollution and emission control. A theoretical approach to the design of emission control equipment is taken by Crawford (1984). The issues of air quality criteria are addressed in the WHO publication European Series No. 23 (1987). UK legal practice and management issues are covered annually in the handbook edited by Loveday Murley, who has also edited a book on clean air around the world (Murley, 1991).

Other books relating to different aspects of air pollution effects and control include:

Cheremisinoff, P. N. (ed.) (1993) *Air pollution control and design for industry* (Marcel Dekker).
Buonicore, A. J. and Davis, W. T. (eds) (1992) *Air pollution engineering manual* (Van Nostrand-Reinhold).
Licht, W. (1988) *Air pollution control engineering*, 2nd ed. (Marcel Dekker).
Theodore, L. and Buonicore, A. (1988) *Air pollution control equipment*, 2 vols (CRC Press).
Bennett, G. (ed.) (1991) *Air pollution control in the European Community: implementation of the European Community Directives in the twelve member states* (Graham and Trotman).
Muezzinoglu, A. and Williams, M. L. (eds) (1992) *Industrial air pollution: assessment and control* (Springer-Verlag).
Mellanby, K. (ed.) (1988) *Air pollution, acid rain and the environment* (Elsevier).
Wellburn, A. (1988) *Air pollution and acid rain: the biological impact* (Longman).
Open University Environmental Control and Public Health PT272 course team (1975) *Air pollution* (Open University).

Journals

The main journals dealing with atmospheric research topics are:

Air and Waste: Journal of the Air and Waste Management Association
Atmospheric Environment Part A and Part B
Journal of Atmospheric and Terrestrial Physics

Odour

The control of odours is a particular aspect of air pollution which is receiving attention, especially in the areas of solid and liquid waste handling and treatment. An excellent reference manual in this field is that edited by Valentin *et al.* (1980) which provides a sound basis for the measurement and prediction of odours. Cheremisinoff (1992) comprehensively reviews the control problems in industry. Other books on odour generation and control are:

Bowker, R. P. G. et al. (1989) *Odour and corrosion control in sanitary sewerage systems and treatment plants* (Noyes Data Corporation).

Dragt, A. J. and van Ham, J. (eds) (1992) *Biotechniques of air pollution abatement and odour control policies: proceedings of an international symposium*, Maastricht, 27–29 October, 1991 (Elsevier).

Connor, E. S. and Bruce, A. M. (eds) (1984) *Stabilization, disinfection and odour control in sewage sludge treatment: an annotated bibliography covering the period 1950–1983* (Ellis Horwood).

Noise

A large number of texts have been written dealing with various aspects of noise and vibration generation and control.

The following selection covers a range of issues on the subject:

Adams, M. S. and McManus, F. (1994) *Noise and noise law* (Wiley).

Sharland, E. (1986) *Woods practical guide to noise control*, 4th ed. (Woods Acoustics).

Bies, D. A. and Hansen, C. H. (1988) *Engineering noise control* (Unwin Hyman).

Harris, C. M. (ed.) (1991) *Handbook of acoustical measurements and noise control*, 3rd ed. (McGraw-Hill).

Health and Safety Executive (1990) *Noise at work, noise assessment, information and control: noise guides 3–8*, Health and Safety Guidance Booklets (HMSO).

Anderson, J. S. and Bratos, Anderson, M. (1993) *Noise: its measurement, analysis, rating and control* (Avebury).

Sound Research Laboratories (1991) *Noise control in industry*, 3rd ed. (Spon).

Saenz, A. L. and Stephens, R. W. B. (eds) (1986) *Noise pollution: effects and control* (Wiley).

OECD Centre for Development Studies (1986) *Fighting noise: strengthening noise abatement policies* (OECD).

Roberts, J. and Fairhall, D. (eds) (1989) *Noise control in the built environment* (Gower).

McMullan, R. (1991) *Noise control in buildings* (BSP Professional).

Barber, A. (1993) *Handbook of noise and vibration control*, 6th ed. (Elsevier).

Journals

Journal of Sound and Vibration

Transactions of the ASME. Journal of Vibration and Acoustics.

Environmental impact assessment

The assessment of environmental impact features strongly in virtually all new public or private development projects. This is particularly true for projects connected with solid and liquid waste management. Public perception tends to view such projects as generating waste rather than treating waste in an environmentally acceptable manner. The Institute of Environmental Management keeps a library of environmental assessment case studies associated with recent infrastructure developments. The following texts cover the general requirements for producing effective environmental statements associated with new development projects.

Cheremisinoff, P. N. and Morresi, A. C. (1977) *Environmental assessment and impact statement handbook* (University of Michigan Press).
Canter, L. W. and Hill, L. G. (1980) *Handbook of variables for environmental impact assessment* (University of Michigan Press).
O'Riordan, T. and Hey, R. D. (eds) (1976) *Environmental impact assessment* (Saxon House).
Wood, D. A (Chairman) (1992) *Assessing the environmental impact of road schemes: the Standing Advisory Committee on Trunk Road Assessment* (HMSO).
Canter, L. W. (1977) *Environmental impact assessment* (McGraw-Hill).
Petts, J. and Eduljee, G. (1994) *Environmental impact assessment for waste treatment and disposal facilities* (Wiley).
Gilpin, A. (1994) *Environmental impact assessment (EIA): cutting edge for the 21st century* (Cambridge University Press).
Biswas, A. K. and Agarwala, S. B. C. (1992) *Environmental impact assessment for developing countries* (Butterworth-Heinemann).
Hyman, E. L. and Stiftel, B. (1988) *Combining facts and values in environmental impact assessment: theories and techniques* (Westview Press).
Turnbull, R. G. H. (ed.) (1992) *Environmental and health impact assessment of development projects: a handbook for practitioners* (Elsevier).
Glasson J. et al. (1994) *Introduction to environmental impact assessment* (UCL Press).
Morris, P. and Therivel, R. (eds) (1994) *Methods of environmental impact assessment* (UCL Press).
Economic Commission for Europe (1991) *Policies and systems of environmental impact assessment* (United Nations).
Economic Commission for Europe (1990) *Post project analysis in environmental impact assessment* (United Nations).
Wood, C. and Jones, C. (1991) *Monitoring environmental assessment and planning* (HMSO).
Department of the Environment (1989) *Environmental assessment: a guide to the procedures* (HMSO).
Smith, D. A. (1993) *Being an effective expert witness* (Thames Publishing).

References

American Water Works Association (1990) *A handbook of community water supplies* 4th ed. New York: McGraw-Hill.

Cheremisinoff, P. N. (1992) *Industrial odour control.* Oxford: Butterworth-Heinemann.
Crawford, M. (1984) *Air pollution control theory.* India: Tata McGraw.
Greenberg, A. E., Clesceri, L. S. and Eaton, A. D. (eds) (1992) *Standard methods for the examination of water and wastewater* 18th ed. Washington, DC: American Public Health Association.
Hammer, M. J. (1986) *Water and wastewater technology* 2nd ed. SI version. New York: Wiley.
James, A. (ed.) (1992) *An introduction to water quality modelling* 2nd ed. Chichester: Wiley.
Klein, L. (1959) *River pollution* Vol. 1 chemical analysis. London: Butterworths.
Klein, L. (1962) *River pollution* Vol. 2 causes and effects. London: Butterworths.
Klein, L. (1966) *River pollution* Vol. 3 control. London: Butterworths.
Mason, C. F. (1991) *Biology of freshwater pollution* 2nd ed. Harlow: Longman.
Metcalfe and Eddy Inc. Staff and Tchobanglous, G. (1990) *Wastewater engineering, treatment, disposal and re-use* 3rd ed. New York: McGraw-Hill.
Methods for the examination of waters and associated materials (1976–) London: HMSO.
Murley, L. (ed.) (annual) *NSCA pollution handbook.* Brighton: National Society for Clean Air.
Murley, L. (ed.) (1991) *Clean air around the world* 2nd ed. Brighton: International Union of Air Pollution Prevention Associations.
Operation of municipal wastewater treatment plants (1990) Vol. 1 management and support systems; volume 2 solid processes. Alexandria, VA: Water Pollution Control Federation.
Parsons, T. R. et al. (1984) *A manual of chemical and biological methods for seawater analysis.* Oxford: Pergamon.
Pescod, M. B., Anderson, G. K. and Elliott, D. J. (eds) (1991) *Urban solid waste management.* Geneva: World Health Organization.
Reynolds, T. D. (1982) *Unit operations and processes in environmental engineering* 3rd ed. New York: McGraw-Hill
Stern, A. C. (1968–1986) *Air Pollution* 8 vols. New York: Academic Press.
Tebbutt, T. H. Y. (1992) *Principles of water quality control* 4th ed. Oxford: Pergamon.
Twort, A. C. et al. (1993) *Water Supply* 4th ed. London: Arnold.
Valentin F. H. H. et al. (eds) (1980) *Odour Control: a concise guide.* Stevenage: Warren Spring Laboratory.
World Health Organization (1987) *Air Quality Guidelines for Europe* World Health Organization Regional Publications European Series No. 23. Geneva: World Health Organization.

CHAPTER TWENTY-NINE

Transportation and traffic planning and engineering

P. W. BONSALL

Scope

This chapter is concerned with the planning, design and management of road and rail-based passenger transport systems and with the behaviour of the users of these systems. It does not attempt to cover air or water modes nor to deal with freight distribution.

General background

The transport sector is a significant consumer of non-renewable energy resources and a significant contributor of greenhouse gases. Its activities contribute in no small way to the gross national product and represent a significant item of expenditure by the average citizen. Current concern is focused, not only on the consequences for safety and environmental quality, but on the problems of congestion which is seen as a major source of inefficiency in industrial/commercial activity. Traffic levels have tended to rise, along with economic activity, faster than any increase in system capacity and this has inevitably led to congestion. The traditional 'rush hours' have expanded to cover the whole working day and congestion has spread from the urban areas onto the inter-urban routes. Improvements in the management of the infrastructure (notably through increasingly sophisticated traffic control systems) have managed to squeeze out some extra capacity from the existing infrastructure but the general consensus is that the emphasis must now switch from capacity enhancement to demand management using regulations or pricing mechanisms and the encouragement of public transport where possible. There is currently a very lively debate on the merits of pricing for urban and inter-urban roads, on the role of new technologies in this area and on the need for an integration of land use and transport planning.

The most dramatic increases in demand have been for inter-urban journeys and off-peak travel. This reflects the fact that the capacity for further growth in urban travel at peak periods is no longer available and that the patterns of economic activity and lifestyle are changing – perhaps partly in response to the constraints and opportunities offered by the transport system.

The passenger transport sector is now dominated by private vehicles but public transport modes play a crucial role in large metropolitan areas and are clearly essential for those without access to a car. Car ownership continues to rise with the main growth now being in multiple car ownership.

The importance of walking has often been overlooked but there is some evidence that the needs of pedestrians and other vulnerable road users are beginning to receive more attention. Cyclists, who are also part of the vulnerable road user group, continue to be a minority group despite some improvement in specialist facilities in recent years.

Guidance on sources

General

The practice of transport and traffic planning and engineering is eclectic. It draws on a diverse range of skills including civil engineering (for the design, construction and maintenance of infrastructure), computing and operations research (for the design and operation of traffic control systems) and economics and law (for the evaluation of investments and the design of effective regulatory and fiscal management strategies). It requires a sound understanding of the interaction between transport and land use systems, the factors influencing demand for travel and the behaviour of drivers. Current concerns for safety and environmental quality make a knowledge of these factors particularly important. Against this background it is inevitable that the would-be student or practitioner has to consider a wide range of information sources for their work.

There are many ways of tracking down information in this field. The choice will, as always, depend on the nature of the information sought, the searcher's level of knowledge and the time/resources available for the search.

If someone new to the field desires to achieve a broad and in-depth knowledge of the field, it would obviously be desirable if they could spend time working in the industry, attending one of the excellent degree courses or undertaking research. Assuming time is not available for such luxuries, the would-be student will have to be content with reading some of the available textbooks, attending some carefully selected short courses and capitalizing on the expertise of the teachers/ lecturers.

If the searcher already has the necessary background and wishes to improve their knowledge of a specialist field, then a consultation with a relevant expert, reading of selected articles and attendance at a relevant short course or conference will be the best course of action.

If the answers to specific questions or detailed guidance on matters of fact are sought, then the solution will be to identify and approach the relevant organization or specialist expert or to consult the relevant guidelines or official circulars. If access to statistical data is sought, it may be possible to find required information in one of the many published sources or, failing that, to seek it directly from the relevant organization.

In any event it will first be necessary to identify the relevant organizations, literature and databanks. This chapter is laid out to facilitate this.

Identifying the correct organizations and the individuals within them

The quickest way to identify the correct individual or organization is to invest in a few phone calls or e-mail messages. Letters or even face-to-face meetings may be appropriate to make more detailed enquiries but much time will be wasted if the appropriate individual is not first identified. A guide to some of the most relevant organizations is provided later in this chapter.

Some organizations will regard provision of information and advice as part of their function and may have a designated officer whose prime function is to provide such information. In my experience the Department of Transport (and its subsidiary organizations) and the various membership organizations are very well set up in this respect. In other organizations where there is no designated information officer the enquirer will be dependent on the goodwill of individual staff and should not be surprised if they are reluctant to devote much time to a query from an 'outsider'. Information is very definitely a tradable commodity and one should not always expect free access!

Keeping abreast of the field

General

Because transport and traffic issues are of interest to a large lay audience, newspapers, magazines, television and radio provide a fairly good coverage of developments in the field via news items, special reports and documentaries. The standard of presentation in the serious press and broadcast media is generally good and even those working in the industry should not feel they will learn nothing from them. However, there are of course occasions when the facts are, deliberately or otherwise, somewhat distorted and anyone with a professional interest in the subject would do well to double check before acting on information from these sources.

We are fortunate in having several non-academic publications targeted at the professional audience which provide a relatively painless way of staying abreast of developments. One, *Local Transport Today*, is particularly good for its coverage of news and developments relating to roads and public transport planning. Others include: *Traffic Engineering and Control* (which has good coverage of technical developments in traffic planning and operations and includes in-depth articles, as well as product and service information); *Transport; Traffic Technology International*; *Highways and Transportation*; *Modern Railways*; and *Bus and Coach*, whose coverage will be apparent from the titles. Two recently introduced newsletters whose contents will not immediately be obvious to the uninitiated are *Inside IVHS* and *RTI Focus News* (to be retitled as *Inside ITS* and *ITS Focus News* respectively) which have been spawned by the growth of interest in the role of new technology in the planning and operation of traffic and transport.

These and other more academic publications are listed below. It will be noted that several are produced by professional institutions or other membership organisations. As will be discussed later in this chapter, membership of the appropriate institution or association is clearly an important means of keeping abreast of one's chosen field.

Accident Analysis and Prevention (Elsevier) 1969–.
Environment and Planning (Pion) 1969–.
Highways (formerly *Highways and Public Works*) (Faversham House Group) 1985–.
Highways and Transportation (Institution of Highways and Transportation) 1983–.
International Journal of Transport Economics (Revista Internazionale di Economica dei Traporti) 1974–.
ITE Journal (Institute of Transportation Engineers) 1978–.
IVHS Journal (IVHS America) 1993–.
Journal of Advanced Transportation (Institute for Transportation Inc) 1967–.
Journal of Regional Science (Regional Science Research Institute) 1958–.
Journal of Transport Economics and Policy (University of Bath) 1967–.
Journal of Transportation Engineering (American Society of Civil Engineers) 1969–.
Journal of Transport Geography (Elsevier) 1992–.
Journal of Transport History (Manchester University Press) 1953–.
Local Transport Today (LTT Ltd) 1989–.
Modern Railways (Ian Allan) 1946–.

Passenger Transport (American Public Transit Association) 1943–.
Proceedings of the Chartered Institute of Transport (CIT London) 1991–.
Progress in Planning (Pergamon) 1974–.
Public Transport International (formerly UITP Revue) (UITP) 1952–.
Railway Gazette International (Reed Business Press) 1835–.
Regional Science and Urban Economics (North Holland) 1971–.
Regional Studies (Carfax Publishing) 1966–.
Traffic Engineering and Control (Printerhall) 1960–.
Traffic Technology International (UK and International Press) 1995–.
Transactions of IBG (Institute of British Geographers) 1935–.
Transport Policy (Elsevier) 1993–.
Transport, Proceedings of the Institution of Civil Engineers (ICE) 1992–.
Transport Reviews (Taylor and Francis) 1981–.
Transportation (Kluwer) 1972–.
Transportation Journal (American Society of Transportation and Logistics) 1961–.
Transportation Planning and Technology (Gordon and Breach) 1972–.
Transportation Quarterly (formerly *Traffic Quarterly*) (Eno Foundation) 1947–.
Transportation Research (A, B and C), (Elsevier) 1967–.
Transportation Research Record (formerly *Highway Research Record*) (Transportation Research Board) 1973–.
Transportation Science (Operational Research Society of America) 1967–.

For those wanting to update themselves on recent developments in a given topic area there are a number of useful sources of review material. These include the reports of special enquiries or studies conducted for the Department of Transport (e.g. by the Standing Advisory Committee on Trunk Road Assessment (SACTRA)) or by/on behalf of professional bodies or pressure groups (selected examples are listed below). Many review studies may not be published as free standing documents and may be difficult to locate. Two useful sources, however, are the articles appearing in the journal *Transport Reviews* and the *State of the Art Reviews* produced by the Transport Research Laboratory (TRL) which contain detailed studies of individual topics by 'acknowledged experts' in the field. Other publishers who specialize in review material include the VNU Science Press (via their Topics in Transportation series) and the major newspaper publishers (via special supplements or free standing reports; the *Financial Times* in particular has produced a number of very informative reviews relevant to transport and traffic planning). Specific examples include the following:

Armitage Report, *Report of the inquiry into lorries, people and the environment* (1980) (HMSO).
Buchanan Report, *Traffic in Towns* (1963) (HMSO).
Leitch Report, *Report of the Advisory Committee on Trunk Road Assessment* (1977) (HMSO).
SACTRA Reports (HMSO):
- on *Urban road appraisal* (1986).

500 Transportation and traffic planning and engineering

- on *Environmental evaluation* (1992).
- on *Traffic generation* (1994).

Highway capacity manual (TRB special report 209) (1994) (Transportation Research Board).
Traffic appraisal manual(1993) (HMSO).
Traffic signs manual (1982) (DOT).
Design manual for roads and bridges, vols 1–10 (1994) (HMSO).
Planning policy guidance note 13 (1994) (DOE).
Microprocessor based traffic signal controller for isolated linked and urban traffic control installations (1991) (DOT).
COBA9 manual (1981 with subsequent updates to individual sections) (DOT).
ROSPA road safety engineering manual (1993) (ROSPA).
Reports of the House of Commons Select Committee on Transport (HMSO) (Various).
Royal Commission on Environmental Pollution. Eighteenth report; *Transport and the environment* (1994) (HMSO).

Attendance at conferences is regarded by many as the best way to keep abreast of developments. If the conference is well organized and the delegate takes the opportunity to ask questions and mix with others 'in the bar' then the time can be well justified. Simply scanning the proceedings (if available) is quicker but can leave points ambiguous and important nuances missing. Attendance at the annual conference of one's professional institution is certainly a useful way of keeping up with professional as well as technical matters.

It is perhaps worth mentioning here four regular conferences which cover a wide range of topics: the PTRC European Transport Forum (previously known as the PTRC Summer Annual Meeting) has become the premiere European event for updating on current issues and research in transport; the annual US Transportation Research Board (TRB) Conference is the world's biggest and most wide ranging conference on transport and traffic planning and engineering. Unlike PTRC and TRB the triennial World Conference on Transport Research is primarily for academics, it is an ambitious undertaking whose success is dependent on the skill of the organizers in any given year. It has traditionally tended to concentrate on 'soft' issues rather than engineering. The annual Universities Transport Studies Group Conference is restricted to academics and provides a valuable forum for the exchange of ideas.

A number of other conferences are held on a regular basis but they differ from those listed above in that they have specialist themes. Examples include: the International Association for Travel Behaviour Research Conference, the Survey Methods in Transport Conference, the International Symposia on Transport and Traffic Theory, the Vehicle Navigation and Information Systems Conference (VNIS), the Permanent International Association of Road Congress Conference (PIARC),

Transportation and traffic planning and engineering 501

the International Symposium on Automotive Technology and Automation (ISATA).

Numerous other conferences are held on an *ad hoc* basis, often sponsored by one or other of the professional institutions. Lists can be found in the professional journals. A list of organizations responsible for such conferences is given below.

IATB Conference, IATBR, (current Secretary) Prof K Axhausen, Institut für Strassenbau und Verkehrsplanung, Leopold-Franzens-Universität, Technikerstr. 13, Innsbruck 6020, Austria.

IEE Symposia, IEE, Savoy Place, London, WC2R 0BL, UK.

International Symposium of Transportation and Traffic Theory (organizer of most recent (1993) event) Institute of Transportation Studies, University of California, Berkeley, USA.

ISATA Symposium, ISATA, 42 Lloyd Park Avenue, Croydon, Surrey CR0 5SB, UK.

PTRC Conferences, PTRC, Glenthorne House, Hammersmith Grove, London W6 OLG, UK.

Public Transport Symposium, Transport Operations Research Group, University of Newcastle upon Tyne, Newcastle upon Tyne NE1 7RU, UK.

Survey Methods in Transport (co-organizer of several conferences in the series) Liz Ampt, c/o Steer Davies Gleave, 32 Upper Ground, London SE1 9PD, UK.

TRB Conference, TRB, 2101 Constitution Avenue NW, Washington DC, 20418, USA.

UTSG Conference (members only).

VNIS Conference, SAE Inc, 400 Commonwealth Drive, Warrendale, PA 15096, USA.

WCTR Conference, WCTRS (secretariat) Laboratorie d'Economie des Transports, MRASH–14, Avenue Berthelot, 69363, Lyon Cedex 07, France.

Attendance at specialized short courses organized by universities or consultants can be a useful way of keeping abreast of the field. Many such courses are run on a one-off basis in response to a topical issue (e.g. rail privatization), while others are run on a regular basis or as part of a modular degree or diploma course. Some of the most popular courses are oriented to the use of a particular technique or the conduct of a particular procedure. The professional press (e.g. *Local Transport Today* and *Traffic Engineering and Control*) contain lists of forthcoming courses while, for longer term planning, it is also possible to get oneself put on the mailing lists of the main course organizers such as PTRC and the Universities of Leeds and Nottingham.

Networked or electronic communication is fast becoming an important mode of communication between individuals and institutions researching in transport and traffic. Electronic mail is used as an efficient medium via which messages and text files can be exchanged, thereby making cooperative projects and document production more feasible across large distances than ever before. Electronic bulletin boards, newsgroups and discussion lists are used to exchange views and disseminate results among the community of interest in such a way as to spawn new ideas and cooperative ventures that would previously not have been imaginable. Some of the ideas are at the leading edge of research while others are more political or issue related. Some of the most popular of the transport-oriented user-networks are listed below, they range from amateur newsgroups such as TRANSPORT.UK, general purpose mailing lists such as UTSG Mailbase and TRANSP-L to more specialist networks such as IVHS-L which deals with transport telematics issues. These networks are home for cruisers of the 'information superhighway' and are as yet un-policed; this means that there is no quality control on the information content or language – you have been warned!

UTSG (UK-based transport research issues network) mailbase@mail base.ac.uk.
TRANSP-L (US-based transportation planning issues network) listproc@gmu.edu.
IVHS-L (US-based transport telematics network) majordomo@mail hub.ornl.gov.
TRANSIT (US-based public transport issues network) listserv@gitvm1.bitnet.
TRANSPORT.UK (UK-based Usenet Newsgroup dominated by amateur enthusiasts!) available on most computers via the netscape browser.
MISC.TRANSPORT.URBAN-TRANSIT (US-based Usenet Newsgroup dominated by amateur enthusiasts!) available on most computers via the netscape browser.

World Wide Web (WWW) is another fast growing system available via the Internet. Unlike the previously mentioned networks, WWW is

a multimedia hypertext system which is a natural medium for the transfer of graphics, digital images and voice data as well as text data. WWW is made up of a series of linked 'sites' each of which contain a structured data tree. In addition to bibliographic data and research abstracts, WWW contains a variety of potentially useful information and services ranging from online data on traffic flows at certain junctions in Southern California to editorial comment on current issues provided by the Friends of The Earth organization.
Useful pages include the following:

http://www.worldbank.org/html/extpb/Publications.char.html (World Bank publications)
http://www.its.leeds.ac.uk (a UK entry site for transport pages on the worldwide web)
http://www.itsonline.com/(Journal of the intelligent transport systems society)
http://www.dialog.com/cgi-bin/taos_doc.pl? databases+790+tris (TRIS database)
http://www.sor.princeton.edu/~dhb/TSS/ (transportation science section of INFORMS database)
http://www-mctrans.ce.ufl.edu/mct.htm (information about transport software)
http://gatekeeper.unicc.org/unctad/ship/mt-home.dir/mt-head1.htm(multimodel section of UNCTAD)

Publications

The main journals carrying articles relevant to transport and traffic planning and engineering have been listed above. Even before adding on all the various conferences that are held each year, the list is far too long for anyone to hope to keep up-to-date with all the new articles and there is no single journal to which one can safely devote all one's attention. The simplest way to keep a watching brief on new publications and conference proceedings is to subscribe to one of the specialist abstracting services or bulletins of new publications. The *Department of Environment/Department of Transport Library Bulletin* used to be a very useful monthly summary of new publications but it is unfortunately no longer publicly available. TRL's series *Current Topics in Transport* (which brings together the abstracts of papers on a given theme from the IRRD database) is a useful source, provided of course that one's particular interest has featured as a theme in the recent past. Other abstracting and current research services are:

CURRENT RESEARCH

CRIB, *Current research in Britain* (Longman Cartermill).
IRF research and development (IRF).
UTSG current research listing (UTSG).

ABSTRACTS AND INDEXES

ASCE Publications Abstracts (ASCE).
Current Literature in Traffic and Transportation (Northwestern University).
Department of Transport/Department of the Environment Publications (DOT/DOE).
The ENDS Report (Environmental Data Service).
Government Publications (HMSO).
Highway Research Information Service (HRIS) Abstracts (TRB).
HMSO Transport Sectional List (HMSO).
SAE Technical Literature Abstracts (SAE).
Transportation Research Abstracts (TRB).
TRB Publications (TRB).
TRL Headstart (TRL).
TRL Current Topics in Transport (TRL).
UITP Index (UITP).
Urban Transportation Abstracts (TRB).

If one has the time, the best tactic is undoubtedly to conduct one's own periodic search of recent additions to one or more of the computerized publications databases available online via networks such as Internet or offline via regularly updated CD-ROM discs. Whether for monitoring new publications or for conducting *ad hoc* literature searches on specific topics, these 'live' databases have undoubtedly become the most efficient way of trawling the published literature. By using a combination of publications database and citation index it is possible to select an article on the basis of its author, its topic, its date or place of publication, its inclusion or exclusion of specified words in the title, or its having referenced, or having been referenced in, another specified article. The titles thus located can be displayed with or without the accompanying abstracts and filed for further action as required. Online or CD-ROM catalogues can then be used to discover whether the full article can be found in one's chosen library. If one has access to the online networks or CD-ROM databases, one not only has the freedom to conduct one's own searches but one can receive the results in a file on one's own computer for further processing.

The relevant databases include: OECD's IRRD (International Road Research Documentation) which contains over 200 000 abstracts of research related to all modes of transport in OECD member countries; the US Transportation Research Board's TRIS (Transportation Research Information Service) which contains some 250 000 bibliographic citations subsuming the English language content of IRRD along with additional abstracts relating to US research in transport; TRANSDOC which is supported by ECMT and contains information from ICTED (International Cooperation in Transport Economics Documentation) and IUR (International Union of Railways), TRANSDOC contains some 30 000 research digests together with bibliographic records and is particularly strong on policy and socio-economic aspects of transport planning; ACOMPLINE which is derived from *Urban Abstracts* and concentrates on matters relating to the urban and regional planning of transport; COMPENDEX* PLUS which concentrates on the engineering aspects of transport; NTIS which contains report literature; and PASCAL which is multi-disciplinary and multilingual. Transport and traffic make up a very small part of ISTP (the Index to Scientific and Technical Proceedings) which has very good coverage of conferences and of the SCIENCE CITATION INDEX and the SOCIAL SCIENCES CITATION INDEX, but the structure of these databases makes them particularly valuable. CORDIS (the EU's Community Research and Development Information Service) is another multi-sector database which has useful coverage of transport; it contains digests of results of EU-funded research as well as information about current and upcoming research programmes.

Access to these databases from one's own desk can be obtained via networks such as JANET or Internet using a service such as NISS (National Information Software System) or hosts such as BIDS, ESA-IRS, DIALOG or WWW. Access to some databases is free and unrestricted while others may require a subscription or search fee. Help screens are available to assist new users but it is quite easy to misspecify a search and end up with mountains of irrelevant material, so novice users are well advised to seek advice from an experienced colleague or librarian. Technical libraries such as that at TRL will, on payment of a fee, conduct specified searches to order.

If one makes substantial use of these databases the access fee and communications costs can be quite substantial and it may be more economic to purchase an offline database such as Silver Platter's TRANSPORT CD-ROM. This CD is the result of cooperation between OECD, TRB and ECMT to produce a single database containing all the information in IRRD, TRIS and TRANSDOC. The database contains over 475 000 entries and is available on 2 disks updated quarterly. The first release in January 1995 contained all publications from 1968 to November 1994. Other databases, such as CORDIS are also now being made available on CD-ROM.

Despite all this information about computerized databases and online access, it would be quite wrong to assume that developments in computerized literature searching have made traditional libraries redundant for the student of transport and traffic planning and engineering. It is still necessary to use the library to read the full articles, to borrow books and search the vast archives of material which is not yet available via electronic means. Also, the knowledge and expertise of subject specialist librarians is extremely valuable and can save hours of otherwise frustrating searches. Technical collections and specialist librarians are to be found in many of the key organizations listed elsewhere in this chapter.

If one wishes to keep fully up to date with useful techniques for transport and traffic planning it will be necessary to scan the journal articles and conference proceedings as described above. As has been mentioned already, attendance at conferences and discussion with presenters can also be useful provided that the conference is well targeted and/or sufficiently prestigious to draw important speakers. Textbooks are, of course, useful if they happen to cover the topic one is interested in. They generally relate to established techniques and procedures rather than to the very latest developments and this may or may not be a problem.

The books detailed below are generally acknowledged to be useful, but the inclusion of a particular title should not be taken as an unreserved recommendation nor should exclusion be taken to indicate dismissal – there are many widely used books which are now somewhat out of date just as there are many very good texts which are too specialist to ever appeal to a general audience. The list includes evergreen classics such as Webster and Cobbe (1966) which is still something of a bible for traffic engineers, Bruton (1985) which continues to provide a sound introduction to underlying principles; O'Flaherty (1986 and 1987) which still provide a good coverage of highway engineering practice; and *Roads and traffic in urban areas* (IHT/DOT, 1987) which is a good source of technical advice and has the particular merit of the seal of approval from the Department of Transport and the Institution of Highways and Transportation. More recent texts include Ortuzar and Willumsen (1994) which provides a fairly advanced coverage of the use of models in transport planning and the revised edition of *The highway capacity manual* (TRB, 1994) which is widely acknowledged as a standard international reference on state of the art transport analysis methods. New books include a multi-authored text edited by O'Flaherty outlining the principles of transport planning and engineering, and a book by Taylor, Young and Bonsall designed to aid the reader's understanding of traffic systems.

Bruton, M. J. (1985) *Introduction to transportation planning*, 3rd ed. (Hutchinson).
Cooper, J. (ed.) (1990) *Logistics and distribution planning* (Kogan Page).
Croney, D. and Croney, P. (1991) *Design and performance of road pavements*, 2nd ed. (McGraw-Hill).
Cullingworth, J. B. (1988) *Town and country planning in Britain*, 10th ed. (Unwin Hyman).
Dickey, J. W. et al. (1983) *Metropolitan transportation planning*, 2nd ed. (McGraw-Hill).
Fowkes, A. S. and Nash, C. A. (eds) (1991) *Analysing demand for rail travel* (Avebury).
Gerlough, D. L. and Huber, M. J. (1976) *Traffic flow theory*, TRB Special Report SR 165 (Transportation Research Board).
Highway capacity manual (1994) TRB special report SR 209 (Transportation Research Board).

Institution of Highways and Transportation/Department of Transport (1987) *Roads and traffic in urban areas* (Department of Transport).
Jane's Urban Transport Systems (1993) (Jane's Information Group).
Jones, P., Dix, M., Clarke, M. and Heggie, I. (1983) *Understanding travel behaviour* (Gower).
Mannering, F. L. and Kilareski W. P. (1990) *Principles of highway engineering and traffic analysis* (Wiley).
Nash, C. A. (1982) *Economics of public transport* (Longman).
O'Flaherty, C. A. (1986) *Highways*, Vol. 1 Traffic Planning and Engineering, 3rd ed. (Edward Arnold).
O'Flaherty, C. A. (1987) *Highways*, Vol. 2 Highway Engineering, 3rd ed. (Edward Arnold).
O'Flaherty, C. A. (ed.) (1996) *Transport planning and engineering* (Edward Arnold).
Ortuzar, J. Dios and Willumsen, L. (1994) *Modelling transport*, 2nd ed. (Wiley).
Salter, R. J. (1989) (1989) *Highway traffic analysis and design*, 2nd ed. (Macmillan).
Starkie, D. (1982) *The motorway age* (Pergamon).
Stopher, P. R., Meyburg, A. H. and Brog, W. (1981) *New horizons in travel-behavior research* (Lexington Books).
Taylor, M., Young, W. and Bonsall P. (in press) *Understanding traffic systems* (Avebury/Gower).
Thomas, R. (1991) *Traffic assignment techniques* (Avebury).
Van Vuren, T. and Van Vliet, D. (1992) *Route choice and signal control: the potential for route guidance* (Avebury).
Watson, J. P. (1989) *Highway construction and maintenance* (Longman).
Webster, F. V., Bly, P. H. and Paulley, N. J. (eds) (1988) *Urban land-use and transport interaction: policies and models* (Avebury).
Webster, F. V. and Cobbe, J. (1966) *Traffic signals*, Road Research Technical Paper 56 (HMSO).
White, P. R. (1986) *Planning for public transport* (Hutchinson).
Wistrich, E. (1983) *The politics of transport* (Longman).
Wohl, M. and Hendrickson, C. (1984) *Transportation investment and pricing principles* (Wiley).

Accessing official regulations and guidance

Government departments produce a series of guidance notes and circulars to assist practitioners in understanding the relevant legislation and procedures and to promote good practice. These notes and circulars are sent as a matter of course to local authorities and other public sector bodies but are also available to other interested parties. Most of the relevant documents come, as would be expected, from the Department of Transport (DOT) but the Department of the Environment is involved in matters relating to the link with land use planning and environmental appraisal. Current lists of notes and circulars are available from the Departments concerned.

Several circulars and guidance notes are produced each year and they typically cover a fairly narrow field (e.g. on the design of roundabouts, or on public consultation procedures). From time to time more comprehensive advice is published in the form of manuals. Current examples of important DOT publications include the multi-volume *Design manual for roads and bridges*, the *Traffic appraisal manual (TAM)*, and *The COBA manual*. The *Design manual for roads and bridges* comprises current technical and guidance notes and so covers a variety of topics ranging from the design of grade separated junctions to recommended procedures for environmental appraisal. *TAM* is a loose-leaf document first produced in 1981 and subsequently updated from time to time, it provides guidance on the procedures and tools for forecasting travel demand. *The COBA manual* describes the approved usage of the Department of Transport's Cost Benefit Appraisal software and is updated to reflect new versions of that software.

Other documents which indicate the Department's thinking on current issues and which, while not having the force of law, should obviously be taken into account by practitioners, include the Department's response to reports such as those by SACTRA and various consultation papers including draft circulars/guidance notes, White and Green papers – again lists are available from the relevant Departments. Also of interest are papers presented at conferences by Department personnel – even when they contain the standard disclaimer, they can give a good insight into insiders' thinking.

It is useful to keep an eye on the reports produced by parliamentary select committees – the House of Commons Select Committee on Transport is an obvious source but *ad hoc* reports are produced by others such as the House of Lords Committee on Science and Technology. Some relevant titles have been detailed above.

The importance of an international dimension in certain aspects of transport planning and engineering has long been recognized. In many areas (e.g. driver and vehicle licensing, traffic signing standardization) cooperation has been achieved through voluntary agreement or international conventions (e.g. the Geneva and Vienna conventions on the design of traffic signs). More recently the European Union (EU) has begun to issue directives relating to environmental and safety matters (e.g. Directive 85/337/EEC, 27 June 1985, on the assessment of the effects of certain public and private projects on the environment). Current discussions within ECMT and EU on the harmonization of traffic signs and the 1993 EU White Paper on transport policy are indicative of the fact that the EU is likely to become an increasingly important source of standards, protocols and perhaps even policy direction. The EU Commission has a directorate (DGVII) specifically related to transport and its Information Technology Directorate (DGXIII) has been very active in promoting the adoption of advanced telematics in the transport sector.

Sources of statistical information

As a general rule it is easier to access national and international statistics than to find information about a given site and this is due in no small part to the efforts of various national agencies which have compiled data from a range of local sources. Perhaps the most generally useful of the 'official' UK documents is the annual *Transport statistics Great Britain* which brings together a wide range of statistics and is consciously designed to assist informed discussion of transport developments and policies. It comprises over 200 pages of statistical tabulations and graphs along with commentary and background information. Other reports deserving special mentioned are *Road accidents Great Britain*, which is produced annually, the *Report of the national travel survey*, which is produced about once every three years and the monthly *New motor vehicle registrations*. A complete list can be obtained from the Department of Transport (DOT – on 0171 276 8513). Official publications typically disaggregate the data to regions or counties but DOT can sometimes respond, at cost, to requests for more detailed data or special tabulations. Equivalent documents and services are available for some other countries on application to the appropriate government departments and, as can be seen from the list below there are a number of sources of international statistics.

Selected statistical publications by DOT and HMSO

Annual vehicle census Great Britain (DOT) annual
Bus and coach statistics Great Britain (HMSO) annual
Continuing survey of road goods transport (HMSO) annual
Heavy goods vehicles in Great Britain (HMSO) annual
International road haulage by UK registered vehicles (HMSO) annual
International comparisons of transport statistics (HMSO) occasional
Local road maintenance expenditure in England and Wales (HMSO) annual
Merchant fleet statistics (HMSO) annual
National road maintenance condition survey (DOT) annual
National travel survey (HMSO) occasional
New motor vehicle registrations Great Britain (DOT) monthly
Quarterly road casualties Great Britain (DOT) annual
Quarterly transport statistics (DOT) quarterly
Road accidents Great Britain. The casualty report (HMSO) annual
Road lengths in Great Britain (HMSO) annual
Road traffic statistics Great Britain (DOT) annual
Road accident statistics – English regions (HMSO) annual
Seaborne trade statistics of the United Kingdom (HMSO) annual
Traffic in Great Britain (DOT) quarterly
Transport of goods by road in Great Britain (HMSO) annual
Transport statistics for London (HMSO) occasional
Transport statistics Great Britain (HMSO) annual

United Kingdom shipping industry: international earnings and expenditure (HMSO) annual
Vehicle speeds in Great Britain (DOT) occasional

International statistical publications

Annual bulletin of transport statistics for Europe (ECE) annual
Census of motor traffic on main international traffic arteries (ECE) 1993
International statistical handbook of public transport (UITP) 1985
International railway statistics (International Union of Railways) annual
Statistical report on road accidents (ECMT) annual
Statistical trends in transport (ECMT and OECD) annual
Statistics of road traffic accidents in Europe (ECE) annual
Transport accident reports/briefs (NTIS/US DOT) annual and occasional
Transport: annual statistics (EUROSTAT, Stats. Office of European Communities) annual
US DOT national transportation statistics (US DOT) annual
World transport data (IRTU) irregular
World road statistics (IRF) irregular

In addition to the official published documents there is, of course, a considerable volume of statistical data held in archives and databases. Some of the publicly available archives/databases are listed below. Certain of these are produced on a commercial basis (e.g. ABC's timetables, or the various trip rate databanks), while others are designed primarily for the research community (e.g. the UK's ESRC DATA ARCHIVE, or the US's NATIONAL TECHNICAL INFORMATION SERVICE). Some sources are freely available while others are provided only for research or private use and may not be made available to commercial organizations. A fuller list of these sources, together with requisite names and addresses, is contained in the Transport Statistics User Group's *Directory of sources in transport statistics*. This very useful document, which is available in hard copy or on disk, contains a cross reference table to enable users to discover which organizations are willing to provide data of a particular type. TSUG is jointly sponsored by the Chartered Institute of Transport and the Statistics Users Council as a forum for information on the whereabouts of transport statistics.

Worldwide public transport timetables (ABC)
Basic road statistics (British Road Federation)
Motorists' opinions – survey of 1500 UK motorists (Lex Service PLC)
Public transport statistics (London Regional Transport)
Greater London transportation studies (London Research Centre)
London area transport surveys (London Research Centre)
Motoring statistics (Royal Automobile Club)
Vital travel statistics (Transport 2000)
Bus registration index (Transport Research Laboratory)
Thomas Cook European and worldwide timetables (Thomas Cook)
TSUG *Directory of sources in transport statistics* (available through TSUG

Chairman (currently Mr P. Clifford c/o Ove Arup, 13 Fitzroy Street, London W1 6BQ))

ESRC DATA ARCHIVE contains data from national surveys such as the National Travel Survey and the Family Expenditure Survey as well as databases compiled on ESRC sponsored research projects. Contact: via University of Essex.

TRICS (Trip Rate Information Computer System) compiled by various local authorities and consultants. Contact: JMP Consultancy, 172 Tottenham Court Road, London W1P 9LG.

NTIS (National Technical Information Service). Contact: TRB, Washington DC.

GENERATE (Trip rate data bank) compiled by West Midlands local authorities. Contact: West Midlands Joint Data Team, PO Box 1777, Clarendon House, High Street, Solihull B91 3RZ.

TRANSPORT CD-ROM available from Silver Platter Information Inc. via e-mail: davidb@silverplatter.com.

Key organizations

The Department of Transport and its agents

The Secretary of State for Transport has ultimate responsibility for all transport and traffic planning within the UK. Some of his/her functions are executed directly by the Department of Transport (DOT) but much of the work is done by agents and agencies with various degrees of autonomy. Typical examples would be the Highways Agency, which deals with the construction and maintenance of trunk roads and motorways, and the TCSU (Traffic Control Systems Unit) which is the DOT's agent for the management of traffic control in London. Given its central role in UK transport and traffic planning it is not surprising that DOT should be the pre-eminent source of information and advice. In addition to its production of advice notes and circulars it produces or sponsors periodic statistics, reports and design guides, maintains a very good library and issues monthly bulletins of relevant literature.

Recent years have seen a number of changes in the nature and responsibilities of the various bodies involved in transport and traffic planning in the UK. One trend has been the gathering in to DOT of responsibilities for aspects of transport previously dealt with elsewhere (notably taxis, aircraft and shipping); the only major function that is still outside DOT is enforcement of traffic regulations which is a police responsibility and so comes under the Home Office. Another trend in recent years has been reduction in the effective power of locally elected strategic planning bodies (notably via abolition of the Greater London Council and the Metropolitan County Councils, with their functions now being exercised by the Department of Transport, the London

Traffic Control Systems Unit, the Director of Traffic for London and via a variety of collaborative arrangements between unitary authorities). Another important trend has been the increased role for private sector organizations. This has come about through deregulation of the bus industry and increased competition in the airline industry, and, more recently, through encouragement of private sector involvement in road planning, financing and operation, and the (currently envisaged) privatization of British Rail. Another example of increased commercialization is the current intention to transfer the Transport Research Laboratory (TRL) to the private sector.

Given these trends the traditional role of various organizations has been changing. Thus, for example, the Passenger Transport Executives (PTEs) are now primarily concerned with acting as an information source for registered bus services, administering any concessionary fares scheme and identifying the need for service enhancements and administering tenders for these, rather than with attempting to plan and operate an integrated passenger transport system. Notwithstanding this change of emphasis, some PTEs have been active in the promotion of innovations such as guided buses and 'supertrams'. Local authorities, while recognizing the need for the development of integrated transport strategies, often find themselves too small to find the necessary skills in-house and so they increasingly find it necessary to seek strategic planning advice from transport planning consultants. The currently contemplated reorganization of local authority functions in the shire counties into smaller unitary authorities will no doubt accelerate this trend.

The changing role of TRL ought also to be mentioned here, TRL began life as the Road Research Laboratory and undertook research on highways vehicles and traffic matters initially under the Department of Scientific and Industrial Research and then under the Department (or Ministry) of Transport. Subsequently renamed the Transport and Road Research Laboratory (TRRL) to reflect a wider remit including public transport and forecasting and management, it produced a series of informative and authoritative reports on a wide variety of subjects. These reports were issued free to interested parties and undoubtedly served to improve the general standard of transport and traffic planning in the UK. The TRL also contributed to the discipline by providing services such as maintenance of IRRD database. Increasing commercialization of TRL seems likely to lead it to target its resources on activities for which it can earn a financial return and this would almost certainly reduce its role as an independent source of background information, advice and services for the industry at large. It is not yet clear how the role of TRL will evolve but, as of now, it remains a very useful source of information and expertise through its library and report series.

Institutions, associations and universities

Practitioners in the field of transport and traffic planning may belong to one or more of a number of professional institutions. Those from a civil engineering background are likely to belong to the Institution of Civil Engineers but those who have chosen to specialize in transport or traffic rather than other aspects of civil engineering are more likely to be members of the Institution of Highways and Transportation or, particularly if they are involved in freight or passenger transport operations, of the Chartered Institute of Transport. Personnel involved in traffic signals planning and operation may be members of the Institution of Electrical Engineers or the US-based Institute of Electrical and Electronics Engineers. Staff who have a background in land use planning may well be members of the Royal Town Planning Institute. Other specialist institutions include the Association of Public Lighting Engineers, the Institute of Transportation Engineers (based in the USA), the Institute of Road Safety Officers and the Association of Transport Coordinating Officers. Each of these professional bodies has specialist libraries and organizes periodic conferences on specialist topics. Some of them have also sponsored special reports or studies. Their annual yearbooks and/or membership lists are a useful source of information on companies and individuals active in the field and their newsletters and journals provide a good way of keeping up to date with the field. A list of professional institutions is given below.

Chartered Institute of Transport, 80 Portland Place, London W1N 4DP, UK. Tel: +44 171 636 9952.

Institution of Civil Engineers, 1–7 Great George Street, Westminster, London SW1P 3AA, UK. Tel: +44 171 222 7722.

Institution of Electrical Engineers, Savoy Place, London WC2R 0BL, UK. Tel: +44 171 240 1871.

Institute of Electrical and Electronics Engineers, 345 East 47th Street, New York, NY10017, Tel: +1 212 705 7900 and 445 Hoes Lane, Piscataway, NJ 08854, USA. Tel: +1 908 981 0060.

Institution of Highways and Transportation, 3 Lygon Place, Ebury Street, London SW1W 0JS, UK. Tel: +44 0171 730 5245.

Institute of Road Transport Engineers, 1 Cromwell Place, London SW7 2JF, UK. Tel: +44 171 589 3744.

Institute of Transportation Engineers, 525 School Street SW, Suite 410, Washington DC 20024, USA. Tel: +1 202 554 8050.

514 Transportation and traffic planning and engineering

Another useful source of information and literature are the various associations whose prime function is to serve the needs of their members or to advance a particular cause. Examples include: The Road Haulage Association; The British Road Federation; The International Road Federation; The Bus and Coach Council; The Automobile Association; Royal Automobile Club, Transport 2000 and the Pedestrians' Association. Contact details for these organizations are listed below.

Automobile Association, Fanum House, Basingstoke, Hants RG21 2EA, UK. Tel: +44 1256 20123.

British Parking Association, 7 Hillside, Portbury, Bristol BS20 9UD, UK. Tel: +44 1275 374078.

British Road Federation, Pillar House, 194–202 Old Kent Road, London SE1 5TG, UK. Tel: +44 171 703 9769.

Bus and Coach Council, Sandinor House, 52 Lincoln's Inn Fields, London WC2A 3LZ, UK. Tel: +44 171 831 7546.

Royal Automobile Club, 14 Cockspur Street, London SW1Y 5BL, UK. Tel: +44 171 389 8900.

Pedestrians Association, 126 Aldersgate Street, London EC1A 4JQ, UK. Tel: +44 171 450 0750.

Transport 2000, Walkden House, 10 Melton Street, London NW1 2EJ, UK. Tel: +44 171 388 8386.

Relevant associations active in the local government sector include the Association of Metropolitan Authorities and the County Surveyors Society. Although such bodies do not have the resources to support a large staff or substantial libraries in-house, they are a useful source of contacts and do occasionally support specialist studies, conferences or seminars. Contact details for public sector organizations are listed below.

Association of County Councils, 66A Eaton Square, London SW1W 9BH, UK. Tel: +44 171 235 1200.

Association of District Councils, 9 Buckingham Gate, London SW1, UK. Tel: +44 171 828 7931.

Association of Metropolitan Authorities, 35 Great Smith Street, Westminster, London SW1P 3BJ, UK. Tel: +44 171 222 8100.

Association of Metropolitan District Engineers, City of Salford, Civic Centre, Chorley Road Surbiton, Swinton M27 6BW, UK. Tel: +44 161 788 8282.

County Surveyors Society (Hon Sec (1994–97) D. T. Gardner), Wiltshire County Council, County Hall, Trowbridge, Wiltshire BA14 8AD, UK. Tel: +44 1225 713301.

Department of Transport, 2 Marsham Street, London SW1P 3EB, UK. Tel: +44 171 276 3000.

Economic and Social Research Council and Science Research Council, Polaris House, North Star Road, Swindon, Wilts SN2 1UJ and SN2 1ET, respectively, UK. Tel: +44 1793 413000 and 1793 411000.

London Boroughs' Association, Westminster City Hall, Victoria Street, London SW1E 6LB, UK. Tel: +44 171 834 6788.

London Planning Advisory Committee (LPAC), Eastern House, 8–10 Eastern Road, Romford RH1 3PN, UK. Tel: +44 1708 24515.

London Research Centre, Parliament House, Black Prince Road, London SE11, UK. Tel: +44 171 222 5600.

London Regional Transport, 55 Broadway, London SW1H 0BB, UK. Tel: +44 171 222 5600.

London Traffic Control Systems Unit, Kings Building, Smiths Square, London SW1P 3HQ, UK. Tel: +44 171 821 6744.

Transport Research Laboratory, Old Wokingham Road, Crowthorne, Berks RG11 6AU, UK. Tel: +44 1344 773131 (currently (1995) still in the public sector)

UTSG (current (1995) Secretary; Dr N. Hounsell, Transportation Research Group, University of Southampton, Highfield, Southampton SO17 1BJ, UK.

The Engineering and Physical Sciences Research Council and the Economic and Social Research Council, as sponsors of research and occasional seminars in transport, are a useful source of information about current research, but more detailed information about current research in the university sector can be obtained through the Universities Transport Study Group (UTSG) or the *Current research in Britain* (CRIB) directory which is also available on CD-ROM. A number of university groups are active in transport and traffic research and it would be invidious to single out only a few from the 70 or so institutions currently belonging to UTSG. Very useful annual summaries of research underway in the main academic groups are published annually in *Traffic Engineering and Control*.

Organizations outside the UK

A number of organizations outside the UK have already been mentioned but it is worth drawing attention to the fact that international bodies are increasingly influential, particularly with respect to standards. The current trend is for an increasing proportion of transport research to be funded at an international level and for international collaboration in research to be the norm. UK universities and consultancies have a well-established tradition in transport research and are currently making an important contribution to research programmes funded by the EU and other international bodies. Some of the relevant bodies are listed below.

AASHTO, American Association of State Highway and Transportation Officials, 444 North Capitol Street, NW, Suite 225, Washington, DC 2001, USA.

APTA, American Public Transit Association, 1201 New York Avenue NW, Suite 400, Washington, DC 20005, USA.

ARRB, Australian Road Research Board, 500 Burwood Highway, Vermont South, Victoria 3133, Australia.

ASCE, American Society of Civil Engineers, 345 East 47th Street, New York, NY 10017, USA.

Commission of the European Union;

DGVII (Transport Directorate) Rue de la Loi 200, B1049 Brussels, Belgium.

DGXIII (Informatics Directorate – DRIVE Office) Rue de Treves 61, B1040 Brussels, Belgium.

ECE, United Nations Economic Commission for Europe, Palais des Nations, CH 1211 Geneve 10, Switzerland.

ECMT, European Conference of Ministers of Transport, 19 Rue de Franqueville, F–75775 Paris, Cedex 16, France.

Eurostat, Batiment Jean Monet, Plateau du Kirchberg, PO Box 1907, L–2920 Luxembourg.

INRETS, Institut National de Recherche sur les Transports et leur Securité, 2 Avenue du General Malleret Joinville, BP 34, 94114 Acruiel-Cedex, France.

Institute for Road Safety Research SWOV, PO Box 170, 2260 AD Leidschendam, Netherlands.

International Bank for Reconstruction and Development (The World Bank), 1818 H Street, NW Washington, DC 20433, USA.

IUR, International Union of Railways, 14 Rue Jean Rey, 75015 Paris, France.

IRF, International Road Federation, 63 Rue de Lausanne, Geneve 1202, Switzerland.

IRTU, International Road Transport Union, Centre Internationale, 3 rue de Vareombe, 1211 Geneve 20, Switzerland.

Laboratoire Central des Ponts et Chaussées, 28 Rue des Saints-Peres, F 75007, Paris.

National Swedish Road and Traffic Research Institute, S–581 01 Linköping, Sweden.

OECD, Organisation for Economic Cooperation and Development, 19 Rue de Franqueville, F–75775 Paris, Cedex 16, France.

PIARC, 27 Rue Guénégaud, 75006 Paris, France.

SAE, Society of Automotive Engineers, 400 Commonwealth Drive, Warrendale, PA 15096, USA.

The ENO Foundation, PO Box 2055, Saugatuch Station, Westport, CT 06880, USA.

TNO, Institute of Spatial Organisation, Postbus 6041, 2600 JA Delft, Netherlands.

Transportation Association of Canada, 2322 St Laurent Boulevard, Ottawa, Ontario, K1G 4K6, Canada.

Transportation Research Forum, 1600 Wilson Boulevard, Suite 905, Arlington, VA 22209, USA.

TRB, Transportation Research Board, 2101 Constitution Avenue NW, Washington, DC USA.

UITP, International Union of Public Transport, Avenue de l'Uruguay 19, B–1050 Brussels, Belgium.

US DOT, US Department of Transport (and Federal Highway Administration, National Highway Traffic Safety Administration, etc.), 400 7th Street SW, Washington, DC 20590, USA.

Acknowledgements

I am pleased to have this opportunity to thank my colleague Howard Kirby for his useful comments on an earlier draft of this chapter, and colleagues Roy Clarke, Frank Montgomery, Ken Fox and Yim-ling Siu for their advice on access to online databases and networks. I must however remain responsible for any errors that have crept in.

CHAPTER THIRTY

Construction engineering

W. ADDIS AND M. BUSSELL

Introduction and scope

The term construction engineering is a rather tautologous heading which comprises those aspects of engineering which concern the construction industry: in its broad connotation it is less often used than specific fields such as civil, structural, hydraulic, bridge and geotechnical engineering. In the present context it is taken to cover those aspects of engineering, not covered elsewhere in this volume, which concern the main materials employed in the construction industry and matters relating to manufacture and production using those materials. For this reason, and since it reflects how the industry is structured, the information is presented under three main headings: general information, material-related information, aspects of the construction process.

While some of the cited sources of information cover design issues, these are mainly dealt with in other specialist chapters (structural engineering (Chapter 31), transport engineering (Chapter 29) etc.). Construction management issues are also covered in another chapter (Chapter 32).

The material which follows is of two distinct types – sources of information (publications, databases, etc.) and where to find it (institutions, libraries, etc.). Furthermore, the construction industry is categorized in several overlapping ways – by profession (civil engineer, builder, etc.), by material (steel, concrete, etc.) and by building element (wall, foundation, etc.). It has therefore not always been possible to present the information according to a single structure. It has also been judged helpful to order the different data in a variety of ways which are appropriate to their context. For instance, the more general information and references are presented before the more specific; where works are of roughly equal rank, they are arranged alphabetically; British publications and organizations precede those based elsewhere. Under each construction material, there first appear organizations providing guidance, then specific technical guidance and finally books which range between the general and the specific.

Professional institutions, learned societies, research institutes, etc.

Information related to all aspects of the construction industry closely reflects the structure of the industry itself. It is, therefore, appropriate to begin by covering the main organizations which create, handle and hold construction information.

The Building Centre, 26 Store Street, London WC1E 7BT, UK. Tel: +44 171 637 1022. Fax: +44 171 580 9641.

The Building Centre is an important and unique centre for information in construction. It houses a permanent display of construction products and materials and changing exhibitions in its gallery. In its Information Exchange it has an unrivalled library of building product data as well as information about materials, standards and regulations. The European Construction Centre gives access to both computer and hard-copy databases on technical, product and legislative information and links to a network of specialist European information centres, consultancies and business organizations. The Centre for Construction Market Information provides marketing information to subscribing companies and produces marketing reports for sale to all. Last but not least there is the Building Bookshop which offers perhaps the widest selection of construction publications in Britain. The Building Centre Trust sponsors a variety of educational projects such as videos, exhibitions, slide packages and a database of training material. There are regional Building Centres in Strathclyde, Newcastle, Manchester and Bristol.

Building Research Establishment (BRE), Bucknalls Lane, Garston, Watford, Herts WD2 7JR, UK. Tel: +44 1923 894040. Fax: +44 1923 664010.

This organization is the former British government research establishment and for many decades has undertaken fundamental research in most aspects of construction related to buildings. In addition to publications emanating from this work, the BRE is one of the leading centres for information about construction and now has its own online database. The BRE publishes information to suit the full range of needs from client, house occupier and architect to the results of highly specialized engineering research. Recently there has been a growing body of work devoted to 'green' issues and energy conservation. The publications themselves appear either as bespoke reports or as one of a number of continuing series which include the following:

BRE Digests
BRE Reports
BRE Newsletter
BRE Information Directory
Good Building Guides (which have superseded the Defect Action Sheets)
Overseas Building Notes
Information Papers

Chartered Institute of Building (CIOB), Englemere, Kings Ride, Ascot, Berkshire SL5 8BJ, UK. Tel: +44 1344 23355. Fax: +44 1344 23467.

The Institute has an Information Resource Centre the contents of which are disseminated through its Construction Information Digests and covers mainly management, contractual and procurement aspects of construction. The Institute also publishes a dozen or so Construction Papers every year which cover technology, management and best practice in construction. An annual list of publications is available from the Institute's bookshop.

Construction Industry Computing Association (CICA), Guildhall Place, Cambridge CB2 3QQ, UK. Tel: +44 1223 311246. Fax: +44 1223 62865.

The CICA is a source of information and advice about the use of computers and software in construction. For several years in February it has organized Britain's leading exhibition devoted to computers in construction which acts as a valuable source of information itself.

Construction Industry Research & Information Association (CIRIA) 6 Storey's Gate, London SW1P 3AU, UK. Tel: +44 171 222 8891. Fax: +44 171 222 1708.

This independent organization undertakes, commissions and publishes reports and research findings notes across the full range of issues relating to construction with the needs of professionals particularly in mind. They issue a publications list annually which currently mentions around 250 Reports, Special Publications, Technical Notes, Books, Guides and training videos. A selection of these are also available on microfiche.

522 Construction engineering

Institution of Civil Engineers (ICE), 1–7 Great George Street, London SW1P 3AA, UK. Tel: +44 171 222 7722. Fax: +44 171 222 7500.

This is the oldest and still the leading professional institution for the construction industry. It hosts and organizes a large number of conferences, seminars, meetings and CPD events. Meetings of its members are reported in the Proceedings of the Institution, and have been for over 150 years – these publications are a major source of information for those wishing to pursue historical investigation. The proceedings are now published by Thomas Telford, in six specialist journals. In addition it publishes two periodicals *Advances in cement research*, and *Géotechnique*, the world's leading journal on geotechnical engineering.

The library, for the use of members (and others, by arrangement), is one of the world's best in its field and houses some 90 000 volumes and hundreds of periodicals (many going back well into the last century) on all aspects of construction engineering, both practical and academic. For historical research it is unsurpassed in Britain. The library subscribes to all the major CD-ROM and online databases.

Institution of Structural Engineers (IStructE), 11 Upper Belgrave Street, London SW1X 8BH, UK. Tel: +44 171 235 4535. Fax: +44 171 235 4294).

This professional institution was founded in 1908 (as the Concrete Institute) and commissions and publishes Technical Reports, design and appraisal guidance, conference and symposium papers related to structural engineering. In addition it organizes and hosts conferences, seminars and CPD events and publishes *The Structural Engineer* bi-monthly which combines academic papers with institution news. There is a library for the use of members (and others by arrangement) which has a particularly wide selection of periodicals and a good collection from the early years of reinforced concrete.

The Institution is also the British base of two important international organizations, the International Association for Bridge and Structural Engineering (IABSE) and the Fédération Internationale de la Précontrainte (FIP).

International Association for Bridge and Structural Engineering (IABSE) ETH-Hönggerberg, CH-8093 Zürich, Switzerland. Tel: +41 1 377 2647. Fax: +41 1 371 2131.

IABSE organizes regular conferences throughout the world on many aspects of engineering and their proceedings are valuable for both their depth and breadth. The British branch can be contacted at the Institution of Structural Engineers, see above.

Society for the Protection of Ancient Buildings (SPAB), 37 Spital Square, London E1 6DY, UK. Tel: +44 171 377 1644. Fax: +44 171 377 5296.

Transport Research Laboratory (TRL), (formerly the Transport & Road Research Laboratory, originally the Road Research Laboratory), Old Wokingham Road, Crowthorne, Berks RG11 6AU, UK. Tel: +44 1344 773131. Fax: +44 1344 770356.

Most countries have their own equivalents of the British professional institutions and research institutes. These can usually be tracked down in directories such as:

CIRIA Guide to European Community and international sources of construction information. 2nd ed. (1991)(CIRIA) International Directory of Engineering Societies and Related Organisations (1993) (American Association of Engineering Societies).

A few organizations of international significance are:

American Society of Civil Engineers (ASCE), 345 East 47th Street, New York.

Their publications are distributed in Europe, including Britain, by:

Clarke Associates – Europe Ltd, 4th Floor, The Rackhay, Queen Charlotte Street, Bristol BS1 4HJ, UK. Tel: +44 117 268864/225864. Fax: +44 117 226437.

École Nationale des Ponts et Chaussées, 28 rue des Saint-Pères, 75007 Paris, France. Tel: +33 1 42 60 34 13.

International Council for Building Research Studies and Documentation (CIB), PO Box 1837, NL 3000 BV Rotterdam, Netherlands. Tel: +31 10 411 0240. Fax: +31 10 433 4372.

Réunion Internationale des Laboratoires d'Essais et de Recherches sur les Matériaux et Constructions (RILEM), Avillon du Crous, 61 Avenue de Président Wilson, F–94235, Cachan, CEDEX France. (Tel: +33 1 47 40 08 59. Fax: +33 1 47 40 01 13.

Verein Deutscher Ingenieure (Association of German Engineers), Graf-Recke Strasse 84, Postfach 1139, 4000 Düsseldorf 1, Germany. Tel: +49 211 62140. Fax: +49 211 6214–575.

Standards organizations

British Standards Institution (BSI), 389 Chiswick High Road, London W4 4AL. Tel: +44 181 996 9000. Fax: +44 181 996 7400.

(also the contact point for other international and national standards organizations, including:

- International Standards Organization (ISO),
- Comité Européen de Normalisation (CEN, European Committee for Standardization),
- those of European Union nations including France (AFNOR) and Germany (DIN),
- American Society for Testing and Materials (ASTM),
- Standards Association of Australia (SAA),
- Standards Council of Canada (SCC), and other bodies)

It publishes a monthly periodical *BSI News*.

European Committee for Standardization (CEN), rue de Stassart 36, B–1050 Bruxelles, Belgium. Tel: +32 2 519 6811. Fax: +32 2 519 6819.

Trade associations

Those related to construction materials are covered under the relevant material. Others are too numerous to mention and can be found e.g. in the CIRIA Information Guide (see **General** section below).

Companies

Research in the private sector of the construction industry is rather rare. Nevertheless a few companies and their research arms are large enough to be worthy of special mention and represent valuable points of contact. They are usually fully commercial and do work for outside clients. Some of this work may lead to reports which are generally available, for instance as CIRIA Reports or private publications. The following are some of the better-known:

Arup Research and Development, 13 Fitzroy St, London W1P 6BQ, UK. Tel: +44 171 636 1531. Fax: +44 171 465 3669.

W. S. Atkins, Woodcote Grove, Ashley Road, Epsom, Surrey KT18 5BW, UK. Tel: +44 1372 726 140. Fax: +44 1372 740 055.

British Steel Structural Steel Advisory Service (BSC), Steel House, Redcar, Cleveland, TS10 5QW, UK. Tel: +44 1642 474111. Fax: +44 1642 489466.

Davis, Langdon and Everest, 39 Kingsway, London WC2B 1TP, UK. Tel: +44 171 497 9000. Fax: +44 171 379 3030.

John Laing plc, 133–139 Page Street, Mill Hill, London NW7 2ER, UK. Tel: +44 181 959 3636. Fax: +44 181 906 5297.

Pilkington Glass Ltd, Prescot Road, St Helens, Merseyside WA10 3TT, UK. Tel: +44 1744 692000. Fax: +44 1744 692882.

Redland Bricks Ltd, PO Box 3, Brampton Hill, Newcastle under Lyme, Staffs ST5 0QU, UK. Tel: +44 1782 715500. Fax: +44 1782 715511.

George Wimpey plc, 26 Hammersmith Grove, London W6 7EN, UK. Tel: +44 181 748 2000. Fax: +44 181 748 0076.

Universities

Universities in Britain are the focus of a wide variety of government and privately sponsored research, and consultancy of both a theoretical and practical nature. Every civil and structural engineering and building department undertakes this variety of work, though each tends to have its specialisms according to the particular staff and facilities it fosters.

Bodies concerned with regulatory compliance

Health and Safety Executive, Broad Lane, Sheffield S3 7HQ, UK. Tel: +44 114 892 345. Fax: +44 114 892 333.

Institute of Building Control, 21 High Street, Ewell, Epsom, Surrey KT17 1SB, UK. Tel: +44 181 393 6860. Fax: +44 181 393 1083.

National House Building Council (NHBC), Buildmark House, Chiltern Avenue, Amersham, Bucks HP6 5AP, UK. Tel: +44 1494 434 477. Fax: +44 1494 728 521.

Access to information

The construction industry is so complex that it has long been necessary to provide assistance with finding information and a number of specialist organizations have developed to serve these needs. Many of them are also concerned with originating information. Although most of

526 Construction engineering

these organizations are for practising professionals they also give a direct route to the work of academics and research institutes.

The *Construction Industry Information Group* (CIIG), 26 Store Street, London, WC1E 7BT, UK. Tel: +44 171 637 1022. Fax: +44 171 580 9641 is an organization specifically for those concerned with information sources in construction. Membership includes institution, academic, commercial and company libraries and publishers. It has regular meetings and produces a newsletter about six times a year.

In general, the libraries housed by most professional institutions, notably the Institution of Structural Engineers, Institution of Civil Engineers, the Building Services Research and Information Association (BSRIA) and the various research institutes, such as the Building Research Establishment, serve most people seeking information. Some of these libraries are open only to members; most now make a charge for some or all of their services.

Initial searches

Before beginning a search for bibliographic information it might be wise to consult the following to assess the most effective places to search, although it does not fully cover recent developments in IT such as online databases and CD-ROM:

Cox, J. (1991) *Building, construction, architecture databases: 1991* (Aslib).

Various organizations produce guides and directories which all provide good starting points and usually lead quickly to the appropriate specialists.

General

Richardson, B. G. (1989) *CIRIA UK construction information guide* (Spon).

A comprehensive directory of British sources of construction information including trade associations, research institutes, societies and other organizations (including all British local authorities). It is widely available though currently out of print; a new edition is imminent.

Sebestyén, G. and Pollington C. E. (eds) (1986) *International directory of building research, information and development organisations* 5th ed. (Spon).

Building regulations and design standards

Both the following are useful starting points, though neither is up-to-date:

A survey of building regulations worldwide 3rd ed. (1989) (BYGGDOK – the Swedish Institute of Building Documentation) in English.

Structural design standards: selected national and international titles and Sources (1985) (CIRIA).

Specification

National building specification (NBS Services Ltd) periodically updated. Library of standard clauses covering materials and workmanship for structural and other building trades.

Specification (EMAP Business Publishing). This annual volume is aimed mainly at specifiers of materials and products for construction though is effectively a good introduction for anyone. It comprises three volumes – Products, Technical and Specification Clauses. The technical volume gives data about all the main materials and elements used in building construction as well as relevant standards and Codes of Practice and a short bibliography. *Specification* is in its 88th edition (1994) and earlier editions, back to 1898, are a good source of information about old building technologies. From 1995, specification is available only as a CD-ROM.

Catalogues

Many organizations which produce publications collate their material in catalogues:

- HMSO
- British Standards Institution
- Building Research Establishment
- Construction Industry Research and Information Association (CIRIA)
- professional institutions (e.g. ICE, IStructE, CIOB)
- research institutes and trade associations (e.g. BCA, SCI, TRADA, BDA)
- Health and Safety Executive
- commercial publishers (e.g. Blackwell, Butterworths, Longman, Macmillan, Spon, Thomas Telford, Wiley, etc.).

Directories, indices and electronic databases

These are increasingly becoming available on online databases, CD-ROM or microfiche versions. Some systems offer a licence agreement allowing the user to obtain hard-copy of stored images, partially obviating the need for a paper library. University library catalogues and an increasing number of databases are accessible on the JANET and Internet systems and are easily accessible within universities. The major databases and indices are as follows:

ARCHITECTURE DATABASE (1978–)

Published by the Royal Institute of British Architects, this online database incorporates the *Architectural Periodicals Index* since 1978 and architectural books held by the RIBA library since 1984.

BARBOUR INDEX

This well known microfiche index continues to grow and been repackaged under four names: Building Technical Microfile, Building Product Microfile, Engineering Technical Microfile, and Engineering Product Microfile.

BCQ AND ON DEMAND INFORMATION

This new online database gives access to virtually all the regulations, standards and technical information for construction professionals including the full contents of publications by BRE, HMSO, BSI (British Standards, Codes of Practice), British Board of Agrément, many of the major trade associations (e.g. BCA, BDA, TRADA, NHBC) and many similar organizations, as well as a wealth of trade and product data. At present there is no overall index.

BUILDING TECHNICAL MICROFILE AND BUILDING PRODUCT MICROFILE

These two microfiche directories are aimed mainly at architects though there is some overlap with the Engineering Technical and Product Microfiles from the same publisher. The Technical Microfile serves as a compact copy of many publications of relevance to the construction industry including British Standards, Reports by BRE and CIRIA, and so on. The Product Microfile fiches are updated twice yearly and contain a list of all known manufacturers of a product; many companies catalogues are also contained on the fiche.

CITIS

CD-ROM systems for *International Civil Engineering Abstracts* (journal articles, 1972–) and *Software Abstracts for Engineers* (1984–).

CONCQUEST

A search service for publications relating to concrete design and construction from the British Cement Association.

CONSTRUCTION INFORMATION SERVICE (TECHNICAL INDEXES LTD)

This contains full copies of all the British Standards relating to construction and is, in effect, a selection from the same publisher's more

Construction engineering 529

comprehensive database WORLDWIDE STANDARDS SERVICE which covers all branches of engineering throughout the world. (TECHNICAL INDEXES LTD)

CONSTRUCTION & CIVIL ENGINEERING INDEX

This contains a wide range of literature and information about products relevant to the construction industry. It has been regularly updated for over 25 years and is available both on microfiche and CD-ROM.

CURRENT TECHNOLOGY INDEX (1981–) (BOWKER-SAUR)

Covers British periodicals in engineering and technology. Between 1962 and 1980 it was published as the British Technology Index. Available on CD-ROM and online.

EI COMPENDEX* PLUS DATABASE (1970–)

An online database surveying some 4500 journals and conference proceedings as well as selected reports and books. It covers publications worldwide in all fields of engineering, though, in any one field, is not comprehensive. It is the electronic version of the hard-copy *Engineering Index* published by the same organization (Engineering Information).

ENGINEERING INDEX (1884–) (ENGINEERING INFORMATION)

This index currently surveys, worldwide, some 4500 journals, conference proceedings and selected reports and books in all fields of engineering. The *Engineering Index* is still available as hard copy but is now also available both online and on CD-ROM. However, it has become so large that an edited version is now available devoted to civil and structural engineering only. Since 1970 it is available as an online database called EI COMPENDEX* PLUS.

ENGINEERING TECHNICAL MICROFILE AND ENGINEERING PRODUCT MICROFILE

These two microfiche directories are aimed mainly at civil, structural and construction engineers though there is some overlap with the Building Technical and Product Microfiles from the same publisher (Barbour Index). The Technical Microfile serves as a compact copy of many publications of relevance to the construction industry including British Standards, Reports by BRE and CIRIA, and so on. The Product Microfile fiches are updated twice yearly and contain a list of all known manufacturers of a product; many companies' catalogues are also contained on the fiche.

530 Construction engineering

ICE PUBLICATIONS DATABASE (1836–)

Some 25 000 records of the *Proceedings of the Institution of Civil Engineers, Géotechnique, Municipal Engineer, Transactions*, books and most ICE conferences.

ICONDA – INTERNATIONAL CONSTRUCTION DATABASE (1974–)
(FRAUNHOFER-GESELLSCHAFT INFORMATIONZENTRUM RAUM UND BAU(IRB)

Excellent on European coverage but poor on British coverage since c.1990 due to lack of input. Available on CD-ROM from SilverPlatter.

INSTRUCT (1923–)

This covers a wide range of books, periodicals and related publications (e.g. BRE Digests) held in the library of the Institution of Structural Engineers. The database includes all books, all articles in *The Structural Engineer* (the Institution journal) from 1923–date, all standards and Codes of Practice held, all reports and articles from other journals added since 1986, and many reports and series published before 1986.

Other specialist or specific databases include BRIX for BRE publications, HSELINE for health and safety sources (Health and Safety Executive), and STANDARDLINE for BSI standards.

Conference proceedings

Generally published by the organizing bodies such as ICE, IStructE, the British Masonry Society, the Council on Tall Buildings and the Urban Habitat (New York), and the Fédération Internationale de Précontrainte (FIP). The British Library produces regular lists of conference proceedings received in the British Library. The Ei Engineering Meetings, which is part of the EI COMPENDEX* PLUS database lists individual conference papers.

University PhD theses

The British Library publishes a quarterly index *British Reports, Translations and Theses*. University Microfilms International has copies of PhD theses on microfilm and these can be purchased on film or as hard-copy reprints. They cover US theses from 1861 and British theses from about 1970 (depends on the university); there is also a separate database for British theses (BRITS – the British Thesis Service). Canadian and European theses have been held since 1989 and many theses from other countries are available but this is not a comprehensive holding. University Microfilms have recently (since about 1990) started to include Masters theses in their holding, though this depends on the University. The DISSERTATION ABSTRACTS database is available online, on CD-ROM or in a printed version. Further services include a reprint service for many technical journals which are held on microfilm or microfiche, and Books on Demand which can provide reprints of some 120 000 out of print books.

Specialist publishers

The construction industry is served by a number of specialist publishers who deal variously with academic and professional books, guides, directories and yearbooks, conference proceedings, and both academic and professional serial publications. The following are probably the most significant:

Blackwell
Butterworth-Heinemann Architecture
Civil-Comp Press. Publishers of the proceedings of the main international conferences devoted to the use of computers in civil and, indeed, many other branches of engineering.
Longman
Macmillan
Spon
Thomas Telford. Originally grew out of the publications arm of the Institution of Civil Engineers and now forms probably the largest construction engineering imprint in Britain.

Periodicals

A large number of journals contain material relevant to construction and structural design. In London the libraries of the Institution of Structural Engineers (IStructE) and the Institution of Civil Engineers (ICE) receive and hold many relevant periodicals from throughout the world. Both institutions have a useful list of periodcal holdings. Back issues of the main periodicals are kept in good technical libraries.

In Britain the professional and academic interest is served principally by the two main institutions. The ICE has published a record of its proceedings since 1836. Currently these appear as the *Proceedings of the Institution of Civil Engineers* (ICE) which, since 1992, have been published in six parts – *Civil Engineering, Geotechnical Engineering, Municipal Engineer, Structures and Buildings, Transport and Water, Maritime and Energy*. The Institution of Structural Engineers publishes *The Structural Engineer* bimonthly; it dates back to 1923.

Among the more specialist periodicals, the following are widely known:

Building (The Builder Group)
Building Research and Information (Spon)
Building Services (Chartered Institution of Building Services Engineers) (Building Services Publications)

Bulletin of the International Association for Shell and Spatial Structures (IASS)
Composites. The International Journal of the Science and Technology of Reinforced Materials (Butterworth-Heinemann)
Computers and Structures (Pergamon)
Concrete (Concrete Society)
Construction and Building Materials (Butterworth-Heinemann)
Construction Computing (Englemere Services Ltd)
Construction News (Construction Publications)
Earthquake Engineering and Structural Dynamics (International Association for Earthquake Engineering, (Wiley)
Géotechnique (ICE)
International Journal of Solids and Structures (Pergamon)
International Journal of Space Structures (Multi-Science Publishing)
New Civil Engineer (EMAP)
New Steel Construction (Steel Construction Institute/BCSA) (Kingslea Press Ltd)
Structural Design of Tall Buildings (Wiley)
Structural Engineering International (International Association for Bridge and Structural Engineering)
Structural Engineering Review (Pergamon)
Structural Safety (Elsevier)
Structural Survey (MCB University Press).

In the USA the American Society of Civil Engineers (ASCE) publishes journals as hard copy and on CD-ROM on most aspects of construction engineering including:

Journal of Computing in Civil Engineering
Journal of Construction Engineering and Management
Journal of Geotechnical Engineering
Journal of Performance of Constructed Facilities
Journal of Structural Engineering.

Widely read periodicals from other countries include:

ACI Materials Journal (American Concrete Institute)
ACI Structural Journal (American Concrete Institute)
Annales de l'Institut Technique du Bâtiment et Travaux Publics (Institut Technique du Bâtiment et Travaux Publics) (France)
Annales des Ponts et Chaussées (SPPIF – Masson Service) (France)
Architectural Science Review (University of Sydney)
Bauingenieur (Association of German Engineers – VDI) (Springer Verlag)
Bautechnik (Ernst und Sohn) (Germany)
Beton Armé (Societe des Editions Andre Guerrin) (France)
Beton und Stahlbetonbau (Ernst und Sohn) (Germany)
Canadian Journal of Civil Engineering (National Research Council of Canada)
Civil Engineering (American Society of Civil Engineers)
Civil Engineering in Japan (Japan Society of Civil Engineers)
Concrete International (American Concrete Institute)
Engineering Journal (American Institute of Steel Construction)

ENR (Engineering News Record) (McGraw-Hill) (USA)
Giornale del Genio Civile (Instituto Poligrafico e Zecca dello Stato) (Italy)
Materials and Structures (RILEM) (France)
Stahlbau (Ernst and Sohn) (Germany)
Structure (Japan Structural Consultants Association).

Specialist bookshops

There are three specialist bookshops of note (the latter two of which produce useful annual catalogues):

Institution of Civil Engineers Bookshop, ICE, 1–7 Great George St, London SW1P 3AA, UK. Tel: +44 171 222 7722. Fax: +44 171 222 7500.

The Building Bookshop, Building Centre, 26 Store St, London WC1E 7BT, UK. Tel: +44 171 637 3151. Fax: +44 171 636 3628.

RIBA Bookshop, 66 Portland Place, London W1N 4AD, UK. Tel: +44 171 251 0791. Fax: +44 171 255 1541. RIBA has regional outlets in Belfast, Birmingham, Cambridge, Leeds, Manchester and Nottingham.

General reference works

Terminology

With international work and legal contracts becoming more common, precise terminology is essential. The standard work of reference (624 pages) for the English language is:

British Standards Institution (1993) *Glossary of building and civil engineering terms (BS 6100)* (Blackwell for BSI).

More general guidance might be found in:

Maclean, J. H. and Scott, J. S. (1993) *Penguin dictionary of building*, 4th ed. (Penguin).
Scott, J. S. (1991) *Penguin dictionary of civil engineering*, 4th ed. (Penguin).

General reference

The following is a small selection of general books which are all readily available. They all refer the reader to more specialist sources where appropriate:

Black, L. (ed.) (1991) *Builder's reference book*, 12th ed. (Spon).
Blake, L. S. (ed.) (1989) *Civil engineer's reference book*, 4th ed. (Butterworth-Heinemann).

Groàk, S. (1992) *The idea of building* (Spon).
Harvey, R. C. and Ashworth, A. (1993) *The construction industry of Great Britain* (Butterworth-Heinemann).
Housing and construction statistics 1981–1982: Great Britain (1993) (HMSO, Department of the Environment).
Merritt, F. S. (ed.) (1983) *Standard handbook for civil engineers*, 3rd ed. (McGraw-Hill).

Yearbooks and directories

Thomas Telford publishes a range of annual directories and yearbooks which cover most aspects of construction engineering. Each comprises a general introduction including an extensive bibliography, information about relevant British Standards and Codes of Practice, lists of relevant institutes and associations both in Britain and abroad, materials and products data, and addresses of many firms offering consulting and contracting services and those concerned with construction plant and equipment. The following are a selection:

Concrete yearbook
Ground engineering yearbook
NCE road construction & traffic directory
Steel construction yearbook
Underground directory
Waste, recycling and environmental directory
Water directory.

The European steel industry has published its own directory:

ECCS steel construction international directory (1992) (Kingslea Press).

Construction history

Compared to other branches of engineering, history plays a very important role in the work of the construction engineer – from the protection of remains of archaeological importance and preservation of buildings several hundred years old to the adaptation and rehabilitation of old buildings for modern use or building demolition. These all require an understanding of the materials and construction techniques of the past.

The main organizations concerned with old buildings themselves are the Panel for Historic Engineering Works (PHEW), based at the ICE, and the Society for the Protection of Ancient Buildings (SPAB). These seek to record works of historic importance and play a role in their responsible treatment. They can also be a useful guide towards getting appropriate professional assistance in dealing with such buildings. As with other areas of construction, expertise and information is gathered sometimes according to material, sometimes building type, sometimes the date of construction.

General information about construction history can be found in the annual periodical *Construction History* produced by the Construction History Society which is based at the CIOB offices in Ascot. Information is best sought through personal recommendation which can be found by consulting persons associated with PHEW, the CHS or the History Study Group based at the IStructE.

For general construction and building history, the following are recommended and all contain references to other more specialized works.

Balcombe, G. (1985) *History of building* (Batsford).
Bowyer, J. (1993) *History of building*, 3rd ed. (Attic Books).
Cowan, H. J. (1977) *An historical outline of architectural science*, 2nd ed. (Applied Science).
Davey, N. (1961) *A history of building materials* (Phoenix House).
Elliott, C. D. (1992) *Technics and architectural sciences* (MIT Press).
Graefe, R. (ed.) (1989) *Zur geschichte des konstruierens* (Deutsche Verlags-Anstalt).
Pannell, J. P. M. (1964) *An illustrated history of civil engineering* (Thames & Hudson).
Straub, H. (1952) *A history of civil engineering* (Leonard Hill).
Strike, J. (1991) *Construction into design: the influence of new methods of construction on architectural design 1690–1990* (Butterworth-Heinemann).

Old and obsolete technical publications are often held by the original publishers (e.g. British Standards by BSI) or by the ICE or IStructE libraries, in reserve stock on limited access. Such material, and much old product literature, is also held by the Science Reference and Information Service of the British Library.

Construction materials (general)

Construction materials include very nearly all materials and so a brief comprehensive general work is a contradiction in terms. Nevertheless, between them the following give most of what is needed and all should be consulted before delving deeper:

Addleson, L. and Rice, C. (1991) *Performance of materials in buildings: a study of the principles and agencies of change* (Butterworth-Heinemann).
Doran, D. K. (1992) *Construction materials reference book* (Butterworth-Heinemann).
Illston, J. M. (ed.) (1993) *Construction materials: their nature and behaviour*, 2nd ed. (Spon).
Jackson, N. and Dhir, R. K. (eds) (1988) *Civil engineering materials*, 4th ed. (Macmillan).

In the construction industry there is probably more experimentation (sometimes inadvertent!) in using materials in new combinations than in most other industries. It is an unfortunate fact that materials are often specified by those who do not fully understand the consequences of their selections. It must also be said that buildings are often expected to last longer than many other artefacts and long-term performance is often difficult to predict. When in doubt (and perhaps also when not in doubt) experts must be consulted. The trade associations for different materials are of invaluable help. With such difficulties in mind, there is one organization which collects all types of engineering materials data and makes them available to members in the manner which best suits the needs of design and research engineers:

ESDU International plc (formerly Engineering Sciences Data Unit), 27 Corsham Street, London N1 6UA, UK. Tel: +44 171 490 5151. Fax: +44 171 490 2701.

Concrete

Organizations providing guidance

The following bodies all provide technical information to members and publish a wide range of material from the introductory to the latest research:

British Cement Association (BCA), Century House, Telford Avenue, Crowthorne, Berkshire RG11 6YS, UK. Tel: +44 1344 762 676. Fax: +44 1344 761 214.

The British Cement Association, formerly the Cement and Concrete Association, publishes a large number and range of *Guides, Construction guides, Interim technical notes,* bibliographies, reprints, *R.c. reviews, Project profiles* and videos.

British Precast Concrete Federation, 60 Charles Street, Leicester LE1 1FB, UK. Tel: +44 1533 536 161. Fax: +44 1533 514 568.

Concrete Society, Framewood Road, Wexham Springs, Slough, Bucks SL3 6PJ, UK. Tel: +44 1753 662226, Fax: +44 1753 662126.
   Publishes many technical reports, guides and digests.

Fédération Internationale de la Précontrainte (FIP) (International Prestressing Federation), c/o Institution of Structural Engineers (IStructE), 11 Upper Belgrave Street, London SW1X 8BH, UK. Tel: +44 171 235 4535. Fax: +44 171 235 4294.
   An international organization for the development of structural concrete; it publishes many recommendations, guides to good practice, reports, state of the art reports, technical reports and special reports on both reinforced and prestressed concrete.

American Concrete Institute, 22400 West 7 Mile Road, Detroit, MI 48219–1849, USA. Tel: +1 313 532 2600. Fax: +1 313 538 0655.

Technical guidance

Technical guidance about concrete design and production is in abundance. The following is a selection of the standard works:

BS 1881: 1970–1992 Testing concrete
BS 5075: 1982–1985 Concrete admixtures
BS 5328: 1990–1991 Concrete
BS 5975: 1982 Code of practice for falsework
BS 6073: 1981 Precast concrete masonry units
BS 6089: 1981 Guide to assessment of concrete strength in existing structures

BRE Digest 325 Concrete part 1 materials (1987) (BRE).
BRE Digest 326 Concrete part 2 specification design and quality control (1987) (BRE).

The BCA is home to the Concrete Information Service which publishes *Concrete Current Awareness*, a monthly periodical giving bibliographic details of all the information about concrete it receives. The BCA itself publishes many specialist guides and reports of which the following is a selection:

Appearance matters (parts 1–9) Ref 47.101–9
Concrete mixes for general purposes Ref 45.031
Concreting in hot weather Ref 45.013
Reinforcement Ref 45.105
Winter concreting Ref 45.007

Books

Good technical libraries will have many books on concrete, a large number of which are unfortunately now out of print. The following should all be currently available:

Allen, R. T. L. *et al.* (eds) (1992) *Repair of concrete structures*, 2nd ed. (Spon).
Bungey, J. H. (1989) *Testing of concrete in structures*, 2nd ed. (University of Surrey Press).
Clarke, J. L. (ed.) (1993) *Structural lightweight aggregate concrete* (Spon).
Cooke, T. H. (1990) *Concrete pumping and spraying* (Thomas Telford).
Dewar, J. D. and Anderson, R. (1992) *Manual of ready-mixed concrete* (Blackie/Chapman and Hall).
Fordyce, M. W. and Wodehouse, R. G. (1983) *GRC and buildings* (Butterworths).
Hewlett, P. C. (ed.) (1988) *Concrete admixtures*, 2nd ed. (Longman).
Kohlhass, B. (1983) *Cement engineers' handbook*, 4th ed. (Bauverlag).
Lauritzen, E.K. (ed.) (1993) *Demolition and reuse of concrete and masonry* (Spon).

Mays, G. H. (ed.) (1992) *Durability of concrete structures: investigation, repair and protection* (Spon).
Neville, A. M. (1986) *Properties of concrete*, 3rd ed. (Longman).
Neville, A. M. and Brooks, J. J. (1987) *Concrete technology* (Longman).
Perkins, P. H. (1986) *Repair, protection and waterproofing of concrete structures* (Elsevier).
Richardson, J. G. (1977) *Formwork construction and practice* (Viewpoint).
Richardson, J. G. (1991) *Quality in precast concrete: design, production, supervision* (Longman).
Tattersall, G. H. (1991) *Workability and quality control of concrete* (Spon).
Taylor, H. P. J. (ed.) (1992) *Precast concrete cladding* (Edward Arnold).
Waddell, J. J. and Dobrowolski, J. A. (1993) *Concrete construction handbook*, 3rd ed. (McGraw-Hill).

Steel

Organizations providing guidance

The following organizations all provide technical information to members and publish a wide range of material to help designers and constructors:

British Cast Iron Research Association (BCIRA), Cast Metals Technology Centre, Alvechurch, Birmingham B48 7QB, UK. Tel: +44 1527 66414. Fax: +44 1527 585070.

British Constructional Steelwork Association Ltd (BCSA), 4 Whitehall Court, London SW1A 2ES, UK. Tel: +44 171 839 8566. Fax: +44 171 976 1634.
Publications include

design guides
Handbook of structural steelwork (1990) (with SCI)
Historical structural steelwork handbook (1984)
National structural steelwork specification for building construction, 3rd ed. (1994).

It also handles the publications of other organizations such as:
American Institute of Steel Construction (AISC)
European Convention for Constructional Steelwork (ECCS).

BSC General Steels, Commercial Division – Sections, PO Box 24, Steel House, Redcar, Cleveland TS10 5QL, UK. Tel: +44 1642 474 111. Fax: +44 1642 489 466.
Provides a wide range of information such as the guide to BSC Sections: Structural Sections and corrosion protection guides.

Construction engineering

Steel Construction Institute (SCI), Silwood Park, Ascot, Berks SL5 7QN, UK. Tel: +44 1344 23345. Fax: +44 1344 22944.

A research institute (which incorporates the work of the former CONSTRADO) which produces design guides, commentaries on codes of practice and manuals on all matters relating to the use of steel in buildings. Much is also published in association with mainstream publishers.

American Institute of Steel Construction, One East Walker Drive, 400 North Michigan Avenue Ste. 3100, Chicago, IL 60601 2001, USA. Tel: +1 312 670 2400. Fax: +1 670 5403.

Technical guidance

Technical guidance about steel design and production is in abundance. The following is a selection of the standard works:

BS 4: Structural steel sections Part 1: 1993 Specification for hot-rolled sections. Partially replaces BS4: Part 1: 1980 Specification of structural steelwork.
BS 4848: 1980–1991 Specification for hot-rolled structural steel sections.
BS 5493: 1977 Code of practice for protective coating of iron and steel structures against corrosion.

Of very practical use to the full range of professionals involved in steel construction is a series produced by the BCSA usually referred to by their colour, and updated frequently:

- *The red book (designing)* – Handbook of structural steelwork
- *The green book (connecting)* – Joints in simple construction, Vol. 2: Practical applications
- *The black book (specifying)* – National structural steelwork specification for building construction, 3rd ed.
- *The orange book (contracting)* – The contractual handbook
- *The blue book (erecting)* – Erectors' manual

Books

Good technical libraries will have many books on steel. The following is a selection of useful titles:

Angus, H. T. (1976) *Cast iron: physical and engineering properties*, 2nd ed. (Butterworths)
Blanc, A., McEvoy, M. and Plank, R. (eds) (1993) *Architecture and construction in steel* (Spon).
Chandler, K. A. and Bayliss, D. A. (1985) *Corrosion protection of steel structures* (Elsevier).
Davies, B. J. and Crawley, E. J. (1980) *Structural steelwork fabrication* (British Constructional Steelwork Association).

Dowling, P. J., Owens, G. W. and Knowles, P. (1988) *Structural steel design* (Butterworths).
Hart, F., Henn, W. and Sontag, H. (1985) *Multi-storey steel framed building* English edition G. B. Godfrey (ed.) (Collins).
Smith, R. C. and Lawrence, L. (eds) (1992) *Fire protection for structural steel in buildings*, 3rd ed. (Association of Specialist Fire Protection Contractors and Manufacturers and Steel Construction Institute).
Twelvetrees, W. N. (1900) *Structural iron and steel* (Fourdrinier).

Natural stone, masonry and brickwork

The term 'masonry' includes load-bearing elements and facades of both natural stone and bricks; it is sometimes used to describe mass concrete (without reinforcement), and the general term 'ceramic' is also sometimes used.

Organizations providing guidance

Brick Development Association (BDA), Woodside House, Winkfield, Windsor, Berks SL4 2DX, UK. Tel: +44 1344 885651. Fax: +44 1344 890129.

The BDA is the principal focus of information about brickwork. It publishes many design guides, engineers' file notes, design notes and technical information papers.

Other relevant organizations include:

British Masonry Society, c/o CERAM Research (see below).
This society publishes much useful information, including, especially:

Publication 92/1, sources of technical information for the design, construction and maintenance of masonry structures (1992) A substantial reference work in itself.

CERAM Research (formerly British Ceramic Research Limited), Queens Road, Penkhull, Stoke-on-Trent ST4 7LQ, UK. Tel: +44 1782 45431. Fax: +44 1782 412331.

CERAM publishes many research papers and Special Publications, and specifications such as the *Model specification for clay and calcium silicate structural brickwork* (1988).

Stone Federation of Great Britain, 82 New Cavendish Street, London W1M 8AD. Tel: +44 171 580 5588. Fax: +44 171 631 3872.

Technical guidance

BRE Digest 362 building mortar (BRE) (1991).

Natural stone is of particular concern to restorers of buildings and the following are all of use:

Hart, D. (1988) *The building magnesian limestones of the British Isles* (HMSO).
Hart, D. (1991) *The building slates of the British Isles* (HMSO).
Leary, E. (1983) *The building limestones of the British Isles* (HMSO).
Leary, E. (1986) *The building sandstones of Great Britain* (HMSO).
Schaffer, R. J. (1972) *The weathering of natural building stones* (HMSO).

Those considering new stone will find help in:

The selection of natural building stone BRE Digest 269 (1991) (HMSO).

Books

Books on design guidance are listed in Chapter 31. Two general works are:

Brick Development Association (1994) *Brick Development Association guide to successful brickwork* (Edward Arnold).
Sowden, A. M. (ed.) (1990) *The maintenance of brick and stone masonry structures* (Spon).

Timber

Organizations providing guidance

Timber Research & Development Association (TRADA), Stocking Lane, Hughenden Valley, High Wycombe, Bucks HP14 4ND, UK. Tel: +44 1494 563091. Fax: +44 494 565487.

TRADA publishes a number of books and a vast range of detailed information in the form of Wood Information sheets.
 Both the BRE and CIRIA also publish much useful information on timber.
 Many other countries have institutes equivalent to TRADA, for example:

American Institute of Timber Construction, 11818 S.E. Mill Plain Blvd, Vancouver, WA 98684–5092, USA. Tel: +1 206 254 9132. Fax: +1 206 254 9456.

Construction engineering

Technical guidance

TRADA publications provide useful coverage.

The specification of timber is addressed by a short BRE information paper (*How to specify structural timber* BRE IP 5/79) and in many British Standards, of which three important ones are:

BS 4978: 1988 Specification for softwood grades for structural use.
BS 5756: 1985 Specification for tropical hardwoods graded for structural use.
BS 6446: 1984 Specification for manufacture of glued structural components of timber and wood-based panel products.

Books

The following books give a good overview and provide reference to other more specialist works:

Baird, J. A. and Ozelton, E. C. (1989) *Timber designer's manual*, 2nd ed. (BSP Professional).
Sunley, J. and Bedding, B. (eds) (1985) *Timber in construction* (Batsford/TRADA).
Timber construction manual, 4th ed. (1994) (American Institute of Timber Construction).
Yeomans, D. (1992) *The architect and the carpenter* (RIBA Heinz Gallery).

Glass and glass composites

See Chapter 31 on structural engineering.

Aluminium and its alloys

See Chapter 31 on structural engineering.

Sealants and adhesives

Organizations providing guidance

Poor or inappropriate use of sealants and adhesives probably accounts for more than half of all problems found in buildings. Good guidance can be obtained from:

British Adhesives and Sealants Association, 33 Fellowes Way, Stevenage, Herts SG2 8BW, UK. Tel: +44 1438 358514. Fax: +44 1438 742565.

Technical guidance

BRE Digest 340, choosing wood adhesives (1989) (BRE).
TN128 Civil engineering sealants in wet conditions – review of performance and interim guidance on use (1987) (CIRIA).

Books

The following books are all currently available:

Klosowski, J. M. (1989) *Sealants in construction* (Marcel Dekker).
Mays, G. C. and Hutchinson, A. R. (1992) *Adhesives in civil engineering* (Cambridge University Press).
Panek, J. R. and Cook, J. P. (1991) *Construction sealants and adhesives*, 3rd ed. (Wiley).

Construction plant and equipment

Technical guidance

BS 5228: 1984–1992 Noise control on construction and open sites
BS 5531: 1988 Code of practice for safety in erecting structural frames
BS 5744: 1979 Code of practice for safe use of cranes
BS 6031: 1981 Code of practice for earthworks
BS 6187: 1982 Code of practice for demolition
BS 8000: 1989–1994 Workmanship on building sites
BS 8004: 1986 Code of practice for foundations
BS 8103 Structural design of low-rise buildings. Part 1: 1986 Code of practice for stability, site investigation, foundations and ground floor slabs for housing.

It is inevitable in a very practical area of the engineer's work that up-to-date books going into great detail are rare. Trade catalogues and manufacturers' information provides the latest detail, usually with cost information, while a wide range of general and introductory books serve the educational and training market:

Building and Construction Index (1993) (Benn) A comprehensive guide to construction plant and their suppliers. It comprises the former *Sell's Building and Construction Index* and *Construction Weekly Plant Directory*.
Spon's Plant and Equipment Guide (Building Materials Market Research) (annual, loose-leaf) Gives price, size and performance data for new and used plant.

Books

Building Advisory Service (1993) *Construction safety* (Building Employers Confederation) Updated regularly in 32 sections.
Davies, V. J. and Tomasin, K. (1990) *Construction safety handbook* (Thomas Telford).

Day, D. A. and Benjamin, N. B. H. (1991) *Construction equipment guide*, 2nd ed. (Wiley).
Edmeades, D. H. (1972) *The construction site* (Estates Gazette).
Harris, F. (1989) *Modern construction and ground engineering equipment and methods* (Longman).
Harris, F. and McCaffer, R. (1991) *Management of construction equipment*, 2nd ed. (Macmillan).
Illingworth, J. R. (1982) *Site handling equipment* (Thomas Telford).
Illingworth, J. R. (1993) *Construction methods and planning* (Spon).
Whyte, I. L. (ed.) (1988, 1990, 1992) *Decommissioning and demolition. Proceedings of an international conference 1, 2 and 3* (Thomas Telford).

On a smaller scale the Institution of Civil Engineers publishes a series of *Works construction guides* (Thomas Telford) on most aspects of site practice including temporary works, excavation, formwork, piling and safety.

Temporary works

Temporary works comprise all those structural and construction equipment and plant which are removed from the site after a project is complete. It is in its very nature that it touches on many different aspects of the construction process and the many professionals and subcontractors involved. Its successful management and control is essential to safe and cost-effective work: unfortunately this ideal is not always achieved and failures of temporary works, usually as a result of poor organization and planning, still account for many deaths each year.

Technical guidance

BS 5531: 1988 Code of practice for safety in erecting structural frames
BS 5973: 1993 Code of practice for access and working scaffolds and special scaffold structures in steel
BS 5974: 1990 Code of practice for temporarily installed suspended scaffolds and access equipment
BS 5975: 1982 Code of practice for falsework
BS 6100: section 6.5: 1987 Formwork

Books

A number of excellent books and guides should help to reduce the number of accidents, but only if their advice is heeded:

Concrete Society and Institution of Structural Engineers (1986) *Formwork: a guide to good practice* (Concrete Society).
Illingworth, J. R. (1987) *Temporary works: their role in construction* (Thomas Telford).

Irvine, D. J. and Smith, R. J. H. (1992) *Trenching practice* CIRIA report 97, rev. ed. (CIRIA).
Irwin, A. W. and Sibbald, W. I. (1984) *Falsework, a handbook of design and practice* (Granada).
Mackay, E. B. (1986) *Proprietary trench support systems* technical note 95, 3rd ed. (CIRIA).
Timber in excavations, 3rd ed. (1990) (Thomas Telford/TRADA).

Civil engineering

This term usually includes all branches of construction except those relating to buildings. Although there is no room here to deal with any field comprehensively, the following general works are worth mentioning:

Organizations providing guidance

HMSO publishes works for various departments, including:

Department of Transport: *Design manual for roads and bridges (1993)A* multivolume work providing essential guidance.
Manual of contract documents for highway works (1992).
Transport Research Laboratory: Contractor reports, research reports, state of the art reviews, supplementary reports.

Books

Department of Transport design manual for roads and bridges (1993) (HMSO).
Golze, A. R. (1977) *Handbook of dam engineering* (Van Nostrand Reinhold).
Graff, W. J. (1981) *Introduction to offshore structures: design, fabrication and installation* (Gulf Publishing).
Jansen, R. B. (ed.) (1988) *Advanced dam engineering: for design, construction and rehabilitation* (Van Nostrand Reinhold).
Metcalf and Eddy Inc. (1991) *Wastewater engineering: treatment disposal and reuse*, 3rd ed. (McGraw-Hill).
Megaw, T. M. and Bartlett, J. B. (1981–1982) *Tunnels: planning, design and construction* (Ellis Horwood).
Podolny, W. and Scalzi, J. B. (1986) *Construction and design of cable stayed bridges*, 2nd ed. (Wiley).
Salter, R. J. (1983) *Highway traffic analysis and design*, 2nd ed. (Macmillan).
Salter, R. J. (1988) *Highway design and construction*, 2nd ed. (Macmillan).
Whittaker, B. N. and Frith, R. C. (1990) *Tunnelling: design, stability and construction* (Institution of Mining and Metallurgy).
Wignall, A., Kendrick, P. S. and Ancill, R. (1991) *Roadwork: theory and practice*, 3rd ed. (Newnes/Butterworth-Heinemann).

CHAPTER THIRTY-ONE

Structural engineering

M. BUSSELL AND W. ADDIS

Introduction and scope

Structural engineering has been defined comprehensively (if not succinctly) by the Institution of Structural Engineers as: 'the science and art of designing and making, with economy and elegance, buildings, bridges, frameworks, and other similar structures so that they can safely resist the forces to which they may be subjected'.

Structural engineering can be viewed as a major branch of civil engineering, standing alongside hydraulic, ground, transportation and environmental engineering. It is also an essential element within construction engineering, which is covered in chapter 30. That chapter deals with general information on construction, as well as with information related to construction materials and aspects of the actual construction process.

This present chapter concentrates mainly on information relating to structural design, an activity usually performed by structural engineers but invariably requiring collaboration with other relevant disciplines. For buildings and other structures these disciplines include architecture, environmental and ground engineering, and both quantity and building surveying. In addition there are the building and civil engineering contractors and suppliers responsible for turning a design into the built reality and, not least, the regulatory bodies (building control authorities, Health and Safety Executive, etc.) established by statute to ensure safety both during construction and in the completed work.

To match the diversity of disciplines there is often a corresponding fragmentation of organization. At its most fundamental this is epitomised by a design, prepared for a client by a design team, being tendered for and subsequently constructed by a contractor who had no involvement in the design process. (Even the currently fashionable and growing approach of 'design and build' will usually involve an *ad hoc* conjunction of several organizations, each contributing its particular skills.) Consequently the structural designer needs to be aware of the work of related disciplines and practicalities of construction, as well as understanding the management of the design process. This is not overlooked in the sources cited below, although contractual matters, procurement, site organization and other construction matters are dealt with under construction engineering (chapter 30) and construction management (chapter 32).

The present chapter has been compiled jointly with the chapter on construction engineering and duplication has been avoided wherever possible. It is therefore recommended that readers always consult the relevant section in the other chapter.

The information is not presented according to a single structure. It has also been judged helpful to order the different data in a variety of ways which are appropriate to their context. For instance, the more general information and references are presented before the more specific; where works are of roughly equal rank, they are arranged alphabetically. Under each construction material, there first appear organizations providing guidance, then relevant standards followed by specific technical guidance and books which range between the general and the specific.

Information has been arranged in several categories; after an overview of structural design and work on existing structures, specific design guidance is presented on

- structural loadings and fire engineering
- construction materials, subdivided into the design of new structures and the appraisal and refurbishment of existing structures
- building elements
- building and structure types.

Information specialists, periodicals, professional institutions, research organizations, learned societies, trade associations, etc.

Details of organizations and general reference works are covered in Chapter 30 on Construction Engineering.

Structural design – general

Structural engineers are usually guided by the so-called Codes of Practice (now British Standards e.g. BS 8110: 1985) which cover loads acting on structures, the properties of the various structural materials and the design and construction of structures using those materials. The codes are not legal documents; however, if an engineer chooses not to follow them, very good reasons will be sought. Nowadays the Codes are becoming enshrined in computer programs, a move which many engineers fear will encourage the avoidance of original and critical thought.

With an eye to the future, it is worth noting that an entirely new set of Codes is soon to be adopted by all the members of the European Union. Drafts of the following Eurocodes are currently available or will soon be available from British Standards Institution as ENVs (European pre-standards) and will all become current within the next 3–4 years:

EC1 – Basis of Design and Actions on Structures
EC2 – Design of Concrete Structures
EC3 – Design of Steel Structures
EC4 – Design of Composite Steel and Concrete Structures
EC5 – Design of Timber Structures
EC6 – Design of Masonry Structures
EC7 – Geotechnical Design
EC8 – Earthquake Resistant Design of Structures
EC9 – Design of Aluminium Alloy Structures

REGULATORY GUIDANCE

The scope and context of structural design in Britain is outlined in the Building Regulations and the forthcoming regulations that will cover construction, design and management. For guidance see:

Approved document 'A' of the Building Regulations 1991: Structure (1991) (HMSO).
Davis, L. *Guide to the Building Regulations 1991: for England and Wales* (1992) (Butterworth Architecture).
N.B. Separate regulations apply in Scotland and in Northern Ireland.

The *Construction (design and management) regulations 1994* (1995) (HMSO) require all parties including clients, designers and contractors to give formal consideration to health and safety of all involved in construction and maintenance.

GENERAL GUIDANCE

Structural engineers need to be able to deal with a great many different materials and types of construction – both modern and old, maybe several hundred years old. An ability to find and use the appropriate specialist design and construction guidance is thus an essential skill for structural engineers. A number of organizations publish a wide range of publications to assist designers of building structures. The Building Research Establishment (BRE) and CIRIA produce both general and specific information relating to materials, existing buildings and the

550 Structural engineering

design and construction of new buildings. The Institution of Structural Engineers (IStructE) has produced a range of valuable overviews of key aspects of the engineer's job:

Aims of structural design 2nd ed. (1987)
The achievement of structural adequacy in buildings (1990)
Stability of buildings (1988)
Communication of structural design (1974)
Structural failures in buildings (1980)
Qualitative analysis of structures (1989)
Guide to good management practice for engineering design offices (1991).

Specific technical guidance on structural design is covered in later sections. The following are of more general relevance:

Addis, W. (1994) *The art of the structural engineer* (Ellipsis, formerly Artemis).
Alexander, S. J. and Lawson, R. M. (1981) *Technical note TN107. Design for movement in buildings* (CIRIA).
Hodgkinson, A. J. (ed.) (1980) *Handbook of building structure* 2nd ed. (Architectural Press) Good general overview but becoming elderly.
Report of the Standing Committee on Structural Safety (IStructE) every 3–5 years.

Structural design must interact with the other functional needs of a building, such as weathertightness, thermal and acoustic performance, etc. The Building Regulations and their Approved Documents are a natural starting point for guidance. Numerous reports and other guidance are published, in particular by BRE and CIRIA.

This chapter does not set out to cover the analysis of structures, which is an increasingly mathematical and computer-dependent activity for which there is a vast and growing specialist literature, not all of which is directly useful to the designer. It is, however, worth mentioning one series of long-established and authoritative texts by the late S. P. Timoshenko, published by McGraw-Hill and part of the engineering experience of many practitioners in Britain and the USA. The series, frequently reprinted, includes:

Theory of structures
Strength of materials (2 vols)
Theory of elasticity
Theory of elastic stability
Theory of plates and shells

In a world becoming more reliant on computers, it is now well recognized how essential it is to understand the behaviour of structures without relying wholly on software. Essential checks which should be carried out continuously by the wise engineer are outlined in:

Macleod, I. A. (1988) *Technical Note TN133. Guidelines for checking computer analysis of building structures* (CIRIA).

Work on existing structures

The structural designer is increasingly concerned with existing structures and their investigation, appraisal, testing, repair and maintenance, and alteration and strengthening for re-use. This demands additional skills and knowledge. Building design and construction techniques change with time, so it is essential to have an understanding of the approach and methods of the original designers (if indeed there were any), and of the original builders. Secondly, the existing structure is a working entity that does not necessarily conform to modern codes of practice. So, to determine its adequacy calls for engineering judgement and pragmatism, coupled with a systematic approach. And, finally, any alteration and strengthening work must involve the designer in considering the stability of the structure while the work is done.

GUIDANCE ON INSPECTION, APPRAISAL, REPAIR AND STRENGTHENING

Primary guidance:

Appraisal of existing structures (1980) (IStructE).
BRE Digest 366, Structural appraisal of existing buildings for change of use (1991) (BRE).
Health and Safety Executive (1990) *Guidance Note G58, evaluation and inspection of buildings and structures* (HMSO)
Guide to surveys and inspections of buildings and similar structures (1991) (IStruct).
Report 111, structural renovation of traditional buildings (1994) (CIRIA).

Other guidance:

Beckmann, P. (1995) *Structural aspects of building conservation* (McGraw-Hill)
Brebbia, C. A. et al. (eds) (1989, 1991, 1993) *Structural repair and maintenance of historical buildings I, II and III* (Computational Mechanics Publications) Conference proceedings.
Health & Safety Executive (1992) *Guidance note GS51, façade retention* (HMSO).
Highfield, D. (1991) *The construction of new buildings behind historic facades* (Spon).
Holland, R., Montgomery-Smith, B. E. and Moore, J. F. A. (eds.) (1992) *Appraisal and repair of building structures* (Thomas Telford) Further volumes are planned for concrete, steel and iron, masonry and timber.
Load testing of structures and structural components (1989) (IStructE).
Robson, P. (1991) *Structural appraisal of traditional buildings* (Gower Technical).
Structural Survey (MCB University Press) quarterly.

The Society for the Protection of Ancient Buildings has published a number of Technical Pamphets and Information Sheets on causes of deterioration in old buildings and on suitable remedial measures.
Particular building types have generated the need for systematic

appraisal as a result of performance failures. Sources on concrete system buildings and steel-framed housing are referred to where relevant below. Sports grounds also have been subject to attention:

Appraisal of sports grounds (1990) (IStructE).

SOURCES FOR BUILDING CONSERVATION

Ashurst, J. and Ashurst, N. (1988) *Practical building conservation*, 5 vols (Gower Technical).
Brereton, C. (1991) *The repair of historic buildings: advice on principles and methods* (English Heritage).
Conservation Unit, Museums and Galleries Commission (1991) *Conservation sourcebook* (HMSO) Lists many organizations involved in conservation, including technical specialists.
Smith, J. F. (1978) *A critical bibliography of building conservation* (Mansell) Not up-to-date, but useful.
Strengthening of historic monuments and structures (1986) (ICOMOS – the International Council on Monuments and Sites) A bibliography, again not fully up-to-date, but supplementing the previous source.

SOURCES ON ORIGINAL BUILDING CONSTRUCTION METHODS

Much information on earlier forms of building construction is to be found in contemporary textbooks. A few have been reprinted recently and are listed in the relevant sections below. Others such as earlier editions of *Mitchell's building construction* and *Dorman Long safe load tables for steel* – both originating a century or so ago – are now collectable items and scarce, although major institutional libraries (ICE, IStructE, and the Royal Institute of British Architects, as well as the British Library) are valuable sources for such material.

The following works contain much general and particular information concerning a variety of existing building types:

Ackermann, K. (1991) *Building for industry* (Watermark).
Fitzgerald, R. (1988) The development of the cast iron frame in textile mills to 1850 *Industrial Archaeology Review*, **10**, 127–145.
Giles, C. and Goodall, I. H. (1992) *Yorkshire textile mills: the buildings of the Yorkshire textile industry 1770–1930* (HMSO).
Hay, G. D. and Stell, G. P. (1986) *Monuments of industry – an illustrated historical record* (Royal Commission on the Ancient and Historical Monuments of Scotland).
Melville, I. A. and Gordon, I. A. (eds) (1992) *Structural surveys of dwelling houses* 3rd ed. (Estates Gazette).

Locating information on a particular existing building is time-consuming but essential. Sources to be considered include previous owners, the original designers, the local authority building control department, local archives, and the following more general sources:

Structural engineering 553

- Journals publishing articles and papers on individual buildings are usually indexed by the publisher; for the present century the *Architects' Journal* and *Architectural Review* (EMAP Publishing), *The Structural Engineer* (IStructE), and the *Proceedings of the Institution of Civil Engineers (PICE)*; for the last century the *PICE*, the *Civil Engineers and Architects' Journal*, and *The Builder*
- British Architectural Library at the Royal Institute of British Architects
- National Register of Archives for possible archive locations.

STRUCTURAL HISTORY

In addition to the historical references given in Chapter 30, the following are examples of historical studies that include useful information on design thinking and construction methods of the past:

Addis, W. (1990) *Structural engineering: the nature of theory and design* (Ellis Horwood).
Collins, A. R. (ed.) (1983) *Structural engineering – two centuries of British achievement*, 2nd ed. (Tarot Print for IStructE).
Mainstone, R. (1975) *Developments in structural form* (Allen Lane).
Timoshenko, S. P. (1953) *History of strength of materials* (McGraw-Hill).
Todhunter, I. and Pearson, K. (1886–1893) *A history of the theory of elasticity and of the strength of materials*, 2 vols (Cambridge University Press).

Professional institution interest groups concerned with structural history include:

Study Group for the History of Structural Engineering (IStructE)
Panel for Historic Engineering Works (ICE)
Construction History Society (c/o the Chartered Institute of Building)

Specific design guidance – structural loadings

STANDARDS

In Britain, Building Regulations refer to BS 6399 and CP 3: Chapter V: Part 2 for minimum loadings to be applied in building design. Handbooks extend the code guidance:

BS 6399: Part 1: 1984, Code of practice for dead and imposed loads
BS 6399: Part 2: 1995, Code of practice for wind loads
BS 6399: Part 3: 1988, Code of practice for imposed roof loads

Eurocode 1 for the basis of design and actions on structures is in preparation. Part 2.1: Actions on structures deals with loads and should be available as a prENV in 1995–6.

An additional loading consideration is seismic, from earthquakes. At present this is only significant in Britain for large or high-risk structures (nuclear installations, offshore platforms, major bridges, etc.), but for design overseas it may be a dominant factor. There is no British seismic design standard for general use at present.

Eurocode 8 on earthquake resistant design of structures is in preparation. Part 1.1: Seismic actions and general requirements for structures, and Part 1.2: General rules for buildings, should be available as prENVs in 1995–6.

GUIDANCE

BRE Digest 346, the assessment of wind loads (in 8 parts).
There are various other BRE Digests and Reports on wind effects.
Currie, D. M. (1993) *Handbook of imposed roof loads: a commentary on British Standard BS 6399: loading for buildings*, part 3, BRE Report BR 247 (BRE).

BOOKS

Booth, E. (ed.) (1994) *Concrete structures in earthquake regions: design and analysis* (Longman).
Cook, N. J. (1986) *The designer's guide to wind loading of building structures*, part 1: Background, damage survey, wind data and structural classification (Butterworths).
Cook, N. J. (1990) *The designer's guide to wind loading of building structures*, part 2: Static structures (Butterworths).
Dowrick, D. J. (1987) *Earthquake resistant design for engineers and architects*, 2nd ed. (Wiley).
European recommendations for steel structures in seismic zones (1988) (European Convention for Constructional Steelwork, available via BCSA).
Study on design of steel buildings in earthquake zones (1986) (ECCS).

The effects of earthquakes on existing construction (sadly but mercifully, to date, in other countries) are described and illustrated in a series of reports published by EEFIT, the Earthquake Engineering Field Investigation Team, obtainable from the Institution of Structural Engineers.

Specific design guidance – fire engineering

The term fire engineering is comparatively new and represents a changing attitude to the need to protect buildings and the people inside them from the ravages of fire. Fire is now seen as a 'load' which a structure must withstand rather as it might have to withstand the loads imposed by an earthquake.

The structural designer must first address the performance of a structure in a fire in terms of the period of fire resistance required (in hours), and then the means of ensuring this. This may entail providing new fire protection or enhancing the existing fire resistance. An additional requirement for a fire-damaged structure is to assess whether the fire has permanently affected or weakened the structure. The following sources give guidance:

Buchanan, A. H. (eds.) (1994) *Fire Engineering Design Guide* (Centre for Advanced Engineering, University of Canterbury, New Zealand)

Stollard, P. & Johnston, L. (eds.) (1994) *Design against fire: an introduction to fire safety engineering designs* (Spon).

GENERAL

Approved document B of the Buildings Regulations 1991: fire safety (1991) (HMSO).

Guidelines for the construction of fire-resisting structural elements, BRE Report BR 128 (1988) (HMSO)

CONCRETE

BS 8110: Structural use of concrete, Part 1: 1985, Code of practice for design and construction, and Part 2: 1985, Code of practice for special circumstances

Assessment and repair of fire damaged concrete structures, technical report CSTR33 (1990) (Concrete Society).

Harmathy, T. Z. (1993) *Fire safety design and concrete* (Longman).

Lawson, R. M. (1985) *Fire resistance of ribbed concrete floors*, CIRIA Report R107 (CIRIA).

STEEL

Lawson R. M. and Newman, G. M. (1990) *BS 5950: Structural use of steelwork in building, Part 8: 1990, Code of practice for fine resistant design of steel structures – a handbook to BS 5950: Part 8* (SCI).

Smith, R. C. and Lawrence, L. (1992) *Fire protection for structural steel in buildings*, 3rd ed. (ASFPCM/SCI).

The reinstatement of fire-damaged steel and iron-framed structures (1986) (British Steel Corporation, Swinden Laboratories).

MASONRY

BS 5628: Code of practice for use of masonry, Part 3: 1985, Materials and components, design and workmanship

Sowden, A. M. (ed.) (1990) *The maintenance of brick and stone masonry structures* (Spon).

TIMBER

BS 5268: Structural use of timber, Part 4: 1978, 1989 Fire resistance of timber structures (in 2 sections)

Increasing the fire resistance of timber floors Digest 208 (1988) (BRE).
Schneider, U. and Nagele, E. (eds) (1989) *Repairability of fire-damaged structures* W14 report (CIB).
Smith, R. B. (1993) *Fire protection of timber floors* (ASFPCM).

Specific design guidance – by material

Concrete

Technical guidance on structural design in concrete is available in abundance and the selection presented here is of standard works. Various organizations and trade associations provide specialist information and guidance to assist structural designers. Details of these are given in the relevant part of Chapter 30 on construction engineering. In particular, the publications of the following organizations should be consulted:

British Cement Association
Concrete Society
Fédération Internationale de la Précontrainte (c/o IStructE)

STANDARDS

BS 8110: 1985 Structural use of concrete (3 parts)

Rowe, R. E. *et al.* (1987) *Handbook to British Standard BS 8110: 1985 structural use of concrete* (Palladian).

Eurocode 2 for design of concrete structures is in preparation. EC2 Part 1.1: General rules and rules for buildings, is available as a DD ENV 1992–1–1 (European pre-standard) from BSI.
The British Cement Association and Thomas Telford are publishing guides and concise versions to assist understanding and use of EC2.

GENERAL GUIDANCE

Deacon, R. C. (1986) *Concrete ground floors: their design, construction and finish*, 3rd ed. (BCA).
Elliott, K. S. and Tovey, A. K. (1992) *Precast concrete frame buildings: design guide* (BCA).
Guide to the design of anchor blocks for post-tensioned prestressed concrete CIRIA guide G1 (1975) (CIRIA).
Irwin, A. W. (1984) *Design of shear wall buildings* CIRIA report 102 (CIRIA).
Kong, F. K. and Evans, R. H. (1987) *Reinforced and prestressed concrete*, 3rd ed. (Chapman and Hall).
Manual for the design of reinforced concrete building structures (1985) (IStructE/ICE).
Ove Arup & Partners (1977) *Design of deep beams in concrete* CIRIA guide G2 (CIRIA).

Reynolds, C. E. and Steedman, J. C. (1988) *Reinforced concrete designer's handbook*, 10th ed. (Spon).
Standard method of detailing structural concrete (1989) (IStructE/Concrete Society).
Structural use of lightweight aggregate concrete (1988) (IStructE).
Taylor, H. P. J. (ed.) (1992) *Precast concrete cladding* (Edward Arnold).
Whittle, R. T. (1994) *Design of reinforced concrete flat slabs to BS 8110* CIRIA report 110, 2nd ed. (CIRIA).
Wilby, C. B. (1993) *Concrete dome roofs* (Longman).

GUIDANCE ON EXISTING CONCRETE STRUCTURES

Technical guidance on existing concrete structures is available covering investigation, appraisal, testing and repair. The BRE (in particular) has in addition published a very substantial amount of material on precast building systems, to assist designers in recognition and understanding of the systems and in identifying repair needs. The Concrete Society also produces useful guidance, and back copies of periodicals such as *The Structural Engineer*, and the now ceased *Concrete and Constructional Engineering* (monthly, 1908–66). The following is a selection of specific guidance:

Bate, S. C. C. (1984) *High alumina cement concrete in existing building superstructures* BRE report BR235 (HMSO).
Campbell-Allen, D. and Roper, H. (1991) *Concrete structures: materials, maintenance and repair* (Longman).
Concrete core testing for strength technical report CSTR11 (1976) (Concrete Society).
Guidance note on the security of the outer leaf of large concrete panels of sandwich construction (1989) (IStructE).
Guide for making a condition survey of concrete in service ACI 201.1R–68 (1984) (American Concrete Institute).
Kay, T. (1992) *Assessment and renovation of concrete structures* (Longman).
Non-structural cracks in concrete technical report CSTR22, 2nd ed. (1992) (Concrete Society).
Post-tensioning systems for concrete in the UK: 1940–1985 CIRIA report 106 (1985) (CIRIA).
Pullar-Strecker, P. (1987) *Corrosion damaged concrete: assessment and repair* (Butterworths, for CIRIA).
Structural effects of alkali-silica reaction: technical guidance on appraisal of existing structures (1992) (IStructE).

Steel

Technical guidance on structural design in steel is available in abundance and the selection presented here is of standard works. Various organizations and trade associations provide specialist information and guidance to assist structural designers. Details of these are given in the relevant part of Chapter 30 on construction engineering. In particular, the publications of the following organizations should be consulted:

British Constructional Steelwork Association
British Steel
Steel Construction Institute

STANDARDS

BS 5493: 1977 Code of practice for protective coating of iron and steel structures against corrosion
BS 5950: 1982–1994 Structural use of steelwork in building (9 parts)

Lawson, R. M. (1990) *Commentary on BS 5950: part 3: section 3.1 composite beams* (SCI).
Steelwork design guide to BS 5950 part 1 (1990). Vol. 1: Section properties and member capacities, 3rd ed. (1992); Vol. 2: Worked examples, rev. Edn (1991); Vol. 3: Commentary (1991); Vol. 4: Essential data for designers (1991) (SCI).

Eurocodes 3 and 4 for design of steel and composite steel and concrete structures respectively are in preparation. EC3 Part 1.1: General rules and rules for buildings, and EC4 Part 1:1: General rules and rules for buildings, are available as European pre-standards DD ENV 1993–1–1 and 1994–1–1 from BSI.

The Steel Construction Institute and Thomas Telford are publishing guides and concise versions to assist understanding and the use of these Eurocodes. One example is:

Johnson, R. P. and Anderson, D. (1994) *Designers' handbook to Eurocode 4, part 1.1: design of composite steel and concrete structures* (Thomas Telford).

GENERAL GUIDANCE AND BOOKS

Blanc, A., McEvoy, M. and Plank, R. (eds) (1993) *Architecture and construction in steel* (Spon).
Burgan, B. A. (1993) *Concise guide to the structural design of stainless steel*, 2nd ed. (SCI).
Cheal, B. D. (1980) *Design guidance notes for friction grip bolted connections* CIRIA technical note TN98 (CIRIA).
Dowling, P. J., Harding, J. E. and Bjorhovde, R. (eds) (1992) *Constructional steel design: an international guide* (Elsevier).
Hart, F., Henn, W. and Sonntag, H. (1985) *Multi-storey steel framed buildings* English ed. G. B. Godfrey (ed.) (Collins).
Hayward, A, and Weare, F. (1992) *Steel detailers' manual*, rev. ed. (Blackwell).
Joints in simple construction Vol. 2: Practical applications (1991) (BCSA/SCI).
Manual for the design of steelwork building structures (1989) (IStructE).

Owens, G. W. and Knowles, P. R. (eds) (1992) *Steel designers' manual*, 5th ed. (Blackwell).
Randerson, K. (1991) *The choice of steels for building and bridges* (British Steel Technical).

GUIDANCE ON EXISTING STEEL AND IRON STRUCTURES

An excellent introduction to this subject is the two chapters devoted to cast and wrought iron by R. J. M. Sutherland in Doran, D. K. (ed.) (1992) *Construction materials reference book* (Butterworth-Heinemann).
The British Cast Iron Research Association can advise on existing cast iron, but may charge for this.
Practical guidance on existing steel structures is limited if adequate. BRE has published a substantial amount of material on steel-framed and steel-clad (prefabricated) housing, to assist designers in recognition and understanding of the systems and in identifying repair needs.

Bates, W. (1984) *Historical structural steelwork handbook* (BCSA).
Bussell, M. (In press) *Appraisal of existing iron and steel structures* (SCI).

Natural stone, masonry and brickwork

Technical guidance on structural design in masonry is widely available and the selection presented here is of standard works. Various organizations and trade associations provide specialist information and guidance to assist structural designers. Details of these are given in the relevant part of Chapter 30 on construction engineering. In particular, the publications of the following organizations should be consulted:

Brick Development Association
British Masonry Society
British Ceramic Research Limited
British Cement Association
Building Research Establishment
Stone Federation

Technical guidance on structural design is fairly widely available, although less so than for concrete and steel. The following are standard works.

STANDARDS

BS 5628: 1985–1992 Code of Practice for Use of Masonry (3 parts)

Haseltine, B. A. and Moore, J. F. A. (1981) *Handbook to BS 5628: structural use of masonry, part 1: unreinforced masonry* (BDA).
Haseltine, B. A. and Tutt, J. N. (1991) *Handbook to BS 5628: part 2, section 1: background and materials* (BDA).

Haseltine, B. A. and Tutt, J. N. (1992) *Handbook to BS 5628: part 2, section 2: reinforced masonry* (BDA).
Roberts, J. J., Edgell, G. J. and Rathbone, A. J. (1986) *Handbook to BS 5628: part 2: structural use of reinforced and prestressed masonry* (Spon).

Eurocode 6 for design of masonry structures is in preparation. Part 1.1: Rules for reinforced and unreinforced masonry should be available as a prENV in 1995–1996.

GENERAL GUIDANCE AND BOOKS

Curtin, W. G., et al. (1987) *Structural masonry designers' manual*, 2nd ed. (BSP Professional).
Curtin, W. G., et al. (1984) *Structural masonry detailing* (Granada/BSP Professional).
Heyman, J. (1982) *The masonry arch* (Ellis Horwood).
Heyman J. (1995) *The stone skeleton: structural engineering of masonry architecture* (Cambridge University Press).
Hendry, A. W. (1990) *Structural masonry* (Macmillan).
Orton, A. (1992) *Structural design of masonry*, 2nd ed. (Longman).
Reinforced and prestressed masonry (1981) (IStructE).
Roberts, J. J. et al. (1983) *Concrete masonry designer's handbook* (Spon).

GUIDANCE ON EXISTING MASONRY STRUCTURES

Bray, R. N. and Tatham, P. R. B. (1992) *Old waterfront walls: management, maintenance and rehabilitation* (Spon).
Macgregor, J. E. M. (1971) *Outward leaning walls* technical pamphlet 1 (SPAB).
Sowden, A. M. (ed.) (1990) *The maintenance of brick and stone masonry structures* (Spon).
Thiel, M. J. (ed.) (1993) *The conservation of stone and other materials* (Spon).
Warland, E. G. (1986) *Modern practical masonry*, 2nd ed. (Stone Federation).
Williams, G. B. A. (1976) *Chimneys in old buildings* technical pamphlet 3 (SPAB).

Timber

Technical guidance on structural design using timber is not as widely available as for steel and concrete. Various organizations and trade associations provide specialist information and guidance to assist structural designers. Details of these are given in the relevant part of Chapter 30 on construction engineering. In particular, the publications of the following organizations should be consulted:

Timber Research & Development Association
Building Research Establishment.

Further guidance can be obtained from the national timber organizations in countries which use the material more often than in Britain, especially Canada, Finland, Sweden, Switzerland and the USA.

STANDARDS

BS 5268: 1978–1991 Structural use of timber (7 parts).

Eurocode 5 for design of timber structures is in preparation. Part 1.1: General rules and rules for buildings is available as European pre-standard DD ENV 1995-1-1 from BSI.

GENERAL GUIDANCE AND BOOKS

Approved document: timber intermediate floors for dwellings (excluding compartment floors) (1992) (TRADA).
Baird, J. A. and Ozelton, E. C. (1989) *Timber designers' manual*, 2nd ed. (BSP Professional).
Bull, J. W. (1994) *The practical design of structural elements in timber*, 2nd ed. (Gower).
Desch, H. E. (1981) *Timber, its structure and properties*, 6th ed. Macmillan).

GUIDANCE ON EXISTING TIMBER STRUCTURES

Boutwood, J. (1991) *The repair of timber frames and roofs* technical pamphlet 12 (SPAB).
Charles, F. W. B. and Charles, M. (1995) *Conservation of timber buildings* (Donhead).
Hewett, C. A. (1980) *English historic carpentry* (Phillimore).
Macgregor, J. E. M. (1985) *Strengthening timber floors* technical pamphlet 2 (SPAB).
Newlands, J. (1990) *The carpenter's assistant: the complete practical course in carpentry and joinery* (Studio Editions) Reprint of the 1857 edition.
Swindells, D. J. and Hutchings, M. (1993) *A checklist for the structural survey of period timber-framed buildings* (RICS Books).
Yeomans, D. T. (1992) *The trussed roof: its history and development* (Scolar Press).

Aluminium alloys

Despite wide application in other fields of engineering, aluminium alloys remain a little-used structural material; this is reflected in the limited availability of design guidance.

The main trade association for aluminium, which can provide potential users with guidance is:

Aluminium Federation Ltd, Broadway House, Calthorpe Road, Birmingham B15 1TN, UK. Tel: +44 121 456 1103. Fax: +44 121 456 2274. It publishes its own excellent handbook. *The properties of aluminum and its alloys* 9th ed. (1993).

Structural engineering

STANDARDS

The two principal British Standards concerning aluminium are:

BS 1161: 1991 Specification for aluminium alloy sections for structural purposes.
BS 8118: Structural use of aluminium, Part 1: 1991, Code of Practice for Design.
Eurocode 9 for design of aluminium alloy structures is in preparation.

GUIDANCE AND BOOKS

Bull, J. W. (1994) *The practical design of structural elements in aluminium* (Avebury).
European recommendations for aluminium alloy structures fatigue design (1992) (European Convention on Constructional Steelwork) Available via BCSA.
Lane, J. (1992) *Aluminium in building* (Gower).
Mazzolani, F. M. (1985) *Aluminium alloy structures* (Pitman).
Sharp, M. L. (1993) *Behaviour and design of aluminium structures* (McGraw-Hill)

Glass and glass composites

ORGANIZATIONS PROVIDING GUIDANCE

For advice on the use of glass, especially in unusual applications, the following should be consulted:

British Glass Manufacturers Confederation, Northumberland Road, Sheffield S10 2UA, UK. Tel: +44 114 686201 Fax: +44 114 681073.

Glass & Glazing Federation, 44 Borough High Street, London SE1 1XB, UK. Tel: +44 171 403 7177. Fax: +44 171 357 7458.

Pilkington Glass Advisory Service, Prescot Road, St Helens, Merseyside WA10 3TT, UK. Tel: +44 1744 692000. Fax: +44 1744 692882.

TECHNICAL GUIDANCE

Glass has recently entered its third age. Traditionally it was seen mainly as a material for glazing. This aspect of glass use has long been guided by a British Standard:
BS 6262: 1982 Code of practice for glazing for buildings.
Since the 1950s glass in fibre form, working compositely with plastics and cements, has gained increasing use as a structural material.

There is no code governing this usage, but some guidance for the construction industry is given in:

BRE Digest 331, GRC (glass-reinforced cement) (1988) (BRE)

Recently the increasing use of glass in glazed walls and atria has encouraged the glass industry to look at the potential of using glass as a structural material. As yet, no official guidance is available but a valuable guide has been produced by the company which knows the material best:

Button, D. and Pye, B. (eds) (1993) *Glass in building*: (Butterworth Architecture, with Pilkington Glass Ltd.).

BOOKS

Hollaway, L. (1978) *Glass fibre reinforced plastics in construction: engineering aspects* (Surrey University Press).

Majumdar, A. J. and Laws, V. (1991) *Glass fibre reinforced cement* (BSP Professional).

Maunsell Structural Plastics is the leading consultant in the field of GRP; they can be contacted at 154 Croydon Road, Beckenham, Kent BR3 4DE, UK. Tel: +44 181 663 6565. Fax: +44 181 659 5568.

The first major use of structural (load-bearing) glass was in the glazed facade at La Villette, in Paris; it was designed by Peter Rice and RFR. This work has been described in great detail in:

Rice, P. and Dutton, H. (1995) *Structural glass* 2nd ed. Spon.

Although not written as a handbook, this book conveys the essential philosophy behind using glass as a load-bearing material.

Specific design guidance – building elements

The building envelope

The envelope comprises walls, cladding and roofing. Its design is not just a structural matter, but must also take account of weathertightness, thermal and acoustic performance, etc. It is therefore a multi-disciplinary activity.

A recent large work (7 vols) considers most aspects of walls (not only structural) and is an excellent place to start in this field of interest:

Wall technology, Special Publication No. SP87 (1992) (CIRIA).

Cladding materials are generally concrete or masonry, glass, or proprietary metal, plastics or lightweight fabric. The Centre for Window and Cladding Technology located at the University of Bath is active in this field. The Architectural Cladding Association is one arm of the British Precast Concrete Federation. They can be found at 60 Charles

Street, Leicester, LE1 1FB, UK. Tel: +44 1162 536 161. Fax: +44 1162 514 568.

Roofing materials depend on whether the roof is pitched or flat, but include tiles, metal sheet, asphalt, felt, and synthetic membranes. Information on the use of these is available from the tile manufacturers, the Copper/Lead/Zinc Development Associations, the British Flat Roofing Council, the Mastic Asphalt Contractors and Employers Federation, and the felt and membrane manufacturers.

The following is a selection of published information sources for cladding:

STANDARDS

BS 5628: Code of practice for use of masonry, Part 3: 1985, Materials and components, design and workmanship
BS 6262: 1982 Code of practice for glazing for buildings
BS 8200: 1985 Code of practice for design of non-loadbearing external vertical enclosures of buildings
BS 8217: 1994 Code of practice for built-up felt roofing
BS 8297: 1995 Code of practice for design and installation of non-loadbearing precast concrete cladding
BS 8298: 1994 Code of practice for design and installation of natural stone cladding and lining
CP 143: 1958–1988 Code of practice for sheet roof and wall coverings (covers aluminium, zinc, galvanized steel, copper, bitumen sheet)
CP 144: 1970 Roof coverings (covers bitumen felt and mastic asphalt)

GUIDANCE AND BOOKS

Arup Research and development (1993) *Flat roofing: design and good practice* (CIRIA/British Flat Roofing Council).
Brookes, A. J. and Grech, C. (1990) *The building envelope: applications of new technology cladding* (Butterworth Architecture).
Saxon, R. (1994) *The atrium comes of age* (Longman).
Standard and guide to good practice for curtain walling (1993) (Centre for Window and Cladding Technology).
Vanderberg, M. (ed.) (1974) *AJ handbook of building enclosure* (Architectural Press) Becoming elderly.

For the investigation and appraisal of existing cladding and roofing systems, there are numerous reports and other guidance published, in particular, by BRE and CIRIA. The sources listed above generally give information on maintenance.

Foundations

The relevant British Standard is BS 8004: 1986 Code of practice for foundation and there are three standard texts worth mentioning:

Curtin, W. G. et al. (1994) *Structural foundation designers' manual* (Blackwell).
Tomlinson, M. J. (1995) *Foundation design and construction*, 6th ed. (Longman).
Tomlinson, M. J. (1994) *Pile design and construction practice*, 4th ed. (Spon).

Fixings

Fixings for securing into concrete and masonry are essentially proprietary elements, so that their performance and load-carrying capacity is determined from tests and published in manufacturers' technical literature. Fixings to steel and timber are covered by the material codes of practice.

The Construction Fixings Association can provide details of its members. It is located at

Light Trades House, Melbourne Avenue, Sheffield S10 2QJ, UK. Tel: +44 114 663 084 Fax: +44 114 670 910.

STANDARDS

BS 5080: 1993 Structural fixings in concrete and masonry (2 parts to date).

GUIDANCE AND BOOKS

Performance, selection, and testing of fixings are covered by the following publications:

Comité Euro-Internation du Béton Bulletin d'Information 206, Fastenings to Reinforced Concrete and Masonry Structures 2 parts (1991) (CEB).
CIRA Technical Note 137, Selection and Use of Fixings in Concrete and Masonry: Interim Update to CIRIA Guide 4 (1991) (CIRIA).

Specific design guidance – building and structure types

Published material on types of buildings and structures ranges from a review of the form in general (including buildings) to standards for design of particular functional types.

At the most general level, information on spatial forms and requirements of particular building types is available in two books intended for architects but with much of interest and value to the structural engineer:

Neufert, E. (1980) *Architects' data* English 2nd ed., V. Jones (ed.) (Granada).
Tutt, P. and Adler, D. (eds) (1992) *New metric handbook*, 5th ed. (Butterworth Architecture) Currently under revision.

Many other books intended for architects contain a useful review of completed buildings and can serve to increase the structural engineer's vocabulary and knowledge of precedent. Three examples are:

Structural engineering

Brookes, A. J. and Grech, C. (1992) *Connections: studies in building assembly* (Butterworth Architecture).
Orton, A. (1988) *The way we build now: form, scale and technique* (Van Nostrand Reinhold).
Wilkinson, C. (1991) *Supersheds: the architecture of long-span large volume buildings* (Butterworth Architecture).

Bridges

Within a chapter devoted to structural engineering, only a small selection of information relating to bridges can be given. The structural design of bridges is well codified in Britain. Guidance embraces loadings, bearings, and fatigue considerations as well as basic design in concrete or steel. Less codified guidance is available for masonry and timber, although a revival of interest in masonry arch construction is stimulating new work and publication in this area.

HMSO publishes a wide range of material to assist bridge designers. This emanates mainly from the Department of Transport, e.g. the *Design manual for roads and bridges* (1993), which is multi-volume and gives essential guidance and the *Manual of contract documents for highway works* (1992), and the Transport Research Laboratory which publishes Contractor Reports, Research Reports and State of the Art Reviews.

Many sources of information applicable to buildings apply also to bridges, e.g. structural use of materials. The following is of more specific value:

STANDARDS

BS 5400: 1978–1990 Steel, concrete and composite bridges (10 parts).

Steel Construction Institute (1991) *Commentary on BS 5400: part 3 (1982): code of practice for the design of steel bridges* (SCI).

GUIDANCE AND BOOKS

Assessment of reinforced and prestressed concrete bridges (1988) (IStructE).
Clark, L. A. (1983) *Concrete bridge design to BS 5400* (Construction Press) Supplement issued 1984.
Iles, D. C. (ed.) (1992) *Replacement steel bridges for motorway widening* (SCI).
Leonhardt, F. (1982) *Bridges: aesthetics and design* (Architectural Press).
Page, J. (ed.) (1993) *Masonry arch bridges* (HMSO) A TRL state of the art review.
Troitsky, M. S. (1988) *Cable-stayed bridges: theory and design*, 2nd ed. (BSP Professional).

Technical guidance on the inspection, maintenance, repair and assessment of bridges is published by HMSO in the *Department of*

Transport design manual for roads and bridges, comprising Vol. 3 (1993). Although prescribed only for trunk road structures, this guidance is widely applied to other existing bridges, whether public or private.

Offshore structures

As a relatively new field in structural design in Britain, offshore construction has generated much research, particularly on the estimation of wave loading and on the durability of concrete and steel. This is gradually being codified. The Institutions of Civil and Structural Engineers and CIRIA have published a range of material on general issues, with the institutions also publishing various papers on particular offshore installations. The following is a selection of standard works:

Coastal structures and breakwaters (1992) (Thomas Telford).
P112 – Interim guidance notes for the design and protection of topside structures against explosion and fire (1992) (SCI) essential reading for all engineers and managers concerned with safety of platform design.
Blain, W. R. (ed.) (1992) *Marina technology* (Thomas Telford/Computational Mechanics).
Department of Energy (1990) *Offshore installations: guidance on design, construction and certification*, 4th ed. (HMSO).
Offshore structures (1975) (Thomas Telford)
Recommendation of the Committee for Waterfront Structures 5th ed. (1986) (Ernst).

Silos and tanks

Anchor, R. D. (1992) *Design of liquid-retaining concrete structures*, 2nd ed. (Edward Arnold).
Reimbert, M. L. and Reimbert, A. M. (1987) *Silos: theory and practice*, 2nd ed. (Lavoisier).
Safarian, S. S. and Harris, E. C. (1985) *Design and construction of silos and bunkers* (Van Nostrand Reinhold).
Selective bibliography on the design of hoppers and silos CIRIA special publication 1 (1970) (CIRIA).

Towers and masts

BS 4485: Water cooling towers, Part 4: 1975, Structural design of cooling Towers.
BS 8100: 1986–1995 Lattice Towers and Masts (2 parts).

Heinle, E. and Leonhardt, F. (1989) *Towers: a historical survey* (Butterworth Architecture).
Hill, G. B., Pring, E. J. and Osborn, P. D. (1990) *Cooling towers*, 3rd ed. (Butterworths).

Miscellaneous

All manner of specialist books exist for particular types of buildings and are immediately recognizable by their titles! A few miscellaneous examples to end this chapter include:

The design of air-supported structures (1984) (IStructE)
Design recommendations for multistorey and underground car parks. Report of a joint committee of the Institution of Structural Engineers/ Institution of Highways and Transportation 2nd ed. (1984) (IStructE).
Safety considerations for the design and erection of demountable grandstands (1989) (IStructE).

CHAPTER THIRTY-TWO

Construction management

D. A. LANGFORD and A. RETIK

Introduction

Information is the lifeblood of many managers. Managers in the construction industry are no exception to this shibboleth. However the type of information that construction managers demand will vary according to their position within the organization, the purpose of the information and the level of detail required. This chapter will consider the sources of information that construction managers may use from these perspectives:

The level at which the information is used may be broken down into three parts:

1. information necessary for an understanding of the construction industry, its structure, organization and future direction;
2. the information necessary for the planning and control of firms operating within the industry – be they consultants contributing to design or contractors undertaking construction work;
3. the information necessary to ensure effective project management including coordinated project information.

The range of information that a construction manager is likely to come into contact with is vast and this reflects the wide range of disciplines upon which construction managers draw. So, information rooted in management, technology, engineering, economics, law (contract, environmental and health and safety), personnel management, operational research and statistics, to name but a few, will be pertinent to the work of construction managers. Not only are the sources diverse but the requirement for information is separated by time. Information may be provided continuously, periodically and occasionally. The differences are delineated by Newcombe, Langford and Fellows (1990). 'Continuous inputs are produced frequently with short time intervals between each provision e.g. site labour productivity measures; FT share indices.' Periodic information occurs regularly at prescribed intervals. External information such as the Budget and the *Housing and Construction Statistics* appear periodically as do internal information such as companies' annual report and accounts. Occasional information occurs irregularly and may be generated externally by, for example, reports on the industry's performance or internally as a reaction to events.

In the face of such a barrage of information, powers of selectivity are vital. Indeed the very scope of available information will mean that this chapter is necessarily selective. It will focus upon main sources of information at each level in the world of construction management. Comprehensiveness is impossible, discrimination of the information sources quoted has been judiciously applied. No apology is offered for this selectivity.

Data and information in construction management

So far the text has used the word 'information' to convey the material used by construction managers, but other concepts are visible. The differences between data and information needs to be noted; they are not the same. Data are raw facts which organizations, be it the firm or the project, collect. Data can arise from within the organization and typical data would be the likely costs of different construction elements. This is useful in pricing the next project. Equally, data may be provided from external sources – for example, levels of output costs, etc.

Newcombe, Langford and Fellows (1990) have argued that information is data which have been pressed into a form which is useful to construction managers. Inevitably there will be some ambiguity regarding when data become information. External information encompasses reports and similar documents which have been produced using raw data. Examples would be the NEDO reports *Faster building for industry* (1983) or *Faster building for commerce* (1988), reports on how the national Budget will influence the construction industry, etc. This type of information is of necessity general, and individuals and organizations will need to contextualize the information for their particular purposes.

In contrast, internal information is organization specific. It is information generated by and used by organizations and its managers. At the corporate level information such as the strategic plan may provide internal information which is at the pinnacle of the information pyramid but this will frequently be turned into secondary information such as marketing plans, etc. At the project level the master plan or milestone chart may be broken down into monthly, weekly and daily short term programmes.

The role of information in a construction manager's job

One of the challenges of construction management is to align the different sources of information so that a coherent picture is obtained; this is difficult, for construction management can be viewed according to Fellows *et al.* (1983) 'in two dimensions – project management and business management – which are in practice interdependent'; but the information required by construction managers will draw upon different sources and the synthesis of business data and project management information will be vital.

The tone of the discussion in many texts tends to see information as that which is commonly accessible – published sources, etc. However, for managers, information is drawn from wider sources which is consistent with the nature of managerial work. Mintzberg (1973, 1983) has noted that managerial work is characterized by variety, brevity and discontinuity and much of a manager's time is spent transmitting and receiving information in an informal way. Consequently construction managers are likely to combine formal information sources with the informal, spoken rather than written, methods of data collection and distribution.

Whatever preferences an individual construction manager may have for the sources of information one precept may be held constant – each construction manager will have a role which requires the use of information. As Fryer (1990) notes, managers have informational roles which require the manager to act as 'monitor, disseminator and spokesman . . . reviewing data and events, giving and receiving information, and passing information from the group to others outside'.

This role necessitates the construction manager to develop information-handling skills. As businesses and projects have become more complex, so construction managers need a combination of human, technical and conceptual skills (Katz, 1974) to locate and interpret information. Making judgements about the accuracy and utility of the myriad information sources will be an important component of the construction manager's job.

The information available to construction managers is presented in a variety of forms and the chapter will address the delivery of this information – much of it will be in the format of paper reports, journals, texts, etc. Increasingly, the science of information technology and computer-based information sources will be the source of construction information.

Information sources for the industry

The primary sources of information about the nature of the construction industry are provided by the Government. Foremost is the *Housing and construction statistics*.

The housing and construction statistics

These are published quarterly and trace the economic and employment trends in the industry. These are broken down into the various regions. Tables 32.1 and 32.2 illustrate the range of information available from the data produced within the Department of the Environment, analysed by the Central Statistical Office and produced by HMSO. Part 1 of the quarterly report focuses upon housing, and the range of information available is shown in Table 32.1. Table 32.2 shows the data which may be obtained from part 2 of the *Housing and construction statistics* and

Table 32.1. Information carried by the Housing and Construction Statistics.

	Contact point for further information
Housing performance, seasonally adjusted	01272-218050
Permanent dwellings started, under construction and completed: United Kingdom: by country	01272-218050
Permanent dwellings started and completed England: by region	01272-218050
Estimated time lag: start to completion	01272-218050
Specialized dwellings	01272-218050
Houses and flats completed: by number of bedrooms: England	01272-218050
Permanent dwellings: tenders accepted (net): local authorities and new towns	01272-218050
Mortgages: main institutional sources: United Kingdom	0171-276 3504
Banks: mortgage advances: United Kingdom	0171-276-3504
Insurance companies: dwelling prices and mortgage advances: United Kingdom	0171-276 3504
Building societies: dwelling prices and mortgage advances: United Kingdom	0171-276 3504
Building society mortgage advances and commitments: United Kingdom	0171-276 3504
Building societies: dwelling prices, mortgage advances and income of borrowers: United Kingdom	0171-276 3504
Rent registration: average registered rents and change on previous rents	0171-276 3505
Housing benefit referrals to rent officers: mean referred and determined rents	0171-276 3505
Building materials and components production, deliveries and stocks	0171-276 4761

Table 32.2. Type of information available from part 2 of the Housing and Construction Statistics.

	Contact point for further information
Construction cost and price indices	0171-276 3460
Housing costs and prices	0171-276 3460
Value of output	0171-276 3526
Value at current prices of contractors output: by type of work	0171-276 3526
Value at current prices of contractors output: by region	0171-276 3526
Value of new orders obtained by contractors	0171-276 3526
Value at current prices of new orders obtained by contractors: by type of work	0171-276 3526
Value at current prices of new orders obtained by contractors: by region	0171-276 3526
Construction manpower: employees in employment	0171-276 3526
Sales of housing land and disposal of land and dwellings under licence	0171-276 3505
Right to buy: progress of applications	0171-276 3505
Sale and transfer of dwellings	0171-276 3505
Value of dwellings sold by local authorities and new towns	0171-276 3505
Houses and flats sold by local authorities and new towns	0171-276 3505
Renovations: England	01272 218456
Renovations: grants for private owners and tenants	01272 218456
Renovations: grants under the Local Government and Housing Act 1989	01272 218456
Renovations: grants paid to private owners to tenants	01272 218456
Renovations: grants to private owners and tenants for disabled facilities	01272 218456
Renovation of specialized dwellings: work completed for local authorities	01272 218456
Grants paid for HMSO, common parts and minor works	01272 218456
Renovations: by region	01272 218456
Stock of dwellings: by tenure	01272 218078
Stock of dwellings: by region, metropolitan county and conurbation	01272 218078
Stock of dwellings: by tenure and region	01272 218078
Housing loans by local authorities to private persons	0171-276 3505
Housing loans by local authorities to housing associations	0171-276 3505
Dwellings demolished or closed and unfit dwellings made fit: England	01272 218456
Dwellings demolished or closed and unfit dwellings made fit: by region	01272 218456

as can be seen the emphasis is upon value of output, employment, project durations as well as housing data.

New earnings survey

Secondly, but none the less useful, government data can be found in publications such as the *New earnings survey*. This is produced by the Department of Trade and Industry and charts the changes in hours worked, basic pay and earnings across a wide range of crafts and professions. This enables construction managers to check for any differences between the wages and salaries a company is paying from that typical of the industry, or other industries, in a region.

Those companies that are sufficiently sophisticated as to engage in formal corporate planning, and many are, will frequently seek information which is indicative of the business behaviour of potential clients. Gathering data about political, economic, social and technological trends (PEST analysis) can assist a construction organization to direct its marketing and align the firm's resources to growing, future markets, not current and perhaps dormant ones. In order to gather information on trends in society the triennial publication *Social Trends* will be a useful source of information. This volume is based upon mass surveys of consumer behaviour as well as economic and social trends in terms of how disposable income is spent, lifestyle preferences, etc.

The use of such data is manifold; for example, some firms detecting the growing number of elderly but wealthy homeowners launched (temporarily) successful developments of sheltered housing for the elderly. Others, reading the increasing lifestyle choices in sport and leisure, speculatively built golf courses, sports complexes, etc. Careful analysis of such information can help construction organizations to fit with the environment from which they draw their clients. When a firm is not synchronized with its environment then it is in difficulty. The Darwinist theory of adaptability and survival lives on in the world of business!

Corporate information

The big picture provided by the government statistics is supplemented by a wide range of information sources which seek to be either futuristic, current or historical in emphasis. Information sources useful to a company can:

1. forecast trends in particular construction markets;
2. describe the current state of trade;
3. provide a historical account of performance on a particular facet of the construction industry.

Forecasting information

JFC Construction Forecasts and Research is an organization which grew out of the 'privatization' of the construction forecasting arm of the National Economic Developments Office's (NEDO's) committee for building and civil engineering. As NEDOs were abandoned in the early 1990s, the forecasting service was expected to fall, but fortunately for the construction industry a 'management buyout' enabled the service to survive. The forecasts provide a quarterly commentary and projections for the market sectors shown in Table 32.3.

This forecasting service draws upon known economic barometers to predict future demand and a continuous extrapolation of previous trends is its principal method of deriving a forecast. It uses historical data for the previous seven years and prepares a three-year forecast with a commentary.

Periodically, regular forecasts are supplemented by occasional commentaries upon the future of the industry. Most notable has been the Cambridge Econometrics foray into forecasting demand for construction work into the early part of the twenty-first century. On a broader basis the output from the Centre for Strategic Studies in Construction (CSSC) (1988) at the University of Reading has considered the possible generic shape of the industry in the year 2001. Part of this scenario planning rested upon market forecasts. Fellows and Langford (1992) also produced a document which addressed likely growth areas in construction. Other academic institutions have also tackled the thorny problem of marketing in construction, the Macroeconomic Modelling Bureau at the University of Warwick being among them.

Other organizations have an interest in providing industry-wide forecasts. For example, the Centre for Construction Market Information, based at the Building Centre in London, produces market forecasts for the industry. The professions also have an interest in forecasting, certainly the Royal Institution of Chartered Surveyors regularly produces forecasts of house prices by region.

Table 32.3. JFC Market Sector Projections.

1	Housing	– private
		– public
2	Infrastructure	– private
		– public
3	Public non-housing	
4	Private non-residential	– industrial
		– commercial
5	Repair and maintenance	

The financial institutions also trade in information. The stockbrokers who have an interest in broking in construction industry stock will use their analysts to forage around and produce reports on the likely prospects for individual firms. However, these reports are limited to clients who are investors, nonetheless they are a potent source of information for construction managers. Firms noted for this kind of report are:

James Capel & Co
Centre for Interfirm Comparison
Financial Times Information Services
Mintel Publications
Morgan Grenfell Securities
Dun and Bradstreet
Savory Milne

All of this forecasting data is by its nature general and locates growth or decline in sub-markets. For more specific marketing information, construction managers may usefully consult a quarterly publication entitled *Construction Leads*. This document sold by Camargue Communications, a market research organization, provides construction firms with information about the building intentions of major construction clients.

Not only will construction firms want to know whether a client has intentions to build, but also what is the creditworthiness of the client. Will the contractor get paid on time? A vital question in credit-sensitive times. Information provided from Trade Indemnity Plc enables firms to examine the track record of firms when it comes to paying their bills on time. Such information may well influence the tender submitted for particular clients.

The current state of trade

Employers' associations, professional institutions and leading journals serving the construction industry conduct surveys which describe the current state of trade. Those emanating from the employers' association have a twofold purpose. The most obvious is to inform the members of what is 'around the corner' and this enables members of the employers' association to gauge the market for construction work in the next six months, and so inform the adjudication of tenders being submitted. Less obvious, but sufficiently transparent, is to create publicity which, hopefully, prompts the Government to 'do something' in support of the construction industry.

The employers' associations who publish bi-annual reports are the Building Employers Confederation (BEC), the Federation of Civil Engineering Contractors (FCEC) and the Federation of Masters Builders (FMB).

In addition to the employers' pressure groups, the professions will also provide 'state of the industry' reports. The most prominent, since it affects a large number of people, is the Royal Institution of Chartered Surveyors (RICS) survey of price movements in the housing market. Less well known but perhaps more predictive of future trends in the construction industry is the Royal Institution of British Architects (RIBA) survey of workload in the architectural profession. This samples architectural practices of the advance commissions they have secured and is a direct indicator of construction work in the pipeline.

By gathering in such information, the construction manager is better prepared to prepare and plan the direction of the individual enterprise within the industry. Information about relevant organizations is given later in the chapter.

Historical account of performance

Information which records an aspect of the industry's performance is a useful benchmark to compare an individual firm against an industry standard. This can be financial data, size by turnover, number of employees, PE (profit/earnings) ratios, etc. This is typically produced annually in *Building*. Other information, sombrely received by construction managers and workers alike is the Health and Safety Executive's report on safety performance in the industry (Factory Inspectors Annual Report, HMSO). This will assist in identifying hazards and so help eliminate them.

Project-based information

The construction project information system

In 1971 the HMSO published a report entitled *An informative system for the construction industry – final report of the Working Party on Data Co-ordination*, which identified how the use of information was a key issue in the industry. It noted that each of the professions contributing to the industry had their own way of handling project information. Moreover, the information presented by each professional needed to be coherent with that presented by others, but each member of the professional team had to contribute to several projects, each with its own system for presenting and handling project information. This resulted in a poorly coordinated and therefore unsatisfactory way of handling project information. The report also argued that contractors were often uncertain about where (or whether) a piece of information was located within the project documentation, and designers could not always find relevant research or product information to meet their design needs. This latter finding is supported by work by Mackinder and Marvin (1982) who found that architects specify from experience rather than from searches through product information or basic data.

Project information may be layered according to Day and Langford (1990) into three levels:

1. General information relating to the project: e.g. building regulations, standard method of measurement, product information.
2. Specific information on firms involved in the project: e.g. office procedures, manufacturing techniques.
3. Project information specific to the project site: e.g. brief, scheme design, production drawings, bill of quantities or other vehicle for determining price, construction programme, as built drawings, design and other warranties, etc.

Following the *Data coordination* report, several of the principal professional and trade organizations (RIBA, RICS, ACE and BEC) established a body entitled the Coordinating Committee for Coordinated Project Information. After some eight years' work, the concept of coordinated project information (CPI) was launched. The documentation on CPI included guidance on the layout of information. Central to this concept was the Common Arrangement of Works Section (CAWS). This sought to organize construction project information into 300 work categories based upon four source documents; the Standard Method of Measurement, the National Building Specification, the Price Adjustment Formula categories and the Property Services Agency General Specification. However the coordinating committee for CPI found it difficult to reconcile these documents, and so based their 300 work sections upon observations of how work is carried out on site. The common arrangement of works information is structured in three levels. Level 1 is a group of activities, e.g. groundwork, air conditioning/ventilation, Level 2 is a sub-grouping which typically defines the work of a specialist contractor, Level 3 is a description of an operation. These three levels may be illustrated as follows:

Level 1	Group	Section D Groundwork
Level 2	Sub Group	Section D3 Piling
Level 3	Work	Section D30 Cast in place concrete piling.

Supporting CAWS is the Production Drawings Code. This advises designers about the arrangement and presentation of drawn information including sheet sizes, annotation, titling and numbering as well as the structuring of information. The Project Specification Code aims to make project specification comprehensive, precise, practical, specific and clearly written.

The specification relies upon libraries of clauses based upon the

Construction management 579

library of standard descriptions, written by authors from the PSA, RICS and BEC.

Under the CPI system the specification is an integral part of the design, not something prepared by the QS and semi-detached from the architect's design work. The project specification is organized in accordance with the methodology established for CAWS. Finally, CPI is completed by the Standard Method of Measurement 7 (SMM7). This is an integrating document, set out in accordance with CAWS, to forge coherent links between design information and that needed to build.

The SMM7 enables bills of quantities to be prepared in SMM order and thus be coordinated with other project information and introduce a coding system creating an appropriate environment for computerization.

Project information sources for designers

Most designers will have access to catalogues of trade literature compiled and updated by the premier construction project information source Barbour. The compendium produced by *Barbour Index* holds all the trade literature relevant to buildings. The *RIBA.ti Technical Information Microfile* (produced jointly by RIBA Services and Technical Indexes) provides a similar index of various technical information.

Project information sources for contractors

Much of the project information held by contractors arises from information generated by other participants in the project. However, contractors will draw upon standard databases for cost and price data and production information to build up estimates for work and construction programmes.

These databases are discussed in greater depth in the next section.

Information from journals relevant to construction management

Institutional and commercial journals

Architects' Journal (Royal Institute of British Architects) Issues relevant to the architectural profession. Architectural reviews, practice management, etc.

Chartered Builder (Chartered Institute of Building) Journal of the Chartered Institute of Building covers managerial and technical issues facing the building industry.

Chartered Quantity Surveyor (Royal Institution of Chartered Surveyors) Journal of the QS Division of the RICS. Covers managerial and economic issues facing the construction industry.

580   Construction management

Proceedings of the Institution of Civil Engineers (Institution of Civil Engineers).

Construction Computing (Chartered Institute of Building) Developments of computing to the design and construction of buildings.

Civil Engineering ASCE (American Society of Civil Engineers) A monthly publication which carries a mixture of themes, opinion and job adverts.

Structural Engineer (Institution of Structural Engineers) Mainly technical papers on the theme of structural engineering but occasionally carries management papers.

New Civil Engineer (Institution of Civil Engineers) A weekly publication which carries a mixture of themes, opinion and job adverts.

Construction News (Publications) A newspaper which carries news and views from the industry. Widely read for its jobs section.

Building (Builder Group) A weekly publication which carries a mixture of business news, project reports, opinion, job adverts, etc.

Academic journals

Construction Management and Economics (Spon) Construction management, construction economics within an academic context.

Journal of Construction Engineering and Management (ASCE) Construction engineering case; construction management and economics, research products, presented in an academic style.

Journal of Professional Issues in Engineering Education and Practice (ASCE) Matters of professional concern to engineers, coverage include ethics, education, professional development, etc.

International Journal of Information Technology in Construction (University of Salford) Application of information technology, robotics, computer applications to the design and construction of buildings.

International Journal of Project Management (Butterworth-Heinemann) A quarterly journal from the International Project Management Association with coverage of project management issues ranging from project control to strategic issues.

Automation in Construction An International Journal for the Building Industry (Elsevier) The use of information technologies in architecture, engineering construction technology and maintenance and management of construction facilities.

Architecture, Construction and Engineering Management (University of Loughborough). A journal launched in 1994 to extend the coverage of academic reporting of Construction Management issues.

Computer driven information sources

It is a common view that all information sources will be computerized in the very near future (see *Building IT 2000*). Computer networks and telecommunication systems will increase accessibility of data banks and provide user friendly retrieval facilities.

It does not necessarily mean that the current information sources will be abandoned and, instead, new electronic ones will be created. There are two main trends that have been developed concurrently in information management:

- to enhance existing sources using computers in order to facilitate collecting, input, maintenance (update) and retrieval of the information.
- to create new tools in order to increase effectiveness of information management (knowledge-based systems, multimedia approach, visualization, telecommunication and networking, and others are among these tools).

Today's DBMS (Database Management System) allows one to produce in-house databases tailoring the information use and management to the organization's format and convenience. Usually an access to such information denied to the 'rest of the world'. This section will describe the information sources produced by public and commercial agencies for the external users. These sources (usually called databases or data banks) can be delivered to the user's computer offline (on disk, tape, CD-ROM, etc.) or online (via computer or telephone networks). The latter method requires special devices to establish communication (a computer card for the network option and a modem for connection through the telephone lines). The major advantage of having an online service is the almost instant access to the latest information. Online information retrieval is carried out using a computing terminal to search for data (textual or numerical) stored on computers around the world. Search terms can be combined in various ways online to give a more specific search than is possible using printed volumes. The cost of use depends on the database searched, the amount of time spent on line and the number of data items printed.

Selected offline information sources and services

Offline information is usually a part of the task-oriented commercial software (mainly on diskettes) or large online databases (mainly on CD-ROM). The most updated information (usually local or national) about these sources can be found in software catalogues and directories, professional journals, computer exhibitions, etc. Nevertheless, there are several directories of databases publicly available worldwide. *Directory of portable databases* and *Computer-readable databases* (both by Cuadra/Gale) are among the most comprehensive, and have recently been merged with the *Directory of online databases* to form the *Gale directory of databases* vol. 1: Online databases.

RIBA.ti (Construction Information Service) Is available on CD-ROM or network version. The database includes technical Information and the Building Supplement (equivalent to more than 150 000 pages of

582 Construction management

text) such as British Standards, codes of practice, legislations, agreement certificates, advisory material from government and public associations, and so on. Annual subscription with update three times per annum available from RIBA Services, UK.

BCQ/ODI (On Demand Information) CD-ROM construction information system available at an annual subscription. It includes product literature and technical information from manufacturers, as well as technical information from different government and public information providers. Online access, now available from PC, will require an ISDN telephone line and software provided by ODI.

CSSP/Thomas Telford Software (Institution of Civil Engineers) Develops and markets commercial database management systems for the construction industry. Standard libraries for quantities measurement and price databases are integrated in the software and can be altered by the user.

CITIS Very comprehensive bibliographic database of international civil engineering abstracts and software abstracts of construction computer programs worldwide. Published by CITIS in Blackrock, Dublin, Ireland.

MEANS (R.S. Means Company, Inc.) A very large and comprehensive source for US construction data. The company publishes over 20 annual cost guides and about 50 reference books. All cost data are available in Lotus 1-2-3 format or through different commercial estimating software packages (Data Source, Timberline, Astro II, Pulsar, and others).

PERINORM (Technical Indexes Ltd) A CD-ROM produced by BSI, AFNOR and DIN containing detailed bibliographic entries on all the current, full and draft standards and specifications from all three bodies plus all European and international standards produced by ISO, IEC and CEN/CENELEC.

WESSEX (Wessex Software UK Ltd) Commercial software for estimating, BOQ pricing, etc. It integrates construction cost databases from price books published by Wessex Group. The software enables the databases to be altered either globally or by estimate, to allow users to adjust data to suit local conditions and individual prices.

Selected online information sources

There are over 7000 (seven thousand!) online databases available worldwide today. Several commercial companies provide current information on databases accessible through online services.

Gale Research Inc. publishes as a part of the Electronic Information Series several directories covering various aspects of the information industry. Among them is the *Gale directory of databases*. It provides detailed entries with database type, subject, producer, online services, product description, language, geographic coverage, time span covered and frequency of updating. Separate sections list contact information for database producers and online services, as well as master, subject and geographic indices.

Aslib, The Association for Information Management, has launched the Aslib Online Series. The series includes more than ten rather subject-oriented directories. Cox (1991) and Pilkington and Rhodes (1992) are the most relevant to construction management.

BCIS ON-LINE (Building Cost Information Service, owned by the Royal Institution of Chartered Surveyors) has been designed to allow sections of data from the BCIS data banks for storage on the user's own computer. The data bank now has over 10 000 cost analyses from UK construction projects available to subscribers. Analyses covers a period of 20 years from 1973 but the majority (73 per cent) are from the last 10 years. Indices, construction economic indicators, background information as well as Approximate Estimate Package software program are also available to the subscribers. These are updated on a weekly basis.

BRIX/FLAIR Produced by the Department of the Environment, this source comprises of two databases (BRIX and FLAIR) which contain more than 180 000 citations, with abstracts to the worldwide journal articles, reports and books relating to building sciences.

BUILDING ONLINE Commercial information services mainly for the UK construction industry. BUILDING ONLINE provides access to a library of reference information including the Building Regulations, lists of manufacturers, standards, statistics, several price databases (among them Wessex Prices and average building prices from BCIS) and others. The service also provides electronic mail between users (annual subscription, can be accessed from microcomputer with an appropriate modem and communications software).

COMPENDEX (Computerised Engineering Index) (Engineering Index Inc.) Contains more than 2.5 million bibliographic records including all aspects of civil and related engineering.

DITR (Deutsches Informationszentrum für Technische Regeln) German technical rules and standards with English abstracts.

DODGE CONSTRUCTION ANALYSIS SYSTEM Produced by several American companies through McGraw-Hill, Construction Information Group. The database includes 4 million time series for construction projects. Gives information on contract value, square footage, etc. Information by country, structure type, ownership, type of construction, storey height, framing cost is also presented.

ICONDA The CIB International Construction Database, covering almost all aspects of civil engineering. The bibliographic section alone contains about 300 000 references. Available online and on CD-ROM.

PLANEX Produced by The Planning Exchange (Glasgow), covers European and some international urban development and planning, housing transportation, employment, statistics, etc.

SCAN-A-BID Produced by United Nations, Development Business/ Development Forum contains details of major construction projects worldwide by international organizations.

TED Tenders Electronics Daily Produced by the Office for Official Publications of the CEC (Commission of the European Communities). It is a directory of invitations to tender for public works and supply contracts from the 12 member states of the European Community, plus African, Caribbean and Pacific countries associated with the EEC. All community languages except Greek.

NON-ENGLISH SELECTED DATABASES

Dutch: VROMDOC
Finnish: SFS
French: BATIC, BATIMENT, CIMBASE, PASCAL
German: DITR, BAU LITDOK, RSWB
Swedish: BODIL (also in Danish, English, French, German, Norwegian), BYGGFO, BVR (Building Commodity File)

Electronic networks and major online hosts

There are many online databases which are accessible through the educational and commercial computer networks. The examples of the educational networks are JANET, Bitnet and Internet.

Internet is actually a 'supernetwork' through which all other networks are interconnected.

JANET is the acronym for the Joint Academic NETwork which connects campus networks in the UK. There is a variety of services available over JANET including: access to bulletin boards (BUBL for services resources available on networks), specialist databases (bibliographic BIDS, software NISSPAC, HENSA) computerized catalogues of academic institutions (UK, US/Europe) and e-mail.

Information services companies (called network hosts) provide online directories of the databases available from them. Data Star, the largest European online host system, provides users with access to over 250 public databases. DIALOG is a huge database host offering access to over 400 public databases. These two services have now merged into Knight-Ridder Information. Other hosts where construction information can be found are OVID Technologies (formerly BRS and CDP Online), ESA-IRS, FIZ TECHNIK, COMPUSERVE INFORMATION SERVICES – KNOWLEDGE INDEX, ORBIT (now Questel.Orbit), and others. Remember that the same database may be available through different hosts at different charges.

Research organizations

Construction managers may have occasion to draw upon the outputs of research organizations which serve the construction industry. The information produced by the organizations will vary in the immediacy of its use and some research information will be presented in an immediately digestible form, for example, a new code of practice, while others' output will be more speculative and require development by the user so that the information generated is useful. The list below addresses the main suppliers of construction research information.

Construction Industry Research and Information Association (CIRIA)

This body is funded by subscription from firms within the construction industry and carries a small government grant. The research carried out is largely contract research and consultants or academic institutions are invited to undertake a piece of research on behalf of the Association.

It has several themes to its research programme and is particularly strong in hydrology, geotechnics and structural behaviour. More recently however it has sponsored projects which relate to construction management issues. Among them has been contributions to materials management, quality assurance for construction and procurement systems. Each piece of research is presented as a report which is distributed free to CIRIA members, but may be purchased by non-members. The annual report lists the publications available and the *CIRIA guide to sources of construction information* is an invaluable source document.

Building Services Research and Information Association (BSRIA)

The work of this organization is similar to that of CIRIA but its emphasis is on the field of building services. Occasionally it will venture into the field of construction management, particularly in respect of the coordination of building services with other construction elements and trades. Its reports are available from the organization.

Building Research Establishment (BRE)

The BRE produces information concerning technical issues, but since the demise of the production and building economics division, its value to construction managers has diminished. It produces a catalogue of publications each year of conference organizers. This section catalogues some of the regular conferences, although obviously other more *ad hoc*, specialist events are frequently held.

International Council for Building Research Studies and Documentation (CIB; Council Internationale du Bâtiment)

This organization was founded in 1953 to encourage research and foster international cooperation in building and construction related research. The Council has its headquarters in Rotterdam and, as it has grown, has subdivided its activities into working commissions. There are now over 100 working commissions under the umbrella of the CIB. The ones most relevant to construction management are:

Working Commission (W)65 Organization and Management of Construction
W55 Building Economics
W70 Maintenance Management and Modernization of Building Facilities
W92 Procurement Systems

Each working commission holds seminars and conferences, and the proceedings constitute literature which most libraries with an interest in construction will maintain.

Internet

This organization grew out of the American Association of Project Managers and now is an international organization which organizes major conferences on the theme of project management. It meets alternatively in the USA and Europe. Originally its focus was the control of time and cost on projects but its scope is now considerably widened to incorporate strategic aspects of project management. Again, its proceedings are available in the public literature.

The Bartlett International Summer School

This organization is based around the Bartlett School of Architecture at University College London, and holds an annual conference (usually) in Europe during August/September of each year.

The theme of the conference is the production process involved in the built environment with a particular emphasis upon labour process issues and the political economy of the building industry. It presents a refreshing alternative to the output of many conferences since it attracts contributors who are economists, sociologists, anthropologists, etc., who cast a fresh eye on the workings of the construction industry. The proceedings are published each year and are available from the Bartlett School.

Association of Researchers in Construction Management

This organization is one which seeks to coordinate academic research in the field of construction management and assist in disseminating the products of research to industry. It maintains a database of current research, produces newsletters describing current research activity and holds highly successful conferences each year.

Construction Industry Computing Association (CICA)

The association was set up in 1973 to promote the effective use of computers in design, construction and property. Today, CICA is a company owned by its members. Among the services provided are regular bulletins, software directories and publications, exchange of information and experience, information on current and future markets, and others.

Source books for information

Bradfield, V. J. (ed.) (1983) *Information sources in architecture* (Butterworths).
Richardson, B. G. (compiler) (1989) *The CIRIA UK construction information guide* (Spon).
Sebestyen, G. and Pollington, C. E. (eds) (1986) *International directory of building research, information and development organisations. Council for Building Research Studies and Documentation*, 5th ed. (Spon.).

References

Building IT 2000: a study by experts on the future of building and information technology presented on a hypertext database. (1991) London: CERCI.
Centre for Strategic Studies in Construction (CSSC) (1988) *Building Britain 2001.* Reading: CSSC.
Cox, J. (1991) *Building, construction, architecture databases.* London: Aslib.
Day, A. and Langford, V. (1990). The implementation of CPI in the UK building industry. Report to the Building Centre Trust, University of Bath, Bath: University of Bath.
Department of the Environment (1971) *Data coordination and information flow in the construction industry.* London: HMSO.
Fellows, R. and Langford, D. (1992) *Marketing and the construction client.* Ascot: Chartered Institute of Building.
Fellows, R. et al. (1983) *Construction management in practice.* Harlow: Longman.
Fryer, B. (1990) *The practice of construction management* 2nd ed. Oxford: BSP Professional.
Harris, R. and McCaffer, R. (1989) *Modern construction management* 3rd ed. Oxford: BSP Professional.

Katz, R. L. (1974) Skills of an effective administrator. *Harvard Business Review*, **52**(5). 90–100.
Mackinder, M. and Marvin, H. E. (1982) *Design decision making in architectural practice*. BRE Information Paper 11/82.
Mintzberg, H. (1983) *Power in and around organisations*. Englewood Cliffs, NJ: Prentice Hall.
Mintzberg, H. (1973) *The nature of managerial work*. New York: Harper & Row.
National Economic Development Office (1988) *Faster building for commerce*. London: HMSO.
National Economic Development Office (1983) *Faster building for industry*. London: HMSO.
Newcombe, R., Langford, D. and Fellows, R. (1990) *Construction management*. London: Batsford.
Pilkington, S. and Rhodes, R. (1992) *Online engineering databases*. London: Aslib.

CHAPTER THIRTY-THREE

Mineral process engineering

M. R. SMITH

Introduction

Mineral process engineering is a multi-disciplinary subject requiring knowledge of geology, mineralogy, the properties of natural materials and material science, chemistry, process engineering and extractive metallurgy. The realization of a mining project, including the mineral processing plant or mill, also requires that the engineer is familiar with the practices of mechanical, civil and electrical engineering. Thus, as perhaps with no other branch of engineering, it is likely that useful sources of information will often include textbooks and journals directed towards other engineering disciplines, especially chemical engineering. Since the mineral industry is a large, basic wealth-creating activity which contributes significantly to the economies of most nations and dominates the economies of others it is also necessary for the mineral processing engineer to be closely concerned with finance, economics and business management.

Brief history

Mineral processing is an ancient skill. Indeed, the development of mankind has been chronicled by the attainment of skills to fashion tools first from flint (stone age), then from an alloy of copper and tin (bronze age) and subsequently from iron. Ancient civilizations plus the Egyptians, Greeks and Romans accumulated substantial elementary knowledge of mineral sources and the processing of ores in order to obtain base and precious metals, ceramics, glass, coloured pigments and gemstones. Several metals e.g. gold, silver, tin, copper and iron were extracted by direct smelting of ores but concentrates were also produced by manual sorting and selection of the minerals and by processes using water to separate minerals which would be recognized today as 'gravity concentration'. One of the earliest printed books *De re metallica* was written by Georgius Agricola (Hoover and Hoover, 1950) in 1556 who recorded and illustrated the practices of mining and mineral processing at that time in the Harz Mountains of Germany.

It is doubtful whether the knowledge at that time was truly a science and mineral processing has been described much more recently as an 'art'. With such a long history, mineral processing has obtained several synonyms including; mineral dressing, ore dressing, beneficiation, mineral concentration and milling. The frequent use of water has led to the use of the term 'washing', still prevalent in the coal mining industry to describe coal preparation. In North America and elsewhere the mineral processing engineer may be known as a 'metallurgist' with reference to the part that mineral processing plays in the extraction of metals (Gilchrist, 1989) rather than the technology of forming, working and using metals.

The technology of mineral processing developed rapidly during the Industrial Revolution particularly with mechanization of coal preparation replacing hand-sorting at coal mines, which were called upon to provide fuel for steam engines, coke for ironmaking, tar and domestic coal. During the first few decades of the twentieth century significant advances took place including the development of new separation processes such as froth flotation for the recovery of many base metals from their ores and cyanidation leaching of ores of gold and silver. Several legendary textbooks were printed during this period including Richards and Locke (1940), Truscott (1923), Taggart (1927) and Gaudin (1939). Much of their content remains relevant and is certainly of interest but, clearly, a great many advances have taken place in the understanding and practice of mineral processing since that time that were not included. Unfortunately, with the exception of Pryor (1965), it is only recently that comprehensive textbooks have been printed to replace them.

Minerals

Modern civilization has created a huge demand for the mineral products upon which it depends. Obvious examples are fuels such as coal, oil and uranium, construction materials such as steel, sand, glass and cement, metals for engineering and a huge variety of chemicals including fertilisers. A useful guide to the source, processing and use of mineral products is Johnstone and Johnstone (1961). The elements and compounds needed to meet this demand are derived from 'minerals'; a mineral being a naturally occuring chemical compound or 'native' element of approximately fixed composition and having a given crystal structure or form. Chemically, many minerals are relatively simple but are known by historical names, frequently Greek, Roman or medieval. This name is often unrelated to the chemical composition but determined by a physical property, locality of occurrence or first discovery or even the name of the geologist who first identified the mineral. Basic texts on mineralogy such as Gribble (1988), Dana (1949) or Klein and Hurlbut (1993) are useful if not essential.

Minerals are formed by geological processes and assemblages of minerals form rocks. These geological processes may create anomalously high concentrations of certain potentially valuable minerals within a rock which becomes a 'mineral deposit'. If the deposit can be exploited at a profit the rock becomes 'ore'. Some examples are provided in Park and MacDairmid (1975) and Jensen and Bateman (1981).

Mineral liberation and separation

The valuable element or compound is contained within a mineral which occurs as single crystals or units comprising clusters of crystals called 'grains' dispersed throughout the ore. The remainder of the rock mass may have no value and is known as waste or gangue. The proportion of value varies enormously; typical proportions being greater than 50 per cent for coal, 30 per cent for many iron ores, less than 2 per cent for copper ores and 10 parts per million (ppm) for gold ores. The size of the valuable mineral grains or unit also varies over orders of magnitude from coal seams several metres thick to grains of base and precious metal sulphide minerals of size less than 0.050 mm.

In order to be able to separate one mineral type from another, without the physical destruction of either through solution, melting or volatilization, it is necessary for the ore to be transformed into a state in which all of the valuable mineral exists as discrete, mono-minerallic particles. The mineral is then said to be liberated and the state is achieved by comminution (reduction of particle size) of the ore. Only in such a state can the maximum differences between the physical or chemical properties of the individual mineral particles be realized. This is a fundamental requirement, since these differences form the basis of the separation process. In addition, unless the valuable mineral is liberated it is not possible to produce a mineral concentrate of high grade or achieve a high recovery of the value from the ore. The liberation of mineral ores has been the subject of a textbook by Barbery (1991).

Even in the case of chemical processing, say by leaching, it is necessary for the surface of the desired mineral to be exposed to the liquid or gaseous reagent. Again, this is achieved through the crushing and grinding of the ore to a particle size which is of the same order as that of the grains of valuable mineral.

Therefore, it can be seen that comminution and sizing processes of industrial scale are extremely important with respect to mineral process engineering.

Mineral process engineering

In essence, mineral process engineering encompasses (i) the mineralogical appraisal of an ore; (ii) the design of a process flowsheet to separate the valuable mineral by appropriate combination of unit processes; (iii) selection and specification of equipment within the flowsheet; and (iv) commissioning, operation and control of the process plant. However, it may also extend to the supervision of construction and the transport and marketing of mineral concentrates or metals. Regrettably, few complete accounts or case histories have been published and none recently, but see Scobie and Wyslousil (1968), Abu Rashid and Smith (1982), Lewis and Martin (1983).

By its very nature, the mineral industry is often associated with large areas of land use and the general public will be concerned that the impact upon the environment is reduced to a minimum. Therefore, it is most probable that mineral processing engineers will need to consider the environmental aspects of any process at the design stage and to be familiar with technologies such as water treatment, gas cleaning and dust collection/suppression.

A notable feature of the mineral industry is the generation of waste mineral, gangue or tailings, which usually greatly exceeds the quantity of concentrate produced. The selection and design of an appropriate, safe and environmentally acceptable method of disposal in tips or tailings dams is an exercise in civil engineering for which the mineral processing engineer must accept some responsibility.

Finally, it is not uncommon for the plant manager to be responsible for the restoration and revegetation of exhausted mine workings and waste tips, a requirement of law in many countries. Some knowledge of techniques employed in agriculture and forestry is, therefore, useful.

The principal unit processes and subjects of relevance to mineral process engineering are;

- Mineralogy
- Comminution
- Sizing
- Mineral separation
- Process modelling, simulation and control
- Solid-liquid separation
- Bulk materials handling.

Therefore, it can be appreciated that information sources will include publications intended for chemical engineers, and mechanical engineers, in particular Coulson and Richardson (1991).

Mineral industry

Although several multinational mining companies exist with a wide range of mineral interests, clear divisions can be recognized within the mineral industry. The major divisions are between (a) the production of metals, (b) the production of coal, (c) the production of construction aggregates and non-metalliferous industrial minerals, and (d) the production of gemstones. Interest in metals can be sub-divided into (i) iron ore and steel alloy metals, (ii) base metals, copper, lead, zinc, etc., (iii) precious metals, gold and silver, (iv) aluminium, (v) nickel, and (vi) platinum group metals. This situation has promoted the publication of textbooks and journals catering for specific interests such as *Coal*, *Industrial Minerals* and *Rock Products*.

Information sources

In general there is remarkably free exchange of technical information within the mineral industry. The paucity of modern textbooks is compensated for by verbal exchange of information, meetings of institutions and societies and symposia and conferences. The journals, in particular, assume great importance as sources of information upon current practice and new developments. However, the aluminium, nickel, platinum and gemstone, especially diamond, industries tend to be vertically integrated from mining through processing to marketing of the finished product, and technical information is often confidential and difficult to obtain.

Learned societies, authorities and organizations

The principal professional institutions to which mineral processing engineers employed or educated in the UK belong are:

Institution of Mining and Metallurgy (IMM)
Institution of Mining Engineers (IMinE)

The first represents the interests of all those working in the mining industry other than coal and certain stratiform deposits e.g. clays. The latter represents mining engineers and those employed in coal preparation. Both offer members the opportunity to register with the Engineering Council (EC) including the status of Chartered Engineer. The Minerals Engineering Society (MES) (formerly the Coal Preparation Society), was created within the UK coal mining industry to specifically represent the interests of coal preparation engineers. Until recently it was a nominated body of the EC through IMinE.

The interests of those employed in the quarry industry both in UK and abroad, particularly in Australia, New Zealand, South Africa and the Far East, e.g. Hong Kong, are served by the Institute of Quarrying (IQ). This body does not belong to the Engineering Council. Internationally, emigrant mining engineers have created institutions offering similar benefits and services in the important mining districts of the world that once formed part of the British Empire. The more important are:

Canadian Institute of Mining and Metallurgy (CIM)
South African Institution of Mining and Metallurgy (SAIMM)
Australasian Institute of Mining and Metallurgy (AusIMM)

The Society of Mining, Metallurgy and Exploration (SME) and The Metallurgical Society (TMS-AIME) represent the interests of engineers employed both in the mining and extractive industries of the USA and internationally. The TMS and SME form part of the American Institute of Mining, Metallurgical and Petroleum Engineers (AIME).

Research centres

A number of research centres exists that are funded by national, federal or local government to conduct research and development in the fields of mining and mineral processing in order to support an important wealth creating or export industry. These include:

United States Bureau of Mines (USBM), Washington DC which has a number a research centres in the USA such as those located at Reno, Twin Cities (Minneapolis–St. Paul) and Salt Lake City.
Commonwealth Scientific and Industrial Research Organization (CSIRO), Division of Mineral Process Engineering, Australia.
Warren Spring Laboratory (WSL), Harwell.
MINTEK (National Institute for Metallurgy, NIM).
Canadian Centre for Mineral and Energy Technology (CANMET).
Ontario Research Foundation, Canada.
Bureau de Recherches Geologiques et Minières (BRGM), France.
Institute of Geology and Mineral Exploration (IGME), Greece.
Mineral Research and Exploration General Directorate (Maden Tetkik ve Arama Genel Mudurlugu), Turkey.
Centro de Investigacion Minera y Metalurgica, Chile.

Until privatization in 1994, British Coal maintained a research and development centre, Technical Services and Research Executive (TSRE), previously known as the Mines Research and Development Establishment (MRDE), at Bretby, near Burton upon Trent. It is now a private organization, Technical Engineering Services, TES.

Much of the work of these organizations is published in journals and proceedings of conferences.

Commercial research laboratories

There are many commercial research laboratories some of which have been or continue to be partly funded by government including, for example, AEA Technology (formerly WSL). Two well known research centres are attached to universities: Mineral Resources Research Center (MRRC), University of Minnesota, USA and the Julius Kruttschnitt Mineral Research Centre (JKMRC), University of Queensland, Australia.

A large number of fully commercial laboratories is located near the major mining fields in USA, Canada, Australia and South Africa, e.g. Lakefield Research, Ontario; Kilborn Engineering, Toronto and Mountain State, Colorado, Robertson Research went into liquidation. Several major mining companies have established research laboratories which may also undertake external research, e.g. RTZ, Anglo-American/ DeBeers, Noranda and BHP. However, the results of their research are not generally available.

Research organizations

Two organizations which exist to identify the research needs of the mineral industry and manage collaborative research projects between industrial partners and both commercial and university research laboratories are:

Mineral Industry Research Organisation (MIRO), UK.
Australian Mineral Industries Research Association (AMIRA).

Universities

While the universities containing departments of, say, Civil or Chemical Engineering, are too numerous to mention, relatively few contain departments teaching degrees in mining, mineral engineering or mineral processing. The following list, which is not exhaustive, presents sources of expert opinion and information contained in libraries.

UK:

Department of Earth Resources Engineering, Royal School of Mines, Imperial College, London
Department of Mining and Mineral Engineering, University of Leeds
Department of Mineral Resources Engineering, University of Nottingham
Department of Chemical Engineering, University of Birmingham
Camborne School of Mines, University of Exeter

Europe:

Austria: Leoben
Finland: Helsinki
France: Paris, Ales and Nancy

Germany: Aachen, Berlin, Clausthal and Freiburg
Greece: National Technical University, Athens
Holland: Delft Technical University
Italy: Cagliari, Milan, Rome and Trieste
Norway: Trondheim
Portugal: Lisbon and Oporto
Spain: Madrid and Oviedo
Sweden: Lulea and Goteborg
Turkey: Middle East Technical University, Ankara and Istanbul Technical University

International:

USA: Colorado School of Mines, Michigan Technological University, Henry Krumb School of Mines, Colombia University, New York, Massachusetts Institute of Technology, and the Universities of Utah, Penn State, Arizona, California, Missouri-Rolla and Minnesota
Canada: British Colombia, Vancouver; McGill University, Montreal; Laval University, Quebec City, Queens University, Kingston and the Universities of Alberta and Nova Scotia
South America: Brazil, Sao Paulo and Rio de Janiero; Chile, University of Chile, Santiago, Guyana, Georgetown
Africa: Ghana, Tarkwa School of Mines; South Africa, Universities of the Witwatersrand, Johannesburg, Stellenbosch and Cape Town; Zambia ,University of Zambia, Lusaka; Morocco, Rabat
Australasia: Western Australian School of Mines, Kalgoorlie, South Australian Institute of Technology, Adelaide, Western Australia Institute of Technology and the Universities of Queensland and Western Australia, Perth

Journals

Foremost sources of technical information are the journals published by the learned societies and institutions.

Transactions of the IMM (TransIMM), Section C, 'Mineral processing and extractive metallurgy'.
Mining Engineering.
Journal of the SAIMM (JSAIMM).
Bulletin of the CIM (CIMBulletin).
Bulletin of the Aus.IMM.

The SME also publishes *Minerals and Metallurgical Processing* and *Mining Engineering.*

Commercial publications enjoying wide circulation and containing articles or papers upon mineral processing include:

Mining Magazine (Mining Journal Ltd).
Engineering and Mining Journal (E/MJ) (Maclean Hunter).
International Journal of Mineral Processing (Elsevier).
Minerals Engineering (Pergamon).

Papers upon the chemical processing of ores by leaching and hydrometallurgy are contained in *Hydrometallurgy* published by Elsevier.

The journal *World Mining* no longer appears, although past copies are available for reference in most mining libraries. It has been replaced by *World Mining Equipment* (WME) (Metal Bulletin) which primarily promotes equipment and services to the industry.

Information and news upon the production of construction materials (aggregates), cement, open-cast coal and other minerals by quarrying is published in *Quarry Management* by the Quarry Managers Journal. The same interest in the USA is served by the journal *Rock Products* (Maclean Hunter).

A principal journal of the British coal mining industry is *Colliery Guardian* published by FMJ International. The American journal *Coal Age*, previously published by McGraw-Hill is now contained in *Coal* (Intertec). The title *World Coal* has been merged into *World Mining Equipment* (WME) (Metal Bulletin).

A specialist journal for the non-metalliferous mining industry is *Industrial Minerals* published by Metal Bulletin.

A number of newsletters and journals exists that provide information upon new developments and events within the mineral industry. The most famous of these is the weekly *Mining Journal* published in London by Mining Journal Ltd. The USBM publishes *Minerals Today* and the IMM circulates *Minerals Industry International* to its members. Others include:

Canadian Mining Journal (Southam Business Communications).
Australian Mining Monthly (Thompson Publications Australia).

The concern with the environmental impact of mining has stimulated publications dedicated to this subject and it can be anticipated that their number will increase. An example is:

Mining Environmental Management (Mining Journal).

Textbooks

Comprehensive

There are only a few modern, comprehensive textbooks upon mineral processing namely: Pryor (1965), Kelly & Spottiswood (1982) and Wills (1992). The last is a popular introductory text, available in paperback, recommended to undergraduate students in the UK. The most recent comprehensive text comprising many chapters written by experts was edited by Weiss (1985) and published by the SME. This huge volume effectively updates and replaces the earlier version of Taggart (1927).

598 Mineral process engineering

Texts that are comprehensive with respect to particular mineral commodities or divisions within the mineral industry are Leonard (1991), Lefond (1983) and Littler (1990).

The design of a mineral processing plant was comprehensively discussed as a number of papers or chapters edited by Mular and Bhappu (1980) and published by the SME.

Additional texts

MINERALOGY

Gribble (1988) and Dana (1949) provide introductory information upon the identification, composition, occurrence and properties of minerals. A more comprehensive text is also available: Klein and Hurlbut (1993).

COMMINUTION

The theoretical and practical aspects of the design of process plant for crushing and grinding of mineral ores are discussed in Mular and Jergensen (1982) and Komar Kawatra (1992). More specific information is provided by McQuiston and Shoemaker (1978) and Austin, Klimpel and Luckie (1984).

The design of crushing plant to produce aggregates is the subject of Mellor (1990).

Additional information can be found in relevant chapters of texts upon chemical engineering (Coulson and Richardson, 1991, Perry and Green, 1984).

SIZING

An introduction to the concepts and measurement of particle size is given by Allen (1990). The selection and application of vibrating screens is the subject of a chapter in Mular and Bhappu (1980). The hydrocyclone classifier is in widespread use in the mineral industry as a sizing device and is discussed by Svarovsky (1984).

MINERAL SEPARATION

Mineral separation processes can be categorized as physical or chemical processing. One very important process, however, does not fit easily into either category and that is froth flotation. Mular and Anderson (1986) discuss the design of the mineral separation section of the plant.

PHYSICAL SEPARATION PROCESSES

The process exploits differences of physical properties, e.g. density, magnetism and electrical conductivity to separate the minerals. Separation on the basis of density includes processes known as dense medium separation and gravity concentration. Gravity concentration processes, plant design and flowsheets were described in Burt and Mills (1984). Svoboda (1987) discusses magnetic methods of mineral separation and electrical separation is discussed by Inculet (1984).

CHEMICAL SEPARATION PROCESSES

Also referred to as hydrometallurgy, these processes include leaching; solution purification by precipitation, solvent extraction or ion exchange; carbon adsorption for gold recovery, electro-winning and electro-refining. The most important applications are the production of alumina (aluminium), the recovery of gold by cyanidation and the leaching of copper and uranium from their ores.

The basic chemistry is discussed in Burkin (1966) and in American Cyanamid (1958). No single textbook has been devoted to the subject of hydrometallurgy, not even to the cyanidation of ores of gold and silver. However, much useful information can be obtained from collections of papers upon specific aspects of leaching practice. Examples are McQuiston and Shoemaker (1975), Brent Hiskey (1983) and Van Zyl, Hutchinson and Kiel (1988).

FROTH FLOTATION

There is probably more written upon froth flotation than any other aspect of mineral processing but, in part, this reflects the importance of the separation process to the modern industry. Most base metals, with the exception of tin and tungsten, and a significant proportion of precious metals are concentrated by froth flotation. In addition, it is applied to iron ores, coal and a wide variety of industrial minerals.

Although there has been substantial development, particularly of flotation reagents, and the unit size of flotation machines has increased dramatically in attempts to obtain economies of scale and energy consumption, the basic principles are unchanged. Therefore, many of the early texts remain useful and include Gaudin (1957), Sutherland and Wark (1955), and Glembotski, Klassen and Plaksin (1963). One of the most comprehensive reviews of the theory and application of flotation, undertaken to mark 50 years of industrial application of flotation, was edited by Fuerstenau (1962) and published by the SME. The conference held as a memorial to Professor Gaudin produced another important reference, Fuerstenau (1976). The most recent textbook with the title *Flotation* has been written by Crozier (1992).

An understanding of the surface chemistry of minerals and the function of flotation reagents to alter the nature of this surface is fundamental to flotation. The science is discussed by Leja (1982). Discussion of flotation reagents constitutes a major part of Jones and Oblatt (1984) and Somasundaran and Moudgil (1987). Manser (1975) specifically reviewed the use of reagents for the flotation of silicate minerals. Other than the introduction of specific reagents the most significant, recent development has been that of column flotation which is the subject of the textbook by Finch and Dobby (1990) and a symposium edited by Sastry (1988).

SOLID-LIQUID SEPARATION

Comprehensive reviews have been given by Svarovsky (1990) and Purchas (1981). Greater detail of the design procedures has been provided by Purchas (1986). The subject is also dealt with in textbooks of chemical engineering (Coulson and Richardson, 1991, Perry and Green, 1984).

Handbooks, data sources and directories

The most comprehensive, readily available data source is Weiss (1985) and much useful data are also contained in Mular and Bhappu (1980).
 The Nordberg Machinery Co. produced a *Reference manual* (Nordberg, 1983) containing information upon the application of crushing, grinding, screening and handling equipment manufactured by that company. Pegson Ltd (1953) produced a similar publication for the assistance of producers of crushed rock, sand and gravel. The Allis Chalmers Corp., now Allis Mineral Systems Ltd, reprinted a paper by Bond (1954) on crushing and grinding calculations. The Denver Equipment Company (DECO) (1965) has printed a handbook detailing typical flowsheets for separation of minerals and providing advice upon application of their wide range of process equipment.
 Manufacturers of flotation reagents, flocculants and other mining chemicals provide extensive technical literature of which American Cyanamid (1989) and Dow Chemical Co. (1976) are well known and much used examples.
 Several compendia of mineral processing flowsheets and accounts of operating experience provide useful comparative information for design purposes including Pickett (1978) and Woodcock (1980). The publishers McGraw-Hill compiled three handbooks from articles selected from *Coal Age* and *E/MJ* (Merritt, 1978, Thomas, 1977 and White, 1980).

Conferences and symposia

In 1952 the IMM hosted a conference entitled Recent developments in mineral processing which became the first of the International Mineral

Processing Congresses (IMPC) which are held every two or three years in international locations; the eighteenth in 1993. The proceedings constitute a substantial information source. Similar events are organized by the coal industry, the International Coal Preparation Congresses, and the industrial minerals industry, the Industrial Minerals International Congresses.

The TMS-AIME and SME(AIME) hold regular quarterly and annual meetings, the proceedings of which are sometimes published as separate works, for example Sastry (1988) and Zunkel *et al.* (1985). The SME symposium held to honour Nathaniel Arbiter was published as *Advances in mineral processing* (Somasundaran, 1986).

A short course organized at the University of Arizona in 1961 with the title Computers and computer applications in the mineral industry became the first of a series of conferences variously known as Computers in mining and Computers in the mineral industry, and generally as Application of computers and operations research in the mineral industry or APCOM.

The Queensland branch of the AusIMM occasionally organizes symposia for the exchange of operating experience as the Mill operator's conference. A specific symposium for operators of dense medium plants took place in 1987.

The British Hydraulics Research Association (BHRA), Cranfield, organizes international conferences with the title *Hydrocyclones* which are relevant to mineral processing engineers.

Finally, the Society of Chemical Industry (SCI), London has organized the International Solvent Extraction Conferences (ISEC) and the *Hydromet* conferences which include discussion of hydrometallurgical processing of mineral ores.

Reviews and directories

Several journals contain annual reviews of the mineral industry on a basis of commodity, metal or mineral product, rather than technology and engineering. However, Mining Journal produces the more comprehensive *Mining annual review*.

The USBM publishes production statistics for the mining industry worldwide in the *Minerals yearbook*.

There are several directories of mining companies and mineral operations:

Financial Times mining international yearbook (Longman).
E/MJ International directory of mining (Maclean Hunter).
The mining directory (Don Nelson Publications).
Industrial minerals directory (Metal Bulletin).
Guide to the coalfields (Colliery Guardian).

Directory of mines and quarries (British Geological Survey).
Directory of quarries, pits and quarry equipment (Quarry Management).

Patents

The Noyes Data Corporation has published a review of patents with respect to flotation (Ranney, 1980).

Abstracts and indexes

Indexes are available for the journals *Mining Journal, Mining Magazine* and *Industrial Minerals. Coal Abstracts* is prepared by the International Energy Agency (IEA) in printed form. However, the principal service is provided by the IMM through *IMM Abstracts*.

Computerized literature search

The following are available:

IMMAGE (Institution of Mining and Metallurgy).
ECOMINE (Bureau de Recherches Geologique et Minière) (BRGM).
STIMLINE (Deutsches Bundesanstalt für Geowissenschaften und Rohstoffe) (BGR).
AUSTRALIAN EARTH SCIENCES INFORMATION DATABASE (Australian Mineral Foundation).
COAL DATA BASE (IEA Coal Research).
NTIS (US National Technical Information Service).

References

Abu Rashid, A. R. and Smith, M. R. (1982) *Development of a selective flocculation-froth flotation process to beneficiate a non-magnetic iron ore of the Kingdom of Saudi Arabia* Proceedings of 14th International Mineral Processing Congress. Toronto: CIM.
Allen, T. (1990) *Particle size measurement* 4th ed. London: Chapman and Hall.
American Cyanamid Co. (1989) *Mining chemicals handbook*. Wayne, NJ: American Cyanamid.
American Cyanamid Co. (1958) *Chemistry of cyanidation*. New York: American Cyanamid.
Austin, L. G., Klimpel, R. R. and Luckie, P. T. (1984) *Process engineering of size reduction: ball milling*. Littleton, CO: SME(AIME).

Mineral process engineering 603

Barbery, G. (1991) *Mineral liberation*. Quebec: Les Editions GB.
Bhappu, R. and Harden, R. (1989) *Gold forum on technology and practices–World Gold '89*. 1st international meeting between SME and AusIMM. Littleton, CO: SME(AIME).
Bond, F. C. (1954) Crushing and grinding calculations. *CIM Bulletin*, **47**, 466–472.
Burkin, A. R. (1966) *The chemistry of hydrometallurgical processes*. London: Spon.
Burt, R. and Mills, C. (1984) *Gravity concentration technology*. Amsterdam: Elsevier.
Coulson, J. M. and Richardson, J. F.(1991) *Chemical engineering* Vol. 2: particle technology and separation processes, 4th ed. Oxford: Pergamon.
Crozier, R. D. (1992) *Flotation*. Oxford: Pergamon.
Dana, E. S. (1949) *Minerals and how to study them* 3rd ed. New York: Wiley.
Denver Equipment Co. (1965) *Modern mineral processing flowsheets* 2nd ed. Denver, CO: Denver Equipment Co.
Dobby, G. S. and Roa, S. R. (eds) (1989) *Processing of complex ores* Proceedings of an International Symposium; 28th Annual Conference of Metallurgists of CIM, Halifax, Nova Scotia, August 20–24, 1989. New York: Pergamon.
Dow Chemical Co. (1976) *Flotation fundamentals and mining chemicals*. Midland: Dow Chemical Co.
Finch, J. A. and Dobby, G. S. (1990) *Column flotation*. Oxford: Pergamon.
Fuerstenau, D. W. (ed.) (1962) *Froth flotation 50th anniversary volume* Rocky Mountain Fund Series, Littleton, CO: SME(AIME).
Fuerstenau, M. C. (ed.) (1976) *Flotation; A.M. Gaudin memorial volume* 2 vols. Littleton, Co: SME(AIME).
Gaudin, A. M. (1939) *Principles of mineral dressing*. New York: McGraw-Hill.
Gaudin, A. M. (1957) *Flotation* 2nd ed. New York: McGraw-Hill.
Glembotski, V. A., Klassen, V. I. and Plaksin, I. N. (1963) *Flotation* (translated from Russian). New York: Primary Sources.
Gilchrist, J.D. (1989) *Extraction metallurgy* 3rd ed. Oxford: Pergamon.
Gribble C.D. (1988) *Rutley's elements of mineralogy* 27th ed. London: Unwin Hyman.
Hiskey, J. Brent (ed.) (1983) *Heap and dump leaching practice* proceedings of the 1993 SME Fall Meeting, Salt Lake City, Utah, October 19–21, 1993. Littleton, CO: SME(AIME).
Hoover, H. C. and Hoover, L. H. (1950) *De re metallica* by Georgius Agricola 1556 (translated from Latin). New York: Dover.
Inculet, I. I. (1984) *Electrostatic mineral separation*. Letchworth: Research Studies Press.

Jensen, M. L. and Bateman, A. M. (1981) *Economic mineral deposits* 3rd ed. New York: Wiley.

Johnstone, S. J. and Johnstone, M. G. (1961) *Minerals for the chemical and allied industries* 2nd ed. London: Chapman and Hall.

Jones, M. J. and Oblatt, R. (eds) (1984) *Reagents in the mineral industry.* London: IMM.

Kelly, E. G. and Spottiswood, D. J. (1982) *Introduction to mineral processing.* New York: Wiley.

Klein, C. and Hurlbut, C. (1993) *Manual of mineralogy* (after J. D. Dana) 21st ed. New York: Wiley.

Komar Kawatra, S. (ed.) (1992) *Comminution, theory and practice.* Littleton, CO: SME(AIME).

Lefond, S. J. (ed.) (1983) *Industrial minerals and rocks* 5th ed. 2 vols. New York: SME(AIME).

Leja, J. (1982) *Surface chemistry of froth flotation.* New York: Plenum.

Leonard, J. W. (ed.) (1991) *Coal preparation* 5th ed. Littleton, CO: SME(AIME).

Lewis P. J. and Martin G. J. (1983) Mahd adh Dhabah gold-silver deposit, Saudi Arabia: mineralogical studies associated with metallurgical process evaluation *Transactions. Institution of Mining and Metallurgy,* **92**, 63–72.

Littler, A. (1990) *Sand and gravel production.* Nottingham: Institute of Quarrying.

Lynch, A. J. et al. (1981) *Mineral and coal flotation circuits: their simulation and control.* Amsterdam: Elsevier.

Manser, R. M. (1975) *Handbook of silicate flotation.* Stevenage: Warren Spring Laboratory.

McQuiston, F. W. and Shoemaker, R. S. (1975) *Gold and silver cyanidation plant practice.* Littleton, CO: SME(AIME).

McQuiston, F. W. and Shoemaker, R. S. (1978) *Primary crushing plant design.* Littleton, CO: SME(AIME).

Mellor, S. H. (1990) *An introduction to crushing and screening.* Nottingham: Institute of Quarrying.

Merritt, P. (ed.) (1978) *Coal Age operating handbook of coal preparation.* New York: McGraw-Hill.

Mular, A. L. and Bhappu, R. B. (eds) (1980) *Mineral processing plant design.* Littleton, CO: SME(AIME).

Mular, A. L. and Jergensen, G. V. (1982) *Design and installation of comminution circuits.* Littleton, CO: SME(AIME).

Mular, A. L. and Anderson, M. A. (1986) *Design and installation of concentration and dewatering circuits.* Littleton, CO: SME(AIME).

Nordberg (UK) Ltd. (1983) *Reference manual.* Coalville: Nordberg Machinery Co.

Park, C. F. and MacDairmid, R. A. (1975) *Ore deposits* 3rd ed. San Francisco: Freeman.
Pegson Ltd (1953) *Aggregate producers handbook*. Coalville, Leicester: Pegson Ltd.
Perry, R. H. and Green, D. (1984) *Perry's chemical engineers handbook* 6th ed. New York: McGraw-Hill.
Pickett, D. E. (ed.) (1978) *Milling practice in Canada*. Montreal: CIM.
Pryor, E. J. (1965) *Mineral processing* 3rd ed. Amsterdam: Elsevier.
Purchas, D. (1981) *Solid/liquid separation technology*. Croydon: Uplands Press.
Purchas, D. (1986) *Solid/liquid separation equipment scale-up* 2nd ed. Croydon: Uplands Press.
Randol International (1992) *Randol gold forum*. Golden: Randol International.
Ranney, M. W. (1980) *Flotation agents and processes*. Park Ridge, NJ: Noyes Data Corporation.
Richards, R. H. and Locke, C. E. (1940) *Textbook of ore dressing* 3rd ed. New York: McGraw-Hill.
Sastry, K. V. S. (ed.) (1988) *Column flotation 88* Proceedings of the SME Annual Meeting, Phoenix, Arizona, January 25–28, 1988. Littleton, CO: SME.
Scobie, A. G. and Wyslousil, D. M. (1968) *Metallurgical testing, design, construction and operation of Lake Dufault treatment plant* CIM Bullet, **61**, 482–488.
Somasundaran, P. (ed.) (1986) *Advances in mineral processing: a half century of progress in application of theory to practice* Arbiter Symposium. Littleton, CO: SME(AIME).
Somasundaran, P. and Moudgil, B. M. (1987) *Reagents in mineral technology* Vol. 27, Surfactant science series. New York: Marcel Dekker.
Sutherland, K. L. and Wark, I. W. (1955) *Principles of flotation*. Melbourne: AusIMM.
Svarovsky, L. (1984) *Hydrocyclones*. London: Holt, Rinehart and Winston.
Svarovsky, L. (ed.) (1990) *Solid-liquid separation* 3rd ed. London: Butterworths.
Svoboda, J. (1987) *Magnetic methods for the treatment of minerals*. Amsterdam: Elsevier.
Taggart, A. F. (1927) *Handbook of ore dressing*. New York: Wiley.
Thomas, R. (ed.) (1977) *E/MJ operating handbook of mineral processing*. New York: McGraw-Hill.
Torma, A. E. and Gundiler, I. H. (eds) (1989) *Precious and rare metal technologies*. Amsterdam: Elsevier.
Truscott, S. J. (1923) *Textbook of ore dressing*. London: Macmillan.

Weiss, N. L. (ed.) (1985) *SME mineral processing handbook.* Seeley Mudd Series. New York: SME.
White, L. (ed.) (1980) *E/MJ second operating handbook of mineral processing.* New York: McGraw-Hill.
Wills, B. A. (1992) *Mineral processing technology* 5th ed. London: Pergamon.
Woodcock, J. T. (ed.) (1980) *Mining and metallurgical practices in Australasia* Monograph Series Volume 10. Parkville: AusIMM.
Van Zyl, D. J. A., Hutchinson, I. P. G. and Kiel, J. E. (1988) *Introduction to evaluation, design and operation of precious metal heap leaching projects.* Littleton, CO: SME.
Zunkel, A. D. et al. (1985) *Complex sulphides: processing of ores, concentrates, and by-products* Proceedings of a Symposium held at the TMS-AIME Fall Extractive Meeting, San Diego, CA, November 10–13, 1985. Warrendale, PA: TMS-AIME.

CHAPTER THIRTY-FOUR

Mining

C. T. SHAW

Introduction

Mining is here defined as the recovery of minerals and hydrocarbons from the earth's crust for the benefit of mankind. As such it can claim to be probably the oldest profession, certainly the oldest engineering profession. It predates farming (as opposed to hunting and gathering) by millennia and it is arguable that one of the distinguishing features of the human species is the use of mining to recover the materials required to ensure humanity's comfort and survival. Mining then goes back to pre-history, to the start of human activity. Indeed the various ages of human history have been named after the dominant mined materials used, such as the stone age, the copper age, the bronze age, the iron age, etc.

Initially very few materials were mined, but over time the number has increased until now uses have been found for virtually all the elements in the periodic table and they are all recovered from mines. In addition, there is a large number of industrial minerals, gemstones and fuels which are mined for themselves rather than for their contained elements. Mining is therefore an extremely broad discipline and covers a wide area of engineering.

This chapter will however only deal with the mining of metals and industrial minerals as the mining of coal and other hydrocarbons has been dealt with in *Information sources in energy* (1988) (Butterworths).

Organizational sources

Mining industrial organizations

There is a vast number of mining operations worldwide and the majority of them are privately owned. Even where there has been state ownership there is a move towards privatization in most countries. However, mining tends to be dominated by a number of very large multinational organizations which own operations in a large number of countries. The mining companies can be researched in a number of publications particularly in the yearbooks which list all the most important mining organizations. The *Financial Times mining international yearbook* (Longman) is an excellent source of information as is the *E & MJ international directory of mining*. A third reference volume is the *Mining directory* 7th ed. (1994) (Don Nelson Publications) which lists mining companies and mining equipment manufacturers and suppliers worldwide.

There are also specialist directories for many sectors of the mining industry. Notable among these are The *Industrial minerals directory* 2nd ed. (1991) (Metal Bulletin Books) which covers all industrial minerals and, other more local publications such as *Spon's quarry guide to the British hard rock industry* (1991) (Spon) which lists all the quarries in the UK. There is also a wide range of publications dealing with coal but, as indicated before, they are not covered in this chapter as they have already been dealt with in *Information sources in energy*.

In the USA the major mining companies have joined together to form a lobbying organization based in Washington, the American Mining Congress. This organization holds regular meetings and publishes a journal *The Mining Congress Journal* (1915–). The governmental organization associated with mining is the United States Bureau of Mines (USBM). The USBM has been responsible for a vast amount of mining related research and publishes three regular series of research monographs. There is the Bureau of Mines Information Circular/19xx series, of which there have now been close to 10 000 produced. Secondly, there is the Bureau of Mines Report on Investigations/19xx of which again some 10 000 have been issued. The USBM also produces a regular *United States Department of the Interior; United States Bureau of Mines; minerals yearbook* (annual). This deals with the production and statistics of the various mineral commodities, commodity by commodity. They also issue for each commodity a regular Mineral Commodity Profile which deals with the sources, producers, uses and statistics of the commodity concerned.

In the UK the mining companies have formed a number of industrial organizations. There is the Mining Association of the UK, the Mining Equipment Manufacturers Association and the Sand and Gravel Association. On the research side there is the Mining Industry Research Organization. None of these organizations issues regular journals or publications in the public domain.

In South Africa, the majority of the mining companies are members of the Chamber of Mines of South Africa. The Chamber of Mines reports regularly on the industry and publishes a *Chamber of Mines*

Mining 609

Newsletter (bi-monthly). This organization also runs a research operation for mining and publishes research reports and a regular *COMRO Bulletin* giving news of research results. It also produces an annual report which contains a detailed statistical summary of the operations of the major gold, coal and uranium mines in South Africa. The Chamber of Mines also deals with safety matters and until 1992 published a monthly safety magazine, *The Reef*.

Professional societies

Mining is served by a large variety of professional societies. In the UK there are three which deal directly with the mining industry. The Institution of Mining and Metallurgy (IMM) at 44 Portland Place in London is the institution which deals with metal mining, industrial minerals, metallurgy and petroleum engineering. The Institution of Mining Engineers (IMinE), headquartered at Danum House in Doncaster, deals with tabular deposits, specifically the mining of coal and iron ore. Finally, the Institute of Quarrying (IQ) deals, as its name implies, with the quarry industry and is based in Nottingham. There is a fourth institution which serves the non-mining engineers in the mining industry, the Institution of Mining Electrical and Mining Mechanical Engineers (IMEMME), but this Institution is about to merge with the IMinE.

The first two of the above institutions have royal charters and are recognized by the Engineering Council as institutes able to grant chartered status to their members. The Institute of Quarrying has not yet achieved that status. Each of them publishes journals of one kind or another. The IMM publishes its *Transactions* (1892–) in three sections, *Section A – Mining Industry*, *Section B – Applied Earth Science* and *Section C – Mineral Processing and Extractive Metallurgy*, each of which is published three times each year. In addition it publishes a regular bi-monthly journal which contains news of the Institute activities and short technical contributions called *Minerals Industry International*. The *Transactions* are fully refereed journals. Finally, the IMM runs an abstracting service which it markets both in hard-copy, the *IMM Abstracts* and as a computer-based service called IMMAGE, currently available on floppy disks and directly by modem connection. This is probably the best computer abstracting service available to the industry as the vast majority of the abstracted material is also available in hard-copy from the IMM. The IMM also runs a continuing series of conferences and publishes the proceedings of those conferences. Finally the IMM commissions books on various mining engineering topics and publishes these.

The IMinE only publishes a monthly journal the *Mining Engineer* (1960–) and the Institute of Quarrying publishes its activities in *Quarry Management* (1918–, Quarry Managers Journal Ltd.) monthly. However, both the latter two organizations also run regular conferences, but the proceedings are not always published. IMEMME publishes *Mining Technology* (1920–, Marylebone Press) 10 per annum, which deals with coal mining equipment. It is not clear what will happen to this journal after the merger.

There are equivalent institutions in all of the other countries of the English-speaking world. The Australasian Institute of Mining and Metallurgy, the Canadian Institute of Mining and Metallurgy and the South African Institute of Mining and Metallurgy are all similar in style to the IMM and all publish transactions. The Australasian Institute publishes both *AusIMM Bulletin* (1983–) 6 per annum, and *The AusIMM Proceedings* (1983–) biannually, the Canadian Institute publishes the *CIM Bulletin* (1898–) monthly, and the South African Institute *The Journal of the SAIMM* (1984–) monthly. In South Africa there is also the Association of Mine Managers of South Africa. This organization prints the papers delivered at its meetings on a regular basis as *AMM Circular No. x/yy*. On a biennial basis these papers are published in book form as the *Association of Mine Managers of South Africa – papers and discussion*. Normally each volume covers the previous two years papers and discussions.

In the USA there is the Society for Mining, Metallurgy and Exploration (SME) which is an independent institution but associated with and a part of the American Institute of Mining, Metallurgical and Petroleum Engineers (AIME). The SME publishes its proceedings in a monthly journal called *Mining Engineering* (1949–). This journal is a mix between a professional journal and a monthly magazine, but does publish refereed papers. The SME also commissions and publishes an extensive list of books, to the degree that many of the standard texts in mining are the publications of the SME. The book list of the SME, as for the IMM, also includes a large number of conference proceedings.

In the European Union there is Eurominerals, a combination of the professional institutions of the UK and their equivalents in the other European countries. There is also the Society of Mining Professors/ Societät der Bergbaukunde, currently with its secretarial office in London at the Royal School of Mines, which has a largely European membership at present and deals with the European Mining industry. The Society publishes the journal *Mineral Resources Engineering* which has been out of publication for a while but which is due to reappear.

There are also specialized, usually international, institutions and societies dealing with various aspects of mining. The International Society for Rock Mechanics and The International Society of Ventilation Engineers are probably two of the most important such associations.

Documentary sources

Journals

There is a large number of periodicals and serials which deal with various aspects of mining. Many of these are commercial publications and these are supplemented by the publications of the wide variety of institutions and societies a lot of which have been mentioned above.

In the UK there is the weekly publication *The Mining Journal* (1835–, Mining Journal plc). This is a newspaper style journal which reports on developments within the industry. This is probably the leading weekly mining newspaper in the world and is extensively read worldwide. A sister publication of the *Mining Journal* is the *Mining Magazine* (1909–, Mining Journal plc) monthly. This has articles in depth on various mining operations around the world as well as reporting on technological developments in the industry. In addition to the above in the UK there are *World Mining Equipment* (1976–), *Mining World* (1915–) and *Industrial Minerals* (1967–, Metal Bulletin) monthly. *Mine & Quarry* (1924–, IML Group) monthly, is the official journal of the UK Minerals Engineering Society.

Mining is relatively poorly served for independent refereed journals. The main such journals are the transactions of the various institutions mentioned above. There are, however, good ones in the area of rock mechanics notably *The International Journal of Rock Mechanics and Mining Sciences* (1964–, Pergamon) 7 per annum, and *Rock Mechanics and Rock Engineering* (1929–, Springer-Verlag) quarterly but otherwise there are few to serve the rest of the discipline. *Mineral Resources Engineering*, which is not currently in publication but restarts in 1995 will partially fill the gap. There are of course the publications of the smaller societies such as *The Journal of the South African Mine Ventilation Society* (1948–) which publishes mine environmental material.

In the USA, other than the journals previously mentioned published by the professional and industrial associations, the leading mining journal is *E&MJ (Engineering and Mining Journal,* (1866–, Maclean Hunter) monthly.

From other countries there are *Australia's Mining Monthly* (1980–) monthly, the *Canadian Mining Journal* (1879–, Southam Magazine Group), bi-monthly, *The Northern Miner* (1915–, Southam Magazine Group) Canada's weekly mining newspaper, and the *South African Mining, Coal, Gold and Base Minerals* (1891–, Thomson Publications) monthly.

Statistical sources

There is a large variety of statistical sources of information on the mining industry available. Both the *Engineering and Mining Journal (E&MJ)* and the *Mining Magazine* produce annual review issues which cover the developments in the mining industry during the preceding year and give statistics on the production of the various mineral commodities. In addition to these there is a variety of mining yearbooks and annual reviews produced. Probably the most comprehensive, although it normally appears about 2 to 3 years after the period it is covering, is the *United States Department of the Interior; United States Bureau of Mines; minerals yearbook* (USGPO). These are usually in three volumes and cover the mining and minerals industry of the world, commodity by commodity. They give greatest prominence to the US industry but do cover it all. The *Mining annual review* (Mining Journal) is another good review of the world's mining industry. There is also the *World mineral statistics* (British Geological Survey) annual.

Most countries and particularly countries with a significant mining industry produce annual statistics of their mineral production. Thus there are the *United Kingdom minerals yearbook* (British Geological Survey), *Australian mineral industry annual review* (Bureau of Mineral Resources, Geology and Geophysic), *Canadian minerals yearbook* (Department of Energy, Mines and Resources) and *South Africas mineral industry* (South African Department of Mineral and Energy Affairs, Minerals Bureau).

Conferences

The mining industry is very well served for conferences. The most important body of the mining literature is probably the volumes of the proceedings of the various conferences held. Certainly, there is more new material published at conferences than anywhere else. There are two major international general mining conferences. The World Mining Congress is held every two years and involves mining engineers from the entire world. The proceedings are published by the organizing country. The other is the Council of Mining and Metallurgical Institutions conferences. These are held once every four years in countries which are or were members of the British Commonwealth and are major mining countries. The proceedings of these conferences are published by the host organization and include papers on most mining engineering topics.

Then there are specific conferences for various aspects of the mining discipline. For example, there are the APCOM (the Application of Computers and Operations Research in the Minerals Industry). These

Mining 613

are held every 18 months alternately in the USA and in a country outside the North American continent. Again, the proceedings, which are state of the art in the use of computing in mining, are published by the organization hosting the particular conference, *16th, 17th and 23rd Application of Computers and Operations Research in the Mineral Industry* (1979, 1982 and 1992), are available from the SME for example, and *APCOM '84 – Application of Computers and Mathematics in the Minerals Industries: 18th International Symposium* (1984) (Chapman and Hall/IMM). Similarly there are regular conferences in other areas of mining such as the International Mine Ventilation Conferences, e.g. *2nd International Mine Ventilation Congress* (1980) (Society of Mining Engineers), *Mine Ventilation, Third International Congress, Harrogate 1984* (1984)(Chapman and Hall/IMM), *International Mine Ventilation Congress* (4th: 1988: Brisbane) (Balkema), *Proceedings of the 6th US Mine Ventilation Symposium* (1993) (Society of Mining Engineers), Surface Mining Conferences e.g. *International Symposium on Continuous Surface Mining* (2nd: 1988: Austin, TX) (Balkema), Rock Mechanics conferences e.g. *International Congress on Rock Mechanics* (7th: 1991: Aachen) (AusIMM) and so on.

As stated, the proceedings of these and other conferences form a very important part of the body of the literature in the field of mining engineering.

Abstracting and indexing services

As indicated in a previous section, the most important abstracting service in the field of mining engineering is that run by the IMM, which offers a regular bi-monthly volume of abstracts and also offers the abstracts on the IMMAGE database for computers. This can be accessed online or can be purchased on floppy disks.

The *Mining Journal* also runs a computer-based information service.

Reference books, monographs and reports

Mining engineering is such a broad field that a large number of books exists in the general area. That said, there are areas of mining engineering which do not have a widely accepted text. The three best introductory texts in mining engineering are Hartman (1987), Stocks *et al.* (1979) and Thomas (1978).

The SME is the generally accepted source for the main general texts such as Cope and Rice (1992); Crickmer and Zegeer (1981); Hartman (1992); Hustrulid (1982); Kennedy (1990); and Stefanko (1983).

For rock drilling there is a number of texts, many of them produced by the equipment manufacturers such as Atlas Copco (1982), Gardner-Denver (1976), Sandvik Coromant and Atlas Copco (1977a and 1977b), Tamrock (1983 and 1984). Other texts in this area include the works by the Australian Mineral Foundation listed under blasting below and McGregor (1967).

Mining has a wide range of machinery specially designed for the purpose of mining. This machinery and equipment, especially that designed for use underground, is different from that used in any other discipline. There is no really modern text which covers the field of mining machinery but there have been conferences in this area. Probably the best modern source is Almgren et al. (1993) and there is also the *International Conference on Mining Machinery* (1979). Other sources of information in this area are Buchanan (1966), Bartholomae et al. (1983) which deals with machinery noise, Steele (1969), Martin (1982) and Stack (1982). As always there are also the materials supplied by the equipment manufacturers to consult as well.

In the area of blasting the texts tend to be produced by the explosives manufacturers Atlas Powder Company (1987), Du Pont de Nemours & Co. (1977), Nobel's Explosives Co. Ltd. (1972), although much good work has been done in Australia as well in this area by the Australian Mineral Foundation (1977 and 1983). Other important aspects of blasting are covered in Bollinger (1980), Dowding (1985), Gustafsson (1973), Health and Safety Commission (1988), Hemphill (1981), Konya and Walter (1990), Langefors and Kihlstrom (1978).

Mining does affect the environment and there is a number of texts on the effects of mining on the environment such as Down and Stocks (1977). There is even more available on the clearing up of the environment after mining as Anon (1987), Department of the Environment (1986, 1988 and 1991), Littler (1990) and RMC Group (1987).

Mining does have its management, and this is covered in Sloan (1983) and National Coal Board (1979). There are the usual economic concerns and thus there is a branch of studies called Mineral Economics. Gentry and O'Neil (1984), Gocht et al. (1989) and Storrar (1981) deal with various aspects of this topic.

Specialized topics

The following specialized topics have been chosen to show the wide range of subject areas covered by the mining engineering discipline.

Surface mining

The vast bulk of the mineral recovered by mining is recovered from surface mines, some 70 per cent or more of all material mined. This includes the mining of metals, fuels such as coal and the industrial minerals and structural materials required by civil engineering and other sectors. Surface mining of coal has been covered elsewhere in this series.

For metal mining the surface mining techniques are somewhat different and are described well in Kennedy (1990). Surface mining equipment is specialized equipment and can best be researched by using the material supplied by the various manufacturers such as Caterpillar Inc. (1991) Most of these mines use large off-highway haulage systems and these have been covered in the proceedings of a conference on off-highway haulage (Golosinski and Srajer, 1989).

Many surface mines recover material that is already loose, such as river gravels or beach sands. When this is done to recover a valuable metal such as tin or gold, it is called placer mining. These methods of mining are covered by Cope and Rice (1992). However, similar deposits are worked for the sand and gravel themselves to provide building materials, and the methods used are treated in Department of the Environment (1988) and Littler (1990).

Underground mining

Underground mining methods are very different from those used on the surface and form a study in themselves. The first major difference is of course access, which has to be through tunnels or shafts, and will be through shafts for all the deeper operations. The sinking of shafts is a study in itself and the various aspects of shaft sinking and operations can be referenced firstly through Institution of Mining and Metallurgy (1989), Lutgendorf (1986), Bennett et al. (1985) and Health and Safety Executive (1985).

Mining engineers distinguish between rock which can be cut by heavy duty cutting machinery which is called 'soft rock', the typical example of which is coal, and rock which needs to be drilled and blasted, or 'hard rock', which is the type of rock usually mined in the recovery of metals. The mining methods for the two types of rock differ mostly in the kinds of equipment which can be used, which has an effect on the layouts of the mines. For the machinery used in underground mines the material produced by the equipment manufacturers is again a good source such as Atlas Copco (1982), and Tamrock (1983 and 1984).

For the mining methods used in hard rock mining there are a few good references such as Dravo Corporation (1974), Hustrulid (1982), and Stewart (1981). Soft rock mining is mainly for coal, although other materials such as potash and salt are mined by these methods. This type of mining has been covered to an extent in the previous book under solid fuels, but in the context of these mining methods texts worth consulting are Crickmer and Zeeger (1981) and Stefanko (1983).

Rock mechanics

The engineering material with which mining engineers work and in which they design their structures is rock. Rock is a non-homogeneous material, and in addition the engineer is constrained to work with the rock as it is found and is not able to design the material to its purpose. It is therefore important for the mining engineer to have a fundamental understanding of the behaviour of the material with which he will work, and this is rock mechanics. Over the last three decades a great deal has been learned in this area and some understanding of rock behaviour under varying stress conditions is now available. However, this remains one of the main areas of further research in the mining discipline. A good introductory text is Goodman (1988) For more advanced studies in rock mechanics consult Brady and Brown (1993), Franklin & Dusseault (1989), Hoek & Brown (1980), Hudson (1992 and 1989), Hudson et al. (1993) and Whittaker and Frith (1990).

The stability of the slopes in open pit mining is a critical factor in the design of these mines. It is critical both for the safety of the operation and for the optimization of the extraction of the ore through the minimization of the waste mined with it. A very good text for this area of rock mechanics is Hoek and Bray (1981).

The interaction of the openings with the rock and with the preexisting stress field within which the opening has been cut is a very complex interaction. A variety of mathematical models has been, and is being, developed to help in the analysis of these interactions. Finite element, boundary element and other mathematical and numerical techniques are used in the development of these models. The use of these methods in rock mechanics is covered by Pande et al. (1990).

Other texts of interest to the study of rock mechanics deal with some of the techniques used (Price and Cosgrove, 1990, Priest, 1985). As mentioned there is a number of conferences held in this area and the proceedings of these conferences form a major portion of the literature. The subject of support systems for underground openings is covered in the general texts mentioned above, but the latest developments can also be gleaned from Kaiser and McCreath (1992).

Ventilation

The maintenance of an environment which is acceptable for the workforce in an underground mine is of critical importance. This has been traditionally called ventilation but with the steady increase of concern, not just for the air within the mine, but also for the other working conditions such as light, sound, etc. the discipline has tended to become more broad and is called Mine Environmental Engineering.

There is a number of texts which deal with all aspects of airflow in mines and the control of contaminants in that airflow. These topics are covered in Bossard (1983), Hall (1981), Hartman *et al.* (1982), Mine Ventilation Society of South Africa (1982), National Coal Board (1978) Rabia (1988), Sengupta (1990) and Vutukuri and Lama (1986).

In addition to these more general texts there are numerous specific aspects of the ventilation of mines which tend to have their own texts. Thus, the drainage of methane in coal mining operations is comprehensively dealt with in Boxho *et al.* (1980). The movement of the air in mines is controlled by the installation of fans and thus information on fans is very important. A good text in this area is Fan Manufacturers' Association (1981) Ower and Pankhurst (1977) deal with the techniques and equipment for the measurement of airflow in mines in detail. Radchenko (1976) on the other hand covers the problem of dust, and the use of ventilation to assist in the elimination of dust in underground mines. Saxton (1986–1987) is an historical text which covers the history of ventilation in coal mines over the last 400 years or so.

Also in the area of ventilation, due to many of the hazards associated with mining, and particularly with coal mining, as a result of the pick-up of gas by the mine air it is now required that there be constant monitoring of gas levels in mines. This problem leads on to the general area of safety in mines and the need to plan for rescue should an accident happen. These aspects of mining are dealt with in Strang and Mackenzie-Wood (1987).

As mentioned previously there is a number of international and national Mine Ventilation Congresses held regularly. The proceedings of these are the best source of the most up-to-date thinking in the area of mine ventilation. However, in addition to these general conferences there are other conferences which cover specific topics in the area such as the First International Conference on Radiation Hazards in Mining, Golden, edited by Gomez (1981).

References

Almgren, G. *et al.* (eds) (1993) *Mine Mechanisation and Automation, Proceedings of the 2nd International Symposium on Mine Mechanisation and Automation*, Lulea, Sweden, 7–10 June 1993. Rotterdam: Balkema.

Anon (1987) *A practical guide to restoration*. Egham: RMC Group.

Atlas Copco (1982) *Atlas Copco manual*. 4th ed. Stockholm: Atlas Copco.

Atlas Powder Company (1987) *Explosives and rock blasting*. Baltimore, MD: Atlas Power Co.

Australian Mineral Foundation (1977) *Drilling and blasting* 2 vols. Sydney: AMF.

Australian Mineral Foundation (1983) *Drilling and blasting in open pits and quarries. Soft materials* 2 vols. Sydney: AMF.

Bartholomae, R. C. *et al.* (1983) *Mining machinery noise control guidelines*. Washington, DC: USBM.

Bennett, R. D. *et al.* (1985) *State-of-the-art construction technology for deep tunnels and shafts in rocks*. Technical Report WES/MP/GL-85-1. Vicksburg: US Army Engineer Waterways Experimental Station.

Bollinger, G. A. (1980) *Blast vibration analysis*. Carbondale, IL: Southern Illinois University Press.

Bossard, F. (1983) *Manual of mine ventilation design practices* 2nd ed. Butte, MO: Floyd Bossard and Associates.

Boxho, J. *et al.* (1980) *Firedamp drainage: handbook* Essen: Verlag Gluckauf.

Brady, B.H. G. and Brown, E. T. (1993) *Rock mechanics for underground mining* 2nd ed. London: Chapman and Hall.

Buchanan, W. (1966) *Hydraulics applied to underground mining machinery* London: Pitman.

Caterpillar Inc. (1991) *Caterpillar performance handbook* 22nd. ed. Peoria, IL: Caterpillar Inc.

Cope, L. W. and Rice, L. R. (eds), (1992) *Practical placer mining*. Littleton, CO: Society of Mining Engineers.

Crickmer, D. F. and Zegeer, D. A. (eds) (1981) *Elements of practical Coal Mining*. Littleton, CO: Society of Mining Engineers.

Department of the Environment (1986) *Landfilling wastes* Waste Management Paper No. 26. London: HMSO.

Department of the Environment (1988) *Marine dredging for sand and gravel*. London: HMSO.

Department of the Environment (1991) *The control of landfill gas* rev. ed. Waste Management Paper No. 27, London: HMSO.

Department of the Environment (various dates) *Minerals planning guidance*. London: HMSO.

Dowding, C. H. (1985) *Blast vibration monitoring and control* Englewood Cliffs, NJ: Prentice Hall.

Down, C. G. and Stocks, J. (1977) *Environmental impact of mining*. London: Applied Science.

Dravo Corporation (1974) *Analysis of large-scale non-coal underground mining methods*. Springfield, VA: National Technical Information Service, sponsored by the US Bureau of Mines.

Du Pont de Nemours & Co. (1977) *Blasters handbook* 16th ed. Wilmington, DE: Du Pont de Nemours & Co.
Fan Manufacturers' Association (1981) *Fan application guide* 2nd ed. Middlesex: Hevac Association.
Franklin, J. A. and Dusseault, M. B. (1989) *Rock engineering*. New York: McGraw-Hill.
Gardner-Denver (1976) *Rock drilling data* Gardner-Denver.
Gentry, D. W. and O'Neil, T. J. (1984) *Mine investment analysis*. Littleton, CO: Society of Mining Engineers.
Gocht, W. R. *et al.* (1989) *International mineral economics*. Berlin: Springer-Verlag.
Golosinski, T. S. and Srajer, V. (eds) (1989) *International Symposium on Off-Highway Haulage in Surface Mines*, Canada, 15–17 May 1989 Edmonton, Canada. Amsterdam: Balkema.
Gomez, M. (ed.)(1981) *First International Conference on Radiation Hazards in Mining*, Golden, CO. Littleton, CO: Society of Mining Engineers.
Goodman, R. E. (1988) *Introduction to rock mechanics* 2nd ed. New York: Wiley.
Gustafsson, R. (1973) *Swedish blasting technique*. Gothenburg: SPI.
Hall, C. J. (1981) *Mine ventilation engineering*. Littleton, CO: Society of Mining Engineers.
Hartman, H. L. *et al.* (1982) *Mine ventilation and air conditioning*. 2nd ed. New York: Wiley.
Hartman, H. L. (1987) *Introductory mining engineering*. 2nd ed. New York: Wiley.
Hartman, H. L. (ed.) (1992) *SME mining engineering handbook* 2nd ed. Littleton, CO: Society of Mining Engineers.
Health and Safety Commission (1988) *Explosives at quarries*. London: HMSO.
Health and Safety Executive (1985) *Safe manriding in mines: supplement and corrigenda to the first and second reports of the National Commission for Safety of Manriding in Shafts and Unwalkable Outlets*. London: HMSO.
Hemphill, G. B. (1981) *Blasting operations* New York: McGraw-Hill.
Hoek, E. and Brown, E. T. (1980) *Underground excavations in rock*. London: IMM.
Hoek, E. and Bray, J. W. (1981) *Rock slope engineering* 3rd ed. London: IMM.
Hudson, J. A. (1989) *Rock mechanics principles in engineering practice* London: Butterworths.
Hudson, J. A. (1992) *Rock engineering systems: theory & practice*. Chichester: Ellis Horwood.
Hudson, J. A. *et al.* (eds) (1993) *Comprehensive rock engineering: principles, practice and projects* 5 vols. Oxford: Pergamon.

Hustrulid, W. A. (ed.) (1982) *Underground mining methods handbook*. Littleton, CO: Society of Mining Engineers.

Institution of Mining and Metallurgy (1989) *Shaft Engineering*. London: Chapman and Hall.

International Conference on Mining Machinery (1979) Brisbane: Institution of Engineers.

Kaiser, P. K. and McCreath, D. (eds) (1992) *Rock Support in Mining and Underground Construction*. International Symposium on Rock Support Proceedings. Sudbury, Canada, 16–19 June, 1992. Rotterdam: Balkema.

Kennedy, B. A. (ed.) (1990) *Surface mining* 2nd ed. Littleton, CO: Society of Mining Engineers.

Kolsky, H. (1963) *Stress waves in solids*. New York: Dover.

Konya, C. J. and Walter, E. J. (1990) *Surface blast design*. Englewood Cliffs, NJ: Prentice Hall.

Langefors, U. and Kihlstrom, B. (1978) *Modern technique of rock blasting* 3rd ed. New York: Wiley.

Littler, A. (1990a) *Sand and gravel production*. Nottingham: Institute of Quarrying.

Littler, A. (1990b) *Sand and gravel planning and restoration*. Nottingham: Institute of Quarrying.

Lutgendorff, H. O. (1986) 30 years of sliding shaft lining; a retrospect. *Gluckauf*, **122**, 310–316.

McGregor, K. (1967) *Drilling of rock*. London: C. R. Books.

Martin, J. W. (1982) *Surface mining equipment*. Golden, CO: Martin Consultants.

Mine Ventilation Society of South Africa (1982) *Environmental engineering in South African mines*. Marshalltown: Mine Ventilation Society of South Africa.

National Coal Board (1978) *Ventilation in coal mines*. London: NCB, Mining Dept.

National Coal Board (1979) *Work study in mines*. London: NCB, Mining Dept.

Nobel's Explosives Co. (1972) *Blasting practice* 4th ed. Stevenson, Ayrshire: Nobel's Explosives Co.

Ower, E. and Pankhurst, R. C. (1977) *Measurement of air flow* 5th ed. Oxford: Pergamon.

Pande, G. N. et al. (1990) *Numerical methods in rock mechanics*. Chichester: Wiley.

Price, N. J. and Cosgrove, J. W. (1990) *Analysis of geological structures*. Cambridge: Cambridge University Press.

Priest, S. D. (1985) *Hemispherical projection methods in rock mechanics* London: Allen & Unwin.

Rabia, H. (1988) *Mine environmental engineering*. Newcastle upon Tyne: Entrac Software.

Radchenko, G. A. (1976) *Ventilation for dust elimination in underground mines*. New Delhi: Amerind Publishing Co.
RMC Group (1987) *A practical guide to restoration*. Egham: RMC Group.
Sandvik Coromant and Atlas Copco (1977a) *Rock drilling manual: theory and technique* Sweden: Sandvik.
Sandvik Coromant and Atlas Copco (1977b) *Rock drilling manual: theory and technique* Sweden: Sandvik.
Saxton, I. (1986–1987) *Coal mine ventilation – from Agricola to the 1980s*. Published in serial form in *Mining Engineer*.
Sengupta, M. (1990) *Mine environmental engineering* 2 vols. Boca Raton, FL: CRC Press.
Sloan, D. A. (1983) *Mine management*. London: Chapman and Hall.
Stack, B. (1982) *Handbook of mining and tunnelling machinery*. Chichester: Wiley.
Steele, D. J. (1969) *Modern developments in mining machinery*. London: Virtue & Co.
Stefanko, R. (ed.) (1983) *Coal mining technology – theory and practice*. Littleton, CO: Society of Mining Engineers.
Stewart, D. R. (ed.) (1981) *Design and operation of caving and sublevel sloping mines*. Littleton, CO: Society of Mining Engineers.
Stocks, J. et al. (1979) *Mining and mineral processing*. Milton Keynes: Open University Press.
Storrar, C. D. (ed.) (1981) *South Africa mine valuation* rev. ed. Johannesburg: Chamber of Mines of South Africa.
Strang, J. and Mackenzie-Wood, P. (1987) *Manual on mines rescue, safety and gas detection* 2nd ed. Golden, CO: Colorado School of Mines Press.
Tamrock (1983) *Tamrock handbook of underground drilling*. Tampere: Tamrock.
Tamrock (1984) *Tamrock handbook on surface drilling and blasting*. Tampere: Tamrock.
Thomas, L. J. (1978) *Introduction to mining* rev. ed. Sydney: Methuen.
Vutukuri, V. S.and Lama, R. D. (1986) *Environmental engineering in mines*. Cambridge: Cambridge.
Walker, S. C. (1988) *Mine winding and transport*, Amsterdam: Elsevier.
Whittaker, B. N. and Frith, R. C. (1990) *Tunnelling: design, stability, and construction*. London: IMM.

CHAPTER THIRTY-FIVE

Biomedical engineering

S. C. HUGHES AND P. R. RICHARDS

Introduction

Information sources in biomedical engineering are myriad, scattered and occasionally volatile. Among the reasons for this is the fact that it is a young discipline, despite its 50 years or so of professional activity. It is also a relatively small discipline – even although it serves one of the biggest professional disciplines – Medicine.

That it is also a multi-speciality discipline which overlaps, and sometimes jostles with, many of the other paramedical areas, adds both challenge and fun to information sourcing in biomedical engineering. Even in these days of computerized 'easy access' library systems, lateral thinking and serendipity are important tools in searching or researching biomedical engineering sources.

Because the medical technology 'airwaves' are crowded and overlapping, the first difficulty to overcome when searching biomedical engineering literature is to decide what biomedical engineering is. The answer to this question can be interpreted in a variety of ways by its participating professionals and certainly by authors. Each interpretation can make it essential to start the search in different source libraries or from a different viewpoint.

It is appropriate here initially to adopt a broad definition and to set it in context with the competing and complementary techniques in the hope of providing a framework in which a new researcher will find it easier to navigate. The more experienced researcher may disagree with this approach but should find it easy to adapt the framework to suit their own view of the profession. We accept that not every practitioner of the subject will agree with our views, we would be surprised if they did, but we follow Plato's teaching that everything is opinion.

Definitions

Biomedical engineering is defined, in general terms, as the application of engineering principles to the study, diagnosis, treatment and the enhancement of human health and performance.

At first glance, this appears a simple and straightforward definition, but some contemplation reveals that for a subject so defined, the literature spreads across all of engineering, all of medicine, many of the major life sciences and a number of related topics such as sports science, computing and even some social sciences. Challenging though this range of opportunities for research and study may be, the very breadth of topic coverage makes the choice of literature, from which to source information, a very problematic and occasionally random task.

When the authors were starting their careers in the 1960s, there were three generally accepted uses of the word 'engineer'. Academically, someone who studied engineering was involved with the scientific study and design of the mechanics or function of systems and materials. He (seldom in those days she) could be further classified as a mechanical or civil or electrical or marine or chemical or aeronautical engineer according to the nature of the system or materials worked with. Sometimes the boundaries were blurred, but these divisions mainly were sacrosanct. They had been decided earlier (in the late nineteenth and early twentieth centuries in Britain) and noble institutions had been set up to promote the professional welfare of those involved.

The second was, and is, in fact a misuse of the term when it is applied to the person who repairs or maintains equipment. They should more correctly be called a mechanic or technician, both valuable and honourable employments but different from the professional engineer. This comment has the same relevance for all engineers, but in medicine, or medically related areas, this blurred use of the term can lead to real misunderstanding of the roles and responsibilities of individuals and of the information likely to be found in the literature. Most researchers in biomedical engineering are not interested *a priori* in 'hospital engineering', a term usually reserved for the buildings support and supplies equipment within a hospital and which really means the maintenance and installation of such equipment as gas, power, lighting, etc.

The third, and more colloquial, use of the word engineer was 'to enable or to contrive or to arrange or manipulate'. Thus a person might 'engineer a promotion or a position on a committee' or just 'engineer a change'.

Biomedical engineering uses 'engineering' in what we have described above as the academic or professional sense. Its engineering is most often taken from the mechanical, civil or electrical branches of engineering.

The structure of the related disciplines

Biomedical engineering forms part of the 'umbrella' discipline known as bioengineering in which the same principles are applied to the totality of biological systems of which the human system is but a part. As one subset of biomedical engineering, the discipline of biomechanics should also be noted as a route to extensive literature. Biomechanics is the application of mechanics to the study and description of the human body and its function.

It is relevant to highlight some of the subject areas which biomedical engineering helps to serve. These are relevant, because most of these subject areas are outwith the scope of this book but may nevertheless afford a door into parallel literature sources, closely overlapping the more explicit biomedical engineering sources.

Biology

Biology is the scientific study of living organisms. Until this century, there was great debate as to whether living organisms had some 'vital' property which could not be explained in terms of the physical sciences (mechanics). This 'vitalist' versus 'mechanist' debate has largely fizzled out with the mechanists considered the victors (see Savory, 1971, Medawar and Medawar, 1985, Mayr, 1988). The vitalist/mechanist debate occasionally reappears, particularly when considering the vexed question of the 'soul' (Changeaux, 1986). 'When dead, that is, when the soul stops working, the animal machine remains inert and immobile' (Borelli 1989).

In biology, the use of the word 'engineering' has been more often in what we have termed its colloquial role. Thus genetic engineering 'enables' the transfer of a gene from one species to another. The engineering in this context has little if nothing to do with our mechanical or civil or electrical engineers.

There are branches of genetic engineering which relate to a more traditional academic division, namely chemical engineering. This relationship is mainly in the area of process engineering which can contribute to problems associated with the large-scale culture of the micro-organisms used in gene transfer work. These subjects (genetic and process engineering) combine in the fashionable and new (in name at least) field of 'biotechnology' which is often confused with bioengineering and sometimes with its sub-group biomedical engineering by the layman (and companies that send advertising mailshots!). Bioengineering should be clearly and carefully discriminated from biotechnology.

The distinction is important because although biotechnology could be thought to fit into our definition of biomedical engineering (biotechnology gains its fashionability from its medical potential), the two subjects have traditionally been quite separate and there is seldom overlap between their respective literature.

Medicine and biology

Medicine and biology are intimately linked. Many of the early biologists were medical men (Harvey dissected sixty other species besides man in his studies and this eclecticism was not unusual), and in a way, medicine could be seen as a branch of biology which centres around one species, *homo sapiens*.

In the *Oxford concise dictionary*, medicine is defined as 'the art of restoring and preserving health'. This implies not only diagnosis and treatment but a concern with the long-term course of a disease, its prognosis. This latter has sometimes to be considered as the rehabilitation or adaptation of a patient to their altered physical condition (the disease may not be 'curable'). From our information aspect this extends the range of 'medicine' beyond areas covered by the traditional medical practitioner into physiotherapy, occupational therapy and other paramedical fields. It follows that the field of rehabilitation engineering, in which engineering is used to provide equipment and methods to assist in the rehabilitation of patients, should be included in our quest for information.

Biomedical engineering in practice

Biomedical engineering (BME), in the framework we have used, is seen as a partner and a servant of modern medicine and health care. Three emergent strands can be identified

- BME in direct clinical application as in a routine hospital or clinical setting;
- BME in an industrial and commercial role in the design and product development of medical equipment;
- BME research and development conducted within research organizations, universities, or research in the other two strands.

The areas of most intimate overlap between BME and other paramedical professions are also the areas of most overlap in the literature. This overlap occurs with medical physics and with physiological (or clinical) measurement where electronic engineering, instrumentation, ergonomics and mechanical engineering merge to create and develop clinical measurement and treatment equipment.

In the main overlap area, the medical physics and BME professions work in close liaison to provide and support such facilities as, for example, blood handling equipment, cardiac monitoring, cardiovascular measurement in theatre and recovery suites, intensive care respiratory function, ventilation and anaesthetic equipment, nerve conduction assessment, evoked response measurement, electroencephalography. Away from this contested zone, a distinctive boundary at which technology becomes definitively medical physics is identified by accepting that the imaging and ionizing radiation technologies – X-rays, MRI, CAT scanning, radiotherapy, nuclear medicine – are firmly the province of the medical physics profession. It is equally possible to identify clearly those areas which are definitively BME.

One model which can be used effectively to illustrate the differing areas of responsibility, is obtained by classifying, somewhat arbitrarily, the needs of human health into four 'elements':

1. an internal element primarily concerned with the behaviour and function of the human body 'beneath' the skin;
2. an external element representing the social/educational/employment structures in which the individual lives;
3. an interface element primarily concerned with the actual physical contact between the individual and functionally useful devices;
4. an interface element relating to environmental and occupational influences on health.

This model, illustrated in Figure 35.1, is used by selecting a topic item, for example 'pressure sores' and projecting a radial line from the centre circle (the patient) outwards. The intersection of the radial line with the outer rings indicates the field in which information on the chosen topic is likely to be found, in this case pressure sores is in the domain of clinical engineering, rehabilitation engineering and biomedical engineering.

This model is obviously more useful in delineating the differences and overlaps between BME and medical physics in the 'internal element' and the 'functional interface element', than for the other two 'elements' in which these disciplines have a more limited role.

Nevertheless, this model has been found useful by a few hundred postgraduate aspiring biomedical engineers in guiding their interests through the choices and opportunities offered in the diversity of these two disciplines and in their ardent search for information from the range of sources. We hope that these preliminary thoughts will help to indicate some directions for lateral thinking which we believe is the essence of effective literature searching, and to raise the everpresent worry that there may be 'another route' to the information you seek.

Organizations and societies

It is in the nature of a young discipline that the organizations which seek to represent it, promote it and inform the world about it, are them-

628 *Biomedical engineering*

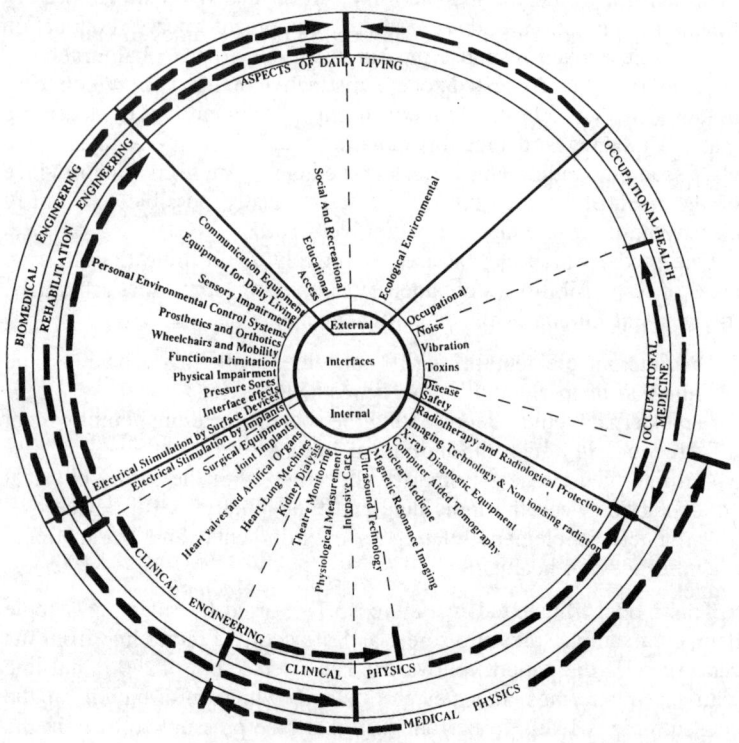

Figure 35.1. Model of Interdisciplinary Overlap in Bioengineering and Medical Physics.
Key – The topics connected on a radial from the centre are within the Discipline(s) shown in the outer circles.

selves young and are multi-faceted, occasionally to the point of confusion. In the older disciplines, for example, the parent engineering mainstream areas, the representation has had time to grow, merge and develop a focused maturity in which a few major institutions can be readily identified. In the biomedical engineering field the situation is evolving and further changes may well occur in the decades surrounding the 'millennium' year 2000.

In medical engineering, or biomedical engineering the first international awakenings were in the late 1950s when the International Federation of Medical Electronics (IFME) was established as the result of conferences in 1958 and 1959, both in Paris. An excellent and brief history of the development of the IFME into what is now the International Federation for Medical and Biological Engineering (IFMBE) has been written by Mallard (1994). This development evolved by including biomechanics – the application of mechanics to study of the human body – in 1965. Between 1977 and 1980 the IFMBE agreed to link cooperatively with the International Organization for Medical Physics (IOMP) under the umbrella of a new body whose name was chosen to be the International Union for Physical and Engineering Sciences in Medicine (IUPESM).

The IFMBE issues news sheets, *MBEC News*, throughout the year, highlighting recent and forthcoming conferences, news of its Council meetings, and items of worldwide interest in science and technology including standards, government legislation and new initiatives.

The IOMP, its sister organization within the IUESPM had itself evolved after considerable and occasionally difficult debate due to the interests of members of the Hospital Physicists Association (UK) which was a professional body for employees of the National Health Service in the UK. The early difficulties are a good illustration of the overlaps which are and will probably remain an inherent feature of biomedical engineering, in that the difference between medical electronics, which had already created its IFME, and medical physics had first to be clarified and agreed. However, the IOMP was, in fact, set up remarkably quickly and inaugurated in 1963. Its bulletin *Medical Physics World* is issued about twice per year.

These international bodies and their umbrella organization organize major international conferences at which state of the art papers are presented, and so form a valuable source of current information in biomedical engineering and medical physics.

A European Federation of Medical Physics (EFOMP) also exists with 25 member states, not restricted to the European Union. It publishes policy statements on best practice in medical physics, and on education and training. It organizes symposia and is a member organization of the IOMP. The European Society for Engineering and Medicine (ESEM) also releases information (*ESEM News*) in bulletins to member organizations.

In the UK and USA, the older institutions also recognize this 'young' field. Biomedical engineering UK activities can be found in the Institution of Electrical Engineers (IEE), and the Institution of Mechanical Engineers (IMechE). The former publishes relevant papers in its journal and the latter has a dedicated journal in its *Proceedings. Part H, Journal of Engineering in Medicine*. A third member of the older UK institutes, the Institute of Physics (IoP) includes articles of a general and review nature in its monthly bulletin, *Physics World*.

There are two specialist UK organizations which deal with biomedical engineering as a main part of their activities. These are the learned society – the Biological Engineering Society (BES) whose members encompass such disciplines as medicine, physiology, biological engineering biomedical engineering and medical instrumentation. Its journal was the *Journal of Biomedical Engineering* which in 1993 changed its title to *Medical Engineering and Physics*. Alongside the BES, the Institute of Physical Sciences in Medicine (IPSM) is a professional and learned institute which grew out of the Hospital Physicists Association. The HPA largely had an employee representational (and professional association) role. Its members supported the creation of the IPSM as a body, intimately linked to the HPA, to reflect the learned activities in which the members of the discipline were involved. The IPSM has a number of publications ranging from its bulletin *Scope*, published quarterly, through *Physics in Medicine and Biology* (PMB) published monthly, to *Physiological Measurement*, published quarterly. *Physiological Measurement* is an official journal of the IPSM, it represents also the EFOMP, IOMP and the Deutsche Gesellschaft für Medizinische Physik.

In 1994, a merger of the BES and IPSM was approved in principle and negotiation over the detailed arrangements began between the councils of both organizations. The outcome of an 'Institute or Council of Engineering and Physics in Medicine' was predicted, offering a larger and more influential platform for medical engineers, medical physicists and other members with related interest. The merger also emphasizes the multi-layered overlapping nature of this field, a nature which strongly influences the routes which have to be chosen by a researcher to ensure the best chance of searching out the fullest possible information from a confusing range of literature. In 1995 this merger was formalized and the Institution of Physics and Engineering in Medicine and Biology (IPEMB) was established.

In the USA, two main bodies are sources of information about aspects of biomedical engineering. These are the American Society of Mechanical Engineers (ASME) and the Institute of Electrical and Electronics Engineers (IEEE). The IEEE has formed a sub group called the IEEE Engineering in Medicine and Biology Society. Their *Transactions on Rehabilitation Engineering* and *Transactions on Biomedical Engineering* are a major source of high quality published material, with a natural tendency to concentrate on research and development work in which instrumentation and/or control engineering plays a predominant part. Conferences and symposia organized by the Rehabilitation Society of North America (RESNA) remain a source of rehabilitation engineering specific data.

There are other organizations which specialize in more restricted areas of biomedical engineering and whose conferences are valuable additional sources. These include for example, the UK Tissue Viability Society, the European Society for Movement Analysis in Children, and in the USA, the International Society of Electrophysiological Kinesiology (ISEK).

The paramedical professions have their own professional bodies, such as the Chartered Society of Physiotherapy and the British Association of Occupational Therapists in the UK. It is increasingly common to find articles published in their society publications (e.g. *British Journal of Occupational Therapy*) in which biomedical engineering methods have been used by the research to study or resolve a problem related to their own profession. At the very least, these journals offer the professional biomedical engineer an insight into what are the problem areas of current concern in patient therapy.

Contact details for key societies and organizations are given below.

American Society for Artificial Internal Organs (ASAIO), PO Box C, Boca Raton, Florida 33429, USA.

Biological Engineering Society (BES) [merged with Institute of Physical Sciences in Medicine in October 1995 to form Institution of Physics and Engineering in Medicine and Biology, Royal College of Surgeons, Lincoln's Inn Fields, London WC2A 3PN, UK.

Biomedical Engineering Society (BES), PO Box 2399, Culver City, California 90231, USA.

European Society for Artificial Organs (ESAO), c/o Centre Medical Universitaire (GETAM), 1 Rue Michel-Servet, CH–1211 Geneva, Switzerland.

European Society for Biomaterials (ESB), c/o Laboratoire de Recherche Orthopedique, Faculté de Medecine, Lariboisiere St Louis, 10 Ave de Verdun, F75010 Paris, France.

European Society of Biomechanics, (MEM) Institute for Biomechanics, PO Box 30, CH–3010 Berne, Switzerland.

European Society for Engineering and Medicine (ESEM), contact: Dr J. Vander Sloten, Katholieke Universiteit Leuven, Division of Biomechanics and Engineering Design, Celestijnenlaan 200A, B–3001 Heverlee, Belgium.

IEEE Engineering in Medicine and Biology Society (EMBC), c/o Institute of Electrical and Electronic Engineers (IEEE), 345 East 47th Street, New York, NY 10017–2394, USA.

Institute of Physical Sciences in Medicine (IPSM) [merged with Biological Engineering Society in October 1995 to form Institution of

Physics and Engineering in Medicine and Biology], 4 Campleshon Road, York YO2 1PE, UK.

Institute of Physics (IOP), 47 Belgrave Square, London SW1X 8QX, UK.

Institution of Physics and Engineering in Medicine and Biology, 4 Camplesham Road, York YO2 1PE.

Institution of Electrical Engineers (IEE), Savoy Place, London WC2R 0BL, UK.

Institution of Mechanical Engineers (IMechE) (Medical Engineering Group), 1 Birdcage Walk, London SW1H 9JJ, UK.

International Federation for Medical and Biological Engineering (IFMBE), contact: IFMBE Secretary General, Department of Medical Physics, Faculty of Medicine, University of Amsterdam, AMC–Meibergdreef 15, 1105 AZ Amsterdam, The Netherlands.

International Organization for Medical Physics (IOMP), contact: Dr. CG Orton, Gershenson Radiation Oncology Center, Harper Hospital and Wayne State University, 3990 John R, Detroit, Michigan 48201, USA.

International Society for Prosthetics and Orthotics (ISPO), 317 E 34 Street, New York, NY 10016, USA.

Books

There are few more obvious ways of revealing one's prejudices and shortcomings than recommending books for others to read. If a book is an indication of the knowledge, wit, and experience of the author, must it not also have something in common with the thinking of those who like and recommend it? The shyness that this thought engenders is reinforced by the fact that in a wide and fast-changing field, recommendation is a difficult task. There is some possibility of shifting the onus onto the reader by claiming that much will depend on what they seek: whether it is a scholarly reference encyclopedia to use as a data source, or bedtime reading to relax into, or just something which will be a window into the field. What follows will show all three types of book, but our emphasis is on the 'window'.

The most commonly levelled criticism about this field is that no 'definitive' textbooks are available. The authors hope that the description of the field outlined above convinces the reader that this 'ideal' aspiration is neither achievable nor desirable, since it is unlikely that the enormous range of topics could be adequately encompassed in one text.

It is a paradox, in view of the newness of the subject and our high-tech age, that the first book on biomedical engineering is very old, but yet only recently generally accessible (it was translated from the original Latin in 1989). This is Borelli's *On the movement of animals* which was first published in 1680, a year after the author's death. While not recommending it as the most up-to-date source (the mechanics are pre-Newtonian!), it is in many ways the most comprehensive single volume work on biomedical engineering. It reflects our problems of definition. Its concern is not bone and muscle mechanics alone, but the heart and circulation, fevers, even the growth and development of plants from seeds. Borelli took the broad view. A philosopher/historian interested in the connections between all the scientific disciplines would be hard pressed to find a better starting point.

Engineers are not all philosophers and historians however, and the usefulness in the modern context lies not so much in this new translation of Borelli, but other translations (this time from German into English) by the translators of Borelli (Maquet and Furlong) published in a series by the same publisher (Springer-Verlag). These are late nineteenth century works on gait by Braune and Fischer (1985, 1987 and 1988); another late nineteenth century work by Wolff (1986); and a more recent work by Pauwels (1980).

It is regrettable that many scientists do not like an historical approach; there is a tendency to think that old is wrong and research will be better if it has been done using new technologies. Recently some colleagues in a collaborating group thought they had made a very interesting discovery about human balance as a result of their use of a state of the art force platform linked to the latest signal analysis software. Their pre-publication literature survey revealed however that Braune and Fischer had described the same phenomenon (admittedly not with good quantitative data) a century ago from results from their 'sealing wax and string' apparatus. As a rider to this, almost every year some student will tell us that there is nothing in the literature on their research project topic (sometimes even when it is in a much researched area); more often than not, questioning will reveal that their search covered the current CD-ROM only. Clearly a compromise on old versus new literature must be made. The word 'research' can after all mean re-search, or search again.

In this propaganda for not forgetting history, you may have noticed a referencing habit that biologists use a great deal but physical scientists use much less. Biologists name-drop! This probably stems from the fact that biological arguments rely much more on the presentation of observed facts; there are few things that can be worked out from first principles. Engineers may look up books of formulae and then use their mathematical abilities to manipulate them for the task in hand. The biologist has great difficulty in doing this, each step in a biological chain of reasoning must be supported by a reference; the step is someone's discovery, not a mathematical process whose instigator has been forgotten. Biologists become adept therefore at acknowledging this or that person (it can also help shift the blame when it is not possible to check!).

This seeming intellectual obsequiousness, although fairly objectionable if carried over into real life, is a useful reference-finding trick worth fostering. If you know that 'Bloggins' works/worked on a particular subject, then this name fed into a library or abstracting index can provide a useful starting point. Gradually we can build in our minds a picture of who works with Bloggins and where they work: centres often build up a tradition of working on particular subjects. It may emerge that Bloggins is the mouthpiece of a group rather than the most brilliant researcher, but by that time you will be well into the field and the name has served a useful purpose.

We hope that in the lists that follow the authors quoted will consider themselves honoured and not insulted by being thought of as useful 'Bloggins's'!

Introductory biological texts

Some introductory books on biology are listed on the grounds that in an engineering text, this is the area where the majority of the readership may feel they require some instruction or vocabulary. Standard texts with 'biology' as the title tend to be of less use as they are usually aimed at training people to be biologists. They may contain a lot of cell biology or 'systematic' biology (which to a biologist implies classification or taxonomy or phylogeny) which is of great import in biology as a whole, but is of less immediate relevance to the prospective biomedical engineer. Books on anatomy and/or physiology are the biological subject area of most concern to them. The text we recommend to engineering students is Tortora and Grabowski (1993), but there are a number of equally good introductory texts of this style. There is a tendency for students to try and start at the top and learn *Gray's anatomy* or one of the heavy physiology tomes off by heart, but these give information overkill and will leave you depressed or bemused rather than enlightened.

A little further discussion is required on some of the other texts; titles do not always reflect the reason for their selection. Thus, the book by Aiello and Dean (1990) is included because a considerable amount of fossil investigation of human evolution involves biomechanical investigation and this text has an extensive reference list. The works by Gray (1953 and 1968) and Alexander (1975 and 1983) are a source of biomechanical information on animals with some human extrapolation; Gray was the doyen of British experimental zoologists for the first half of this century, Alexander is a prolific modern worker. The volumes by Bastian (1993) are a cheap and simple guide to the terminology of the anatomy and physiology of the systems they cover e.g. the skeletal and muscular systems, the urinary system, the cardiovascular system, the nervous system and the respiratory system. Frost (1967) is a classic simple introduction to biomechanics as seen from the viewpoint of an orthopaedic surgeon. The books by Denny (1993) and Leyton (1975) are both good introductions to biological fluid dynamics, although less medically-oriented than the text by Caro *et al.* (1978). The books by Vincent (1993) and Wainwright *et al.* (1976) are excellent introductions to biomaterials. Berne and Levy (1990) is a good introductory physiology text which takes things rather deeper than Tortora and Grabowski (1993).

Although the emphasis in this chapter is laid primarily on the texts which provide a 'window' on the subject, there are also a number of books which stand out as useful reference sources, both for quick introduction to topic areas and for their contained reference lists for further reading.

Engineering and technology texts

The CRC series of texts offer easily accessible, if occasionally brief, descriptions of many techniques, principles and concepts in a wider range of sciences. Included in this series, of special interest to the biomedical engineer is Fleming and Feinberg (1976), in which encyclopedic data across several areas of the discipline can be found. In this there are some areas which may be debated as being biomedical engineering but it contains a substantial reference list for further study. Feinberg (1980) is a splendid example of the potential for confusion between hospital engineering and the more direct type of clinical engineering which would meet with a UK definition. Webster (1988) is a valuable source.

For the new biomedical engineer who wishes to access quickly broad data on rehabilitation engineering there are a number of helpful texts. For an insight into the general nature of diseases which affect human physical and sensory abilities see Goldenson, Dunham and Dunham (1978), Greenwood *et al.* (1993) and Letts (1990).

Study of human and animal movement is probably the oldest application of mechanics to living systems as witnessed by the 'up-to-date' seventeenth century reference above. This may be because it is an immediate and rather obvious form of behaviour to study and because it is one which is accessible to the simplest of measuring systems – observation and the stopwatch. More recently the use of accurate and reliable measuring systems have raised the status of the study to a level at which it forms a major component in any biomedical engineer's professional armoury of skills.

Winter (1990) gives a useful and comprehensive overview of the role of technology in analysing movement. It includes descriptions of mathematical and engineering models which may be used to describe and predict patterns of motion. For researchers interested in dysfunction of movement, other valuable sources are those of Gage (1991), Kralj and Badj (1989) and Martin (1990).

Orthopaedic research and development extends from fracture reduction and the choice of pins, plates, nails, etc., to the design of prosthetic joint implants for hip, knee and elbow replacement. While much of the detailed design information is retained for commercial reasons in the research departments of the implant manufacturers, there is a wealth of information in the scientific literature on implant wear, materials, strength and performance. A few books which might offer useful insight to the principal obstacles and successes include Waugh (1991) which gives an entertaining recount of the first (and some would say still the most successful) hip joint development, at the hands of a innovative and creative surgeon, not an engineer! Bone cement and its application is covered by Saha (1990); and the earlier – but still useful – Black and Dumbleton (1981). For an introduction to the interplay between engineering and medicine, Goel and Weinstein (1990) provides a host of useful references and clinical perspectives.

In cardiovascular research, the text by Caro et al. (1978) will open the door to the use of computational fluid dynamics in describing, or attempting to describe, and modelling the human circulatory behaviour.

It would be wrong to leave this brief snapshot of texts to be considered by the researcher without introducing some important sources in biomaterials. The study of biomaterials is the most critical subject for biomedical engineering applied to devices and equipment which is for intimate contact with the human body and especially for those devices which are to be implanted, whether they are hard implants – as in hip or knee implants, soft implants – as in blood vessel replacement or bypass, or powered implants – such as in functional stimulation of muscle and nerve tissues.

No information guide would be complete without mention of Williams (1981–). He has published many scientific articles and edits a series for CRC Press. Other valuable texts are those by Fung (1993) and Duck (1990), for use as a references' compilation of published research, Marks and Payne (1981) and Millington and Wilkinson

(1983). Related to the physical properties of skin and soft tissue and the effects of their interactions with the physical world is a book which reviews recent work in decubitus ulcers' prevention. Decubitus ulcers, or 'pressure sores' are a major problem in healthcare, costing tens of millions of pounds in the UK (and proportionately more in dollars in the USA). The work which is edited by Bader (1990) provides an essential starting point for engineers who wish to tackle this most challenging of research areas.

Journals

We have made a very broad subdivision of the journals into headings which we consider represent the most common coverage of each category. It is entirely consistent with the entire ethos of our 'overlapping' argument that these divisions are artificial and that subjects using, say ergonomics, will appear in biomedical engineering journals, and engineering papers in medical journals. However the subdivisions we have chosen will at least 'shorten the odds' for the searching reader. Where thought necessary, further comment about the journals is given in the same order as the listings category.

Medical journals

Medical journals are an important source of two types of information for the biomedical engineer. Firstly, they serve to highlight the areas of greatest concern to the medical profession and therefore offer a potentially rich stimulus to the researcher wishing to select a field of interest, or to determine the state of the art in the clinical or medical applications of biomedical engineering. Secondly, they provide a mixed source of 'case study' research which may concentrate on studies in which a small number of subjects or patients are involved or large clinical trials in which the full rigour of the 'scientific method' is applied. Both of these types of research contain useful information although the latter is often propounded as more scientifically credible.

In all cases the published work usually serves to emphasize that any studies in which humans or living systems are the subject contain large variability in the measured data. This is not due to the researchers' inaptitude but to the inherent variability of the living process. Not only are there large differences between individuals in the study, but also the individual may respond to the test stimulus by adaptation from what would otherwise be the representative behaviour of the individual in the absence of the study.

For that reason alone, the researcher needs to become aware of the difficulties in designing medically- or clinically-related research studies. There is a number of useful texts which may help the researcher in planning the work. These are generally outside the scope of this text, however a small, low cost monograph which has been found valuable by postgraduate students in 'reminding themselves' of the precautions is that by Lowe (1993).

It is possible to group the medical journals, indeed all of the journals mentioned here, by adopting a systems approach to the human body; one which falls neatly into the three main types of engineering which apply in biomedical engineering. The musculo-skeletal system in which the principal disciplines which apply are those of mechanics, the description of more or less rigid objects in motion or of their reactions when they come into contact. The body fluids systems and the integumentary system, in which fluid mechanics and the mechanics of soft materials (rheology) are used to describe the behaviour of, for example, the cardiovascular system and its blood flow, the urinary system, or the behaviour of skin and the overlying tissues.

The third system is that of the neurological system and the senses, in which the techniques of control engineering and of instrumentation engineering are most directly applicable.

Obviously, this division is as arbitrary as the overall grouping we have chosen. However, provided one remembers that in this description, as always, overlaps occur, then the division is a convenient and useful one. Many postgraduate students and researchers for over a decade have been using it as an initial doorway into discovering the delights of biomedical engineering.

Journals on the musculo-skeletal include:

American Journal of Surgery
Archives of Orthopaedic and Traumatic Surgery
Bone
British Journal of Rheumatology
Foot
International Orthopaedics
Journal of Bone and Joint Surgery
Spine
The Knee

Clinical orthopaedics and related research journals:

Acta Orthopaedica Scandinavia
Arthritis and Rheumatism
Acta Anaesthesiologica Scandinavica
Acta Cardiologica
American Heart Journal
American Journal of Cardiology
Anaesthesia
Annales de Cardiologie et d'Angeiologie
Annals of Plastic Surgery

British Heart Journal
British Journal of Anaesthesia
British Journal of Plastic Surgery
British Journal of Urology
Decubitus

Fluid and integumentary system journals:

Journal of Tissue Viability
Journal of Urology
Urological Research
Wound and Skin Care (formerly Wound Management)

Neurological systems and the senses:

Acta Neurochirurgica
Age and Ageing
British Journal of Audiology
British Journal of Neurosurgery
Clinical Electroencephalography

General and overlapping areas:

American Journal of Medicine
Annales de Chirurgie
Annals of Surgery
Aviation, Space and Environmental Medicine
Australian and New Zealand Journal of Medicine
British Medical Journal
Clinical Physiology
European Journal of Surgery – Acta Chirurgica
Geriatric Medicine
International Surgery

The main medical physics journals include:

Anaesthesia and Intensive Care
Electrocardiology
IEEE Transactions on Medical Imaging
Image Processing
International Journal of Clinical Monitoring & Computing
Journal of Ambulatory Monitoring
Journal of Cardiovascular Electrophysiology
Journal of Cardiovascular Technology
Journal of Clinical Monitoring
Journal of Clinical Ultrasound
Journal of Ultrasound in Medicine
Medical Physics
Physica Medica
Physiological Measurement
Ultrasound in Medicine and Biology

Biological/life science journals

There are literally thousands of biological/life science journals to choose from. Some regularly publish research of medical significance, others seldom. We know of no journal in this general area which could be said to contain a biomedical engineering article in every issue. Rather than subscribe, the most useful approach may be to use a system which publishes contents pages (e.g. *Current Contents*) and scan certain types of journal.

Thus, physiological journals and those with 'experimental' in their title may be a useful source. The *Journal of Physiology*, for example, has long been a primary outlet, in Britain, for physiological research of medical significance. The fact that the research may be on an octopus, a fish, a cockroach, or a rat or dog, is of little consequence; these are the physiologists' models. Physiology is very much involved with the application of the physical sciences in biology, and the engineer will feel at home with the equipment and analysis, if not the experimental material. As with clinical journals the concern will be with the results from the living organism, and papers on the design of apparatus occur less frequently.

Sometimes a journal may build an association with a particular author (here the 'name dropping' philosophy comes into play). For example, Alexander and his co-workers often publish animal gait articles in the *Journal of Zoology* and it is worth scanning its contents page for this reason; however, other subjects of a biomedical engineering nature appear seldom in this journal.

The inclusion of ergonomics and sports science journals is for want of a better slot; many biologists would not class them within their field. Conversely, many general biological journals which biologists would deem essential (for example, the *Journals of the Linnean Society*, the *Journal of Ecology* and journals on genetics, biochemistry, cell biology, immunology, taxonomy and evolution) are excluded; biomedical engineering articles appear in them very rarely.

The most useful journals in the biological/life sciences include:

Acta Physiologica Scandinavica
American Journal of Physiology
American Journal of Sports Medicine
Applied Ergonomics
Ergonomics
Experimental Physiology
Journal of Applied Physiology
Journal of Experimental Biology
Journal of Experimental Zoology
Journal of Physiology
Journal of Zoology

Philosophical Transactions of the Royal Society: Biological Sciences
Physiological Reviews

Biomedical engineering journals

By now the reader will be all too familiar with the difficulty in structuring confusion of information sources and of the dangers in over arbitrary classification, but in broad terms it is possible to divide these journals into convenient groupings. The medical and materials group offer a mix of applied and clinical data. Of these, the journal *Biomaterials* offers quick sound information on current thrust areas, *Clinical Biomechanics* the insight to applications of biomechanics to 'real problems' and there is a number of journals relating to human movement of which *Gait and Posture* is the newest. A reminder of the fact that all problems have not been solved is readily seen in the *Journal of Long Term Effects of Medical Implants*, a timely reminder that the biomedical engineers' efforts can sometimes only be measured in decades not months.

The second group is self-explanatory and contain good access sources for researching the application of electronics and computing to patient care. In particular, the institute journals (IEE, IEEE and IMechE) are well regarded in the field.

A catch-all group of journals is chosen as the last group, containing a very wide range of work and providing well-referenced reviews. The *Journal of Medical Engineering & Technology* has the useful feature of being a source of 'current technology news' plus it contains the contents lists of some of its sister journals.

Journals on medical and materials include:

Advanced Hospital Technology: Orthopaedics
Advanced Materials
Artificial Organs
Biomaterials
Clinical Biomechanics
Gait & Posture
Human Movement Science
Journal of Applied Biomaterials
Journal of Applied Biomechanics
Journal of Biomaterials Applications
Journal of Biomechanics
Journal of Human Movement Studies
Journal of Long Term Effects of Medical Implants
Journal of Materials Science: Materials in Medicine
Journal of Prosthetics and Orthotics
Proceedings. Institution of Mechanical Engineers. Part H, Journal of Engineering in Medicine

642   Biomedical engineering

Rheology
Transactions of the ASME. Journal of Biomechanical Engineering

Instrumentation and electronics:

Advances in Bioengineering & Instrumentation
Australasian Physical and Engineering Sciences in Medicine
Biomedical Instrumentation and Technology
Biomedical Sciences Instrumentation
Biomedizinische Technik
Communication Outlook
Computers in Biology and Medicine
Computers in Biomedical Research
Health Equipment Information
IEEE Engineering in Medicine and Biology Magazine
IEEE Transactions on Biomedical Engineering
Innovation et Technologie en Biologie et Medecine
International Journal of Biomedical Computing
Japanese Journal of Medical Electronics and Biological Engineering
Journal of Clinical Engineering
Medical and Biological Engineering and Computing
Medical Electronics

General:

Annals of Biomedical Engineering
Critical Reviews in Biomedical Engineering
Frontiers of Medical and Biological Engineering
Journal of Biomedical Engineering
Journal of Medical Engineering and Technology
Medical Progress Through Technology

Rehabilitation and rehabilitation engineering journals have been grouped as one list, emphasizing the close integration of the applied and the theoretical work. It would be fair to expect that the main topics were clinically and patient-directed rather than engineering theory-directed. The most useful journals include:

Ability
American Journal of Occupational Therapy
American Journal of Physical Medicine and Rehabilitation
American Journal of Physical Therapy
Archives of Physical Medicine & Rehabilitation
Assistive Technology
Augmentative and Alternative Communication
Biofeedback and Self Regulation
British Journal of Occupational Therapy
Care of the Elderly
Clinical Rehabilitation
Critical Reviews in Physical and Rehabilitation Medicine

Disability Now
Disability and Rehabilitation
Elderly Care
IEEE Transactions on Rehabilitation Engineering
International Journal of Rehabilitation Research
Journal of Manipulative and Physiological Therapeutics
Journal of Visual Impairment & Blindness
Paraplegia
Physical Medicine and Rehabilitation Clinics of North America
Physical Therapy
Physiotherapy
Prosthetics and Orthotics International
Rehabilitation Digest
Technology and Disability

Additionally, charities and general news publications can be of interest:

AFRC News
ESEM News
Hospital Equipment & Supplies
MBEC News
Medical & Biological Engineering & Computing News
Medical Technologist and Scientist
Physics World
Scope
See Hear
Therapy Weekly

Abstracting and indexing services

Of the abstracting services, probably *Index Medicus* is the most widely used source for published work in medicine and in the medical technologies. It is valuable in several ways. Its classification system of defined literature headings gives a ready made structured list of potential search keywords, which is critical to efficient searching on those computerized databases which use the *Index Medicus* MeSH system for their classifications. It contains good cross referencing which can illuminate a new pathway to otherwise overlooked data. It affords a 'citation' accessing route to those workers who publish most prolifically, not necessarily the best works but at least the most widely circulated ones. There are equivalent engineering abstracts. There are also annually published *Advances in* ... and while some of the material obviously overlaps with *Index Medicus*, much of its does not.

One caveat in connection with *Index Medicus* is that it applies very vigorous rules about the size (measured by circulation), the internationality and the longevity of journals which it selects for inclusion in its abstracting process. It is therefore not a complete source of information. Young and specialist journals may be overlooked in its selection, a danger which increases the chances of the new researcher missing data on the latest emergent interest group. For example, *Gait and Posture* and *Clinical Rehabilitation* are examples of journals awaiting or perhaps only recently awarded inclusion in its service.

Specialist information services exist. In the UK, BECAN is an abstracting service which concentrates on biomedical engineering and issues monthly hard-copy of recent publications. An attractive feature of this is that it includes selected conferences and less well-known sources in its coverage, alongside the more general journals. A less attractive feature is the not totally systematic listing method. Some journals such as the *Journal of Medical Engineering and Technology* include sections on forthcoming meetings, extracted contents pages from other journals, and can be an easily read source of smaller items of interest.

Computerized databases

The automated medical information service of essential use to the biomedical engineer is MEDLINE – which is in fact MEDlars onLINE. The current MEDLINE file covers material from 1977 and is supplemented by backfiles to 1966. MEDLINE is structured on *Index Medicus* but has the advantage that text word searching opens up the access success possibilities. However care in good selection of keywords is a vital element in effective searching.

It follows that the user needs access to good medical and technical dictionaries to check ambiguities. This is particularly important when many more references than expected are counted by the system; an example which comes immediately to mind is 'fixation' which is used in orthopaedic implant surgery, in the preparation of tissue for microscopy and in agriculture – as a process by which certain plant organisms change nitrogen to nitrates – it should not therefore be found in MEDLINE in this context but may occur in biological abstracting systems. A similar example is 'capillary bed' which can turn up thousands of references, the minority of which are related to the human cardiovascular system.

There is rapid expansion of the availability of electronic databases, some operate in CD-ROM format, others through non-refereed bulletin boards on international e-mail. There are several biomedical engineering related bulletin boards, which are great fun and highly effective sources of 'hot' items. Since they are not refereed or assessed in any way their use should be cautious.

Standards: national and international information

As with many disciplines there are an agreed set of standards and guidelines on which the researcher can draw in the early stages of using, developing or designing a new technique or device. The British Standards (BS) are rarely mandatory but contain well-tried guidelines as to best practice. The American equivalent are the American Society of Mechanical Engineers (ASME) Standards and, worldwide, the International Standards Organization (ISO) complements or parallels these national ones. Of particular importance are the Electrical Safety for Medical Equipment Standards of which no researcher should be ignorant.

ASME Standards concentrate on general mechanical engineering and some aspects which may be relevant to medical engineering equipment. Most medical equipment in the USA is required to reach standards set by the Food and Drug Administration (FDA).

The British Standards Institution issues a range of standards which relate to medical engineering in headings such as:

aids for the disabled, e.g. crutches, hearing aids, stair lifts, walking frames, wheelchairs (BS 6935: 1988(1993), 6936: 1988(1993), 6937: 1988(1993))
medical electrical equipment (BS 5724: 1979–1993)
patient transport equipment (BS 896: 1960–1965)
manufacture, quality assurance (BS EN 46001: 1994, 46002: 1994).

The International Standards Organization issues standards which in many areas overlap with the British Standards and can be accessed via the catalogue of standards in the UK, USA, Australia, etc. or through various computer-based indexes.

Governments of different countries issue advice from time to time, or reports on equipment performance. The UK Department of Health, through its Medical Devices Directorate (MDD) is the most likely source of this information, and equivalent notices are issued by the German Government and by the Nordic countries. An approach to HMSO in the UK or to the appropriate commercial or scientific attachés at the embassy of a country of interest will give access to the sourcing routes of such national government publications.

Conclusions

The authors have sought to demonstrate and perhaps even overemphasize the challenge, difficulties, dangers and excitement of searching in a field which encompasses such a wide range of disciplines. In doing so we may have omitted useful or favoured referencing sources. Our choice has been a pragmatic one based, inevitably, on personal usage and prejudice. We believe it is one which at the very least will guarantee a fascinating entry into the literature and more importantly will give a landscape map in which a few clear features are to be found so that the researcher can be 'rescued' from the byways into which the too diverse literature options may take them.

References

Aiello, L. and Dean, C. (1990) *An introduction to human evolutionary anatomy*. London: Academic Press.
Alexander, R. M. (1975) *Biomechanics*. London: Chapman and Hall.
Alexander, R. M. (1983) *Animal mechanics* 2nd ed. Oxford: Blackwell.
Bader, D. L. (ed.) (1990) *Pressure sores, clinical practice and scientific approach*. Basingstoke: Macmillan.
Bastian, G. F. (1993) *An illustrated review of anatomy and physiology*. New York: HarperCollins.
Berne, R. M. and Levy, M. N. (1990) *Principles of physiology*. London: Wolfe.
Biomechanics in sport (1988) Bury St Edmunds: MEP.
Black, J. and Dumbleton, J. H. (eds) (1981) *Clinical biomechanics*. New York: Churchill Livingstone.
Black, J. (1992) *Biological performance of materials: fundamentals of biocompatibility* 2nd ed. New York: Marcel Dekker.
Borelli, G. A. (1989) *De motu animalium (On the movement of animals)*, trans. Maquet, P. New York: Springer-Verlag.
Braune, W. and Fischer O. (1985) *On the centre of gravity of the human body*, trans. Maquet, P. and Furlong, R. New York: Springer-Verlag.
Braune, W. and Fischer O. (1987) *The human gait*, trans. Maquet, P. and Furlong, R. New York: Springer-Verlag.
Braune, W. and Fischer O. (1988) *Determination of the moments of inertia of the human body and its limbs*, trans by Maquet, P. and Furlong, R. New York: Springer-Verlag.
Caro, C. G. et al. (1978) *The mechanics of the circulation*. Oxford: Oxford University Press.
Changeaux, J.-P. (1986) *Neuronal man*. Oxford: Oxford University Press.
The changing role of engineering in orthopaedics (1989) Bury St Edmunds: MEP.
Denny, M. W. (1993) *Air and water: the biology and physics of life's media*. Princeton, NJ: Princeton University Press.
Dowson, D. and Wright, V. (eds) (1981) *Introduction to the biomechanics of joints and joint replacement*. Bury St Edmunds: MEP.

Duck, F. A. (1990) *Physical properties of tissue: a comprehensive reference book*. London: Academic Press.
Feinberg, B. N. (1980) (ed.) *Handbook of clinical engineering*. Boca Raton, FL: CRC Press.
Fleming, D. G. and Feinberg, B. N. (eds) (1976) *Handbook of engineering in medicine and biology*. Cleveland, OH: CRC Press.
Frost, H. M. (1967) *An introduction to biomechanics*. Springfield, IL: Charles Thomas.
Fung, Y. C. (1993) *Biomechanics: mechanical properties of living tissues* 2nd ed. New York: Springer-Verlag.
Gage, J. R. (1991) *Gait analysis in cerebral palsy*. Oxford: MacKeith Press.
Goel, V. K. and Weinstein, J. N. (eds) (1990) *Biomechanics of the spine: clinical and surgical perspective*. Boca Raton, FL : CRC Press.
Goldenson, R. M., Dunham, J. R. and Dunham, C. S. (eds) (1978) *Disability and rehabilitation handbook*. New York: McGraw-Hill.
Gray, J. (1953) *How animals move*. Cambridge: Cambridge University Press.
Gray, J. (1968) *Animal locomotion*. London: Weidenfeld and Nicolson.
Greenwood, R. et al. (eds) (1993) *Neurological rehabilitation*. Edinburgh: Churchill Livingstone.
Kralj, A. and Bajd, T. (1989) *Functional electrical stimulation: standing and walking after spinal cord injury*. Boca Raton, FL : CRC Press.
Letts, R. M. (ed.) (1990) *Principles of seating the disabled*. Boca Raton, FL: CRC Press.
Leyton, L. (1975) *Fluid behaviour in biological systems*. Oxford: Oxford University Press.
Lowe, D. (1993) *Planning for medical research: a practical guide to research methods*. Middlesbrough: Astraglobe.
Mallard, J. (1994) The birth of the international organisations – with memories. *Scope*, **3**(2), 25–31.
Marks, R. and Payne, P. A. (1981) *Bioengineering and the skin*. Lancaster: MTP Press.
Martin, W. R. W. (1990) *Functional imaging in movement disorders*. Boca Raton, FL: CRC Press.
Mayr, E. (1988) *Toward a new philosophy of biology*. Cambridge, MA: Belknap.
Medawar, P. B. and Medawar, J. S. (1985) *Aristotle to zoos: a philosophical dictionary of biology*. Oxford: Oxford University Press.
Millington, P. F. and Wilkinson, R. (1983) *Skin*. Cambridge: Cambridge University Press.
Nose, Y. (1973) *Manual on artificial organs* vol. 2 the oxygenator. St Louis: Mosby.

Pauwels, F. (1980) *Biomechanics of the locomotor apparatus*, trans. Maquet, P. and Furlong, R. New York: Springer-Verlag.
Saha, S. (1990) *Bone cement and its application in orthopaedic surgery*. Boca Raton, FL: CRC Press.
Savory, T. H. (1971) *The principles of mechanistic biology*. Watford: Merrow.
Smith, R. V., Leslie, J. H. Jr. (1990) *Rehabilitation engineering*, Boca Raton, FL: CRC Press.
Tortora, G. J. and Grabowski, S.R. (1993) *Principles of anatomy and physiology*, 7th ed. New York: HarperCollins.
Vincent, J. F. V. (1993) *Structural biomaterials* 2nd ed. London: Macmillan.
Wainwright, S. A. *et al.* (1976). *Mechanical design in organisms*. London: Edward Arnold.
Wallace, W. A. *et al.* (eds) (1992) *Joint replacement in the 1990s*. Bury St Edmunds: MEP (Mechanical Engineering Publications Ltd). MEP also publish *Advances in Bioengineering* annually, about a year in arrears.
Waugh, W. (1991) *John Charnley: the man and the hip*. London: Springer-Verlag
Webster, J. G. (ed.) (1988). *Encyclopaedia of medical devices and Instrumentation* (4 volumes). New York: Wiley.
Williams, D. F. (1981–), *Series on biocompatibility*. Boca Raton, FL: CRC Press.
Winter, D. A. (1990) *Biomechanics and motor control of human movement* 2nd ed. New York: Wiley.
Wolff, J. (1986) *The law of bone remodelling*, trans. Maquet, P. and Furlong, R. New York: Springer-Verlag.

CHAPTER THIRTY-SIX

Concurrent engineering

D. F. RADCLIFFE

Specialization and departmentalization in corporate structures in many Western companies have led to an artificial segmentation and serialization of the product development process into stages, viz. market research, concept development, product planning, product engineering or design, production planning, production development and production. This segmentation encourages an 'over the wall' approach to information transfer within companies with consequential misunderstandings, delays and inefficiencies in getting new products to market. The term concurrent engineering emerged in the early 1990s to describe the process of carrying out these stages of the product development cycle in a predominantly parallel fashion to overcome the shortcomings of this serial approach. Concurrent engineering is a radical departure from what has become the conventional management of the product development process. It is a systematic approach to the integrated, simultaneous design of both products and their related processes, including manufacturing, test and support (Turino, 1992); a fundamentally different way of developing products through the timely and open sharing of product information between company functions. It is worth noting that most Japanese companies do not even have a name for this approach, it has been so embedded in the company culture for so long (Hartley, 1992).

Interest in the concept of concurrent engineering has been spurred on by increasing world competition in manufactures with the consequential need to reduce the time-to-market, the imperative of increased product quality, and a focus on customer-oriented design (Clark and Fujimoto, 1991). It is not merely the current fad in manufacturing or engineering design but rather the overarching paradigm that embeds techniques such as quality function deployment, design for manufacture and assembly, and computer integrated manufacture into a business strategy. Concurrent engineering fosters a collaborative approach that enables concepts like 'right-first-time' to be realized. The persuasive logic and compelling common sense of concurrent engineering recommends it, but its implementation requires more than just common sense (Linton et al., 1992). Concurrent engineering is one of several terms that have emerged to describe the shift from a serial to a parallel consideration of issues in part or all of the product development process. Andreasen and Heim (1987) coined the term integrated product development to describe a global business strategy that brings together three prime concerns of a manufacturing company – marketing, product design and development and production – in parallel. Simultaneous engineering (Hartley and Mortimer, 1991) pre-dates concurrent engineering as a term but has a broadly similar meaning. While simultaneous design would seem to imply a more restricted meaning than concurrent engineering, it is used synonymously with concurrent engineering in some writing (e.g. National Research Council, 1991a) as is concurrent design (National Research Council, 1991b). The term parallel engineering is also used although a distinction is sometimes drawn; parallel engineering being applied to the parallel conduct of uncoupled stages in product development while concurrent engineering is applied to those stages that are highly coupled or interdependent.

Concurrent engineering encompasses a broad raft of techniques, tools and technologies. Consequently the literature on the implementation of concurrent engineering draws on at least four themes; (1) principles of concurrent engineering, (2) cross-functional teams, (3) tools and techniques, and (4) enabling technologies. Work on the principles of concurrent engineering is emerging in the new journals and monographs. In contrast there is an extensive literature on tools and techniques such as quality function deployment (QFD), design for manufacture and other design for 'x' methodologies, enabling technologies including CAD/CAM/CAE, information systems, multimedia, etc. and on teamwork. As a consequence sources of information on concurrent engineering are widely scattered.

Learned societies

As concurrent engineering spans activities traditionally seen as the province of engineering design and manufacturing (including production engineering) numerous professional bodies have an active interest in concurrent engineering. Three international bodies, ISPE, CIRP and IFIP, concerned with automation and information technology in manufacturing with representation from Europe, Asia and North America sponsor activities in concurrent engineering. Of the three, the Institute of Concurrent Engineering of the International Society for Productivity Enhancement (ISPE) is the most centrally focused on concurrent engineering. CIRP (in French: Collège International pour l'Etude Scientifique des Techniques de Production Méchanique; in English: International Institution for Production Engineering Research) encourages international cooperation on a broad range of fundamental issues in manufacturing; it has sponsored several workshops and conferences on concurrent engineering and other aspects of the integration of people and technology in the product realization process. IFIP (International Federation for Information Processing) has a technical committee on Computer Applications in Technology that is concerned with the technologies that will be implemented in computer integrated manufacture. Several Working Groups (including 5.2, 5.3 and 5.8) that form part of this committee have sponsored workshops on concurrent engineering.

In the USA, the Society of Concurrent Engineering (SCE), Society for Computer-Aided Engineering (SCAE) a technology association of the Fabricators and Manufacturers Association, International and the Computer and Automated Systems Association (CASA) of the Society of Manufacturing Engineers (SME), the Design Theory and Methodology and the Design Automation Committees of the American Society of Mechanical Engineers (ASME), The Institute of Electrical and Electronics Engineers (IEEE) have all sponsored meetings on concurrent engineering and their archive journals and magazines contain numerous papers and articles on the topic.

The same holds for the Institution of Electrical Engineers (IEE), the Institution of Production Engineers (IProdE), before amalgamation with the IEE, and the Institution of Mechanical Engineers (IMechE) in the UK and the VDI-Gesellschaft Produktionstechnik in Germany. The WDK (Workshop Design-Konstruktion), an international society for the science of engineering design based in Switzerland, incorporates concurrent engineering in its biennial conference series. In Japan, the Japanese Society of Mechanical Engineers (JSME) has sponsored several joint meetings with ASME on design and concurrent engineering.

Reference books and monographs

Chapter 2 of Bakerjian (1992) provides an excellent introduction to and overview of concurrent engineering. It introduces the ten key principles of concurrent engineering: (1) understand your customer, (2) use product development teams, (3) integrate process design, (4) early involvement of suppliers and subcontractors, (5) use digital product models, (6) integrate CAE, CAD and CAM tools, (7) simulate product performance and manufacturing processes electronically, (8) use quality engineering and reliability techniques, (9) create an efficient development approach and (10) improve the design process continuously. Each of these principles is explained and discussed and the interrelationships between them articulated. The remainder of this extensive handbook is devoted to the many designs for manufacture techniques on which concurrent engineering depends.

The earliest monographs on concurrent engineering were those of Andreasen and Heim (1987) on integrated product development and Nevins and Whitney (1989) on concurrent design of products and processes. Both offer broad strategies for integrating design and manufacturing methods and the tools and techniques they use. Andreasen and Heim (1987) is richly illustrated with conceptual models and metaphors for the product development process and cartoons of practice that engage the reader in the breadth of the topic. However, since 1991 there has been something of an explosion in the number of titles on concurrent engineering. This trend seems set to continue as the field matures.

Hartley and Mortimer (1991) and Hartley (1992) present a general overview of concurrent engineering aimed at informing management of the principles, techniques and implementation of the concept. Both books are sprinkled with anecdotal evidence and simple product examples drawn from major corporations in North America, Europe and Japan. These examples show the opportunities and benefits presented by concurrent engineering. Carter (1992) also provides a hands-on management overview of concurrent engineering and identifies five forces for change: technology, tools, tasks, talent and time.

Susman (1992) explores the design-manufacturing interface through a collection of scholarly papers. The book identifies some of the strategic capabilities necessary to support competitive product development and highlights the significance of the social, political and cultural context of design for manufacturing. An ergonomic or human factors perspective of design for manufacture and concurrent engineering is presented in the book edited by Helander and Nagamachi (1992). Teamwork is a key to concurrent engineering and numerous books on management consider this topic. Many of the key issues specific to concurrent engineering are brought together in the round table discussion edited by Beaumariage and Shunk (1990). A more complete analysis of the transition of an enterprise to cross-functional teams is presented in Dimancescu (1992).

A number of books approach concurrent engineering from the perspective of CAD/CAM, CAE or computer integrated manufacturing, the themes which dominated much of the manufacturing debate in the 1980s. Twigg and Voss (1992) and Stark (1992) show how these technologies can support concurrent engineering with emphasis on information management. Kusiak (1993) is a collection of papers on design for manufacture and design automation related to concurrent engineering. It has a strong emphasis on the use of AI techniques in manufacture.

The literature on concurrent engineering is beginning to differentiate in line with specific industry sectors. For instance, the manufacture of electronic products with their unique need for assembly, testability and serviceability is addressed by Shina (1991), Turino (1992) and Classon (1993). In contrast, Haug (1993) is aimed at the application of concurrent engineering to mechanical systems design in the context of the automotive, aerospace and naval systems industries. SAE (1991) is concerned with the product-process interface in the aerospace industry while SAE (1992) is focused on the particular requirements of the automobile industry. These latter two are also an example of the trend by learned societies to issue guidelines on concurrent engineering tailored to the needs of their members. Another example is the information pack from the IEE (Coupland, 1992). These trends will continue as the principles of concurrent engineering are adapted to particular industries and professional groupings (Miller, 1993).

In addition to the monographs that deal with concurrent engineering principles and practices, there are numerous reference texts that specialize in the specific tools and techniques upon which concurrent engineering depends. Akao (1991) is the accepted authority on the principles and application of quality function deployment. The most comprehensive monograph on design for manufacture is the handbook edited by Bralla (1986), while one of the most widely accepted sources on design for assembly is the books and software of Boothroyd and Dewhurst (1987). Jacobs (1991) has edited the first monograph on the principles and case study applications of stereolithography and rapid prototyping.

Journals

A wide range of major international journals publishes research papers, reviews and conference reports in the area of concurrent engineering. Several new journals that focus specifically on this topic have commenced publication in recent years. *Concurrent Engineering: Research and Applications (CERA)*, the official journal of the Institute for Concurrent Engineering of the International Society for Productivity Enhancement (ISPE), is concerned exclusively with the principles, techniques and enabling technologies of concurrent engineering. It is published four times per year with the first issue in January 1993. An earlier journal with a similar title, *Concurrent Engineering*, commenced in Jan/Feb 1991 but apparently ceased publication in December 1991. The *Journal of Design and Manufacturing* incorporates concurrent engineering with a primary focus on the development of design principles for capturing and transmitting designer intent from design through to manufacture.

A number of journals in the field of engineering design places emphasis upon concurrent engineering. The *Journal of Engineering Design* presents research into the improvement of design process and practices in industry and the creation of advanced engineering products. It deliberately embraces both the industrial and the academic engineering design communities across a wide range of engineering disciplines. This journal carries regular papers on concurrent engineering with an emphasis on the management of the process. *Research in Engineering Design* places more emphasis on analytical and computational aspects of engineering design research including concurrent engineering. It has the sub-title 'Theory, applications and concurrent engineering' and an editor-in-chief for concurrent engineering. *Artificial Intelligence for Engineering Design, Analysis, and Manufacturing* has a similar emphasis with a stronger AI flavour.

Design Studies, a journal on design research and design theory of long standing, presents an eclectic range of papers on the nature of design that draw on traditions of engineering, architecture, planning and industrial design. It tends to be descriptive and contains many papers on the social and organizational aspects of design practice relevant to the establishment of concurrent engineering.

While the journals in engineering design have arguably taken a lead in publishing work on concurrent engineering, the traditional journals in production and manufacturing systems increasingly contain original work in this area. The principal examples are *International Journal of Production Research, Manufacturing Engineering, Proceedings of the Institution of Mechanical Engineers. Part B, Journal of Engineering Manufacture, IEEE Transactions on Components, Packaging and Manufacturing Technology, Part A: Transactions on Manufacturing Technology*. In addition several new journals specializing in aspects of manufacturing place particular emphasis on concurrent engineering. *Journal of Electronics Manufacturing* includes concurrent engineering of electronic products as one of its principal areas.

In recent years an increasing number of articles on teamwork, the adoption of computer based design and manufacturing engineering techniques that are part of concurrent engineering have begun to appear in engineering and technology management journals including *IEEE Transactions on Engineering Management*. An interdisciplinary forum on the theoretical, practical and social aspects of the information technologies that will enable team processes especially in concurrent engineering can be found in *Computer Supported Cooperative Work* (CSCW). *Harvard Business Review* often contains topical articles on new trends in manufacturing management pertinent to concurrent engineering (e.g. Katzenbach and Smith, 1993).

Concurrent engineering 655

As the concept of concurrent engineering has become more widely discussed, a number of industry journals have run special issues that explore the implications of concurrent engineering. *Aerospace America* (April 1993) published a special issue entitled 'CE: engineering a change in the design process'. It contains a series of case studies from major aerospace contractors that demonstrates the application of concurrent engineering principles. Similarly, the editors of *Manufacturing Engineering* (January 1992) presented a series of articles under the collective title *Future View* that outlined the status and future trends in manufacturing organization and management in the USA. While not always explicitly stated, concurrent engineering principles underscored much of the discussion.

Conferences

A clear pattern of conferences specifically on concurrent engineering is yet to be established. One exception is the annual conference on concurrent engineering and computer-aided acquisition and logistics support (CE & CALS) held in Washington, DC in early June, sponsored by the Society for Computer-Aided Engineering (SCAE) and the CERC at the University of West Virginia. This series has been running since the early 1990s and is strongly associated with the US Department of Defense initiative in concurrent engineering. The SCAE and the Society of Concurrent Engineering are beginning to co-sponsor conferences such as Concurrent Engineering: Practical Applications for Superior Performance, Costa Mesa, California, March 1993.

The first CIRP International Workshop on Concurrent Engineering for Product Realization was held in June, 1992 at Chou University in Tokyo. It was organized jointly by CIRP and IFIP. IFIP also sponsored a Working Conference on Manufacturing in the Era of Concurrent Engineering (Halevi and Weill, 1992).

The Design Engineering Division of the ASME hold an annual series of Design Technical Conferences in September. Two of these, Design Theory and Methodology and Design Automation, have included papers on concurrent engineering and related technologies in recent years. The Annual Winter Meeting of ASME has also included sessions on concurrent engineering (e.g. Sharon, 1990). ICED '93 held in The Hague, the most recent biennial International Conference on Engineering Design (ICED) included papers and workshops on concurrent engineering and teamwork in design. JSME-ASME Joint Workshop on Design '93: Frontiers in Engineering Design, June, 1993 included papers on cooperative product development and concurrent engineering.

A number of meetings organized by the AI (artificial intelligence) community is including aspects of concurrent engineering technology. The American Association for Artificial Intelligence (AAAI) held a workshop on AI in Collaborative Design as part of AAAI '93 and co-sponsored with IEEE workshops on Enabling Technologies Infrastructure for Collaborative Enterprises (WET-ICE). Similarly the Association for Computing Machinery (ACM) meetings such as SIGGRAPH and CHI have sessions and workshops on group technologies. The AI in Design series of conferences also contain some relevant papers.

Research and industry reports

The Concurrent Engineering Research Centre (CERC) at West Virginia University is a (US) national research centre established to promote and facilitate the adoption of concurrent engineering technology by US industrial users and suppliers. Funding for CERC and related programmes is provided by the Defence Advanced Research Projects Agency (DARPA) Initiative in Concurrent Engineering (DICE). The CERC offers an extensive range of technical reports in its role as National Repository for Concurrent Engineering (srobin@cerc.wvu.wvnet.edu). Research reports on work in progress and theses related to concurrent engineering can be obtained from the Engineering Design Research Centre, Carnegie Mellon University and the Centre for Design Research at Stanford University.

The Task Group on Electronic Systems of the CALS/Concurrent Engineering Industry Steering Group, composed primarily of US aerospace contractors, prepared a comprehensive report that outlined the proposed eleven first principles of concurrent engineering (Linton *et al.*, 1992). In addition, this report presents a self-assessment procedure to assist an organization in understanding and improving its concurrent engineering environment, including a series of best practice templates. The enabling tools and technologies and possible obstacles to implementation are also discussed.

Government reports in the USA and the UK highlight the importance of adopting concurrent engineering principles in manufacturing industry. The (US) National Research Council report (National Research Council, 1991a) identifies and analyses research needs in five critical areas of manufacturing. Concurrent engineering principles are cited as crucial to two of these, equipment reliability and maintenance and product realization. A complementary report (National Research Council, 1991b) considers the state of engineering design under three headings: designing for competitive advantage; improving engineering design education and a national agenda for engineering design research. This research agenda focuses on three broad themes: developing scientific foundations for design models and methods, creating and improving design support tools and relating design to the business enterprise. It identifies concurrent design as one of the modern practices that should be adopted. Although the use of concurrent design concepts has met success, little is known about how to organize and manage concurrent processes and cross-functional teams effectively. Department of Trade and Industry (1989) identifies the drivers that will impact upon UK manufacturing industry in this decade and proposes responses to the opportunities and threats that are presented. These responses, new product process, the rational factory, integrated logistics, integrated organization and integrated information echo the principles of concurrent engineering.

Current awareness sources

CERC at West Virginia University produces *Concurrent Engineering Research in Review* (two issues per year). It is an excellent source of information that contains overview articles, reports of meetings, workshops and conferences, a calendar of forthcoming conferences and recent reports published by the CERC. This review is also available electronically (carriger@cerc.wvu.wvnet.edu). In addition the CERC produces a bibliography on concurrent engineering. A similar one was produced by Corlett (1992).

International Network of Engineering Design Researchers publishes a newsletter that contains a list of researchers and their primary areas of interest, reports of meetings, book reviews, conference announcements and topical issues.

Computerized information sources

There are numerous informal electronic mail groups with interests in aspects of concurrent engineering or the tools and enabling technologies on which it depends. One point of initial access to these is via the CERCnet at the CERC at West Virginia University. CERCnet is a free electronic subscription service that provides access to concurrent engineering related abstracts, a bulletin board of events and conferences. Register via e-mail to sysop@cerc.wvu.wvnet.edu.

An electronic bulletin board called mailbase in the UK includes a user group in engineering design. To join, e-mail the following; subscribe engineering-design your-first-name your-surname to the e-mail address mailbase@mailbase.ac.uk. For more information, including a user guide, send the message; send mailbase overview send mailbase userguide lists to mailbase.

References

Akao, Y. (1991) *Quality function deployment*. Cambridge, MA: Productivity Press.
Allen, C. W. (ed.) (1990) *Simultaneous engineering: integrating manufacturing and design*. Dearborn, MI: Society of Manufacturing Engineers.
Andreasen, M. M. and Heim, L. (1987) *Integrated product development*. Bedford: IFS (Publications).
Bakerjian, R. (ed.) (1992) *Tool and manufacturing engineers handbook* 4th ed., Vol. 6 Design for manufacturability. Dearborn, MI: Society of Manufacturing Engineers.
Beaumariage, K. and Shunk, D. (eds) (1990) *Issues in migrating to teamwork*. Dearborn, MI: Society of Manufacturing Engineers.
Boothroyd, G. and Dewhurst, P. (1987) *Product design for assembly*. Wakefield, RI: Boothroyd Dewhurst Inc.
Bralla, J. G. (ed.) (1986) *Handbook of product design for manufacturing*. New York: McGraw-Hill.
Carter, D. E. (1992) *CE, concurrent engineering: the product development environment for the 1990's*. Reading, Mass: Addison-Wesley.
Chappell, C. and Stevenson, I. (1992) *Concurrent engineering: market opportunity*. London: Ovum.
Chase, T. R. (ed.) (1992) Engineering data management: key to integrated product development. *Proceedings of the ASME International Computers in Engineering Conference*, San Francisco, 1992. New York: ASME.
Clark, K. B. and Fujimoto, T. (1991) *Product development performance: strategy, organization and management in the world auto industry*. Boston, MA: Harvard Business School Press.
Classon, F. (1993) *Surface mount technology for concurrent engineering and manufacturing*. New York: McGraw-Hill.
Corlett, J. (ed.) (1992) *Concurrent engineering: a selected bibliography*. London: BKT Information Services.
Coupland, J. W. (1992) *Concurrent engineering: an information pack*. London: IEE Technical Information Unit.
Department of Trade and Industry (1989) *Manufacturing into the late 1990s*. London: HMSO.
Dimancescu, D. (1992) *The seamless enterprise: making cross functional management work*. New York: HarperBusiness.
Dwivedi, S. N., Verma, A. K. and Sneckenberger, J. E. (eds) (1991) *Proceedings 5th international conference on CAD/CAM, robotics and factories of the future. 2 vols*. Berlin: Springer-Verlag.
Halevi, G. and Weill, R. D. (eds) (1992) *Manufacturing in the era of concurrent engineering* revised papers and discussions from IFIP

Technical Committee 5 / Working Group 5.3 / Working Group 5.2 Working Conference, Herzilya, Israel. Amsterdam: North-Holland.
Hartley J. R. and Mortimer, J. (1991) *Simultaneous engineering,* 2nd ed. Dunstable: Industrial Newsletters Ltd.
Hartley, J.R. (1992) *Concurrent engineering.* Cambridge, MA: Productivity Press.
Haug, E. J. (ed.) (1993) *Concurrent engineering: tools and technologies for mechanical system design.* Berlin: Springer-Verlag.
Helander, M. and Nagamachi, M. (eds) (1992) *Design for manufacturability: a systems approach to concurrent engineering and ergonomics.* London: Taylor and Francis.
Jacobs, P. F. (1991) *Rapid prototyping and manufacturing; fundamentals of sterolithography.* Dearborn, MI: Society of Manufacturing Engineers.
Katzenbach, J. R. and Smith, D. K. (1993) The discipline of teams, *Harvard Business Review,* **71**, 111–120.
Kuo, W. and Pierson, M. M. (eds) (1993) *Quality through engineering design.* Proceedings of the Conference, Bangalore, India, 11–14 January, 1993. Amsterdam: Elsevier.
Kusiak, A. (ed.) (1993) *Concurrent engineering: automation, tools, and techniques.* New York: Wiley.
Linton, L., et al. (1992) *First principles of concurrent engineering: a competitive strategy for product development.* Washington DC: CALS/CE Working Group – Electronic Systems.
Mayer, R. J., Keen, A. and Wells, M. S. (eds) (1993) *Information integration for concurrent engineering.* Wright-Patterson Air Force Base, OH: Armstrong Laboratory, Air Force Systems Command.
Miller, L. C.G. (1993) *Concurrent engineering design: integrating best practices for process improvement.* Dearborn, MI: Society of Manufacturing Engineers.
Mortimer, J. and Hartley, J. (eds) (1991) *Simultaneous engineering: an executive guide.* London: DTI.
National Research Council (1991a) *The competitive edge: research priorities for US manufacturing.* Washington, DC: National Academy Press.
National Research Council (1991b) *Improving engineering design: designing for competitive advantage,* Washington, DC: National Academy Press.
Nevins, J. L. and Whitney, D. E. (eds) (1989) *Concurrent design of products and processes: a strategy for the next generation in manufacturing.* New York: McGraw-Hill.
Parsaei, H. R. and Sullivan, W. G. (eds) (1993) *Concurrent engineering: contemporary issues and modern design tools.* London: Chapman and Hall.

SAE (1991) *Aerospace product/process design interface*. Warrendale, PA: Society of Automotive Engineers.

SAE (1992) *Simultaneous engineering in automotive development*. Warrendale, PA: Society of Automotive Engineers.

Sharon, A. (ed.) (1990) *Issues in design/manufacture integration*, Winter Annual Meeting of ASME, Dallas, Texas Nov 25–30, 1990. New York: American Society of Mechanical Engineers.

Shina, S. G. (1991) *Concurrent engineering and design for manufacture of electronics products*. New York: Van Nostrand Reinhold.

Stark, J. (1992) *Engineering information management systems: beyond CAD/CAM, to concurrent engineering support*. New York: Van Nostrand Reinhold.

Susman, G. I. (ed.) (1992) *Integrating design and manufacturing for competitive advantage*. Oxford: Oxford University Press.

Turino, J. L. (1992) *Managing concurrent engineering: buying time to market; a definitive guide to improve competitiveness in electronics design and manufacturing*. New York: Van Nostrand Reinhold.

Twigg, D. and Voss, C. A. (1992) *Managing integration in CAD/CAM and simultaneous engineering: a workbook*. London: Chapman and Hall.

Index

Main treatments are indicated by **bold** page numbers. Figures and tables are indicated by *fig* or *tab* following the page number. Entries are arranged in letter-by-letter order rather than word-by-word; also names beginning Mc are treated as if they began with Mac: e.g. McGreavy/Macgregor/machines/machine tools/MacKellan/McKinney/mackintosh. Most common nouns are in the plural form (e.g. 'machines' not 'machine').

The following abbreviations may be found:

Amer.	American	fed'n	federation
Ass'n	Association	inst.	institute
Bd	Board	inst'n	institution
Brit.	British	internat.	international
Cncl	Council	Min.	Ministry
co.	company	nat.	national
conf.	conference	org'n	organization
confed'n	confederation	Proc.	Proceedings
corp'n	corporation	pub.	published by
dept	department	soc.	society
ed.	edited by	trans.	translated by/ translation
Eur.	European		
fed.	federal	Transac.	Transactions

A

AA (Automobile Ass'n) **514**
AAAI *see* Amer. Ass'n for Artificial Intelligence
Aachen
   Colloquia [fluid power] 292
   Technical University 360

AAES *see* Amer. Ass'n of Engineering Socs.
AAR *see* Ass'n of Amer. Railroads
AASHTO (Amer. Ass'n of State Highway and Transportation Officials) **516**

Abaqus [program] 148
Abbott, D., *Biographical dictionary* ... 126–7
Abbott, H. [*safety/design*] 334
Abbott, J.A. [*fire hazard*] 321
abbreviations, dictionaries of **124**

662 Index

ABC [co.]
 timetables 510
 Worldwide public transport timetables 510
A.B. Chance [co.] 366
Abel, P.D. [*water quality*] 487
Aben, H. [*photoelasticity*] 145
Abercrombie, S.A. 316
Ability 642
Abramowitz, M. *and* Segun, I.E. 454
abrasion 173, 358
Absorption heat pumps congress (pub. EC) 208
abstracting/indexing services 97–103, 224, 227, 354, 643–4
 nature and use 98–9, 472
 author-/editor-supplied abstracts 64, 65, 78
 currency 99, 227
 directories of 99, 227
 duplication in 102
 INSPEC 421–2
 keywords 56, 59, 98, 109, 485
 languages, non-English 101–2
 printed formats 97–8
 selective lists of 100–3, 109–13
 books 103
 conferences 63
 Proceedings 76
 contents pages 102, 103
 current awareness 102–3
 journals 63, 64, 65, 74–5, 97–103
 house journals 70, 74
 trade journals 74–5
 newsletters 70
 newspapers 70, 74–5
 patents 379
 reports 102
 software 113, 528
 theses 63, 78–9
 see also databases; indexes and specific topics
Abstracts of Science and Technology in Japan 101–2
Abu Rashid, A.R. *and* Smith, M.R. 592
ac (alternating current) 320, 432, **436**
academic journals *see* journals
academic organizations *see* organizations
Academic Press
 Academic Press dictionary of science and technology [ed. Morris] 123
Advances in chemical engineering 104, **272**
Accident Analysis and Prevention 327, 498
Accident prevention manual (pub. NSC) 315–16
accidents 181, 315–16, **317**–18, **323**–4
 terminology 316
 CIMAH 318

journals 281, 327, 498
mining 617
near-miss reporting 318
nuclear 266
road **509**–10
 and Standards 21
 see also safety
accumulators, hydraulics 294
ACE *see* Ass'n of Consulting Engineers
Achievement of structural adequacy . . . (pub. IStructE) 550
ACI
 ACI Materials Journal 532
 ACI Structural Journal 532
 see otherwise Amer. Concrete Inst.
acidity, pH control 280
acid rain 430, 491
Ackermann, K. 552
ACM (Ass'n for Computing Machinery) 656
 ACM GUIDE TO COMPUTING LITERATURE 400, 401
 ACM Guide to Computing Literature and Computing Reviews 113
 ACM-NS Info Flash 115
 Computer Graphics 339
ACOMPLINE database **505**
acoustics 119, **492**
acronyms, dictionaries **124**
ACS *see* Amer. Chemical Soc.
Acta Anaesthesiologica Scandinavica 638
Acta Cardiologica 638
Acta Chirurgica [*Eur. Journal of Surgery*] 639
Acta Mechanica 459
Acta Neurochirurgica 639
Acta Orthopaedica Scandinavica 638
Acta Physiologica Scandinavica 640
Acta Press 379
actuators, electrohydraulic 286, 286*fig*, 290, 292–3
AD *see* Astia Document
ADAMS [program] 170*tab*
Adams, C.E. [*gears*] 165
Adams, D.P. *et al. see under* Hain
Adams, L.F. [*measurement*] 367
Adams, M.S. *and* McManus, F. [*noise*] 492
Adams, R. W. [*trade associations*] 125, 340
Addis, W. 550, 553
Addison Wesley Publishers Ltd 423
Addleson, L. *and* Rice, C. 535
Adept Scientific Micro Systems [co.] 160
adhesives 362, **542**–3
 databases **112**, 314, 355
 expert system 312
 journals 357
Adhesives Abstracts 112
adiabatic processes **200**, 201
Adkins, B. *and* Harley, R.G. [*electricity*] 433

Adkins, C.J. [*thermophysics*] 205
Adler, U. 16
advanced gas-cooled (AGR) reactors 261–2
Advanced Hospital Technology: Orthopaedics 641
Advanced Manufacturing Engineering, former 356
Advanced Manufacturing Technology conference 359
Advanced Manufacturing Technology Research Institute (AMTRI) **371**
Advanced Materials 641
Advanced Materials and Processes 357
advanced procurement and logistics systems (APLS) 311
Advanced Robotics 378
Advances in Applied Mechanics 104
Advances in Applied Probability 460
Advances in Automobile Engineering 104
Advances in Bioengineering 648
Advances in Bioengineering and Instrumentation 642
Advances in [biomedical engineering] 643
Advances in Chemical Engineering 104, **272**
Advances in Computers 104
Advances in Concrete and Cement Research 522
Advances in Corrosion Science and Technology 104
Advances in Cryogenic Engineering 104
Advances in the economics of energy and resources (pub. JAI) 232
Advances in Electrochemistry and Electrochemical Engineering 105
Advances in Electronics and Electron Physics 105
Advances in Energy Systems and Technology 233
Advances in Engineering Software 105
Advances in Environmental Science and Technology 105
Advances in Heat Transfer 105
Advances in Human Factors Ergonomics 339
Advances in Materials Research 105
Advances in Nuclear Science and Technology 105, 266
Advances in Radio Research 105
Advances in Robotics 381
Advances in Robotics and Automation conference 382
Advances in robotics and automation (pub. IASTED) 382
Advances in Solar Energy 244
Advances in Space Science and Technology 105

Index 663

Advances in Welding Processes
  conference 360
advertisements 64, 65, **66**, **69**,
  71, 82
AEA [1]
  AEA Technology [orgn] **177**,
    179, **262**, 266, 595
  National Centre of
    Tribology 177
  see otherwise Atomic Energy
    Authority
AEA [2] (Amer. Electronics
  Ass'n) 414
AEC see Atomic Energy
  Commission
AECB see Canadian Atomic
  Energy Control Bd
AECL (Atomic Energy of
  Canada Ltd) **263**
AEI (Agence Européenne
  d'Informations) 210
aerodynamics see aerospace
  engineering
Aeronautical Journal **143**
aeronautics see aerospace
Aerospace America 655
Aerospace Daily 72
AEROSPACE DATABASE 112
aerospace engineering
  aircraft
    DOT and 511
    engines 208
  airlines, deregulation 512
  aviation Standards 16
  and biomedical engineering
    639
  combustion/aerodynamics 235
  components 91
  and concurrent engineering
    653, 654, 656
  conferences 204
  databases 91–2, 112
  design 155, 653
  dictionaries 123
  handbooks 653
  hydraulics 286, 292
  Internet resources 119
  journals 7, 67, 68, 72, 112,
    143–4, 639, 655
  reviews 105
  measurement 367
  organizations 7, 138, 210–11,
    421, 475
  product information 91
  reliability 178
  reports 4, 116
  solar energy 243–4
  spacecraft
    thermal systems 204
  tribology 176
  Standards 16, 23, 91, 421
  AIAA **26**
  BSI Aerospace series 21
Aerospace product/process design
  interface (pub. SAE) 653
AESJ (Atomic Energy Soc. of
  Japan) **265**
aesthetics **336**
Aesthetics in engineering design
  ... (pub. SEED) 336
AFNOR see Ass'n Française de
  Normalisation

AFRC AFRC News 643
AFREPREN (African Energy
  Policy Research Network)
  242
AFRI (Ass'n Française de
  Robotique Industrielle) 381,
  388
AFRI Liaison 378
Africa
  books-in-print lists 104
  construction 584
  energy options 242
  minerals 596
  patents 46
African Energy Policy Research
  Network (AFREPREN) 242
After Dark service [of BRS] 107
AGA see Amer. Gas Ass'n
Age and Ageing 639
Agence Européenne
  d'Informations (AEI) 210
Aggregate producers handbook
  (pub. Pegson) 600
aggregates 593, 597, 598, 600
agitation 275
AGMA (Amer. Gear
  Manufacturers Ass'n) 164
AGR (advanced gas-cooled
  reactors) 261–2
Agricola, Georgius
  De re metallica 590
  ed. Bauer 154
agriculture
  abstracts 101
  databases 111
  and energy 256
  experimentation 179
  hydraulics and 286
  and mineral processing 592
  mobile machinery 286
  solar energy 244
  waste conversion 249–50
AGV (automated guided vehicle)
  363, 385, 386, 470tab
AI
  AI Magazine 402
  see otherwise artificial
    intelligence
AIAA see Amer. Inst. of
  Aeronautics and
  Astronautics
AIChE
  AIChE Journal 281, 407
  see otherwise Amer. Inst. of
    Chemical Engineers
AIDA design system 348
AI in Design conferences 656
AIEDAM see Artificial
  Intelligence for Engineering
  Design ...
Aiello, L. and Dean, C. 634–5
AIME see Amer. Inst. of Mining,
  Metallurgical ...
Aims of structural design (pub.
  IStructE) 550
air
  air-supported structures 568
  compressed see compressors;
    pneumatics etc.
  conditioned see air
    conditioning

fluidized beds 430
  and hydraulic fluids 288–9
internal combustion engines
  200
mines 616–17
polluted see air pollution
thermodynamic properties
  202–3, 253–4
see also atmosphere; gases;
  wind etc.
air conditioning 204–5, **207**, 244,
  253–5
journals 67
organizations 210–11, 403
Air coolers, cooling towers and
  evaporative coolers (pub.
  NEL) 207
air pollution 486, **490–2**
journals 491
odour **491–2**
Air pollution (pub. OU) 491
Air quality guidelines ... (pub.
  WHO) 491
air transport see aerospace
Air Transport World 68
Air and Waste Management
  Association, Air and Waste
  491
AISC see Amer. Inst. of Steel
  Construction
AJ see Architects' Journal
Akao, Y. 653
Akzo [co.] 305
Alabaster, J.S. and Lloyd, R. 487
Alawi, H. et al. 243
Alban, L.E. 165
Al-Chalabi, F.J. 239
alcohols, and energy 250
Aleksander, I. et al. [artificial
  intelligence] 397
alerting see selective
  dissemination ...
Alexander, D.C. [ergonomics]
  336
Alexander, J.M. et al.
  [manufacture] 361
Alexander, R.M. [biomechanics]
  635, 640
Alexander, S.J. [design/buildings]
  550
Alford, M.H.T. and Alford, V.L.
  123
Algar, P. [safety/management]
  325
algebra see under mathematics
Alger, J.R.M. and Hays, C.V.
  [design] 332
Alger, P.L. [induction motors]
  435
Algor Europe Ltd 170tab
Aliabadi, M.H. and Brebbia,
  C.A. 455
alkalis, alkali-silica reactions 557
Allan R.N. et al. [power systems]
  433
Allen, A.J. et al. [stress analysis]
  147–8
Allen, C.W. [concurrent
  engineering] 658
Allen, P.W. et al. [rubber] 169
Allen, R.T.L. et al. [concrete]
  537

664 Index

Allen, T. [particles] 598
Allis Mineral Systems Ltd [Allis Chalmers] 600
Allmendinger E.E. 476
alloys 593
aluminium 313, 549, **561–2**, 564
casting 357
defence uses 307
equivalent 26, 302–3, 313
expert system, proposed 312
handbooks 307
inter-metallics 353
magnetic materials 419
Standards 26, **302–3**
All-Union Institute of Scientific and Technical Information (VINITI) 101
Almgren, G. et al. 614
Alphanumeric reports series index (ARPI) (pub. BL) 5
Altan, T. et al. 364
alternating current (ac) 320, 432, **436**
Alternative Energy 242
Alternative Energy Digests 242
alternative/renewable energy sources **209–10**, 223, **241–50**, 253–7, **430–1**
biomass **249–50**
books 209–10, **241–2**
conferences 209–10, 242, 243, 246, 248
costs 430
geothermal **245–6**
hydroelectricity see hydroelectric power
hydrogen 241
journals 209, **242**, 244–5, 246, 247, 249–50, 431
nuclear **250–1**, **259–67**
organizations 210, 225
solar **242–5**
synthetic fuels 241
water 210, **247–9**, 430
wind see wind
see also solar etc.
alumina 599
aluminium 593, 599
abstracts/indexes 57, 110
alloys 313, 549, **561–2**, 564
handbooks 313, 561
organizations 303, **561**
Standards 549, 562
Aluminium Federation Ltd **561**
Aluminum Association of America 303
Ambrose, E.R. 208
AMC see Amer. Mining Congress
America, see also United States etc.
American Association for Artificial Intelligence (AAAI) 656
AI Magazine 402
American Association of Engineering Societies (AAES) 128
International directory ... 125, 523

American Association of Project Managers, and 'Internet' org'n 586
American Association of State Highway and Transportation Officials (AASHTO) **516**
American Bureau of Shipping, Rules for building and classing of steel ships 477
American Ceramic Society, Ceramic Abstracts 228
American Chain Association, Chains for power transmission ... 164
American Chemical Society (ACS) 239, 324
Annual Reviews of Industrial and Engineering Chemistry 105
Chemical Abstracts see Chemical Abstracts
Chemical and Engineering News 282
conferences 324
Energy and Fuels 234
Industrial and Engineering Chemistry Research 282
Journal of Chemical Engineering Data 282
Mineral matter and ash in coal [ed. Vorres] 238
American Concrete Institute (ACI) 532, **537**, 557
ACI Materials Journal 532
ACI Structural Journal 532
American Council for an Energy Efficient Economy 257
American Cyanamid Co.
Chemistry of cyanidation 599
Mining chemicals handbook 600
American Electronics Association (AEA) 414
American Gas Association (AGA) 226–7
Energy Analysis 234
Gas Energy Review 240
TERA Analysis 235
American Gear Manufacturers Association (AGMA) 164
American Geological Institute Bibliography and Index of Geology 228
GEOREF 229, 230
American Heart Journal 638
American Heat Transfer and Fluid Mechanics Institute 205
American Institute of Aeronautics and Astronautics (AIAA) and AEROSPACE DATABASE 112
conferences 204
and magnetohydrodynamics 252
publications
International Aerospace Abstracts (IAA) 7, 112
Journal of Thermophysics and Heat Transfer 205
Progress in Astronautics and Aeronautics 105

Standards 26
Future of aerospace standards 26
and thermodynamics 210
American Institute of Chemical Engineers (AIChE) **270**
conferences 270
journals **281**
AIChE Journal 281, 407
Biotechnology Progress 281
Chemical Engineering Progress **281**
International Chemical Engineering 240, 281
Process Safety Progress 281, 328
publications **270**, 272, 274, 279, 317, 324
Fire and explosion index ... 317, 322
Heat transfer 277
online 110
Symposium Series 270
and safety 279, 317, 318, 324, 325
and software 281
American Institute of Mining, Metallurgical and Petroleum Engineers (AIME) **594**, 601, 610
Metallurgical Society (TMS-AIME) **594**, 601
see also Soc. of Mining, Metallurgy and Exploration (SME-AIME)
American Institute of Steel Construction (AISC) 532, 538, **539**
American Institute of Timber Construction 541
American Journal of Cardiology 638
American Journal of Medicine 639
American Journal of Occupational Therapy 642
American Journal of Physical Medicine and Rehabilitation 642
American Journal of Physical Therapy 642
American Journal of Physiology 640
American Journal of Sports Medicine 640
American Journal of Surgery 638
American Library Association, publications 131
American Machinist 356
American Mathematical Society (AMS) **113**, **453**
preprints, electronic 116
American Mining Congress (AMC) 227, **608**
American National Standards Institute (ANSI) 16, **25**, 123, 449
ANSI Reporter 25
ANSI Standards Action 25
Catalogue of American national standards 25

Index 665

American National Standards
  Institute (ANSI) (contd)
  European office, former 20
  Industrial engineering
    terminology 123
American Nuclear Society (ANS)
  265
  Fusion Technology 251
  Nuclear News 68, 267
  Nuclear Science and
    Engineering 251, 266
  and Nuclear Technology 251,
    266
  Transactions 251
American Petroleum Institute
  (API) 226, **270–1**, **283**
  APIBIZ **229**, 230
  APILIT **229**, 230, 231
  APIPAT **229**, 231
  Literature Abstracts 228, 229
  literature indexes 283
  Management of process
    hazards . . . 323, 325
  patent indexes 229, 283
  Technical data books 273
American Public Health
  Association 494
American Public Transit
  Association (APTA) **516**
  Passenger Transport 499
American Society for Artificial
  Internal Organs (ASAIO)
  631
American Society of Civil
  Engineers (ASCE) **144**, **516**,
  523
  and energy 247
  ocean energy 248
  and environmental engineering
    488
  Journals **532**
    Civil Engineering 532, **580**
    . . . of Aerospace
      Engineering 144
    . . . of Computing in Civil
      Engineering 532
    . . . of Construction
      Engineering and
      Management 532, **580**
    . . . of Energy Engineering
      255
    . . . of Environmental
      Engineering 487, 488
    . . . of Geotechnical
      Engineering 532
    . . . of Performance of
      Constructed Facilities
      532
    . . . of Professional Issues
      in Engineering
      Education and Practice
      580
    . . . of Structural
      Engineering **144**, 532
    . . . of Transportation
      Engineering 498
  publications 523
  ASCE Publications Abstracts
    504
  Civil engineering guidelines
    for . . . hydroelectric
    developments 247

Guidelines for rehabilitation
  of . . . hydroelectric
  plants 247
indexes to 102
and training 580
and transport 498, 504, 516
American Society for
  Engineering Education,
  Chemical Engineering
  Education 282
American Society for
  Experimental Stress
  Analysis, former 133
American Society of Heating,
  Refrigerating and Air
  Conditioning Engineers
  (ASHRAE) 210, 227, 254
publications
  ASHRAE handbooks **207**
  Terminology of heating,
    ventilation . . . 253
American Society of Mechanical
  Engineers (ASME) **144**, **152**
  and alternative energy 210
  and automation/robotics 359,
    381
  and biomedical engineering
    630, 645
  and concurrent engineering
    651, 655
  conferences **162–3**, 173, 206,
    317, 319
  Design Technical
    Conferences 319, 342,
    655
  Flexible Assembly 359
  Flexible Automation 359
  Heat Transfer 317
  Heat transfer . . . Solar
    Thermal Systems 210
  ICED 162, 342
  Japan/US symposia 359
  Mechanisms 163
  Robotics and Manufacturing
    381
  Winter Annual Meetings
    293, 317, 319, 381, 655
  World Congress on
    Machines . . . 162
and control systems 401
and design 162–3, 319, 339,
  342, 359, 651, 655
Design Automation Committee
  651
Design Engineering Division
  342, 359, 655
Design Theory and
  Methodology Committee
  651
and energy 210, 228, 236, 252
and environmental engineering
  492
and fluid mechanics 210, 287,
  290, 293, 296, 297
and Journal of Fluids
  Engineering 115
Journal of Manufacturing
  Systems 356, 378
and machines 152, 162–3
and manufacture 355, 356, 359
Manufacturing Engineering
  350, 356, 654, 655

and marine technology 475
and materials 309
and measurement 401
Ocean Engineering Section
  475
publications 297, 372
  Advances in Robotics 381
  Applied Mechanics Reviews
    102, 153, 228, 355
  ASME boiler and pressure
    vessel code 25, 206
  ASME steam tables 236
  electronic 115
  Graphic symbols for fluid
    power . . . 287
  Instructional aid for
    occupational safety and
    health . . . 325
  Mechanical Engineering 64,
    157, 296, 338
and reliability 38, 319
and risk analysis 317
and safety 317, 318, 319, 325
Standards 17, 22, **25–6**, **206**,
  645
and stress analysis 144
and thermodynamics **206**, 208,
  209, 317
Transactions **144**, 339, 409
  Journal of Applied
    Mechanics 144
  Journal of Basic
    Engineering 297, 405
  Journal of Biomechanical
    Engineering 642
  Journal of Dynamic
    Systems, Measurement
    and Control 292, 298,
    401
  Journal of Engineering for
    Gas Turbines and
    Power 208, 209
  Journal of Engineering for
    Industry 356
  Journal of Engineering
    Materials and
    Technology 144, 309
  Journal of Fluids
    Engineering **115**, 290
  Journal of Mechanical
    Design 157, 160, 163,
    168, 338
  Journal of Pressure Vessel
    Technology 144
  Journal of Solar Energy
    Engineering 244
  Journal of Tribology 174,
    175
  Journal of Vibration and
    Acoustics 492
and tribology 173, 174, 175
American Society for Metals
  (ASM) **313**
  databases 305, 314
  and EMA 355
  MAT.DB 314
  and METADEX 111, 355
  and Engineered Materials
    Abstracts 228, 229
  and Materials Information
    service 305, 309, 355

American Society for Metals
  (ASM) (contd)
  and Metals Abstracts 309, 355
  publications 125, 302, 371,
    374
  ASM handbooks 121, 125,
    125, 302, 312, 340
  ASM International 313
  ASM materials engineering
    dictionary [ed.Davies]
    123
  Engineered materials
    handbook 121, 312
  Guide to materials... data
    ... 121
  World-wide guide to
    equivalent irons and
    steels 26
  and safety 317
  Standards 26
American Society for Precision
  Engineering (ASPE)
  conferences 359
  Precision Engineering 357
American Society for Quality
  Control 453
  conferences 359–60
  Quality progress 358
American Society of Safety
  Engineers 327
  Directory of safety-related
    computer resources 328
  Professional Safety 328
American Society for Testing
  and Materials (ASTM) 16,
    25, 524
  CHETAH programme 324
  conferences 303
  and databases 304, 313
  Proceedings 303
  publications/Standards 302,
    304, 313
  Annual book of ASTM
    standards 25, 302, 314,
    340
  ASTM Standardization News
    25
  Five[/Fifty]-Year Index...
    102
  Special technical
    publications 321
American Society of
  Transportation and Logistics,
  Transportation Journal 499
American Standards Association,
  former (ASA) 25
American Statistical Association
  113, 453
American Water Works
  Association
  Handbook of community water
    supplies 489
  Journal... 489
American Wind Energy
  Association (AWEA)
  Standard performance testing
    ... 246
  Wind Energy Weekly 247
AMIRA see Australian Mineral
  Industries...
AMM (Ass'n of Mine Managers
  [of S. Africa]) 610

AMM Circulars 610
ammonia, thermodynamic
  properties 202–3
amperes 28
amplifiers 146, 286
  operational 341, 423
Amrine, H.T. et al. 365
AMS see Amer. Mathematical
  Soc.
Amstead, B.H. et al. 360
AMTRI (Advanced
  Manufacturing Technology
  Research Inst.) 371
anaesthesia 638–9
Anaesthesia 638
Anaesthesia and Intensive Care
  639
analogue signals 412, 423, 441–2
analysis
  methodologies 432–3
  see also numerical analysis;
    stress analysis and
    specific topics
Analysis of large-scale
  non-coal... mining...
  (pub. Dravo) 615
anatomy 634–5
Anchor, R.D. 567
Anderson, B.D.O. et al. [control
  systems] 396
Anderson, B.D.O. and Moore,
  J.B. [control systems] 392,
  394, 397
Anderson, H.H. [pumps] 208
Anderson, J.S. and Bratos, M.
  [noise] 492
Anderson, L.L. and Tillman,
  D.A. [fuels] 237, 241
Anderson, R.O. [petroleum] 239
Anderson, V. [energy] 233
Andreasen, M.M. et al.
  [concurrent engineering]
  337, 361
Andreasen, M.M. and Heim L.
  [concurrent engineering]
  650, 652
Andrew, W.G. et al. 279
anemometry, Doppler 298
Angeles, J. and Lopez-Cajun,
  C.S. 168
Anglo-American/De Beers [co.'s]
  595
Angrist, S.W. and Hepler, L.G.
  201
Angus, H.T. 539
animals, as machines 625
Annales de Cardiologie et
  d'Angéiologie 638
Annales de Chirurgie 639
Annales de l'Institut Technique
  du Bâtiment et Travaux
  Publics 532
Annales des Ponts et Chaussées
  532
Annals of Biomedical
  Engineering 642
Annals of the Institute of
  Statistical Mathematics 460
Annals of Nuclear Energy 251
Annals of Plastic Surgery 638
Annals of Surgery 639

Annual book of ASTM standards
  25, 302, 314, 340
Annual bulletin... for Europe
  series (pub. UN/ECE)
  ... coal statistics ... 237
  ... electric energy statistics
    ... 252
  ... general energy statistics
    ... 232
  ... transport statistics ... 510
Annual Energy Review (pub.
  USDOE) 232
Annual Instrumentation and
  Measurement Technology
  Conference (IMTC) 360
Annual Quality Congress 359–60
annual reports 82, 83, 93, 478,
  481
Annual Review of Energy 232
Annual Review of Energy and the
  Environment 232
Annual Review of Fluid
  Mechanics 105, 205
Annual Review of Materials
  Science 105
Annual Reviews of Industrial and
  Engineering Chemistry 105
Annual Technical Report: Energy
  Materials... (pub. USDOE)
  233
Annual vehicle census Great
  Britain (pub. DoT) 509
ANS see Amer. Nuclear Soc.
ANSI
  ANSI Reporter 25
  ANSI Standards Action 25
  see otherwise Amer. Nat.
    Standards Inst.
Ansys [program] 148
antennas 447, 448
Anthony, L.J. 121
anthropometrics 336
  journals 339
Aoki, M. 392
APAET see Associação
  Portuguesa de l'Análise...
APCOM see Application of
  Computers... Mineral
  Industry
API
  APIBIZ database 229, 230
  APILIT database 229, 230, 231
  APIPAT database 229, 231
  see otherwise Amer. Petroleum
    Inst.
APLS (advanced procurement
  and logistics systems) 311
Appearance matters (pub. BCA)
  537
Appliance 67
Appliance Manufacturer 67
Application of Computers... in
  the Mineral Industry
  (APCOM) conferences 601,
  612–13
Application of process
  integration to utilities...
  (pub. ESDU) 204
Applications in Manufacturing
  and Robotics conference 382
Applications of Mathematics 459

Applied Earth Science (IMM Transac.) 609
Applied Energy 209, 234
Applied Ergonomics 339, 640
Applied Mathematical Modelling 293
Applied Mechanics Reviews 102, 153, 228, 355
Applied Motion [program] 170tab
Applied Science and Technology Index (ASTI) 100, 102, 110, 380
  online 110, 400, 401, 402
Applied Statistics 460
Appraisal of existing structures (pub. IStructE) 551
Appraisal of sports grounds (pub. IStructE) 552
Approved Documents [construction] 549, 550, 555, 561
APTA see Amer. Public Transit Ass'n
AQUALINE database 485
Aquatechnic International 68
Arbiter, Nathaniel 601
archaeological aspects
  construction 534
  industrial archaeology 552
Archer, J.S. and Wall, C.G. [petroleum] 239
Archer, L.B. [innovation/design] 332
arches 560, 566
Architects' Journal (AJ) 553, **579**
  publications 564
Architectural Cladding Association 563
Architectural Periodicals Index 528
Architectural Review 553
Architectural Science Review 532
architecture 467, 547–68
  Bartlett School of Architecture 586
  books 533, 565–6
  handbooks 565
  databases 528
  'design-and-build' 548
  energy aspects 253–5
    solar energy 243–4
    wind energy 246
  history 534–5
  information co-ordination 578
  journals 528, 532, 553, 579, 580, 654
  naval see under marine technology
  organizations 521, 533
  safety 326–7
  and timber 542
  see otherwise structural engineering; construction
Architecture, Construction and Engineering Management **580**
ARCHITECTURE DATABASE 527, **528**
Archive for Rational Mechanics and Analysis 459
archives 93–4, 510–11, 552–3

National Register 553
Archives of Orthopaedic and Traumatic Surgery 638
Archives of Physical Medicine and Rehabilitation 642
Arden, M.E. et al. 243
Argonne National Laboratory [US] 251
Aris, R. 275
Arizona, University 596, 601
ARL see Ass'n of Research Libraries
Armed Services Technical Information Agency [US] 5
Armitage Report (1980): ... Lorries, people and the environmment 499
Armor, M. 208
arms see defence
Armstead, H.C.H. 245
Arnell, R.D. et al. 173
Arnold, L. 458
Aronson, E.J. 5
Arora, C.P. [cooling systems] 207
Arora, J.S. [design] 334
ARPI see Alphanumeric reports
ARRB (Australian Road Research Bd) **516**
Arrillaga, J. 432
Arrillaga, J. and Arnold, C.P. 432
Arrowsmith, D.K. and Place, C.M. 457
Arthritis and Rheumatism 638
artificial intelligence (AI) 280, 378, **396–7**
  abstracts/indexes 400
  and concurrent engineering 355, 653, 655–6
  conferences 382, 403, 655–6
  and control systems 390–1, 392, **396–7**
  databases 400
  and design 335, 339
  expert systems see expert systems
  journals 339, 356, **357**, 379, 397, 400, **402–3**, 654
  knowledge-representation 402–3
  machine vision 363, 377, 382, 384, 385, 386
  neural networks 390–1, 396, **402**
  organizations 402, 656
  pattern-recognition 390
  qualitative models **397**
  and robotics 378, 385
  and simulation 392
Artificial Intelligence 356, 400, 402
ARTIFICIAL INTELLIGENCE database 400
'Artificial intelligence ... ' [from IEE Proc.] 397
Artificial Intelligence in Engineering 357
Artificial Intelligence for Engineering Design, Analysis and Manufacturing (AIEDAM) 339, 654

artificial organs 631, 641
Artificial Organs 641
Arup, Ove, & Partners 556
  Research and Development **524**, 564
ASA (American Standards Ass'n, former) 25
Asai, K. and Takashima, S. 363, 386
ASAIO see Amer. Soc. for Artificial Internal Organs
ASC directory 86
ASCE
  ASCE Publications Abstracts 504
  see otherwise Amer. Soc. of Civil Engineers
Ascher, H. and Feingold, H. 178
Asfahl, C.R. 383
ASFPCM, publications 555, 556
ash 235, 238
Ash, N.I. 159
Ashby, M.F. 304, **308**, 337
Ashby, M.F. and Jones, D.R.H. 337
Ashford, F.C. 336
ASHRAE
  ASHRAE handbooks 207
  see otherwise Amer. Soc. of Heating ... Engineers
Ashurst, J. and Ashurst, N. 552
Asia
  construction 584
  electronics 418
  energy 232
  minerals 594
  productivity 350
Asian Productivity Organization 350
Asimov, Isaac 377
Asimow, M. 332
ASIS see Associazione Italiana per l'Analisi ...
Aslib (Ass'n for Information Management), publications 587, 588
  Directory of information sources in the UK 303
  Index to theses accepted ... 79
  Online Series 583
ASM/ASM International **313**
  ASM handbooks 121, **125**, **302**, 313, 340
  ASM materials engineering dictionary [ed. Davies] 123
  see otherwise Amer. Soc. of Metals
ASME
  ASME boiler and pressure vessel code 25, 206
  ASME steam tables [ed. Meyer] 236
  see otherwise Amer. Soc. of Mechanical Engineers
Asociación Española y Certificación 27
UNE 27
ASPE see Amer. Soc. for Precision Engineering
asphalt, roofing 564

assembly 354, 355, 357, 358, **361–2**, 369
automated 357, 359, 361, 369, 379, 385
design 337, 653
electronics 361
flexible 359
journals 379
part-feeding 369
welding/joining **360–1**
see also manufacture
Assembly Automation 357, 379
Assessment of reinforced... concrete bridges (pub. IStructE) 566
Assessment and repair of fire-damaged concrete structures (pub. CS) 555
Assistive Technology 642
Associaçáo Portuguesa de l'Análise Experimental de Tensoes (APAET) **141**
Associated Spring SPEC Ltd
Engineering guide to spring design 187
software 169
Association of American Railroads (AAR)
publications 324
Emergency handling of hazardous materials... 323
Association of British Plywood and Veneer Manufacturers 303
Association for Computing Machinery see ACM
Association of Consulting Engineers (ACE) 578
Consulting engineer's... yearbook 126
Association of Control and Automation Manufacturers 403
Association of County Councils **514**
Association of District Councils **514**
Association of Energy Engineers, *Energy Engineering* 234
Association Entropie, *Entropie* 234
Association française de normalisation (AFNOR) **23–4**, 524
Catalogue of English translations 23
Catalogue des normes françaises 23
Catalogue des ouvrages 22
Catalogue des recueils 24
Normes françaises 23
and PERINORM 22, 152, 582
Association Française de Robotique Industrielle (AFRI) 381, **388**
AFRI Liaison 378
Association for Information Management see Aslib
Association of Metropolitan Authorities **514**

Association of Metropolitan District Engineers **515**
Association of Mine Managers of South Africa (AMM) 610
Circulars **610**
Association of Nuclear Energy Producers (OPEN) see Organisation des Producteurs...
Association for the Prevention of Steam Boiler Explosions, former 21
Association of Public Lighting Engineers 513
Association of Researchers in Construction Management **586–7**
Association of Research Libraries (ARL), *Directory of electronic journals...* 114, 115, 116
associations see organizations
Association of Specialist Fire Protection Contractors... 540
Association Technique Maritime et Aéronautique [France] 475
Association of Transport Co-ordinating Officers 513
Associazione Italiana per l'Analisi delle Sollecitazioni (ASIS) **141**, 145
Associazione Italiana di Tecnica Navale 475
ASTI see *Applied Science and Technology Index*
Astia [Armed Services Technical Information Agency] Document (AD) 5, 11
ASTM
ASTM Standardization News 25
see otherwise Amer. Soc. for Testing and Materials
Astro II [program] 582
Astrom, K.J. 391, 392, 395
Astrom-Hagglund approaches 392–3
Astrom, K.J. *et al.* 396
Astrom, K.J. *and* Wittenmark, B. 393, 395
astronautics see aerospace engineering
asynchronous transfer mode (ATM) 444
AT & T (American Telephone and Telegraph Company), *AT & T Technical Journal* 417
Atherton, D.P. *and* Borne, P. 454, 455
Atkins, W.S. [co.] **524**
Atlas Copco, *Atlas Copco Manual* 614, 615
Atlas Powder Co., *Explosives and rock blasting* 614
ATM (asynchronous transfer mode) 444
atmosphere
databases 229

flammable 320, 321
pollution **490–2**
psychrometry 253–5
thermodynamics 243–5
see also air
Atmospheric Environment 491
Atom 266
Atomic Energy Authority (AEA [UK]) 11, 226
AEA [Industrial] Technology 177, **179**, **262**, 266, 595
Condition Monitoring and Sound Assessment Club 179
National Centre of Tribology 177
publications 347
Atom 266
Chernobyl accident... [ed. Gittus *et al.*] 266
see also Harwell
Atomic Energy of Canada Ltd (AECL) **263**
Atomic Energy Clearing House 251
Atomic Energy Commission (AEC [Japan]) **263**
Atomic Energy Society of Japan (AESJ) **265**
Atomic Weapons Research Establishment [UK] see under Harwell
atoms
atomic energy see nuclear energy
atomic weights 203
fission process 259–60
fusion process 261
Atoms in Japan 266
atria 563, 564
Attia, Y.A. 237
audiology 639
audio transmission, Internet 503
audits
design 179, 333
energy 254, 257
quality 178, 179
safety 325
Audubon House: building the... energy-efficient house (pub. Nat. Audubon Soc.) 254
Auger, C.P. 14, 61, 121, 122
Augmentative and Alternative Communication 642
Aureille, M. 245
AusIMM
AusIMM Bulletin 596, 610
see otherwise Australasian Inst. of Mining...
Austin, L.G. *et al.* 598
Australasia, electronics 418
Australasian Institute of Mining and Metallurgy (AusIMM) **594**, 601, 610, 613
AusIMM Bulletin 596, 610
Proceedings 610
Australasian Physical and Engineering Sciences in Medicine 642
Australia
Bureau of Mineral Resources... 612

Index 669

Australia (contd)
  construction 532
  earth sciences 602
  energy 232, 245
  electricity 427
  Institution of Engineers 173, 614
  journals 73
    electronic newsletter 115
  medicine 639, 642
  minerals/mining 593–4, 595, 596, 597, 601, 602, 610, 611, 612, 613, 614
  newspapers 75
  patents 45
  Perth, Curtin University 115
  process engineering 73
  Queensland University 595
  research 594
  Standards 113, **524**
  transport 516
  tribology 173
AUSTRALIAN EARTH SCIENCES INFORMATION DATABASE 602
Australia and New Zealand Journal of Medicine 639
Australian Mineral Foundation 602
  Drilling and blasting ... 614, 618
Australian Mineral Industries Research Association (AMIRA) 595
Australian mineral industry annual review (pub. Aus. Bureau of Mineral Resources) 612
Australian Mining Monthly 597
Australian and New Zealand Solar Energy Society 245
Australian Road Research Board (ARRB) 516
Australia's Mining Monthly 611
Austria 143
  Leoben mining school 595
  minerals 595
  patents 45, 46
  Standards 22
  stress analysis 143, 144–5
  transport 501
autoclaves 205
Autofact '92 conference 382
automated guided vehicles (AGVs) 363, 385, 386, 470tab
Automated Imaging Association 382
Automatica 400, 403, 407, 408
Automatic Control and Computer Sciences 400
automation 353–4, 360, 366, 369, **377–88**
  abstracts/indexes **380**, 399–403
  assembly 357, 359, 361, 369, 379, 385
  automotive industry 377, 378, 379
  books **382–6**
  chemical engineering 279, 280
  and competitiveness 383–4
  and concurrent engineering 653

  conferences 291, 359, **381–2**, 401
  construction 378, 382, 580
  databases **380**, 399–403
  design 383, 401, 653
  economic aspects 386
  ergonomics and 384
  flexible systems see flexible manufacturing
  hydraulics 285
  journals 357, **378–9**, **400–1**, 580
  organizations **380–2**, **387–8**, 403, 650–1
  pneumatics 291
  and quality 366
  robotics **377–88**
  safety 279, 325
  Standards **383**, 420
  teleoperation 383
  test equipment 379
  updating 385
  vehicles (AGVs) 363, 385, 386
  see also control; robotics
Automation in Construction 580
Automation and Control 400
Automation and Remote Control 400
Automation and Robotics in Construction conference 382
Automobile Association (AA) **514**
automobiles see automotive engineering; cars etc.
Automobile series [Standards (pub. BSI)] 21
automotive engineering
  abstracts 228
  automation 377, 378, 379
  batteries 252
  and concurrent engineering 653
  conferences 501
  design 179, 338, 341
  journals 68, 104, 144, 252, 338, 379
  organizations 20, 24, 25
  quality 367
  robotics 377, 388
  Standards 16, 20, 24–5, **26**
  BSI Automobile series 21
  SAE handbook 26, 208
  statistics 509–11
  vehicles see vehicles
  see also transport; machines etc.
Automotive handbook [Adler] 16
auto testing 379
Avallone, E.E. and Baumeister, T. III, Mark's standard handbook ... 160
Avery, W.H. and Wu Chi 247
aviation see aerospace
Aviation, Space and Environmental Medicine 639
Aviation Week and Space Technology 68
Avitzur, B. 364
AWEA see Amer. Wind Energy Ass'n

B
BAC (Business Archives Cncl) 94
Bacha, J.D. et al. 239
Backe, W. and Hoffmann, W. 296
Backhurst, J.R. and Harker, J.H. 271–2
Bacon, D.H. [thermodynamics] 203, 205
Bacon, J. [concurrent systems] 398
Bader, D.L. [decubitus] 637
Badin, E.J. [coal] 238
Badr, O. et al. [thermodynamics] 203
BAIE (British Ass'n of Industrial Editors) 83
Bailey, J.E. and Ollis, D.F. [biochemical engineering] 278
Bailey, R.L. [creativity] 332
Bain, R.W. 235
Baird, J.A. and Ozelton, E.C. 542, 561
Bajpai, A.C. et al. 455
Baker, A.C. [tidal power] 247
Baker, A.K. [brakes/clutches] 164
Baker, W.E. and Tang, M.J. [explosions] 321
Bakerjian, R. 124, **368**, 651–2
balanced bridge circuits 135
balancing 161tab, 166, 167, **168–9**
Balcombe, G. 535
Ball Bearing Journal 176
Ball and Roller Bearing Engineering 176
Ball and Roller Bearing Manufacturers Association **177**
Balmer, D.W. and Paul, R.J. 392
Bamford, W.H. et al. 318, 325
Banerjee, P.K. and Butterfield, R. 455
Banerjee, P.K. and Marino, L. 455
Banerjee, P.K. and Mukherjee, S. 455
Banerjee, P.K. and Watson, J.O. 455
Banerjee, P.K. and Wilson, R.B. 455
Banieghbal, M.R. 297
Banister, M. 255
Banks, D.D. and Banks, D.S. 287
Bansal, N.K. 253
Barber, A. [pneumatics] 285
Barber, A. [noise/vibration] 182, 492
Barber, M.J. [oil hydraulics] 288
Barbery, G. [minerals] 591
Barbour index (microform) **181**, **528**, **529**
  Building Product Microfile **528**, **529**
  Building Technical Microfile **528**, **529**
  Engineering Product Microfile **528**, **529**

Barbour index (microform) (contd)
 Engineering Technical Microfile 528, **529**
 Environmental Health, Fire and Safety Microfile **181**, 183, 334
bar coding 385
Bardeen, John 411
Bardeen, J[ohn] and Brattain, W[alter] E. 423
Barents Sea 482
Baril, R. [*manufacture*] 363
Barin, I. [*thermochemistry*] 272
barium, *as* fission product 259
Barker, T.B. [*Taguchi/design*] 179, 334
Barkman, W.E. [*in-process quality*] 367
Barltrop N.D.P. *et al*. [*wind energy*] 246
Barltrop N.D.P. and Adams, A.J. [*marine engineering*] 473
Barnard, J.A. and Bradley, J.N. 235
Barnett, S. and Cameron, R.G. 456
Barry, T.I. and Reynard, K.W. 313
bars 137
Bartholomae, R.C. *et al*. 614
Bartknecht, W. 321
Bartlett International Summer School 586
Barton, J. and Rogers, R. 323
Bartz, W. 164, 165
Barwell, F.T. 175
BASF [org'n] 305
Basic road statistics (pub. BRF) 510
Bass, H.G. [*safety/electricity*] 320
Bass, L. [*product liability*] 318
Bassani, R. and Piccigallo, B. 175
Bastian, G.F. 635
Bastress, E.K. 235
Basu, P. 235
batch control/processing 279, 280, **432**
Bate, S.C.C. [*concrete*] 557
Bateman, G. [*magnetohydrodynamics*] 251
Bates, W. [*steelwork*] 559
Bath, University
 Building Centre Trust 587
 Centre for Window and Cladding Technology 563, 564
 Design Group 344
 Fluid Power Workshops 297
 Information and Data Services *see* BIDS
 Journal of Transport Economics and Policy 498
Bathe, K.-J. 455
BATIC database 584
BATIMENT database 584
batteries, vehicles 252
Baughman, G.L. 239, 241

Bauingenieur 532
BAU LITDOC database 584
Baumeister, T., *Mark's standard handbook* 124
Bautechnik 532
Bautista, E. *et al*. 162
Baxter, B. 473
Bayer AG [co.] 313
Baylet, V. *et al*. [*fluid power*] 297
Bayley, S. [*design/taste*] 336
BCA *see* British Cement Ass'n
BCIRA *see* British Cast Iron Research Ass'n
BCIS [Building Cost Information Service] database **583**
BCQ [Building Centre/Quantarc] databases 96
BCQ-CONSTRUCTION INDUSTRY INFORMATION SYSTEM 95
BCQ/ODI [(On Demand Information] **528, 582**
BCSA *see* British Constructional Steelwork Ass'n
BDA *see* Brick Development Ass'n
beams, structural 137, 556, 558
Beano [comic paper] 34
Bear, D. 443
bearings 154, **163**, 174, **175–7**
 books 161*tab*, 173, 176, 341
 flexure-pivots 177
 gas 177
 journals 176
 magnetic 177
 organizations 177
 software 161*tab*, 173, **175–6**
Beasy [program] 149
Beauchamp, K.G. 446
Beaumariage, K. and Shunk, D. 652
BEC *see* Building Employers Confedn
BECAN abstracting service 644
Becker, A.A. 149, 455
Beckmann, P. 551
Beer, F.B. and Johnson, E.R. [*vectors/kinematics*] 158
Beer, J.M. and Chigier, N.A. [*combustion*] 235
Belfast, RIBA bookshop 533
Belgium, fluid mechanics 297
Bell System Technical Journal, former 417
Belter, P.A. *et al*. 278
belts 161*tab*, **164**, 173, 341
BEM *see* boundary element methods
Benchmark 139
benchmarks 139, 433
Benedict, M. *et al*. 265
beneficiation 590
Benford, H. 476
Benham, P.P. and Crawford, R.J. 141
Benn Business Information Services [co.] 72, 85
Benn's media 72
Building and construction index 543
Process engineering directory 86

Bennet, D.J. and Thomson, J.R. [*nuclear power*] 250, 265
Bennett, G. [*air pollution*] 491
Bennett, G.F. *et al*. [*hazardous substances*] 324
Bennett, P. [*safety/computers*] 326
Bennett, R.D. *et al*. [*mining*] 615
Bennett, S. and Virk, G.S. [*control systems*] 393
Benson, R.S. 203, 235
Benson, R.S. and Whitehouse, N.D. 235
Bentley, J.P. 423
Benuzzi, A. and Zaldivar, J.M. 324
Berghmans, J. 208
Berk, A.A. 335
Berkowitz, N. 238
Berliand, G.T. 243
Bermacsek, G.M. 247, 255
Berman, E.R. 245
Berne, R.M. *and* Levy, M.N. 635
Bernoulli equations 291
Berridge, G.L.C. 208
Bertain, L. 363
BES [1] *see* Biological Engineering Soc.
BES [2] *see* Biomedical Engineering Soc.
Besant, C.B. and Lui, C.W.K. [*CAD/CAM*] 335, 362
Bessant, J. *and* Lamming, R., *Macmillan dictionary of production*... 123
BEST database 152
Best's loss control engineering manual 316
Beton Armé 532
Beton und Stahlbetonbau 532
Bettess, P. 455
Beuth Verlag [publishers)] 24
Bever, M.B., *Encyclopedia of materials science*... 122
Beyer, R. [*kinematics*] 185
Beyer, W.H. [*mathematics*] 454
Bezdek, J.C. 457
BFPA *see* British Fluid Power Ass'n
BGMA (Brit. Gear Manufacturers Ass'n) 154
BGR *see* Deutsches Bundesanstalt für Geowissenschaften...
BGS *see* Brit. Geological Survey
Bhappu, R. *and* Harden, R. 603
BHP (Broken Hill Pty) 595
BHRA *see* Brit. Hydromechanics Research Ass'n
BHR Group [co.] **101**, 112, **291**
Bhusan, B. *and* Gupta, B.K. 176
bibliogaphic databases *see* databases
bibliographic details, defined 98
Bibliographic Index 104
Bibliographie Internationale 101, 227, 230, 584
online 101, 231
Bibliographie Nationale Française 103
bibliographies 99, **103–4**

bibliographies (contd)
  in journals, author-supplied 64
  library-catalogues 103
  national 103
  nature and use 99
  online 103
  Pascal 101
  specialized 104
  standard guides **121–2**
*Bibliography and Index of
  Geology* 228
BIDS (Bath Information and
  Data Services) 77, **108–9**,
  505
BIDS database **354**, 584
Bies, D.A. and Hansen, C.H.
  181, 492
Biggs, N.L. 457
Billett, M. 175
Billinton, R. 428
Billinton, R. and Allen R.N. 428
binary notation, digital signals
  442
bins 278
biochemical engineering 269,
  271, **278–9**, 282–3
  control systems 391
  journals 281–2, 488, 490
  see also biotechnology *etc.*
bioenergy *see* biomass
bioengineering **625–6**
  see otherwise biomedical
    engineering
Bioengineering Abstracts 100
*Bioengineering and
  Biotechnology Abstracts*,
  former 100
Biofeedback and Self Regulation
  642
biofuels *see* biomass
biographical data **126–7**
*Biographical dictionary of
  scientists* [ed. Abbott] 126–7
BIOLOGICAL ABSTRACTS 486
biological engineering 629–30
Biological Engineering Society
  (BES) 629–30, **631**
  *Medical Engineering and
    Physics* 630
biology 278–9, **625–6**, **633–5**
  databases 111, 486
  and environmental engineering
    486, 487, 488, 489, 491
  evolution 634
  journals 639, **640–1**, 642
  and medicine 626, 629–30
  microbiology 486
  and vitalism 625
  and water treatment 489
biomass, energy from 210, 225,
  241, **249–50**
  IEA Technical Information
    Service 225
Biomass and Bioenergy 249
Biomass Bulletin 249
Biomass Energy Directory 249
biomaterials 631, 635, **636–7**,
  638, 639, 641–2
Biomaterials 641
biomechanics 133, **625**, 634–6
  journals 641–2

organizations 629, 631
and vitalism 625
*Biomechanics in sport (pub.
  IMechE))* 646
biomedical engineering (BME)
  623–48
  nature and role **623–7**, 635,
    636, 645–6
  applications 626–7, 628*fig*,
    638
  abstracts/indexes 100, 102,
    112, 486, **643–4**
  biology **625–6**
  biomaterials 631, 635, **636–7**,
    638, 639, 641–2
  books **632–7**
  dictionaries 644
  conferences 628, 629, 630, 644
  databases 486, **644**
  journals 112, 629–31, **637–43**
  organizations **627–32**
  rehabilitation 626, 627, 628*fig*,
    630, 635, 642–3
  Standards **645**
  see also biomechanics
Biomedical Engineering Society
  (BES [US]) **631**
*Biomedical Instrumentation and
  Technology* 642
*Biomedical Sciences
  Instrumentation* 642
Biomedizinische Technik 642
bioproducts, journals 281
Bioresource Technology 249
*Biotechnologie-Verfahren,
  Anlagen, Apparate, former*
  283
biotechnology **625–6**
  abstracts 112, 283, 400
  biomass energy **249–50**
  chemical engineering and 112,
    269, 625
  databases 112, 283
  environmental management
    492
  journals 281–2, 283
  see otherwise biomedical *etc.*
*Biotechnology Apparatus, Plant
  and Equipment* 283
Biotechnology Progress 281
BIPM *see* Bureau Internat. des
  Poids et Mesures
Birley, A.W. *et al.* 307
Birmingham
  RIBA bookshop 533
  University, chemical/mineral
    engineering 595
B-ISDN (integrated services
  digital network: broadband)
  442, **445**
Bishop, D.N. 279
Biswas, A.K. *and* Agarwala,
  S.B.C. 493
Bitnet [network] 584
Bittencourt, J.A. 250
bitumen, cladding/roofing 564
BL *see* Brit. Library
Black, J. [*biomechanics*] 646
Black, J. *and* Dumbleton, J.H.
  [*biomechanics*] 636
Black, L. [*building*] 533

Black, R.M. [*design*] 337
Black & Decker [co.] 366
'Black book' [1] [*Metals black
  book* (ed. Bringas)] 312
'Black book' [2] [steel structure
  specifications (*pub.* BCSA)]
  539
Blackburn, J.O. 209
Blackwell Publishers 239, 527,
  531
Blain, W.R. 567
BLAISE [Brit. Library Automated
  Information Service]
  database 110, 226
BLAISE-LINE database host 10,
  77, 78
Blake, L.S. 125, 533
Blakstad, M. 334
Blanc, A. *et al.* 539, 558
Blasters handbook (pub. Du
  Pont) 614
blasting **614**, 615
Blasting practice (pub. Nobel)
  614
BLDSC *see* Brit. Library:
  Document Supply Centre
Bleazard, G.B. 446
Blix, H. 250
BLLD (Brit. Library Lending
  Division) 8
Bloch, H.P. 274
Bloch, H.P. *and* Geitner, F.K.
  179
Blockley, D.I. 315, 326
BLSRIS *see* Brit. Library:
  Science Reference and
  Information . . .
'Blue book' [1] [electrical/
  electronics directory (*pub.*
  Peregrinus] **418**
'Blue book' [2] [steel structure
  erecting (*pub.* BCSA)] 539
'Blue book' [3]
  [telecommunications
  Standards (*pub.* CCITT)]
  449
Blums, E. *et al.* 251
Bly, J.H. 364
BMFT [German Fed. Min. of
  Research and Technology]
  355
BMHB *see* Brit. Materials
  Handling Bd
BMT (Brit. Maritime
  Technology) 478
BMT Abstracts 480
BNB (*Brit. Nat. Bibliography*)
  103
BNES *see* Brit. Nuclear Energy
  Soc.
BNF (Brit. Nuclear Forum) 263
BNFL (Brit. Nuclear Fuels plc)
  262
Bodsworth, C. *and* Appleton,
  A.S. 202
Bodurtha, F.T. 321
Boiler operator's handbook (pub.
  NIFES) 206, 236
boilers 205, **206**, **235–7**, 277
  controls 279
  and electricity generation 252,
    428–9

boilers (contd)
  fluidized beds 235–7
  journals 237
  nuclear reactors 260
  safety 21, 327
  Standards 21, 22, **25–6, 206**, 235
  steam generators 260
Boilers and pressure vessels ... survey (pub. BSI) 235
boiling 204, 206
boiling water reactors (BWRs) 260, 261–2
Bolin, B. *et al.* 255
Bollinger, G.A. 614
Bolz, R.E. *and* Tuve, G.L. 124, 232
bomb disposal 378
Bond, F.C. [*minerals*] 600
Bond, J. [*fire hazard*] 321
bonding 362
bone [materials] 308
Bone [clinical] 636, 638
books 103–4, **121–32**, 151–2, **231–5**, 435, **470–3, 632**
  abstracts 103, 111, 112
  bibliographies **103–4**
  books-in-print lists 104
  catalogue problems 269–70
  content/title 316
  disks accompanying *see under* disks: floppy
  obsolescence denied 452
  online 110, 281, 323, 354
  out-of-print 535, 552
  publishers, selected 422–3, 527, **531**
  reprints 552, 561
  reviews/notices 65, 104
  and theses 78
  as 'windows' 632
Books on Demand service 531
Bookseller 104
Books in print [UK; *pub.* Whitaker] 104, 224
Books in print [US] 104
Boon, P.J. *et al.* 487
Boos, B. 255
Booth, E. [*structures*] 554
Booth, J., *IEEE standard dictionary* ... 123
Booth, K.M. [*cooling systems*] 207
Boothroyd, G. 361
Boothroyd, G. *et al.* 361
Boothroyd, G. *and* Dewhurst, P. 337, 653
Boothroyd, G. *and* Knight, W.A. 363
BOQ 582
Borelli, G.A. 625, **633**
Borer, J. 279
boron
  in magnetic materials 435
  as neutron-absorber 260
Bose, B.K. 436
Bossard, F. 617
Boston Spa
  BOSTON SPA CONFERENCES ON CD-ROM 110
  see otherwise British Library: Document Supply

Bothe, H.-H. 457
Bott, T.R. 277
Bottema, O. *and* Roth, B. 158
Boudart, M. 275
boundary element methods/analysis (BEM) 134, **137**, 149, 451, **455**, 457, **458**, 459, 616
Bouska, V. 238
Boutwood, J. 561
Bowden, F.P. *and* Tabor, D. 173
Bowker A&I [co.] ROBOTICS 380
Bowker, R.P.G. *et al.* 492
Bowker, R.R. [publishers], *Ulrich's periodicals directory* 73, 74, 99, 227, 400
Bowker-Saur [publishers] *and Current Technology Index* 380
  online 529
  Guides to Information Sources 122
  Information sources in metallic materials 303
  World databases in patents 57
Bowyer, J. 535
Boxer, G. 202
Boxho, J. *et al.* 617
Boyen, J.L. 206
Boyle, J.M. *and* Maciejowski, J.M. 397
Boylston, R. 320
Bozic, S.M. 392
BP (Brit. Petroleum Co. plc) 210
BP Statistical Review of World Energy 232
District Heating and ... Developments 209
Horizon 481
BRA *see* British Robot Ass'n
Bradfield, V.J. 587
Bradley, D.A. 172
Brady, B.H.G. *and* Brown, E.T. [*rock mechanics*] 616
Brady, G.S. *and* Clauser, H.R. [*Materials handbook*] 122
brakes 161tab, 164, 173, 177
Bralla, J.D. 337, 340, 653
Brandes, E.A. *and* Brook, G.B., *Smithells metals reference book* 125, 302
brand names 88
  see also trademarks
Branover, H. 251
Brattain, Walter E. 411
Brauer, R.[L.] 315, 328
Braune, W. *and* Fischer, O. 633
Braunkohle 239
Bray, R.N. *and* Tatham, P.R.B. 560
Brazil, minerals 596
brazing 362
BRE
  BRE Digests 207, 207, 208, **521**, 530, 537, 541, 543, 551, 554, 556, 563
  BRE Information directory **521**
  BRE Newsletter **521**
  BRE Reports **521**, 528, 529, 554, 555, 557

see otherwise Building Research Establishment
breakwaters 567
Brebbia, C.A. 455, 458
Brebbia, C.A. *et al.* 551
Brebbia, C.A. *and* Chaudouet-Miranda, A. 455
Brereton, C. 552
Bretherick, L., *Bretherick's handbook of reactive chemical hazards* 323
Brewster, R.L. 443, 444, 445, 446
BRF *see* Brit. Road Fed'n
BRGM *see* Bureau de Recherches Géologiques ...
Brichta, A.M. *see under* Gobel
Brick Development Association (BDA) 527, 528, **540**, 559
  ... Guide to successful brickwork 541
  publications 560
bricks/brickwork **540–1, 559–60**
  organizations 95, 559
  Standards 555, 559–60
  stress analysis 137
  see also masonry
Bridger, R.S. 336
bridges 519, 545, 547, **566–7**
  design/Standards 500, **508**, 545, 566
  organizations 522–3
  seismic loading 554
bridge structures
  hydraulic systems 292
  welding 292
Bridgwater, A.V. 249
Bridgwater, A.V. *and* Grassi, G. 249
Briggs, R.A. 20
Brin, A. 248
Bringas, E.A. 312
Bristol, Building Centre 520
Britain *see* United Kingdom
British Adhesives and Sealants Association **542**
British Architectural Library 528, 552, **553**
British Association of Industrial Editors (BAIE) 83
British Association of Occupational Therapists 631
British Board of Agrément 528
British Cast Iron Research Association (BCIRA) **538**, 559
British Cement Association (BCA) 528, **536**, 556, 559
CONQUEST database 528
Concrete Information Service 537
Concrete mixes for general purposes 537
publications **536**, **537**, 556
Appearance matters 537
Catalogue 527
Concreting in hot weather 537
British Ceramic Research Ltd, *former* 540, 559

Index 673

British Chartered Institution of
 Building Services Engineers,
 conferences 206
British Coal 54, 226
 privatized 594
British Commonwealth *see*
 Commonwealth
British Compressed Air Society
 208
British Computer Society 319,
 413, **416**
British Constructional Steelwork
 Association (BCSA) **538**,
 554, 558, 562
 publications **538, 539**, 558,
 559
 New Steel Construction 532
 'Red/green *etc.* books' 539
British Cryogenics Council,
 Cryogenics safety manual
 327
British Electricity International
 [co.] 251–2, 429
British exports 92
British Flat Roofing Council 564
British Fluid Power Association
 (BFPA) **287, 296**
 *Fluid power engineers data
 book* 285, 287
British Gas [co.] 226, 238
 *British Gas directory of
 energy-saving equipment*
 (*pub.* Energy Information
 Centre) 255
British Gas Scotland [co.],
 Gaslife 69
British Gear Association 164,
 165
British Gear Manufacturers
 Association (BGMA) 154,
 164
British Geological Survey (BGS)
 *Directory of mines and
 quarries* 602
 *United Kingdom minerals
 yearbook* 612
 World mineral statistics 612
British Glass Manufacturers
 Confederation 562
British Heart Journal 638
British Hydraulics Research
 Association (BHRA) 601
British Hydromechanics Research
 Association (BHRA) 101,
 210, 297, 298, 601
 BHR Group Ltd **101**, 112, **291**
 conferences 208, **291**, 296
 *Large scale applications of
 heat pumps* 208
British Institute of Management
 210
 Energy Managers' Workshops
 205
British Journal of Anaesthesia
 639
British Journal of Audiology 639
British Journal of Neurosurgery
 639
*British Journal of
 Non-Destructive Testing* 358
*British Journal of Occupational
 Therapy* 631, 642

*British Journal of Plastic
 Surgery* 639
British Journal of Rheumatology
 638
British Journal of Urology 639
British Library (BL)
 Automated Information Service
 see BLAISE
 conference *Proceedings* 77,
 110, 530
 Document Supply Centre
 (BLDSC) 8, **142, 152**
 BOSTON SPA CONFERENCES ON
 CD-ROM 110
 *British National
 Bibliography* (*BNB*)
 103
 British Thesis Service 78,
 142, 530
 BRITS 530
 conferences 77, 110, 142
 Focus Bulletins 142
 grey literature 10, 78
 newsletters 70
 reports 5, 7, 8, **9**, 11, 142,
 226
 Subject Search 142
 theses 9, 78, 142, 530
 translations 9
 and energy 224, **226**
 Lending Division (BLLD),
 former 8
 machines 152
 Patent Information Network
 60–1
 publications
 *Alphanumeric reports series
 index (ARPI)* 5
 *British National
 Bibliography* (*BNB*)
 103
 *British Reports, Translations
 and Theses (BRTT)* **9**,
 78, 530
 *Conference proceedings
 index* 77, 110, 530
 Guide to directories ... 125
 *Inventory of abstracting and
 indexing* ... 227
 *Scientific abstracting and
 indexing* ... *in the
 British Library* 227
 Trademarks ... *guide* [ed.
 Newton] 121, 130
 Science Reference and
 Information Service
 ([BL]SRIS) 93–4, 152,
 153
 abstracts 153
 Business Information
 Service 83, 93, **94**
 databases 153
 directories 125
 and energy 224, **226**
 house journals 83
 Japanese Information
 Service 102
 obsolete publications 535,
 552
 patents information 60–1,
 341

 product information **93–4**,
 95, 535
 trade journals 67–8
 trade literature 93–4, 95
 trademarks 121
British Maritime Technology Ltd
 (BMT) 478
 BMT Abstracts 480
British Masonry Society 530,
 540, 559
British Materials Handling Board
 (BMHB), *Draft code* ...
 for ... *silos, bins, bunkers
 and hoppers* 278
British Medical Journal 639
*British National Bibliography
 (BNB)* 103
British Nuclear Energy Society
 (BNES) **265**
 Nuclear Energy 251, 266
British Nuclear Forum (BNF)
 263
British Nuclear Fuels plc (BNFL)
 262
British Parking Association **514**
British Petroleum Co. plc *see* BP
British Plastics Federation
 *Guidelines for the safe
 production of phenolic
 resins* 325
 Thermosetting Material Group
 325
British Precast Concrete
 Federation **536**, **563**
British Quality Foundation **371**
British Rail [org'n], privatization
 502, **512**
*British Reports, Translations and
 Theses (BRTT [pub. BL])* **9**,
 78, 530
 online 78
British Road Federation (BRF)
 514
 Basic road statistics 510
British Robot Association (BRA)
 381, **388**
 Robot facts 382
British Safety Council **183**
 Safety Management 328
British Scientific Instrument
 Research Association,
 former 415
British Society for Strain
 Measurement (BSSM) **137**,
 138, **144**
 *BSSM strain measurement
 reference book* 141
 Strain **138**, **142**, 145
 BRITISH STANDARDS database 90
British Standards Institution
 (BSI) **21–2**, 89, **153**, 210,
 342, **524**
 and BSS 29
 codes of practice *see* codes of
 practice
 and international Standards
 21–2, 524
 and PERINORM 22, 152, 582
 principles (*BS 0*) 152
 publications **21–2**, 182
 Automobile series 21

674 Index

British Standards Institution
  (BSI)
publications (contd)
  Boilers and pressure
    vessels... survey 235
  BS 4940: ...
    Recommendations
    for... technical
    information... 81, 82
  'BSI bestsellers' 22
  BSI handbooks 21, 368
  BSI News 152, 157, 524
  BSI standards catalogue 21,
    22, 152, 154, 340, 367,
    527
  BSI Update 152
  Codes of Practice 21, 182,
    528
    on database 528
  Drafts for Development
    (DD) series 177
  external Standards 21–2
  General Series (GS) 21, 182
  glossaries 123, 177, 533
  guidance documents 182
  Legislation (L) 182
  Management of design...
    337
  Manual of British Standards
    in engineering drawing
    and design [ed. Parker]
    337
  Manual of British Standards
    in engineering
    metrology [ed. Brooker]
    368
  Marine series 21
  Microfiche index to British
    standards 22
  Plant and Machinery (PM)
    182
  Quality assurance handbook
    368
  Quality management
    handbook 177–8
  special series 21
  Sales and Accounts 153
  STANDARDLINE database 530
  see otherwise Standards
British Standards Society (BSS)
  29
British standards yearbook,
  former 21
British Steel [Corporation] (BSC)
  538, 558, 559
  Structural Steel Advisory
    Service 524–5
British Technology Index, former
  341, 529
British Telecom [co.] (BT), BT
  Technology Journal 417,
  449
British Telecommunications
  Engineering 414
British Thesis Service see under
  British Library: Document
  Supply
British Timken [co.] 176
British Wind Energy Association
  (BWEA) 242, 246
  Wind turbines... 246

BRITS (British Thesis Service)
  database 530
brittle lacquer stress analysis 136,
  147
brittle materials 359
BRIX/FLAIR databases 230, 530,
  583
broadband communications 442,
  444, 445
Broadbent, G. 332
broadcasting
  information from 497–8
  International Convention 417
brochures see trade literature
Brodowicz, K. and Dyakowski,
  T. 277
bronze 589, 607
Brooker, K., Manual of British
  Standards in engineering
  metrology 368
Brookes, A.J. and Grech, C.
  [structures] 564, 566
Brookes, L.G. and Motamen, H.
  [nuclear energy] 250
Brown, F.T. [fluid power] 297
Brown, G.C. and Skipsey, E.
  [energy] 233
Brown, M.E. [thermal analysis]
  205
Brown, R.G. and Hwang, P.Y.C.
  [control systems] 392
Brown & Root Ltd 94
Browning, R.L. 317
browsers, Internet 118
BRS/BRS ONLINE database
  host, former 107, 108, 422,
  584
  After Dark service 107
BRTT see Brit. Reports...
Bruce, M. 334
Bruges, E.A. 204
Brundle, C.R. et al. 122
Bruton, M.J. 506
Bryers, R.W. 235
BS (Brit. Standard) 16, 21
BSC see Brit. Steel [Corp'n]
BSI
  'BSI best sellers' 22
  'BSI bestsellers' 22
  BSI handbooks 21, 368
  BSI News 152, 157, 524
  BSI standards catalogue 21,
    22, 152, 154, 340, 367,
    527
  BSI Update 152
  see otherwise Brit. Standards
  Inst'n
BSRIA see Building Services
  Research and Information
  Ass'n
BSS (Brit. Standards Soc.) 29
BSSM
  BSSM strain measurement
    reference book [ed. Pople]
    141
  see otherwise Brit. Soc. for
    Strain Measurement
BT (British Telecom) [co.], BT
  Technology Journal 417,
  449
bubble-columns 275

BUBL bulletin board 584
Buchanan, A.H. [fire
  engineering] 555
Buchanan, Prof. Sir Colin,
  Report (1963): Traffic in
  towns 499
Buchanan, W. [mining
  hydraulics] 614
Buchhändlervereinigung [org'n]
  103
Buckley, J.D and Stein, B.A.
  [assembly] 362
Buckley, P.S. et al. [distillation/
  controls] 279
Builder, The, former 553
Building 531, 577, 580
Building Advisory Service 543
Building Bookshop 520, 533
Building Britain 2001 (pub.
  CSSC) 575, 587
Building Centre [London] 88, 95,
  520
Building Bookshop 520, 533
Centre for Construction Market
  Information 520, 575
databases 95, 520
see also BCQ
European Construction Centre
  95, 520
Information Exchange 95, 520
training material 521
Trust 520, 587
Building Centres, regional 520
Building Centre Trust 520, 587
Building Commodity File see
  BVR
Building and construction index
  (pub. Benn) 543
Building Cost Information
  Service see BCIS
Building Employers
  Confederation (BEC) 543,
  576, 578, 579
Building Energy Management
  Systems Centre 403
Building IT 2000 (pub. CERCI)
  580
Building Management South and
  West Ltd 94
BUILDING ONLINE database 583
Building Product Microfile 528,
  529
Building Regulations 95, 549,
  550, 583
  Approved Documents 549,
    550, 555, 561
Building Research Establishment
  (BRE) 210, 520–1, 549–50,
  585
  BRIX 230, 530, 583
  and concrete 537, 557
  and conferences 585
  and conservation issues 521
  databases 520
  library 526
  and masonry 541
  publications 254, 521, 549–50,
    557, 559, 560, 564
  BRE Digests 207, 208, 521,
    530, 537, 541, 543,
    551, 554, 556, 563

Index 675

Building Research Establishment (BRE)
  publications (contd)
    BRE Information directory **521**
    BRE Newsletter **521**
    BRE Reports **521**, 528, 529, 554, 555, 557
    catalogues 527, 585
    Condensing boilers 206
    on database 528, 529, **530**
    Defect action sheets, former 521
    Good building guides **521**
    Increasing the fire resistance of . . . floors 566
    Information papers (IPs) **521**, 542, 588
    Overseas building notes **521**
  and timber 541, 543
Building Research and Information 531
buildings
  building envelope **563–4**
  building types **565–8**
  energy issues 206, 207, 208, 228, 230, **253–5**, 256, 403
    see also specific fuels
  façades 551, 563
  foundations 122, 543, **564–5**
  historical **534–5**, **551–3**
  industrial 552
  noise 492
  'passive' 253
  safety 316
  'supersheds' 566
  see otherwise structural engineering; construction
Buildings Energy Technology 254
Building Services 207, 531
Building Services Contractor 67
Building Services Engineer, former 207
Building Services Research and Information Association (BSRIA) 210, **585**
IBSEDEX 230
International Building Services Abstracts 228
  library 526
  Refrigeration and the environment 207
Building Technical Microfile **528**, 529
Bulgaria
  machines 161
  nuclear energy 261
bulk effects 134
bulk modulus, of fluids 288–90
Bull, J.W. 561, 562
bulletin boards, electronic **117–18**, 341–2, 502, 584, **644**
Bulletin of the Geothermal Resources Council 246
Bulletin of the International Peat Society 239
Bulletin Signalétique, former 101, 227, 230
  online 231
Bungay, H.R. *and* Belfort, G.

[*biochemical engineering*] 278
Bungey, J.H. [*concrete*] 537
bunkers 278, **567**
Buonicore, A.J. *and* Davis, W.T. 491
Burden, R.L. *and* Faires, J.D. 455
Burditt, M.F. 315
Bureau of Indian Standards, *Standards India* 27
Bureau International des Poids et Mesures (BIPM) 27–8
  publications
    CGPM/CIPM *Proceedings* 27–8
    Metrologia 28
    Système international des unités [SI units] 28
Bureau de Recherches Géologiques et Minières (BRGM [France]) 594, 602
Burgan, B.A. 558
Burgess, J.A. 332
Burghardt, M.D. 202
Burkin, A.R. 599
Burnham, L. *and* Johannsson, T.B. 242
Burns, T.R. *and* Midttun, A. 248, 255
Burt, R. *and* Mills, C. 599
Burton, R. 227
Buschart, R.J. 320
Bus and Coach 498
Bus and Coach Council **514**
Bus and coach statistics Great Britain (pub. HMSO) 509
Busemann, F. *and* Casson, W. 433
buses 498, 509, 510, **512**, 514
  deregulation 512
  guided 512
  see also transport
Business Archives Council (BAC) 94
Business Automation 400
business information 571, **574–7**
Business Alerts 306, 309
  creditworthiness 576
  current trends 576–7
  databases **229**, 230, 306, 422
  directories 84–6
  energy 229, 230
  finance 576
  Financial Times Business Information 70
    see also *Financial Times*
  forecasts 575–6, 575*tab*
  journals 65, 400
  marine 469*tab*
  materials 306, 309
  multimedia 342
  New earnings survey 574
  performance statistics 577
  PEST analysis 574
  Business Information Service *see under* British Library; *Financial Times*
  Bus registration index (pub. TRL) 510
Bussell, M. 559

Butcher, D. 126
Butterworth Heinemann [publishers] 527
Butterworth Architecture 531
  conferences 359
  Information sources in energy 607
Button, D. *and* Pye, B. 563
buyers' guides 71, 85, 86, 340, 414
Buyers' guide to uninterruptible power supplies (pub. ERA Technology) 414
Buzzi, R.A. 324
BVR [Building Commodity File] database 584
BWEA *see* Brit. Wind Energy Ass'n
BWR (boiling water reactors) 261–2
BYGGDOK [Swedish Institute of Building Documentation] 527
BYGGFO database 584
Bylanski, P. *and* Ingram, D.G.W. 443, 446
Byte 72, 402

C
C [computer language] 455
CAA (Civil Aviation Authority) 421
cables
  submarine 481
  telecommunications 441
CAD *see* CAD/CAM
CADalog [program] 176
CAD/CAM (computer-aided design [and manufacture]) **155**, **295**, **334–5**, **362–3**, **390–1**, **397–8**
  abstracts/indexes 355
  and automation 383, 385
  books, dictionaries 370
  and communications 362
  and concurrent engineering 650, 651, **652–3**, 654
  conferences 339, **359**
  data transfer 309–11
  design **334–5**, 339, 341, 342
  expert systems 396, 397
  machines **151–2**, 158–9
  systems design **390–1**, **395**, 397–8
  exhibitions 342
  fuzzy logic 391
  historical aspects 155, 390
  hydraulics **294–6**
  journals **339**, 357, 654
  manufacture **362–3**
    see also CIM
  marine technology 474
  materials 310–11
  mathematical aspects 455
  product information 310–11
  robotics 383
  stress analysis 137
  see also control systems
CADCAM 169, 339
CAD/CAM International 357

676 Index

CAD/CAM, Robotics and Factories of the Future conference 359
CADDET
   CADDET Newsletter 234
   see otherwise Centre for the Analysis and Dissemination...
cadmium, *as* neutron-absorber 260
CAE (computer-aided engineering) 650, 651, **652–3**
   see also CAD/CAM; automation *etc.*
   CAE (Computer-Aided Engineering) 339
CA FILE *see* Chemical Abstracts: online
Caines, P.E. 391
Cairncross, S. *and* Feachem, R.G. 488
Cakmak, A.S. *et al.* 454
calculations, handbooks 124
calculators, electronic 412, **452**
calculus 451
Caldwell, J.B. *and* Ward, G. 474
California, University of 257, 596
   Institute of Transportation Studies 501
CALS (computer-aided acquisition and logistic support) [program] **310–11**, 314, 655
   Steering Group 656
CAM (computer-aided manufacturing) *see* CAD/CAM
Cam-1 Organization 353
Camargue Publications [co.] 576
Camatini, E. *and* Kester, T. 207
Camborne School of Mines 595
Cambridge, RIBA bookshop 533
Cambridge dictionary of science and technology [ed. Walker] 123
Cambridge Econometrics [co.], construction forecasts 575
Cambridge Materials Selector [co.], CAMBRIDGE MATERIALS SELECTOR database **308**, 313
Cambridge Scientific Abstracts [co.] 100, 111
   databases 380
   Pollution Abstracts 228, 230
Cambridge University Press
   Cambridge dictionary of science and technology [ed. Walker] 123
   software 452
cameras, patent dispute 43
Camlinks [program] 168, 170*tab*
Campbell, J. 364
Campbell-Allen, D. *and* Roper, H. 557
CAMPUS database project **304–5**, 309, 313
cams 156, 159, 160, 161*tab*, 166, **167–8**, 173

Canada
   books-in-print lists 104
   chemical engineering 282
   conferences 76, 162
   construction 532, 541
   control systems 291
   energy 234, 253
      nuclear energy 261–2, 263, 264, 265
      power engineering 426
   fluid power 291
   journals 72–3
   marine technology 478
   mechanical engineering 162
   minerals/mining 594, 595, 596, 597, 610, 611, 612
   National Research Council 253, 478, 532
   Ontario Research Foundation 594
   research 253, 478, 532, 594
   Standards 27, **524**
   theses 530
   timber 560
   transport 517
   see also North America
Canadian Atomic Energy Control Board (AECB) 264
Canadian Centre for Mineral and Energy Technology (CANMET) 594
Canadian Energy Research Institute, *Geopolitics of Energy* 234
Canadian Institute of Mining and Metallurgy (CIM) **594**, 610
   Bulletin 596, 610
   CIMBASE database 584
Canadian Journal of Chemical Engineering 282
Canadian Journal of Civil Engineering 532
Canadian minerals yearbook 612
Canadian Mining Journal 597, 611
Canadian Nuclear Association (CNA) 263
Canadian Nuclear Society (CNS) **265**
Canadian Society of Chemical Engineering, *Canadian Journal of Chemical Engineering* 282
Canadian Standards Association 27
   Catalogue 27
   Info update 27
candelas 28
Candu (Canadian deuterium-uranium reactors) 261–2
CANMET (Canadian Centre for Mineral and Energy Technology) 594
Cannon, A.G. *and* Bendell, A. 179–80
CAN/OLE database host 422
cans, manufacture 364
Canter, L.W. 493
Canter, L.W. *and* Hill, L.G. 493
CAOLD database 231

Caoutchouc [rubber Standard] 24
Capeheart, B.L. *et al.* 255
C[v]apek, Karel 377
Capel, James, & Co. 576
carbon
   adsorption, gold processing 599
   carbon dioxide (CO2
      as coolant 260, 261
      lasers 364
   composites 307
   and fission process 260
   hydrocarbons 239, 240, 273, 282, 467, 607
   taxes on 233
Cardew, M.H. 275
cardiology 638–9
cardiovascular system 636
Care of the Elderly 642
cargo, marine 469*tab*, 480, 482
   safety 477
Cargo Handling Abstracts 480
Caribbean countries
   construction 584
   energy 232
Carlson, A.B. [*communications*] 443
Carlson, H. [*springs*[169
Carnegie Mellon University [US] 656
Carnot cycles **201**, 203
Caro, C.G. *et al.* 635, 636
car parks 514, 568
carpentry 542, 561
Carraro, C. *and* Siniscalco, D. 233
cars 16, 496
   organizations 514
   parking 514, 568
   see also automotive; traffic *etc*
Carson, P.A. *and* Mumford, C.J. 323, 324
   Newnes hazardous chemicals pocket book 323
Carter, A.D.S. [*reliability*] 178
Carter, C. *and* De Villiers, J. [*solar energy*] 243
Carter, D.E. [*concurrent engineering*] 652
Carter, D.E. *and* Baker, B.S. [*concurrent engineering*] 312
Carter Bennan Co. 172*tab*
Cartwright, M. 457
Carvill, J. 124
Cary, H.B. 362
CASA *see* Computer and Automated Systems Ass'n
CA Search *see* Chemical Abstracts: online
case studies 64, 493
   companies 356, 357, 358, **363**, **365–6**, 383, 386
Casey, H. *and* Panish, M.B. 448
Cashdollar, K.L. *and* Hertzberg, M. 321
Cassella, P. *et al.* 474
Cassidy, K.A. 322
CASSIS databases 58, **59**
casting **364–5**, 369
   die-casting, automated 377
   journals 68, 357

casting (contd)
  precision casting 364–5
  residual stress 134
Catalogue of American national
  standards (pub. ANSI) 25
Catalogue of the Canadian
  Standards Ass'n (pub. CSA)
  27
Catalogue of English translations
  (pub. AFNOR) 23
Catalogue des normes françaises
  (pub. AFNOR) 23
Catalogue des ouvrages (pub.
  AFNOR) 22
Catalogue des recueils (pub.
  AFNOR) 24
catalogues 66, **82–3**, 85, 87, 93,
  340, 579
  on CD-ROM 90, 306
  of databases 504
  on disk 306
  exhibition catalogues 84, **87–8**
  government publications 224
  house journals as 69
  of information sources see
    directories etc.
  library 13, **103**, 223, 269–70,
    584
  in microform 90, 157
  organizations' 476, 527
  publishers' 104, 527
  and software 151
  see also product information
catalysis 273, 275
catastrophe theory **457**
Catchword and Trade Name
  Index (CATNI) 100, 341
Caterpillar Inc., Caterpillar
  performance handbook 615
cat flaps, and patents 34
CATNI see Catchword and
  Trade Name Index
Catsim [program] 296
Caulfield-Browne, M. et al. 168
caustics [light] stress analysis
  136, 147
cavitation 174
CAWS (Common Arrangement
  of Works Section) **578–9**
CBD Research [co.]
  Current British directories 86,
  125
  Directory of European
    industrial and trade
    associations 370
CBI (Confed'n of Brit. Industry)
  157, 342
CCIR [Internat.
  Radiocommunications
  Consultative Committee]
  449
CCITT [Internat. Telegraph and
  Telephone Consultative
  Committee] 449
CD [1] (compact disc) 413
  see also CD-ROM
CD [2] (Committee Draft, of
  ISO) 18
CDM (Construction design and
  management regulations)
  549

CD-PLUS database host 107, 108
CDP Technologies [co.] **108**, 110
CD-PLUS database host 107,
  108
CDP Online database host,
  former **422**, 584
CD-ROM (compact disc:
  read-only memory) 89,
  90–1, 107, **108**, 231, 267,
  306
  access to 224, 305, **422**, **427**
  access speed 108
  costs 306, 505
  directories of **231**, 504
  e-mail 511
  fee-basis 103, 108
  jukeboxes 111
  LANs 108
  networked **108**, 111
  options **108**
  software 426, 427
AEROSPACE DATABASE 112
ASTI 110
automation 380
BCQ/ODI **582**
bibliographies 103
and biomedical engineering
  644
buyers' guides 157
CASSIS 58, **59**
catalogues 90, 306
chemical engineering 271, 354
chemicals 323
citations 109
CITIS **113**, **528**, **582**
COMPENDEX **110**, 354, 380
computing 426
conference Proceedings 77,
  110
construction 113, 527–30,
  581–2, 583
contents pages 111
CORDIS 505
CURRENT CONTENTS 111
databases **109–13**, 580
design 157, 340, **341**
directories 73, 74, 400
electrical/electronics **111**, 422,
  426, **427**
encyclopaedias 122, 271
energy 231, 267
ENERGY SCIENCE AND
  TECHNOLOGY 112
environmental engineering
  485–6
FIRST 59
ICONDA 113
INSPEC **111**, **422**, **426**, 450
ISMEC 111
manufacturing 354
materials 112, **306**
mathematics 113, 460
mechanical engineering 111
METADEX COLLECTION 112
newspapers 71, 75
patents 55, 57, 59, 89, 341
product information 89, **90–1**,
  92, 95, 96, 529
research 515
RIBA/TI **581–2**
robotics 380

Index   677

safety 323
science 109
Standards 22, 89, 113, 152,
  340
  BRITISH STANDARDS 90
  NORMIMAGE 22
  PERINORM **22**, 152, **582**
theses 78, 530–1
transport 504, **505**, **511**, 515
WATER RESOURCES ABSTRACTS
  113
see also databases
CE [1] (Comité Européen) Marks
  19–20
CE [2] see concurrent
  engineering
CEA see Commissariat à
  l'Énergie . . .
CEA (Commissariat à l'Énergie
  Atomique) **263**
CEB see Comité
  Euro-International du Béton
CEBA see CHEMICAL ENGINEERING
  AND BIOTECHNOLOGY . . .
Cebon, D. and Mitchell, C.G.B.
  316, 317
CEC
  CEC Joint Research Centre
    313
  see otherwise Europe:
    European Commission
CE & CALS Conference 655
CECC (CENELEC Electronic
  Components Committee) 91,
  420
CEDOCAR database host 422
CEE [1] see Europe: European
  Commission
CEE [2] (Internat. Commission
  for Conformity Certification
  of Electrical Equipment) 19
CEGB see Central Electricity
  Generating Bd
cellular materials, handbook 312
cement 536–8, 557, 590
  bone cement 636
  database 528
  glass-reinforced (GRC) 537,
    562–3
  handbooks 537
  journals 522, 597
  organizations 536, 556
  see also concrete
Cement and Concrete
  Association, former 536
CEN see Comité Européen de
  Normalisation
CENELEC [Eur. Committee for
  Electrotechnical
  Standardization] 19, 113,
  420
  Electronic Components
    Committee (CECC) 91,
    420
  and PERINORM 23, 582
  Standards for access to the EC
    markets (CEN/CENELEC)
    20
Census of motor tarffic on main
  international traffic arteries
  (pub. UN/ECE) 510

Center for Chemical Process
    Safety [US]
  publications
    *Chemical reactivity
      evaluation* . . . 324
    Guidelines series *see*
      Guidelines
    Safe automation . . . 279
Center for Coal Science [US]
  239
Center for Design Research [US]
  656
Center for Renewable Resources
  [US], *Renewable energy in
    cities* 210
Central Electricity Generating
  Board (CEGB), *former
    Modern power station practice*
  251–2, **429**
  test results 433
Central Fuel Research Institute,
  Indian coals 238
Central Product Information
  Body, proposed 93
Central Statistical Office (CSO)
  572
  Guide to official statistics 126
  Social trends 574
Centre for the Analysis and
  Dissemination of
  Demonstrated Energy
  Technologies (CADDET),
  Newsletter 234
Centre for Construction Market
  Information **520**, 575
Centre for Environmental
  Management and Planning,
  *Environmentally sound
    development* . . . 255
Centre Européen de Recherches
  Nucléaires *see* CERN
Centre for Interfirm Comparison
  [org'n] 576
Centre National de la Recherche
  Scientifique (CNRS) 101
  Bibliographie Internationale
  101, 227, 230, 231, 584
  Bulletin Signalétique 101, 227,
  230, 231
  and NTIS 355
  PASCAL **230**
Centre for Strategic Studies in
  Construction (CSSC),
  Building Britain 2001 575,
  587
Centre for Window and Cladding
  Technology (CWCT) 563,
  564
centrifuges, safety 327
Centro de Investigación Minera y
  Metalurgica [Chile] 594
*CERA (Concurrent Engineering:
  Research and Applications)*
  653
Ceramic Abstracts 228
Ceramic Industry 68
ceramics 302, 307, 353, 540–1
  abstracts/indexes 112, 228,
    309, 355
  bricks *see* bricks
  conferences 359

databases 112, 306, 308, 355
  early uses 589
  and electronics 419
  handbooks 302, 307, 312
  and hydraulics 288
  Internet resources 119
  journals 68
  organizations 140, 303, 327,
    359
  safety 327
  Standards 302
  see also engineered materials
Ceramics International 68
CERAMITEC symposium 359
CERAM Research [co.] **540**
CERC (Concurrent Engineering
  Research Center) 654, **656**,
  657
CERCnet 657
CERCI 587
Building IT 2000 580
CERN (Centre Européen de
  Recherches Nucléaires [Eur.
  Centre for Nuclear
  Research])
  Preprint Server 116
  World Wide Web 118, 119
certification, of components 414
CETIM database 314
CETOP (Eur. Oil Hydraulic and
  Pneumatic Committee) 287
CFCs (chlorofluorocarbons) 202
CFD (computational fluid
  dynamics) **294**
CGPM *see* Conférence Générale
  des Poids et Mesures
Chadderton, D.V. 253
Chadwick, A.T. 243
chain reactions **260**, 266
chains 161*tab*, **164**
*Chains for power
  transmission* . . . (pub. Amer.
  Chain Ass'n) 164
'Challenges to control . . . ' (pub.
  IEEE) 398
Challoner, J. 233
Chamber of Mines of South
  Africa 608–9
  COMRO Bulletin 609
  Newsletter 608–9
Chambers [publishers], *Chambers
  materials science and
  technology dictionary* [ed.
  Walker] 123
Champeney, D.C. 454
Chance, A.B. [co.] 366
Chandler, K.A. *and* Bayliss, D.A.
  539
Chandrasekhar, S. 243
Changeaux, J.-P. 625
*Changing role of engineering in
  orthopaedics* (pub. MEP)
  646
Chang -Kuo Ho *and* Jui-Tien
  Chi 253
chaos theory **457**
Chapman & Hall Ltd [publishers]
  423
Charles, F.W.B. *and* Charles, M.
  [*timber*] 561
Charles, J.A. *and* Crane, F.A.A.
  [*materials/design*] 337

Charlier, R.H. 248
Charlier, R.H. *and* Justus, J.R.
  248, 255
Chartered Builder **579**
Chartered Engineer [qualification]
  593
Chartered Institute of Building
  (CIOB) **521**
  catalogue 521, 527
  and Construction History
    Society 535, 553
  journals
    Chartered Builder **579**
    Construction Computing
      532, **580**
    Construction History 535
  publications 587
    *Construction information
      digests* 521
    Construction papers 521
Chartered Institute of Transport
  (CIT) 510, **513**
  Proceedings 499
  and TSUG 510
Chartered Institution of Building
  Services Engineers (CIBSE)
  210
  CIBSE guide 207
  Journal, former 207, 531
*Chartered Mechanical Engineer,
  former* 356
Chartered Quantity Surveyor **579**
Chartered Society of
  Physiotherapy 631
charts, maritime 481
Chase, M.A. *and* Sheth, P.N.
  [*design*] 170*tab*
Chatenever, R. 207
Chatfield, C. 458
Cheal, B.D. 558
ChE (chemical engineer[ing]),
  ChE Electronic Newsletter
  115
CHEMABS *see* Chemical Abstracts:
  online
Chemical Abstracts (CA) 57, 98,
  228, 283
  CAOLD 231
  online **230**, 231
Chemical Engineer **281**
chemical/process engineering
  269–83
  nature and role 269–70
  abstracts/indexes 100, 112,
    228, **283**, 400
  automation 279, 280, 325
  biochemical engineering *see*
    biochemical engineering
  books 269, **271–3**, 274–80,
    317
  bibliographies 270
  data 272–3, 282
  dictionaries *etc.* 122–4, 271
  directories 86, 270, 272
  handbooks 124, **271–3**
  conferences **270**, 272, 281
  controls **279–80**, 282, 391, 400
  databases 90, 112, 229, 230,
    231, 271, **283**, 354, 400
  data transfer 310
  design aspects *see under*
    design applications

Index 679

chemical/process engineering (*contd*)
  electrochemical processes 105, 274, 357, 363
  environmental aspects 281, 486, 489, 490
  fluid mechanics aspects 269, **276**
  and genetic engineering 625
  Internet resources 119, **281**
  journals 72, 104–5, 280, **281-3**, 358
    electronic newsletter 115, 281
    trade journals 65, 67–8, 87
  mass transfer **273**
  mineral processing **589–606**
  organic 273
    see also bio-
  organizations 270–1, 281–2
  particle technology **278**
  patent searches 283
  physical chemistry **275–6**
  process analysis/synthesis 269, 279
  product information 310
  properties data 272–3, 282
  reactors **275**
  safety aspects 270, 279, 281–2, 283, 315–19, **322–4**, **325–6**
  software 273, **281**
  Standards 26, 271, 279
    specifications 310
  thermal processes **276–7**
  thermodynamic aspects **199**, **275–6**, 282
  training 270, 282
  transport processes **273**
  unit operations 269, 271–2, **273–4**, 488
  water treatment 489
  see also chemicals; processes; manufacturing
Chemical Engineering 72, 282, 406
Chemical Engineering Abstracts, former 283
CHEMICAL ENGINEERING AND BIOTECHNOLOGY ABSTRACTS database (CEBA) 112, 229, 230, 231, **283**, 400
  print versions 283
Chemical Engineering Education 282
Chemical and Engineering News 282
Chemical Engineering Progress **281**
Chemical Engineering Research and Design (IChemE Transac.) 281
Chemical Engineering Science 282
Chemical and Engineering Technology 282
Chemical Hazards in Industry 283, 327
Chemical Information Conference 61

Chemical Manufacturers Association (CMA), *Process safety management* . . . 325
Chemical Patents Index 59
Chemical and Petroleum Engineering 240
Chemical Plants and Processing 282
Chemical reactivity evaluation and application to process design (pub. Center for Chemical Process Safety) 324
Chemical Rubber Co., *Handbook of chemistry and physics* 272
Chemical safety data sheets (pub. RSC) 323
chemicals industries 269–70
  abstracts/indexes 228, **230**
  books 271–3, 323–4
    materials handbooks 308
  conferences 61, 601
  contamination 322–3
  journals 67, 87, 104, 281–2
  and mineral resources 590, 601
  organizations 270–1, **270**, 325, 601
  patents 56, **57**, 58, **59**
  process integration 204
  properties data 323
  reactions 322–3
  safety 270, 317–21, **322–4**, 326
    emergencies 323–4
  Standards 271
  synthetic chemicals 203
  thermal processes 205
  toxicity 322, 323
  see also chemical engineering
Chemie-Ingenieur-Technik 282
Chemiker-Zeitung 67
chemistry 322–3
  abstracts/indexes 228, **230**, 283
  biochemistry **278–9**, 281–3, 488, 490
    see also biochemical engineering
  books 269
  dictionaries 271
  handbooks 272–3
  and chemical engineering 269, **275**
  chemical thermodynamics **199**
  constants 232
  databases 111, 229, **283**
  and electricity generation 252, **429**
  and environmental engineering 488, 489, 490
  equilibrium 199
  geothermal systems 245
  journals 105, 281–2, 283
  mineral processes 238, 589, 599–600
  physical **275–6**
  reactions/reactors 271, **275**, 278–9, **322–4**
  safety 322–4
  thermodynamic aspects 199, 202–3, 272–3, 282

  and water treatment 489
  welding 362
  see also chemical engineering
Chemistry of cyanidation (pub. Amer. Cyanamid) 599
Chemistry data series (pub. DECHEMA) 272
Chemistry and Industry 281
Chen, F.Y. 168
Cheong, V.E. 446
Cheremisinoff, N.P. [*heat/mass transfer; fluid mechanics*] 275, 276, 277
Cheremisinoff, P.N. [*air quality*] 491–2
Cheremisinoff, P.N. and Cheremisinoff, N.P. [*heat transfer*] 277
Cheremisinoff, P.N. and Morresi, A.C. [*geothermal energy*] 245
Cheremisinoff, P.N. and Morresi, A.C. [*environmental assessment*] 493
Chernobyl accident and its consequences [ed. Gittus *et al.*] 266
CHETAH hazards programme 324
CHI [IT] conferences 656
Chiang, C.H. 159
Chile, minerals 594, 596
chimneys 560
China, Nanyang Technological University 359
Chin, W.W. and El-Masri, M.A. 204
China
  automation 404
  energy 234, 241, 253
  fluid power 290, 297
  see also Chinese language
China Energy Report 234
Chinese Journal of Fluid Power Engineering 290
Chinese language
  dictionaries 454
  patents 47
chips, electronic 395, **412**
Chironis, N.P. 156
Chisholm, D. 206
Chopey, N.P. and Hicks, T.G. 272
Chorlton, A. and Shackshaft, G. 433
Chow, W.W.-C. 158, 337
CHP *see* combined heat and power
CHPA (Combined Heat and Power Ass'n) 209, 210
CHRIS hazardous chemicals data manual (pub. US Coast Guard) 323
Christensen, J.J. *et al.* 272
Christou, A. 367
Christy, D.P. *and* Watson, H.J. 392
Chryssolouris, G. 364
CHS (Construction History Soc.) 535, 553
Chui, C.K. 457

680 Index

Churchill, R. and Ulfstein, G.
  [marine disputed areas] 482
Churchill, S.W. [viscous flows]
  276
CIB see Internat. Cncl for
  Building Research Studies
Ciba Geigy [co.] 305
CIBSE
  CIBSE guide 207
  see otherwise Chartered Inst'n
    of Building Services
    Engineers
CICA (Construction Industry
  Computing Ass'n) 521, 587
CIE [Internat. Commission on
  Illumination] 22
CIGRE see Conférence
  Internationale...
CIIG (Construction Industry
  Information Group) 526
CIM [1] (computer integrated
  manufacturing) 362–3
  case studies 363
  CIMBASE database 584
  and concurrent engineering
    650, 652–3
  conferences 359, 360
  FMS 362–3
  journals 356–7, 379, 651
  management 362, 363, 369
  patents 379
  robotics 379, 383, 385, 386
  see also CAD/CAM
CIM [2] (Canadian Inst. of
  Mining and Metallurgy)
  594, 610
  CIMBASE database 584
  CIM Bulletin 596, 610
CIMAH (control of industrial
  major accident hazards) 318
CIMBASE database 584
CIMIO Ltd 172tab
CIOB see Chartered Inst. of
  Building
CIPM see Comité Internat. des
  Poids et Mesures
circuits
  abstracts 421–2
  analysis 433
  conferences 417
  cycloconvertors 436
  drives 436
  handbooks 418–19
  integrated (IC) 412, 419
  VLSI 415, 417
  journals 415–16, 421
  solid-state 416, 417, 433
Circuits, Devices and Systems
  (IEE Proc.) 415
CIRIA
  CIRIA Guides 521, 556, 565
  ... guide to European ... and
    international ...
    information 523
  ... guide to sources of
    construction information
    585
  ... UK construction
    information guide 524,
    526, 587
  see otherwise Construction

Industry Research and
  Information Ass'n
CIRP
  CIRP annals 358, 371
  see otherwise Collège Internat.
    pour l'Étude...
CIS (Commonwealth of
  Independent States) see
  Russia; Soviet Union
CISPR [Internat. Special
  Committee on Radio
  Interference] 22
CIT see Chartered Inst. of
  Transport
citations 102, 109, 224, 400, 461,
  504
CITIS Ltd 113, 582
  CITIS databases (CD-ROM)
    113, 528, 582
Civil Aviation Authority (CAA)
  421
Civil-Comp Press [publishers]
  531
civil engineering 269, 519, 545,
  547
  abstracts 100, 113
  and biomedical engineering
    624
  books
    dictionaries etc. 122, 123
    handbooks 125
    yearbooks 126
  databases 90, 113, 527–30,
    582–4
  data transfer 310
  and environmental engineering
    486, 489, 490, 492–3
  historical aspects 94
  hydraulics and 286
  hydroelectricity 430
  Internet resources 119
  journals 67
  mineral processing 589, 592
  mobile plant 286
  organizations 576
  product information 310
  safety 315, 318, 326
  Standards, specifications 310
  structural aspects see structural
    engineering
  surface mining and 615
  trade literature 90, 94
  and transport 496, 513
  water engineering 247–9, 489,
    490
  see also construction; roads
    etc.
Civil Engineering (ICE Proc.)
  531
Civil Engineering (pub. ASCE)
  532, 580
Civil engineering guidelines
  for... hydroelectric
  developments (pub. ASCE)
  247
Civil Engineering in Japan 532
Civil engineering work: a
  compendium of occupational
  safety practice (pub. ILO)
  326
Civil Engineers' and Architects'
  Journal, former 553

Civil and Structural Engineering
  Abstracts 100
  cladding 557, 563–4
  CLAIMS database 58
Clare, R. 248
Clark, A.M. [product liability]
  334
Clark, A.P. [telecommunications]
  446
Clark, K.B. and Fujimoto, T.
  [concurrent engineering]
  649
Clark, L.A. [bridges] 566
Clarke, E. [electricity] 433
Clarke, J.L. [concrete] 537
Clarke Associates [co.] 523
Clason, W.E. [chemical/
  dictionary] 271
classification societies [maritime]
  472
classification systems
  INSPEC 421
  MeSH system 643
  patents 48–9, 56–7
  reports 5–6, 8, 9
  Standards
    ICS 18
    SIC 85, 92
    UDC 102, 121
Classon, F. [concurrent
  engineering] 653
clays 593
Clayton, B.R. 246
Clegg, A.J. 365
Clevett, K.J. 279
clients [of networks]/
  client-servers 117–18
climate 233, 234, 253, 255, 256
  global warming 256
  and hydroelectricity 430
  and wastewater 488
  see also greenhouse effect
Clinical Biomechanics 641
Clinical Electroencephalography
  639
clinical engineering/physics
  626–7, 628fig, 629–32, 635
Clinical Physiology 639
Clinical Rehabilitation 642, 644
clutches 161tab, 163–4, 173,
  177, 341
CMA see Chemical
  Manufacturers Ass'n
CMOS (complementary
  metal-oxide semiconductors)
  424
CNA (Canadian Nuclear Ass'n)
  263
CNAA see Cncl for Nat.
  Academic Awards
CNC (computer numerical
  control) 385
CNRS see Centre Nat. de la
  Recherche...
CNS (Canadian Nuclear Soc.)
  265
coaches see buses
coal 226, 237–9, 607, 615
  abstracts/indexes 112, 228
  alternatives to 209
  books 235–8, 240

Index 681

coal (contd)
 coke 590
 conferences 601
 databases 602
 derived fuels 241
 environmental effects 238
 flotation 599
 gasification 237–9, 240
 historical aspects 226
 industrial importance 590, 593
 journals **239**, **593**, 597, 611
 liquefaction 225, 237–9
 open-cast 597
 organizations 20, 225, 226, 227, 239, 609
 research 594
 seam sizes 591
 South Africa 609
 statistics 237
 surface mining 615
 transport 238
 'washing' 590
 waste proportions 591
 see also mining etc
Coal 593, **597**
Coal Abstracts 112, 228, 602
Coal Age, former 597, 600
Coal Calendar 239
Coal conversion systems technical data book (pub. USDoE) 237
Coal data (pub. NCA) 237
COAL DATA BASE 602
Coal and the environment (pub. Commission on Energy and the Environment) 238
Coal Industries [co.] 54
Coal Industry Advisory Board [of IEA] 238
Coal Information 239
Coal liquefaction.. (pub. IEA/OECD) 238
Coal Outlook 239
Coal Preparation 239
Coal Preparation Society, *former* 593
Coal Research Establishment (CRE) 226
Coal Science and Technology 239
Coal transport infrastructure . . . (pub. IEA/OECD) 238
Coal Week 239
Coal Week International 239
coastal structures 567
Coastal structures and breakwaters (pub. Telford) 567
coating see surface treatments
Coating 68
COBA (cost benefit appraisal) [program] 508
COBA manuals (pub. DoT) 500, **508**
Coburn, A.W. and Spence, R.J.S. 327
codes of practice (CPs) 15, 25, 181, **182**, 471, **548–9**
 BSI 21, **182**, 528
 health and safety/HSE 151, **181**, 182, 315, 477

Lloyds Register 151
 online 528
 see also Standards
Codes of Practice (pub. BSI) 21
coding
 of information **445**, **457**
 see also classification
Codus Ltd 91
CODUS databases **91**, 418
co-generation see combined heat and power
Cogeneration 209
Cohen, A.M. [*mathematics*] 455
Cohen, H. *et al.* [*turbines*] 235
coke, ironmaking 590
Collège International pour l'Étude Scientifique des Techniques de Production Mécanique (CIRP) **370–1**, **650–1**, 655
CIRP Annals 358, 371
Collier, J.G. 235
Colliery Guardian 597
 Guide to the coalfields 601
Collin, R.E. [*radio*] 447
Collins, A.R. [*structures history*] 553
Colorado
 Mountain State [co.] 595
 School of Mines 595
colour, early pigments 589
COMADEM (Condition Monitoring and Diagnostic Engineering Management) conferences 296
Comar Instruments [co.] 146
combinatorics 451
Combined heat and electrical power generation . . . (pub. DoE) 213–14
combined heat and power (CHP)/ co-generation 204, **208–9**, 235, 238, **253–5**, 429
 journals 209, 255
 organizations 209, 225
Combined Heat and Power Association (CHPA) 209, 210
Combined heat and power 209
combustion **199**, 204, 207, **235–7**
 combined-cycle processes **429**
 and electricity generation 428, **429–30**
 fluidized beds **430**
 internal combustion engines **200**, 203, 225–7
 journals 105, **237**, 282
 see also thermodynamics
Combustion Institute, *Combustion and Flame* 237, 282
Combustion Science and Technology 237
Comer, P. 186
Comisión Panamericana de Normas Técnicas (COPANT) 20
Comité Euro-International du Béton (CEB), *Bulletins* 565
Comité Européen de Normalisation (CEN) [Eur. Committee for

Standardization] **19–20**, 23, 113, **524**
 CE Marks 19–20
 and PERINORM 23, 582
 publications, *Standards for access to the EC markets* (CEN/CENELEC) 20
Comité International des Poids et Mesures (CIPM) 27–8
 Procès-verbaux des séances [*Proceedings*] (pub. BIPM) 28
 Sessions des comités consultatifs [*Proceedings of consultative committees*] (pub. BIPM) 28
Commercial Motor 68
comminution **591–2**, **598**, 600
Commissariat à l'Énergie Atomique (CEA [France]) **263**
Commission on Energy and the Environment, *Coal and the environment* 238
Commission of the European Communities (CEE/EC) see under Eur. Union
Committee on Information Hang-ups [of NTIS] 13
Committee on Science and Technology [House of Lords] 508
Committee for Waterfront Structures, *Recommendations* 567
commodities, statistics 75
Common Arrangement of Works Section (CAWS) **578–9**
Common Market see Europe
Common standards for enterprises (pub. EU) 20
Commonwealth of Independent States (CIS) see Soviet Union
Commonwealth Regional Renewable Energy Resources Information Service (CRRERIS) 225
Commonwealth Scientific and Industrial Research Organization (CSIRO [Australia]) 594
Commonwealth universities yearbook (pub. Gale) 126
communication **441**
 and biomedical engineering 642
 multimedia 342, 503
 'open systems' (OSI) 450
 skills of, *and* design **338**, 550
 speech
  and Internet 503
  and telephony 441
 see otherwise telecommunications
Communication in design . . . (pub. SEED) 338
Communication Outlook 642
Communications in Applied Numerical Methods **458–9**

682 Index

Communications (IEE Proc.) 448
Communications Networks 449
Communications News 449
Communications in Numerical Methods in Engineering **143**
Communications Speech and Vision (IEE Proc.) 415
Communication of structural design (pub. IStructE) 550
compact disc *see* CD; CD-ROM
companies
 annual reports 82, **83**, 93, 478, 481
 case studies 356, 357, 358, **363**, 365–6
 house journals *see under* journals
 patents *see under* patents
 and product information 82–3
 trademarks/trade names *see* trademarks
 see also organizations
Companies and their brands (pub. Gale) 88
compatibility [Standards] *see* harmonization
COMPENDEX [*PLUS]
 (Computerized Engineering Index) databases 92–3, **110**, 354, **399**, **400**–2, **422**, 485, **529**
 automation 380
 conferences 77, **400**, 529, 530
 construction 529, 530, 583
 control systems 399, **400**–2, **422**
 design 341
 electronics 422
 energy **229**, 230, 231
 environmental engineering 485
 Internet access 120
 machines 153
 manufacturing 354
 product information 92–3
 robotics 380
 transport 505
 water engineering 485
Compendium of fire safety (pub. FPA) 321
Competitive edge (pub. Nat. Research Cncl) 650, 656
competitiveness 336, **337**, **353**, 361, 366, 383–4, 385, 649, 650
components **156**–**7**, **159**, 160, **163**–**9**, 337–8
 abstracts/indexes 421–3
 books 124, 161*tab*, 163–9, 340
 databases 91, 418, 421–3
 data interpretation 288–9, **291**
 data transfer 310
 directories 370, 418
 electronics *see under* electronic engineering
 failure 177
 hydraulics *and* 286–7, **288**, **291**
 journals **163**, 165, 168, 169, 416, 654
 miniaturization 287
 optoelectronics 448

reliability 368
research 291
software 163
'standard parts' 337–8
Standards 91, 163–4, 368, 414, 420
 certification 414
 preferred lists 91, 418
 see also equipment; product information *and specific topics*
Components in Electronics 416
composite materials 133, 302, 307
 abstracts 309, 355
 databases 306, 308, 355
 encyclopaedias 122
 glass composites 537, 562–3
 handbooks 142, 302, 307, 313, 361
 journals **143**, 357, 531
 manufacture 355, 357, 362
 organizations 303
 products 355
 properties 355
 Standards 302
 and stress analysis 133
 see also engineered materials
Composites **143**, **531**
Composites Manufacturing 357
Compressed air and energy use (pub. EEO) 208
Compressed Gas Association, *Handbook of compressed gases* 325
compression, fluid 199–201, 204, **208**, **288**–**90**, 325
compressible flow functions 202
compressors 200, 204–5
 refrigeration 205
CompuMath Citation Index **461**
 online 402, 403
COMPUSERVE list-server 584
computational fluid dynamics (CFD) **294**
Computational Intelligence 406
Computer 403
computer-aided acquisition and logistic support *see* CALS
Computer-aided acquisition and logistic support (CALS) . . . guide (pub. USDoD) 314
computer-aided design [and manufacture] *see* CAD[/CAM]
Computer-Aided Design **169**, 309, 339
Computer-Aided Design Report 339
computer-aided engineering *see* CAE
Computer-Aided Engineering (CAE) 339
Computer-Aided Engineering Journal 309
Computer-Aided Geometric Design 339
computer-aided [design and] manufacture *see* CAD[/CAM]

Computer and Automated Systems Association (CASA [of SME]) 378, 381, 382, 651
Computer and Control Abstracts 101, 341, 380, **399**
 online 101, **111**, **230**, 380, 399
COMPUTER DATABASE **422**
computer graphics *see* graphics: computer
Computer Graphics 339
Computer Graphics Forum 339
Computer and Information Systems Abstracts 100
computer integrated manufacturing *see* CIM
Computer Integrated Manufacturing 356
Computer Integrated Manufacturing conference 359
Computer Integrated Manufacturing Systems 356–7
Computerized Engineering Index *see* COMPENDEX
computer numerical control (CNC) 385
computer programs *see* software
Computer-readable databases (pub. Cuadra/Gale) 581
computers
 nature and role 335, **335**, 377, 391, **432**–**3**
 abstracts/indexes 100, 101, 111, **113**, 341, **399**–**403**, **421**–**2**
 access/availability 151–2
 artificial intelligence *see* artificial intelligence
 and biomedical engineering 624, 639, 642, 643
 books 423
 buyers' guides 422
 dictionaries 124
 CAD/CAM *see* CAD/CAM
 calculators, pocket 412, **452**
 and chemical engineering 273
 and concurrent engineering 335
 conferences 319, 360, 601, 655–6
 APCOM 601, 612–13
 CE & CALS 654
 Civil-Comp Press 531
 EWICS 319
 and construction 521, 531, 532, 580, 587
 cpI 579
 and control systems 205, **391**, **393**–**4**, 398, **432**–**3**
 cybernetics 336
 databases 111, **113**, **228**–**9**, 355, **399**–**403**, **421**–**2**, 426, 580–1
 data compression 115, **457**
 and design 155, **334**–**5**, **339**, **397**–**8**
 see otherwise CAD/CAM
 ergonomic aspects **336**, 339
 exhibitions 342, 521

Index 683

computers (contd)
  expert systems 312, 396
  graphics see graphics:
      computer
  historical aspects 377, 411,
      412
  and hydraulics 285–7, 286fig
      see also fluid mechanics
  icons 295
  images see image processing;
      imaging
  journals 282, 295, 339, 356–7,
      399–403, 415–16, 415–17,
      448, 532, 580, 639, 642,
      643
    current awareness 399
    electronic 115
    newsletters 72, 115
    reviews 104–5
    trade journals 67
  kinematics 158
  languages
    'C' 455
    Fortran 335, 455
    Pascal 455
    'logic machines' 377
  machine drives 434
  and manufacture 355, 362–3,
      377–88
      see also CAD/CAM; CIM
  materials 419
  mineral processing 601
  mining 601, 612–13
  modelling see modelling
  mouse 295
  networks see networks,
      computer
  numerical analysis 134
  organizations 113, 413, 521,
      587
  parallel processing 398
  personal 151–2, 306
    CD-ROM 422
    CAD/CAM 165, 168, 169,
        170–2tab, 295
    and CDs 224
    and control systems 393
    databases on 108, 111, 305–
        6, 355, 422, 504, 581–
        2, 609, 613
    and Internet 114–20
    and measurement 367
    and networks 446
    system analysis 432
        see also disk: floppy
  and pneumatics 285
  product information 422
  programmable logic controllers
      285, 383, 385–6
  read-write heads 419
  remote 118
      see also networks: computer
  and robotics 377, 378
  and safety 319, 320, 326
  Standards 420
  and structures 549, 550
  system analysis 432–3
  technology see information
      technology
  telecommunications 442, 445–
      6

and transport 496
  see also computer-aided etc.;
      software; networks;
      information technology
Computers in Biology and
    Medicine 642
Computers in Biomedical
    Research 642
Computers and Chemical
    Engineering 282
Computers and Digital
    Techniques (IEE Proc.) 415,
    448
Computers and Graphics 339
Computers and Industrial
    Engineering 357
Computers in Mechanical
    Engineering 295
Computer Society [of IEEE] 381,
    403
Computers and Structures 532
Computer Supported Cooperative
    Work (CSCW) 654
Computer Technology in
    Welding conference 360
Computer Weekly 67, 70
Computing 67
Computing and Control
    Engineering Journal 400,
    416
COMPUTING REVIEWS database 403
COMRO Bulletin 609
Concordia, C. 433
CONCQUEST database 528
concrete 536–8, 556–7
  alkali-silica reactions 557
  bridges 566–7
  cladding 557, 563–4
  corrosion 557
  database 528
  design 528, 538, 556–7
  domes 557
  falsework 537
  and fire hazard 555
  fixings 565
  formwork 538
  handbooks 534
  journals 68, 522, 537
  masonry work 560
  offshore structures 567
  organizations 140, 536–7, 556
  quality 538, 552
  safety 544
  and seismic hazard 554
  Standards 537, 549, 555, 556,
      564
  tanks 567
Concrete 68, 532
Concrete and Constructional
    Engineering, former 557
Concrete core testing for
    strength (pub. CS) 557
Concrete Current Awareness 537
Concrete Information Service 537
Concrete Institute, former 522
Concrete International 532
Concrete mixes for general
    purposes (pub. BCA) 537
Concrete Society (CS) 536, 556
  Assessment and repair of
      fire-damaged . . .
      structures 555

publications 544, 555, 557
Concrete 532
Concrete core testing for
    strength 557
Technical reports (CSTRs)
    555, 557
Concrete yearbook (pub. Telford)
    534
Concreting in hot weather (pub.
    BCA) 537
concurrent engineering (CE)
    649–60
  nature and role 649–50
  checklists 651–2
  abstracts 657
  books 274, 312, 651–3
  bibliographies 657
  reports 656–7
  bulletin boards 657
  and CAD/CAM 650, 651,
      652–3, 654
  conferences 651, 655–6
  and control systems 398
  databases 657
  and design management 333,
      335, 337
  industry variation 653
  journals 356–7, 653–5, 657
  organizations 650–1, 653
  National Repository [US]
      656
  self-assessment 656
  terminology 649, 650
  thermodynamic 204
Concurrent Engineering 653
Concurrent Engineering:
    Research and Applications
    (CERA) 653
Concurrent Engineering Research
    Center (CERC) 654, 656,
    657
CERCnet 657
Concurrent Engineering Research
    in Review 657
condensing 204, 206, 235, 237,
    277
  refrigeration 207
Condensing boilers (pub. BRE)
    206
condition monitoring see under
    reliability
Condition Monitoring and Sound
    Assessment Club [of DTI]
    178–9
conduction, thermal 199, 202,
    277
conductors, electrical 419
  superconductors 100, 312, 420
      see also semiconductors
Cone, C. 206
Confederation of British Industry
    (CBI) 157, 342
Conférence Générale des Poids et
    Mesures (CGPM) 27–8
Comptes rendus des séances
    [Proceedings] (pub.
    BIPM) 28
conference-groups [Internet] 305
Conférence Internationale des
    Grands Réseaux Électriques
    (CIGRE) 225, 431

684 Index

Conférence Internationale des Grands Réseaux Électriques (CIGRE) (contd)
  Electra **431**, 437
Conference papers index (pub. BL) 77
Conference proceedings index (pub. BL) 77, 110, 530
Conference on Radiation Hazards in Mining 617
conferences 63, **75–7**, 142, 309, 358, 471, **500**
  abstracting/indexing 76, 103, 110, 400, 529
  calls for papers 64, 76
  computer-mediated see discussion lists
  currency 75, 76
  directories 76–7, 267, 281
  e-mail announcements 116
  and grey literature 10
  joint sponsorship 417
  keynote addresses 471
  official delegates 508
  Proceedings 64–5, 72, 75, **76**, 103, **109–10**, 111, 153, 161–3, 358–60, **471**, 474, 530
    abstracts/indexes of 77, 111, 112, 354, **400**
    directories 77, 530
    'posters' 76
    publishers, selected 530, 531
    special subject series 474
    see also under specific topics and organizations
CONFERENCES in ENERGY, PHYSICS, MATHEMATICS AND CHEMISTRY database 110
Conference on the Stability of Ships and Ocean Vehicles 474
Congressional Information Bureau [US] 251
Conlon, P.C. and Mason, A.M. 324
Connor, E.S. and Bruce, A.M. [sewerage] 492
Connor, F.R. [antennas] 447
Conseil International du Bâtiment (CIB) see Internat. Cncl for Building Research Studies
conservation
  of energy see under energy
  environmental see environmental engineering
  of structures 551–3
Conservation sourcebook (pub. Museums and Galleries Commission) 552
Considine, D.M. [instruments/control] 125, 398
Considine, D.M. and Considine, G.D. [instruments/control] 398
Considine, D.M. and Considine, G.D., Van Nostrand's scientific encyclopedia 122
Constantinescu, V.N. et al. 175
constants
  physical/chemical 203, 232

thermodynamic 203
CONSTRADO, former 539
constraint 158
Construction and Building Materials 532
CONSTRUCTION AND CIVIL ENGINEERING INDEX database **529**
Construction Computing 532, **580**
Construction design and management regulations (CDM) 549
construction engineering **519–45**
  nature and role 519, 545, 547
  abstracts/indexes 100, 113, 228, **527–30**
  automation 378, 382, 580
  books
    bookshops, specialist 520, **533**
    catalogues 527, 528, 529, 533, 543
    dictionaries 123, 533–4
    directories 85–6, 126, 475, 523, **527–30**, **534**, 543, 585, 587
    handbooks 125, 326, 508, **533–4**, 537–8
    publishers, selected **531**
    theses **530–1**
  civil engineering **545**
  computers and 521, 531, 532, 580, 587
  cpI 579
  conferences 382, 474, 522–3, **530**, 585–7
  Interbuild 87, 89
  contractors 67, 547–8, 569, 576, 579
  databases 95, 113, 230, **527–31**, **581–4**
    catalogues 90
  demolition 543, 544
  design see under design applications
  energy services see electricity; heating etc.
  environmental impact **492–3**
  exhibitions 520, 521
  historical **534–5**
    archives 93, 94
    see also under structural
  journals 64, **67**, 68, 105, 522, **531–3**
  management **569–88**
    marine 469tab, 472, 474
    marine see under marine technology
  materials see under materials
  organizations 95, 210–11, 403, 511, **520–6**, 536–7, 538–9
    CIIG **526**
    ECC 95, **520**
    international **523–4**
    regulatory **525**
    trade/companies **524–5**, 528
  plant 67, **543–4**
  decommissioning 544
  power issues see buildings; energy and specific fuels

product information 85–6, 520, 527, 528–9
quality 585
research 230, 520–5, 530–1
robotics 378, 382
safety 326, 525, 530, 532, 543–4, 569
  HSE reports 577
  temporary works 544
Standards 95, 472, **524**, 526–7, 528–30, 534, **537**, 543, **544–5**, 582, 583
project co-ordination 578–9
Regulations 95, **549**, 550, 583
specification 95, **527**, 538, 540, 542, 578
technical information 82
statistics 534, 570, **572–4**
temporary works **544–5**
trade literature 93–5
trademarks 88
and transport 496
see also structural engineering; construction management; building
Construction Equipment 67
Construction Europe 69
Construction Fixings Association **565**
Construction History 535
Construction History Society (CHS) 535, 553
Construction Industry Computing Association (CICA) **521**, 587
Construction Industry Information Group (CIIG) **526**
Construction Industry Research and Information Association (CIRIA) **521**, **549–50**, 585
  publications **521**, 541, 549–50, 557, 558, 564, 567
  catalogue 521, 527, 585
  CIRIA guide to European ... and international ... information 523
  CIRIA guide to sources of construction information 585
  CIRIA UK construction information guide 524, 526, 587
  Guides **521**, 556, 565
  Reports **521**, 524, 528, 529, 545, 551, 555, 556, 557, **585**
  Special Publications **521**, 563, 567
Structural design standards ... **527**
Technical notes (TNs) **521**, 543, 545, 550, 558, 565
and sealants 543
training videos 521
Construction information digests (pub. CIOB) 521
CONSTRUCTION INFORMATION SERVICE database **528–9**, 581–2

Index 685

Construction leads (pub.
  Camargue) 576
construction management 519,
  550, **569–88**
  nature and role 569–71
  abstracts/indexes **582–4**
  CAWS **578–9**
  conferences **585–7**
  consultants 569
  contractors 67, 547–8, 569,
    576, 579
  costing 583
  current trends information
    576–7
  databases **581–4**
  economics issues 570, 580,
    583, 586
  financial information 576
  forecasts 575–6, 575*tab*
  handbooks 125
  information issues 569–71
  project information **577–9**
  journals 67–8, 69, **579–80**
  maintenance 586
  market adaptability 574
  organizations 521, 576–7,
    585–7
  planning 584
  procurement 585, 586
  Production Drawings Code 578
  project management 569, 586
    information co-ordination
    577–9, 587
  Project Specification Code
    578–9
  software **581–2**
  statistics 569, 570, **572–4**,
    572–3*tab*
  temporary works **544–5**
  tenders 584
  see otherwise construction
    engineering
*Construction Management and
  Economics* **580**
Construction News 532, **580**
Construction papers (pub. CIOB)
  521
Construction safety (pub.
  Building Advisory Service)
  543
*Construction Weekly plant
  directory, former* 543
consultants
  construction industry 569
  transport 511, **512**
  yearbook 126
*Consulting engineer's who's who
  and yearbook* (pub. ACE)
  126
Consumer Association, *Which?*
  105
consumer magazines 72, 105
containers [transport], Standards
  23
Contemporary Ergonomics 339
contents pages **102, 103**, 111,
  460, 640
*Continuing survey of road goods
  transport* (pub. HMSO) 509
Contol Theory and Applications
  (IEE *Proc.*) 401

Contractor reports (pub. TRL)
  545, 566
contractors
  construction industry 333,
    547–8, 569, 579
  journals 67
  organizations 576
  electronics 361
  transport 545, 566
  contracts 569
Contractual handbook (pub.
  BCSA) 539
Control and Dynamic Systems
  401
*Control and Dynamic Systems –
  Advances...* 105
control engineering *see* control
  systems
Control Engineering (pub.
  Cahners) 400
Control Engineering series (pub.
  IEE) 401
control of industrial major
  accident hazards (CIMAH)
  318
Control and Instrumentation 280,
  402
Control of landfill gas (pub.
  DoEnv) 614
Control of Pollution Act (1974)
  487
Controls and Systems 379, 401
Control of Substances Hazardous
  to Health (COSHH),
  Regulations 181
control systems 285–7, 286*fig*,
  389–409, 435–7
  nature and role 279, 285–7,
    286*fig*, **389–90**
  abstracts/indexes 101, 111,
    341, **399–403**
  adaptive 363, 390, 393, **395–6**,
    398, 401
  aerospace engineering 286
  and artificial intelligence
    396–7, **402–3**
  Astrom-Hagglund approaches
    392–3
  automation *see* automation
  batch control 279, 280, **432**
  biochemical engineering 278
  biomedical engineering 630,
    638, 639
  books 279–80, 383, **391–9**,
    423
  sources summarized 399
  dictionaries *etc.* 154, 398–9
  directories 400
  handbooks 125, 288, 327,
    398, **399**
  chemical engineering **279–80**
  chips, commercial 395
  computer control 205, **391,
    393–4**, 398, **432–3**
  computer-aided design
    397–8
  see also CAD/CAM
  and concurrent engineering
    398
  conferences 291–2, 296, 298,
    319, **399–403**

continuous-time control 392–3
Coon-Cohen approaches 392
cross-couplings 394
databases 111, **399–403**, 421–2
design of *see under* design
  applications
deterministic 390, 393
digital systems **393–4**
discrete-time control 393
electric drives 434–7
and electricity generation 252,
  429, 435–7
electrohydraulic 285–99
expert systems 164, 167, **312,
  396, 397, 402–3**
feedback 286, 286*fig*, **389**,
  393, 394, 397
field-oriented 437
frequency-responses 392, 394
fuzzy logic 391, **396**, 403
Gaussian (LQG) regulators
  393, 394, 397, 398
H2/H-infinity 390, **397–8**
hierarchical 393–4
historical aspects 389–90, 395
hydraulics *and* **285–7**, 286*fig*
identification/estimation 390,
  391–2, 393, 395
instrumentation **398**
Internet resources 119
journals 86, 87, 280, 282, 290,
  292, **356–7**, 378–9,
  399–403, 415–17, **459**,
  460
  electronic newsletter 115
  reviews 105
Kalman filters **391–2**, 394
knowledge-based 312, 393,
  396, 402–3
learning systems 396, **402–3**
linear 389, 391, **392–4, 456**
linear-quadratic (LQ[G])
  approaches 393, 394,
  397, 398
non-linear 391, 393, 394,
  457, 460
linguistic models 396
loops 391
manufacture 354, 356, 358
marine 469*tab*
mineral processing 592
minimum-variance regulators
  393
modelling 389, **390–8**, 459
  linguistic 396
  qualitative 397
monitoring 390
multivariable 286, **394–5**
numerical 383, 386
on-line tuning 390, **395–6**
optical methods 298
optimality **390**, 393, 394–5,
  398, 401
organizations 279, 291–2, **371,
  400–3**, 413, 414–15
parallel processing 392, **398**
parameter drift/perturbation
  395, 397
phase margins 392
PI[D] (proportional, integral
  and [derivative]) 389–90,
  392–3, 397

Index

control systems (contd)
  pneumatics 285
  process control *see under* processes
  proportional, integral [and derivative] (PI[D]) **292–3**, 389–90, 397
  real-time systems 393–4, 398, 403
  reliability 319
  and robotics 378
  robust **390**, 393, 394–5, 396, **397**
  safety 279, **319**, 320
  relief systems 327
  self-learning **402–3**
  self-organizing **396**
  self-tuning 390, **395–6**, **402–3**
  signal processing **286–7**, 286*fig*, 392, 393, 398
  simulation 390, **392**
  software 398
  Standards 279
  state space approaches 390, 391, 393, 394
  stochastic control **390**, 391, 393, 395–6
  supervisory 391
  system analysis **432–3**
  system transfers 390, 393
  thermal systems 205
  traffic 495–6, 498–9
  training 399
  vector control 437
  verification **390**
  VME control 393
  Ziegler-Nichols approaches 392–3
  see also automation; mechatronics; systems *etc.*
Control Systems Magazine (pub. IEEE) 401
convection 199, 202, 235
Convention du Mètre (1875) 27
conveyors
  automated 385
  particles 278
Cook, N.J. [*wind/structures*] 554
Cooke, P. *et al.* [*design*] 337
Cooke, T.H. [*concrete*] 537
cooking 203
cooling/cooling systems 199–201, 204–5, **207**, 253–5
  coolant pressures 261
  cooling towers 207, **567**
  district 209, 253
  nuclear reactors 260–2
Coombe, R.A. 252
Coon-Cohen control methods 392
Cooper, B.R. *and* Ellingson, W.A. [*coal*] 238
Cooper, J. [*traffic*] 506
Cooper, P.F. *and* Findlater, B.C. [*water*] 488
Cooper, W. Fordham [*safety/electrical*] 320
co-ordinated project information (CPI) **578–9**, 587
Co-ordinating Committee 578
COPANT *see* Comisión Panamericana ...

Cope, L.W. *and* Rice, L.R. 613, 615
copper 593, 607
  bronze 589
  cladding/roofing 564
  early uses 589
  leaching 599
  organizations 303, 564
  waste proportions 591
Copper Development Association 303, 564
copyright 36
  electronic publishing 119
  licensing 41
  of patent documents 48
Corbett, J. *et al.* 337
CORDIS database **505**
Corduneanu, C. 456
Corfield, K.G. 329
Corlett, J. 657
Cornish, E.H. 312, 337
corrosion 539, 558
  and concrete 557
  corrosives 323
  expert systems 312
  journals 104
  thermal processes 235, 236, 238
  water treatment/sewerage 492
Cortada, J.W. *and* Woods, J.A. 334
COSATI (Committee on Scientific and Technical Information) 6, 9
COSHH [control of substances hazardous to health] Regulations 181
Cossalter, V. *et al.* 186
Costanzo, L. 336
cost benefit appraisal [program] *see* COBA
Costing in design ... (pub. SEED) 337
Cote, E.A. *and* Linville, J.L. 322
Coughanowr, D.R. 279
Coulson, J.M. *and* Richardson, J.F. **271–2**, 592, 598, 600
Council for Building Research Studies ... *see* Internat. Cncl ...
Council of Mining and Metallurgical Institutions, conferences 612
Council for National Academic Awards (CNAA) 79
Managing design ... 329
Council for Scientific and Industrial Research, Feilden Report 329
Council on Tall Buildings and the Urban Habitat [US] 530
Counihan, M. 232
county councils *see* local authorities
County Surveyors Society 514, **515**
Coupland, J.W. 653
couplings 155, 161*tab*, **163–4**, 341
Court, A.W. *et al.* 340
Cover, T.M. *and* Thomas, J.A. 457

Cowan, H.J. 535
Cox, B. *and* Webb, R. [*costing*] 337
Cox, J. [*building databases*] 526, 583
Cox, S.J. *and* Tait, N.R.S. [*safety*] 325
cp *see* Codes of Practice
CPA (critical path analysis) 334
CPD (contact potential difference) 522
cpI *see* co-ordinated project information
Crabb, J.A. 248
cracking,
  environmentally-assisted 140
crack-tip stresses 136
Crafer, R.C. *and* Oakley, P.J. 364
Craig, J.J. 394
Crane, F.A.A. *and* Charles, J.A. 308
cranes 436, 543
Cranfield Unit for Precision Engineering 359
cranks 155
Cravalho, E.G. *and* Smith, J.L. 202
Crawford, M. 491
'Crawler' Internet list 312
CRC Press [publishers] 635, 636
CRE (Coal Research Establishment) 226
Creasey, D.J. 443
creativity 156, 330, **332**
creditworthiness 576
CRIB see Current research in Britain
Crickmer, D.F. *and* Zegeer D.A. 613, 616
criticality, nuclear reactions **260**
critical path analysis (CPA) 334
Critical Reviews in Biomedical Engineering 642
Critical Reviews in Physical and Rehabilitation Medicine 642
Croatia, stress analysis 144–5
Croft, A. *et al.* 454
Croner Publications
  Croner's health and safety at work **181**, 182, 334
  Croner's Health and Safety Briefing 181, 183
  Croner's manual handling operations 325
  Trade directories of the world 86
Croney, D. *and* Croney, P. [*roads*] 506
Croney, J. [*anthropometrics*] 336
Croome, D.J. *and* Roberts, B.M. 207
Cross, N. [*design*] 156, 331
Cross, N. *et al.* [*design*] 156
Crossley, E. [*design*] 166
Crossley, F.R.E. [*IFToMM*] 152
Crowder, M.J. *et al.* 178
Crowl, D.A. *and* Louvar, J.F. 325
Crozier, R.D. 599
CRRERIS *see* Commonwealth Regional Renewable ...

Cruse, T.A. 319
crushing, mineral processing 591, 598, 600
cryogenic engineering 104, 327
Cryogenics safety manual (pub. Brit. Cryogenics Cncl) 327
crystals **590–1**
  crystallites 136
  crystallization 274
  grains 136, 591
  lattices 136
CS *see* Concrete Society
CSCW (*Computer Supported Cooperative Work*) 654
CSIRO *see* Commonwealth Scientific and Industrial...
CSO *see* Central Statistical Office
CSSC *see* Centre for Strategic Studies in Construction
CSSP/Thomas Telford Software databases 582
CTI PLUS [*Current Technology Index*] database 104, 153, 400, 401, 402, **486**, **529**
Cuadra [publishers] 581
Cubbon, R.C.P. 327
Cullen, J. *and* Hollingum, J. 179
Cullen Report (1990) [Piper Alpha disaster] 477
Cullingworth, J.B. 506
Cumo, M. *and* Naviglio, A. 325
Cumulative Book Index (pub. Wilson) 103
current
  alternating (ac) 320, 432, **436**
  direct (dc) 320, 435, **436–7**
  HVDC 432
current awareness services **102–3**
  books 104
  contents pages 102
  and databases 93, 229, 421–2
  and journals 65, 87
  'ready-made' 96
  trade literature 96
  see also selective dissemination of information
Current Bibliography on Science and Technology 227
Current Biotechnology 283
Current Biotechnology Abstracts, former 283
Current British directories (pub. CBD) 86, 125
Current Contents (CC) [series] 102, 111, 640
CURRENT CONTENTS on DISKETTE 111
CURRENT CONTENTS SEARCH 111
... *Engineering, Technology and Applied Sciences* **102**, **111**, 228, **380**
  online versions **111**, **230**
... *Physical, Chemical and Earth Sciences* **111**, 228
Current Index to Statistics 113, **460–1**
Current Literature in Traffic and Transportation 504
Current Mathematical Publications 113, **460**

Current Papers in Computing and Control 399
Current Research in Britain . . . (CRIB) (pub. Longman Cartermill) 78, 126, 503, 515
Current Technology Index (CTI) **100**, 102, 153, 227, 341, **380**
  online/[CTI PLUS] 104, 153, 400, 401, 402, **486**, **529**
Current Topics in Transport **503**, 504
Curriculum for design . . . (pub. SEED) 332, **333**
Currie, D.M. 554
Curtin, W.G. *et al.* 560, 565
Curtin University of Technology [Australia] 115
curves, mathematics of 454, **455**
Cusumano, J.A. *et al.* 238
Cuthbert, L.G. *and* Sapanel, J.-C. 444
cutting processes 353, 358, 359, 360, 362, 363–4
  laser cutting 363
CWCT (Centre for Window and Cladding Technology) 563, 564
cyanidation 590, **599**, 600
cybernetics
  and ergonomics 336
  journals 401
  and robotics 378
  see also computers; robotics etc
Cybernetics and Systems 401
cyclic processes 200
cyclists 496
cycloconvertors 436
cylinders, power 288, 290
Czech Republic
  machines 162
  stress analysis 144–5

D
DABS compendium (pub. MBR) 85–6
DADS [program] 170*tab*
Dai, S.-H. *and* Wang, M.-O. 178, 179
Dale, B.G. *and* Oakland, J.S. 178
Dalgleish, D.I. 444, 447
Dally, J.W. *et al.* 367
dams 247–8, 255, 545
  tailings dams 592
Dana, E.S. 591, 598
danger, *defined* 316
... *Dangerous goods Regulations (pub.* HSE) 477
Daniel Industries 276
Danish language, databases 584
Danish Ship Research Institute 478
Dansk Standardiseringsraad 27
Dansk standard 27
Danubia-Adria Symposium **144–5**
Darlow, M.S. 167
DARPA *see* Defense Advanced Research . . .

data
  compression of 115, **457**
  co-ordination concept (CPI) 577–9
  Working Party 577
  and 'information' 570
  manufacturers' *see under* components
  networks *see* networks, computer
  security of 362
  source books **124–5**, 272–3
  telecommunications 412, 443, 444
  transfer *see* data transfer
  see also information; statistics etc.
data banks *see* databases
database management system (DBMS) 581
databases **107–20**, 224, **228–9**, 304, **504–5**, 527, **580–1**, 582
  CD-ROM *see* CD-ROM
  abstracts/indexes 100–3
  bibliographic 63, **107–13**
  bibliographies 103
  of catalogues 90
  catalogues of 504
  citations 102, **109**, 224, 400, 461, 504
  conferences 76–7
  Proceedings 103, 109–10, 421
  contents pages 111
  currency 306
  data transfer *see* data transfer
  directories of 57, 328, 581
  on diskette *see* disks: floppy
  encyclopaedias 271
  export/report *see* data transfer
  fee basis of 53–4, **93**, 103, **108**, 581
  connect-time 91, **93**, 108
  free/subsidized 304, **305**
  licences 527
  'quality time' 108
  subscription 103, 108
  trip-rates 510, **511**
  variations 108, 584
  general, selective list **109–11**
  host services **107–9**, 229, **422**, 505, **584**
  INSPEC **421–2**
  journals 71–2, 73, **74–5**
  newsletters 70
  keywords 98
  materials **304–7**
  multilingual 505
  and networks **108–9**, **117–20**, 527
  newspapers 72–5
  numerical data 306–7, 310–12
  offline services 581–2
  'oldest' 58
  patents 48, 53–6, 57–60, 341
  product information 89, **90–3**
  reports 10–12
  as sales aids 91, 93, 304–5
  searching approaches 55–6, **98–9**, **421**, 504, 505
  coding systems 56, 58

databases
  searching approaches (contd)
    decision-trees 312
    full-text 56, 72
    indexes 90, 99, 354, **421**, 461
    keywords 56, 59, 98, 109, 485
    keywords, ambiguous 644
    librarian-/specialist-mediated 98, 107, 505
    menus 72
    thesauri 354, 421
    user friendliness 98, 107, **108**
    see also indexes
  SIGLE 10–11
  software accompanying 59–60
  specialist, selective lists **111–13, 229–31**
  statistics 511
  tape versions **108, 422**
    leasing 108, 450
  trade directories 91–3
  trademark directories 88
  updates 306
  see also abstracting services and specific topics
Data book for tyres and rims (pub. ETRTO) 20
Datacom 449
Data co-ordination . . . in the construction industry (pub. DoEnv.) 577–8
Data digests (pub. D.A.T.A. International Inc.) 418–19
data networks *see* networks, computer
DATAPLAS database 313
Data Resources Inc. (DRI), *DRI Energy Bulletin* 234
data sheets 82, **83**, 93, 291, 302, 330
Data Source [program] 582
DATASTAR database host 107, 108, **109–12**
  CEBA 112
  citations 109
  COMPENDEX 110
  construction 584
  contents pages 111
  CURRENT CONTENTS 111
  and DIALOG 107
  electrical/electronics 111, 422
  INSPEC 111, 422
  journals 75
  and Knight-Ridder 108
  materials 112
  METADEX 112
  newspapers 71
  NTIS 109
  product information 92, 93
  RAPRA ABSTRACTS 112
  science 109
  theses 78
data transfer **309–11, 445**
  expert systems 311–12
  Standards/protocols 118, 305
  CALS 310–11, 314
  IGES 335

STEP 307, **309–10**, 314
  see also networks
data trees 503
DAT (digital audio tape) 108
DAtF (Deutsches Atomforum) **263**
Daubert, T.E. *and* Danner, R.P. 272
Davey, N. [*building materials*] 535
Davidson, J. [*reliability*] 319
Davidson, J.F. *et al.* [*fluidization*] 278
Davies, A.C. [*welding*] 362
Davies, B.J. *and* Crawley, E.J. [*steel*] 539
Davies, C. [*furnaces*] 206
Davies, D.W. *and* Barber, D.L.A. [*networks*] 446
Davies, V.J. *and* Tomasin, K. [*construction safety*] 543
Davinson, D. [*theses*] 78
Davis, G. [*org'ns/who's who*] 125, 126
Davis, G.H. [*water energy*] 248
Davis, J.R., *ASM materials . . . dictionary* 123
Davis, L. [*Building Regulations*] 549
Davis, M.D. [*decision theory/design*] 332
Davis, M.H.A. *and* Vinter, R.B. [*control systems*] 391
Davis, Langdon & Everest [co.] 525
Dawson top 3000 directories and annuals 86, 125
Day, A. *and* Langford, V. [*cpi*] 578
Day, D.A. *and* Benjamin, N.B.H. [*construction equipment*] 544
DBMS (database management system) 581
DC [1] *see* Design Cncl
dc [2] *see* direct current
DD (BSI Drafts for Development series) 177
Deacon, R.C. 556
dead loads 553
Dean, A.E. *and* Tower, K. 322
Deasington, R.J. 445, 446
De Beers [co.] 595
De Bono, E. 332
De Bremaecker, J.C. 245
decay 199
DECHEMA eV **273**
  and CEBA 283
  Chemistry data series **272**
  data compilations 273
Decision and Control conference 403
decision-making/decision theory 155–6, 308, 332, 333–4, 358, 401, 403
decision-support systems 393
decision-trees 312
Deckwer, W.D. 275
DECO (Denver Equipment Co.), *Modern mineral processing flowsheets* 600

Decommissioning of radioactive facilities (pub. IMechE) 250
Decubitus 639
decubitus [pressure sores] 637, 639
Defence Helicopter 67
defence industries
  databases 91–2
  data transfer 311
  handbooks 307
  journals 67, 460
  reports 4, 8–9
  robotics 378
  software **310–11**
  Standards **23**, 91, 421
  component specifications **91**, 418
  US 22, **26–7**, 91–2, **310–11**
  see also aerospace; marine etc.
Defence Research Agency (DRA) 8
Defence Research Information Centre (DRIC) 8–9, **23**
  DRIC Abstracts 9
  Guide to services 23
Defence specifications (DEF/SPEC) 23
Defence standards (DEF/STAN) 23
Defense Advanced Research Projects Agency (DARPA [USDoD]), DICE 656
deformation
  forming 364
  and strain/stress **134**, 176, 306–7
DeFrancis, J. 454
DEF/SPEC (*Defence specifications*) 23
DEF/STAN (*Defence standards*) 23
DeGarmo, E.P. *et al.* 360
degradation
  energy 201, 204
  materials 199
Dekker [*publishers*], *Encyclopedia of chemical processing and design* 271
DE/Mec [program] 171*tab*
Deming, 367
demolition 543, 544
Dempa Publications, *Japan electronics buyers' guide* 418
De Nevers, N. 276
Denmark
  design 162
  marine technology 478
  Standards 20, **27**
  wind energy 247
  see also Danish language
Denn, M.M. 276
Dennis, G.H., *Guide to official statistics* 126
Denno, K. 210
Denny, M.W. 635
dense medium separation 599, 601
density, *and* thermodynamics 202
Denver Equipment Co. (DECO), *Modern mineral processing flowsheets* 600

Index 689

Departments/Ministries *see under*
United Kingdom; United
States *etc*.
Derby, B. *et al.* 337
De Renzo, D.J. 235
Derive [program] 452
Derrick, A. *et al.* 243
Derry, T.K. and Williams, T.I.
154
Derwent Information/Derwent
Publications [co.] 56, **59–60**
WORLD PATENTS INDEX 55, 58,
59, 380
desalination 203, 204–5, **243–4**
Desch, H.E. 561
Deshpande, P.B. 279
design [1], procedures **155–6**,
166, 301, 308, **329–51**
nature and role 98, **155–8**,
329–32, 331*fig*
interdisciplinary aspects 329,
330, 332, 333
strategic role 329, 330, **333**
abstracts/indexes 156–7, **341**
aesthetics **336**
analysis/stress analysis **157–8**,
159, **160**, 179, 273, 291,
330, 338
needs analysis 333
sensitivity analysis 333, 334,
335, **337**
synthesis 166–7, **273**
value analysis 337
anthropometrics **336**, 339
artificial intelligence 335, 339
assurance 332
audits/reviews 179, 333
books 156–7, 158, 274, 308,
312–13, 329–51
buyers' guides **157**, **340**
case studies 332, 337
catalogues 90, 156–7, 340,
579
directories **340**, 370
handbooks **124–5**, 156, 159,
160, 313, **339–41**, 446,
653
reports **329**, 330, 341, 656–7
sales literature 304–5
communication 178, 330, **338**
complexity 330
components 156, **163–000**,
291, **337–8**
computer-aided **334–5**, 339,
341, 342, 362, **397–8**
properties data 273
software 158, 159, 160–1,
161*tab*, **170–2***tab*, 281,
653
see also CAD
concepts **155–6**, 301
and concurrent engineering
333, 335, 337, **649–60**
continuous development 652
conferences 156, 309, 322,
330, **342**, 358–9, 401,
474, 655–6
costs/costing 330, 333–4,
336–7, 369
creativity 156, 330, **332**
customer-orientation 649, 651

databases 90, **341**, 528
CD-ROM 157, 340, **341**
data transfer 310
see also materials
data comparison **160–1**,
161*tab*, 291
decision-making 155–6, 308,
332, **333–4**
'design-and-build' 548
details 333, **337–8**
dimensions 337
dimensionless parameters
156
three-dimensional 336
discipline 330
education for *see* training
electrical aspects 331, 340
engineering drawing 310,
337–8
environmental aspects 430
ergomonics **329**, 330, **335–6**,
339
evaluation 333
exhibitions **342**
expert systems 396, 397
'features' 310
fundamental design 332
historical aspects 155, 535
interior design 94
industrial aspects **336–7**
'industrial design' 336
information sources **339–42**
information technology 125
innovation 156, **332**
integration aspects 330, **335**,
337
interdependencies 332
Internet resources 119, **341–2**
iteration 330, 332, 335
journals 139, 142–4, **157**, **163**,
169, 254, 266, 281, 330,
338–9, 356–7, 416, 481,
653–5
kinematics **158–9**
legislation 333, **334**
linear systems 334
management *and* 329, 330,
333–4, **336–7**, 338, 365–6
and marketing 304–5, 331*fig*,
332, 335–6
mass transfer 273
materials 330, 337–8, 341
see also materials
mathematical aspects **334**, 455
mechanical aspects 331, 340
methodologies 155–6, 166,
330, **331–2**, 333–4, 397–8
modelling **334–5**
needs analysis 333
noise control 181
Nyquist Stability technique
397
optimization 156, 166, 167,
169, 178, 333, **337**, 397
organizations 152, **342**, 587,
656, 657
reliability 178–9
patents 330, **341**
PDS 333
precision 354
problem-solving 155–6, 159,
330

PABLA 156, 332
process analysis/synthesis **273**
process-based 330, 331–2
product-based 330, 331, **331***fig*
as product development 336
product information 81, 94,
340, 416
catalogues 90
product liability 334
profitability 337
prototyping, rapid 653
psychological aspects 332
quality 177–83, 333, **334**, 337
reliability 178–9, 318, 319,
333, **334**, 337
reviews/audits 179, 333
'right first time' 333
and risk 333
'robust design' **179–80**, 280,
334
and safety 182, 315, 318, 319,
322, **325–7**, **334**, 368
searches 334
sensitivity analysis 333, 334,
335, **337**
sequencing 332, 331*fig* 332,
334
simulation 335
simultaneous design 650
specifications **155–7**, 178,
331*fig*, **333**, **577–9**
standard parts 337–8
Standards 160, 161*tab*, 330,
333, **337**, **339–40**, 343–4,
348–9, 526–7
EU Directives 341
EU Eurocodes 549
safety **182**, 368
stress analysis **133–49**
sub-processes 273
synectics 346
syntheses 166–7, 273
system analysis 115, 273–4,
331–2, **432–3**
Taguchi system 177, **179–80**
tolerances 337
'total activity' 330
total design 331, 333
trade names 431
training 329, 332, **334**, **336**,
338
Design Council review 336
OU courses 321, 334
SEED publications *see*
SEED
tribological aspects 173, 174
trouble-shooting 160, 166
unit operations **273–4**, 488
variables 334
see also design applications;
materials
design [2], applications 269, 329,
336–7
aerospace 155, 653
alternative energies 245, 246
and assembly 337, **361**, 653
automation 383, 401, 653
automotive industry 179, 338,
341
biochemical engineering **278–9**
chemical engineering 269, 271,
272, **273–4**

design [2], applications
  chemical engineering (contd)
    chemical reactors 275
    physical chemistry 275–6
    unit operations 273–4
    construction 508, 519, 522,
      526–7, 528, 535, 587
    bridges 500, **508**, 545, 566
    'design and build' 548
    energy systems 253–5
    Production Drawings Code
      578
    structures **547–68**
    see also architecture
  control systems 279–80, 341,
    379, **390–1**, 393, **396**,
    397–8, 423
  curves 455
  electrical/electronics 156, 331,
    340, 341, 420, 423,
    434–5, 446, 448
  energy systems 205, 209, 236,
    252, 253–5, 429, 430, 434
    nuclear engineering 156,
      250–1, 259–61, 266
    solar energy 243–4
    water energy 247–8
  environmental engineering 430,
    488, 489, 491
  fibre optics 448
  fluid mechanics **276**, **295**
  heating, district 209
  industrial aspects **326–7**
  machines **151–84**, 313, 331,
    338, 340, 341, 368
    electric 434
    machine tools 360
  manufacturing aspects 36,
    329–51, **333–4**, **336–7**,
    338, 356–7, 358, 359,
    360, **365–6**, 369, 384
  marine 240, 467, 469tab, 473,
    474, 477, 480
  mineral processing 598–600
  mining 616
  offshore installations 240
  particle| technology **278**
  power engineering 429, 434
  roads 500, **508**, 545, 566
  robotics 383
  ships/shipping **467**, 474
  solar energy 243–4
  surfaces 455
  of systems 331–2, 356, **390–1**,
    395, 397–8, 415, 423, 448
  telecommunications 446, 448
  thermal processes 210, **276–7**
  transportation 338, 500, **508**,
    511
  water energy 247–8
  see also design procedures
*Design of air-supported
  structures* (pub. IStructE)
  568
Design Automation Conference
  401
Design Council (DC) 336, 342
  publications 130, 303, 313,
    342–7, 350–1
    *Industrial design
      education . . .* 336

Design Engineering 163, 176
Design Engineering (pub.
  Maclean Hunter) 338
Design Engineering (pub.
  Morgan Grampian) 157, 338
Design Handbooks (pub.
  Associated Spring) 187
Design Management Institute,
  Design Management Journal
  338
*Design manual for roads and
  bridges* (pub. DoT) 500,
  508, 545, 566–7
Design and Manufacturing
  Systems Conference **342**,
  359
Design News 338
*Design and operation of
  industrial compressors* (pub.
  IMechE) 208
Design procedural guides (pub.
  SEED) **160**, 161*tab*, 163,
  341
Design Products and
  Applications 157, 338
*Design recommendations for . . .
  car parks* (pub. IStructE/
  IHT) 568
Design Studies 338, **654**
Desktop Engineering Ltd 171*tab*
desk-top publishing 14
De Sola, R. 124
Detail design phase . . . (pub.
  SEED) 333, **337**
deterministic processes **451**
deuterium
  and fission process 260
  and fusion process 261
Deutsche Gesellschaft für
  Medizinische Physik 630
Deutsche Informationszentrum
  für Technische Regeln
  (DITR) 24
  DITR DATENBANK 24, 583, 584
Deutsche Nationalbibliographie
  103
Deutschen Keramischen
  Gesellschaft, conferences
  359
Deutsches Atomforum eV
  (DAtF) **263**
Deutsches Bundesanstalt für
  Geowissenschaften und
  Rohstoffe (BGR) 602
Deutsches Institut für Normung
  eV (DIN) 16, **24**, 524
  and BSI 22
  *DIN Katalog für technische
    Regeln* 24
  metrology 28
  and PERINORM 22, 152, 582
Devaney, R.L. 457
developing countries
  patents 45–6
  wastes, solid 490
  wastewater 488
  water treatment 489
*Development of alternative
  sources of energy* (pub.
  DOE) 210
Devol, George C. 377

DeVor, R.E. et al. [quality] 334
De Vos, A. [solar energy] 243
De Vries, L. and Herrmann,
  T.M. 123
Dewar, J.D. and Anderson, R.
  537
Dewis, M. and Murdoch, J.,
  *Tolley's health and
  safety . . . handbook* 181
Deyell, John, Co. [publishers]
  400
DG[I/II etc] (Directorates of Eur.
  Commission) see under
  Europe
DHA see District Heating Ass'n
diagnostic engineering
  conferences **296**
  journals 327, 358
  organizations 327
  see also fault diagnosis;
    condition monitoring
Diagnostic Engineering 327
Dial engineering 85
Dial industry 92
DIALOG database host 107,
  107–8, **229–30**, 231
  AEROSPACE DATABASE 112
  automation 380
  biotechnology 112
  CEBA 112
  chemical engineering 112
  citations 109
  COMPENDEX 110, 153, 354, 380
  conference papers 77
  construction 584
  contents pages 111
  CURRENT CONTENTS 111
  and DataStar 107
  directories 73
  electrical/electronics 111, 422,
    426
  energy **229–30**, 267
  ENERGY SCIENCE AND
    TECHNOLOGY 112
  FLUIDEX 112
  INSPEC 111, 380, 422, 426
  ISMEC 111
  journals 75
  and Knight-Ridder 584
  KNOWLEDGE INDEX 107
  machines 111, 153
  manufacture 354
  materials 112
  MATHSCI 113
  METADEX 112
  newspapers 71
  NTIS 109
  patents 58, 59, 380
  product information 92, 93
  RAPRA ABSTRACTS **112**
  robotics 380
  science 109
  Standards **113**, 368
  statistics 113
  theses 78
  transport 505
  WATER RESOURCES ABSTRACTS
    113
Dialog Information Services **229**,
  426
Diamant, R.M.E. and Kut, D.
  209, 253

Index 691

diamonds 593
  journals 357
DICE (DARPA Initiative in Concurrent Engineering) 656
Dickey, J.W. *et al.* 506
Dickson, E.M. and Loperena, G.A. 246
dictionaries/glossaries/encyclopaedias **121–4**, 232
  abbreviations 124
  acronyms **124**
  aerospace 123
  bi-/multilingual **123**, 124, 152, 154, 233, 267, 271, 370, 454
  CAD/CAM 370
  chemistry/chemical engineering 122, 271
  civil engineering 123
  computers/information technology 124
  construction 123, 533–4
  control systems 154, **398–9**
  cooling processes 207
  electrical/electronics 123, 124, 421
  encyclopaedias **122**
  energy 232, 233, 256, 267
    nuclear energy 267
  environment 123
  eponyms 122
  ergonomics 123
  fluid mechanics 276, 287
  glossaries **123**
  information processing 124
  languages **123**, 124, 152, 154, 233, 267, 271, 370, 454
  machines 123, 152, **154**, 370
  machine tools 370
  magnetic materials 312
  management 123
  manufacturing 123, **370**
  materials 122, 123, 309, 312, 313, 370
  mathematics 453, 454
  measurement/instruments 154, 398
  physics 122
  process control systems 398–9
  production management 123
  quality/reliability 154, 177, 179, 180
  report series codes 5
  robotics 154
  safety 316, 334
  statistics 453, 454
  symbols **124**, 287
  thermodynamics/thermal systems 201, 207, 253
  translating **123**
  tribology 154
  see also directories
die-casting, automated 377
dielectric materials 419
Dienes, L. *et al.* 233
diesel oil, *and* wind energy 246
Dieter, G.E. 156, 308, 332
Dietrich, A. 169
differentials, mechanical 156
diffraction, stress analysis **136**, 147–8

Diffrient, N. *et al.* 336
diffusion 275
Digest of United Kingdom Energy Statistics (pub. HMSO) 232
Digital Signal Processing 416
digital signals **412**, 419, **423**, **441–2**, 443
  audio tape (DAT) 108
  computers 412
  control systems **393–4**
  digital buses 125
  telecommunications 412, 441–2, 443
  see also signal processing
digitization, images 503
Dijksman, E.A. 158
Dimancescu, D. 652
Dimarogonas, A.D. 158, 163
DIMDI 77, 109, 111
DIN [German Standard (/organization)]
  DIN Katalog für technische Regeln 24
  see otherwise Deutsches Institut für Normung
Din, F. 240
Dinter, F. *et al.* 205
Dirac delta 457
direct current (dc) 320, 435, **436–7**
  HVDC 432
Direction de la Sûreté des Installations Nucléaires (DSIN [France]) **264**
Directives [of Eur. Commission]
  see under Europe
Directorate of Technical Development (DTD [of Min. of Aviation]) 23
directories **84–6**, 92, **125–6**, 303
  CD-ROMs 231, 504
  of abstracting/indexing services 99, 227
  biographical/who's who **126–7**
  books 104
  chemical engineering 86, 270, 272
  chemicals 323
  components 370, 418
  conferences 76–7, 267, 281
    Proceedings 77, 530
  construction 85–6, 126, 475, 523, **527–30**, **534**, 543, 585, 587
  data 142, 272
  of databases 304, 504, 526, 581, 582–3
  design **340**, 370
  of directories 72, 84, **86**, **125**, 340
  discussion lists 116
  on diskette 418
  electrical/electronics 85, **418**
  energy 232, 233, 242, 249, 255
    nuclear energy 233, 261, 267
  Europe **126**, 340, 418
  exports 92
  hazardous substances 323

heat recovery 206
  indexes to 72–3, 76, 77, 84
  information sources 303
  instruments 418
  Internet information 118–19
  journals 70, **72–3**, 74, 99, **400**
    electronic journals 114, 115, 116
    online journals 74
  local authorities 526
  machine tools 85
  manufacturers 340
  manufacturing 370
  marine technology **474–5**, 478
  materials 86, 303, 323, 370
  metals 303
  minerals/mining 600, 601–2, 608, 612
  newspapers 72–3
  offshore engineering 474–5
  organizations 85, **125–6**, 303, **340**, 370, 523, 526, 552
  ports 475
  process engineering 86
  product liability 334
  quarrying 602
  research 123, 370
  roads/transport 510–11, 534
  safety 323
  shipping 474–5
  software 328, 587
  Standards organizations 18, 25
  statistics 510
  thermal systems 206
  theses 79
  trade directories 65, **84–6**, 87, 92, 370
  trademarks/-names **88**, 92, 100, 341
  transport/roads 510–11, 534
  universities 126
  yearbooks 72, 73
Directories in print (pub. Gale) 84, **86**
Directory of databases (pub. Gale) 581, 582–3
Directory of Directors 92
Directory of EEC information sources (pub. Euroconfidentie) 126
Directory of electronic journals (pub. ARL) 114, 115, 116
Directory of European industrial and trade associations (pub. CBD) 370
Directory of information sources in the UK (pub. Aslib) 303
Directory of mines and quarries (pub. BGS) 602
Directory of online databases (pub. Gale) 581
Directory of periodicals online ... (pub. Fed. Document Retrieval) 74, 400
Directory of portable databases (pub. Quadra/Gale) 581
Directory of published Proceedings 77
Directory of quarries, pits and ... equipment (pub. Quarry Management) 602

Directory of safety-related computer resources [ed. Brauer] 328
Directory of sources in transport statistics (pub. TSUG) 510–11
DIS (Draft Internat. Standard [of ISO]) 18
Disability Now 643
Disability and Rehabilitation 643
Discrete Applied Mathematics 460
discussion lists, electronic 116–17, 502
 directory 116
 moderated lists 116
Diskette of world electronics data (pub. Elsevier) 418
disks
 CD-ROM 306
 floppy/diskettes
  data on 108, 111, 305–6, 355, 418, 422, 504, 581–2, 609, 613
  with books 152, 158, 159, 160, 167–8, 178, 183–4, 281
 see also software; computers
dispersion 199
display screens 182
disposal 310
 see also waste
Dissertation abstracts 78
 online 78, 530–1
dissertations see theses
distillation 273, 274, 279
 control systems 398
distress 174
district engineers, organizations 515
district heating 208–9, 210, 225, 253–5
District Heating Association (DHA) 209, 210
 Practical aspects of district heating... 209
District heating combined with electricity generation... (pub. DoE) 209
District Heating and Cooling 209
District heating handbook (pub. IDHA) 209
District heating and technological developments (pub. BP) 209
DITR
 DITR DATENBANK database 24, 583, 584
 see otherwise Deutsche Informationszentrum...
Dixon, A.E. and Leslie, J.D. 243
Dobby, G.S. and Roa, S.R. 603
DODGE CONSTRUCTION ANALYSIS SYSTEM database 583
DoE/DOE (Dept of Energy) see under United Kingdom; United States
DoEnv (Dept of Environment) see under United Kingdom
Dohr, G. 240
Dokumentation Electrotechnik see ZDE

domes, concrete 557
Doppler effects, flow analysis 298
Doran, D.K. [construction materials] 535, 559
Doran, P. [hazards] 317
Dorf, R.C. and Nof, S.Y. 125
Dorling, A.R. 453
Dorman Long safe load tables for steel 552
Dormer, P. 336
Dossett, D. 16
DoT/DOT (Dept of Transport) see under United Kingdom; United States
Doughty, S. 159
Douglas, J.M. 272, 274
Dow Chemical Co. 305
 Fire and explosion index... (pub. AIChE) 317, 322
 Flotation fundamentals and mining chemicals 600
Dowding, C.H. 614
Dowling, P.J. et al. 540, 558
Down, C.G. and Stocks, J. 614
downstream processes 278
Downton, A. 336
Dowrick, D.J. 327, 554
Dowson, D., [tribology] 154, 173
Dowson, D. et al. [tribology] 174
Dowson, D. and Wright, V. [biomechanics] 646
Doyle, J.C. et al. 397
DRA (Defence Research Agency) 8
Drabble, G.E. 158
Draft code... for... silos, bins, bunkers and hoppers (pub. BMHB) 278
drafting, Standards 16
Drafts for Development (DD) series (pub. BSI) 177
Drago, R.J. 164
Dragt, A.J. and van Ham, J. 492
Dravo Corp'n, Analysis of large-scale non-coal... mining... 615
dressing, minerals/ore 590
Drewry Shipping Consultants [co.], Shipping Statistics and Economics 479
DRI (Data Resources Inc.), DRI Energy Bulletin 234
DRIC
 DRIC Abstracts 9
 see otherwise Defence Research Information Centre
driers/drying 205, 244, 274, 321
drilling
 metals 363
 mineral extraction 614, 615
 Drilling and blasting... (pub. Australian Mineral Foundn) 614, 618
Driskell, L.R. 279
drives 163–5, 195, 433–7
 switched-reluctance 435
Drives and Controls 165
Druce, G. and Watson, A.C. 167
Dryden, I.G.C. 205, 232

Dryden, I.G.C. and Griffith, M. 232
drydocks 478
drying/driers 205, 244, 274, 321
DSH program 296
DSIN see Direction de la Sûreté...
DTD (Directorate of Technical Development [of Min. of Aviation]) 23
DTI (Dept of Trade and Industry) see under United Kingdom
Duan, G.R. 397
Duan, G.R. et al. 397
Duck, F.A. 636
Ductile Iron Pipe Association 303
ductility 303, 364
Duderstaadt, J.J. and Hamilton, L.J. 265
Dudley's gear handbook [ed. Townsend] 164
Duffie, J.A. and Beckman, W.A. 210, 243
Dukelow, S.G. 279
Dun & Bradstreet [co.] 576
Dun and Bradstreet Europa 85
Key British Enterprises 85
Dunlop, J. and Smith, D.G. 443, 447
Dunn, I.J. [biochemical engineering] 278
Dunn, P.D. [alternative energies] 210
Du Pont/Du Pont de Nemours [co.'s] 305
 Blasters handbook 614
 Elastomers' notebook 69
Dupraz, J. 458
Durelli, A.J. and Parks, V.J. 147
Durelli, A.J. and Riley, W.F. 145
dusts see powders
Dutch language
 databases 584
 dictionaries 154, 271
 journals 458
 Standards 20
Dwivedi, S.N. et al. 362
Dyke, P. and Whitworth, R. 158
Dym, C.L. 335
dynamics 151, 159–60, 183
 analysis 158
 books 152, 154, 159
 casting processes 364
 CFD 294
 conferences 161–3
 generators 433
 hydrodynamics 251–3, 276, 276, 469tab
 journals 159–60, 378, 379
 magnetohydrodynamics 251–3
 marine 469tab
 and robotics 378, 379
 solidification 364
 systems 389, 391, 395
 thermodynamics 199–200
 transmission lines 292–3
 see also machines; fluid power etc.
Dynapak [program] 170tab

Index 693

E
EAGLE (Eur. Ass'n for Grey Literature in Europe) 10
Ealing Electro-Optics plc 146
Earle, J.H. 335
Earney, F.C.F. 482
Earth Energy 250
Earthquake Engineering Field Investigation Team (EEFIT) 554
Earthquake Engineering and Structural Dynamics 532
earthquakes 327, 532, 549, **553–4**
  seismic loadings 553–4
earth sciences
  databases 111, 229, 602
  journals 609
  see also geology *etc.*
earthworks 543
East, S. 366
Easterby, R. 336
Eastern Bloc Energy 234
Easthope, C.E. [*kinematics*] 159
East Kilbride, NEL 139
Eastop, T.D. *and* Croft, D.R. [*thermodynamics*] 255, 277
Eastop, T.D. *and* McConkey, A. [*thermodynamics*] 202
Ebert, K. *and* Von Ammon, R. 324
EBSCO [co.] 74, 400
  Index and abstract directory 99
EBU (Eur. Broadcasting Union) 417
EC [1] (Eur. Commission) *see under* Europe
EC [2] (Eur. Community, *former*) *see* Europe
EC [3] *see* Eurocodes
EC [4] *see* Engineering Cncl
ECC *see* Eur. Construction Centre
ECCS
  ECCS steel construction international directory 534
  see otherwise Eur. Convention for Constructional Steelwork
ECE (Economic Commission for Europe) *see under* United Nations
EC Energy Monthly (*pub.* Financial Times) 234
ECIF (Electronic Components Industry Fed'n, *former*) 414
Eckhoff, R.K. 322
ECMT *see* Eur. Conference of Ministers of Transport
École Nationale des Ponts et Chaussées **523**
ecology 640
  databases 229
  ecosystems 255
  water 487
  see also environment; environmental engineering
ECOMINE database 602
econometrics, construction forecasts 575

Economic Commission for Europe *see under* United Nations
economics
  and automation 386
  Budget information 570
  and construction 570, 580, 583, 586
  and energy 205, 208–9, 210, 229, **231–5**, 239, 255–6, **425, 429–31**
    alternative energies 242–3, 245, 255, 430–1
    'energy economics' 223, 234
    environmental pressures 430
    nuclear energy 250, 252, 429
    power systems **425**, 428
  family expenditure 511
  financial information 84, **576**, 589
  forecasts 575–6, 575*tab*
  and geology 229
  and information 574
  journals 75, 234–5, 358, 498–9, 580
    trade journals 66
  and management 358, 569
  and marine technology 479, 482, 510
  and mineral processing 589, 594
  and mining 614
  and production 358
  and quality 368
  and telecommunications 443
  and transport 495, 496, 498, 501, 503, 504–5, 507
  urban 499
  see also statistics
Economic and Social Research Council (ESRC)
  ESRC 510, **511, 515**
  and Science Research Council] 515
Economic use of gas-fired boilers (*pub.* DoE) 206
Economides, M.J. *and* Ungemach, P.O. 245
Economist, The 75
Economist Intelligence Unit, *Quarterly Energy Review*s 232–3
ECSC (Eur. Coal and Steel Community) 20
EDCs (Eur. Documentation Centres) 313, **341**
Edel, D.H. [*design*] 332
Eder, W.E. [*design Proceedings*] 162
EDF (Électricité de France), EDF-DOC database 230
Edgar, T.F. *and* Himmelblau, D.M. 274
Edge, K.A. *and* Johnston, D.N. 293
EDI *see* electronic data interchange
EDIFACT Standard **311**

Edinburgh
  'Edinburgh Engineering Virtual Library' ('EEVL') internet site 120
  Heriot Watt University 120
Edison, Thomas Alva 411
Edmeades, D.H. 544
Edmonds, D. 96
education, of engineers *see* training
Educational Resource Information Center (ERIC [US]) 11
Edwards, L. *and* Endean, M. 312, **360**
EEA (Electronic Engineering Ass'n, *former*) 414
EEC (Eur. Economic Communities, *former*) *see* Europe
EEF
  EEF directory... 85
  see otherwise Engineering Employers Fed'n
EEFIT (Earthquake Engineering Field Investigation Team) 554
EEMUA *see* Engineering Equipment and Materials Users Ass'n
EEO *see* Energy Efficiency Office
'EEVL' ('Edinburgh Engineering Virtual Library') Internet site 120
efficiency, thermal **200, 203–4**, 205, 277, **428**
Efficient use of energy [ed. Dryden/*pub.* DoE] 205
effluent *see* wastewater
EFOMP (Eur. Fed'n of Medical Physics) **629**, 630
EFTA *see* Eur. Free Trade Ass'n
EGF (Eur. Group on Fracture, *former*) 140
Eggleston, D.M. *and* Stoddard, F.S. 246
Ehringer, H. *et al.* 207, 209
EHV (extra-high voltage) 252, **429**
EI [1] (Engineering Index)
  EI COMPENDEX *see* COMPENDEX
  EI MANUFACTURING database 354
Ei [2] (Engineering Information)
  'Ei Engineering Meetings' 530
  EI PAGE ONE database 229
  'Ei Village' Internet site 120
  see otherwise Engineering Information
EIA [1] (Electronics Industries Ass'n) 414
EIA [2] (Energy Information Administration [US]) 226
  EIA data index (*pub.* USDOE) 232
EIA [3] (environmental impact assessment) **492–3**, 500
eigenfunctions 397, **456**
Eindhoven University of Technology [Netherlands] 115

694 Index

EIS (Engineering Integrity Soc.) **137**
Eisenreich, G. and Sube, R. 454
Eisenstadt, M. and O'Shea, T. 397
elasticity 134, 550, 553
  elastic emission machining 357
  photoelasticity 135, 145–6
  thermoelasticity **136**, 147
  viscoelasticity 134
elastohydrodynamics 174, 176
elastomers 69, 169
Elastomers' notebook (pub. Du Pont) 69
Elderly Care 643
Electra **431**, 437
Electrical Communication 449
Electrical and Electronics Abstracts **101**, 341, 380, 399, **421–2**, 426
  indexes to 99, 421
  online **111**, 230, 380, 399, **421–2**, 426
  see also INSPEC
Electrical and electronics trades directory (pub. Peregrinus) 418
Electrical Engineer 427
electrical engineering 223–57, 425–39
  abstracts/indexes **101**, 111, 341, 354–5, 380, 399
  and biomedical engineering 624, 629, 639, 645
  books
    'Blue book' **418**
    buyers' guides 414
    dictionaries 123, 124
    directories 418
    handbooks 125, 340, 341, 431
  conferences 434
  controls *see* control systems
  control systems for 391
  databases 59, 111, 230, **421–2**, 426
    see also INSPEC
  design 156, 331, 340, 341
  electrochemical processes 105, 274, 357, 363
  electrohydraulics **285–99**
  electromagnetism 419–20, 434
  electro-optics 146
  electrophysical processes 358
  historical aspects 94
  Internet resources 119
  journals 67, 87, 415–17, 426–7, 434, 435
  machine drives **433–7**
  magnetohydrodynamics **251–3**
  manufacturing 354–5, 356, 357, 363
  materials 419–20
  mineral processing 589
  mining 609
  organizations 24, 138, 320–1, 371, **413–14**
  patents 59
  power systems **429**
  product information 414
  safety 182, **320–1**

'intrinsic' 320, 321
Standards 16, **18–19**, 24, **26**, 123, **320**, 645
Electricity at Work Regulations **182**
IEE wiring regulations **320**, 321
  telecommunications **441–50**
  trade literature 94
  water engineering 247–9
  wiring 320, 321
  see also electricity; electronics; electrotechnical *etc.*
electrical machines
  nature and role 425, **433–5**
  power-conditioning 426, **435**
Electrical Patents Index 59
electrical power **425–39**
Electrical Research Association, *1967 steam tables* . . . 235
electrical resistance strain guages 135
Electrical Review 427
Electrical Times 427
Electrical Wholesaler 67
Electrical World 252, 427
Électricité de France (EDF), EDF-DOC 230
electricity 320, **425**, **428**
  current 320, **436–7**
    alternating 320, 432, **436**
    direct 320, 435, **436–7**
    HVDC 432
  energy efficiency 203
  generation of *see* energy technology; nuclear energy *etc.*; power engineering
  legislation 320
  safety **320–1**, 322
  Standards 27–8, **320**
  static **320**, 321, 322
  and telecommunications 441–2
  and telephony 441
  terminals 444
  voltage **320**
    EHV 252, 429
    HVDC 432
    motors 435, 436
  see also electrical engineering; electricity generation *etc.*
Electricity Council, *Power system protection* 320
Electricity at Work Regulations **182**
Electric Machines and Power Systems 252
Electric Power Applications (IEE Proc.) 252, 434
ELECTRIC POWER DATABASE (EPD) **229**, **426**
Electric Power Research Institute (EPRI [US]) 227, 264, **264**
  bulletin board 117
ELECTRIC POWER DATABASE **229**, 426
  Journal 252
  publications 238, 246
Electric Power Systems Research 252
Electric Vehicle-Battery Technology 252

Electrocardiology 639
electrochemical processes 105, 274, 357, 363
electro-discharge processes 357, 363
electroencephalography 639
electrohydraulics **285–99**
  see also fluid mechanics
electromagnetism 419–20, 434
electro-magnetic compatibility (EMC) 420
Electronic Components Industry Federation, *former* (ECIF) 414
electronic data interchange (EDI) 16, **311**
Electronic Design 416
electronic engineering **411–24**
  nature and role **411–13**, 422
  analogue/digital division 412, **423**
  abstracts/indexes 100, 101, 102, 111, 341, 355, 380, 400–3, **421–2**, 426
  applications information **419**
  and biomedical engineering 626–7, 628, 629, 630, 639, 641, 642
  books **417–20**, 421, 422–3
    'Blue Book' **418**
    data books 418, 420
    dictionaries 123, 124, 421
    directories 85, **418**
    handbooks 418–19, **421**
    publishers, selected 422–3
    buyers' guides 414, 418
  circuits
    applications **419**
    integrated **412**
    components 91, 360, 412–13, 414, 416, **418–19**, **420**, 421–2, 448
    equivalents 418
    preferred lists 91, 418
    Standards 91, 420
  and concurrent engineering 653, **654**
  conferences **417**, 421
  consumer goods 412–13, 416–17
  contract manufacturing 361
  control systems **389–409**
  databases 90, 91, 111, 355, 380, 400–3, 418, **421–2**, 426
  data transfer 310
  design *see under* design applications
  and electrical drives **435–7**
  equipment 146–7
  exhibitions 414
  hydraulics and 285–7
  instrumentation 415
  journals 361, 413–14, **415–17**, 421–2, 426–7, 435, 448–9, 654
  electronic journals 115
  reviews 105
  trade journals 67, **416**
  machine drives 434, **435–7**
  manufacture 353, 355, 360, 367, 369

Index 695

electronic engineering
  manufacture (contd)
  assembly 361
  materials 417, **419–20**
  mechatronics 166, **169–73**, 292
  medical 102, 629
  microelectronics 353, 367, 424
  nucleonics 267
  optoelectronics 353, 415, 423, **447–8**, 448
  organizations 24, **413–15**, 628
  product information 310, **414**, 416, **417–20**
  reliability 319, 412
  safety **319**, 320
  signal processing *see* signal processing
  Standards **18–19**, 24, 25, **91**, 123, 414, **420–1**, **449–50**
  specifications 310, **418**, **420–1**
  training 413, 414, 415, 420
Electronic Engineering 67, **416**
Electronic Engineering Association, *former* (EEA) 414
Electronic Industries Association (EIA [US]) 414
Electronic Libraries (Elib) programme 120
electronic mail *see* e-mail
Electronic Packaging and Production 361
Electronic Parts Information Centre/CODUS EPICTM 91
electronic publishing **71–2**, **114–16**
  newsletters **115**
  preprints 115–16
  see also journals: electronic
Electronic Research Centres, *former* 126
Electronics 416
Electronics and Communications Abstracts 100
Electronics and Communications Engineering Journal **416**, 449
Electronics and instruments directory (*pub.* Morgan Grampian) 418
Electronics Letters 115, **415**, 449
  online version **115**
electronic speckle pattern interferometry (ESPI) **135–6**, 146–7
Electronics Today International 417
Electronics Weekly 416
Electronics World and Wireless World 417
Electronic Times 416
electrons **411**
  electron-beam processes 364
  electron tubes 411
electro-optics 146
electrophysical processes 358
electro-refining 599
electrostatics **320**, 321, 322
electrotechnical engineering abstracts 101

Standards 17, **18–19**, 24
  see otherwise electric; electronic *etc.*
electro-winning, mineral processes 599
E-Letter on Systems, Control and Signal Processing [Netherlands] 115
Elgerd, O.I. 428
Elia, F.A., Jr 317
Elib (Electronic Libraries) programme 120
Elliott, C.D. [*architecture*] 535
Elliott, K.S. *and* Tovey, A.K. [*concrete*] 556
Ellis, A.J. *and* Mahon, W.A.J. 245
Elmahdy, A.H. 253
El-Masri, M.A. 204
Elsevier Science Ltd [publishers] 85, **151**, 209
  dictionaries, multilingual 123
  Diskette of world electronics data 418
  Elsevier materials selector databook 132, 308, 313
  Elsevier's dictionary of chemical engineering [ed. Clason] 271
  European electronics directory 85, 418
  Handbook of industrial materials 313
  International electronics directory 418
EMA/EMA see Engineered Materials Abstracts
e-mail (electronic mail) **114–17**, **502**, 584
  and biomedical engineering 644
  bulletin boards **117–18**, 341–2, 502,, 584, **644**
  and concurrent engineering 657
  and databases 229, 511, 583
  design 341–2
  discussion lists **116–17**, 502
  filtering 117
  newsletters **115**
  news reader programs 117
  and transport 502
Emanuel, G. 203
EMBS see Engineering in Medicine and Biology Soc.
EMC (electro-magnetic compatibility) 420
emergencies 317–18, **323–4**, 325–6
  see also safety
Emergency handling . . . in surface transportation (*pub.* AAR) 323
Emergency power systems at nuclear power plants . . . (*pub.* IAEA) 325
El/[&]MJ (Engineering and Mining Journal) 596, 600, 611, 612
El/[&]MJ international directory of mining 601, 608

employers' associations *see under* organizations
employment, statistics **574**
EMS Chemie [co.] 305
EN [1] Standards 19
En [2] (English translation of Standard) 24
encyclopaedias **122**
  see otherwise dictionaries
Encyclopedia of chemical processing and design (*pub.* Dekker) 271
Encyclopedia of hydraulics, soil and foundation engineering [ed. Vollmer] 122
Encyclopedia of materials science and engineering [ed. Bever] 122
ENDS (Environmental Data Service), *ENDS Report* 504
energy 199–201, 223–57
  theory of **199–201**, 223
  alternative sources *see* alternative energies
  atomic *see* nuclear energy
  audits of 254, 257
  availability **201**, **203–4**
  biomass 210, 225, **241**, **249–50**
  coal and 590
  and combustion 199, 428
  conferences 110
  Energy Managers' Workshops 205
  conservation [thermodynamics] **199–200**, 204, 223, 521
  conservation [economical use]
    see under energy technology
  conversion of **199**, **425**, 428, 434
  power engineering **425**, **429–30**
  current trends **209–10**, **429–30**
  databases 110, 225
  degradation 201
  economics aspects *see under* economics
  efficiency **200**, **203–4**, 205, 277, **428**
  electricity *see* electricity
  electromagnetic 434
  enthalpy **202–3**
  entropy *see* entropy
  exergy **201**, **203–4**
  geothermal **245–6**
  heat *see* heat
  historical aspects 94
  journals 7–8, 67, 209
  kinetic 256
  legislation **233**, 240, 320, 430
  low/high grade **204**
  and manufacture **363–4**
  nuclear **259–62**
  and process 269
  quality/quantity **204**
  recovery of **206–7**
  redistribution of 277
  renewable *see* alternative energy
  solar *see* solar energy
  stable 199, 203

energy (contd)
  storable 199, 203
  storage systems 205, **206–7**, 210, **255–7**
  technology *see* energy technology
  thermal **199**
  total energy resource analysis (TERA) 235
  trade literature 94
  and transport industry 495
  transport of 202–3
  water 210, **247–9**, **430**
  wind *see* wind
  and work 199–201
  see otherwise energy technology
Energy 234
ENERGY database 225, 230
Energy Abstracts 227
Energy Analysis 234
Energy balances of OECD countries (*pub*. OECD/IEA) 232
energy-beam processes 357, **363–4**
Energy from Biomass and Wastes 250
Energy Conservation Digest 257
Energy conservation in IEA countries (*pub*. IEA) 256, 257
Energy Conservation News 257
Energy Conversion and Management 234, 257
Energy Daily 234
Energy database (*pub*. USDOE)
  ... *subject categories and scope* 8
  ... *subject thesaurus, permuted listing* 232
Energy Design Update 254
Energy Economics 234
Energy Economics and Climate Change 234
Energy Economist 234
Energy efficiency in buildings... (*pub*. EEO) 253
Energy Efficiency Office (EEO) 208, 210
  Compressed air and energy use 208
  Energy efficiency in buildings... 253
Energy Engineering 234
Energy and Environment 242
Energy and the environment (*pub*. Friends of the Earth) 255
Energy Exploration and Exploitation 234
Energy and Fuels 234
Energy Information Administration (EIA [US]) **226**
  EIA data index 232
Energy Information Centre, British Gas *directory of energy-saving...* 255
Energy in Japan 234
Energy Journal 234

Energy law... (pub. IBA) 233, 240
Energy Letters 234
ENERGYLINE database 230
Energy manager's workbook (*pub*. Energy Publications)) 205
Energy Managers' Workshops 205
Energy maps (*pub*. Petroleum Economist)
  ... *of the Middle East* 232
  ... *of Europe* 232
Energy Papers (*pub*. HMSO) 234
Energy Policy 234
Energy Prices and Taxes 234
Energy Publications [co.], *Energy manager's workbook* 205
Energy recovery in process plants (*pub*. IMechE) 207
Energy Research 234
Energy Research Abstracts (ERA) **7–8**, 11, 112, 227, 267
ENERGY SCIENCE AND TECHNOLOGY database **112**, 229
Energy Sources 234
Energy statistics and balances of non-OECD countries (*pub*. OECD/IEA) 232
Energy statistics of OECD countries (*pub*. OECD/IEA) 232
Energy statistics yearbook (*pub*. UN) 232
Energy Systems and Policy 234
Energy Systems Trade Association 403
energy technology **223–57**, **425–39**
  nature and scope **223–4**, 231–2
  abstracts/indexes 101, **112**, **227–8**
  alternative energy *see* alternative energy
  automation 403
  books **121**, **231–3**, 235–57
  bibliographies 227, 228
  dictionaries 232, 253, 256, 267
  directories 232, 233, 242, 249, 255
  general/multidisciplinary **231–5**
  handbooks 231–3, 235, 256–7
  maps 232
  business information **229**, 230
  coal processes **237–9**
  combustion processes **235–7**
  conferences 224, 226, 233, 242
  conservation [economical] systems 209, 223, **253–7**
  databases **112**, 226, 227, **228–31**
  design *see under* design applications
  efficiency **255–7**
  energy audits 254, 257
  environmental issues *see under* environment

gas **239–41**
Internet resources 119
journals 112, 226, 227, 231–2, **233–5**, 237, 239, 240–1, 242, 254, 257, 426–7
legislation 233, 240
management systems **255–7**
nuclear *see* nuclear energy
oil **239–41**
organizations 210–11, **224–7**, 234, 403, 594
  international **224–5**
  national **225–7**
policies 223, 225–6, **231–5**
renewable energy *see* alternative energy
reports 224, 225, 227
statistics 225, **231–5**, 239, 252
storage systems **255–7**
taxes 233
total energy approaches 204, 235
see also thermal systems; power engineering *etc*.
Energy Technology Support Unit (ETSU) **226**, 243, 245, 246, 247
Energy terminology... dictionary (*pub*. World Energy Conf.) 233
Energy Today 234
Energy for tomorrow's world (*pub*. WEC) 233
Energy Trends (*pub*. DTI) 234
Energy World 234
Engelberger, Joseph 377
Engineer 67, 103, **157**, 338
Engineer buyers' guide **85**, **157**, 340
engineered materials 112, 124, 228, **302**, 303, 307–8, 312
  abstracts/databases 228, 229, **306**, 341, **355**
Engineered Materials Abstracts (EMA) 228, 341
  online 112, **229**, 230, 231, **306**, **355**
Engineered materials handbook (*pub*. ASM) 121, 312
engineering **1–2**, 119, **624**
  biographical data **126–7**
  concurrent **649–60**
  history of 66, 93–4, 99, **154–5**, 624, **633**
    archives **93–4**, 510–11, 552–3
    biographical 126–7
    buildings 93, 522, 527, **534–5**, 538, 549, **551–3**
    civil engineering 94
    computers 377, 411, **412**
    control systems 389–90, 395
    databases 58
  design 155, 535
    CAD/CAM 155, 390
    interior design 94
  electronics 411–13, 433–4, 435
  fuels 58, 590
  gas industry 94, 226
  gears 154

Index

engineering
  history of (contd)
    gemstones 589
    ICE 522
    IEE 413
    interior design 94
    iron 589, 607
    IStructE Study Group 553
    linkages 155
    machines/mechanisms
      154–5, 159, 166, 173,
      433–4
    materials 93, 535, 551–3
    metals 589–90
      smelting 589
    minerals/mining 154, 226,
      589–90, 607, 617
    patents **31–2**, 43, 51, 58, 59
    petroleum industry 94
    pigments 589
    power generation 226, 429,
      431, 433–5
    privatization 226, 429, 502,
      512, 594
    product information 93–4
    robotics 378
    silver 154, 589
    Standards 17, **20–1**
    steel 538
    stone/rock 589, 607
    stress analysis 133
    system analysis 432–3
    telecommunications 441–2,
      448
    television 411
    thermal systems 206
    tin 589
    trade literature 93–4
  interdisciplinary aspects 269,
    353, 623–4, 633
  parallel 650
  professional issues 580
  qualifications 371, 593, 609
  Reports, official **329**
  simultaneous 650
  technology transfer 371, 415
  topicality 99
  training *see* training
Engineering 67, 103, 157, 163,
  169, 338
*Engineering Applications of
  Artificial Intelligence* 357
Engineering and Automation 401
Engineering Council (EC)
  affiliated bodies 593–4, 609
  and design 342
  *Engineers and risk issues:
    code* . . . 315
  and qualifications 593, 609
  Risk guidelines for engineers
    315
engineering design *see* design
Engineering design [Feilden
  Report] 329
Engineering Designer 157, 163,
  338
*Engineering Design Graphics
  Journal* 338
*Engineering Design and
  Manufacturing Index (pub.
  TI)* **156–7**, 165, 370

Engineering Design Research
  Center [US] **656**
Engineering Distributor 68, 86
engineering drawing 337–8
  journal 338
  Standards 310, 337
Engineering Employers
  Federation (EEF)
  EEF directory 85
  Health and Safety Newsline
    328
  Practical risk assessment 317
Engineering Equipment and
  Materials Users Association
  (EEMUA), *Safety-related
  instrument systems* . . . 319
Engineering Failure Analysis 327
Engineering Fracture Mechanics
  143
*Engineering guide to spring
  design* 187
Engineering Index (*EI*) **99**, **100**,
  153, 228, 341, **355**, 380,
  529
  online **110**, 402, 485, **529**
  see otherwise COMPENDEX
  patents 58
Engineering Index Inc. 100, 583
Engineering Information Inc. (Ei)
  93, 100, **110**, 314, 400
  EI PAGE ONE 229
  'Ei Village' Internet site 120
  and MECHANICAL ENGINEERING
    ABSTRACTS 380
Engineering Integrity Society
  (EIS) **137**
Engineering Journal 532
Engineering Management 65
Engineering and Mining Journal
  (*E/[&]MJ*) 596, 600, 611,
  612
*E&MJ international directory
  of mining* 601, 608
Engineering News Record (ENR)
  72, 533
Engineering our future [Finniston
  Report] 329
Engineering and Physical
  Sciences Research Council
  (EPSRC) 415, **515**
Engineering Product Microfile
  528
  CD-ROM version **529**
Engineering Research Centres
  126, 370
Engineering Sciences Data Unit
  see ESDU
Engineering Standards
  Committee, *former* 21
Engineering Technical Microfile
  528
  CD-ROM version **529**
Engineering World 87
*Engineers and risk issues: code
  of professional practice*
  (*pub.* Engineering Cncl) 315
engines 205, 208, 235–7
  cyclic processes 200
    Rankine cycles 203, 207
  design 159
  internal combustion **200**, 203,
    235–7

steam 200
tribology 174
see also turbines *etc*.
England *see* United Kingdom
Englefield, M.J. 454
English Heritage [org'n],
  publications 552
ENIAC computer 412
Enichem [org'n] 305
ENO Foundation [US] **517**
  Transportation Quarterly 499
  E-Notes (*pub*. Internat. Inst. for
    Energy Conservation) 257
ENR (*Engineering News Record*)
  72, 533
ENS *see* Eur. Nuclear Soc.
enthalpy 202–3
Entropie 234
entropy **201**, 202–3, 204
  and 'civilization' 224
  constant 289
  organization/journal 234
ENV [Eur. pre-Standard]/
  Eurocodes **549**
ENVIROLINE database 229
environment **485–94**
  and chemical engineering 281,
    486, 489, 490
  and construction 569
  databases **229**
  and design 430
  and energy technology 223,
    228, 229, 230, 231–5,
    236, 238, **241–2**, **253–7**
  buildings 253–5
  electricity generation 255–7,
    430
  nuclear energy 250, 266
  and ergonomics 336
  Flowers Report 266
  and fluid mechanics 287
  and human factors 336
  journals 247, 257, 281, 498,
    504
  and manufacturing 361, 371
  and marine technology 470*tab*,
    487
  and materials 355
  and mineral processing 592,
    597
  and mining *see under* mining
  organizations 503
  PEST analysis 574
  restoration 592
  revegetation 592
  Royal Commission 500
  and transport industry 493,
    495, 496, 498, **500**, 507–8
  waste 493
  and water engineering **486–9**,
    493
  see also environmental
    engineering
Environment 257
Environmental assessment . . .
  (*pub*. DoEnv) 493
ENVIRONMENTAL BIBLIOGRAPHY
  database 229
Environmental Data Service
  (ENDS), *ENDS Report* 504
environmental engineering
  485–94, 547

environmental engineering (contd)
 nature and role 485, **486**, 489–91, 492–3
 abstracts 100, 228, **485–6**
 air **490–2**
 and biomedical engineering 639
 biotechnical approaches 492
 books
  bibliographies 492
  dictionaries 123
  handbooks 487, 489, 490, 491, 492, 493, 534
 case studies 493
 conferences 492
 construction research 521
 databases 111, **229**, **230**, **485–6**
 and design 488, 489, 491
 developing countries 488, 489, 490
 examinations **486–7**
 impact assessment (EIA) **492–3**, 500
 Internet resources 119
 journals 105, **257**, **487–92**, 597
 legislation 247, 486–7, 492
 measurements 486–7, 491, 492
 mine environmental engineering 617
 noise **492**
 odour **491–2**
 organizations 255, 487–93
 Standards **486–7**, 490, 492
 training 491
 wastes **488**, **489–90**
  see also waste
 water 485, **486–9**, 493
  see also environment; alternative energy
Environmental Engineering Abstracts 100
Environmental engineering in South African mines (pub. Mine Ventilation Soc. of S.A.) 617
Environmental evaluation (pub. SACTRA) 500
environmental health *see* pollution *etc.*; safety
environmental impact assessment (EIA) **492–3**, 500
Environmental Impact Assessment Review 257
Environmentally sound development in . . . energy and mining . . . (pub. Centre for Environmental Management and Planning) 255
Environmental Management 257
Environmental Pollution 487
Environmental Protection Act (1990) 487
Environmental Protection Agency (EPA [US])
 Extremely hazardous substances . . . 323
 and NTIS 486
Environmental Protection Bulletin 281

ENVIRONMENTAL RESOURCES TECHNOLOGY (ERTH) database 230
Environmental Science and Technology 487
Environmental Studies Institute, ENVIRONMENTAL BIBLIOGRAPHY 229
Environment and Planning 498
Environmetrics 257
EPA *see* Environmental Protection Agency
EPC *see* Eur. Patent Convention
EPCEM *see* Eur. Permanent Committee for Experimental Mechanics
EPCSA *see* Eur. Permanent Committee for Stress Analysis
EPD (ELECTRIC POWER DATABASE) **229**, **426**
EPIDOS database 58, **59**
EPO *see* Eur. Patent Office
EPOS database 313
EPRI
 EPRI Journal 252
 see otherwise Electric Power Research Inst.
EPSRC (Engineering and Physical Sciences Research Cncl) 415
equilibrium, thermal **199**
equipment **269–84**
 capital 369
 journals 67, 87
 reliability 368
 safety **182–3**, **186–7**, **326–7**
  interlocking devices 368
  'intrinsically safe devices' 320
 selection of 269, **273**, 274, 279
 Standards 368
 stress analysis 145–6
 Use of Work Equipment Directive **183**
 see also machines; plant; work equipment; components
ERA *see* Energy Research Abstracts
ERA Technology [co.] 414
 Buyers' guide to uninterruptible power supplies 414
Erdman, A.G. 155
Erdman, A.G. *and* Sandor, G.N. 160, 166, 167
Erectors' manual (pub. BCSA) 539
ergonomics 329, 330, **335–6**
 abstracts/indexes 100
 anthropometrics 336
 and automation 384
 and biomedical engineering 626–7, 640
 and concurrent engineering 652
 conferences 339
 dictionaries 123
 journals **339**, 637, 640
 organizations 339
 Standards 348–9

Ergonomics 339, 640
Ergonomics in Design 339
Ergonomics in engineering design . . . (pub. SEED) 336
Ergonomics Research Society, *Ergonomics* 339, 640
Ergonomics Society, *Proceedings* 339
ERIC (Educational Resource Information Center) 11
Erickson, W.D. 164
Ernst, R. 232, 370
erosion 173
Ertas, J./A. *and* Jones, J.C. 156, 331
ERTH (ENVIRONMENTAL RESOURCES TECHNOLOGY) database **230**
ESA *see otherwise* Eur. Space Agency
ESA/IRS database host 71, 108, 153, **230–1**
 automation 380
 conference papers 77
 construction 584
 directories 73
 electrical/electronics 111, 422
 energy **230–1**, 267
 INSPEC 111, 422
 machines 153
 MATHSCI 113
 newspapers 71
 NTIS 109
 RAPRA ABSTRACTS 112
 reports 7, 111
 robotics 380
 Standards **113**
 transport 505
ESAO (Eur. Soc. for Artificial Organs) 631
ESB (Eur. Soc. for Biomaterials) 631
Eschmann, P. *et al.* 176
ESDU (Engineering Sciences Data Unit)/ESDU International plc **142**, 155, 171*tab*, **273**, **536**
 publications
  Application of process integration to utilities . . . 204
  chemical/process engineering 273, 276, 277
  data compilations 273, 276, 277
  design procedures 273, 340
  Fluid mechanics: internal flow 276
  'Mechanisms' subseries 165, 167, 168, 169
  Physical data: chemical engineering 272
  Process integration 204
  'Stress and Strength' subseries 188
  'Tribology' subseries 173, **175–6**, 177, **187**, 188
  Validated engineering data index 142
 Rolling bearings 188
 software **161**, 161*tab*, 163, 173
  ESDUpacs 168, **171***tab*, **175–6**

Index 699

ESDU (Engineering Sciences
  Data Unit)/ESDU
  International plc
  software (contd)
  ESDUview interface 169
ESEM (Eur. Soc. for Engineering
  and Medicine) 629, **631**
  ESEM News 629, 643
  *Eshbach's handbook of
    engineering fundamentals*
    [ed. Tapley/Poston] 124
ESIS (Eur. Structural Integrity
  Soc.) **140**
ESPI *see* electronic speckle...
ESRC (Economic and Social
  Research Cncl), ESRC DATA
  ARCHIVE database 510, **511**
Essex, University, *and* ESRC 511
Estates Gazette, publications 552
ETRTO (Eur. Tyre and Rim
  Technical Org'n), *Data book
    for tyres and rims* 20
ETS/NET [Eur.
  Telecommunications
  Standard] 449
ETSI (Eur. Telecommunications
  Standards Inst.) 421, **449**
ETSU *see* Energy Technology
  Support Unit
Ettlie, J.E. *and* Stoll, H.W. 333,
  365–6
EU (Eur. Union) *see* Europe
EU directive on machinery safety
  19–20, 181, **183**
Euler, Leonhard 154–5, 159
EUNS (Ex-USSR Nuclear Soc.)
  265
EuReDatA (Eur. Reliability Data
  Banks Ass'n) 180
*Eureka: Engineering Materials
  and Design* 157, 309, 338
Euro Abstracts (*pub*. EU) 227
Eurocodes/ENVs [Eur.
  pre-Standards] **549**
Euroconfidentie [co.], *Directory
  of EEC information sources*
  126
Eurominerals [co.] **610**
EURONORM standards 20
Europa [co.]
  Europa world yearbook 126
  World of learning 126
Europe/European Union [EU]
  abstracts/indexes 227
  Agence Européenne
    d'Informations (AEI) 210
  automation 381, 382
  bibliographies 103–4
  buyers' guides 157
  CEC Joint Research Centre
    314
  chemical engineering 281
  committees 480
  Community Research and
    Development Information
    Service (CORDIS) **505**
  computer graphics 339
  conferences 77
  construction 69, 95, 584
  contractors
    Contractors' Meetings 235,
      244

electronics 361
databases 584
  CORDIS **505**
  see also ESA/IRS
and design 341
Directives [of EC] 19–20,
  180–1, **183**, 341, 508
Electro-Magnetic
  Compatibility 420
  environment 486, 491, **508**
  Machinery Directive **19–20**,
    181, **183**
  see also Standards *below*
Directorates *see under*
  European Commission
  below
directories 340
documentation **126**, **341**
Eastern Europe
  energy 225, 234
  patents 46
electricity *see* energy *below*
electronics 361, 417, 418,
  420–1
energy **225**, 227, 232–5, 241,
  252
  alternative sources 243–4,
    245, 247
  commercial aspects 252
  Energy Research and
    Development
    Programme 209, 225
  Eurowin database 247
  nuclear energy 262, 265,
    266
  Petroleum Science and
    Technology Institute
    (PSTL) 240
  environment 486, 487, 491,
    493, 508
European Commission (CEE/
  EC/CEC)
  Absorption heat pumps
    congress 208
  conferences 359
  contractors' meetings 207,
    209
  databases published 584
  Directives *see* Directives
    above
  Directorates (DGs) **508**, **516**
  Joint Research Centre 314
  publications 232, 252
  and Standards 19, 180–1
  Statistical Office 232
fluid power 287
gas industry 480
graphics 339
information sources 125, **126**,
  505
and information technology
  508, **516**
journals 72, 75
machines 19–20, 162–3, 181,
  183
manufacturing 357, 360, 361
  CIM 359
marine technology 474, 480
materials 304
minerals 480, **595–6**
newspapers 71, 75

Office of Official Publications
  20, 232, 584
official publications [*of* EU]
  126
offshore engineering 480
organizations **125**, 126, 340,
  370
patents 37, **46**, 54, 58–9, **341**
  Convention, proposed 46
  petroleum 480
product information 85, 93
product liability 334
publications
  Euro Abstracts 227
  Fluidized bed systems...
    235
  research 314
  current **126**, 505
  Research and Development
    Programmes [EU] 225,
    304, 305, 516
  robotics 382
  safety 180–1, **183**, 508
  Standards **19–25**, 420–1, 449,
    524
  *Common standards for
    enterprises* 20
  Directives *see* Directives
    above
  electronics 420
  ENVs/Eurocodes **549**
  harmonization 19–20, 27–8,
    420, 508
  machines **19–20**, 162–3,
    181, **183**
  regional, EU **19–20**
  *Standards for access to the
    EC markets* 20
  statistics 225, 232, 382, 510,
    516
  Annual Bulletins 252
  see also Eurostat
stress analysis 138, **139–40**
telecommunications 449
and thermal systems 207, 208,
  209
theses 78, 530
trade directories 85
trade literature *see* product
  information *above*
and transport 500, 504, **508**,
  510, **516**
UN Economic Commission for,
  see under United Nations
welding 360
WHO series for 491
European Association for
  Computer Graphics,
  Computer Graphics Forum
  339
European Association for Grey
  Literature in Europe
  (EAGLE) 10
European Broadcasting Union
  (EBU) 417
European Bulletin (*pub*. IChemE)
  281
European Centre for Nuclear
  Research *see* CERN
European Coal and Steel
  Community (ECSC) 20

European Commission (EC) *see under* Europe
European Committee for Electrotechnical Standardization *see* CENELEC
European Committee for Standardization *see* Comité Européen de Normalisation
European Community[ies] (EC) *see* Europe
European Conference on Automated Manufacture 381
European Conference on Fracture **140**
European Conference of Ministers of Transport (ECMT) 508, **516**
and databases 504, 505
Statistical report on road accidents 510
Statistical trends in transport 510
and TRANSDOC 504
European Conference on Power Electronics and Applications 417
European Construction Centre (ECC) 95, **520**
European Convention for Constructional Steelwork (ECCS) 538, 554, 562
ECCS steel construction international directory 534
European recommendations for aluminium alloy structures... 562
European recommendations for steel structures in seismic zones 554
Study on design of steel buildings in earthquake zones 554
European Documentation Centres (EDCs) 313, **341**
European Economic Area, safety 181
European electronics directory (*pub.* Elsevier) 85, 418
European Energy Report 234
European Environmental Law Review 487
European Federation of Medical Physics (EFOMP) **629**, 630
European Free Trade Ass'n (EFTA)
construction 69
Standards 19, 420
European Group on Fracture, *former* (EGF) 140
European Information Association, publications 132
European Journal of Non-Destructive Testing 357
European Journal of Surgery – Acta Chirurgica 639
European Nuclear Forum (FORATOM; Forum Atomique Européenne) **262**

European Nuclear Society (ENS) **265**
and Nuclear Engineering and Design 266
and Nuclear Technology 266
European oil and gas demonstration project... (*pub.* PSTI) 240
European Oil Hydraulic and Pneumatics Committee [CETOP] 287
European Patent Convention (EPC) **46**, 49
European Patent Office (EPO) 37, **46**, 54
databases 58–9
EPIDOS 58, **59**
European Permanent Committee for Experimental Mechanics (EPCEM) 133, **139–40**, 145
European Permanent Committee for Stress Analysis, *former* (EPCSA) **133**, 138, 139
European Power News 67
European recommendations for aluminium alloy structures... (*pub.* ECCS) 562
European recommendations for steel structures in seismic zones (*pub.* ECCS) 554
European Reliability Data Banks Association (EuReDatA) 180
European research centres (*pub.* Longman) 126
European Society for Artificial Organs (ESAO) **631**
European Society for Biomaterials (ESB) 631
European Society of Biomechanics [MEM] 631
European Society for Engineering and Medicine (ESEM) **629**, **631**
ESEM News 629, 643
European Society for Movement Analysis in Children 630
European sources of scientific and technical information (*pub.* Longman) 126
European Space Agency (ESA) 7, 210
conferences 204
online services *see* ESA/IRS
Spacecraft thermal... systems 204
European Structural Integrity Society (ESIS) **140**
European Telecommunications Standards Institute (ETSI) 421, **449**
ETS/NET [European Telecommunications Standard] 449
European Transactions on Electrical Power Engineering 252
European Transport Forum **500**, 501
European Tyre and Rim

Technical Organization (ETRTO), *Data book for tyres and rims* 20
European Union *see* Europe
European Workshop on Industrial Computer Systems (EWICS) 319
Europe Energy 234
Eurostat bureau 232, 510, **516**
Eurostat Energy Statistical Yearbook 232
Transport: annual statistics 510
EuroTech Direct conference 162–3
Eurotrans [co.], *Glossary of transmission elements* 154
EUROWELD conference 360
Eurowin project 247
evaluation, PERT 334
Evans, B.G. 443, 444
evaporation 277, 280
cooling systems 207
Everett, J.L. 444
Everyday with Practical Electronics 417
Evett, J.B. *and* Liu, C. [*fluid mechanics/hydraulics*] 205
Evetts, J. [*materials*] 312
evolution 634
Ewalds, H.L. *and* Wanhill, R.J.H. 141
EWICS *see* Eur. Workshop on Industrial Computer...
Ewing, J.A. [*engines*] 235
Ewing, W.M. *et al.* [*petroleum*] 240
excavation 544–5
excimer lasers 364
'Excite' Internet list 119
exergy **201**, **203–4**
Exeter, University, Camborne School of Mines 595
exhibitions 75, 87–8, 89
Building Centre 95
catalogues 84, **87–8**
see also conferences *and specific topics*
exothermic systems 259–60
expanders 205, **208**
expansion, fluid **199–201**, 204
experimental mechanics **133–4**, 174
see otherwise stress analysis
Experimental Mechanics 141, **142**
Experimental Physiology 640
Experimental Techniques 141, **142**
Expert: Intelligent Systems and their Applications (*pub.* IEEE) 402
expert systems **312**, **396**
component design 164, 167
design 397
journals **402–3**
Expert Systems: the International Journal of Knowledge Engineering and Neural Networks 402
exploration, geophysical 240

explosions
 hazard 236, 317, **321-2**, 323
 nuclear 261
explosives, mining/quarrying **614**
Explosives at quarries (pub.
 HSC) 614
Explosives and rock blasting
 (pub. Atlas Powder) 614
extraction processes
 minerals 589
 see also mining *etc.*
extra-high voltage (EHV) 252,
 429
Extremely hazardous
 substances . . . (pub. EPA/
 NTIS) 323
extrusion processes 364
Ex-USSR Nuclear Society
 (EUNS) 265
Eyre, E.C. 338

F
fabric, cladding 563-4
Fabricators and Manufacturers
 Association International,
 SCAE 651, 655
façades 551, 563
factories
 construction of 552
 'rational' 657
 see also workplaces
Factory Automation 400
Factory Inspectors' annual
 report (pub. HSE) 577
Factory Mutual Engineering
 Corp'n, *Loss prevention*
 data . . . 316
FAF (Forum Atomique
 Française) 264
FAG (UK) [co.] 176
 Ball and Roller Bearing
 Engineering 176
failure
 of structures 316
 of systems **178**, 315-21, 325-8
 failure-rates/modes analysis
 (FMEA/FMECA) 91,
 178, **179-80**, 318, **327**,
 367
 and parallel processing 398
 see also fault diagnosis;
 reliability
Faires, J.D. *and* Burden, R.L.
 455
fairground rides 286
Fairplay Publications [co.],
 Fairplay International
 Shipping Weekly 480
falsework 537, 544, 545
Faltinsen, O.M. 473
Family Expenditure Survey 511
Fan, L.S. 276
Fan application guide (pub. Fan
 Manufacturers Ass'n) 617
fans 205, **208**, 617
FAO (Food and Agriculture
 Org'n) 248, 255
Faraday Transactions [of RSC*]*
 282
Farag, M.M. 337
Far East *see* Asia

Farmer, F.R. [*nuclear energy*]
 265
Farmer, I.W. [*coal*] 238
Farr, R.E. [*telecommunications*]
 443
Farr, S. [*Standards*] 20
Farrashkhalvat, M. *and* Miles,
 J.P. 456
fastening processes 22, 340, 362,
 369
 see also joints
Faster building for commerce
 (pub. NEDC) 570
Faster building for industry (pub.
 NEDC) 570
fatigue, human **335**
fatigue, materials 140, 318
 bearings 176
 journals 143
Fatigue and Fracture of
 Engineering Materials and
 Structures **143**
Faulkner, E.A. 236
fault diagnosis
 control systems *and* 288, 291,
 397
 fault trees 179, 180, 317
 machines **178-80**
 power engineering 432-3
 Standards 180
 structures **133-49**
 see also safety; condition
 monitoring
Fault tree handbook (pub. NRC)
 317
Fawcett, H.H. *and* Wood, W.S.
 324, 326
FCC (Fed. Communications
 Commission) 421
FCEC (Fed'n of Civil
 Engineering Contractors)
 576
FEA *see* finite element analysis
Featherstone, R. 172*tab*
Federal Communications
 Commission (FCC [US])
 421
Federal Council for Science and
 Technology [US], COSATI
 6, 9
Federal Document Retrieval
 service [US], *Directory of*
 periodicals online 74, 400
Federal Highway Administration
 [US] **517**
Federation of Civil Engineering
 Contractors (FCEC) 576
Federation of the Electronics
 Industry (FEI) 414
Fédération Internationale de la
 Précontrainte (FIP) 522,
 530, **536**, 556
Federation of Master Builders
 (FMB) 576
feedback, control systems 286,
 286*fig*, **389**, 393, 394, 397
FEI (Fed'n of the Electronics
 Industry) 414
Feigenbaum, A.V. 366
Feilden, G.B.R., Report
 [*Engineering design*] 329

Feinberg, B.N. 635
Felder, R.M. *and* Rousseau,
 R.W. 271
Fellows, R. *et al.* 571
Fellows, R. *and* Langford, D.
 575
felt, roofing 564
FEM *see* finite element methods
Fenn, J.B. 201
Ferguson, C.R. [*engines*] 232
Ferguson, J.M. *et al.* [*marine*
 safety] 478
Ferodo [co]., design manual 164
Ferrero, G.L. *et al.* 249
fertilizers 590
Fettweis, G.B. 238
fibre-optics *see* optical fibres
fiche *see* microform
Field, R.W. 271
Fifty-Year Index to ASTM
 Technical Papers and
 Reports (pub. ASTM) 102
Fighting noise . . . (pub. OECD)
 492
film, photographic 25
film industry 286
filtration 282
Filtration Society, *Filtration and*
 Separation 282
Final environmental impact
 statement for . . .
 [OTEC] . . . (pub. Office of
 Ocean Minerals and Energy)
 248, 256
financial information 84, **576**,
 589
financial institutions **576**
Financial Times (FT) **70**, 103,
 499
 Financial Times Business
 Information 70, 325, 576
 EC Energy Monthly 234
 North Sea Letter 480
 Financial Times mining
 international yearbook
 601, 608
 FT PROFILE database host
 71, 75, 108
 FT Share Index 570
Finch, J.A. *and* Dobby, G.S. 600
finishing processes 87, 357, 359,
 369
 see also surface *etc.*
finite element analysis/methods
 (FEA/FEM) 134, **136-7**,
 451, **455**, 616
 books 139, 148, 364, **455**, 456
 hydraulics 294
 infinite elements 455
 journals 139, 143, 459
 manufacturing 364
 multi-dimensional elements
 137
 NAFEMS **139**, 148
 software 148-9, 306-7
 stress analysis 134, **136-7**,
 148-9
 wind turbines 246
Finite element primer (pub.
 NAFEMS) 148
Finite Elements in Analysis and
 Design **143**

Fink, D.G. and Beaty, H.W. 125
Fink, D.G. and Christiansen, D.
    421
Finland
    construction 584
    minerals 595
    Standards 20
    timber 560
    see also Finnish language
Finlay, M. 231
Finnish language
    databases 584
    dictionaries 154
Finniston, M., Report
    [Engineering our future] 329
FIP see Fédération Internationale
    de la Précontrainte
Fire and explosion index hazard
    classification guide (pub.
    Dow/AIChE) 317, 322
fire hazard 316, **321–2**, **554–6**
    books 235–7, 317, 321–2
    conferences 322
    databases 230
    electrical 320
    exhibitions 87–8
    fire engineering 554–6
    fire resistance 320–1
    fluids 288
    flame 235, 236, 237
    flammability 320, 322, 323
    hydraulics 288
    intrinsic safety 320, 321
    journals 327–8
    nuclear power stations 325
    offshore engineering 475
    organizations **321**, **327**
    Standards **321**
    steel 540
    see also safety; combustion
Fire hazard properties of
    flammable liquids, gases and
    volatile solids (pub. NFPA)
    322
Fire and Materials 327
Fire Protection Association
    (FPA) 321, **327**
    FPA Compendium of fire
    safety 321
Fire protection and fire fighting
    in nuclear installations (pub.
    IAEA) 325
Fire Research Station, FLAIR 230
Firesafe exhibition 87–8
Fire Safety Engineering 328
Fire Surveyor, former 328
FIRST database 58, **59**
First Gear [program] 165
First Search service [of OCLC]
    103
Fisher, R.A. [design/experiments]
    179
Fisher, T.G. [controls] 279, 319
fisheries
    abstracts 101
    and environmental protection
        487
    and hydroelectricity dams
        247–8, 255
fission, nuclear 250–1, **259–61**,
    266

fission products 259–60
Fitch, A.A. [petroleum] 240
Fitch, E.C. [maintenance] 178
Fitt, P.W. and Moses, R.T. 208
Fitzgerald, A.E. et al. [electric
    motors] 434
Fitzgerald, R. [factory building]
    552
Five language technology
    dictionary (pub. Gale) 123
Five-Year Index to ASTM
    Technical Papers and
    Reports (pub. ASTM) 102
fixings, construction 565
FIZ Karlsruhe [co.] 110
database hosts
    FIZ CHEMIE 283
    FIZ TECHNIK 584
    STN INTERNATIONAL
        231
FLAIR database 530, 583
    see also BRIX/FLAIR
flame see combustion
Flamm, J. and Luisi, T. 180
flammability 320, 322, 323
Fleagle, R.C. and Businger, J.A.
    243
Fleming, D.G. and Feinberg,
    B.N. [bioengineering] 635
Fleming, Sir John Ambrose 411
Fleming, P.J. [control systems]
    398
Fletcher, A. [electronics] 361
Fletcher, L.S. [thermal systems]
    204
Flexible Assembly conference
    359
Flexible Automation conference
    359
flexible manufacturing systems
    (FMS) 355, 357, **362–3**,
    365, 386
    conferences 359
    journals 379
    management 363, 386
    quality management 366
flexure-pivots 177
Flight International 68
flint 589
flocculants 600
Flood, J.E. 443, 444
Flood, J.E. and Cochrane, P. 444
floors 555, 556, 561
flotation
    column flotation 600
    froth flotation 590, **599–600**,
        602
Flotation fundamentals and
    mining chemicals (pub.
    DOW Chemicals) 600
flow see fluid
flowcharts 152, 159, 167, 176
Flowers, Sir Brian, Report,
    Nuclear power and the
    environment 266
Flow Measurement and
    Instrumentation 280
flowsheets, mineral processing
    592, 599, 600
FLUCOME (fluid control,
    measurement and

visualization) conferences
    290, **292**, 297
Fluent Inc./Fluent Europe Ltd
    294
Fluent program 294
Fluid Abstracts 297
    online 400
fluid dynamics
    biological 635, 636
    casting 364
    see otherwise fluid mechanics
fluid engineering see fluid
    mechanics
Fluid Engineering Abstracts 112
    online see FLUIDEX
FLUIDEX database **101**, 205, **229**,
    400
fluid film lubrication see
    lubrication
fluidization 276, 278
Fluidized bed combustion of coal
    (pub. NCB) 236, 238
fluidized beds **235–7**, 238, **430**
    journals 235
Fluidized bed systems ... (pub.
    EU) 235
fluid mechanics 204, **205–6**, 269,
    276, **285–99**
    nature and role 276, **285–7**
    abstracts/indexes 101, 205
    biological 638–9
    books 269, 271–2, 274, 276
    dictionaries 276, 287
    chemical engineering and 269,
        276
    conferences 205–6, 210
    controls 280
    databases 205, **229**, 400
    design 276, 295
    dynamics 364, 635, 636
    electricity generation 428
    environment and 287
    hydraulics see fluid power
    journals **205**, 280, 281–2, 459
    electronic 115
    reviews 105
    trade journals 67
    organizations 205, 210–11,
        281–2
    Standards 112, 206
    thermal systems 204, **205–6**,
        207–11
    training 288
    see also fluid power
Fluid mechanics: internal flow
    (pub. ESDU) 276
fluid power **285–99**
    nature and role 276, **285–7**,
        286fig
    terminology/symbols 285,
        287
    abstracts/indexes 297
    books **287–8**
    dictionaries 276, 287
    handbooks 285, **287–8**
    CFD modelling **294**
    components data 288–9, **292**
    computer simulation/(CAD)
        290–6, 291, 297
    conferences 289, **291–2**, 293,
        296–7, 298

Index 703

fluid power
  conferences (contd)
    Aachen Colloquia 292
    Bath *Workshops* 297
    Bath Workshops 297
    BHRA series 291, 296
    COMADEM 296
    FLUCOME 290, 292, 297
  continuity equations 290–2
  and control systems 280,
    285–7, 286*fig*
  exhibitions 296
  flow forces 290–2
    flow reaction force 291
  flow geometry 294, 297
  flow rates 290–2, 294
  fluid properties 288–90
  integrated systems 287
  journals 290, 292, 296–7
  leakage 290–1
  noise/vibration 287, 291, 292
  organizations 285, 287, 291–2,
    296–7
  power systems 286–7, 286*fig*
  safety 288
  software 294–6
  Standards 287, 288
  theoretical studies 297–8
  transmission lines 292–4
  vizualization 290, 292, 298
  see also fluid mechanics;
    hydraulic
Fluid Power 296
Fluid power engineers data book
  (*pub.* BFPA) 285, 287
*Fluid power systems and
  components . . .* (*pub.* ISO)
  287
fluids
  bulk modulus 288–90
  compressibility 200, **288–9**
  fire resistant 322
  heaters for 205
  hydraulic 288–90
  organic 203
  properties 273, **288–90**
  fire resistant 288
  thermodynamic 202–3
  synthetic 288
  see also fluid mechanics; fluid
    power
fluid transfer, hydraulics 286*fig*
Flurscheim, C.H. 336
FMB *see* Fed'n of Master
  Builders
FMEA/FMECA (failure modes,
  effects [and criticality]
  analysis) 180, 367
  see otherwise under failure
FMS *see* flexible manufacturing
  systems
foams, databases 308
Foden, C.R. *and* Weddell, J.L.
  324
Fogler, H. Scott 275
Folland, G.B. 457
Follett, *Report* [on university
  libraries] 120
Food and Agriculture
  Organization (FAO) 248,
  255

*Food and Bioproducts
  Processing* (IChemE
  Transac.) 281
food industries 269
  environment and 256
  freezing 207
  journals 281
  nutrition database 229
  thermodynamics and 205
Foot 638
FORATOM *see* Forum Atomique
  Européenne
force
  hydraulic systems 290–2
  moment 134
  and stress 133, 134
  viscous 290, 291
  see also load
Fordham Cooper, W. 320
Ford Motor Company 377
Fordyce, M.W. *and* Wodehouse,
  R.G. 537
forests/forestry
  abstracts 101
  as energy source 249–50
  and hydroelectricity 256
  and mineral processing 592
  tropical 256
forging 286, 292, 364, 369
*Formal methods in safety-critical
  systems* (*pub.* IEE) 319
forming 355, 358, 359, 360–1,
  363–5, **369**
high-energy-rate 364
marine technology 469*tab*
formulae, mathematical/physical
  124, **452, 454**
formwork 538, 544
Forschungsberichte **9–10**
Forschungsgemeinschaft
  Ultrapräzisionstechnik eV
  (UPT [Germany]) 359
Fortran [computer language] 335,
  455
Forum Atomique Européenne
  (FORATOM; Eur. Nuclear
  Forum) 262
Forum Atomique Française
  (FAF) **264**
Foster, J.E. 246
Foulds, L.R. 457
fouling, heat exchangers 277
Foumeny, E.A. *and* Heggs, P.J.
  277
foundations
  encyclopaedias 122
  structures 543, **564–5**
  Standards 564
*Foundations of Computing and
  Decision Science* 401
Foundation for Woodstove
  Dissemination (FWD) 242
founding processes **363–5**
  journals 68
  see also forming *etc.*
Fourier analysis 454, **456, 457**
Fowkes, A.S. *and* Nash, C.A.
  506
FPA (Fire Protection Ass'n) 321,
  327
FPA compendium of fire safety
  321

Fraas, A.P. 277
fractals 457
fracture 133–4, **140**, 141, 142,
  143, 318
frameworks, structural 544, 547,
  556, 558, 561
France
  abstracts/indexes 101
  aerospace engineering 475
  automation 378, 381
  Bibliothèque Nationale,
    *Bibliographie Nationale
    Française* 103
  books-in-print lists 104
  CNRS *see* Centre National de
    la Recherche . . .
  concurrent engineering 651
  construction 523, 532, 584
  École Polytechnique 154–5
  electricity 230, 253
  energy 235
  fluid mechanics/hydraulics 67,
    297
  Institut National des Sciences
    Appliquées [Lyon] 173
  Laboratoire d'Economie des
    Transports 501
  Laboratoire Central des Ponts
    et Chaussées **517**
  library, national 103
  marine technology 475
  materials 305
  medicine 638, 642
  minerals 594, 595, 602
  Minitel system 305
  nuclear energy 251, 260,
    261–2, 263, 264, 265
  Phénix reactor 260
  patents 58
  research 230, 355
  robotics 378, 381, 388
  Standards 22, **23–4**, 113, 524
  stress analysis 141
  transport 501, 516, 517
  tribology 173
  see also French language
Francis, B.A. 398
Franco, S. 423
Franke, M.E. *and* Drzewiecki,
  T.M. 293
Frankena, F. 249
Franklin, G.F. *et al.* [*control
  systems*] 393
Franklin, J.A. *and* Dusseault,
  M.B. [*rock mechanics*] 616
Fraunhofer Gesellschaft [Society]
  113
IRB 530
Freeman, R.L. 125
freezing *see* cryogenic;
  refrigeration
French, D.N. [*boilers*] 236, 327
French, M.J. [*design*] 169, 332
French language
  bibliographies 103
  conference *Proceedings* 359
  databases 231, 584, 602
  dictionaries **123**, 154, 267,
    271, 454
  journals 234, 235, 251, 253,
    431, 532, 638, 639, 642

704   Index

French language
 journals (contd)
  abstracting 101, 227, 231
  Standards 20, 23–4
freons
 CFCs 202
 thermodynamic properties
  202–3
frequency, Standards 27
frequency domain analysis 293,
  294, 390, 394
Freris, L.L. 246, 430
Freris, L.L. *and* Sasson, A.M.
  433
fretting 173
Fretwell-Downing Data Systems
  [co.] 418
friction 173, 174
 and control systems 395
 and forming 364
 hydraulic systems 290, 291,
  293
 stiction 290
 and thermal processes 201
Friedman, A. 455
Friends of the Earth [org'n] 256,
  503
 Energy and the environment
  255
fringes, optical 135
 moiré fringes **135**, 147
Fritz, J.J. 248
Frocht, M.M. 145
Froment, G.F. *and* Bischoff, K.
  275
*Frontiers of Medical and
 Biological Engineering* 642
Frost, B.R.T. [*nuclear energy*]
  250
Frost, H.M. [*biomechanics*] 635
froth flotation 590, **599–60**, 602
Fryer, B. 571
FT
 FT PROFILE database host
  71, 75, 108
 FT Share Index 570
 see otherwise Financial Times
Fuel 239, 240, 282
Fuel Efficiency Booklets (*pub.*
  DoE) 206, 208
Fuel and Energy Abstracts 227,
  283
Fuel Processing Technology 282
fuels 223, **237–42, 429–30**
 nature and role 224
 abstracts/indexes 227–8
 alternative **241–57**
 biofuels 210, 225, **241, 249–50**
 derived *see* synthetic *below*
 efficiency 203, **428, 429–30**
 fossil
  combustion **199**, 200, 203,
   428, 429–30
  environmental damage **430**
  non-renewability 428,
   429–30
 historical aspects 590
 journals 239, 240–1, 282
 mineral sources 590
 nuclear 250–1, **259–60**, 266,
  429

enriched 260
listed 261
organizations 205, 224–7
renewable **430–1**
synthetic 203, 237–40, **241**
 journals 241
 terminology 223
 see also energy technology
  and specific topics
*Fuel Science and Technology
 International* 239, 240
Fuerstenau, D.W. 599
Fuerstenau, M.C. 599
Fuller, D.D. 175, 177
fundamental design method 332
Fung, Y.C. 636
Furlong, R. 633
furnaces 205, **206**
fusion, nuclear 250–1, **261**
Fusion Engineering and Design
  251
*Fusion Power Program
 Quarterly Progress Report*
  (*pub.* USDOE) 251
Fusion Power Report 251
Fusion Technology 251
Future of aerospace standards
  (*pub.* AIAA) 26
fuzzy logic systems **391, 396**,
  403, **457**, 460
Fuzzy Sets and Systems 403, 460
FWD (Found'n for Wovestone
  Dissemination) 242
Fylde Electronic Laboratories Ltd
  146

G

Gage, G.R. 636
Gaggioli, R.A. 204
Gaillard, M.L. *and* Quenzer, A.
  364
Gait and Posture 641, 644
Gale Research Inc.
 *Commonwealth universities
  yearbook* 126
 Companies and their brands
  88
 Directories in print 84, **86**
 Gale directory of databases
  581, 582–3
 World business directory 84
Galer, I.A.R. 336
Galileo Galilei 133
Gall, T. 312
Gallant, R.W. *and* Yaws, C.L.
  273
Gallick, E.C. 240
Galton, A. 457
Galyer, J.F.W. *and* Shotbolt,
  C.R. 367
GAMAC *see* Groupement pour
  l'Avancement . . .
games
 electronic 412
 patentability 36
 game theory 344
gangue 591, 592
Gans, R.F. 158, 167
Gantmacher, F.R. 456
Gardiner, P. *and* Rothwell, R.
  332

Gardner-Denver [co.], *Rock
 drilling data* 614
Garner, S. 336
Garrad Hassan & Partners Ltd
  246
Garside, R. 320
Garzon, C.E. 248, 256
Gas Abstracts 228
Gas Energy Review 240
*Gas Engineering and
 Management* 240
gases
 cleaning of 592
 combustion 200, 202
 compressed 205, 325
 desulphurization 430
 engines **200**
 environmental effects *see*
  greenhouse effect *etc.*
 expansion 200
 flammable 322
 gas bearings 177
 journals 282
 and landfill 490
 liquefaction 202, 205, 206, 236
 mineral processing 591, 592
 in mines 617
 power generation *see* gas
  industry; power
  engineering
 processes 273–4
 properties 273
 purification 282
 reactor coolants 260–2
 safety 321, 325
 thermodynamic properties
  202–3, 240, 273
 turbines *see* turbines
 see also air *etc.*; fluid
gasification, *and* energy
 production 237–9, 249
gas industry 226, **239–41**
 abstracts/indexes 228
 books 235–9
 properties data 273
 controls 279
 historical aspects 94, 226
 journals 67, 69, 208, 209, 234,
  240–1, 282, 480
 natural gas 429
 offshore operations 467–8
 organizations 225, 226–7, 480
 safety 321, 325
 thermal systems 204
 trade literature 94
 see also offshore engineering
Gaslife 69
Gas Separation and Purification
  240, 282
Gaston, F.M.F. *and* Irwin, G.W.
  392
gas turbines *see under* turbines
Gas Turbine World 209
Gaudin, A.M. 590, 599
gauges/gauging 367
 strain gauges **135**, 136, 146
 see also measurement
Gaussian regulators 393, 394,
  397, 398
Gavalas, G.R. 238
Geankoplis, C.J. 274

Gear design . . . manual (pub.
  SAE) 164
gears 155, 159, 161*tab*, **164–5**,
  173, 177, 341
gearboxes 341
historical 154
organizations 154, 164, **165**
GEC (General Electric [Co.])
  305, 366
GEC Measurements [co.] 432
GEC Review 83
  Protective relays . . . 432
Gemeinschaft für Experimentelle
  Spannungsanalyse (GESA)
  141
gemstones 593, 607
  early uses 589
General Electric [co.] *see* GEC
General Motors [co.] 366
General Series (GS) publications
  [of BSI] 21, 182
GENERATE database **511**
generation *see under* power
  engineering *etc.*
*Generation, Transmission and
  Distribution* (IEE *Proc.*)
  252, 428
generators 252, **428**, **429**, **433**
  steam 260
genetic engineering 625
  patentability 37
Geneva, Convention on traffic
  signs 508
Genta, G. 256
Gentry, D.W. and O'Neil, T.J.
  614
GEOARCHIVE database 229
GEOBASE database **229**, 230
Geodex Retrieval System 99
*Geodex System/s Structural
  Information Service* 99
geography
  databases 229
  journals 498–9
  organizations 498, 499
  regional science 498–9
  and transportation 498–9
geology
  abstracting/indexing 113, 228
  books 240, 612
  databases **229**, 602
  geothermal energy 245–6
  journals 240
  and minerals 240, 589, **591**
  rock workings 615, 616
  see also earth sciences;
    petroleum *etc.*
geometric interference 135
geophysics 229, 240, 245–6
Geopolitics of Energy 234
GEOREF database **229**, 230
Georgius Agricola, *De re
  metallica* 154, 590
geosciences *see* earth sciences;
  geology *etc.*
Geotechnical Abstracts 99
geotechnical engineering 519,
  585
  journals 99, 522, 530, 532
  Standards 549
  see also ground *etc.*

Geotechnical Engineering (ICE
  Proc.) 531
Géotechnique (*pub.* ICE) **522**,
  530, 532
geothermal energy **245–6**
Geothermal Energy 246
*Geothermal Resources Council
  Bulletin* 246
  Transactions 246
Geothermics 246
Geriatric Medicine 639
Gerlough, D.L. and Huber, M.J.
  506
German, R.M. 365
germanium, *and* electronics 411
German language
  abstracts/indexes 101
  bibliographies 103
  books, general 141, 158, 162,
    302–3, 313, 535
  conference *Proceedings* 359
  databases **101**, 400, 583, 584,
    602
  dictionaries **123**, 154, 267,
    271, 370, 454
  journals 9–10, 67–8, 239, 240,
    245, 247, 255, 282, 283,
    297, 459, 461, 481,
    532–3, 642
  Standards 20
Germany
  abstracts/indexes 101
  bibliographies 103
  biomedical engineering 630,
    642, 645
  books-in-print lists 104
  ceramics 359
  chemical engineering 282, 283
  chemicals 67
  concurrent engineering 651
  conferences 110
  construction 532, 584
  databases 110 113
  design 162, 342
  *Deutsche
    Nationalbibliographie* 103
  electronics 421
  energy 239, 240
  alternative energies 245,
    247, 255
  Federal Ministry of Research
    and Technology (BMFT)
    355
  fluid power 292, 297
  Hamburg Ship Model Basin
    478
  journals 73
  kinematics 158
  manufacture 355, 359, 360
    CIM 362
  manufacturing, precision
    engineering 359
  marine technology 475, 478
  materials 68, 303, 359
  metrology 28
  minerals 590, 596, 602
  nuclear energy 261, 263, 265
  process engineering 73
  reports 9–10, 11
  research 230, 355, 359
  Standards 20, 22, **24–5**, 28,
    113, 303, 421, 524

stress analysis 141
Technical University for
  Rhineland and Westphalia
  360
water engineering 68
see also German language
Gertman, D.I. *and* Blackman,
  H.S. 319
Gerwin, D. *and* Kolodny, H. 366
GESA *see* Gemeinschaft für
  Experimentelle . . .
Gess, M.A. *et al.* 272
Ghana, minerals 596
Ghatak, A.K. *and* Thyagarajan,
  K. 448
Giacomo, J.D. 125
Gibbons, J.H. *and* Chandler,
  W.U. 256
Gibilisco, S. 123
Gibilisco, S. *and* Sclater, N.J.
  421
Gibson, J. [*coal*] 238
Gibson, L.J. *and* Ashby, M.F.
  [*materials*] 312
Giddings, J.C. 274
GIDEP [Government-Industry Data
  Exchange Program] database
  91–2
Gilbert, E.G. [*control systems*]
  394
Gilbert, E.G. [*directories/BL*] 125
Gilchrist, J.D. 590
Giles, C. *and* Goodall, I.H. 552
Gillespie, L.K. [*manufacture*]
  361
Gillispie, C.C. [*biography/
  dictionary*] 126
Gilmore, R. 457
Gilpin, A. [*environment*] 493
Gilpin, A. *and* Williams, A.
  [*energy*] 232
Giornale del Genio Civile 533
Gittus, J.H. *et al., Chernobyl
  accident* . . . 266
Glaeser, W.A. 176
Glasgow, DRIC 8–9
glass fibre **562–3**
glass/glazing 562, 563–4, 590
  cladding 563–4
  construction engineering **562–3**
  databases 308
  early uses 589
  GRC *see* cement:
    glass-reinforced
  handbooks 307, 312
  journals 67
  organizations 303, 562, 563
  Standards 562, 564
Glass and Glazing Federation
  562
Glassman, I. [*combustion*] 236
Glasson, J. *et al.* [*environment*]
  493
Glasstone, S. *and* Sesonske, A.
  [*nuclear energy*] 250, 266
Glazman, J.S. *and* Rumble, J.
  [*materials/databases*] 313
Glegg, G.L. 332
Glembotski, V.A. *et al.* 599
Global Automation Information
  Network 382

706　Index

GLOBAL BOOKS in PRINT database [CD-ROM] 103
Glor, M. 322
Gloss, D.S. and Wardle, M.G. 315
glossaries 123
  see otherwise dictionaries
Glossary of building and civil engineering terms (pub. BSI) 123, 533
Glossary of transmission elements (pub. Eurotrans) 154
glycol, hydraulic fluids 288
Glynn, P. 241
Gobel, E.F. [springs] [trans. Brichta] 169
Goble, W.M. [reliability] 319
Gocht, W.R. et al. 614
Goel, V.K. and Weinstein, J.N. 636
Goetsch, D.L. 385
Gohar, R. 175
gold 593
  carbon adsorption 599
  cyanidation 590, 599
  early uses 589
  mining 609, 611
  placer mining 615
  waste proportions 591
Golden [Colorado], conference 617
Goldenson, R.M. et al. 635
Goldin, A. 248
Golosinski, T.S. and Srajer, V. 615
Golze, A.R. 545
Gomez, M. 617
Gonzalez, M.O. 456
Goodall, P.M. 236
Good building guides (pub. BRE) 521
Goodger, E.M. 201, 202, 236
Good Housekeeping 21
Goodman, R.E. 616
Goodwin, G.C. and Sin, K.S. [control systems] 395
Goodwin, M.J. [dynamics] 188
Gopalakrishnan, C. 256
Gopher/Gopher-Plus client-servers 116, 117–18, 305–6
Gordon, G.D. and Morgan, W.L. [satellites] 447
Gordon, M. and Singleton, C. [information acronyms] 124
Gordon, W.J.J. [creativity/design] 332
Gorman, G.E. and Mills, J.J. 99
Gosatomnadzor [Russian nuclear regulatory body] 264
Gosling, C.T. [cooling systems] 207
Gosling, W. [radio] 443
Gosney, W.B. 207
Goulding, J.R. et al. 243
Goult, R.J. 456
Government-Industry Data Exchange Program/GIDEP database [US] 91–2
Government Publications

[abstracts (pub. HMSO)] 504
Government Reports Announcements and Index (GRA&I; US) 6, 7, 11, 27, 109, 341
governments
  conferences 76
  and grey literature 10
  patents 31–2, 42
  publications 72, 126
  see also reports
  Standards initiated by 16
  see also specific countries and topics
Govind, R. and Mocsny, D. 274
Gowar, G. [optical fibres] 447
Gower [optoelectronics] 423
GRA&I see Government Reports ...
Gradshteyn, I.S. and Ryzhik, I.M. 454
Graefe, R. 535
Graff, W.J. 240, 545
Graham & White Instruments Ltd 146?+
Grainger, L. and Gibson, J. 238
grains, of crystals 591
grains, crystalline 136
grandstands 568
Granqvist, C.G. 243
graphics
  computer 335
  conferences 359
  data transfer 310, 335
  Internet 118, 503
  journals 338, 339
  organizations 339
graphical information 481
  see also symbols; engineering drawing
Graphic symbols and circuit diagrams for fluid power ... (pub. BSI) 287
Graphic symbols for fluid power diagrams (pub. ASME) 287
graphite, as reactor moderator 260, 261
graphs/graph theory 451, 452, 454, 455, 457, 460
Grassam, N.S. and Powell, J.W. 177
Grassi, G. et al. 249
gratings, experimental 136, 147
gravel 600, 615
  organizations 608
gravity concentration 590, 599
Gray, J. 635
GRC see cement:
  glass-reinforced
Greater London Council, abolished 511–12
Greater London transportation studies (pub. London Research Centre) 510
Greece, minerals 594, 596
Green, R.E., Machinery's handbook 16, 340
Green, W.G. [machine theory] 158

Green's functions 455, 457
Greenberg, A.E. et al. [water analysis] 486
Greenberg, H.R. and Cramer, J.J. [risk assessment] 317
'Green book' [steel structure connecting (pub. BCSA)] 539
greenhouse effect 233, 255, 430
IEA research programme 225, 226
and transport industry 495
'green' movement see ecology etc.; environment
Greenwald, E.K. [safety/electrical] 320
Greenwood, R. et al. 635
Gregory, S.A. 332
grey literature 10–11, 78, 121
materials 303–4
SIGLE 10–11, 78, 231
Gribble, C.D., Rutley's elements of mineralogy 591, 598
Griffiths, J.M. [telecommunications] 443
Griffiths, J.M. et al. [telecommunications] 445
Griffiths, R.F. [risk] 334
grinding 363
  conferences 359
  diamonds 357
  mineral processing 591, 598, 600
grippers, automated 379
Groàk, S. 534
Groover, M.P. et al. 385
Groover, M.P. and Zimmers, E.W. 335
Grosjean, J. 158, 167
Gross, C.A. [power systems] 432
Gross, R.A. [fusion energy] 250
ground engineering 544, 547
  encyclopaedias 122
  handbooks 534
  safety 543
  see also geotechnical
Ground engineering yearbook (pub. Telford) 534
Groupement pour l'Avancement des Méthodes d'Analyse des Contraintes (GAMAC) 141
group technology 355, 357
Grove, D.M. and Davis, T.P. 179
GRP see plastics: glass-reinforced
Gruber, H.E. [imagination/design] 332
Gruber, R. et al. [mathematics] 458
GS (BSI General Series) 21, 182
guards 368
Guards illustrated ... (pub. NSC) 327
Guenther, R. 448
Guest, P.B. 457
Guidance note on ... concrete panels (pub. IStructE) 557
Guidance notes on ... the wiring regulations (pub. IEE) 321
Guide to the coalfields (pub. Colliery Guardian) 601

Index 707

Guide to the design of anchor-blocks... (pub. CIRIA) 556
Guide to directories at the Science Reference and Information Service [ed. Gilbert] 125
Guide to good management... for engineering design... (pub. IStructE) 550
Guidelines, see also under Standards
Guidelines (pub. Center for Chemical Process Safety)
... auditing process safety management... 325
... chemical process quantitative risk analysis 317
... engineering design for process safety... 325
... hazard evaluation procedures 317
... investigating chemical process incidents 325
Plant guidelines for technical management of chemical process safety... 325
... process equipment reliability data... 318
... safe automation of chemical processes 325
... safe storage and handling of high toxic hazard materials 324
... technical management of chemical process safety... 325
Guidelines for the construction of fire-resisting structural elements (pub. BRE) 555
Guidelines for descriptive cataloging of reports... (pub. XXXX) 13
Guidelines for rehabilitation of... hydroelectric plants (pub. ASCE) 247
Guidelines for the safe production of phenolic resins (pub. Brit. Plastics Fed'n) 325
Guide for making a condition survey of concrete... (pub. ACI) 557
Guide to materials engineering data and information (pub. ASM) 121
Guide to official statistics [ed. Dennis (pub. CSO)] 126
Guide to the offshore installations... Regulations (pub. HMSO) 477
Guide to the safe handling of radioactive wastes... (pub. IAEA) 250
Guides [construction information] (pub. CIRIA)] see CIRIA
Guide to services (pub. DRIC) 23
Guide to surveys and inspections... (pub. IStructE) 551

Gulliver, J.S. and Arndt, R.E.A. 248, 430
Gunn, D. and Horton, F. 206
Gunn, D. and Horton, R. 277
Gupta, P.K. 176
Gustafsson, R. 614
Guyana, minerals 596
GWF: das Gas und Wasserfach. Gas, Erdgas 240
gyroscopes 176

H

H2/H-infinity methods 390, 397–8
Haar, L. *et al.* 236
HAC (Heating and Air Conditioning Journal) 67
Hackleman, M.A. 209
Haeder, W. and Gärtner, E. 28
Haigh, M. 335
Hain, K. 158
[trans. Adams, D.P. *et al.*] 158, 166
Haiping, Z. 295
Hales, C. 333
Halevi, G. and Weill, R.D. 655
Hall, C.J. [mine ventilation] 617
Hall, D. [Standards] 20
Hall, D.O. [biomass] 249
Hall, J.R. [mine safety] 320
Hall, R.W. [manufacturing excellence] 366
Halpin, J.C. 142
Halsall, F. 446
Hamilton, P.H. 163, 164
Hammar, J. [cams] 170–2tab
Hammer, M.J. [water] 488, 489
Hammer, W. [safety] 326
Hamming, R.W. 445
Hamrock, B.J. 175
Hamrock, B.J. and Dowson, D. 176
Handbook of applicable mathematics [ed. Lederman] 454
Handbook of chemistry and physics (pub. Chemical Rubber Co.) 272
Handbook of community water supplies (pub. Amer. Water Works Ass'n) 489
Handbook of compressed gases (pub. Compressed Gas Ass'n) 325
Handbook of industrial materials (pub. Elsevier) 124–5, 313
Handbook for probabilistic risk assessment for nuclear power plants (pub. NRC) 319, 325
handbooks 82–3, 84–6, **124–5**
Handbook of structural steelwork (pub. BCSA/SCI) 538, 539
handling 355
bulk 592
cargos 355, 477, 480
emergencies 323
handbooks 355
hazardous substances 324, 355
radioactive 250
mineral processing 592

organizations 355, 480
safety 325
Handroos, H.M. 296
Hang, C.C. *et al.* 393
Hanover, University, TIB 10, 11
Harmathy, T.Z. 555
harmonization, Standards 16, 17, **19–20**, 26, 324, 508
Harms, A.A. [nuclear energy] 250
Harnby, N. *et al.* 274
Harns-Ringdahl, L. [safety] 317
Harrington, R.L. 476
Harris, C.H. and Billings, S.A. [control systems] 395
Harris, C.M. [noise/vibration] 125, 492
Harris, F. [construction plant] 544
Harris, F. and McCaffer, R. [construction plant] 544
Harris, T.A. [bearings] 176
Harris Publishing, *1992 US electronics industry directory* 418
Harrison, R. *et al.* [geothermal energy] 245
Harrison, R.M. [pollution] 487
Hart, D. [stone] 541
Hart, F. *et al.* [structures/steel] 540, 558
Hart, G.C. [structures/safety] 327
Hart, I.B. [Leonardo] 189
Hartenberg, R.S. and Denavit, J. 154–5, 166
Hartley, J.R. 649, 652
Hartley, J.R. and Mortimer, J. 650, 652
Hartman, H.L. 613
Hartman, H.L. *et al.* 617
Harvard Business Review 351, 587, 654
Harvey, A. *et al.* [hydroelectricity] 248
Harvey, B. [quality] 367
Harvey, R.C. and Ashworth, A. [construction] 534
Harvey, William 626
Harwell [Oxon]
AEA Technology Ltd **177**, 179, **262**, 266, 595
Atomic Weapons Research Establishment [of AEA]
alternative fuels 249, 256
PABLA 156, 332
ETSU **226**, 243, 245, 246, 247
Warren Spring Laboratory (WSL), *former* 494, 594, 595
Harwood, N. and Cummings, W.M. 147
Haseltine, B.A. and Moore, J.F.A. 559
Haseltine, B.A. and Tutt, J.N. 559–60
Hasse, P. 320
Haug, E.J. 653
haulage *see* transport
HAVAC 254
Hawkes, B. and Abinett, R. 331
Hay, G.D. and Stell, G.P. [industrial buildings] 552

Hayes, P. *and* Smith, K.
  [*greenhouse effect*] 233
Haykin, S.S. 446
Hayward, A. *and* Weare, F.
  [*steel*] 558
Hayward, A.T.J. [*hydraulics*] 289
Haywood, R.W.
  [*thermodynamics*] 202, 205, 236
'hazan' 317
Hazardous materials emergency response guidebook pub. USDoT) 326
Hazardous waste treatment processes (*pub*. Water Pollution Control Fed'n) 490
hazards **315–28**
  defined 316
  CHETAH 324
  CIMAH 318
  handbooks 316
  hazardous areas **320**, 321
  hazardous substances **322–4**, 326
  company safety-Standards 28
  COSHH Regulations 181
  explosive substances 322
  flammable substances 320, 322, 323
  radioactive waste 260
  wastes **489–90**
  'hazop' and 'hazan' 317
  human factors 178, **316**, 318, 319, **326**, 336
  journals 283, 327, 490
  machinery 178
  see otherwise safety
Hazewinkel, M. 453
Haznews 490
'hazop' 317
HDTV (high definition television) 413
health
  occupational 183, 316, 320, 335
  public health 486, 491
  see otherwise Health and Safety
Health Equipment Information 642
Health and Safety Commission (HSC), *Explosives at quarries* 614
Health and Safety Executive (HSE) **182, 525**
  area offices 182
  Enquiry Points 152, **182**
  HSELINE database 530
  [HM] Nuclear Installations Inspectorate (NII) 264
  publications 181, 288
  advisory literature **182**, 334
  catalogue 527
  Codes of Practice 151, 182
  Factory Inspectors' annual report 577
  Guidance Notes (HS(G)) **182**, 492, 551
  HSE Books **182**
  Noise at work ... *guides* 492

Programmable electronic systems ... 319
Quantified risk assessment ... **179**, 334
Safe manriding in mines 615
Regulations (HS(R)) **180–2**
Dangerous goods ... 477
Health and Safety Newsline 328
Health and Safety Practitioner 183
Health and Safety at Work 183, 338
Health and Safety at Work Act (1974) 152, **180–1**, 334
  advisory literature **182**
Heap, R.D. 253
heat **199–211**
  theoretical aspects **199–201**
  compression 199–200, 208
  cooling *see* cooling
  exergy 200
  expansion 199–200, 204, 205, 208
  heat and power, cogeneration *see* combined heat and power
  heat production processes, efficiency 203–4
  multiphase flow 274, 282
  processes 28, 276–7, 369
  pyrolysis 249
  specific 202, 203
  storage of **206–7**
  thermal loading 134
  thermal processes **203–5**
  thermal reservoirs/sinks 200
  thermal stress 137
  transfer *see* heat transfer
  transport of 202–3
  tribology 174
  waste **206–7, 208–9**
  see also heat recovery *etc.*; energy
Heat Engineering 255
heat exchangers 205, **206–7**, 270, **276–7**
  fouling 277
heating/heating systems 203, **203–5, 206**, 223, **253–5**, 277
  alternative sources *see* alternative energy
  automated 403
  in buildings 206, **253–5**
  coal 590
  conferences 206
  district heating **208–9**, 210, 225, **253–5**
  journals 67, 254–5
  organizations 206, 210–11, 403
  Standards 206
Heating and Air Conditioning Journal (HAC) 67
Heating, Piping, Air Conditioning 67
Heating, ventilating and air conditioning ... (*pub*. ASHRAE) 207
Heating, Ventilating and Air Conditioning Manufacturers Association 403
Heaton, C.A. 271

heat pumps 204, **207–8**, 225, **253–5, 276–7**
heat recovery systems **206–7**
Heat Recovery Systems and CHP 254
heat transfer **199–201, 205–6**, 207, 269, **276–7**
  books 202–3, 233, 237, 269, 271–2
  conferences 205–6, 210, 277, 317
  databases 205
  journals 105, **205**, 282
  organizations 205, 210
  and process engineering 269, 271–2, 275
  safety 317
  Standards 206
  theory of **199**, 200–1, 275–7
  thermal systems 204–5
  see also fluid mechanics
Heat transfer (*pub*. AIChE) 277
Heat transfer ... *Conference* ... (*pub*. IChemE) 277
Heat transfer (*pub*. ESDU) 207
Heat Transfer and Fluid Mechanics Institute (HTFMI [US]) 205, 210
heat treatments 28, 369
Heavy goods vehicles in Great Britain (*pub*. HMSO) 509
Hedges, A. 256
Heideklang, H.R. 178
Heilich, F. III *and* Stube, E.E. 165
Heinle, E. *and* Leonhardt, F. 567
Heiskanen, K.I. 278
Helander, M. *and* Nagamachi, M. 652
Helgert, H.J. 445
helicopters 67
Heller, E.D. 737
Helm, D. *et al.* 252
Helstrom, C.W. 458
Hemphill, G.B. 614
Henderson, G.P *and* Henderson, S.P.A. [*associations*] 125
Henderson, S.P.A. *and* Henderson, A.J.W. [*associations*] 340
Hendry, A.W. 560
Henley, E.J. *and* Kumamoto, H. 178, 179, 319
Henry, M.F. 324
Henry Crumb School of Mines 596
HENSAdatabase 584
Henstock, M.E. [*recycling*] 337, 490
Hepburn, C. *and* Reynolds, R.J.W. 169
Heriot Watt University, Edinburgh 120
Her Majesty's Stationery Office (HMSO) publications **182**, 645
  anthropometrics 347
  Catalogue 527
  construction 528, 534, 541, 549, 551, 555, 557, 566, 567, 572, 577, 587–8

Her Majesty's Stationery Office
(HMSO) publications
construction (*contd*)
civil engineering 545
on database 528
design 347
Digest of ... energy statistics
232
energy 232, 235, 238, 248,
256
Energy Papers 234
nuclear energy 266
environmental engineering 486,
490, 492, 493
Government Publications
[abstracts] 504
IEA publications 225
marine technology 477
Reports 266, 477, 499–500
safety **182**, 319, 477, 577
statistics 232, 509–10
stress 142
transport/traffic 493, 499–500,
504, 507, 509–10
*HMSO Transport Sectional
List* 504
*United Kingdom Energy
Statistics* 235
Herrick, G. 361
Herridge, S.J. *et al.* 203
Hertz contacts 165, 168
Herzfeld, C.M. 202
Heskett, J. 333, 336
Heurista [publishers] **162**
Hewett, C.A. [*carpentry*] 561
Hewitt, G.F. [*chemical
engineering*] 274
Hewlett, P.C. 537
Heyman, J. 560
HFA/B/C/D fluids 288
Hicks, T.G. [*calculations*] 124,
232, 252
Hickson, D.C. *and* Taylor, F.R.
[*thermodynamics*] 202
high definition television
(HDTV) 413
Highfield, D. 551
Highlands and Islands
Development Board
[Scotland], *Oil map* 481
high power pressure-tube
(RBMK) reactors 261–2
HIGH TEMPERATURE MATERIALS
DATABASE 313
high voltage direct current
(HVDC) 432
Highway capacity manual (pub.
TRB) 500, **506**
highway engineering *see*
transport
*Highway Research Information
Service* (HRIS) *Abstracts*
504
*Highway Research Record,
former* 499
Highways 498
Highways Agency **511**
*Highways and Public Works,
former* 498
Highways and Transportation
498

Hill, G.B. *et al.* [*towers*] 567
Hill, R. [*coding theory*] 457
Hill, T. [*manufacturing/
management*] 366
Hills, M.T. *and* Kano, S.
[*electronics*] 443
Himmelblau, D.M. 272
Hippensteele, S.A. 203
Hirst Business Communications
[co.], *I.C. master* 419
Hiskey, J. Brent 599
*Historical structural steelwork
handbook* (pub. BCSA) 538
history of engineering *see under*
engineering
Hitachi [co.] 363, 386
HM [Her Majesty's] Nuclear
Installations Inspectorate
(NII) **264**
HMSO *see* Her Majesty's
Stationery Office
Hobson, G.D. 240
Hochreiter, L.E. *and* Shiralkar,
B.S. 317
Hodge, B.K. 205
Hodgkinson, A.J. 550
Hoechst [org'n] 305
Hoek, E. *and* Bray, J.W. 616
Hoek, E. *and* Brown, E.T. 616
Hoeltzel, D.A. *and* Chieng,
W.-H. 167
Hoffman, J.M. *and* Maser, D.C.
[*chemical hazard*] 324
Hofmann D.A. *et al.* [*gears*] 164
Hogan, E. *et al.* 249
Holbeche, R.J. 443
Holdsworth, W. *and* Sealey, A.
253
Holland, C.D. *and* Anthony, R.G.
[*chemical reactions*] 275
Holland, F.A. [*fluid mechanics*]
276
Holland, F.A. *et al.* [*heat pumps*]
253
Holland, R. *et al.* [*buildings*] 551
Hollaway, L. 563
Holmberg, K. *and* Folkeson, A.
178
Holmes, J.R. 490
holographic interferometry
135–6, 146–7
Holt, K. 332
Hong, S.I. 204
Hong Kong
American Chamber of
Commerce 234
minerals 594
University 382
Hoover, H.C. *and* Hoover, L.H.
590
hoppers 278, **567**
Horan, N.J. 488
Hord, R.M. 243
Horizon 481
Horlock, J.H. 209, 253
Horlock, J.H. *and* Winterbone,
D.E. 236
Horowitz, P. *and* Hill, W.
[*electronics*] 423
Horowitz, S.H. [*electricity*]
431–2

Horvath, T. 320
Hosford, W.F. *and* Caddell, R.M.
364
hospital engineering 624, 635
Hospital Equipment and Supplies
643
Hospital Physicists Association
(HPA) 629, 630
Hoss, R.J. 448
Hostetter, G.H. *et al.* 393
host services **107–9**, 229, **422**,
505, **584**
see also Dialog *etc.*
Hottell, H.C. 203
Houldcroft, P.T. 303
house journals *see under* journals
housing 543, **572–4**
forecasts 575, 575*tab*
*Housing and construction
statistics* (pub. DoEnv) 534,
570, **572–4**
Howard, J.R. [*fluidized beds*] 236
Howard, R. *et al.* [*buildings/
energy*] 254
Howard-Williams, J. [*marine
technology*] 482
Howe, E.D. 243
Howell, J.R. *et al.* 210, 244
Howell, J.R. *and* Buickius, R.O.
202
How to ... series (pub.
NAFEMS) 139
How to get started with FEA
139
HPA *see* Hospital Physicists
Ass'n
HRD4 failure-rate prediction 91
HRIS (Highway Research
Information Service), *HRIS
Abstracts* 504
Hrones, J.A. *and* Nelson, G.L.
155, 167
HSC *see* Health and Safety
Commission
HSE
HSE Books [co.] **182**
HSELINE database 530
see otherwise Health and
Safety Executive
HS(G) (Guidance) documents
(HS(G)) 182
HS(R) (Regulations) documents
180–2
Hsu, P.H. *and* Wu, Y.S. [*marine
engineering*] 474
Hsu, T.R. *and* Sinha, D.K.
[*CAD/CAM*] 335
HTML (HyperText Markup
Language) 118
Huang, F.F. 202
Hubka, V. *et al.* 162, 332
Hubka, V. 162
Hubka, V. *and* Andreasen, M.M.
162
Hubka, V. *and* Eder, W.E. 162
Hubka, V. *and* Kostelic, A. 162
hubs 341
Hudson, J.A. 616
Hudson, J.A. *et al.* 616
Hufnagel, W. 313
Hughes, F.W. [*electronics*] 361,
419

710 Index

Hughes, O.F. [*marine design*] 476
Hüls AG 305, **313**
human body
  anthropometrics 336, 339
  biomedical engineering **623–48**
  evolution 634–5
  and experimental data 637–8
  'health elements' 627, 628*fig*
  movement 630, 633, 634–6, 641
  pressure sores 637
  systems analysis 638
  and vitalism 625
Human–Computer Interaction 40, 339
human factors
  automation 384
  communication 338
  and computers 339, **339**, 378, 400, 402
  computer-use 336
  and controls design **396–7**
  ergonomics **335–6**, 652
  and experimentation 637–8
  journals 339, 400
  reliability 319
  safety 178, **316**, 318, 319, **326**, 336
  travel behaviour 500, 507, 509
  see otherwise ergonomics; social
Human Factors 339
Human Factors and Ergonomics Society
  Ergonomics in Design 339
  Human Factors 339
  Proceedings 339
Human Factors Society, *former* 339
Human Movement Science 339, 641
Human Systems Management 339
humidity, psychrometry 253–5
Humphreys, K.K. *and* Wellman, P. 337
Hungarian language, dictionaries 454
Hungary
  Central Statistical Office, *Statistical dictionary* 454
  design 162
  stress analysis 144–5
Hunt, K.H. [*kinematics*] 158, 159
Hunt, T.M. [*wear debris/oils*] 179
Hunt, V.D. [*fuels*] 241
Hunter, R. *and* Elliott, G. [*wind energy*] 246
Hunter, T.A. [*safety*] 181, 315, 334
Hurricks, P.L. 190
Hurst, K.S. 169
Huston, R.L. 159
Hustrulid, W.A. 613, 615
HVDC (high voltage electric current) 432
Hycaaf [program] 296
Hycad [program] 295
hydraulic engineering **285–99**, 519, 547

  applications 285–6
  control systems 285–7, 391
  electrohydraulic systems 285–7, 286*fig*
  encyclopaedias 122
  fluids 285, **288–90**
  journals 67
  oil hydraulics, advantages 285–6
  see otherwise fluid mechanics
Hydraulic trainer . . ., (*pub*. Rexroth GmbH) 288
Hydrocarbon Processing 240, 282
hydrocarbons 239, 240, 273, 282, 467, 607
  see also petroleum *etc.*
hydrocyclones 598, 601
hydrodynamics 276
  magnetohydrodynamics **251–3**
  marine 469*tab*
  microhydrodynamics 276
hydroelectric power **247–9**, 255, 256, **430**
  costs 430
  see also dams
hydrogen
  as energy source **241**
  abstracts/indexes 228
  journals **241**
  organizations 225
  and fission process 260
  and fusion process 261
  heavy 260
  'Hydrogen Bomb' 261
  pH control 280
Hydrogen Energy Co-ordinating Committee Annual Report (*pub*. USDoE/NTIS) 241
Hydrogen Energy Quarterly Literature Review 228
hydrology 489, 585
hydromechanics, organizations 210
hydrometallurgy 597, **599**, 601
Hydrometallurgy 597
Hydromet conference 601
hydro-power *see* hydroelectric; water engineering
hydrostatics 467
Hyman, E.L. *and* Stiftel, B. 493
hypertext information 587
  HTML 118
  Internet 118, 503
hypothesis tests 451

I

IAA *see* Internat. Aerospace Abstracts
IABSE *see* Internat. Association for Bridge and Structural Engineering
IAEA
  IAEA Bulletin 266
  IAEA safeguards . . . 250
  see otherwise Internat. Atomic Energy Agency
IAHE (Internat. Ass'n for Hydrogen Energy) 225
IASS *see* Internat. Ass'n for Shell and Spatial Structures

IASTED *see* Internat. Ass'n for Science and Technology for Development
IATB[R] *see* Internat. Ass'n fot Travel . . .
IBA *see* Internat. Bar Ass'n
IBG *see* Inst. of British Geographers
IBM [co.] 306, 366
  IBM Systems Journal 417
  IBM Technical Disclosure Bulletin 69
IBSEDEX database 230
IC (integrated circuits) **412**, 419
ICAO (Internat. Civil Aviation Orgn) 16
ICARCV *see* Internat. Conf. on Automation Robotics . . .
ICAT exhibition 342
ICCAS (Internat. Conf. on Computer Aided Shipbuilding) 474
ICE
  ICE PUBLICATIONS DATABASE **530**
  see otherwise Inst'n of Civil Engineers
ICED *see* Internat. Conf. on Engineering Design
ICEM (Internat. Conf. on Electrical Machines) 434
IChemE *see* Inst'n of Chemical Engineers
ICI (Imperial Chemical Industries) [co.] 305
ICI Engineering **313**
  publications 317
I.C. master (*pub*. Hirst) 419
ICOMOS *see* Internat. Cncl on Monuments and Sites
ICONDA [Internat. Construction Database] 113, **530**, **583**
ICRP (Internat. Commission on Radiological Protection) 262
I & CS (*Instrumentation and Control Systems*) 401
ICS (Internat. Classification of Standards) 18
ICTED *see* Internat. Cooperation in Transport Economics Documentation
IDA (Irish Development Ass'n) 314
Identification of the fire hazards of materials (*pub*. NFPA) 322
IDHA *see* Internat. District Heating Ass'n
IDHCA (Internat. District Heating and Cooling Ass'n) 209
  District Heating and Cooling 209
IEA
  IEA Biomass Conversion Technical Information Service 225
  IEA Coal Research 225, 239, 602
    Coal Abstracts 112, 228, 602

Index 711

IEA (contd)
  IEA Greenhouse Gas
    Programme 225, 226
  see otherwise Internat. Energy
    Agency
IEC
  IEC bulletin 19
  IEC catalogue 19
  see otherwise Internat.
    Electrotechnical
    Commission
IECQ 91
IED see Inst'n of Engineering
  Designers
IEE
  IEE Colloquium Digest 405
  IEE Control Engineering series
    401
  IEE News **416**, 450
  IEE Review **416**, 427
  IEE wiring regulations **320**,
    321
  see otherwise Inst'n of
    Electrical Engineers
IEEE
  IEEE Computer Society Press
    381, 403
  IEEE Control Systems
    Magazine 401
  IEEE Engineering in Medicine
    and Biology Magazine
    642
  IEEE Engineering in
    Medicine and Biology
    Society 630, **631**
  IEEE Expert: Intelligent
    Systems ... 402
  IEEE Journal of Solid-State
    Circuits 416
  IEEE Micro 416
  IEEE Power Engineering
    Review 252
  IEEE Reliability Test System
    433
  IEEE standard dictionary of
    electrical and electronic
    terms [ed. Booth] 123
  IEEE Transactions see under
    Institute of Electrical and
    Electronics Engineers
  see otherwise Institute of
    Electrical and Electronics
    Engineers
IEEE/IEE PUBLICATIONS ONDISC (IPO)
  427
IEEIE (Inst'n of Electronics and
  Electrical Incorporated
  Engineers) 414
IESS (Inst'n of Engineers and
  Shipbuilders in Scotland)
  475
IFAC 381
  conferences 381
IFIP see Internat. Fed'n for
  Information Processing
IFLA (Internat. Fed'n of Library
  Ass'ns) 10
IFMBE see Internat. Fed'n for
  Medical and Biological
  Engineering
IFME see Internat. Fed'n of
  Medical Electronics

IFR
  IFR Robotics Newsletter 378
  see otherwise Internat. Fed'n
    of Robotics
IFS Publications [co.] 38
IFToMM see Internat. Fed'n for
  the Theory of Machines ...
IGES (Internat. Graphics
  Exchange System) 335
IGME see Inst. of Geology and
  Mineral ...
ignition 320, **321–2**
IGU (Internat. Gas Union) 225
IHS (Information Handling
  Services), IHS INTERNATIONAL
  STANDARDS AND
  SPECIFICATIONS database **230**
IHT see Inst'n of Highways and
  Transportation
Iles, D.C. 566
ILI 113
Illingworth, J.R. 544
Illinois Institute of Technology
  289
Illston, J.M. 535
illumination see lighting
ILO see Internat. Labour Office
IMA (Inst. for Mathematics and
  its Applications) **453**
  IMA Bulletin 458
  IMA Journal of Applied
    Mathematics 459
  IMA Journal of Mathematical
    Control and Information
    459
image processing 311, 458, 503,
  639
  compression 457
  image analysis 460
Image Processing 639
imaging
  medical/clinical 627, 639
  stress analysis 136
IMarE see Inst. of Marine
  Engineers
Imarisio, G. and Bemtgen, J.M.
  241
IMC (Irish Manufacturing
  Committee) 359
IMEC (Internat. Machine Tool
  Engineers Conf.) 359
IMechE see Inst'n of Mechanical
  Engineers
IMEKO (Internat. Measurement
  Confed'n) **140**
IMEMME see Inst'n of Mining
  Electrical ...
IMinE see Inst'n of Mining
  Engineers
IMM
  IMM Abstracts 602, 609, 613
  IMMAGE database 602, 609,
    613
  see otherwise Inst'n of Mining
    and Metallurgy
IMMS see intelligent
  manufacturing: management
  systems
IMO (Internat. Maritime Org'n)
  16, **477**
IMPC see Internat. Mineral
  Processing Congress

impedance, hydraulics 289,
  293–4
Imperial Chemical Industries see
  ICI
implants 636, 641
Improving engineering design
  (pub. Nat. Research Cncl)
  650, 656–7
IMTC (Annual Instrumentation
  and Measurement
  Technology Conf.) 360
IN (intelligent networks) 442
incineration 236
Increasing the fire resistance of
  timber floors (pub. BRE)
  556
Incropera, F.P. and Dewitt, D.P.
  205, 277
Inculet, I.I. 599
Index and abstract directory ...
  (pub. EBSCO) 99
Index of conference proceedings
  received (pub. BL)
  [Conference proceedings
    index] 77, 110, 530
  online 77, 110
indexes/indexing **102**, 105, 227,
  643
  CD-ROM 90
  nature and use 97–9
  bibliographic 103
  catalogues 87
  citations 102, **109**, 224, 400,
    461, 504
  conference proceedings 76
  controlled language 56
  currency 99, 227
  to databases etc. 90, 354
    CompuMath Citation Index
      461
    INSPEC 99, **421**
  data sources 142
  directories 72–3, 76, 77, 84
  journals 74–5, **102**, 309, 553
    abstracting journals **99**, 101,
      461
    cumulative 102
    house journals 74
    trade journals 66, 74
  keywords 56, 59, 98, 109, 485
    KWIC [keywords in
      context] 18
  microform 90
  newspapers 70, 74
  patents 47–8, 54, 56, 57
  see otherwise abstracting;
    databases: searching
Index Medicus **643–4**
Index to Scientific Reviews (pub.
  ISI) 105
Index to scientific and technical
  Proceedings (ISTP) (pub.
  ISI) 77
  online 77, **109**, **505**
Index to Statistics and
  Probability 113
Index to theses accepted ...
  (pub. Aslib) 79
India
  coal 238
  energy 255

India (contd)
  machines 161
  Standards 27
Indian coals (pub. Central Fuel Research Inst.) 238
Indian Institute of Technology 255
induction 419–20
  induction motors 434–5, 436
Industrial Archaeology Review 552
Industrial design education... (pub. Design Council) 336
industrial engineering, Internet resources 119
Industrial and Engineering Chemistry Research 282
Industrial engineering terminology (pub. ANSI) 123
Industrial Equipment News 67, 87
Industrial Minerals 593, **597**, 602, 611
Industrial minerals directory (pub. Metal Bulletin) 601, 608
Industrial Minerals International Congress 601
Industrial Press Ltd, *Machinery's handbook* 16, 124, 160, 340, 369
Industrial research in the United Kingdom (pub. Longman) 126
Industrial Robot 379
Industrial trade names (pub. Reed) 88
Industrial Waste Management 488, 490
industry
  companies *see* companies
  construction *see* construction
  journals *see* trade journals *and specific topics*
  statistics 75, 572–7
  see otherwise manufacturing *and specific topics*
Info Globe [publishers] 400
Infonortics [co.] 61
InfoPro Technology [co.] **107**
informatics
  EC Directorate 516
  see otherwise information technology; telematics
information 1–2, 63, 81, **121–32**, 303, 496–8, 569–71
  biographical 126–7
  coding **445**, **457**
  as commodity 497
  and communication 441
  and data **570**
  electronic **107–20**
  see also databases *etc.*
  formal/informal 114, 571
  'grey literature' **10–11**, **231**
  'Hang-ups Committee' 13
  integrated 657
  internal/external 570–1
  journals 459
  major sources **121–32**
and management 354, 569–71
  needs variance 301–2, 569–71
  non-print sources, summarized 89
  online *see* online
  and optimization 354
  primary/secondary 570–1
  printed, advantages 97–8
  project-coordinated 577–9
  searching approaches **468–73**, 623, 627, 634, 640
  selected *see* selective dissemination
  and selectivity 570, **571**
  specificity levels 578
  Standards for
    BS 4940:... presentation of technical information... 81, **82**
    CALS **310–11**, **314**, 655, 656
    STEP 307, **310–11**, 314
  'superhighway' 502
  theory of **445**, **457**
  time aspects **569–70**, **633**
  types of 89, 570–1, 578
  'universal availability' concept 10
  see also product information *and specific topics*
information brokers, patents 53, **61**
information centres, *for* Standards 18, 24
information dissemination services *see* bulletin boards
Information Handling Services (IHS), IHS INTERNATIONAL STANDARDS AND SPECIFICATIONS **230**
information management
  and CIM 362
  and concurrent engineering 649, **652**
  current trends 581
information officers 497
Information Paper (IP [of BRE] **521**, 542, 588
information processing
  books, dictionaries 124
  conferences 655
information professionals *see* information management; libraries
Information retrieval... design teaching (pub. SEED) 339
information science
  organizations 61
  and patents 38, 39, 61
  see also libraries *etc.*
Information Service in Mechanical Engineering *see* ISMEC
information services, commercial patents 53
  see also information brokers; databases *etc.*
Information sources in energy (pub. Butterworth) 607
Information sources in metallic materials (pub. Bowker-Saur) 303
information 'superhighway' 502
information systems
  controls 393
  transport 502, 508, 516
Informationszentrum Raum und Bau (IRB) 530
information technology (IT)
  abstracts/indexes 100, **113**, 355
  books 125
  dictionaries 124
  and concurrent engineering 650, 654, 655–6
  conferences 655–6
  databases **113**, 355, **421–2**
  journals 69, 580, 654
  electronic newsletter 115
  organizations 113, 413, 415, **516**, 650–1, 655–6
  Joint Framework 415
  and scholarly communication 63, **114–20**
  and transport 498, 516
  see also electronic; computers; telecommunications
Informative system for the construction industry (pub. HMSO) 577–8
INFORMS database 503
infra-red, stress analysis 136
INID *see* Internationally agreed Numbers...
INIS
  INIS Atomindex **9**, 228, 267
  online 231, 267
  see otherwise Internat. Nuclear Information System
injection moulding 365
INLA (Internat. Nuclear Law Ass'n) **263**
innovation 156, 332
  see also patents
Innovation et Technologie en Biologie et Médicine 642
INPADOC database 55, **58**, 59, 60
  see also EPIDOS
INPI database 58
INPO (Inst. of Nuclear Power Operations) **264**
inquiries, public 471, 477, 499
  expert witnesses 493
INRETS (Institut National de Recherche sur les Transports et leur Sécurité) **516**
Inside ITS 498
Inside IVHS 498
insolation 243
INSPEC (Information Services for the Physics and Engineering Communities) **230**, **413**, 450
  publications (*INSPEC*)
    Computer and Control Abstracts 101, **111**, **230**, 341, 380, 399, **399**
    Current Papers in Computing and Control 399
    Key Abstracts **399–400**, **421–2**, **426**, **428**
  online *see* INSPEC
  Science Abstracts 228, 355, **380**, **421**

Index 713

INSPEC (Information Services for the Physics and Engineering Communities) publications (*INSPEC*) (*contd*)
    taped versions 422, 450
INSPEC databases 100, **101**, 110, 111, **354–5**, **380**, **399–403**, **421–2**
    CD-ROM formats 111, 422, **426**, 450
    current awareness services 421–2
    indexes to 99, **421**
    journals sourced 99, **111**, 399, **421**
    SDI services 421–2
    taped versions 422, 450
    thesaurus **421**
    automation 380
    computers 101
    control systems 101
    design 341
    · electrical engineering **101**, **421–2**, **426**
    electronics **101**, **421**
    ELECTRONICS AND COMPUTING ONDISC 426
    energy **230**, 231
    INSPEC PERIODICALS ON DISC 111
    instrumentation 421–2
    international access 422
    manufacturing 354–5
    physics 101
    power engineering 421–2
    product reviews 105
    robotics 380
    telecommunications 421–2, 450
INSTA (Internordisk Standerisering) 20
InstE *see* Inst. of Energy
Institute of British Geographers (IBG), *Transactions* 499
Institute of Building Control **525**
Institute of Ceramics 327
Institute of Concurrent Engineering [of ISPE] **650–1**
    CERA 653
Institute of Electrical and Electronics Engineers (IEEE [US]) **387**, **413–14**, **513**
    and automation 381, 387, 401, 403
    and biomedical engineering 630, 639, 641
    and computing 381, 403, 420
    and concurrent engineering 651
    conferences 296, 360, 381, 401, 403, **414**, **417**, 450
    and control systems 296, 398, 401, 402, 403
    and design 342, 401
    and electrical engineering **426–7**, 431–2, 433, 435–6
    and electronics **413–14**, 420, 435–6
    and energy 324
    and fluid power 296

IEEE/IEE PUBLICATIONS ONDISC (IPO) **427**
    and INSPEC 111, 399
    and instrumentation/controls 401, 402, 420
    journals, general **414**, **416**, 641
    Computer 403
    IEEE Control Systems Magazine 401
    IEEE Engineering in Medicine and Biology Magazine 642
    IEEE Expert: Intelligent Systems... 402
    IEEE Journal on Selected Areas in Communications 449
    IEEE Journal of Solid-State Circuits 416
    IEEE Micro 416
    IEEE Power Engineering Review 252
    Journal of Electronic Materials 420
    Spectrum 416, **427**
    and manufacture 360
    and mathematics 460
    and nuclear energy 251
    and power engineering 426–7
    Proceedings 253, 401, 403, **416**, 439
    publications **423**
    'Challenges to control...' 398
    IEEE Computer Society Press 381, 403
    IEEE standard dictionary of electrical and electronic terms 123
    Reliability handbook 319
    Standards 123, 319, 433
    Technology Updates series 439
    and reliability/quality 319, 360, 433
    Reliability Test System 433
    and robotics 378, 381, 387, 401
    and safety 319, 433
    Standards **420**, 449–50
    and system analysis 433
    technical societies **414**, 416, 420
    Computer Society 381, 403
    Engineering in Medicine and Biology Society (EMBS) 630, **631**
    Instrumentation and Measurement Society 360
    Robotics and Automation Society 378, 381
    and telecommunications 449–50
    Transactions 65, **414**, **416**, **426–7**, 431–2, 434, 435–6
    data sets 433
    ... *on Automatic Control* 398, 401, 405
    ... *on Biomedical Engineering* 630, 642

    ... *on Communications* 449
    ... *on Components, Packaging and Manufacturing Technology* 654
    ... *on Energy Conversion* 252, 434, 436
    ... *on Engineering Management* 654
    ... *on Fuzzy Systems* 460
    ... *on Industrial Electronics* 436
    ... *on Industry Applications* 436
    ... *on Instrumentation and Measurement* 402
    ... *on Knowledge and Data Engineering* 402
    ... *on Magnetics* 420
    ... *on Medical Imaging* 639
    ... *on Nuclear Science* 251
    ... *on Power Applications and Systems* 438
    ... *on Power Delivery* 428
    ... *on Power Electronics* 252, 436
    ... *on Power Systems* 252, **428**, 437
    ... *on Rehabilitation Engineering* 630, 643
    ... *on Robotics and Automation* 378, 401
    ... *on Signal Processing* 449
    ... *on Systems, Man and Cybernetics* 378
    ... *on Vehicular Technology* 252
    and transport 513
Institute of Energy (InstE [ofDOE]) 205, 210, **226**
    Efficient use of energy [ed. Dryden] 205
    Energy World 234
    Fuel and Energy Abstracts 227, 283
    Journal 234
Institute of Energy Economics 234
Institute of Environmental Management 493
Institute of Fuel, *former* 205
Institute of Gas Technology 250
    Gas Abstracts 228
Institute of Geology and Mineral Exploration (IGME [Greece]) 594
Institute of Industrial Engineers 326
Institute of Information Scientists
    and patent searches 61
    Searcher 61
Institute of Marine Engineers (IMarE) **137**, **475**
    Marine Engineers' Review 476
    Offshore Technology 476
Institute of Materials **138**, **303**, 313
    databases 229, **305**, 355
    and Engineered Materials Abstracts 228, 229

714 Index

Institute of Materials
databases (contd)
MATERIALS DATABASE FOR
STUDENTS 308
and METADEX 111, 230, 355
and Materials Information
service 305, 309, 355
Materials World 309
and Metals Abstracts 309
Institute of Mathematical
Statistics 113
Institute for Mathematics and its
Applications see IMA
Institute of Measurement and
Control 371
conferences 279
Instrument engineer's
yearbook 86
Transactions 401
Institute of Metals, publications
313, 346
Institute of Nuclear Materials
Management 251
Institute of Nuclear Power
Operations (INPO [US]) 264
Institute of Oceanographic
Sciences 248
Institute for Operational
Research, publications 348
Institute of Petroleum (IP) 270
and International Petroleum
Abstracts 228
IPABASE 230
Model code of safe
practice... 321, 322, 326
Petroleum Review 68, 240
... Quarterly Journal of
Technical Papers 282
Institute of Physical Sciences in
Medicine (IPSM), former
630, 631–2
Physics in Medicine and
Biology (PMB) 630
Physiological Measurement
630, 639
Scope 630, 643
Institute of Physics (IOP/IoP)
138, 374, 632
Institute of Physics Publishing
[co.] 143
Physics World 629, 643
Stress Analysis Group 138
Institute of Quality Assurance
180
Quality Forum 358
Quality News 358
Institute of Quarrying (IQ) 594,
609–10
Quarry Management 597, 602,
609
Transactions 65
Institute of Road Safety Officers
513
Institute for Road Safety
Research [SWOV
(Netherlands)] 516
Institute of Road Transport
Engineers 513
Institute of Science and
Technology [Manchester,
University] (UMIST) 358–9

Institute for Scientific
Information (ISI [US])
CompuMath Citation Index
402, 403, 461
Current Contents 111, 380
see also Current Contents
Index to Scientific Reviews 105
Index to scientific and
technical Proceedings 77,
109, 505
and Science Citation Index
(SCI) 102, 109, 227, 380
Institute of Spatial Organization
[TNO (Netherlands)] 517
Institute of Statistical
Mathematics, Annals... 460
Institute for Transportation,
Journal of Advanced
Transportation 498
Institute of Transportation
Engineers (ITE [US]) 513
ITE Journal 498
Institute of Transportation
Studies [US] 501
Institute of Tribology [Leeds]
173, 177
institutes, general see
organizations
Institution of British
Telecommunications
Engineers, British
Telecommunications
Engineering 414
Institution of Chemical Engineers
(IChemE) 270, 327
and CEBA 283
conferences 270, 277
and energy 226
journals 281
Chemical Engineer 281
Environmental Protection
Bulletin 281
European Bulletin 281
Loss Prevention Bulletin
281, 328
What's On 281
publications 170, 274, 277,
280, 317, 321, 322, 323,
326, 327
data compilations 273
Heat transfer...
.Conference... 277
Symposium Series 270
training 270
User guide on process
integration... 277
and safety 317, 318, 321, 322,
323, 326, 327
Safety and Loss Prevention
Subject Group 327
Transactions 281
Chemical Engineering
Research and Design
281
Food and Bioproducts
Processing 281
Process Safety and
Environmental
Protection 281
Institution of Civil Engineers
(ICE) 138, 513, 522

bookshop 533
conferences 522, 530
journals, general
Advances in concrete and
cement research 522
Géotechnique 522, 530, 532
New Civil Engineer 67, 490,
532, 580
Offshore Engineer 68, 240,
476
library 522, 526, 531, 535, 552
and marine technology 476
Panel for Historic Engineering
Works (PHEW) 534–5,
553
Proceedings (PICE) 522, 530,
531, 553, 580
Civil Engineering 531
Geotechnical Engineering
531
Maritime and Energy 531
Municipal Engineer 530,
531
Structures and Buildings
531
Transport and Water 499,
531
publications 522, 556, 567
Catalogue 527
ICE PUBLICATIONS DATABASE
530
indexes to 102
Works construction guides
544
and stress analysis 138
and Thomas Telford,
publishers 531
Transactions 530
and transport 499, 513
Institution of Diagnostic
Engineers, Diagnostic
Engineering 327
Institution of Electrical Engineers
(IEE) 138, 210, 371, 388,
413, 513, 632
history and role 413
and artificial intelligence 397
and automation 380, 381, 388
and biomedical engineering
629, 632, 641
and Brit. Computer Soc. 413,
416
and computing 413
and concurrent engineering
651, 653
conferences 242, 381, 413,
417, 432, 434, 450, 501
and control systems 381, 397,
398, 399, 401, 413
databases 413, 422
IEEE/IEE PUBLICATIONS ONDISC
(IPO) 427
and INSPEC 101, 111, 230,
354, 380, 399, 421, 426
see also INSPEC
and design 338, 342
and electrical engineering
426–7, 428, 432, 434
and electronics 413, 415–16,
417, 420
and energy 226, 241

Index

Institution of Electrical Engineers
(IEE) (contd)
  and ERA Technology 414
  information services 371
  Information Services for the
    Physics and
    Engineering
    Communities (INSPEC)
    413
    see also INSPEC databases
  and information technology
    413
  and IProdE 651
  journals, general 415–16,
    421–2
    Computer and Control
      Abstracts 101, 341,
      380, 399
    Computing and Control
      Engineering Journal
      400, 416
    Electrical and Electronics
      Abstracts 99, 101, 111,
      230, 341, 380, 399,
      421–2, 426
    Electronics and
      Communication
      Engineering Journal
      416, 449
    Electronics Letters/
      Electronics Letters
      Online 115, 415, 449
    Engineering Management 65
    IEE Colloquium Digest 405
    IEE Control Engineering
      series 401
    IEE News 416, 450
    IEE Review 416, 427
    Key Abstracts 399–400,
      421–2, 426, 428
    Manufacturing Engineer 87,
      356
    Physics Abstracts 100, 101,
      111, 230, 380, 399
    Power Engineering Journal
      253, 427, 435, 438
    Science Abstracts 228, 355,
      380, 421
  library 371, 413
  and manufacturing 356, 371
  and physics 413
  Proceedings 65, 253, 338, 397,
    404–7, 415, 426–7, 437–9
  'Artificial intelligence . . . '
    397
  Circuits, Devices and
    Systems 415
  Communications 448
  Communications Speech and
    Vision 415
  Computers and Digital
    Techniques 415, 448
  Control Theory and
    Applications 401
  Electric Power Applications
    252, 434
  Generation, Transmission
    and Distribution 252,
    428
  IEE Colloquium Digest 405
  Microwaves, Antennas and
    Propagation 448

Optoelectronics 415, 448
Special Issues 397
Production Engineer, former
  356
professional groups
  Computing and Control
    Division 381
  Education Division 242
  Electronics Division 450
  Power and Science Division
    242
  Technology Division 242
  publications 413, 423
  Conference Publications
    series 432
  Formal methods in
    safety-critical systems
    319
  Guidance . . . on . . . wiring
    regulations 321
  IEE wiring regulations 320,
    321
  International Conference on
    Renewable Energy . . .
    242
  [Peter] Peregrinus co. 413
  Safety-related systems series
    319
  'Strategic directions for
    control' 398
  telecommunications series
    443–4
  Wiring regulations/Guidance
    320, 321
  and qualifications 371
  and research 414
  and robotics 380, 381, 388
  and safety 319, 320, 321
  and software 413
  and Standards 420
  and stress analysis 138
  Technical Information Unit
    413, 422
  technology transfer 371
  and telecommunications
    443–4, 448–9, 450
  and thermal systems 210
  and tidal power 247
  and training 420
  and transport 501, 513
Institution of Electronics and
  Electrical Incorporated
  Engineers (IEEIE) 414
Institution of Engineering
  Designers (IED) 152, 153,
    342
  conferences 162
  Engineering Designer 157,
    163, 338
  Official reference book and
    buyers' guide 340
Institution of Engineers
  [Australia] 173, 614
  International Conference on
    Mining Machinery 614
Institution of Engineers and
  Shipbuilders in Scotland
  (IESS) 475
Institution of Gas Engineers 240
Institution of Heating and
  Ventilating Engineers,
  former 207

Institution of Highways and
  Transportation (IHT) 513
  Design recommendations
    for . . . car parks 568
  Highways and Transportation
    498
  Joint Committee with IStructE,
    Report 568
  Roads and traffic in urban
    areas 506, 507
Institution of Mechanical
  Engineers (IMechE) 138,
  153, 180, 371, 388, 632
  Advances in Bioengineering
    648
  and automation 381, 388
  and biomedical engineering
    629, 632, 641
  and concurrent engineering
    651
  conferences 162–3, 169, 173,
    207, 208, 289, 326
  EuroTech Direct 162–3
  ICED 162, 342
  Internat . . . Materials and
    Design Against Fire
    322
  Process Tech 326
  World Congress on the
    Theory of Machines . . .
    162
  and design 162–3, 338, 339,
    342
  and energy 226, 248, 250
  Engineering Manufacturing
    Industries Division 326
  Fluid Machinery Group 208
  and fluid mechanics 206–7,
    208, 210, 289, 292, 293
  information services 371
  journals, general
    Chartered Mechanical
      Engineer, former 356
    Journal of Strain Analysis
      for Engineering Design
      139, 142, 338
    Professional Engineering 65,
      71, 86, 157, 295, 339,
      356
    synopsis journals 71
  Library 152, 371, 413
  and machines 152, 153, 162–3,
    168, 169, 179, 181
  and manufacturing 356, 371
  and materials 138, 303, 309
  Materials and Mechanics of
    Solids Group 138
  and mechatronics 173
  Medical Engineering Group
    629, 632
  Proceedings 65, 144, 162, 292,
    293, 309, 338, 641
  Journal of Aerospace
    Engineering 144
  Journal of Automobile
    Engineering 144
  Journal of Engineering
    Manufacture 356, 654
  Journal of Engineering in
    Medicine 629, 641
  Journal of Engineering
    Tribology 174

Institution of Mechanical
    Engineers (IMechE)
  Proceedings (contd)
    Journal of Mechanical
      Engineering Science
      144, 159, 163, 168, 297
    Journal of Systems and
      Control Engineering
      290, 292
    Process Engineering Group
      206–7
  publications 303, 326
    Biomechanics in sport 646
    Changing role of
      orthopaedics... 646
    Decommissioning of
      radioactive facilities
      250
    Design and operation of...
      compressors 208
    Energy recovery in process
      plants 207
    Machine condition
      monitoring 179
    Management... of...
      radioactive waste 250
    Mechanical Engineering
      Publications Ltd 142,
      144, 173, 179, 303
    Nuclear power plant safety
      standards... 324
    Practical guide to the
      machinery directive 181
    Wave energy 248
    and qualifications 371
    and quality 179, 180, 319
    and robotics 381, 388
    and safety 181, 319, 322, 326
    and stress analysis 138, 144
    technology transfer 371
    and thermodynamics 207, 210
    Thermodynamics and Fluid
      Mechanics Group 206–7
    and tribology 173
    and wave energy 248
Institution of Mechanical
    Incorporated Engineers 153
  Mechanical Incorporated
    Engineer 157
Institution of Mining Electrical
    and Mining Mechanical
    Engineers (IMEMME) 609
  Mining Technology 610
Institution of Mining Engineers
    (IMinE) 593, 609–10
  Mining Engineer 609
Institution of Mining and
    Metallurgy (IMM) 545, 593,
    609, 613
  conferences 600–1
    Mine ventilation...
      congress 613, 617
  IMM Abstracts 602, 609, 613
  IMMAGE 602, 609, 613
  Minerals Industry International
    597, 609
  Shaft engineering 615
  Transactions (TransIMM) 596,
    609
    Applied Earth Science 609
    Mineral Processing and

Extractive Metallurgy
  609
Mining Industry 609
Institution of Nuclear Engineers
  (INucE) 265
  Nuclear Engineer 266
Institution of Occupational Safety
  and Health 183
  Health and Safety Practitioner
    183
Institution of Physics and
  Engineering in Medicine and
  Biology (IPEMB) 630, 631,
  632
Institution of Production
  Engineers (IProdE), former
  344
  and concurrent engineering
    651
Institution of Structural Engineers
  (IStructE) 138, 522, 547
  and car park design 568
  conferences 522, 530
  History Study Group 535
  INSTRUCT database 530
  and international organizations
    536, 554, 556
  Joint Committee with IHT,
    Report 568
  library 522, 526, 530, 531,
    535, 552
  Proceedings 522
  publications 522, 544, 550,
    553, 556, 557, 560, 567
    Achievement of structural
      adequacy... 550
    Aims of structural design
      550
    Appraisal of existing
      structures 551
    Appraisal of sports grounds
      552
    Assessment of reinforced...
      concrete bridges 566
    Catalogue 527
    Communication of structural
      design 550
    Design of air-supported
      structures 568
    Design recommendations
      for... car parks 568
    Guidance notes 522, 557
    Guide to good management
      practice for... design
      offices 550
    Guide to surveys and
      inspections... 551
    Load testing of
      structures... 551
    Qualitative analysis of
      structures 550
    Safety... of demountable
      grandstands 568
    Stability of buildings 550
    Structural failures... 550
    Technical reports 522
  Standing Committee on
    Structural Safety, Report
    550
  and stress analysis 138
  Structural effects of

alkali-silica reaction...
  557
Structural Engineer 522, 530,
  531, 553, 557, 580
Study Group for the History of
  Structural Engineering
  553
institutions, general see
  organizations
Institut National de Recherche
  sur les Transports et leur
  Sécurité (INRETS) 516
Instituto Poligrafico e Zecca
  dello Stato [Italy] 533
Institut für Strassenbau und
  Verkehrsplanung [Austria]
  501
Institut Technique du Bâtiment et
  Travaux Publics [France]
  532
INSTRUCT database 530
Instructional aid for occupational
  safety and health in...
  design (pub. ASME) 325
Instrumentation and Control
  Systems (I & CS) 401
Instrument engineer's yearbook
  (pub. Inst. of Measurement
  and Control) 86
Instrument and Measurement
  Society [of IEEE] 360
instruments/instrumentation
  abstracts/indexes 421–2
  and biomedical engineering
    626–7, 630, 635, 638, 642
  books 86, 125, 423
  dictionaries 398
  directories 418
  handbooks 398
  conferences 360
  current trends 367
  databases 421–2
  and electricity generation 252,
    429
  electronic 411–24
  equipment 146–7
  journals 87, 280, 400–1, 402
  organizations 140, 270, 278,
    279, 360, 414–15
  process control 398
  programmable 420
  reliability testing 319
  and safety 320
    safety-related 319
  Standards 420
  training 415
  see otherwise control systems;
    measurement
Instrument Society of America
  (ISA) 270–1
  publications 278, 279, 280,
    319, 321
  Standards... for
    instrumentation and
    control 279
insulators, electrical 419
insurance industry
  marine insurance 478
  and safety 180, 181, 316, 321
integrated circuits (IC) 412, 419
Integrated Environmental
  Management 487

Integrated Manufacturing Systems 357
integrated processes *see* concurrent engineering
integrated product development 650, 652
integrated services digital network (ISDN) **442**, **445**
broadband (B-ISDN) 442, **445**
integrated systems 287, 366
Integrated Systems Inc. 295
intellectual property
inventions 32
WIPO 45
see also copyright; patent *etc.*
intelligence, artificial *see* artificial intelligence
intelligence, commercial
and patents 60
trade secrets 36
intelligent manufacturing 357, 361, 402
management systems (IMMS) 363, 386
see otherwise artificial intelligence
Intelligent motor control (pub. Quin Systems) 172
intelligent networks (IN) 442
intelligent transport 503
Intelligent Transport Systems Society, *Journal*... 503
Interavia 68
Interbuild [International Building and Construction Exhibition] 87, 89
Intercept [publishers], *Water treatment handbook* 489
Interceram 68
InterDok 153, **163**
Interfaces 404
interference, geometric 135
interferometry
holographic **135–6**, 146–7
moiré 135
speckle-pattern (ESPI) **135–6**, 146–7
Interim guidance notes for... topside structures... (pub. Steel Construction Inst.) 567
inter-metallic materials 353
International Abstracts in Operations Research 355
International Aerospace Abstracts (IAA) 7, 112
international agencies, selective list 224–5
International Association for Bridge and Structural Engineering (IABSE) **522–3**
conferences **522–3**
Structural Engineering International 532
International Association for Earthquake Engineering, *Earthquake Engineering and Structural Dynamics* 532
International Association of Energy Economics, *Energy Journal* 234
International Association for Hydrogen Energy (IAHE) 225
International Association of Science and Technology for Development (IASTED) **387**
Advances in robotics and automation 382
conferences 382, 403
International Journal of Power and Energy Systems 234
International Journal of Robotics and Automation 378–9
International Association for Shell and Spatial Structures (IASS), *Bulletin* 532
International Association for Travel Behaviour Research, conference 500
International Atomic Energy Agency (IAEA) 225, **262**
publications
Emergency power systems... 325
Fire protection and fire fighting... 325
Guide to the safe handling of radioactive wastes... 250
IAEA Bulletin 266
IAEA safeguards... 250
INIS Atomindex **9**, 228, 231, 267
Manual on maintenance of systems... 324
Meetings on atomic energy 267
Nuclear power: status and trends 261, 267
reports **9**
International Bank for Reconstruction and Development *see* World Bank
International Bar Association (IBA), *Energy law*... 233, 240
International brands and their companies 88
International Broadcasting Convention 417
International Building and Construction Exhibition (Interbuild) 87, 89
International Building Services Abstracts (pub. BSRIA) 228
International Cargo Handling Co-ordination Association, *Cargo Handling Abstracts* 480
International CERAMITEC Symposium 359
International Chemical Engineering 240, **281**
International Civil Aviation Organization (ICAO), Standards 16
International Civil Engineering Abstracts 113, 528
online 113, 528
International Classification for Standards (ICS) 18
International Coal Letter 239
International Coal Preparation Congress 601
International Coal Report 239
International Commission for Conformity, Certification of Electrical Equipment (CEE) 19
International Commission on Illumination [CIE] 22
International Commission on Radiological Protection (ICRP) **262**
International comparisons of transport statistics (pub. HMSO) 509
International Conference on Automation Robotics and Computer Vision (ICARCV) 382
International Conference on CAD/CAM (MICAD) 359
International Conference on Computer Aided Shipbuilding (ICCAS) 474
International Conference on Electrical Machines (ICEM) 434
International Conference on Engineering Design (ICED) **162**, **342**, 655
International Conference on Experimental Mechanics 145
International Conference on Materials and Design Against Fire 322
International Conference on Metal Forming 359
International Conference on Mining Machinery (pub. Inst'n Engineers [Australia]) 614
International Conference on Mobile Radio and Personal Communications 417
International Conference on Renewable Energy... (pub. IEE) 242
International Conference on Robotics and Automation 381
International Conference on Solid-State Devices and Materials 417
International Conference on Surface Finishing [ISFEC] 359
International Conference on Ultra-Precision in Manufacturing Engineering (UME) 359
International Congress on Rock Mechanics 613
INTERNATIONAL CONSTRUCTION DATABASE *see* ICONDA
International/Paris Convention [on patents] (1883) 38–9
International Cooperation in Transport Economics Documentation (ICTED) 504
International Copper Research Association 303

International Council for
 Building Research Studies
 and Documentation [CIB]
 523, **586**
  publications 556, 587
  specialist commissions 585
International Council on
 Monuments and Sites
 (ICOMOS) 552
*International directory of
 engineering societies...*
 (*pub.* AAES) 125, 523
*International directory of new
 and renewable energy...*
 (*pub.* Unesco) 242
International District Heating
 Association (IDHA [US])
 211
 District heating handbook 209
International District Heating and
 Cooling Association
 (IDHCA [US]) 209, 211
 District Heating and Cooling
 209
*International electronics
 directory 90'* (*pub.* Elsevier)
 418
International Electrotechnical
 Commission (IEC) 16, 17,
 18–19
 and BSI 21, 22
 information centres 18
 and ISO 17, **18–19**
 and PERINORM 582
 publications 19
  IEC Bulletin 19
  IEC catalogue 19
  Technical guides 19
International Energy Agency
 (IEA) **224–5**
 bioenergy agreement 249
 Coal Industry Advisory Board
 238
 conferences 254
 IEA Biomass Conversion
  Technical Information
  Service 225
 IEA Coal Research 225, 239,
  602
 IEA Greenhouse Gas
  Programme 225, 226
 publications 232, 243
  Coal Abstracts 112, 228,
   602
  Coal Calendar 239
  Coal liquefaction... 238
  Coal quality and ash...
   238
  *Coal transport
   infrastructure*... 238
  *Energy balances of IEA
   countries* 232
  *Energy Conservation in IEA
   countries* 256
  *Energy Conservation in IEA
   Countries* (*pub.* OECD)
   256, 257
  Energy Prices and Taxes
   234
  *Energy statistics and
   balances in non-OECD
   countries* 232

*Energy statistics of OECD
 countries* 232
International Energy Society,
 Energy Letters 234
International Federation for
 Information Processing
 (IFIP) 474, **650–1**, 655
International Federation of
 Library Ass'ns (IFLA), UAP
 10
International Federation for
 Medical and Biological
 Engineering (IFMBE) **628–
 9**, **632**
 MBEC News 629, 643
International Federation of
 Medical Electronics (IFME),
 former 628, 629
International Federation of
 Operational Research
 Societies, *International
 Abstracts in Operational
 Research* 355
International Federation of
 Robotics (IFR) 381, **387**
 IFR Robotics Newsletter 378
 *World industrial robot
  statistics* 382
International Federation for the
 Theory of Machines and
 Mechanisms (IFToMM) 151,
 152, **153**
 Commission 'A' 151, 154
 conferences 152, **161–2**
 *Mechanism and Machine
  Theory* 151, 153, 159,
  163, 166, 168
International Gas Union (IGU)
 225
International Graphics Exchange
 System (IGES) 335
International Information
 Association 372
International Institute for Energy
 Conservation, *E-Notes* 257
International Institute for
 Environment and Safety 248
International Institute for
 Environment and Society
 255
International Institute of
 Technology 272
International Institute of Welding
 371
International Institution for
 Production Engineering
 Research *see* Collège
 International pour la
 Recherche Scientifique...
*International Journal for Adaptive
 Control and Signal
 Processing* 401
*International Journal of Adhesion
 and Adhesives* 357
*International Journal of
 Advanced Manufacturing
 Technology* 356, 379
*International Journal of Ambient
 Energy* 242, 244
*International Journal of
 Biomedical Computing* 642

*International Journal of Clinical
 Monitoring and Computing*
 639
*International Journal of
 Computer Integrated
 Manufacturing* 356–7
International Journal of Control
 401
*International Journal of
 Electronics* 416
*International Journal of Energy,
 Environment, Economics*
 234
*International Journal of Energy
 Research* 234
*International Journal of Expert
 Systems, Research and
 Applications* 402
International Journal of Fatigue
 143
International Journal of Fracture
 143
*International Journal of Global
 Energy Issues* 234
*International Journal of Heat
 and Fluid Flow* 205, 282
*International Journal of Heat
 and Mass Transfer* 282
*International Journal of Human–
 Computer Interaction* 339
*International Journal of Human–
 Computer Studies* 339, 402
*International Journal of Human
 Factors in Manufacturing*
 339
*International Journal of
 Hydrogen Energy* 241
*International Journal of
 Industrial Ergonomics* 339
*International Journal of
 Information Technology in
 Construction* **580**
*International Journal of the
 Japan Society of Mechanical
 Engineers* 292
*International Journal of the
 Japan Society for Precision
 Engineering* 357
*International Journal of Machine
 Tools and Manufacture* 356
*International Journal of
 Mechanical Sciences* **143**
*International Journal of Mineral
 Processing* 239, 596
*International Journal of
 Multiphase Flow* 255,
 282
*International Journal of Neural
 Systems* 402
*International Journal for
 Numerical Methods in
 Engineering* **143**, **458–9**
*International Journal of
 Operations and Production
 Management* 358
*International Journal of Power
 and Energy Systems* 234
*International Journal of Pressure
 Vessels and Piping* **143**
*International Journal of
 Production Economics* 358

Index 719

International Journal of Production Research 356, 654
International Journal of Project Management **580**
International Journal of Rehabilitation Research 643
International Journal of Robotics and Automation 378–9
International Journal of Robotics Research 379, 401
International Journal of Rock Mechanics and Mining Sciences 611
International Journal of Solar Energy 244
International Journal of Solids and Structures 532
International Journal of Space Structures 532
International Journal of Technology and Design Education 338
International Journal of Transport Economics 498
International Journal of Vehicle Design 338
International Labour Office (ILO), Civil engineering work... safety... 326
International Liaison Group on MHD Electrical Power Generation 253
Internationally agreed Numbers for the Identification of Data (INID) 48–9
International Machine Tool Design and Research Conference (MATADOR) 358–9
International Machine Tool Engineers Conference (IMEC) 359
International Maritime Organization (IMO) 16, **477**
International Mathematical News **461**
International Measurement Confederation (IMEKO) **140**
International Mineral Processing Congress (IMPC) **600–1**
International Mine Ventilation Congress 613
International Network of Engineering Design Researchers, newsletter 657
International Nuclear Information System (INIS) **9**
see also INIS Atomindex
International Nuclear Law Association (INLA) 263
International Off-Highway and Power Plant Congress and Exposition 296
International Online Information Meeting, and patents 61
International Organization for Legal Metrology (OIML) 27
International Organization for Medical Physics (IOMP) **629, 632**

Medical Physics World 629
International Organization for Standardization see Internat. Standards Org'n
International Orthopaedics 638
International Patent Classification (IPC) 48, 49, **57**, 58
International Peat Society, Bulletin 239
International Petroleum Abstracts 228, 480
International Power Generation 252, 431
International Precision Engineering Seminar (IPES) 359
International Radiocommunications Consultative Committee [CCIR], former 449
International railway statistics (pub. IUR) 510
International Research Center for Energy and Economic Development, Journal of Energy and Development 234
International Road Federation (IRF) **514, 516**
IRF Research and Development 503
World road statistics 510
International road haulage by UK registered vehicles (pub. HMSO) 509
International Road Research Documentation see IRRD
International Road Transport Union (IRTU) **516**
World transport data 510
International Semiconductor Data Summaries [co.], Semicon index 419
International Shipbuilding Progress 479
International Ship Structures Congress (ISSC) 474
International Society of Electrophysiological Kinesiology (ISEK) 630
International Society of Ocean and Polar Engineering (ISOPE) 475
International Society for Productivity Enhancement (ISPE)
conferences 359
Institute of Concurrent Engineering **650–1**, 653
International Society for Prosthetics and Orthotics ((ISPO) **632**
International Society for Rock Mechanics 610
International Society of Ventilation Engineers 610
International Solar Energy Society (ISES) 225, 243
Sunworld 245
International Solid-State Circuits Conference 417

International Solvent Extraction Conference (ISEC) 601
International Special Committee on Radio Interference [CISPR] 22
International standards index: welding... (pub. Woodhead) 26
International Standards Organization/Org'n for Standardization (ISO) 16, **17–18**, 23, 450, 524
and BSI 21, 22
and databases 113
ISONET 18
PERINORM 582
drafts (DIS) 18
and IEC 17, **18–19**
and IEEE 450
information centres 18
OSI (open systems interconnection) 450
publications 17, **18**
Fluid power systems... 287
ISO Bulletin 18
ISO catalogue 18
STEP see STEP
International statistical handbook of public transport (pub. UITP) 510
International Study Group on Risk Analysis, Risk analysis in the process industries 317
International Surgery 639
International Symposium on Automotive Technology and Automation (ISATA) **501**
International Symposium on Continuous Surface Mining 613
International Symposium on Industrial Robots (ISIR) 381
International Symposium on Semiconductor Manufacturing 417
International Symposium on Transportation and Traffic Theory 501
International Symposium on Transportation and Traffic Theory 500
International Technical Information Institute (ITII) 323
International Telecommunications Conference 450
International Telecommunication Union 113, 449
Telecommunication Standards Sector (ITU-T) 449
International Telegraph and Telephone Consultative Committee [CCITT], former 449
International Tin Association 330
International Union of Air Pollution Prevention Associations 494
International Union for Physical and Engineering Sciences in Medicine (IUPESM) 629

720 Index

International Union of Producers and Distributors of Electrical Energy (UNIPEDE) 225
International Union of Public Transport [Union Internationale des Transports Publics (UITP)] 517
International statistical handbook of public transport 510
Public Transport International 499
UITP Index 504
UITP Revue, former 499
International Union of Railways (IUR) 504, 516
International railway statistics 510
International Union of Theoretical and Applied Mechanics (IUTAM) 140
'Internet' [2] [project management org'n] 586
Internet 107, **114–20**, **305–6**, 341, 502, 584
CD-ROMs and 108
audio transmission 503
browsers 118
bulletin boards **117**
clients and servers **117–18**
conference groups 305
databases on 111, 305–6, 503, 504, **505**, 527, 584
and design 341–2
directory listings 118–19
discussion lists **116–17**
and graphics 114, 118
hypertext 118
informal communication **116–17**
information services **117–20**
'EEVL' 120
'Ei Village' 120
'invisible college' 116–17
journals, electronic **114–15**
and materials 305–6, 312
multimedia information 117, **118**
navigation of **117–18**, 312, **503**
newsletters **115**
preprints 115–16
protocols 118
reports 115, 116
safety information 182
Telnet 118
and transport 502–3
Usenet 305, 502
virtual libraries 119–20
see also e-mail *etc.*
Internordisk Standerisering (INSTA) 20
INucE (Inst'n of Nuclear Engineers) **265**
invention 332
see also innovation; patents
Inventory of abstracting and indexing... in the United Kingdom [ed. Stephens; pub. BL] 227
inverters 435

'invisible college' 116
IOMP *see* Internat. Org'n for Medical Physics
ion-beam processes 364
ion exchange, mineral processing **599**
ionizing radiation, Standards 27
IoP *see* Inst. of Physics
IP [1] *see* Inst. of Petroleum
IP [2] (BRE Information Paper) **521**, 542, 588
IP [3] *see* TCP/IP
IPABASE database **230**
IPC *see* Internat. Patent Classification
IPES (Internat. Precision Engineering Seminar) 359
IPO (IEEE/IEE PUBLICATIONS ONDISC) **427**
IProdE *see* Inst'n of Production Engineers
IPSM *see* Inst. of Physical Sciences in Medicine
IQ *see* Inst. of Quarrying
IRB (Informationszentrum Raum und Bau) 530
Ireland
IDA Enterprise Board 314
manufacturing 359
theses 79
IRF
IRF Research and Development 503
see otherwise Internat. Road Fed'n
Irish Development Association (IDA) 314
Irish Manufacturing Committee (IMC) 359
iron 593
construction industry 538–40, 552, 557–9
early uses 589, 607
and electronics 419
and magnetism 419, 435
mining 609
organizations 303, 538, 559, 609
processing of
coke and 590
flotation 599
Standards 25, 26
waste proportions 591
see also steel
Irons, B. *and* Shrive, N. 455
Iron and Steel Engineer 68
IRRD (Internat. Road Research Documentation), IRRD database 503, **504**, 505, 512
irritant substances 323
IRTU *see* Internat. Road Transport Union
Irvine, D.J. *and* Smith, R.J.H. [*trenching*] 545
Irvine, T.F. *and* Liley, P.E. [*steam/gas*] 236
Irwin, A.W. 556
Irwin, A.W. *and* Sibbald, W.I. 545
ISA *see* Instrument Soc. of America

ISATA *see* Internat. Symposium on Automotive . . .
ISDN *see* integrated services digital network
ISEC *see* Internat. Solvent Extraction Conference
ISEK *see* Internat. Soc. of Electrophysiological Kinesiology
isentropic processes 201
ISES *see* Internat. Solar Energy Soc.
ISFEC [Internat. Conf. on Surface Finishing] 359
ISI *see* Inst. for Scientific Information
ISIR (Internat. Symposium on Industrial Robots) 381
ISMEC(Information Service in Mechanical Engineering), *former* 153
ISMEC Bulletin, former 100
ISMEC database **111**, **153**, **230**, 341
ISO
ISO Bulletin 18
ISO catalogue 18
ISONET [information centres] 18
as term 17
see otherwise Internat. Standards Org'n
ISONET [information standards] 18
ISOPE (Internat. Soc. of Ocean and Polar Engineering) 475
isothermal processes 201
isotropism/anisotropism 134, 136
ISPE *see* Internat. Soc. for Productivity Enhancement
ISPO *see* Internat. Soc. for Prosthetics and Orthotics
ISSC (Internat. Ship Structures Congress) 474
IST&B SEARCH database 77
ISTP *see Index to scientific and technical proceedings*
IStructE *see* Inst'n of Structural Engineers
Italian language
dictionaries 154, 271
journals 297, 498, 533
Standards 20
Italy
construction 533
design 162
fluid power 297
marine technology 475
minerals 596
nuclear energy 261
stress analysis 141, 144–5
see also Italian language
ITE (Inst. of Transportation Engineers) **513**
ITE Journal 498
ITII (Internat. Technical Information Inst.) 323
Ito, K. 453
ITS Focus News 498
ITU/ITU-T *see* Internat. Telecommunication Union

Index 721

IUPESM *see* Internat. Union for Physical and Engineering Sciences in Medicine
IUR *see* Internat. Union of Railways
IUTAM (Internat. Union of Theoretical and Applied Mechanics) **140**
IVHS America
  Inside IVHS 498
  IVHS-L [list] **502**
  IVHS Journal 498

J

Jackson, A.T. [*biotechnology*] 278
Jackson, M. [*energy*] 256
Jackson, N. *and* Dhir, R.K. [*construction materials*] 535
Jackson, P. [*expert systems*] 396, 397
Jacobs, P.F. 653
Jaffe, R.I. 236
Jagoda, A. *and* de Villepin, M. 446
JAIF *see* Japan Atomic Industrial Forum
JAI Press, *Advances in the economics of energy* . . . 232
James, A. [*water quality*] 487
James, A.M. [*thermodynamics*] 201
James, A.M. *and* Lord, M.P., *Macmillan's chemical and physical data* 124
James, D. *fire hazard*] 322
James, G. *and* James, R.C. [*mathematics*] 453
Jamshidi, M. *et al.* 395
Jane's Information Group [publishers]
  Jane's fighting ships 474
  Jane's ocean technology 474
  Jane's urban transport systems 507
JANET (Joint Academic Network) **109**, 505, 527, **584**
Jankowski, D.A. *and* Selover, T.B. Jr 272
Janna, W.S. 205
Jansen, R.B. 545
Japan **101–2**
  abstracting/indexing 101–2, 227
  agriculture/fisheries 101
  chemical engineering 282
  concurrent engineering 649, 651, 654, 655
  construction 532–3
  control systems 400
  design 342, 651
  electronics 101, 418, 642
  energy 101, 227, 234
  fire hazard 323
  fluid power 291–2, 296, 298
  Kyushu University 479
  manufacture 357, 359, **363**, 654
    automation 363, 377–8, 381, **386**

CIM 363
  Japan/US symposia 359
  precision engineering 357
  marine technology 475, 478, 479
  mathematics 458
  mechanical engineering 292
  medicine 642
  Ministry of Internat. Trade and Industry (MITI) 355
  nuclear energy 261, 263, 264, 265
  organizations **263**
  patents 43, 45, 47, 58
  powders 359
  power engineering 46
  research 230, 355, 478
  robotics 377–8, 381, 386, 388
  Robotics Society of Japan 378, 381
  Society of Chemical Engineering 282
  Society of Naval Architects 475
  Society of Powder Technology 359
  Standards **27**, 113
  Technical Institute of Research 478
  telecommunications 101
  Tokyo
    Chou University 655
    Fire Dept 323
    Tokyo University Industrial Institute 479
  trade literature 93
  tribology 175
  see also Japanese
Japan Atomic Industrial Forum (JAIF) **264**
  Atoms in Japan 266
Japan electronics buyers' guide (pub. Dempa) 418
Japanese Industrial Robot Association (JIRA) 381, 382, **388**
  JIRA Robot News 378
Japanese Journal of Medical Electronics and Biological Engineering 642
Japanese language
  conference *Proceedings* 359
  patents 47
Japanese Standards Association (JIS) 27
  JIS yearbook 27
Japan Hydraulics and Pneumatics Society (JHPS) **291**, 292, 298
  Journal . . . 296
Japan Information Center of Science and Technology 227
  Current Bibliography . . . 227
Japan[ese] Society of Civil Engineers 532
Japan[ese] Society of Mechanical Engineers (JSME) 175
  Bulletin 298
  and concurrent engineering 651, 655
  and design 342

and fluid power 298
JSME International Journal 292
  Transactions 298
  Workshops, JSME-ASME 655
Japan Society for Precision Engineering, . . .
  International Journal 357
JAPIO database 58, 400
JBCSA *see* Joint Brit. Committee for Stress Analysis
Jeffrey, A. 454, 456
Jenkins, G. 232, 239
Jensen, J. [*energy*] 256
Jensen, M.L. *and* Bateman, A.M. [*minerals*] 591
Jensen, P.W. [*mechanisms/history*] 156, 160, 163, 168
Jesty, P.H. 446
jet re-attachment 294
Jewell, T.K. 335
JFC Construction Forecasts and Research [co.] 575, 575*tab*
JFIT (Joint Framework for Information Technology) 415
JHPS *see* Japan Hydraulic and Pneumatics Soc.
JICST database 400
JIRA (Japanese Industrial Robot Ass'n) 381, 382, **388**
  JIRA Robot News 378
JIS (Japanese Standards Ass'n) 27
  JIS yearbook 27
JIT *see* just-in-time
JKMRC *see* Julius Kruttschnitt Mineral Research . . .
Joel, R. 202
Jog, M.G. 248, 430
Johanson, N.R. *and* Chapman, J.N. [*magnetohydrodynamics*] 252
Johansson, T.B. [*electricity*] 208
John, V. 370
John Deyell Co. [publishers] 400
Johnson, G.L. [*wind energy*] 246
Johnson, M.A. *and* Grimble, M.J. [*control systems*] 394
Johnson, P. [*human/computers*] 336
Johnson, R.C. [*design*] 163, 334
Johnson, R.P. *and* Anderson, D. [*steel/concrete*] 558
Johnston, D.N. *and* Edge, K.A. [*hydraulics*] 290, 293–4
Johnstone, S.J. *and* Johnstone, M.G. [*minerals/industry*] 590
joinery 561
joining processes *see* joints; welding *etc.*
Joint Academic Network *see* JANET
Joint British Committee for Stress Analysis (JBCSA) **138–9**
  Journal of Strain Analysis . . . **139**, **142**, 338
Joint Framework for Information Technology (JFIT) 415

722 Index

Joint Funding Councils' Libraries Review Group 120
joints, anatomical 638
joints/joining 354, **386**
  construction 539, 558
  couplings 155, 161*tab*, **163–4**, 341
  databases **355**
  handbooks 195, 340, **361–2**, 369
  journals 357
  and materials 303
  mechanical 163
  shaft connections **163–4**, 176, 341
  see also welding
Joints in simple construction (*pub*. BCSA/SCI) 539, 558
Jones, D.A. [*safety terminology*] 316
Jones, F.D. [*mechanisms/design*] 156
Jones, G.R. *et al.* [*electrical engineering*] 125, 340
Jones, J.B. *and* Hawkins [*thermodynamics*] 202
Jones, J.C. [*design/methods*] 331
Jones, M.H. *and* Scott, D. [*tribology*] 173
Jones, M.J. *and* Oblatt, R. [*minerals*] 599
Jones, O.C. *and* Telionis, D.P. [*fluid mechanics*] 205–6
Jones, P. *et al.* [*travel behaviour*] 507
Jones, P.F. [*CAD/CAM*] 335, 362
Jones, R. *and* Wykes, C. [*interferometry*] 146
Jones, T.B. *and* Thomas, B. [*safety/powder*] 321, 322
Jones, V. [*architecture*] 565
Jones, W.B. [*optical fibres*] 448
Jones, W.P. [*air conditioning*] 207
Journal of Abstracts of the British Ship Research Association, former 480
Journal of Advanced Transportation 498
Journal of Aerospace Engineering (*pub*. ASCE) **144**
Journal of Aerospace Engineering (IMechE *Proc.*) **144**
Journal of Ambulatory Monitoring 639
Journal of the American Water Works Association 489
Journal of Applied Biomaterials 641
Journal of Applied Biomechanics 641
Journal of Applied Mathematics (*pub*. IMA) 459
Journal on Applied Mathematics (*pub*. SIAM) 459
Journal of Applied Mechanics **144**
Journal of Applied Physiology 640

Journal of Atmospheric and Terrestrial Physics 491
Journal on Automatic Control [IEEE *Transac.*] 398, 401, 405
Journal of Automobile Engineering (IMechE *Proc.*) **144**
Journal of Basic Engineering 297, 405
Journal of Biomaterials Applications 641
Journal of Biomechanical Engineering 642
Journal of Biomechanics 641
Journal of Biomedical Engineering, former (*pub*. BES) 630, 642
Journal of Bone and Joint Surgery 638
Journal of Cardiovascular Electrophysiology 639
Journal of Cardiovascular Technology 639
Journal of the Chartered Institution of Building Services, former 207, 531
Journal of Chemical Engineering Data 282
Journal of Chemical Engineering of Japan 282
Journal of the Chemical Society (*pub*. RSC) 282
Journal of Chemical Technology and Biotechnology 281
Journal of Chemical Thermodynamics 282
Journal of Clinical Engineering 642
Journal of Clinical Monitoring 639
Journal of Clinical Ultrasound 639
Journal of Coal Quality 239
Journal on Communications [IEEE *Transac.*] 449
Journal on Components, Packaging and Manufacturing Technology [IEEE *Transac.*] 654
Journal of Composite Materials **143**
Journal of Computing in Civil Engineering 532
Journal of Constructional Steel Research 64
Journal of Construction Engineering and Management 532, **580**
Journal on Control and Optimization (*pub*. SIAM) 401, 405, 459
Journal on Control (*pub*. SIAM) 405
Journal of Design and Manufacturing 338, 653–4
Journal of Differential Equations 460
Journal on Discrete Mathematics (*pub*. SIAM) 460
Journal of Dynamic Systems,

Measurement and Control 292, 298, 401
Journal of Ecology 640
Journal of Electronic Materials 420
Journal of Electronics Manufacturing 654
Journal on Energy Conversion [IEEE *Transac.*] 252, 434, 436
Journal of Energy and Development 234
Journal of Energy Engineering 255
Journal of Energy, Heat and Mass Transfer 255
Journal of Engineering Design 157, 338, **654**
Journal of Engineering for Gas Turbines and Power 208, 209
Journal of Engineering for Industry 356
Journal on Engineering Management [IEEE *Transac.*] 654
Journal of Engineering Manufacture (IMechE *Proc.*) 356, 654
Journal of Engineering Materials and Technology **144**, 309
Journal of Engineering Mathematics **459**
Journal of Engineering Mechanics **144**
Journal of Engineering in Medicine (IMechE *Proc.*) 629, 641
Journal of Engineering Tribology (IMechE *Proc.*) 174
Journal of Environmental Engineering 487, 488
Journal of Environmental Management 487
Journal of Experimental Biology 640
Journal of Experimental Zoology 640
Journal of Fluid Control **292**, 293, 295, 298
Journal of Fluids Engineering **115**, 290
Journal of Fusion Energy 251
Journal on Fuzzy Systems [IEEE *Transac.*] 460
Journal of Geotechnical Engineering 532
Journal of Human Movement Studies 641
Journal on Industrial Electronics [IEEE *Transac.*] 436
Journal on Industry Applications [IEEE *Transac.*] 436
Journal of the Institute of Energy 234
Journal of the Institution of Heating and Ventilation Engineers, former 207
Journal on Instrumentation and Measurement [IEEE *Transac.*] 402

Index 723

Journal of Intelligent Manufacturing 357
Journal of the Intelligent Transport Systems Society 503
Journal of the Japan Hydraulics and Pneumatics Society 296
Journal on Knowledge and Data Engineering [IEEE Transac.] 402
Journals of the Linnaean Society 640
Journal of Long Term Effects of Medical Implants 641
Journal of Loss Prevention in the Process Industries 282, 328
Journal on Magnetics [IEEE Transac.] 420
Journal of Manipulative and Physiological Therapeutics 653
Journal of Manufacturing and Operations Management, former 358
Journal of Manufacturing Systems 356, 378
Journal of the Marine Technology Society 476
Journal of Materials Processing and Manufacturing Science 357
Journal of Materials Science 309
Materials in Medicine 641
Journal of Mathematical Analysis and Applications 460
Journal on Mathematical Analysis (pub. SIAM) 459–60
Journal of Mathematical Control and Information (pub. IMA) 459
Journal on Matrix Analysis and Applications (pub. SIAM) 458
Journal of Mechanical Design 157, 160, 163, 168, 338
Journal of Mechanical Engineering Science (IMechE Proc.) 144, 159, 163, 168, 297
Journal of Medical Engineering and Technology 641, 642, 644
Journal on Medical Imaging [IEEE Transac.] 639
Journal of Metals 312
Journal of Nuclear Materials 251
Journal of Nuclear Materials Management 251
Journal on Nuclear Science [IEEE Transac.] 251
Journal on Numerical Analysis (pub. SIAM) 459
Journal of the Operational Research Society 358, 404
Journal of Performance of Constructed Facilities 532
Journal of Petroleum Geology 240
Journal of Petroleum Science and Engineering 240

Journal of Petroleum Technology 240
Journal of Physics 298
Journal of Physiology 640
Journal of Planning and Environment Law 247
Journal on Power Applications and Systems [IEEE Transac.] 438
Journal on Power Delivery [IEEE Transac.] 428
Journal on Power Electronics [IEEE Transac.] 252, 436
Journal on Power Systems [IEEE Transac.] 252, 428, 437
Journal für Praktische Chemie: Chemiker-Zeitung 67
Journal of Pressure Vessel Technology 144
Journal of Process Control 280, 282
Journal of Professional Issues in Engineering Education and Practice 580
Journal of Prosthetics and Orthotics 641
Journal of Regional Science 498
Journal on Rehabilitation Engineering [IEEE Transac.] 630, 643
Journal on Robotics and Automation [IEEE Transac.] 378, 401
Journal of Robotic Systems 401
Journal of the Royal Statistical Society 460
Journal of the SAIMM (JSAIMM) **596**, 610
Journal on Scientific and Statistical Computing (pub. SIAM) 459
Journal of Ship Production 476
Journal of Ship Research 476
Journal on Signal Processing [IEEE Transac.] 449
Journals of the Society for Industrial and Applied Mathematics see under SIAM
Journal of the Society of Naval Architects of Japan [SNAJ] 476
Journal of Solar Energy Engineering 244
Journal of Solar Sciences 244
Journal of Solid-State Circuits (pub. IEEE) 416
Journal of Sound and Vibration 492
Journal of the South African Institution of Mining and Metallurgy (JSAIMM) **596**, 610
Journal of the South African Mine Ventilation Society 611
Journal of Strain Analysis for Engineering Design **139**, **142**, 338
Journal of Structural Engineering **144**, 532
Journal of Systems and Control

Engineering (IMechE Proc.) 290, 292
Journal on Systems, Man and Cybernetics [IEEE Transac.] 378
Journal of Thermophysics and Heat Transfer 205
Journal of Tissue Viability 639
Journal of Transportation Engineering 498
Journal of Transport Economics and Policy 498
Journal of Transport Geography 498
Journal of Transport History 498
Journal of Tribology 174, 175
Journal of Ultrasound in Medicine 639
Journal of Urology 639
Journal on Vehicular Technology [IEEE Transac.] 252
Journal of Vibration and Acoustics 492
Journal of Visual Impairment and Blindness 643
Journal of Zoology 640
journals **63–75**, 309, 471, 473, 640
of abstracting/indexing 97–9, **100–3**, 109–13
abstracts/indexes of 64, 65, 66, 70, **74–5**, **100–3**, 354
academic **64**, 71–2, 74, 114
advertisements in 64, 65, **66**, **69**, 71
business journals 65, 400
conference announcements 64, 76
conference papers 76
consumer magazines 72
'core' journals 63–4
currency 51, 66, 68, 70
and current awareness 65, **103**
databases 71–2, 73, **74–5**
directories of 70, **72–3**, 74, 99, 114, 115, 116, **400**
electronic 70, **71–2**, **114–16**
directory 114, 115, 116
discussion lists 116–17
dual publication 71, 74
Internet **114–16**
refereed 114–15
technical challenges 114–15
historic *see* archives
house journals **69–70**, 72, **73**, 74, **83**, **417**, 481
access to 83, 93
BAIE 83
and product information 82, **83**
illustrations 71
indexes 66, 74–5, **102**, 309, 553
magazines, general 497–8
newsletters **70**, 72
newspapers *see* newspapers
online 74, 111
and patent information **50–1**
preprints 10, 115–16
Proceedings 64–5
professional **64–5**, 71

journals (contd)
  refereeing [vetting] of 64
  research 64, 69, 78, 104–5
  reviews, of books 65, 104
  reviews, product 71, 99, 105
  reviews, research 83, 99, 102, 104–5, 111, 231–5, 499
  synopsis journals 70, 71
  trade journals 65–9, 72, 73, 74–5, 86–7, 473, 480
  buyers' guides 86
  controlled circulation journals 69, 73, 416
  and design 330, 473
  online 71
  and product information 83, 86–7, 309, 473
  selective lists 67–8
  user research 64
  Transactions 64–5
  tutorial elements 427
  user research 63–4
Joy, P.F. and Goliazewski, L. 204
J. and P. switchgear book 431
J. and P. transformer book 431
JSAIMM see Journal of the South African Inst'n ...
JSME
  JSME International Journal 292
  see otherwise Japan Soc. of Mechanical Engineers
jukeboxes 111
Julius Kruttschnitt Mineral Research Centre (JKMRC [Australia]) 595
Junge, H.D. 370
Juran, J.M. and Gryna, F.M., Juran's quality control handbook 369
just-in-time production (JIT) 355, 357, 367
Juvinall, R.C. and Marshek, K.M. 163

K
Kailath, T. 391
Kaiser, P.K. and McCreath, D. 616
Kajdas, C. et al. 154
Kakac, S. 277
Kalman, R.E. 392, 394, 395
Kalman filters 391–2, 394
Kalpakjian, S. 360
Kanury, A.M. 236
Karekezi, S. and Mackenzie, G.A. 242
karnaugh maps 383
Kastanek, F. et al. 275
Katz, D.L. et al. [gas] 240
Katz, R.L. [management] 571
Katzenbach, J.R. and Smith, D.K. 654
Katzev, R.D. and Johnson, T.R. 256
Kaufman, J.G. and Drago, V.J. [materials] 313
Kaufman, J.G. and Glazman, J.S. [materials] 313
Kaufman, M. and Seidman, A.H. [electronics] 424

Kay, T. [concrete] 557
Kaye, D. [patents/business] 61
Kaye, G.W.C. and Laby, T.H. [constants/properties] 232, 419
Kayes, P.J. [hazard] 317
Kays, W.M. [heat exchangers] 206
Kearton, W.J. 236
Kececioglu, D. 178
Keenan, J.H. 202
Keenan, J.H. et al. 202
Keiser, G. 447
Keisoku to Seigyo/Society of Instrument and Control Engineers, Journal 402
Keller, A.G. [machines] 154
Keller, H. and Erb, U. [acronyms/abbreviations] 124
Kelley, D.R. [energy/environment] 256
Kelly, A. [materials] 313
Kelly, D.C. [thermodynamics] 202
Kelly, E.G. and Spottiswood, D.J. [minerals] 597
Kelly, P. and Attree, R. [product liability] 334
Kelly's directories 92
kelvins 28
Kemeny, J.G., Report ... on ... Three-Mile-Island ... 266
Kempe's engineers yearbook [ed. Sharpe; pub. M-G Information] 124, 160, 340
Kenjo, T. and Nagamori, S. 435
Kennedy, B.A. [mining] 613, 615
Kennedy, J.F. and Cabral, J.M.S. [biochemical engineering] 278
Kenney, W.F. 326
Kerntechnische Gesellschaft eV (KTG [Germany]) 265
Kettridge, J.O. 129
Key Abstracts [of INSPEC journals] 399–400, 421–2, 426, 428
Key British enterprises (pub. Dun & Bradstreet) 85
keywords 56, 59, 98, 109, 485
  ambiguous 644
KWIC [keywords in context] 18
Kharbanda, O.P. and Stallworthy, E.A. 317
Kicherer, S. 333
kilograms 28
Kim, J.H. and Stringer, J. [chaos theory] 457
Kim, S. and Karrila, S.J. [microhydrodynamics] 276
Kimbark, E.W. 433
Kimbrell, J.T. [kinematics] 158, 167
kinematics 151, 158–9, 183
  books 152, 154–5, 158–9, 165–8, 176
  conferences 161–3
  historical aspects 154–5, 167
  and robotics 379, 383

  see also machines
Kinepak [program] 170tab
kinesiology 630
kinetics, chemical 275–6, 279
King, C.J. [chemical engineering] 274
King, R.I. [machining] 363
King, R.W. [safety] 326
King, R.W. and Hudson, R. [safety] 326
Kinney, T.B. 392
Kinoglu, F. et al. 295
Kinsky, R. 205
Kirillin, V.A. and Sheyndlin, A.E. 252
Kirk-Othmer encyclopaedia of chemical technology (pub. Wiley) 271
  concise version 271
  online 271
Kister, H.Z. 274
Kittel, C. and Kromer, H. 205
Klaassen, K.B. and van Peppen, J.C.L. [systems failure] 178, 319
Klein, C. and Hurlbut, C. [minerals] 591, 598
Klein, L. [water pollution] 487
Kletz, T.[A.] [safety] 181, 317, 318
Kloomok, M. and Muffley, R.V. 155, 168
Klosowski, J.M. 543
Knee, The 638
Knezevic, J. 191
Knight, I. 336
Knight-Ridder Information[co.] 108, 584
Knott, J.F. 142
Knowledge-Based Systems 402
knowledge engineering 312, 393, 396, 402–3
  see also artificial intelligence
Knowledge Engineering Review 402
KNOWLEDGE INDEX database 107, 584
Knox, C.S. 335
Kobayashi, A.S. [mechanics/materials] 141, 147
Kobayashi, S. et al. [finite elements/forming] 364
Koberg, D. and Bagnall, J. 331
Kobe Steel [co.] 363, 386
Kohlhass, B. 537
Kokoshima, Y. et al. 297
Koloc, Z. and Vaclavik, M. 168
Komar Kawatra, S. 598
Kompass: the authority on British industry 85, 87, 157, 340
  online 91, 92
Kondratyev, K.Y. 244
Kong, F.K. and Evans, R.H. 556
Koninklijke Instituut van Ingenieurs Afdeling Maritieme Technologie [Netherlands] 475
Konya, C.J. and Walter, E.J. 614
Korea, South, Society of Naval Architects 475

Index 725

Koster, M.P., *Vibrations of cam mechanisms* 191
Kotas, T.J. 204
Kotz, S. 454
Kotz, S. *and* Johnson, N.L. 453
Kovach, E.G. [*thermal storage*] 207
Kovach, L.D. [*mathematics*] 454
Kovarik, T. *et al.* 246
Kowalevicz, A. 236
Kozlov, L.V. *and* Nusinov, M.D. 204
Kralj, A. *and* Bajd, T. 636
Krause, F. *et al.* 256
Kraushaar, J.J. *and* Ristinen, R.A. 233
Kreider, K.G. *and* McNeil, M.B. 206
Kreith, F. 205
Kreith, F. *and* Black, W.Z. 205
KR Information OnDisc [co.] 110, 111
Krinitzsky, E.L *et al.* 327
Kron, G[abriel] 433
Kruger, P. *and* Otte, C. 245
Krylov Shipbuilding Research Institute [Russia] 478
krypton, *as* fission product 259
KTG *see* Kerntechnische Gesellschaft
Kucera, V. 393
Kunii, D. *and* Levenspiel, O. 276
Kunststoffe Plastics 68
Kuo, C. [*business/engineering*] 473
Kuo, K.K. [*combustion*] 236
Kurtz, M. [*mathematics*] 453
Kusiak, A. 652-3
Kuske, A.A. *and* Robertson, G. 145
Kut, D. [*energy*] 254, 256
Kutz, M. [*mechanical engineering*] 124
Kwakernaak H. *and* Sivan, R. 394
Kwauk M. 272, 278
KWIC [keywords in context] index of international standards 18
Kyle, B.G. 275

L

L (Legislation) publications [of BSI] 182
Laboratoire Central des Ponts et Chaussées [France] 517
Laboratoire d'Economie des Transports 501
laboratories, national [UK] 27-8
 National Engineering Laboratory 139, 207, 211
 National Physical Laboratory 23, **28**, 147
 National Weights and Measures Laboratory 23, **28**
 Naval Research Laboratory 252
 Sound Research Laboratory 181, 492, 493
 Standards Laboratories **27-8**
 Transport Research Laboratory *see* Transport Research Laboratory
 see also under specific countries
lacquer, brittle lacquer stress analysis 136, 147
Laidler, P. 20
Laing, John, plc 525
Laithwaite, E.R. 435
Lamarsh, J.R. 250, 266
LAN
 LAN Magazine 449
 see otherwise network: local area
Lancaster, D. 419
Landau, Y.D. 395
Landels, J.G. 191
landfill 490, 614
Landfilling wastes (pub. DoEnv) 614
Landsberg, P.T. 202
land use, *and* transport **496**
Lane, G.A. [*solar energy*] 244
Lane, J. [*aluminium/structures*] 562
Lang, K.R. *and* Donohue, D.A.T. [*petroleum*] 240
Lange, K. [*machining*] 364
Langefors, U. *and* Kihlstrom, B. 614
Langley, B.C. 207
languages, non-English *see* dictionaries: multilingual; translations *and specific languages*
Lapin, L.L. 458
Laplace transforms 390, **457**
Large scale applications of heat pumps (pub. BHRA) 208
lasers **363-4**, 448
 anemometry 298
 journals 459
 semiconductors 412, 459
 Standards 27
 suppliers 146-7
 training 364
lateral thinking 344
Latham, R.W. 332
Lathi, B.P. 423
lattices, crystalline 136
Laughton, M.A. 210
Lauritzen, E.K. 537
law *see* legislation
Lawson, R.M. 555, 558
Lawson, R.M. *and* Newman, G.M. 555
Layfield, Sir Frank, *Sizewell-B public inquiry* 266
leaching, minerals processing 597, **599**
lead 564, 593
 in petrol 256
 roofing 564
Lead Development Association 564
Leaney, P.G. 296
learned journals *see* journals
learned societies *see* organizations
Leary, E. 541
Ledermann, W. 454
Lee, J.M. [*biochemical engineering*] 278
Lee, S.M. [*composites*] 122
Lee, W.C.Y. [*telecommunications*] 446
Leech, D.J. 333
Leech, D.J. *and* Turner, B.T. 337
Leeds
 RIBA bookshop 533
 University
  Institute of Tribology 173, **177**
 Leeds-Lyon Symposia 168, 173, **174**
 mineral engineering 595
 transport courses 502
Lees, B. [*corrosion*] 236
Lees, F.P. [*safety*] 318
Lees, F.P. *and* Ang, M.L. [*safety*] 318
Lee Spring Ltd 169
Lefebvre, A.H. 236
Lefond, S.J. 598
Leggett, J.K. 256
legislation/law 471
 BSI 'L' publications 182
 and design 333, **334**
 and electricity 320
 and energy 233, 240, 320, 430
 nuclear energy 250, 263
 and environment 247, **486-7**, 492
 expert witnesses 493
 and management 569
 maritime 469*tab*
 and minerals 233, 240
 and patents 32, 43, 45-6
 product liability 334
 and safety 180, 315, 334
 and trade journals 66
 and transport 496, 507-8
 see also Standards
Le Gourièrés D. 246
Leigh, J.R. [*control systems*] 393
Leinonen, T.E. *et al.* 151, 154
Leitch, Report (1977): ... *Trunk road assessment* 499
Leitch, R.R. [*control systems/AI*] 397
Leja, J. 604
Leliavsky, S. 248
length, Standards 27-8
Lennard, D.E. [*ocean energy*] 248
Leonard, J.W. [*coal*] 598
Leonardo da Vinci 154
Leonhard, W. [*drives*] 406
Leonhardt, F. [*bridges*] 566
Letters Patent 31-2
Letts, R.M. 635
Levenspiel, O. 277
levers 154
Levi, S.T. *and* Agrawala, A.K. 406
Lewins, J.D. *and* Gittus, J.H. 250
Lewis, B. *and* Von Elbe, G. [*combustion*] 236
Lewis, E.V. [*naval architecture*] 476
Lewis, F.L. [*control systems*] 394-5

Lewis, P.J. and Martin, G.J.
  [*mineral processing*] 592
LEXIS database 107
LEXPAT database 56
Lex Service plc, *Motorists'
  opinions..* 510
Leyton, L. 635
Li, N.N. and Strathmann, H. 274
libraries **505**, 526, 531
  archives 94–5
  and BLDSC 152
  catalogues 13, **103**, 223,
    269–70, 584
  union catalogues 103
  database searches by 505
  Electronic Libraries
    programme 120
  Follett Report 120
  IEE 413
  IFLA 10
  national 224
  'obsolescence' 96, 505
  'package libraries' 66
  and patents 33, 38, 39, **47**,
    60–1, **341**
  and reports 10, **12–13**
  Standards 13, 152
  trade literature 94–5
  university 120
  virtual 119–20
  see also British Library *etc.*
Library Association
  *Pugh's dictionary of
    acronyms* . . . 124
  Walford's guide . . . [ed.
    Mullay/Schlicke] 121
Library Bulletin (pub. DoEnv/
  DoT) 503
Library of Congress [US], . . .
  *National Union Catalog:
    Books* 103
licensing
  databases 527
  of intellectual property 41
  patents 41–2
Licht, W. 491
Lichtarowicz, A. 297
Lichtarowicz, A. *et al.* 297
Lickley, R.L. *et al., Report
  [Engineering Design
  Working Party]* 329
life sciences 102, 111, 624
  see otherwise biology;
    medicine *etc.*
light
  lasers **363–4**
  and optical fibres 441, 449
  Standards 27–8
  and telecommunications 447–8
  see also lighting; photo- *etc.*
Lighthill, M.J. 457
lighting
  organizations 22, 513
  safety 320
  solar energy 243–5
  Standards 21, 22
Lightwave 449
lignin 238
Liley, P.E. 202
Limaçon [co.] 170*tab*
Limaye, D.R. 254

limestone 541
Lindberg, A. 361
Lindgren, B.W. 332
Lindley, J. 327
linear quadratic [Gaussian])
  (LQ[G]) approaches 393,
    394, 397, 398
Lingaiah, K. 124, 340
linkages 158, 159, 160, 161*tab*,
    166–7
  historical aspects 155
Linnhoff, B. 204
Linton, L. *et al.* 650, 656
Lipták, B.G. 125
Lipton, S. *and* Lynch, J. 324
liquid-metal-cooled fast breeder
  reactors (LMFBRs) 261–2
liquids
  fire-resistant 288
  flammable 322
  properties 202–3, 273, **288–90**
  see also fluid[s]; water *etc.*
Lissaman, A.J. *and* Martin, S.J.
  360
lists, network *see* discussion lists
Literature Abstracts [petroleum;
  (pub. API)] 228, 229
Littlechild, S.C. 443
Littler, A. 598, 614, 615
Liu, P.I. [*energy*] 256
Liu, Y.A. *et al.* [*chemical
  engineering*] 274
Livermore [Ca], Lawrence
  Livermore Laboratory 314
Ljung, L. 391
Ljung, L. *and* Soderstrom, T.
  395
Lloyd, A.R.J.M. [*maritime/
  weather*] 482
Lloyd's [co.]
  Codes of Practice 151
  Lloyd's List International 480
  Lloyd's ports of the world 475
  Lloyds Register of Shipping,
    Maritime guide 478
  Register of ships 478
LMFBRs (liquid-metal-cooled
  fast breeder reactors) 261–2
LMS Ltd 170*tab*
loads/loadings **133**, **134**, 176
  fluctuating 341
  hydraulic systems 290
  loadflow 432, 433
  power engineering 432, 433
  sinusoidal/random 136
  structural 327, 551, 552,
    553–4, 566
  thermal **134**, 135
  wave loading 567
Load testing of structures . . .
  (pub. IStructE) 551
Lobanoff, V.S. *and* Ross, R.R.
  208
local area network (LAN) *see
  under* network
local authorities
  databases 511
  directory 526
  disempowered 511–12
  organizations **514–15**
  reorganized 512

transport administration 507–8,
  509, **511-12**
local documents, *as* grey
  literature 10
*Local road maintenance
  expenditure in England and
  Wales* (pub. HMSO) 509
Local Transport Today **498**, 502
Lochner, R.H. *and* Matar, J.E.
  179
Loewy, R. 336
Logan, J.G. 217
loggers 146
logic
  fuzzy **391**, **396**, 403, **457**, 460
  'logic machines' 377
  logic systems 334, 411, **412**,
    423
  handbooks 418
  mathematics 451, **457**, 460
logistics 310–11, 460
  computer-aided *see* CALS
  integrated 657
  transportation 506
Logothetis, N. 334, 367
Logothetis, N. *and* Wynn, H.P.
  179
London
  Director of Traffic 512
  Greater London Council,
    abolished 511–12
  Imperial College, Royal
    School of Mines 595, 610
  organizations **515**
  research 510
  Science Museum, Library 94
  statistics 510
  transport 509, 510, **511–12**,
    515
  statistics 509
  University College London
    255
  Bartlett School of
    Architecture 586
London area transport surveys
  (pub. London Research
    Centre) 510
London Boroughs Association
  515
London Planning Advisory
  Committee (LPAC) **515**
London Regional Transport
  (LRT) [org'n] **515**
  Public transport statistics 510
London Research Centre **515**
  *Greater London transportation
    studies* 510
  London area transport surveys
    510
London Traffic Control Systems
  Unit **511–12**, **515**
Long, R.E. 256
Longman Cartermill Ltd
  [publishers] 78, 152, **153**
  BEST 152
Longman Group Ltd [publishers]
  527, 531
  Engineering research centres
    370
  *European sources of scientific
    and technical information*
    126

Index 727

Longman Group Ltd [publishers] (contd)
  Industrial research in the UK 126
  Who's who in science in Europe 126
Look, D.C. and Sauer, H.J. 202
Lorch, W. 489
Lord, E.A. and Wilson, C.B. [*mathematics*] 455
Lord, N.W. et al. [*heat pumps*] 208
Lordan, M. and Thompson, G. 156
Lorenz, C. 333
lorries
  Armitage Report 499
  HGVs 509
Los Alamos National Laboratory, preprints 116
loss prevention 317, 318, 322
  defined 316
  handbooks 316
  journals 281, 328
  organizations 321, 327
  see otherwise safety
Loss Prevention Bulletin **281**, 328
Loss Prevention Council 321
Loss prevention data by Factory Mutual Engineering ... 316
Lotter, B. 369
Lotus 1–2–3 [program] 582
Love, S.F. 332
Lowe, D. 638
Lowenherz, L. 20
Lowrence, W.W. 334
Lowson, M.V. 246
Loyd, S. [*thermal systems*] 208, 209
LPAC (London Planning Advisory Committee) 515
LP Gas Review 67
LQ[G] (linear quadratic [Gaussian]) approaches 393, 394, 397, 398
LRT *see* London Regional Transport
lubrication 164, 168, 173, 174, **175–6**
  and hydraulics 288, 290–1
  oils 179
  software 175–6
Lucas, W. 362
Ludema, K.C. et al. 337
Ludwig, H.F. et al. 488
Lugowski, J. 297
Lumsdaine, E. and Lumsdaine, M. 332
Lunde, P.J. 210
Lusas [program] 148
Lutgendorff H.O. 615
Luttgens, G. and Glor, M. 321
Luyben, W.L. 280
Luyben, W.L. and Wenzel, L.A. 274
Lydersen, A.L. and Dahlo, I. 271
Lynch, A.J. et al. 604
Lyon, Leeds-Lyon Symposia 168, 173, **174**
Lyon, R.H., *Machinery noise* ... 191

M
Maanadens standard (pub. SIS) 27
Mabie, H.H. and Reinholtz, C.F. . 158
Macario, R.C.V. 444
Macbeth, D.K. 365
McCabe, W.L. et al. 274
McCarthy cards 75
McCarthy Information [co.] 75
McCloy, D. and Harris, D.M.J. 383
MacConaill, P.A. et al. 172, 384
McCormick, M.E. 248
Macdonald [publishers], bibliographic series 104
McEwan, J.R. 178
MacFarlane, A.G.J. 391
McGarva, J.R. and Mullineux, G. 172*tab*
McGeough, J.A. [*machining*] 363
McGhee, T.J. [*water*] 489
McGowan, F. 252
McGraw-Hill Co. [publishers] **122**, 422–3
  McGraw-Hill electronics dictionary [ed. Markus/ Sclater] 123
  databases 70, 583
  MCGRAW-HILL ONLINE database 70
  handbooks, minerals 600
  handbooks, stress analysis 141
  Mark's standard handbook for mechanical engineers 124, 160
  McGraw-Hill dictionary of scientific and technical terms 123
  McGraw-Hill encyclopedia of science and technology 122
McGreavy, C. et al. 280
Macgregor, J.E.M. [*structures*] 560, 561
Mcgregor, K. [*drilling*] 614
McGuigan, D. 208, 246, 248
Machine condition monitoring (pub. IMechE) 179
Machine Design 338
Machine Design International 157, 163, 172
machine drives 163–5, **425**, **433–7**
machine-readable services *see* microform
Machinery 157
Machinery and Production Engineering 356
Machinery buyers guide (pub. MachPress) 157, 340, 370
Machinery Design Centre **152**, 153
Machinery Market 67, 86
Machinery's handbook [pub. Industrial Press; ed. Oberg et al.] 124, **160**, 369
machines 151–97, **434**
  nature and role 151–2, 434
  abstracts/indexes 100, 102, **111**, 151, **153**, 156–7, 228, 341

acquisition of 367
animals as 625
artificial machine vision 363
automated *see* automation
books 151–2, 183–4
  buyers' guides 157
  catalogues 156–7, 164, 169
  data 160–1, 161*tab*
  data comparison **160–1**, 161*tab*
  dictionaries 123, 152, **154**, 370
  handbooks 156, 160, 161*tab*, 164, 168–9, 175, 181, 182, 340, 369
chemical engineering 269–83
components aspects 156–7, 159, **163–000**, 340
and concurrent engineering 653
condition monitoring **178–80**
conferences 151, 152, 153, 161–3, **161–2**, 168, 169, **173**, 177, 434
  quality/reliability 179, 180
databases 111, 151, **153**, 179–80, 355
data transfer 310
design aspects **155–8**, 159–84, 338, 340, 341, 434
dynamic aspects 151, **159–60**
electrical drives 163–5, **425**, **433–4**
  electronics **435–7**
  motors **434–5**
  future trends 363
  'goodness' 435
historical aspects **154–5**, 159, 166, 173, 433–4
'hybrid' 166, **169–73**
hydraulics and 286, 288
information sources 152
journals **144**, 151, 153, **157**, **159–60**, 338, 356, 434, 435–6
  components **163**, 165, 168, 173
  mathematics 459
  processes 358–9
  quality/reliability **180**
  reviews 104
  trade journals 67–8, 86–7
  trade literature 94
  tribology 176
kinematic aspects 151, **158–9**
lubrication *see* lubrication
maintenance 178
and manufacture 353–75
marine *see* marine technology; marine engineering
materials 176–7
measurement 367
mechatronics 166, **169–73**
mobile 286, 288
noise **181–2**
organizations 138, 140, **152–3**, 164, 165, 169, 175, 177, 371
safety/reliability 178–9, **180**, 182–3
patents 152

Index

machines (contd)
Plant and Machinery (PM) publications 182
power engineering **425, 428**
product information 310
quality **177–83**, 367
reliability 152, **177–83**, 319
rotating 425, **428**, 434
safety 19–20, 152, 178, **180–4**, 315, 319, **326–7, 368**
software 151, **155**, 159, 161*tab*, **170–2***tab*, 183–4
components 161*tab*, 165, 166, 167, 168, 169, **170–2***tab*
Standards 16, 19–20, 151, **152**, 160, 367–8
components 161*tab*, **163–4**, 169
EU directive on ... safety 19–20
EU Machinery Directive **19–20**, 181, **183**
quality **177–8**
reliability **180**
safety **182–3**, 368
specifications 310
surface treatments **176–7**
testing 355
theoretical aspects 158–9
thermodynamic aspects 202, 204–5
transformation of motion **166–73**
transmission **163–5**
tribology 168, **173–7**
types of 434, 437
see also machine tools; gears etc.
Machine Systems 162–3
machine tools/tooling 353, 354, **357**, 360–1, **363–5**, 364
automated *see* robotics
ball screws 368
brittle materials 359
CNC 385
conferences 358–9
dictionaries 370
directories 85
elastic emission 357
electro-chemical 357
electro-discharge 357
handbooks 360, **363**, 367–8, **369**
hydraulics 292
journals 356, 357
metal-cutting 353
metrology 367
soft tooling 364
Machine Tool Technologies Association (MTTA) **371**, 381, **388**
machine vision 363, 377, 382, 384, 385, 386
machining *see* machine tools
Machining data handbook (pub. Metcut) 369
MachPress Ltd, *Machinery buyers guide* 157, 340, 370
Maciejowski, J.M. 394
Mackay, E.B. 545

Mackenzie-Kennedy, C. 209, 254
Mackinder, M. *and* Marvin, H.E. 577–8
'McKinley Internet Directory' [list] 119
Maclean, J.H. *and* Scott, J.S. 533
Macleod, I.A. 550
McMahon, P.J. 208
McMillan, G.K. [*biochemical engineering*] 278, 280
Macmillan Publishers Ltd 423, 527, 531
Macmillan dictionary of energy [ed. Slesser] 233
Macmillan dictionary of production management and technology [ed. Bessant/Lamming] 123
Macmillan's chemical and physical data [ed James/Lord] 124
World meetings 76–7
McMullan, J.T. *et al.* [*energy*] 233
McMullan, J.T. *and* Morgan, R. [*heat pumps*] 208, 254
McMullan, R. [*noise*] 492
McNeal-Schwendler Corp'n 306–7
McQuiston, F.C. *and* Parker, J.D. [*heating*] 254
McQuiston, F.W. *and* Shoemaker, R.S. [*mineral processes*] 598, 599
Macroeconomic Modelling Bureau [Warwick] 575
Maden Tetkik ve Arama Genel Mudurlugu [Turkey] 594
Madsen, H.O. *et al.* 327
Maeder, G. *et al.* 148
magazines *see* journals
Magison, E.C. 321
Magison, E.C. *and* Calder, W. 321
magnesium, Magnox nuclear reactors 261–2
magnetic materials **419–20**, 434
bearings 177
encyclopaedia 312
permanent magnets 16, 419, 435
Standards 16, 25, **420**
magnetohydrodynamics (MHD) **251–3**
Magnetohydrodynamics: Journal of the Internat. Liaison Group ... 253
Magnox (magnesium non-oxidizing) reactors 261–2
Mahajan, S. *and* Kimerling, L.C. 122
Mailbase client-server 341–2, 657
mailing lists 66, 83, 86, 93
electronic *see* discussion lists
mailshots 91
Mainstone, R. 553
maintenance
data transfer and 310
machines/systems 178, 326, 358

robotics 384
maintainability 178–80, 319, 368
marine technology 467, 470*tab*
power stations 429
roads 509
Majumdar, A.J. *and* Laws, V. 563
Maldonado, T. 336
Maleki, R.A. 363, 385–6
Mallard, J. 629
Malstrom, E.M. 337
management 353–75, 569–71, 649–60
abstracting/indexing 100
and automation 383–6
business management 571, 589
chemical engineering 270
company information 570, **649**
computer-aided processes 362, 363, 369
concurrent engineering 333, **649–60**
guidelines 651–2
construction industry *see* construction management
corporate planning 333, 574
cost justification 369, 384, 386
current trends **365–6**
databases 230
and design 329, 330, **333–4**, **336–7**, 338, **365–6**
energy management 205, 206, 210, 251–2, **255–7**
and environmental issues 255–7, 487, 489, 490, 491, 493
finance 365
FMS 363
group technology 355, 357
and information **569–71**
information issues 365, 569–71, 577–9, 587
information management *and* CIM 362
and concurrent engineering 649, **652**
current trends 581
journals 65, 66, 67–8, 251, 328, 338, 356, 358, 379, 580
linear systems 334
make-or-buy 369
manufacturing 354, 358, **365–6**, 369
marketing 365
materials 122, 251, 585
mineral processing 589, **592**, 597
mining 614
needs analysis 333, 334
network management 442
nuclear energy 251
optimization 333, 334
OPT 365
organizations for 210, 251, 580
performance measurement **366**
personnel management 365, 569
planning 365

Index 729

management (contd)
  power engineering 428
  product development 333–4, 649
  integrated 650
  production management, dictionaries 123
  product liability 334
  product planning 333–4
  project management 569, 571
    and AI 397
    conferences 586
    data transfer 310
    information 570–1, **577–9**, 587
    journals **580**
    organizations 580, 586
    project control 333
    and quality **177–83**, 333, **334**, 354, 365, **366–7**
    total quality management 366–7, 368, 371
    risk management 317, 325, 326, 333, 334
    and robotics 383–6
    and safety 317, 318, 323–4, **325–7**, **328**, 334
    software, CALS **311**, 314
    sub-contracting 333
    teams, cross-functional 650, 651, **652**, 654, 657
    training 366
    types of 571
    version control 310
Management of design for economic production (pub. BSI) 337
Management of process hazards . . . (pub. API) 323, 325
Management . . . of . . . radioactive waste (pub. IMechE) 250
Managing design: an initiative in management education (pub. CNAA) 329
Manchester
  Building Centre 520
  RIBA bookshop 533
  University
    Institute of Science and Technology (UMIST) 358–9
    MATADOR conference 358–9
Manchester Guardian 21
Manchester Steam Users' Assn 21
Mancuso, J.R. 163
Manheimer, W.M. *and* Lashmore-Davies, C.A. 252
'manipulators' 377, 384, 396, 401
  rigid-link 394
Mannering, F.L. *and* Kilareski, W.P. 507
Manser, R.M. 600
Manterfield, R.J. 444
Manual of British Standards in engineering drawing and design (pub. BSI; ed. Parker) 337

Manual of British Standards in engineering metrology (pub. BSI; [ed. Brooker]) 368
Manual of contract documents for highway works (pub. DoT) 566
Manual for the design of . . . concrete . . . structures (pub. IStructE/ICE) 556
Manual for the design of steelwork . . . structures (pub. IStructE) 558
Manual of hazardous chemical reactions (pub. NFPA) 323
Manual on maintenance of systems . . . important to safety (pub. IAEA) 324
manufacturing **353–75**, 649–60
  nature and role 269–70, **353–4**, 365, **649–50**
  abstracts/indexes 100, **354–6**, 357, 380
  advanced processes 353, 356, 357, 359, 361, **362–5**, 366, 371, **377–88**
  machining 363
  artificial intelligence 357
  assembly **361–2**
    see otherwise assembly
  automated 353–4, 357, **359**, 360, 366, 369, **377–88**
    see also robotics *etc.*
  books **360–70**, 382–6
    data cards 360
    dictionaries 123, **370**
    directories 84, 340, **370**
    handbooks 124, 340–1, **360–1**, 368
    reference **368–9**
  case studies 356, 357, 358, 363, **365–6**
  catalogues *see* catalogues
  cells/subsystems 385
  cleaning 361
  competitiveness 336, **337**, **353**, 361, 366, 383–4, 385, 649, 650
  computer-aided processes **359**, **362–3**
    see also CIM; CAD/CAM
  concurrent engineering **649–60**
  self-assessment 656
  conferences 342, **358–60**, 381–2, 655–6
  control systems 354, 356, 358
  costs 330, 333–4, **336–7**, 353, 361, 369
  economics 358
  equipment 369
  process 360
  current trends 649, 656–7
  cutting 353, 358, 359, 360, 362, 363–4
  databases 230, 341, **354–6**, 368, 380
  CAMPUS project **304–5**, 310, 313
  free/subsidized 304, **305** for PCs 356
  data sheets 82, **83**, 93, 291, 302, 330

data transfer, electronic 310, 311–12
decision-making 358
and design 329–51, 356–7, 358, 360, 361, **365–6**, 369, 384
  strategy **333–4**
  value aspects **336–7**
  electronics 353, 355, 360, 361, 367, 369
  environmental aspects 361, 371
  finishing 87, 357, 359, 369
  flexible systems **359**, **362–3**
    see also flexible systems
  forming *see* forming
  group technology 355, 357
  inspection 354, 366–7
  integrated processes 356–7
    see also concurrent engineering
  intelligent manufacturing 357, 361, **363**, 386, 402
  inventories 365
  journals 67–8, 69, 71, 338, **356–8**, 361, 378–9, 653–5
  trade journals 67–8
  logistics 310–11
  machining **363–5**
    see also machining
  maintenance/maintainability 358, 368
  management 354, 358, **365–6**, 369
    see also management
  materials 301, **337**, 353, 355, 360–1, 364, 369, 370
  materials requirement planning (MRP) 367
  measurement 358, 360, **366–7**, 369
  nano-fabrication 355
  operational research 355, 358
  optimization 358, **365**
  organizations 326, 342, 356–60, **370–1**, 380–2, 387–8, 650–1
  parallel processes 649–50
  patents **32**, 33, 36, 379
  performance testing 366, 368
  planning 354, 356, 357, 358, 365, **365**, 369, 384–5
  precision **354**, 357, 358, **359**
  processes *see* process
  procurement 310–11
  product information 82–3, 357
  production control 356
  product liability 334
  products, new 353
  profitability **337**
  quality **357–8**, **359–60**, 360–1, **366–7**, 368, 649
  research 355, 356, **361**, 370
  right-first-time 650
  safety 371
  sales *see* sales
  Standards 19–20, 26, 28, 360, 366, **367–8**, 371, 383
  testing 355, 357–8, 360, 364, 368
    automated 379

manufacturing
testing (contd)
see also testing
thermal processes 203
time aspects 353, 360
just-in-time 355, 357, 365
lead times 365
scheduling 355
tooling see machine tools
training 364, 366, 371
trouble-shooting 361
see also management
Manufacturing Automation 379
Manufacturing automation (pub.
Hong Kong University) 382
Manufacturing Engineer (pub.
IEE) 87, 356
Manufacturing Engineering (pub.
SME) 350, 356, 654, 655
*Manufacturing into the late
1990s* (pub. DTI) 657
Manufacturing phase... (pub.
SEED) 333
*Manufacturing and Process
Engineering Abstracts* 100
Manufacturing Research
conference 359
Manufacturing and robotics
(pub. IASTED) 382
Manufacturing Systems 379
Manzini, E. 336
Maple [program] 452
maps 232, 481
Maquet, P.[trans.] 633
Mara, D.D. 488
Maral, G. and Bousquet, M. 447
Marciniak, Z. and Duncan, J.L.
364
Marcus, H.S. [*marine transport*]
482
Marcus, R.D. et al. [*pneumatics*]
278
Marecki, J. [*heat/power*] 209
Marek, M. and Schreiber, I.
[*chaos theory*] 457
Margolis, D.L. and Brown, F.T.
290
MARIN (Marine Research Inst.
of the Netherlands) 478
marinas 567
*Marine dredging for sand and
gravel* (pub. DoEnv) 615
marine engineering 467, 469*tab*
see otherwise marine
technology
Marine Engineers' Review 476
Marine Log 480
Marine News 481
Marine Pollution Bulletin 487
Marine Research Institute of the
Netherlands (MARIN) 478
marine sciences 467
*Marine Structures: Design,
Construction and Safety* 481
marine technology 467–83
nature and role 467–70,
469*tab*, 470*tab*
information approaches 468–
73
terminology 467–8
abstracts/indexes 472, 479–80

books 470–3, 482
directories 474–5, 478
graphical material 481
maritime guides 478
reports 477–8
commissioning 470*tab*
and concurrent engineering
653
conferences 471, 474, 476
construction work 469*tab*, 472,
474
data transfer 310
decommissioning 470*tab*
design 240, 467, 473, 474,
477, 479*tab*, 480
disputed areas 482
DOT and 511
drydocks 478
economic aspects 479
and electronics 421
environmental aspects 470*tab*,
487
equipment 475
insurance 478
journals 460, 476, 479–80, 487
legislation 469*tab*
marine engineering 467,
469*tab*
nautical studies 467
naval architecture 467, 476,
482
nuclear reactors 261
ocean engineering 468
offshore installations *see*
offshore engineering
organizations 137, 138, 421,
471–2, 475–80
classificatory 472, 477
official 471–2, 477, 511
ports 469*tab*, 475
product information 310, 481
research 477, 478–9
safety 475, 477, 480
disasters 471, 477
ship classing 472, 477
stabilization 474
Standards 16, 421, 472, 475,
477
BSI *Marine Series* 21
codes of practice 477
Regulations 477
specifications 310
statistics 467, 479, 480, 481,
509–10
training 479
Marine Technology 476
Marine Technology Centre
(MARINTECH [Norway])
478
Marine Technology Society
(MTS [US]) 475
Journal... 476
MARINTECH (Marine
Technology Centre
[Norway]) 478
Maritime Abstracts 480
Maritime and Energy (ICE
Proc.) 531
Maritime guide (pub. Lloyd's)
478
Maritime Monitor 480

marketing *see* sales
market research
construction 576
reports 4, 479
Marks, R. and Payne, P.A. 636
*Mark's standard handbook for
mechanical engineers* [ed.
Baumeister] 124
Markus, J. and Sclater, N.,
*McGraw-Hill electronics
dictionary* 123
Marquis [publishers], *Who's who
in science and engineering,
1994–1995* 126
Marriott, F.H.C. 453
Marschall, H.W. [*Standards*] 24
Marsh, W.D. 208
Marshall, W. [*nuclear energy*]
250, 266
Marshall, G.J.
[*telecommunications*] 446
Marshall, V.C. [*safety*] 318
Marshek, K.M. 158
Marshek, K.M. and Kannapan
S.M. 166
Martin, W.R.W. [*biomedical
engineering*] 636
Martin, J.W. [*mining*] 614
Martin, M.C. [*thermodynamics*]
202
Mashburn, W.H. 233
Maskell, B.H. 366
Mason, C.F. 487
masonry/stone 559–60
bridges 540–1, 566–7
cladding 563–4
fire hazard 555
fixings 565
organizations 530, 559
Standards 549, 555, 559–60,
564
weathering 541
see also brick etc.
mass, Standards 27–8
Massachusetts
Massachusetts Institute of
Technology 596
University 254
mass transfer 199, 273, 274, 275,
282
Masters, K. 274
Mastic Asphalt Contractors and
Employers Federation 564
masts 567
MATADOR conference 358–9
Mataré, H.F. 232
MAT.DB database 313
Material Recovery Weekly 490
materials 269, 301–14, 535–6
theoretical aspects 199–200,
301–2
abstracts/indexes 100, 102,
112, 228, 306, 309
analysis 306–7, 310–12
biomaterials 631, 635, 636–7,
638, 639, 641–2
books 121, 301–4, 308,
312–14
data sheets 302
dictionaries etc. 122, 123,
309, 313, 370

Index 731

materials
  books (contd)
    directories 86, 303, 370
    grey literature 303–4
    handbooks 122, **124–5**,
      302–3, **307**, 312–13,
      322, 369, 559
    reports 303, **304**
  brittle 359
  chemical engineering 272
  composite *see* composite
  conferences 303, 309, 322, 359
  construction 86, 519, **532–3**,
    535–43, 550, **563–4**
    cladding 563–4
    mineral sources 590, 593,
      597
    specification 527
    surface mining 615
  databases **111–12**, **303–7**, **308**,
    313–14
    CD-ROM 306
    analysis 306–7, 310–12
    books 303–4
    business 306
    CAMPUS project **304–5**,
      309, 313
    composites/engineered 112,
      229, 306, 341, 355
    disks, floppy 304–5, **306**
    EMA 306, 355
    free/subsidized 304, **305**
    Internet 119, **305–6**, 312
    METADEX 110, **111–12**, 230,
      306, 341, 355
    microform 307
    networks **305–6**, 312
    product information 92
    RAPRA **112**, 341, 355
    selection 308–9
    suppliers 313–14
  data sheets 302
  data transfer 307, **309–11**
    CALS **311**, 314
    STEP 307, **310–11**, 314
  defence 307
  and design 310–11, 330,
    336–8
  electrical 419–20
  electronics 417, **419–20**
  engineered *see* engineered
    materials
  and environment 355
  expert systems **312**
  fatigue 140, 143, 176, 318
  and fire hazard 321–2, 554–6
  flammable 320, 322, 323
  fluid mechanics 287
  hazardous *see* hazard; *and*
    safety *below*
  high-temperature 313
  historical aspects 93, 535,
    551–3
  inter-metallic 353
  joining methods 303, 357
  journals **142–4**, 251, **309**, 327,
    338, 357, 420, 459,
    532–3, 597, 641–2
    house journals 69
    reviews 104–5
    trade journals 68, 87

machines **176–7**
management issues 585
  and manufacturing 301, **337**,
    353, 355, 360–1, 364,
    369, 370
  and manufacturing, materials
    requirement planning
    (MRP) 367
marine 469*tab*
mass transfer **273**
minerals 589–606
  importance 590, 593
  new 133, 353, 357
  see also engineered
    materials
nuclear engineering 250–1,
  259–61, 266
numerical analysis 306–7,
  310–12
  see also numerical analysis
organizations 138, 278, 302,
  303, 304, 325, 524,
  532–3, **536–43**
particle technology **278**
profitability 337
properties 134, 323, 419
  data sources 122, 272–3,
    304, 306, **307–8**, 355,
    369, 419–20, 589
  mathematical aspects 459
recovery of 490
recycling 334, 337, 490, 534
requirement planning (MRP)
  367
rheology 638
safety 317, 322, 323–4, 326–7
selection of 301, **308–12**
software 304–5, 306–7
solar energy and 243–5
standard materials 338
Standards **25**, **26**, **302–3**, 307,
  314, 368, **537**
  CAMPUS 304–5, 310, 313
  company-specific 28
  specifications 302, 310, 368,
    527
  STEP 307, **310–11**, 314
  *World-wide guide to ...
    irons and steels* 26
stress *see* stress analysis
synthetic *see* engineered
  materials
testing 355
tribology 173, **176–7**
MATERIALS BUSINESS FILE database
  306
MATERIALS DATABASE FOR
  STUDENTS 308
Materials and Design 309, 338
*Materials and design against fire
  (pub. Internat. Conf. on
  Materials ... Against Fire)*
  322
Materials Information [co.] **305**,
  306, 309, **313**, **355**
Business Alerts 306, 309
Metals Abstracts 111, 306,
  309, 341, 355
  see also METADEX; EMA
Materials in Medicine 641
Materials Processing conference
  359

materials requirement planning
  (MRP) 367
materials science *see* materials
*Materials Science and
  Engineering Abstracts* 100
Materials and Structures **143**,
  533
Materials World 309
MathCAD [program] 159, **160**,
  452
Mathematica [program] 159, **160**,
  452
*Mathematical Methods in the
  Applied Sciences* 459
Mathematical Reviews 113, 460
  online 113, 403, 460
mathematics **451–66**
  nature and role **451–2**, 459
  abstracts/indexes **113**, **460–1**
  algebra 158, 451, **452**, 455
    linear **456**, **458**
  modern **457**
  analysis 455–6, 458, **459–60**
  books
    bibliographies **453**
    dictionaries/encyclopaedias
      453, **454**
    formulae/tables **452**, **454**
    handbooks 124, **453–4**
    series 454–5
  boundary elements methods
    (BEM) 134, **137**, 149,
    451, **455**, 457, **458**, 459,
    616
  calculations 124
  calculus 451
  catastrophe theory 457
  chaos theory **457**
  coding 457
  combinatorics 451
  complex variables 158, 456
  conferences 110, 453, **458**,
    460–1
  constants 232
  control systems 389, **390**, 394,
    400–3
  curves 454, **455**
  databases 110, **113**, 402, 403
  Dirac delta 457
  discrete 451, **457**, **460**
  distributions 457
  eigenfunctions 397, **456**
  equations 391, 454, 455
    Bernoulli 291
    differential **137**, 389, 390,
      395, **455**, **456**, **458**,
      459–60
  flow rate/force 290–2
  integral 456
  field theory 456
  finite elements *see* finite
    element
  fluid mechanics 294–6
  formulae **452**, 454
  Fourier analysis 454, **456**, **457**
  fractals 457
  functions 454, **456**
    generalized 457
    Green's functions 455, 457
  fuzzy logic **391**, **396**, 403,
    457, 460

732 Index

mathematics (contd)
  geometry 455
  data transfer 310
  flow 294
  motion 151
  graphs/graph theory 451, **452**, **454**, 455, **457**, **460**
  hydraulics 290–2
  hypothesis tests 451
  infinities 390, 397–8
  journals 401, 402–3, **458–60**
    electronic 115
  logic-based approaches 451, **457**, **460**
  logistics 460
  matrices 145, 158, 383, 390, 391, 394, **456**, **458**
  models see modelling
  motion 158–9
  non-linear processes **457**, **460**
  numerical analysis see numerical analysis
  organizations 113, 401, **453**, **458**
  probability 113, **451**, **454**, **457–8**, **459**, **460**
  queueing theory 458
  reliability theory 458
  and robotics 383
  set theory 451, **457**
  software see under software
  statistics see statistics
  stochastic processes 451, **457–8**, **460**
  surfaces **454**, **455**
  tables **452**, 454
  tensors **456**, **458**
  topology, data transfer 310
  training courses 461
  transforms 390, 451, **457**
  trigonometry 451
  variational approaches 456
  vectors 158, **456**, **458**
  wavelets 457
  see also kinematics etc.
Mathematics Abstracts 460
Mathematics of Computation 459
Mathematics of Control Signals and Systems 459
Matheus, C.J. and Hohensee, W.E. 396
MATHSCI database **113**
MathSoft Inc. 452
MathWorks, The, Inc. 452
Matisoff, B.S. 369
Matko, D. et al. 392
Matlab [program] 452
Matousek, R. 332
matrices see under mathematics
Matrixx [program] 295
MATUS database 314
Maunsell Structural Plastics [co.] 563
Mayer, R.J. et al. [concurrent engineering] 659
Mayne, D.Q. 394
Mayr, E. [biology] 625
Mays, G.C. and Hutchinson, A.R. [adhesives] 543
Mays, G.H. [concrete] 538
Mazda, F. 125, 421

Mazda [co.] 363, 386
Mazzolani, F.M. 562
MBEC News (pub. IFMBE) 629, 643
MBR Publications, *DABS compendium* 85–6
Mc, names beginning with, see Mac
MDD (Medical Devices Directorate) 645
Mead Data Central [co.] 56, 107
Meador, R. 209, 254
Meadowcroft, D.B. and Manning, M.L. 238
MEANS database **582**
Means, R.S., Co. Inc. 582
Measurement 357
measurement/metrology **27–8**, **367**
  acoustic 492
  architectural 565
  biochemical 278
  conferences 28, 292, 358, 360
    Convention du Mètre (1875) 27
  control systems 392
  databases 91
  dictionaries 154
  environmental 486–7, 491, 492
  equipment 145–9
  fluid power 290, 292
  and hydraulics 289–90
  journals 28, 143, 280, 292, 357, 401, **402**, 630
  and manufacturing 358, 360, 366–7, 369
  noise 181
  'noisy' 392
  organizations **27–8**, 140, 279, 360, **371**, 401, **414–15**
  physiological/clinical 626, 630, 636, 639
  and quality 354
  and safety 320
  SI units 16, **28**
  Standards **27–8**, **368**, 420
    Standard Methods (SMMs) 578, 579
  temperature 16, 202
  see also instrumentation
Measurement Science and Technology **143**
Measurements Group UK Ltd 146
mechanic [technician] 624
mechanical engineering **151–97**
  and biomedical engineering 624, 626–7, 629, 638
  see also biomechanics
  and chemical engineering 269
  historical aspects 94
  Internet resources 119
  ISMEC **111**
  mineral processing 589, 592
  mining 609
  terminology 151
  see also machines
Mechanical Engineering 64, 157, 296, 338
Mechanical Engineering Abstracts **100**, 153, 341

  online 111, **355**, 380, 401, 402
Mechanical Engineering Publications Ltd (MEP) **142**, **144**
  see otherwise Institution of Mechanical Engineers: publications
Mechanical Incorporated Engineer 157
mechanics **151–97**, 638
  biomechanics see biomechanics
  experimental see stress analysis
  fluid see fluid mechanics
  photomechanics 145
  rock mechanics 610, 611, 613, **616**
  solid 33, 138, **278**
  spatial 159
  see otherwise machines etc.
Mechanism and Machine Theory 151, 153, 159, 163, 166, 168
mechanisms **156–7**, 165, 166–73
  conferences 161–3
  design 341
  historical aspects 154–5
  organizations 151, 152
  planar 341
  software 155, 161*tab*, 166, **170–2***tab*
  Standards 19–20
  thermodynamic 201
  tribology 174
  see otherwise machines
mechatronics 166, **169–73**, 292
Mechatronics 169, 172, 292
Medard, L.A. 322, 323
Medawar, P.B. and Medawar, J.S. 625
media
  information from 497–8
  see also newspapers etc.
Medical and Biological Engineering and Computing 642
Medical and Biological Engineering and Computing News 643
Medical Devices Directorate (MDD) [of Dept of Health] 645
Medical Electronics 642
Medical Engineering and Physics 630
medical physics 626–7, 628*fig*, 629
  journals 629–30, **639**
  organizations 629–30
Medical Physics 639
Medical Physics World **629**
Medical Progress Through Technology 642
Medical Technologist and Scientist 643
medicine 623, **626**, 629–31, 634–5
  and biology 626
  databases 111
  journals **637–9**, 641–2
  organizations 629–31

Index   733

medicine (*contd*)
 and patents 36
 robotics 378, 383
 see also biomedical
   engineering
Medland, A.J. 155, 167, 335
Medland, A.J. and Mullineux, G.
   335
MEDLINE database 644
meetings, general *see* conferences
Meetings on atomic energy (pub.
   IAEA) 267
Megaw, T.M. and Bartlett, J.B.
   545
Meinel, A.B. and Meinel, M.P.
   244
Melaragno, M. 246
Mellanby, K. 491
Melles Griot Photon Control Ltd
   147
Mellor, S.H. 598
Melville, I.A. and Gordon, I.A.
   552
Melvin, A. 240
MEM *see* Eur. Soc. of
   Biomechanics
membranes, synthetic 137, 274,
   564
Menon, H.G. 366-7
menus, network 117-18
MEP *see* Mechanical Engineering
   Publications Ltd
Merchant fleet statistics (pub.
   HMSO) 509
mercury arcs 435
Meredith, D.D *et al.* 331
Meriam, J.L. and Kraige, L.G.
   158
Merkow, M.S. 124
Merrick, D. 240
Merrill, R. and Gage, T. 210
Merritt, F.S. [*civil engineering*]
   125, 534
Merritt, P. [*coal*] 600
MES *see* Minerals Engineering
   Soc.
MeSH classification system 643
METADEX databases 110, **111–12,**
   230, 306, 341, **355**
Metal Bulletin Books [co.] 597
 Industrial minerals directory
   601, 608
Metal Finishing 357
Metallurgia 68
Metallurgical Society, The
   (TMS-AIME [US]) **594,** 601
Metallurgical Transactions 309
metallurgy *see* metals
Metal Powder Industries
   Federation, publications 373
metals/metallurgy 363-5, 590
 nature and role 353, 593
 abstracts/indexes 111, 309,
   341, 355
 alloys *see* alloys
 base/precious metals 589, **593,**
   599, 611
 boilers 235-6
 books
   directories 303
   handbooks 125, **302–3,** 307,
   312

cladding 563-4
concentration, gravity 590
conferences 358-9
corrosion *see* corrosion
 cutting 353
 databases 308
   METADEX 110, **111–12, 230,**
     306, 341, 355
 and electricity generation 252,
   429
 equivalent 26
 extraction/processing **589–606**
   early 589-90
 extractive metallurgy 589
 flotation 590, **599–600,** 602
 grain size 591
 hydrometallurgy 597, **599**
 inter-metallic materials 353
 journals 68, 111–12, 309, 356,
   357
 machining **363–5**
 manufacture 358-9, **363–5,**
   369
 mineral processing 589-606
   terminology 590
 mining 607-21
   surface 615
 nuclear raectors 259-61
 ores *see* ores
 organizations 26, **303, 593–4,**
   609
 patents 57
 powders 369
 properties 323
 roofing 564
 safety 323
 smelting 589
 Standards 26, **302–3**
 waste proportions 591
 see also steel; materials *etc.*
Metals Abstracts 111, 306, **309,**
   341, 355
Metals black book [ed. Bringas]
   312
METALS DATAFILE database 306
Metalworking Production 356
Metcalf & Eddy Inc. 545
Metcalf & Eddy Inc. Staff *and*
   Tchangolous, G. 488
Metcut Research Associates Inc.,
 Machining data handbook
   369
Meteorological Office, *Solar*
 radiation data for the
 United Kingdom... 243
methane
 mines 617
 thermodynamic properties 203
methanol, heat pumps 208
Methods for the examination of
 waters... (pub. HMSO)
   486
Metocean Ltd, *Newsletter* 480
metres 28
Metrologia 28
metrology *see* measurement
Metropolitan County Councils,
 abolished 511-12
Mexico
 automation/AI 382
 Instituto Technologico... de
   Monterrey 382

power engineering 426
Meyer, C.A. *et al.* [*steam*] 236
Meyer zur Capellen, W.
  [*kinematics*] 155
Meyer, E. [*chemistry/hazard*] 324
Meyers, R.A. [*physics/fuels*] **122,**
   241
M-G Information Services *see*
 Morgan Grampian
MHD (magnetohydrodynamics)
   251–3
MICAD [Internat. Conf. on
   CAD/CAM] 359
Michalski, R.S *et al.* 396
Michigan
 Technological University 596
 University of 479
microbiology 486
microcomputers *see* computers:
 personal
microelectronics 353, 367, 424
Microfiche index to British
 Standards (pub. BSI) 22
microfilm **11–12,** 13, 47, **90**
 and CD-ROM 90
 Barbour Microfiles **181,** 183,
   334, **528, 529,** 579
 catalogues 90, 157
 construction information 521,
   527–9, 579
 directories 73, 370
 environment **181,** 183, 334
 materials 307, 309
 package libraries 66, **90**
 patents 47, 55, 57, 59, 341
 product information 89, **90,**
   528, 529
 reports 11–12
 safety **181, 181,** 183, 334
 Standards 22
 theses 78, 530
microhydrodynamics 276
MicroInfo Ltd 7
Microprocessor-based traffic
 signal controller... (pub.
 DoT) 500
microprocessors 412, 416
Microprocessors and
 Microsystems 416
microwaves 447, 448–9
Microwaves, Antennas and
 Propagation (IEE *Proc.*)
   448
Microwaves and RF 449
Middendorf, W.H. 331
Middle East, energy 232, 233,
   243
Midwinter, J.E. *and* Guo, Y.L.
   448
MIL/MIL-HDBK publications
 [US DoD] 91, 307
Miles, L.D. 337
Military Engineer 67
military engineering *see* defence
Mill, R.C. [*human factors/safety*]
   326
Millard, P.M. [*associations*] 125
Miller, C.J. [*product liability*]
   334
Miller, L.C.G. [*concurrent*
 engineering] 653

734 Index

Miller, S.E. and Chynoweth, A.G. [optical fibres] 448
Miller, S.E. and Kaminow, I.P. [optical fibres] 448
Miller, T.J.E. [electric motors] 435, 437
milling/mills
  metals 363
  minerals 589, 590, 598, 600, 601
Millington, P.F. and Wilkinson, R. 636–7
Millman, J. and Grabel, A. 423
Mill Operator's Conference 601
Mills, D. [particle technology] 278
Mills, J.H. et al. [cam design] 167
Milne, R. 397
Milora, S.L. and Tester, J.W. 245
Min, T.C. and Chiou, J.P. 210
MINATOM [Russian Min. of Atomic Energy] 263
mine environmental engineering 617
Mine and Quarry 611
mineral economics 614
Mineral Industry Research Organization (MIRO) 595
mineralogy 589, 591, 592, **598**, 607
  coal impurities 238
  crystals see crystals
  industrial minerals 607–9, 611
  organizations 609
  'mineral deposits' 591
  tabular 609
  mineral oil see oils; petroleum etc.
Mineral planning guidance (pub. DoEnv) 614
mineral processing **589–606**
  nature and role **589–93**, 594
  industry profiles **593**, 601–2
  processes summarized 591–2
  abstracts/indexes **602**
  books 590, 593, **597–600**
  directories **600**, **601–2**
  handbooks **600**
  bulk handling 592
  carbon adsorption 599
  comminution **591–2**, **598**, 600
  concentration 590, 591, 599
  conferences 593, 594, 599, **600–1**
  cyanidation 590, **599**
  databases **602**
  design 598–600
  electro-refining 599
  electro-winning 599
  environmental impact **592**, 597
  exploration 594
  flotation 590, 598, **599–600**, 602
  flowsheets 592, 599, 600
  grains **591**
  historical aspects **589–90**, 591, 607
  industrial, surface mining 615
  ion exchange **599**
  journals 593, 594, **596–7**
  leaching 590, 591, **599**
  legislation 233
  liberation 591
  marine minerals 482
  milling 590
  organizations **593–6**, 602, 607–10
  patents 602
  plant 598
  equipment 597, 600
  precipitation **599**
  prospecting 240
  separation **591–2**, **598–600**
  chemical 591, **599**
  dense medium 599, 601
  solid/liquid 592, **600**
  sizing 591–2, **598**
  solvent extraction **599**
  terminology **590–2**
  waste **591**, **592**
  see also mining etc.
Mineral Processing and Extractive Metallurgy (IMM Transac.) 609
Mineral Research and Exploration General Directorate [Turkey] 594
Mineral Resources Engineering 610, 611
Mineral Resources Research Center (MRRC [US]) 595
Minerals Engineering 596
Minerals Engineering Society (MES) **593**, 611
Minerals Industry International 597, 609
Minerals and Metallurgical Processing 596
Minerals Today 597
Minerals yearbook (USGPO) (pub. USGPO/USBM) 601, 608, **612**
Mines Research and Development Establishment (MRDE), former 594
Mine ventilation ... congress (pub. IMM) 613, 617
Mine Ventilation Society of South Africa
  Environmental engineering in South African mines 617
  Journal 611
miniaturization 287
  and robotics 378
  surgery 378
mining 589–606, **607–21**
  nature and role 589, 593, 607–8, **614–15**
  historical aspects 154, 226, 590, 607, 617
  abstracts/indexes 613
  blasting **614**, 615
  books 237–9, 609, 610, **613–17**
  directories 534, 601–2, 608, 612
  conferences 601, 609, 610, **612–13**, 614, 616, **617**
  databases 602, 609, 613
  drilling 615
environmental issues 255, **592**, 597, 611, **614**, 617
environment restoration 592, 614
work environment **616–17**
equipment 608, 610, 611, **614**, **615**
hydraulics 288, **292**
industry profiles 601–2
journals **239**, **593**, **596–7**, 608–10, **611**
longwall techniques 238
management 614
mechanization 590
noise 614
organizations 227, **593–6**, **607–10**
placer mining 615
rock mechanics 610, **616**
safety 609, 614, 615, 616, 617
'intrinsically safe devices' 320
slopes 616
statistics 608, 609, **612**
surface 613, **614–15**, 616
underground **615–16**
ventilation 610, 611, 613, **616–17**
see also coal etc.; mineral processing
Mining annual review 601, 612
Mining Association of the United Kingdom 608
Mining chemicals handbook (pub. Amer. Cyanamid) 600
Mining, Coal, Gold and Base Minerals 611
Mining Congress Journal 608
Mining directory (pub. Don Nelson) 601, 608
Mining Engineer 609
Mining Engineering **596**, **610**
Mining Environmental Management 597
Mining Equipment Manufacturers Association 608
Mining Industry (IMM Transac.) 609
Mining Industry Research Organization 608
Mining Journal 597, 601, 602, **611**, **613**
Mining Journal Ltd 596, 597, 612
Mining Magazine 596, 602, 611, 612
Mining Monthly 611
Mining Technology 610
Mining World 611
Ministries/Departments see under United Kingdom; United States etc.
Minitab Inc., Minitab [program] 452
Minitel database access system 305
Minnesota, University 596
Mineral Resources Research Center (MRRC) 595
MINTEK [NIM research centre] 594

Index 735

Mintel Publications [co.] 576
Mintzberg, H. 571
MIRA (Motor Industry Research Ass'n), *MIRA Abstracts* 228
MIRO *see* Mineral Industry Research Org'n
MISC.TRANSPORT.URBAN-TRANSIT [list] 502
Mishkin, E. *and* Braun, L. 395
Misra, K.B. 319
Missouri-Rolla, University 596
Mitchell, C.P. *et al.* [*forest biomass*] 249
Mitchell, F.H. [*CIM*] 362
Mitchell, W.J. *and* Slaughter, J.C. [*biochemistry*] 279
Mitchell's building construction 552
MITI ([Japanese] Min. of Internat. Trade and Industry) 355
Mittendorf, W.H. 156
mixing 274
Mobile and Cellular Magazine 449
mobile plant 286
MoD/MOD (Min. of Defence/Defense) *see under* United Kingdom; United States
Model code of safe practice in the petroleum industry (pub. IP) 321, 322, 326
modelling, mathematical **295, 390, 432–3, 451–2**
 algebra 451, 452, **455**
 biomechanics 636
 books **391**, 454, **455, 457–8**
 chemical engineering 280
 and concurrent engineering 335, 651
 control systems 280, 288, 389, **390–8**, 400–3
 design 158, 334, **334–5**
 deterministic 451
 economic 575
 frequency response 390
 hydraulics 288, 291, 293, **294,** 295–6
 journals 293, **459**
 linguistic models 396
 mineral processes 592
 and neural networks 390–1
 parallel computation 392, 398
 performance 446
 qualitative models 397
 robotics 379
 and simulation **392**
 software **452**
 state space methods 390, 391
 stochastic 451, **457–8,** 460
 stress **136–7**
 system analysis **432–3**
 transport 506
 tribology 176
 verification 433
 see also robotics; mathematics; CAD/CAM etc.
Modelling and Simulation in Materials Science and Engineering 313

models [physical]
 control systems 391
 hydraulics *and* 286
 stress analysis 135
Model specification for ... brickwork (pub. CERAM) 540
modems 581, 583
Modern Casting 68
Modern Metals 68
Modern mineral processing flowsheets (pub. Denver Equip. Co.) 600
Modern Plastics [co.] 313
Modern Plastics International 68
Modern power station practice... (pub. CEGB/Brit. Electricity Internat.) 251–2, **429**
Modern Power Systems 253
Modern Railways **498**
Moeckel, W.E. *and* Weston, K.C. 203
Moffat, D.W. 256
moiré fringe analysis **135,** 147
moiré interferometry 135
Molerus, O. 278
moles 28
Molian, S. 167, 172*tab*
moment [of force], *and* load 134
momentum 277, 291
Mond index [chemicals risks] 317
Moniton, L. *et al.* 248
monopolies
 Statute of Monopolies (1623) **32,** 36
 see also patents
Monsanto [co.] 305
Monthly Energy Review 232
Mooney, D.A. 202
Moore, D.F. [*thermodynamics*] 204
Moore, F. [*heating*] 254
Moore, M.J. *and* Sieverding, C.H. [*steam*] 236
Moran, M.J. 204
Morari, M. *and* Zafiriou, E. 280
Moreau, R. 252
Moreno, R. 247
Morgan, J.R. [*design*] 332
Morgan, N. [*marine engineering*] 483
Morgan Grampian [publishers]/M-G Information
Kempe's engineers yearbook [ed. Sharpe] 124, 160, 340
Electronics and instruments directory 418
Morgan Grenfell Securities [co.] 576
Morocco, minerals 596
Morris, C., *Academic Press dictionary*... 123
Morris, P. *and* Therivel, R. [*environment*] 493
mortar 541
Mortier, K.M. *and* Orszulik, S.T. 175
Mosaic software 118

Moser, F. *and* Schnitzer, H. [*heat pumps*] 208
Mosey, D. [*nuclear safety*] 318
MOSFETs (metal-oxide-silicon/semiconductor-field-effect transistors) 435
Moss, B. [*water/ecology*] 487
Moss, M.A. [*design/reliability*] 178
motion
 dynamics 151
 kinematics 151
 spatial 159
Motion program 168, 170*tab*
Motor Industry Research Association (MIRA), *MIRA Abstracts* 228
Motoring statistics (pub. RAC) 510
Motorists' opinions: survey... (pub. Lex) 510
motors **433–5, 436**
 ac 436
 dc 435, 436–7
 design 341
 and hydraulics 290–1, 293
 induction motors 434–5, 436
 servo motors 293, 436–7
 switching *see* switches
 synchronous 435
 torque *and* flux 436–7
Moubray, J. 178
moulding 364, 365
Moulton, A.E. *et al.* 329
MRDE *see* Mines Research and Development...
MRP (materials requirement planning) 367
MRRC *see* Mineral Resources Research Center
MSC/MVision [microform] 307, 309, 314
MSC/PDA Engineering [co.] 306–7, **314**
MTD Ltd, *Newsletter* 480
MTS *see* Marine Technology Soc.
MTTA *see* Machine Tool Technologies Ass'n
Mucci, P. 160, 161*tab*, 337–8
Muezzinoglu, A. *and* Williams, M.L. 491
Mular, A.L. *and* Anderson, M.A. 598
Mular, A.L. *and* Bhappu, R.B. 598, 600
Mular, A.L. *and* Jergensen, G.V. 598
Mullard *see under* Philips co.
Mullay, M. *and* Schlicke, P., *Walford's guide to reference material* 121
Muller, H.W. 165
Mullin, J.W. 274
MULTIMEDIA exhibition 342
multimedia information
 and concurrent engineering 650
 exhibitions 342
 Internet 117, 118, 503
multiphase flow/systems 255, 274, 282

736 Index

Munasinghe, M. 208
Munasinghe, M. *and* Meier, P. 233
Municipal Engineer (ICE *Proc.*) 530, 531
Murley, L[oveday] 491
Murphy *and* Richardson [*control systems*] 391
Murphy, J. [*plastics*] 313
Murphy, J.M.D. *and* Turnbull, F.G. [*ac drives*] 436
Murphy, S. *and* Henderson, C.A.P. [*directories*] 340
Murrill, P.W. 280
Museums and Galleries Commission
Conservation sourcebook 552
Conservation Unit 552
Mustoe, J.E.H. 242
MVision [microform] 307, 309, 314

N

N document [report] 6
NACA *see* National Advisory Committee for Aeronautics
Nacfaire H. 247
NAFEMS [finite elements [org'n] **139**
Benchmark 139
Finite element primer 148
How to... books 139
NAG [program] 452
Nag, P.K. 202
Najim, K. 280
Nakano, N. 298
nano-fabrication/nanotechnology 355
Nanyang Technological University 359
Naples, Università Federico II 482
NASA *see* Nat. Aeronautics and Space...
Nash, C.A. 507
National Advisory Committee for Aeronautics, *former* (NACA) 211
National Aeronautics and Space Administration (NASA; US) *and* AEROSPACE DATABASE 112
Internet bulletin board **117**
and NTIS 486
patents 7
reports **7**, 11, 116, 117, 486
N documents 6
Scientific and Technical Aerospace Reports **7**, 11, 112
Technical Report Service [Internet] 116
and thermal systems 211
National Agency for Finite Element Methods and Standards, *former see* NAFEMS
National Alcohol Fuel Producers Association 250
National Audubon Society, *Audubon House... 254*
National Bureau of Standards, *former* (NBS) 28, 211

National building specification 527, 578
National Center for Supercomputing Applications (NCSA [US]) 118
National Centre of Systems Reliability **180**
National Centre of Tribology [of AEA] 177
National Coal Association (NCA [US]) 227
Coal data 237
National Coal Board (NCB), *former*
Fluidized bed combustion of coal 236, 238
Ventilation in coal mines 617
Work study in mines 614
National Commission for Safety of Manriding in Shafts... 619
National Conference on Artificial Intelligence, *Proceedings* 403
National Economic Development Council/Office (NEDC/NEDO), *former* 344, 575
Faster building for commerce 570
Faster building for industry 570
National Electrical Code [US] 320
National electrical code handbook (*pub*. NFPA) 321
National Electrical Manufacturers Association (NEMA [US]) 26
National Engineering Laboratory (NEL) 211
Air coolers, cooling towers... 207
NAFEMS 139
National Fire Protection Association (NFPA [US]) **321, 327**
publications 322, 324
Fire hazard properties... 322
Identification of the fire hazards... 322
Manual of hazardous chemical reactions 323
National electrical code... handbook 321
National Health Service (NHS) patents, compulsory licences 42
physicists 629
National Highway Traffic Safety Administration [US] **517**
National House Building Council (NHBC) **525**, 528
National Industrial Fuel Efficiency Service (NIFES) 211
Boiler operator's handbook 206, 236
National Information Services Corp'n 113

National Information Software System (NISS)
NISSPAC 584
programs 505
National Institute for Metallurgy (NIM), MINTEK 594
National Institute for Occupational Safety and Health (NIOSH), *NIOSH Pocket guide to chemical hazards* 323
National Institute of Standards and Technology (NIST [US]) **25, 28**
publications **28**, 304
NIST handbook 28
NIST Journal of Research 28
Standards activities organizations in the United States 25
National Maritime Research Center [US], *Maritime Abstracts* 480
National Nuclear Corporation Ltd (NNC) **262**
National Petroleum Council (NPC [US]) 226
National Physical Laboratory (NPL) 23, **28**, **147**
National Power [co.] 226
National Radiological Protection Board (NRPB) **264**
National Register of Archives 553
National Repository for Concurrent Engineering [US] 656
National Research Council [US]
Competitive edge 650, 656
Improving engineering design 650, 656–7
National Research Council of Canada 253, 478, 532
National road maintenance condition survey (*pub*. DoT) 509
National Safety Council (NSC [US]) **327**
Accident prevention manual 315–16
Guards illustrated... 327
Industrial Section, *Newsletters* **328**
National Science Foundation [US], conferences **342**, 359
National Service Robot Association 382
National Society for Clean Air (NSCA) 494
NSCA pollution handbook 494
National Standards Association (NSA [US]) 25, 113
STANDARDS AND SPECIFICATIONS DATABASE 113, 230, 368
Standards and specifications information bulletin 25
National structural steelwork *specification... (pub.* BCSA) 538, 539

Index 737

National Swedish Road and Traffic Research Institute **517**
National Technical Information Service (NTIS [US]) **6–7**, 323
  Committee on Information Hang-ups 13
  database *see* NTIS *Energy database* ... 8, 232
  Geothermal Energy 246
  Nuclear Fuel Cycle 251
  'Rasmussen Report' 266
  and reports 5, **6–7**, 10, 11, 232, 234, 241, 510
  GRA&I **6**, 7, 11, 27, 109, 341
  Guidelines for descriptive cataloguing ... 13
  NTIS Alerts 6–7
  Subject category descriptions 6
  Transport accidents reports/briefs 510
  *USSR Report*s 228
  Solar Thermal Energy Technology 245
National transportation statistics (pub. USDOT) 510
National Travel Survey 511
  National travel survey Report (*pub.* HMSO) **509**
National Weights and Measures Laboratory 23, **28**
NATO *see* North Atlantic Treaty Org'n
Naumann, E.B. 274
Nautical Institute (NI) 475
  Seaways 476
nautical studies **467**
Naval Architect 476
naval architecture *see under* marine engineering
Naval Research Laboratory 252
Naval Research Logistics 460
Nayler, G.H.F. 123, 154
NBS *see* Nat. Bureau of Standards
NCA *see* Nat. Coal Ass'n
NCB *see* Nat. Coal Bd
NCE 534
  NCE road construction and traffic directory (pub. Telford) 534
NCSA *see* Nat. Center for Supercomputing ...
NDT International, former 357
NDT & E International 357–8
Neale, M.J., *Handbook*s [ed.] 161*tab*, **175**, 177
Neale, M.J. *et al.* 163
NEA (Nuclear Energy Agency) *see under* Org'n for Economic Cooperation and Development
NEDC *see* Nat. Economic Development Cncl
Nederlands Normalisatie-Instituut (NNI) 27
  Normalisatie Nieuws 27
NEDO *see* Nat. Economic Development Office

needs analysis 333
Neelamkavil, F. 392
Neely, W.B. 324
NEL *see* Nat. Engineering Laboratory
Nelson, Don [publishers], *Mining directory* 601, 608
NEMA *see* Nat. Electrical Manufacturers Ass'n
Nemetz, P.N. *and* Hankey, M. 256
Neno, H. *et al.* 298
neodymium-iron-boron materials 435
NET/ETS [Eur. Telecommunications Standard] 449
Netherlands
  construction 523, 584
  design 162
  Eindhoven University of Technology 115
  electronic publishing 115
  journals 73
  marine technology 475, 478
  minerals 596
  patents 45
  planning 517
  process engineering 73
  Standards 22, **27**
  and transport 517
  transport 516
  see also Dutch language
Netscape Communications Software [co.] 118
Netscape/Netscape Navigator software **118**, 119, 502
Network 449
network management (NM) 442
networks, neural *see* neural networks
networks, computer/data 114–20, 341, 443, **445–6**
  concurrent engineering 657
  construction **584**
  contents pages 103
  database options **108–9**, **117–20**, 505, 527
  exhibitions 342
  'free and open' 502
  hosts *see* host services
  information 'superhighway' 502
  and integrated services ([B-]ISDN) 442, 445
  Internet *see* Internet
  JANET **109**, 505, 527, **584**
  journals **449**, 459, 460
  local area (LANs) 108, 231, 446, 449
  databases on 108, 231
  LAN Magazine 449
  Standards 420, 450
  packet switching 444–5
  performance modelling 446
  Standards/protocols 420, 445, 450
  X.25 protocol 305, **445**
  TCP/IP 118
  'superhighway' 502
  'supernetwork' 584

wide area (WANs) **109**
networks, communications *see under* telecommunications
Networks – An International Journal 459
NETWORKS exhibition 342
Neufert, E. 565
neural networks 390–1, 396, **402**
Neural Networks 402
neurology 638–9
neutrons
  neutron diffraction stress analysis **136**, 147–8
  neutron-moderators 260, 261
  nuclear fission 259–61
  fission neutrons 260
Neville, A.M. 538
Neville, A.M. *and* Brooks, J.J. 538
Nevins, J.L. *and* Whitney, D.E. 652
New Builder 532
Newcastle-upon-Tyne
  Building Centre 520
  University, Transport Operations Research Group **501**
New Civil Engineer 67, 490, 532, **580**
Newcombe, R. *et al.* 570
New earnings survey (pub. DTI) **574**
Newell, A. *and* Simon, H.A. [*problem-solving*/*design*] 332
Newell, J.A. *and* Horton, H.L. [*mechanisms*/*design*] 156
Newell, R.B. *and* Lee, P.L. [*controls*] 280
Newlands, J. 561
Newman, P.A. *and* Allison, D.O. 203
New motor vehicle registrations Great Britain (pub. DoT) **509**
Newnes hazardous chemicals pocket book [ed. Carson/Mumford] 323
Newport Micro-Controle [co.] 147
New Products 87
New Scientist 103
newsgroups, electronic 281, 502
  see also e-mail
newsletters **70**, 72, **328**, 472, **480**
  electronic mail **115**
  Internet 115
  online 70, 72
newspapers **70**, 72–5, 472, 480, 497–8
  current awareness 103
  databases 422
  directories 72–3
  indexes 70, 74
  local 71
  online 71
  press cuttings services 75
  special reports 499
  news reader programs [e-mail] 117
New Steel Construction (pub. SCI/BCSA) 532

Newton, C.H. [*materials*] 304
Newton, D.C. [*trademarks*] 96, 121
Newton, Sir Isaac 159
New trade names in the rubber and plastics industries (pub. RAPRA) 88
New York [US]
  Columbia University 596
  State University 115
New York Journal of Mathematics 115
New Zealand
  energy 232, 245
  journals 73
  medicine 639, 642
  minerals 593–4
  process engineering 73
  see also Australasia
NEXIS database 107
NF *see Normes françaises*
NFPA *see* Nat. Fire Protection Ass'n
NHBC *see* Nat. House Building Cncl
NI *see* Nautical Inst.
NIBE [wind energy] project 247
Nichols, G.D. 280
nickel 593
Nieuw Archief voor Wiskunde 458
NIFES *see* Nat. Industrial Fuel Efficiency Service
NII (Nuclear Installations Inspectorate) 264
Nikravesh, P.E. 158–9, 167
Nilsson, N.J. 397
NIM *see* Nat. Inst. for Metallurgy
1967 steam tables . . . (pub. Electrical Research Ass'n) 235
1992 US electronics industry directory (pub. Harris) 418
NIOSH (Nat. Inst. for Occupational Safety and Health), *NIOSH pocket guide to chemical hazards* 323
Nippon Kaiji Kyokai (NK), *Annual report . . .* 478
Nisenfeld, A.E. 280
Nishida, S.I. 318
NISS (Nat. Information Software System)
  NISSPACdatabase 584
  programs 505
NIST
  NIST handbook 28
  NIST Journal of Research 28
  see otherwise Nat. Inst. of Standards and Technology
nitrogen, thermodynamic properties 203
NK *see* Nippon Kaiji Kyokai
NM (network management) 442
NNC (Nat. Nuclear Corp'n Ltd) 262
NNI [1] *see* Nederlands Normalisatie-Instituut
NNI [2] *see* Nuclear Installations Inspectorate

Nobel's Explosives Co., *Blasting practice* 614
Nobel Systems Ltd 146
Noether, D. *and* Noether, H. 271
noise [sound] 181–2, **492**
  DTI unit 179
  hydraulics 287, 291, 292, 293
  journals 492
  machine tools 368
  mining 614
  Standards 181–2, 368, 492, 543
  Regulations 181, **182**
Noise control in industry (pub. Sound Research Laboratories) 181, 492
'noise', electronic 441, 443, 458
  control systems 390
Noise at work . . . guides (pub. HSE) 492
Noltingk, B.E. 398
Non-Linearity 460
Non-structural cracks in concrete (pub. CS) 557
Noranda [co.] 595
Nordberg (UK) Ltd, *Reference manual* 600
Norddeutsche Seekabelwerke AG 481
Nordic countries *see* Scandinavia etc.
Normalisatie Nieuws (pub. NNI) 27
Norman, D.A. 336
Normes françaises (NF; pub.AFNOR) 23
NORMIMAGE databases 22
Norrish, J. 362, 386
Norske Veritas, Det [publishers], *Rules for classification: . . . light craft* 477
North America
  energy 233
  mineral processing 590
  newspapers 71, 75
  theses 78
  trade directories 84–5
  see also Canada; United States
North American Manufacturing Research Institution (*of* SME) 359
North Atlantic Treaty Organization (NATO) 211
  Advanced Science Institute 244
  component specifications 91
  Science Committee 207
  and solar energy 244
  and thermal storage 207
Northern Ireland, Building Regulations 549
Northern Miner [Canada] 611
Northern Telecom [co.] 366
North-Holland [publishers] 381, 474
North Sea, offshore development 467–8, 480, 481
North Sea Letter 480
Northwestern University, *Current Literature in Traffic and Transportation* 504

Norton, J.P.[*control systems*] 391
Norton, M.P. [*noise/vibration*] 181
Norton, R.L. [*design*] 159
Norway 72
  hydro-power 248, 255
  journals 72
  marine technology 478, 479
  minerals 596
  Norwegian Technical University 479
  Standards 20, 113
  Norwegian language, databases 584
Nose, Y. 647
Notes on ship slamming (pub. SNAME) 477
Nottingham
  Nottingham Trent University, RAM 380
  RIBA bookshop 533
  University
    mineral engineering 595
    transport courses 502
Nowacki, P. 238
Noyan, I.C. *and* Cohen, J.B. 148
Noyes Data Corp'n 602
NPA *see under* PTS
NPC (Nat. Petroleum Cncl) 226
NPL (Nat. Physical Laboratory) 23, **28**, **147**
NRC *see* Nuclear Regulatory Commission
NRPB (Nat. Radiological Protection Bd) 264
NSA *see* Nat. Standards Ass'n
NSC [1] *see* Nat. Safety Council
NSC [2] (Nuclear Safety Commission [Japan]) 264
NSCA (Nat. Soc. for Clean Air) 494
NSCA pollution handbook 494
NTIS
  NTIS Abstract Newsletter, *former* 7
  NTIS Alerts 6–7
  NTIS database 7, **109**, **230**, **355**, **485–6**, 510
  control systems 400
  energy 226, 230
  environmental engineering 485–6
  manufacturing 355
  minerals/mining 602
  transport 505, 510, 511
  see otherwise Nat. Technical Information Service
Nuclear Electric plc 262
nuclear energy/engineering 250–1, 259–67
  abstracts/indexes 9, 230, **267**
  alternatives to 209
  books 250–1, 261, **265–6**, **267**, 324–5
  dictionaries 267
  directories 233, 261, 267
  handbooks 261, 267, 319, 324–5, **429**
  Reports 266
  chain reactions **260**, 266
  conferences 267, 324

nuclear energy/engineering (*contd*)
  coolants 260–2
  criticality/supercriticality 260
  databases 9, **230**, 231, **267**
  decommissioning 250
  design 156, 250–1, 259–61, 266
  environmental effects, Flowers Report 266
  exhibitions 267
  fission process 250–1, **259–61**, 266
  fusion process **250–1, 261**
    'Hydrogen bomb' 261
  international usage 261–2
  Internet resources 119
  journals 68, 105, **251, 266–7**
  preprints 116
  'manipulators' 377
  military uses 261
  moderators 260, 261
  organizations 225, 226, **262–5**, 267
    international 262–3
    learned/professional 265
    regulatory 264
    UK companies 262
  reactors *see under* reactors
  safety issues **250**, 260, 265–6, 315, 319, **324–5**
    accidents **266**, 318
    international bodies 262, 324–5
    national bodies **264**
    Standards 26, 324
    legislation 250, 263
  submarines 261
  training 267
  wastes management 250, 260
Nuclear Energy 251, 266
Nuclear Energy Agency (NEA) *see under* Org'n for Economic Cooperation/ Development
Nuclear Energy Data 251
Nuclear Engineer 266
nuclear engineering
  safety 262, 265–6, 267, 315, **324–5**
    national bodies 264
    radioactivity 250–1, 260, 266
Nuclear Engineering and Design 251, 266
Nuclear Engineering International 68, 267
Nuclear Engineering International [co.], *World nuclear industry handbook* 261, 267
Nuclear Fuel Cycle (pub. USDoE/NTIS) 251
Nuclear Installations Inspectorate (NII) **264**
Nuclear legislation . . . (pub. OECD) 250
Nuclear News 68, 267
Nuclear Plant Journal 251
Nuclear power and the environment [Flowers Report] (*pub.* HMSO) 266

Nuclear power plant safety standards . . . (*pub.* IMechE) 324
Nuclear power: status and trends (*pub.* IAEA)) 261, 267
Nuclear Regulatory Commission (NRC [US]) 226, **264**
  Fault tree handbook 317
  Handbook for probabilistic risk assessment . . . 319, 325
  and Nuclear Safety 267
  Reactor safety study . . .
    [Rasmussen Report] 266
Nuclear Safety 267
Nuclear Safety Commission (NSC [Japan]) **264**
Nuclear Science Abstracts, online **230**
Nuclear Science Abstracts, former 8
Nuclear Science and Engineering 251, 266
Nuclear Technology 251, 266
nuclei
  and fission 259–60
  and fusion 261
nucleonics 267
Nucleonics Week 267
Numerical Algorithms Group Ltd 452
numerical analysis 451, **455, 458–9,** 616
  books 152
  data transfer 310, 335
  design 158, 334–5
  hydraulics 294–5
  journals 143
  materials 306–7
  software 452
  stress 133–4, **136–7,** 148–9, 616
  tribology 174
  see also finite element *etc*.
numerical approaches, general
  see mathematics
Numerical Recipes [program] 452
Nutbourne, A.N. *and* Martin, R.R. 455
nutrition, database 229
Nyquist Stability design technique 397

O
Oakland, J.S. 334
Oakland, J.S. *and* Porter, L.J. 179
Oakley, M. 333
Oak Ridge National Laboratory [US] 254
OAPEC *see* Org'n of Arab Petroleum . . . Countries
Oates, G.C. 208
Oberg, E. *et al., Machinery's handbook* 124, **160**
Oborne, D.J. 336
O'Callaghan, P.W. 201, 202, 203, 256
O'Callaghan, P.W. *and* Probert, S.D. 204

occupational health 183, 320
  defined 316
  and ergonomics 335
  see also Health and Safety
Occupational Safety and Health (*pub.* RoSPA) 183
Occupational Safety and Health Administration (OSHA [US]) 320
occupational therapy 626, 631, 642
Ocean Engineering 249, 480
oceans
  desalination 203, 204–5, **243–4**
  and energy **247–9**, 255
  estuaries 487
  marine engineering **467–83**
    marine sciences 467
    ocean engineering **468**
    oceanography 470*tab*
    ocean technology **468**
    ocean thermal energy conversion (OTEC) 248
  pollution 16
  resources 468, 469, 470*tab*, 482
  sand recovery 615
  subsea installations *see* underwater
  water analysis 486
  wave loading 567
  see also offshore; water *etc*.
OCLC (Online Computer Library Center [US]) 110, 111
  First Search service 103
O'Connor, P.D.T. [*reliability*] 178, 179, 319
O'Connor, T. [*directories*] 340
Oden, J.T. 456
ODI *see* BCQ/ODI
odour control **491–2**
OECD *see* Org'n for Economic Co-operation and Development
OEM [co.] 361
OEM Design 157, 338
Offenhartz, P.O'D. *et al.* 208
Office of Buildings and Community Systems [US] 254
Office of Energy Technology [US] 237
Office of Ocean Minerals and Energy [US] 248
  Final . . . impact statement for . . . [OTEC] . . . 248, 256
Office of Scientific Research and Development [US] 4
Office of Scientific and Technical Information [US] 254, 267
Official Journal (OJ), of Patent Office 60
official publications **126,** 224, **471**
  as grey literature 10
Official publications in Britain [ed. Butcher] 126
Official reference book and buyers' guide [of IED] 340
offline services **581–2**

Offshore 480
Offshore Abstracts, former 480
Offshore Engineer 68, 240, 476
offshore engineering 467–83
   nature and role 467–8, 470tab
   abstracts 480
   conferences 474, 476
   controls 280
   directories 474–5
   exhibitions 474
   fisheries 101
   journals 68, 240, 476, 480, 481
   oil map 481
   organizations 475–6
   polar engineering 470tab, 475
   product information 481
   robotics 378
   safety 475, **477**, 567
      seismic loading 554
   Standards 477
   structures 240, 545, **567**
   see also marine technology
Offshore Europe conferences 474
Offshore fire safety 475
Offshore installations ... (pub. DoE) 567
Offshore service vessel register 475
Offshore structures (pub. Telford) 567
Offshore Technology 476
Offshore Technology Conference (OTC) 474
O'Flaherty, C.A. **506**, 507
Ogata, K. 391
Ohio State University, Engineering Design Graphics Journal 338
Ohmi, M. et al. 298
Ohmi, M. and Iguchi, M. 298
Oil and energy trends (pub. Blackwell) 239
Oil and Gas Journal 240, 282
Oil and Gas Technology 480
oil industry see petroleum industry
Oilman, former 480
Oil map (pub. Highlands and Islands Development Bd) 481
oils
   derived fuels 238, 240, **241**
   diesel 246
   engine oils 26, 179
   hydraulics 285, **288–90**
      see also fluid mechanics
   internal combustion engines 200
      see also lubrication; petroleum industry
OIML (Internat. Org'n for Legal Metrology) 27
OIN [Organisation Internationale des Normes] 17
OJ see Official Journal
Oki, M. et al. 298
Okrouhlik, M. and Pust, L. 162
O'Leary, B. 244
Oledski, A. 161
Oleodinamica-Pneumatica-Lubrificazione 297

Ölhydraulik und Pneumatik 297
Olins, W. 333
Olivetti [co.] 347
Olivo, C.T. 207
Olson, D.G. 166
Olver, A.D. 447
Ometron Ltd 147
On Demand Information see BCQ/ODI
Online 61
Online Computer Library Center see OCLC
online information 74, 97, **107–20**, 224, **228–9**, 305, 306
   CD-ROM on LANs 231
   access speed 108
   catalogues of 504
   conferences 61
   currency 306
   and current awareness 103
   current trends **107–9**, **580–1**
   options 108–9
   journals 61, 71–2
   materials databases 305–7
   newspapers 71
   'pseudo-online' 231
   updates 306
   see otherwise databases; CD-ROM; networks etc.
Ontario
   Lakefield Research 595
   Ontario Research Foundation 594
op-amps (operational amplifiers) 341, 423
OPEC (Org'n of Petroleum Exporting Countries) 225, 239
OPEC Review 239
OPEN see Organisation des Producteurs ...
Openshaw, S. 250
open systems interconnection/communication (OSI) 450
Open University (OU)
   Air pollution 491
   design courses 331, 334
   environmental control and public health courses 491
   publications 278, 313, 387
operational amplifiers (op-amps) 341, 423
operational research see operations
Operational Research Society, Journal 358, 404
Operational Research Society of America, Transportation Science 499
Operation of municipal wastewater treatment plants series (pub. Water Pollution Control Fed'n) 488
operations/operational research
   abstracts/indexes 100, 355
   amplifiers 341
   and construction management 569
   ergonomics 335–6
   journals **358**, 499

linguistic models 396
manufacturing 355
organizations 355, 358, 499
   and safety **325–6**
   and transport 496, 499
   marine 469tab
OPI (open to public inspection) see under patents
Oppenheim, A.V. and Schafer, R.W. 393
OPT (optimized production technology) 365
optical fibres 146, **445**, **447–8**
   development 412, 420, 441
   and ISDN 445
   journals 449
optics 145–7, 448
Optimal Control Applications and Methods 401, 408
optimality, control systems **390**, 393, 394–5, 398, 401
optimization 269, 274, **334**
   controls for 391, 401
   design 156, 166, 167, 169, 178, 333, **337**, 397
   information and 354
   journals 358, 401, 459, 460
   manufacture 358, 365
   optimized production technology (OPT) 365
   software 137, 149
   thermal systems 205
optoelectronics 353, 423, **447–8**
   components 448
   journals 415, 448
Optoelectronics (IEE Proc.) 415, 448
'Orange book' [steel structure contracting (pub. BCSA)] 539
ORBIT database host, former see QUESTEL.ORBIT
O'Reilly, J.J. 423, 443
ores 589, **591–2**
   dressing 590
Organisation des Producteurs d'Énergie Nucléaire (OPEN [Ass'n of Nuclear Energy Producers]) 263
Organization of Arab Petroleum Exporting Countries (OAPEC) 225
Organization for Economic Co-operation and Development (OECD) 224–5, **517**
   Centre for Development Studies 492
   and energy 225, 232
   IEA see Internat. Energy Agency
   IRRD database 503, **504**, 505, 512
   Nuclear Energy Agency (NEA) 250, 251, **263**
   publications 232, 238
      Coal Information 239
      Coal liquefaction ... 238
      Coal transport infrastructure ... 238
      Energy balances of OECD countries 232

Index 741

Organization for Economic
Co-operation and
Development (OECD)
publications (contd)
  Energy conservation in IEA
    countries 256, 257
  Energy statistics and
    balances of non-OECD
    countries 232
  Energy statistics of OECD
    countries 232
  Fighting noise . . . 492
  Nuclear Energy Data 251
  Nuclear legislation . . . 250
  Publications and Information
    Centre 250
  Quarterly Energy Balance
    232
  Quarterly Oil Statistics . . .
    239
  Statistical trends in
    transport 510
  and transport 504, 505, 510,
    517
Organization of Petroleum
  Exporting Countries (OPEC)
  225, 239
  OPEC Review 239
organizations **471–2**, 498, 624,
  627–8
  access to **497**
  annual reports 478
  classificatory 472, 477
  conferences 76
  directories of 85, **125–6**, 303,
    340, 370, 523, 526, 552
  and electronic publishing 114,
    117
  electronics 414
  employers' 576
  industrial 69
    for Standards 21
  information provision 497
  internal information 570–1
  international **125–6**
  and journals
    academic 64
    house journals 69
    industrial 69
    professional 64–5, 71, 102
  national 263–4
  newsletters 70
  official 471–2, 477–8
  Proceedings **64–5**, 72
  professional 64–5, 71, 76, 102
  publications **470–81**
    indexes to 102
  reciprocity 270
  research 478–9
  Standards initiated by 16, 24–6
  trade associations 340, 370
  and trade names 88
  Transactions 64–5, 72
  user groups 479
  see otherwise under specific
    topics
organs, artificial 631, 641
O'Riordan, T. and Hey, R.D.
  493
Orn, M.K. 325
Ortega, R. and Spong, M. 396

orthopaedics 635, 636, **638**, 641,
  646
orthotics 632, 641, 643
Orthwein, W.C. 159, 164
Orton, A. 560, 566
Ortuzar, J. Dios and Willumsen,
  L. **506**, 507
Osborn, A.F. 332
Osborn Technical Software [co.]
  172tab
Oscam [program] 168, 172tab
OSHA (Occupational Safety and
  Health Administration [US])
  320
Oshima, S. and Tsuneo, I. 298
OSI (open systems
  interconnection) 450
Osmec program 172tab
Österreichische Ingenieur- und
  Architekten-Zeitschrift **143**
OTC (Offshore Technology
  Conf.) 474
OTEC (ocean thermal energy
  conversion) 248
Ott, K.O. and Spinrad, B.I. 250
Ottinger, R.L. et al. 256
OU see Open University
Overseas building notes (pub.
  BRE) **521**
OVID Technologies [co.] 108,
  109, 111, **422**, 584
Owens, G.W. and Knowles, P.R.
  559
Ower, E. and Pankhurst, R.C.
  617
Oxbridge Communications
  [publishers], Standard
  periodical directory 72–3
Oxford University, publications
  407
Oxford University Press, Oxford
  dictionary of abbreviations
  124
oxidation 199
  and hydraulic fluids 288
Oxley, P.L.B. 363
oxygen
  and combustion 428
  thermodynamic properties 203
Ozisik, M.N. 277

P
PABLA (problem solving by
  logical approach) 156, 332
Pace Co., Pace Synthetic Fuels
  Report 241
package libraries 66, **90**
packaging, journals 654
packet switching **444–5**
paddle work 201
Pafec [program] 148
Page, J. 566
Pahl, G. and Beitz, W. 156, 312,
  331
Palmgren, A. 176
PAL (PERMABOND ADHESIVES
  LOCATOR) 314
Palz, W. 243
Pande, G.N. et al. 616
Pande, G.N. and Middleton, J.
  458

Panek, J.R. and Cook, J.P. 543
Panel for Historic Engineering
  Works (PHEW) 534–5, 553
Pang, D. et al. [control systems]
  396
Pang, G.K.H. and MacFarlane,
  A.G.J. [expert systems/CAD]
  397
Pangborn, R.N. et al. [reliability]
  319
Pannell, J.P.M. 535
PANS ('potential advanced
  network services') 444
Pant, P.D. 316
Pantry, S. 181
Panzhauser, E. and Hogland, I.
  254
Papanek, V. 336
parallel engineering/processing
  398, 650
Paraplegia 643
Pareto, Vilfredo 334
Paris
  La Villette 563
Paris/International Convention
  [on patents] (1883) 38–9
Park, C.F. and MacDairmid R.A.
  [ores] 591
Parker, C.C. and Turley, R.V.
  [information sources] 121
Parker, J. [thermodynamics] 202
Parker, M. [design/Standards]
  337
Parker, R.H. [geothermal energy]
  245
Parker, S.P. [McGraw-Hill
  reference-books] 122, 123
Parmley, R.O. 156, 163, 340
Parrish, D.J. 363
Parry, C.F. 327
Parsons, R.M., Co. 238
Parsons, T.R. et al. 486
particles
  mineral 591, 598
  technology of **278**, 282
  wear particles 174
  see also powder
Pascal [computer language] 455
PASCAL [databases] **231**, **505**
  Pascal – Bibliographie
    Internationale **101**, 584
  online 101, 231
Passenger Transport 499
Passenger Transport Executives
  (PTEs) 512
Pasztor, J. and Kristoferson, L.
  249
Patel, R.V. and Munro, N. 394
Patent Cooperation Treaty (PCT)
  45–6, 49, 59
Patent Information Network 60–1
Patent Office (UK) 37, 42, 60,
  61
  Comptroller 41
  Official Journal 60
  Trademarks Registry 88
patents **31–61**
  abstracts/indexes 48, 56,
    57–60, 112, **229**, 231,
    283, 380
  application procedures 33,
    37–42, 49–50, 51

742 Index

patents
  application procedures (contd)
    examination **39**, 40, 41, 43,
      44, **45–6**, 58
    PCT 45
    US 43–5
  assigning [selling] of **41**
  'basic' patents **39**, 50, 55
  books 61, 121
  classification schemes 40,
    48–9, **56–7**, 59
    commercial coding 56, 58–9
    International Patent
      Classification (IPC) 48,
      49, **57**, 58
  company patents **37**, 38, 46,
    52–3
    and name searches 45, 54
  patenting subsidiaries 54
    US 43, **45**, 59
  conferences 61
  Continuations in Part [US] 44
  as contracts 31, 47
  documents **40–1**, 45, **47–50**
    abstracts, applicant-supplied
      38, 40, 45, 48, **49**, 56,
      58–9
    'A' documents 40, 46, 52
    'B' documents **40**, 46, 47
    bibliographic information
      40, **48–9**, 54, **55**, 58–9
    claims 33, **40–1**, **49–50**, 57
    copyright lack 48
    correspondence 54
    drawings 40, 48, 49, 59
    early published applications
      39, 40, 41, 43, 44, **45**,
      46, 47, 52
    non-completed 39, 43, 52
    format 48–50
    granted patents 40, 41, 44
    INID codes 48–9
    languages used 39, 40, 45,
      47, 49, 55
    numbering systems 39, 40,
      46, **47**, **48–9**, 58
    see also classification
      above
  'patentese' 47, 49, 56
  search reports 45, 46
  specifications 31, 33, 37–8,
    40–1, 42, 45, **48–50**, 57
  titles/title pages 44, **48–9**,
    56, 58–9
  US 41, 43–5
  see also searches below
  duration see term below
  'equivalent' patents **39**, 40, 50,
    55
    Non-Convention 38, 55
    US 44
  ethical aspects 36–7
  excluded inventions **36–7**
  expired 48, 53, 58, 593
  'families' **39**, 49, 50
    family searches 54, **55**, 57,
      58, 59, 61
  file wrappers 54
  gazettes 47–8
  government interventions 42
  granted patents 40–3

non-exploited 41–2
historical aspects **31–2**, 43, 58,
  59
historical value 51
industrial applicability 33, **36**,
  41
as information sources 43,
  47–54, **121**
  as best/only sources 50–2
  commercial intelligence 60
  product information 84,
    88–9
  SDI 59, **60**
  see also searches below
infringement **40**, **42–3**, 50,
  52–4
  commercial searches 61
  threatened 43
international aspects **38–9**, 41,
  45–6, 47, 48
  classification 48, **57**
  developing countries 45–6
  European Patent Convention
    (EPC) **46**, 49
  International/Paris
    Convention 38–9
  language of publication 39,
    40, 45, 47, 49, 55
  Patent Cooperation Treaty
    (PCT) **45–6**, 49, 59
  prior art statements 44
  'quality' 43, 53
  regional systems **46**, 49
  searching 45, 52, 54, 55,
    58–60, **341**
  transnational/universal
    patents **46**
  WIPO 45
invalid see validity below
'inventive step' 33, **35–6**
and journals, general 45, **50–5**,
  57, 65
journals, specialist 47–8, 60,
  61
lapsed 53
leasing of see licensing below
legal aspects **31–7**, 40–3
  appeals procedures 39
  'fraud on the Patent Office'
    [US] 44
  information-provision 41,
    44, **47**, **49**, 51
  litigation 40, 41, **42–3**
  monopoly, 'residual' 41, 50
  'negative right' 31
  non-exploited patents 41–2
  patent falsely claimed 43
  Patents Act (1977) 32
  Polaroid v. Kodak 43
  regional systems 46
  statute of limitations 42
  unification 46
legal status 52, **54**, 61
Letters Patent 31–2
'licensing' [leasing] of **41–2**
  compulsory 41–2
literature searches, by patent
  office 37, 52, 53
  centralized 45
  search reports 45
as monopolies 31–2, 41,
  49–50

negative monopoly 31
mortgaged 41
name [author] searches **54**, 55,
  58–9
NASA 7
Non-Convention Applications
  38, 55
novelty **33–5**, 39, 41, 45, 50,
  58
US 44
'obtaining' [stealing] 37
'obviousness' **35**, 37, 39, 41,
  45, 50, 53
online information see
  databases above
Open to Public Inspection
  (OPI) **31**, **39**, 40
patentability **32–7**, 52, **53–4**
  commercial searches 61
patent agents 50
patentee **40**
  see also name searches
Patent Examiners **39**, 40, 49,
  52, 58
patent offices 33, 37–8, 39, 40,
  45–6, 47, 48, 52, 53
  classification schemes 48,
    56–7
  search facilities 45, **54**,
    60?+
  see also Patent Office
Patents Act, 1977 32
prior art **49**, 53, 56
  US 43, **44**
priority dates 33, **37–8**, 40, 41,
  55
  common see families above
  Convention Applications
    38–9
  for litigation purposes 40
  Non-Convention
    Applications 38
  US 33, 43, **44**
prior publication 33–4, 39, 53
  oral 34
prior use[r] 33, **34–5**
as property 37, **41–2**
provisional applications **37–8**
publication 31
  dates of 39, **41**, 45, 51
  US 43, 44
  print delays 47
  see also prior publication
    above
Registers of **54**, 152
renewal **40**, 44, 52–3
reports, applicant-originated 44
reports on 7, 57
revocation of **42**
searches 33, 38, 39, 44, **52–61**,
  341
  books/journals 283
  commercial services **54**, **61**
  databases 48, 53, **54–60**,
    229
  name [author] searching 54
  official services 60–1
  subject searching 54, **55–6**,
    58–61
  types of 52–3, **54–000**
  see also literature search
    above

Index 743

patents (contd)
  selling [assigning] of 41
  software 59–60
  Standards 48–9
  statistics 39, 40, 43, 46, 47,
    50–1, 88
  statistical analysis 59–60
  status of see legal status above
  stealing ['obtaining'] 37
  sufficiency 41
  term of 32, 40, 53, 59
  US 44, 59
  theft of patentable idea 37
  US 41, 43–5, 53, 58–9
  validity 42, 43, 50, 52–3
  validity searches 53
Patents Act (1977) 32
Patent and Trademark Office, US
  (USPTO) 44, 45, 53
Patstat [program] 59–60
Patton, R. et al. 178, 179
Paul, R.J. and Doukidis, G.I. [AI/
  simulation] 392
Paul, R.P. [robotics] 383
Pauwels, F. 633
PB (Publications Bd) document
  5, 11
PBV Consult [co.] 94
PC see computers: personal
PCM (pulse code modulation)
  442
PCN (personal communications
  network) 442
PCT see Patent Cooperation
  Treaty
PDS (product design
  specification) 333
Pearce, R. 364
peat 238, 239
pedestrians 496, 514
Pedestrians' Association 514
Pedley, J.B. et al. 273
Pegson Ltd, Aggregate producers
  handbook 600
Peitgen, H.O. et al. 457
Pelly, B.R. 436
Peltz, L. 256
Pendleton, R.L. and Tuttle, M.E.
  142
Peng, S.S. 238
Peng, S.S. and Chiang, H.S. 238
Penner, S.S. et al. 240, 241
Pennock, G.R. 163
Penn State University 596
PERA (Production Engineering
  Research Ass'n) 342
Peregrinus, Peter [publishers] 413
  Electrical and electronics
    trades directory 418
performance measurement 366,
  368
performance modelling 446
Pergamon [publishers] 61
PERINORM database 22, 152, 582
periodicals see journals
Perkins, J.L. and Rose, V.E.
  [safety] 318
Perkins, P.H. [concrete] 538
Permabond Ltd 312, 314
  PERMABOND ADHESIVES LOCATOR
    (PAL) database 314

Perry, R.H. and Green, D.W.,
  Perry's chemical engineer's
  handbook 271, 272, 598,
  600
Pershagen, B. and Bowen, M.
  250
personal communications network
  (PCN) 442
Perspectives in Energy 234
PERT (programme evaluation
  and review technique) 334
Pescod, M.B. et al. 490
PEST (political, environmental,
  social and technological)
  analysis 574
Peterka, V. 395
Peter Peregrinus see Peregrinus,
  Peter
Peters, M.S. and Timmerhaus,
  K.D. [design] 274
Petersen, D. [safety] 326
Peterson, R.E. [stress] 142
Petrakis, L. and Grandy, D.W.
  238
Petrecca, G. 256
petrol, lead in 256
Petroleum Abstracts 228
  online 230
Petroleum Economist 240
Petroleum Economist [co.],
  Energy maps
  ... of Europe 232
  ... of the Middle East 232
petroleum industry 239–41, 590
  abstracts 228, 480
  books 231–5
  properties data 273
  controls 279
  databases 229, 230, 231
  derived fuels 241
  historical aspects 94
  journals 68, 83, 240–1, 282,
    480
  and lead 256
  legislation 233, 240
  maps 232, 481
  offshore operations 378,
    467–8, 470tab
  organizations 224–5, 225, 226,
    239–41, 477, 479, 480,
    594, 609
  safety 321, 322, 323–4, 325–6
  'intrinsically safe devices'
    320
  Piper Alpha 477
  trade literature 83, 94
  see also mineral; chemicals;
    oil; offshore
Petroleum Review 68, 240
Petroleum Science and
  Technology Institute (PSTI
  [of EU])
  European oil and gas
    demonstration project ...
    240
Oil and Gas Technology 480
Petroski, H. 318
Petrostrategies 241
Petts, J. and Eduljee, G. 493
pH, control of 280
Phadke, M.S. 179

pharmaceuticals, patents 36, 42
Pheasant, S.T. 336
phenolic resins 325
PHEW see Panel for Historic
  Engineering Works
Philips [co.] 346
  data books 418, 420
  Philips Journal of Research
    417
  Philips (Mullard) data book 3
    420
Phillips, J. 159
photoelasticity 135
  photoelastic stress analysis
    135, 145–6
photography
  patent dispute 43
  Standards 25
photomechanics 145
photometry
  equipment 145–7
  organizations/journals 65
  Standards 27
photon-beam processes 364
photovoltaics (PV) 243–5
  PV News 245
  see also solar energy
physical chemistry 275–6
Physical data: chemical
  engineering (pub. ESDU)
  272
Physical Medicine and
  Rehabilitation Clinics of
  North America 643
Physical Review 423
Physical Therapy 643
Physica Medica 639
physics
  abstracts/indexes 100, 101,
    111, 355, 380, 399
  atmospheric see atmospheric
  books 122, 269, 272
  and chemical engineering 275–
    6
  conferences 110
  constants 203, 232
  databases 110, 111, 229, 355,
    380, 399
  electrophysical processes 358
  and environment 491
  geophysics 229, 240, 245–6
  Internet resources 119
  journals 105, 460, 491, 629
    preprints 115–16
  and manufacture 355
  medical/clinical 626–7, 628fig,
    629–32, 635
  nanotechnology 355
  noise 181
  nuclear energy 250–1
  organizations 28, 138, 413,
    415, 629–32
  plasma physics 250
  stress analysis 133
  system failure 178
  thermophysics 203, 205, 269
  welding 362
  see also thermodynamics
Physics Abstracts 101, 380, 399
  online 100, 111, 230, 380, 399
Physics in Medicine and Biology
  (PMB) 630

744 Index

Physics World 629, 643
Physikalisch-Technische Bundesanstalt (PTB) 359
Physiological Measurement 630, 639
Physiological Reviews 641
physiology 626, 629–30, 634–5, 639, **640–1**
physiotherapy 626, 631, 642–3
Physiotherapy 643
PI[D] (proportional, integral [and derivative]) controls 389–90, **392–3**, 397
PIARC 517
Piatier, A. 332
PICE *see* Institution of Civil Engineers: *Proceedings*
pick-and-place devices 383
Pickett, D.E. 600
PID *see* PI
Piesold, D.D.A. 318
pigments, early 589
Pilavachi, P.A. 256
piles/piling 544, 565
Pilkington, S. *and* Rhodes, R. 583
Pilkington Glass Ltd 525, 563
Advisory Service **562**
Pinches, M.J. *and* Ashby, J.G. 287
pinch-points 204
Piotrowski, J. 163
Pipeline and Gas Journal 67
Pipeline and Utilities Construction 68
Piper Alpha disaster 477
pipes/piping
  conferences 319
  directories 475
  handbooks 285
  heating systems 254
  hydraulic 292
  journals 67–8, 143
  marine pipelines 470*tab*
  organizations 303
  reliability 319
  Standards 302
Pippenger, J. *and* Hicks, T.G. 287
Pippenger, J.J. 287
Pisano, A. *et al.* 163
pistons/piston rings 177, 290, 297
Pitt, M.J. *and* Preece, P.E. [*controls*] 280
Pitts, G. [*design*/*PABLA*] 156
Pitts, G. *and* Lewis, S.M. [*design*] 179
placer mining 615
'plain old telephone service' (POTS) 444
planar mechanisms 341
PLANEX database **584**
planing 357
planning
  databases **584**
  environmental 255–7, 506–7
  manufacturing 354, 356, 357, 358, 362, 365, 369, **384–5**
  marine technology 470
  organizations 512, 513, 515, 517

transport **495–6**, 498–9, 502, 505, 506–7, **511–12**, 513, 515, 517
wind power 247
Planning Exchange [org'n] 584
Planning policy guidance notes (*pub*. DoEnv) 500
plant
  BSI 'PM' publications 182
  chemical engineering 274, 282
  chemicals 204, 205, 271
  conferences 296
  construction industry 67, **543–4**
  environmental issues 256–7
  handbooks 256–7
  mineral processing 589, 592
  mobile 286
  power/utilities 204–5, 319, 427–8
  thermal processes 204–5
  see also equipment; machines etc.
Plant guidelines for . . . chemical process safety (*pub*. Center for Chemical Process Safety) 325
Plant and Machinery (PM) publications [of BSI] 182
Plant Manager's Journal 67
Plant and Operations Progress, former 281
PLASCAMSdatabase 314
plasma arc processes 362, 363
plasma physics 250
plasticity 134
plastics 362
  cladding 563–4
  databases **112**, 314, 355
  glass-reinforced 562–3
  handbooks 307, 313, 362
  joining processes 362
  journals 68
  organizations 112, 303, 325, 563
  safety 325
  trade names 88
  see also engineered materials; polymers
plastic surgery 638–9
plates, structural 136, 137, 550
platinum/platinum group 593
plugs, electrical 16
Plumb, H.H. 202
plutonium **259**, 260, 261
plywood 303
PM (Plant and Machinery) publications [of BSI] 182
PMB (Physics in Medicine and Biology) 630
PMC handbooks 307
pneumatics 278, **285**, 293–4
  control systems 285, 391
  journals 285, 297
  organizations 287
  see also fluid mechanics
Podolny, W. *and* Scalzi, J.B. 545
Polak, P. 332
Poland, machines 161
polar engineering 470*tab*, 475
polariscopes 145–6

Polaroid v. Kodak 43
Policies and systems of environmental impact assessment (*pub*. UN/ECE) 493
political information 66, 229, 574
pollution 485–94
  abstracts 228, 230
  atmospheric **490–2**
  electricity generation 430
  books 236, 487
  combustion 199
  databases 230
  journals 487
  marine 16
  Royal Commission 500
  water 486–7, **486–9**
  see also environment
Pollution Abstracts 228
  online 230
Polovin, R.V. *and* Demutskii, V.P. 252
Polydata Ltd 303–4, **314**
polymers 302, 360
  abstracts 112, 309, 355
  databases 112, 306, 355
    CAMPUS **304–5**, 309, 313
  handbooks 302, 307
  organizations 112, 303
  Standards 302
  see also plastics; engineered materials
polytropic processes 200
P112: . . . topside structures . . . (*pub*. Steel Construction Inst.) 567
Pople, J. 141
poppet valves 291, 297, 298
Porges, F. 254
Porteous, A. 123, 490
ports 469*tab*, 475
  drydocks 478
Portugal
  minerals 596
  stress analysis 141
P&O Shipping [co.], *Annual Report* 481
Postelthwaite, I. *et al.* 398
'posters' [conference submissions] 76
Post project analysis in environmental impact assessment (*pub*. UN) 493
Post-tensioning systems for concrete . . . (*pub*. CIRIA) 557
potash, mining 615
'potential advanced network services' (PANS) 444
POTS ('plain old telephone service') 444
Potter, C.D. 155, 170–1*tab*
Poulter Communications plc 95
powders/dusts **278**
  conferences 359
  injection moulding 365
  journals 282
  metals 369
  mineral processing 592
  mines 617
  organizations 359

Index 745

powders/dusts (*contd*)
 safety **321–2**
 see also particles
Powder Technology 282
power
 electrical systems **425–39**
 emergency systems 327
 fluid power **285–99**
 generation of 208–10
 heat and power, co-generation
  see combined heat and
  power
  as mechanical work 285–7,
  286*fig*, **425**, **428**
 rotary 341
 ships 467
 UPS 414
 see also energy; nuclear *etc.*
power-conditioning 426, **435**
power cylinders 288
power density 261
power Engineering **208–9**, 226,
 251–7, **425–39**
 nature and role **425–6**, 427,
  429–30
  economic context **425**, 428
 abstracts/indexes 421–2, **426–7**
 alternative sources *see*
  alternative energy
 books, handbooks 125, **429**,
  431
 buyers' guides 417
 CHP *see* combined heat/power
 'clean power' 242
 co-generation *see* combined
  heat/power
 combined cycles 429
 conditioning 426, **435**
 conferences 242, 296, 417,
  431, 432, 434
 costs 208, 209, 430
 databases **229**, 230, 421–2,
  426–7
 data transfer 310
 desulphurization 430
 distribution 427–8, **431–2**
 district heating **208–9**, 210,
  225, **253–5**
 electronics and 417, **435–7**
 environmental issues 255–7,
  430
 fuels, economy measures
  429–31
 generation process **427–31**
 historical aspects 226, 429
 hydroelectricity *see*
  hydroelectricity
 Internet resources 119
 journals 67, 208, 209, **426–7**,
  431, 434, 435–6
 loadflow calculation 432, 433
 machine drives 163–5, **425**,
  433–7
 magnetohydrodynamics **251–3**
 management 428
 motors **434–7**
 nuclear power *see* nuclear
  energy
 organizations 209, 210–11,
  225, 226, 227, 414,
  426–7, 431

product information 310, **427**
protective relays 431–2
reliability 319, **428**, 431–2,
 433
renewable resources *see*
 alternative energy
safety **431–2**
Standards 433
 legislation 320, 430
 specifications 310
 terminology 223
 thermal systems 204–5,
  208–11, **425**, **428–30**
 analysis **432–3**
 training 429
 transmission 320, 427–8,
  431–2
 EHV 252, **429**
 HVDC 432
 see also energy technology;
  electrical engineering
Power Engineering Journal (pub.
 IEE) 253, **427**, 435, 438
Power Engineering Review (pub.
 IEEE) 252
PowerGen [co.] 226
Power International 67
power stations **251–3**, 427–8,
 429
 CEGB manual 251–2, **429**
 commissioning 429
 cooling towers 207, 567
 energy efficiency 428
 nuclear 250–1, 252, **259–62**,
  429
 see also reactors
Power system protection (pub.
 Electricity Cncl) 320
Power Systems and Applications
 [Key Abstracts] 428
Practical aspects of district
 heating . . . pub. DHA] 209
Practical Design of Ships
 conferences (PRADS) 474
Practical guide to the Machinery
 Directive (pub. IMechE] 181
Practical guide to restoration
 (pub. RMC] 614
Practical risk assessment (pub.
 EEF) 317
PRADS (Practical Design of
 Ships conf.'s) 474
Prasad, M. 207
Pratt, T.H. 322
Prausnitz, J.M. *and* Lichtenthaler,
 R.N. 275
precipitation 274
precision engineering 354, 357
 casting 364–5
 conferences 358, **359**
 journals 357, 358
 organizations 357, **359**
 ultra-precision 359
Precision Engineering 357
Predicasts *see* PTS
Prentice, G. [*electrochemical*
 engineering] 274
Prentice Hall [publishers] 423
Prentis, J.M. [*dynamics*] 194
preprints 115
 electronic 115–16

as grey literature 10
Press, W.H. *et al.* 455
press cuttings services 75
pressing processes 286, **364**
press releases 92
pressure/pressurization **199–201**
 compression *see* compression
 ductility 364
 equilibrium 199
 high-power pressure tube
  reactor (RBMK) 261–2
 and hydraulics **289**, 290, 291,
  292, 293
 pressure ripples 289, 293
 pressure-surges 292
 pressure waves, *and* telephony
  441
 pressurized water reactor
  (PWR) 261–2
 and safety 322
 and thermodynamics 199–201,
  203
 and volume **200–1**
pressure sores [decubitus] 637,
 639
pressure vessels 235–7
 conferences 319
 journals **143–4**
 reliability 319
 Standards 22, **25**, **206**, 235
 see also boilers
Preston, E.J. *et al.* 335, 370
Prett, D.M. *et al.* 280
Prett, D.M. *and* Garcia, C.E. 280
Prett, D.M. *and* Morari, M. 280
Price, B. [*lead/energy*] 256
Price, N.J. *and* Cosgrove, J.W.
 [*geology*] 616
Price Adjustment Formula 578
price lists 82
Priest, J.W., [*design/testing*] 179
Priest, S.D. [*rock mechanics*] 616
privatization 226, 429, 502, **512**,
 594
probability 113, **451**, **454**, **457–8**,
 459, **460**
Probability in the Engineering
 and Informational Sciences
 460
problem-solving
 design 155–6, 159, 330
 kinematics 158
 PABLA system 156, 332
Probstein, R.F. 276
Proceedings see under
 conferences *and specific*
 organizations
Process Biochemistry 282, 490
Process Biochemistry
 International 488
Process and Chemical
 Engineering 228, 283
process engineering *see* chemical
 engineering; mineral
 processes
Process Engineering 402
 online 400
Process engineering directory
 (pub. Benn] 86
Process Equipment News 87
processes 92, **269–84**, 353, 355,
 358

746 Index

processes (contd)
 nature and role 269–70
 advanced see under
  manufacturing
 automated see automation
 and concurrent engineering
  649, 657
 databases 355, 400
 energy conservation 256
 integrated see concurrent
  engineering
 journals 401
 marine technology 470tab
 mineral **589–606**
 parallel processing **398**, 650
 process analysis/synthesis 269,
  273, 279, 362, **398**
 process control 358, 366, **398**,
  401–2
  AI 397
  dictionaries 398–9
  statistical (SPC) 358, **366–7**,
   369
  see also control systems
 process engineering see
  chemical engineering
 process industries **269**
  co-generation 235
  loss prevention 281
  safety 317, 318, **325–6**
  thermal systems 205
 process planning 354, 356,
  357, 358, 362, 365, 369,
  384–5
  Standards **26**
  company-specific 28
  stochastic **451**, **457–8**, 460
  thermal 199–201, **203–5**
  reversibility **201**, 204
  transport processes **273**
  unit operations 269, 271–2,
   273–4, 488
  up/downstream 278
  see otherwise chemical
   engineering
 process industries see under
  processes
 Processing 282
 *Processing of Advanced
  Materials* 357
 Process integration (pub. ESDU)
  204
 *Process Safety and
  Environmental Protection*
  (IChemE Transac.) 281
 Process safety management...
  (pub. CMA) 325
 Process Safety Progress 281, 328
 process see under processes
 procurement 310–11
 Procyk, T.J. and Mamdani, E.H.
  396
 product design specification
  (PDS) 333
 product development **336**
 Product Engineering 168
 product information 63–72,
  81–96, 417–18
  access to 83, **93–6**, 340
  catalogues **82–3**
  Central Body, proposed 93

collections **93–5**
databases 89, **90–3**, 95
 free 304, 305
data interpretation 288–9, 291
data sheets 82, **83**, 93, 291,
 302, 330
data transfer **309**
electronics 414, **417–20**
exhibition catalogues 84, **87–8**
future trends 96
historical 93–4
Internet 312
materials 304–5, 308
patents 84, **88–9**
primary sources **82–3**
product models 310
product reviews 71, 99, 105
sales representatives 89
secondary sources 83–93
Standards, information from 89
Standards for
 BS 4940: ... presentation of
  technical
  information . . . 81
 CALS **311**, 314
 STEP [*ISO 10303*] 307,
  310–11, 314
trade directories 65, **84–6**, 87,
 92, 370
trade journals **66**, **86–7**
 see also under journals
trademarks 84, **88**
verbal 89
see also trade literature etc.
production see manufacture
Production 356
Production Drawings Code 578
Production Engineer, former 356
production engineering 230, 650,
 651
Production Engineering Research
 Association (PERA) 342
Production Planning and Control
 356
productivity 335, 359, 369, 384,
 386
product liability 318
 and design **334**
 Europe 334
 and manufacture 358
'product models' 310
product reviews 71, 99, 105
Professional Engineering 65, 86,
 157, 295, 339, 356
professional organizations see
 organizations
Professional Safety 328
profitability, and design 337
Progetim [co.] 314
*Programmable electronic systems
 in safety-related applications*
 (pub. HSE) 319
programmable logic controllers
 285, 383, 385–6
programme evaluation and
 review technique (PERT)
 334
programs, computer see software
Progress in Aerospace Sciences
 105
*Progress in Astronautics and
 Aeronautics* 105

*Progress in Batteries and Solar
 Cells* 245
*Progress in Combustion Science
 and Technology* 105
*Progress in Construction Science
 and Technology* 105
*Progress in Energy and
 Combustion Science* 237
Progress in Fine Grinding
 Technology conference 359
Progress in Materials Science
 105
Progress in Nuclear Energy 251,
 266
Progress in Planning 499
projects
 project management see under
  management
 Project Specification Code
  578–9
 summaries/evaluations 4
Prokes, J. 287
Promofluid 67
propellers 467
*Properties of aluminium and its
 alloys* (pub. Aluminium
 Fed'n) 561
properties data
 physical 273
 see also under materials and
  specific topics 273
Property Services Agency (PSA)
 94, 578
 Better trade literature 96
 Building Management South
  and West 94
 and CPI 579
 General Specification 578
 library 94
 and PBV Consult 94
proportional, integral [and
 derivative] see PI
propulsion 208
ProQuest [program] 426, 427
prospecting, minerals 240
prosthetics 133, 383, 632, 636,
 641, 643
*Prosthetics and Orthotics
 International* 643
Protective relays: application
 guide (pub. GEC) 432
protocols, data transfer 118, 305,
 309–11, 335, **445**
Pryor, E.J. 590, 597
PSA see Property Services
 Agency
PSTI see Petroleum Science and
 Technology Inst.
psychology
 design and 332
 aesthetics 336
 ergonomics and 335, 336
 see also human factors
psychrometry 253–5
PTB (Physikalisch-Technische
 Bundesanstalt) 359
PTEs (Passenger Transport
 Executives) 512
PTRC **501**
 European Transport Forum
  500, 501

Index 747

PTRC (contd)
  Summer Annual Meeting, former 500
  training courses 502
PTS [Predicasts] databases
  PTS NEW PRODUCTS ANNOUNCEMENTS (NPA) **92**, 93
  PTS NEWSLETTER 70, 402
  PTS PROMPT 75
Publications Board (PB)
  document 5, 11
public health *see* environmental engineering
Public Transport International 499
Public transport statistics (*pub.* LRT) 510
Public Transport Symposium **501**
public works *see* civil engineering; transport
publishing/publishers
  book-lists 104
  desk-top 14
  electronic *see* electronic publishing
  and journals 114
  selective lists 527, **531**
Pugh, E., *Pugh's dictionary of acronyms and abbreviations* . . . 124
Pugh, S. [*design*] 156, 331
Pulfrey, D.L. 244
Pullar-Strecker, P. 557
Pulsar [program] 582
pulse code modulation (PCM) 442
pulse-width-modulation (PWM) 436
pumps 205, **208**, 270
  hydraulic 286, 289, 290, 293, 298
  pressure ripples 289, 293
  variable displacement 286
  see also heat pumps
punch card, patented 51
Purchas, D. 600
purification
  biochemical 278
  gases 282
Purpa Lines 249
Purser, M. 446
PV
  PV News 245
  see otherwise photovoltaics
PWM (pulse-width-modulation) 436
PWR (pressurized water reactor) 261–2
Pye, D. 336
pyrolysis, *and* energy production 249

Q
Qader, S.A. 238, 241
QFD (quality function deployment) 650, 653
QS (quantity surveyor) 579
Qualitative analysis of structures (*pub.* IStructE) 550
Quality assurance handbook (*pub.* BSI) 368

Quality Forum 358
quality function deployment (QFD) 650, 653
quality management 177–83, **366–7**
  audits 178, 179
  automation 366
  awards 371
  books
    dictionaries *etc.* 154, 177, 179, 180
    handbooks **177–8**, **368–9**
    and concurrent engineering 649, 651–2, 653
    conferences 179, 180, **359–60**, 371
  construction 585
  cultural variation 369
  data analysis 366–7
  data transfer 310
  and design 333, **334**, 337
  economic aspects 368
  finite elements 139
  in-process control 367
  journals 139, **180**, **357-8**
  machines **177–84**
  maintenance 368
  manufacturing **357-8**, **359–60**, 360–1, 365, **366–7**, 368, 649
  organizations 139, 178–9, **180**, 358, 359–60, 371, 453
  planning 369
  quality function deployment (QFD) 650, 653
  Standards 19, 22, 154, **177–8**, 366, 368
  testing *see* testing
  total quality management 366–7, 368, 371
  training 334, 371
  and waste 365
  see also reliability *and specific topics*
Quality management handbook (*pub.* BSI) 177–8
Quality News 358
Quality Progress 358
Quality and reliability . . . (*pub.* SEED) 333, **334**
Quality and Reliability Engineering International 180, 339
Quality Today 180, 358
Quantarc [co.] 95
Quantified risk assessment . . . (*pub.* HSE) **179**, 334
quantities, Standards 27
quantity surveying 579
Quantrille, T.E. *and* Liu, Y.A. 280
quarrying 607–21
  blasting 614
  directories 602, 608
  journals 597, 609, **611**
  organizations **593–4**, **609**
  see otherwise mining
Quarry Management 597, 609
Directory of quarries, pits and . . . equipment 602
Quarry Managers Journal [co.] 597

Quarterly of Applied Mathematics 459
Quarterly Coal Report (*pub.* USDoE) 237
Quarterly Energy Balance (*pub.* OECD) 232
Quarterly Energy Reviews (*pub.* Economist Intelligence Unit) 232–3
Quarterly Journal of Mechanics and Applied Mathematics 459
Quarterly Journal of Technical Papers (*pub.* IP) 282
Quarterly Oil Statistics and Energy Balances (*pub.* OECD) 239
Quarterly road casualties Great Britain (*pub.* DoT) 509
Quarterly transport statistics (*pub.* DoT) 509
Queensland [Australia], University 595
QUESTEL.ORBIT database host 58, 59, 75, **107**, **230**, 231, **584**
  CEBA 112
  COMPENDEX 110
  ICONDA 113
  INSPEC 111, 422
  METADEX 112
  NTIS 109
  RAPRA ABSTRACTS 112
  WORLD PATENTS INDEX 380
queueing theory 458
Quinney, D. 456
Quin Systems Ltd, *Intelligent motor control* . . . 172

R
Raask, E. 238
Rabia, H. 617
RAC *see* Royal Automobile Club
radar
  handbook 125
  patented 51
Radchenko, G.A. 617
Rade, L. *and* Westergren, B. 453
radiation
  of heat 199, 202
  ionizing 27
  and semiconductors 91
  Standards 27
  see also radioactivity
radio 441, **447**
  antennas 447, 448
  conferences 417
  early 411
  frequency (rf) 449
  as information source 497–8
  interference 22
  journals 105, 417, 448–9
  mobile 417, 443, **444**, **446**, 447, 449
  organizations 22
  personal 444
  radio tagging 385
  radio telephony 442, **446**, 449
  receivers 443
  spectra 443, 444
  Standards 22, 421

748 Index

radioactivity 250–1, 260, 266, **324–5**, 617
radiometry, Standards 27
Radio Technical Commission for Aeronautics (RTCA [US]) 421
Radio Technical Commission for Maritime Services (RTCM [US]) 421
radio telephony 442, **446**, 449
Radke, M. 337
RAE (Royal Academy of Engineering) 342
Raeder, J. *et al.* 250
Raheja, D. 334
Rahimi, M. *and* Karwowski, W. 384
Railway Gazette International 499
railways 495–517
  books 506–7
  databases 504–5
  journals 68, 497–9
  organizations 323, 511–17
  privatization 502, **512**
  safety 323, 324
  Standards 507–8
  statistics 509–11
  see also transport
Railway Track and Structures 68
Ralph, W.J. [*Standards*] 16
Ralphs, J.D., [*telecommunications*] 443
RAM [Recent Advances in Manufacturing] database 380
Ramakumar, R. 178, 179
RAMB 356
  online 356
Ramo, S. *et al.* 447
R&D *see* research and development
Randerson, K. 559
Randol International [co.], *Randol gold forum* 605
Rankine cycles 203, 207
Ranney, M.W. 602
Rao, B.K.N. [*reliability*] 178
Rao, J.S. *and* Gupta, K.N. [*machines*] 162
Rao, S.S. [*finite elements*] 455
Raper, R. 154
RAPRA (Rubber and Plastics Research Ass'n of Great Britain 303, **314**
  Adhesives Abstracts 112
  New trade names... 88
  PLASCAMS 314
  RAPRA ABSTRACTS **112**, 341, **355**
  RAPRA Abstracts 112
  RAPRA Technology Ltd 112
Rase, H.F. 275
Rashid, M.H. 436
'Rasmussen report' [on nuclear safety] (*pub.* NRC) 266
RASNA UK Ltd 170*tab*
  RASNA [program] 149
Rautenbach *and* Albrecht, R. 274
Rawson, K.J. *and* Tupper, E. 473
Ray, M.S. 270, 337
RBMK [high power pressure-tube reactors] 261–2
R&D (research and development) patents analysis 60
R&D Abstracts 8
RDP Electronics Ltd 146
reactors
  chemical 271, **275**, 279, 323–4
  bubble-columns 275
  nuclear 250–1, 252, **259–62**, 266, **429**
  coolants 260–2
  fast/fast breeder 260, 262
  moderators 260, 261
  Phénix 260
  power density 261
  Reports/accidents **266**
  safety 318
  seismic loading 554
  in ships 261
  thermal 260
  types of **261–2**
Reactor safety study... [Rasmussen report] (*pub.* NRC) 266
Read, G.F. *and* Vickridge, I. 488
Reading, University, Centre for Strategic Studies in Construction (CSSC) 575
Realizing CIM's Industrial Potential conference 359
Real-Time Systems 403
Reay, D.A. 206
Reay, D.A. *and* Dunn, P. 254
Reay, D.A. *and* Macmichael, D.B.A. 208, 254
Recent Advances in Engineering Science 105
Recent Advances in Manufacturing (RAM) database 380
Recovery of waste heat... (*pub.* DoE) 206
rectifiers 435, 436
  choppers 436
recycling 334, 490
  books, handbooks 534
  and design 337
  journals 490
Recycling World 490
'Red book' [steel structure design (*pub.* BCSA)] 539
Reddy, J.N. [*mathematics*] 456
Reddy, J.N. *and* Rasmussen, M.L. [*mathematics*] 456
Reddy, T.A. [*solar energy*] 210, 244
Redford, A.H. *and* Chal, J. 337, 361
Redland Bricks Ltd 525
Redmill, F.J. *and* Valdar, A.R. 444
Reed, E.W. *and* Larman, I.S. [*fluid power*] 287
Reed Elsevier [co.] 107
Reed Information Services [co.], *Industrial trade names* 88
Reef, The, former 609
Rees, J. *and* Odell, P. 240
Referativnyi Zhurnal 101
refereeing
  discussion lists 116
  electronic journals 114–15
  journals 64
reference books 82–3, 84–6, **121–32**
  biographical data **126–7**
  dictionaries **122–4**
  directories **125–6**
  encyclopaedias 122
  handbooks **124–5**
  literature guides **121–2**
  official publications 126
  yearbooks **125–6**
  see also directories *etc*.
Reference manual [mineral processing] (*pub.* Nordberg) 600
refining 599
refrigeration 202, 203, 204–5, **207**, 236, 253–5, **276–7**
  CFCs 202
  organizations 207
  solar energy 244
Refrigeration and the environment (*pub.* BSRIA) 207
Refrigeration systems and applications (*pub.* ASHRAE) 207
regional science 498–9
Regional Science Research Institute, *Journal of Regional Science* 498
Regional Science and Urban Economics 499
Regional Studies 499
Register of ships (*pub.* Lloyd's) 478
Regulations, Health and Safety **180–1, 182**, 477
  Dangerous Goods 477
  Electricity at Work **182**
  Noise at Work 181, **182**
  see also Building Regulations and specific topics
Rehabilitation Digest 643
rehabilitation engineering 626, 627, 628*fig*, 630, 635, 642–3
Rehabilitation Society of North America (RESNA) 630
Reid, C.E. [*chemical thermodynamics*] 275
Reid, R.C. *et al.* [*chemistry*] 273
Reimbert, M.L. *and* Reimbert, A.M. 567
reinforced materials *see* composites; concrete *etc*.
Reinforced and prestressed masonry (*pub.* IStructE) 560
Reinforcement (*pub.* BCA) 537
Reinstatement of fire-damaged steel and iron... structures (*pub.* BSC) 555
Reiter, S. 206
reliability 151, **177–83, 318–19**
  defined 318
  books 152, 154, 177–8, 318–19, 368
  components 368
  and concurrent engineering 652

reliability (contd)
  condition monitoring **178-80**
  conferences 179, 180, **296**
  fluid power 288
  hydraulics 291
  journals 358
  power systems 432-3
  Standards 179
  conferences 179, 180
    COMADEM **296**
  data 178
  databases 91, 179-80
  and design 178-9, 318, 319,
    333, **334**, 337
  electrical 432-3
  electronics 412
  equipment 368
  failure-rates see failure
  human factors 319
  journals 179, **180**
  life prediction 319
  machines 152, **177-83**, 319
  maintainability 178-80, 319,
    368
  organizations 178-80, **180**
  power systems 319, **428**,
    431-2, **433**
  robotics 384
  robots 384
  and safety 177-83, **318-19**,
    325
  Standards 154, **177-8**, 179,
    189, 319, 368
  systems 368, 432-3
  theory of 458
  see also quality; safety
*Reliability Engineering and
  System Safety* 180
Reliability handbook (pub. IEEE)
  319
relief systems 325, 327
reluctance drives 435
remote computer see networks
remote control/teleoperation 383,
  400
renewable energy see alternative
  energy
Renewable Energy 242
Renewable energy in cities (pub.
  Center for Renewable
  Resources) 210
Renewable Energy News Digest
  242
Renewable Energy Resources
  Information Centre (RERIC)
  225
*Renewable energy sources in the
  United Kingdom* (pub. DoE)
  210
Renewable Sources of Energy
  242
repair see maintenance
repetitive strain injury (RSI) 335
reports **3-14**, 121, 471-2, 477-8,
  481
  abstracts/indexes 102, 111, 112
  access services **6-12**, 102, 110,
    116, 142
  currency 5
  databases 115, **116**, 354
  as grey literature 10-11, 121

identification series **5-6**, 8
  in journals 64
  Standards 4, 13
reprints
  books 552, 561
  theses 530-1
RERIC (Renewable Energy
  Resources Information
  Centre) 225
research
  confidentiality issues 472
  construction industry 230,
    520-5, 530-1
  contract 472
  databases 230-1, 355
  guides to 78, **126**, 152
  Internet and 114-17, 120
  journals 64, 69, 78, **503**
  organizations 78, **126**, 472,
    478-9, 515-16, 594
  product information 82
  reviews 83, **99**, 102, **104-5**,
    111, 231-5, 499
  theses 77-9
research and development (R&D)
  patents analysis 60
  R&D Abstracts 8
Research in Engineering Design
  339, **653**
Research Publications Inc. 59
reservoirs 430
resins 307
safety 325
RESNA (Rehabilitation Soc. of
  N. America) 630
Reuleaux, F., *Theoretische
  Kinematik/Kinematics of
  machinery* [trans.] 155, 166
Réunion Internationale des
  Laboratoires d'Essais et de
  Recherches sur les
  Matériaux et Constructions
  (RILEM) **143**, **523**, 533
REUTER TEXTLINE database **422**
Review of Metal Literature 111
reviews
  of books 65, 104
  research 83, **99**, 102, **104-5**,
    111, 231-5, 499
  product 71, 99, 105
  management audits see audits
Reviews in Chemical Engineering
  105
*Reviews of Renewable Energy
  Resources* 242
*Revista Internazionale di
  Economica dei Traporti* 498
Revue de l'Énergie 235
Revue Générale de l'Électricité
  253
Revue Générale Nucléaire 251
Rexroth, G.L., GmbH, *Hydraulic
  trainer* ... 288
Reynard, K. [*materials*] 303
Reynard, K.W. [*materials*] 121
Reynolds, C.E. and Steedman,
  J.C. [*concrete*] 557
Reynolds, Osborne 19
Reynolds, T.D. [*environmental
  engineering*] 488
Reynolds W.C. and Perkins, H.C.
  [*thermodynamics*] 202

rf (radio frequency) 449
rheology 365, 638
Rheology 642
rheostats 435
Rhine, J.M. *and* Tucker, R.J. 206
Rhodes, M.J. 278
Rhône Poulenc [co.] 305
RIBA
  RIBA Bookshops 533
  RIBA product data 86
  RIBA product selector 86
  RIBA Services [org'n] 86, 582
  RIBA/TI database **581-2**
  see otherwise Royal Inst. of
    Brit. Architects
Rice, I.G. 209
Rice, Peter, architect 563
Rice, P[eter] *and* Dutton, H. 563
Rich, E. *and* Knight, K. 335
Richard, C.W. *et al.* [*fluid power/
  software*] 296
Richards, G.E. [*wood energy*]
  249, 256
Richards, R.H. *and* Locke, C.E.
  [*ores*] 590
Richardson, B.G. [*CIRIA* ...
  guide] 526, 587
Richardson, J.F. *and* Peacock,
  D.G. [*chemical engineering*]
  271
Richardson, J.G. [*concrete*] 538
Richardson, M.I. [*substances/
  hazard*] 323
RICS see Royal Inst. of
  Chartered Surveyors
Ridley, J.R. 326
Riedi, P.C. 202
right-first-time manufacturing 650
RIIA see Royal Inst. of Internat.
  Affairs
RILEM see Réunion Internat.[e]
  des Laboratoires ...
Riley, F.J. [*manufacture/
  electronics*] 361
Riley, W.F. *and* Sturges, L.D.
  [*dynamics*] 158
RINA see Royal Inst'n of Naval
  Architects
risk
  defined 316
  best estimates 317
  and design 333
  risk assessment/analysis 181,
    182, 315, **316-17**, 319,
    327
  loss rates 317
  risk management 317, 325,
    326, 333, 334
  *Risk analysis in the process
    industries* (pub. Internat.
    Study Group on Risk
    Analysis) 317
  Risk guidelines for engineers
    (pub. Engineering Cncl) 315
  Riso National Laboratory 242
rivers
  databases 229
  environmental protection 487
  estuaries 487
  gravel recovery 615
  hydroelectricity 247-9, 255,
    256, **430**

750 Index

rivers (contd)
  management 487
  Severn tidal project 248
  see also water; water engineering
Rizzi, E.A. 254
RMC Group, *Practical guide to restoration* 614
Road accidents Great Britain (pub. HMSO) 509
  ...: *the casualty report* 509
Road accident statistics: English regions 509
Road Haulage Association 514
Road lengths in Great Britain (pub. HMSO) 509
Road Research Laboratory, former 512, 523
Road Research Technical Papers (pub. DoT) 507
roads 495–517, 545, 566–7
  accidents 509–10
  books 506–7
    handbooks 534
  construction 534
  databases 504–5
  deregulation 512
  design 500, 508, 545, 566
  environmental impact 493
  international 510
  journals 497–9
  Leitch Report 499
  motorways 507
  organizations 511–17, 523
  pricing policies 495–6
  roundabouts 508
  SACTRA 493, 499–500
    Wood report 493
  safety engineering 316, 317
  Standards 507–8
  statistics 509–11
  users
    behaviour 496
    vulnerable 496
  see also transport; traffic
Roads and traffic in urban areas (pub. IHT/DoT) 506, 507
Road traffic statistics Great Britain (pub. DoT) 509
road transport industry (RTI), *RTI Focus News* 498
Roark, R.J. and Young, W.C. *Roark's formulas for stress and strain* 142, 341
Robb, C. 313
Roberson, R.E. and Schwertasser, R. [dynamics] 159
Roberts, J. and Fairhall, D. [noise] 492
Roberts, J.H. [telecommunications] 443
Roberts, J.J. et al. [masonry] 560
Robertson Research [co.] 595
Robins, W.P.[telecommunications] 443
Robinson, R.N. [chemical engineering] 272
Robot facts (pub. BRA) 382
Robotica 379
Robotic Industries Association 382

robotics 377–88
  nature and role 377–8
  abstracts/indexes 380, 401
  books 382–6
    dictionaries 154
    handbooks 125, 382
  conferences 359, 381–2
  construction 378, 580
  control systems 378, 383–6, 394, 401
  databases 380, 401
  electrohydraulics 285–6
  film industry 286
  intelligence 377, 378, 384
  journals 378–9, 400, 401
  manipulators 377, 384, 396, 401
    rigid-link 394
  marine technology 470tab
  and mechatronics 172
  mobile robots 384
  offline programming 378
  organizations 378–9, 380–2, 387–8
  patents 379
  reliability 384
  safety 384, 386
  sensors 377–8, 384, 402
    sensor-based control 379
    vision facility 363, 377, 382, 384, 385, 386
  simulation 378
  software 379, 385
  Standards 383
    specifications 384
  statistics 382
  task-planning 379
  training 379
  types of robots 383
  see also automation; controls
ROBOTICS database 380
Robotics and Automation Society [of IEEE] 378, 381
Robotics and Autonomous Systems 379
Robotics and Computer-Integrated Manufacturing 379
Robotics and Control 400
Robotics International [of SME] 387
  conference 381
Robotics and Manufacturing conference 381
Robotics Society of Japan, *Advanced Robotics* 378
Robotics Today 378
Robotics World 379
Robots Institute of America 377
Robots and Vision Automation conference 382
Robson, P. 551
robustness
  design 179–80, 280, 334
  systems 390, 393, 394–5, 396, 397
rock/stone
  blasting 615
  drilling 613–14, 615
  early uses 589, 607
  geothermal energy 245

journals 593–4, 611
  as mineral source 591, 593, 600, 616
  rock mechanics 610, 611, 613, 616
  soft/hard 615
Rock drilling data (pub. Gardner-Denver) 614
Rock drilling manual (pub. Sandvik *and* Atlas Copco) 614
Rock Mechanics and Rock Engineering 611
Rock Products 593, 597
Rodriguez, W. 335
Roffel, B. et al. 280
Rogers, G.F.C. and Mayhew, Y.R. 202
Rohner, P. 288
Rohrbach, C. 141
Roland, H.E. and Moriarty, B. 317, 326
Rolling bearings (pub. ESDU) 188
Roman, S. 457
Ronayne, J. 445
Roney, A. 126
Roofing 67
roofs 557, 561, 563–4
  journals 67
  loadings 553, 554
  organizations 564
Rooke, D.P. and Cartwright, D.J. 142
Rooney, J. and Steadman P. 335
Roots, G. 247
Roozenberg N.F.M. [design] 162
Roozenburg N.F.M. and Eekels, J. [design] 332
Rosenberg, P. [alternative energy] 242
Rosenbrock, H.H. 394
Rosi program 172tab
RoSPA
  RoSPA Bulletin 183
  RoSPA handbook of health and safety . . . [ed. Stranks] 181
  RoSPA road safety engineering manual 500
  see otherwise Royal Soc. for the Prevention of Accidents
Ross, A. [robotics] 383–4
Ross, P.J. [quality] 179
Ross, R.B. [materials] 125
Ross, S.M. [probability/models] 458
Rota, G.-C. 454
rotary power 341
Rothbart, H.A. 124, 168, 340
Rothery, B. 178
Rousseau, R.W. 274
Routledge [publishers]
  Routledge French technical dictionary 123
  Routledge German technical dictionary 123
ROV review 475
Rowe, R.E. [concrete] 556
Rowe, W.B. [bearings] 175

Index 751

Rowe, W.B. et al. [bearings] 175
Roy, D.N. [fluid mechanics] 205
Roy, R. and Wield, D. [design] 333
Royal Academy of Engineering (RAE) 342
Royal Aeronautical Society 138
   Aeronautical Journal 143
Royal Automobile Club (RAC) 514
   Motoring statistics 510
Royal Commission on the Ancient... Monuments of Scotland 552
Royal Commission on Environmental Pollution, Transport and the environment 500
Royal Institute of British Architects (RIBA)
   Architects' Journal 553, 564, **579**
   ARCHITECTURE DATABASE 527, **528**
   British Architectural Library 528, 552, **553**
   and CPI 578
   publications 542
      RIBA product data 86
      RIBA product selector 86
   workload survey 577
   RIBA Bookshops 533
   RIBA Services 86, 582
   RIBA/TI **581–2**
   and Technical Information Microfile 579
Royal Institute of International Affairs (RIIA), Energy and Environmental Programme 252
Royal Institution of Chartered Surveyors (RICS) 253, 561, 575, 577
   BCIS 583
   Chartered Quantity Surveyor **579**
   and CPI 578, 579
Royal Institution of Naval Architects (RINA) **138**, **475**
   conferences 476
   Naval Architect 476
   Ship and Boat International 476
   Transactions 476
Royal School of Mines 595, 610
Royal Society, *Transactions*: *Biological Sciences* 641
Royal Society of Chemistry (RSC) 323
   and CEBA 112, **229**, 283
   journals
      Chemical Hazards in Industry 283, 327
      Faraday Transactions 282
      Journal of the Chemical Society 282
      Process and Chemical Engineering 228, 283
   publications 487
      Chemical safety data sheets 323

Royal Society for the Prevention of Accidents (RoSPA) **183**
   Bulletin 183
   Occupational Safety and Health 183
   RoSPA handbook of health and safety... 181
   RoSPA road safety engineering manual 500
Royal Statistical Society (RSS) **453**
   Journal... : *Technometrics* 460
Royal Town Planning Institute, *and* transport 513
RSC *see* Royal Society of Chemistry
RS Catalogues [co.] 340
RS Components Ltd 87
RSI (repetitive strain injury) 335
RSS *see* Royal Statistical Soc.
RSWBdatabase 584
RTCA/RTCM *see* Radio Technical Commisssion...
RTI (road transport industry), *RTI Focus News* 498
RTZ (Rio Tinto Zinc) Corp'n Ltd 595
rubber
   databases **112**, 355
   handbooks 307
   journals 68
   organizations 112, 303
   Standards 25
      Caoutchouc 24
   trade names 88
Rubber and Plastics Research Association of Great Britain *see* RAPRA
Rubber World 68
RuleCAD [program] 167, 172*tab*
Rules for building and classing of steel ships (pub. Amer. Bureau of Shipping) 477
Rules for classification:... light craft (pub. Norske Veritas) 477
Rumpf, H. [trans. Bull, F.A.] 278
Russell, D. 336
Russia
   abstracting/indexing 101
   All-Union Inst. of Scientific and Technical Information (VINITI) 101
   manufacture 357
   marine technology 478
   nuclear energy 261–2, 265
      Gosatomnadzor [national regulatory body] **264**
      Min. of Atomic Energy (MINATOM) **263**
   see also Russian language; Soviet
Russian Castings Technology 357
Russian language
   dictionaries **123**, 154, 267, 454
   journals, abstracting 101
   patents 47
Rutley's elements of mineralogy [ed. Gribble] 591, 598
Ryan, D.L. 335

S
SAA (Standards Ass'n of Australia) 524
Sacarello, H.L.A. 324
Sachs, M. [*associations*] 125
Sacks, E. *and* Joskowicz, L. [*modelling*] 170–1*tab*
SACTRA (Standing Advisory Committee on Trunk Road Assessment), reports 493, **499–500**, 508
Saddler, J.N. 249
SAE
   SAE handbook **26**, 208
   SAE Technical Literature Abstracts 504
   see otherwise Soc. of Automotive Engineers
Saenz, R.L. *and* Stephens, R.W.B. 492
Safarian, S.S. *and* Harris, E.C. 567
SAFE *see* SOFTWARE ABSTRACTS ...)
Safe automation of chemical processes (pub. Center for Chemical Process Safety) 279
Safe manriding in mines (pub. HSE) 615
safety 180–3, **315–28**
   primary considerations 315–16
   terminology **316**, 318, 320
   audits of 325
   books
      dictionaries 316, 334
      directories 323
      handbooks **181**, **315–16**, 317–21
   case studies **317–18**
   chemicals 270, 317–21, **322–4**, 326
   civil engineering 315
   conferences 181, 317, 318, 319, 322, 324, 326
   databases 328
   emergencies/accidents 316, 317–18, **323–4**, 325–6, 509–10
   equipment **182–3**, **186–7**, **326–7**
   explosion 321–2
   failure analysis 318, 327
   fire 320, **321–2**
   fluid mechanics 288
   handling operations 325
   hazardous substances **322–5**
   human factors **316**, 318, 319, **326**, 336
   'industrial' 316
   and insurance 180, 181, 316, 321
   'intrinsic' 320, 321
   journals 180, 181, **183**, 267, 281–2, 283, **327–8**
   legislation **180–1**, **182**, 315
   machines 178, **180–3**, 315, 317–21, 368
   marine 469*tab*
   nuclear industry *see under* nuclear engineering

safety (contd)
  occupational health 316
  offshore engineering 475, 477, 567
  organizations 180, 181, **182–3**, 262, **264**, 279, 315–26, **327**, 371, 513, 516, 517
  and reliability 177–83, **318–19**, 325
  risk assessment 181, **182**, 315, **316–17**, 325
  software 319, 324, **328**
  Standards **180–1**, **182**, 315, 319, 321, 325, 368, 477
    codes of practice **180–1**, 315
    company-specific 28
    Directives of Eur. Commission 19–20, **180–1**
    legislation 315, 320
    origins 20–1
    Regulations **180–1**
  structures 315, 319, **326–7**, 547–8, 549
  workplace 183
  see also Health and Safety; quality *and specific topics*
Safety considerations for . . . demountable grandstands (pub. IStructE) 568
Safety Equipment Institute [US] 327
Safety and Loss Prevention Subject Group [*of* IChemE] 327
Safety Management 328
Safety . . . of demountable grandstands (pub. IStructE) 568
Safety-related computers . . . (pub. Verlag TuV Rheinland) 319
Safety-related instrument systems . . . (pub. EEMUA) 319
Safety-related systems . . . (pub. IEE) 319
Safety and Reliability Society 180, 183
Safety and Reliability 180
Safety at Sea International 480
Saha, S. 636
SAIMM **594**, 610
Saito, T. *and* Igarashi, Y. 254
sales/marketing
  and concurrent engineering 649, 650
  construction industry 520, 574, 575
  databases and 91, 93
    CAMPUS project 304–5, 310, 313
  design and 304–5, 331*fig*, 332, 335–6
  and disks 304–5, 306
  expert systems 312
  global markets 353
  handbooks and 308
  materials 304–5

minerals 592
PEST analysis 574
Price Adjustment Formula 578
reports 4, 422
representatives 89
trade journals 65
see also product information; trade literature; manufacturing
salt
  mining 615
  salt spray testing 16
Salter, R.J. 545
Salvendy, G. **369**, 393
sand 590, 600, 615
  and fuels 239–40, 241
  organizations 608
Sanders, D.A. [*mechatronics*] 172
Sanders, M.S. *and* McCormick, E.J. [*ergonomics*] 336
Sanders, R.E. [*safety*] 318
Sand and Gravel Association 608
Sandler, B.Z. 384
Sandor, G.N. *and* Erdman, A.G. 167, 170*tab*
Sandvik Coromant *and* Atlas Copco, *Rock drilling manual* 614
sanitary engineering
  journals 68
  see also water
Santa Barbara [US], International Academy, *Alternative Energy Digests* 242
Sargent, P.M. 304, 308, 311
Sarkar, A.D. 173
Sassaman, J.F. *et al.* 248
Sastry, K.V.S. [*flotation*] 600, 601
Sastry, S. *and* Bodson, M. [*control systems*] 396
satellites, communication 412, 443–4, **447**
Sauer, H.J. *and* Howell, R.H. 208, 254
Saunders, E.A.D. [*heat exchangers*] 277
Savory, T.H. 625
Savory Milne [co.] 576
Sawyer, C.N. *et al.* [*environment/chemistry*] 486
Sawyer, P. [*controls*] 280
Sax, N.I. *and* Lewis, R.J. Sr 323
Saxon, R. [*structures*] 564
Saxton, I. [*mine ventilation*] 617
Say, M.G. 434
Say, M.G. *and* Taylor, E.O. 434
Sayer, F.P. *and* Bones, J.A. 158–9
Sayigh, A.A.M. 242
Sayigh, A.A.M. *and* McVeigh, J.C. 244
SCAE *see* Soc. for Computer-Aided Engineering
scaffolding 544
SCAN-A-BID database **584**
Scandinavia
  fluid power 296

medicine/biology 638, 640, 645
patents 46
Standards 20
see also Sweden *etc.*
SCC (Standards Council of Canada) 524
SCE *see* Soc. of Concurrent Engineering
Schaffer, R.J. [*masonry*] 541
Scheaffer R.L. *and* McClave, J.T. [*probability/statistics*] 458
Scheer, A.W. 362
Schey, J.A. 360
Schiffbautechnische Gesellschaft [Germany] 475
Schiffstechnik 481
Schiller, S. *et al.* 364
Schilling, R.J. *and* Lee, H. 456
Schipper, L. *et al.* 256
Schmid, J. *and* Klein, H.P. [*wind energy*] 247
Schmidt, F.W. *et al.* [*thermodynamics*] 202
Schmidt, F.W. *and* Willmott, J.A. [*thermodynamics*] 220
Schneider, U. *and* Nagele, E. 556
Schofield, C. *and* Abbott, J.A. 322
Scholium International [co.] 272
Schueman, D. 254
Schugerl, K. 279
Schuler, M.L. *and* Kargi, F. 279
Schultz, K.P. [*Standards*] 16
Schultz, N. [*fire hazard*] 322
Schwartz, M. [*communications*] 443
Schwartz, M.M. [*machining*] 363
Schwarz, H.R. [*finite elements*] 455
Schweitzer, G. [*bearings*] 177
Schweitzer, P. [*chemical engineering*] 274
Schweizer Maschinenmarkt 67
SCI [1] *see Science Citation Index*
SCI [2] *see* Steel Construction Inst.
SCI [3] *see* Soc. of Chemical Industry
Science Abstracts 228, 355, **380**, **421**
Science China/Scientia Sinica 404
Science Citation Index (*SCI*) **102**, **227**
  online 102, **109**, 110, **230**, **380**, **505**
Science and Engineering Research Council (SERC) 11
  Lickley Report 329
  and offshore engineering 468
  and theses 78
Science Museum [London], trade literature archives 94
Science Research Council *see* Science and Engineering Research Cncl
Scientia Sinica/Science China 404

Scientific abstracting and
  indexing ... in the British
  Library [ed. Burton; pub.
  BL) 227
Scientific Computers Ltd 295
Scientific and Technical
  Aerospace Reports (STAR;
  pub. NASA) 7, 11, 112
SCISEARCH database 230, 380,
  400, 401–2
Scobie, A.G. and Wyslousil,
  D.M. 592
Scope (pub. IPSM) 630, 643
Scotland
  Building Regulations 549
  Highlands and Islands
    Development Board, Oil
    map 481
  marine technology 475
  Royal Commission on the
    Ancient ... Monuments
    552
Scott, C.D. [fuels/biotechnology]
  249
Scott, D. and Crawley, F.
  [design/safety] 326
Scott, J.S. [civil engineering]
  123, 533
Scott, W. [communication] 337
Scott Fogler, H. 275
Scottish Nuclear Ltd 262
Scragg, A.H. 279
screens/screening, minerals 598,
  600
screws
  historical aspects 154
  Standards 16, 20, 368
SDI see selective dissemination
  of information
sea see maritime; ocean
Seaborne trade statistics of the
  United Kingdom (pub.
  HMSO) 509
seals/sealants 161tab, 177, 341,
  542–3
  and friction 290
Searcher 61
Sears, F.W. and Salinger, G.L.
  202
Seaton, W.H. et al. 324
Seaways 476
Sebestyén, G. and Pollington,
  C.E. 526, 587
Seborg, D.E. et al. 280, 395
seconds 28
security aspects
  data 362
  patents, prior use 35
  reports 9, 13
  Standards 26
SEED (Sharing Experience in
  Engineering Design) 160,
  341, 342
  Aesthetics in engineering
    design ... 336
  Communication in design ...
    338
  conferences 342
  Costing in design ... 337
  Curriculum for design ... 332,
    333

Design Guides 160, 161tab,
  163, 341
design model 330, 331fig, 333
Detail design phase ... 333,
  337
Ergonomics in engineering
  design ... 336
Information retrieval ... 339
Manufacturing phase ... 333
Quality and reliability ... 333,
  334
Specification phase ... 333
See Hear 643
Seggern, David von 454
seismic see earthquakes
Select-A-Nalysis [program] 176
Select Committee on Science and
  Technology [UK] 210, 508
Exploitation of tidal power in
  the Severn Estuary 248
Select Committee on Transport
  500, 508
Selected Water Resources
  Abstracts 113
Selective bibliography on ...
  hoppers and silos (pub.
  CIRIA) 567
selective dissemination of
  information (SDI) 229
  BL Subject Search 142
  INSPEC 421–2
  patents 59, 60
  synopsis journals 71
Sell's building and construction
  index, former 543
Semiconductor Index of
  Radiation Effects/CODUS SIRE
  91
semiconductors 419, 437
  abstracts 421–2
  books 419, 423
  CMOS (complementary metal
    oxide) 424
  conferences 417
  databases 91
  handbooks 419
  journals 87, 459
  lasers 412, 459
  properties 419
  and radiation 91
  transistors 411
Semicon index (pub. Internat.
  Semiconductor Data
  Summaries) 419
Sen, P.C. 434
Sengupta, M. [mine environment]
  617
Sengupta, S. and Lee, S.S. [heat
  recovery] 206
Senior, J.M. 447
sensitivity analysis 333, 334,
  345, 347
Sensor Review 402
sensors 377–8, 384
  journals 402
  sensor-based control 379
  see also vision; robotics
Sensors and Actuators 402
Sensor Technology 402
separation 274, 282
  biochemical 278

Separations Technology 282
SERC see Science and
  Engineering Research Cncl
SERI
  see otherwise Solar Energy
    Research Inst.
  SERI Materials Branch
    Semiannual Report 245
  SERI Journal 245
serials see journals
Serials directory 74
  online 400
series impedance 293
servers [of networks] 117–18
Service, L.M. et al. [design/
  profit] 337
Service, T.H. [reliability] 319
Service Station 68
servo motors 293, 436–7
servovalves 286, 290, 292–3, 297
Setian, L. 456
Severn, river, tidal project 248
sewerage 488, 489, 492
Seymour, R.J. 248
SFEN see Société Française
  d'Énergie ...
SFS database 584
Shabana, A.A. 159
Shackshaft, G. 433
Shackshaft, G. and Neilson, R.
  433
Shaft engineering (pub. IMM)
  615
shafts, mechanical 195, 341
  connections 163–4, 176, 341
shafts, mining 615, 616
Shah, R.K. et al. 206
Shaheen, E.I. 272
Shahinpoor, M. 362
shales 239–40, 241
Shamlou, P.A. 278
Sharing Experience in
  Engineering Design see
  SEED
Sharland, E. 492
Sharon, A. 655
Sharp, M.L. [aluminium] 562
Sharp, J.J. [fluid mechanics] 205
Sharpe, C., Kempe's engineers
  yearbook 124, 160
Sharples Photomechanics Ltd 145
Shaw, R. 248
Shearer, J.L. et al. 391
shear stress 133
Sheehy, E.P. 121
Sheehy, E.P. and Balay, R. 121
Sheffield, University 143
Shell, R.L. and Simmons, R.J.
  326
Shell [co.] 280
shells, structural 137, 550
Sherman, K. 446
Sherratt, A.F.C. 206, 208
Shigley, J.E. 332
Shigley, J.E. and Mischke, C.R.
  159, 161tab, 163, 336
Shigley, J.E. and Uicker, J.J. 160
Shimizu, S. and Nakosumi, M.
  298
Shina, S.G. 653
Shinskey, F.G. 398

Ship and Boat International 476
Ship design bulletins (pub. SNAME) 477
Shipping statistics and economics (pub. Drewry) 479
ships/shipping 467–83, 470tab
defined 467
conferences 474
directories 474–5
DOT administration 511
stability 474
see otherwise marine technology
shock, handbooks 125
Shock, R.A.W. 245
Shockley, William 411
Shortlist 172tab
Shtipelman, B.A. 165
shunt admittance 293
SI (Système International) units 16, 28
Système International d'Unités (pub. BIPM) 28
SIAM (Soc. for Industrial and Applied Mathematics) 453
SIAM Journal...
on Applied Mathematics 459
... on Control and Optimization 401, 405, 459
... on Discrete Mathematics 460
... on Mathematical Analysis 459–60
... on Matrix Analysis and Applications 458
... on Numerical Analysis 459
... on Scientific and Statistical Computing 459
SIAM Review 458
SIC (standard industrial classification) 85, 92
Siddall, J.N. 332, 333, 334
Siemens [co.], Siemens Review 417
SIG-GRAPH [IT] conferences 656
SIGLE [System for Information on Grey Literature in Europe] 10–11
classification scheme 9, 10
SIGLE database 10–11, 78, 231
signal processing 286–7, 286fig, 393, 398, 441–50
analogue/digital signals 412, 423, 441–2
applications information 419
books 398, 443–4, 457–8
coding 445, 457
common-channel 444
conditioners 146
conferences 417
equipment 146
and hydraulics 286
journals 401, 416, 459
electronic newsletter 115
modulation 443
PCM 442
noise see 'noise'

random signals 392
and safety 320
stress analysis 136
telecommunications 441–50
VLSI 415, 417
silicon
and electronics 411–12
silicates processing 600
and solar energy 244
silos 278, 567
silver 593
cyanidation 590, 599
early uses 154, 589
SilverPlatter Information Inc. 110, 111, 113, 267, 505, 530
e-mail contact 511
networked 108
Simmons, C.H. and Maguire, D.E. [engineering drawing] 337
Simon Books, Who's who of British engineers 126
Simon, H.A. [design] 331, 335
simulation 273, 294, 295, 335, 362, 392
bondgraphs 287
books 454
conferences 403
control systems 280, 288, 291, 294–6, 379, 390, 392, 403
journals 379, 403
organizations 403
robotics 378
Simulation 403
Simulation Symposium 403
simultaneous engineering/design 650
Simultaneous engineering in automotive development (pub. SAE) 653
Singapore
automation/robotics 382
Inst. of Engineers 382
manufacture 359
National University 359
Singer, C. et al. [technology history] 154
Singer, J.G. [combustion] 237
Singer, S. [coal/combustion] 238
Singh, M.G. 398–9
Sinha, N.K. 393
Sinnott, R.K. 272
sintering 365
SIRA Institute 414–15
SIS see Standardiseringkommission i Sverige
Sitkei, G. 237
Sizewell-B public inquiry [ed. Layfield] (pub. HMSO) 266
sizing, of minerals 598
Skaug, R. and Hjelmstad, J.F. 443
SKF [co.] 176
Ball Bearing Journal 176
Skinner, D.G. 237, 238
Skolnik, M.I. 125
Skrotzki, B.G.A. 202
slate 541
Slemon, G.R. 434

Slesser, M., Macmillan dictionary of energy 233
Sloan, D.A. 614
Slotine, J.E. and Li, W. 394
sludge, sewage 488, 492
SME [1] see Soc. of Manufacturing Engineers
SME [2] (SME-AIME) see Soc. of Mining, Metallurgy...
smelting, early 589
Smethurst, G. 489
Smil, V. 256
Smith, C.A. and Corripio, A.B. [controls] 280
Smith, D.A. [environment/law] 493
Smith, D.J. [reliability] 178, 179, 319
Smith, E.H. [mechanical engineering] 124, 160, 340
Smith, G.M. and Jolly, C.B. [stress analysis] 148
Smith, G.N. [probability/statistics] 458
Smith, H.B. [energy] 233
Smith, J.D. [vibration] 165
Smith, J.F. [building conservation] 552
Smith, J.M. [chemical engineering] 276
Smith, J.M. [CALS] 314
Smith, J.M. and Van Ness, H.C. [chemical engineering] 276
Smith, L.H. et al. [hydraulics] 289
Smith, R. [process integration] 274
Smith, R.B. [fire/timber floors] 556
Smith, R.C. and Lawrence, L. [fire/steel structures] 540, 555
Smith, R.V. and Leslie, J.H., Jr [biomedicine] 648
Smithells metals reference book [ed. Brandes/Brook] 125, 302
Smoot, L.D. and Smith, P.J. 238
SNAJ see Soc. of Naval Architects of Japan
SNAME see Soc. of Naval Architects and Marine Engineers
SNE (Soc. of Naval Engineers) 475
Sneddon, I.N. 453
Snow, D.A. 256
Snyder, A.W. and Love, J.D. 448
social information/aspects 229, 574
automation 379
biomedical engineering 624
databases 111, 505, 511
energy planning 255, 256
ergonomics 335
family expenditure 511
organizations 515
PEST analysis 574
and transport 495–6, 505, 509–11
SOCIAL SCIENCES CITATION INDEX database 505

Social trends (pub. CSO) 574
Societät der Bergbaukunde/
  Society of Mining
  Professors 610
Société Française d'Énergie
  Nucléaire (SFEN) 265
societies see organizations
Society of Automotive Engineers
  (SAE [US]) 211, 517
  Aerospace product/process
    design... 653
  Gear... manual 164
  SAE Technical Literature
    Abstracts 504
  Simultaneous engineering...
    653
  Spring... manual 164, 168–9
  Standards 16, 17, 26
  SAE handbook 26, 208
  Universal joint and drive shaft
    manual 195
Society of Chemical Engineering
  Japan, Journal of Chemical
  Engineering of Japan 282
Society of Chemical Industry
  (SCI) 601
  Chemistry and Industry 281
  Journal of Chemical
    Technology and
    Biotechnology 281
Society for Computer-Aided
  Engineering (SCAE [US])
  651, 655
Society for Computer Simulation,
  Simulation 403
Society of Concurrent
  Engineering (SCE [US])
  651, 654
Society for Experimental
  Mechanics (SEM [US]) 133,
  140–1, 142, 144
  Experimental Mechanics 141,
    142
  Experimental Techniques 141,
    142
Society for Industrial and
  Applied Mathematics see
  SIAM
Society of Instrument and
  Control Engineers/Keisoku
  to Seigyo, Journal 402
Society of Manufacturing
  Engineers (SME [US]) 381,
  387
  Computer and Automated
    Systems Association
    (CASA) 378, 381, 382,
    651
  and concurrent engineering
    651
  conferences 359, 381–2
  North American Manufacturing
    Research Institution 359
  publications 370, 373
  Autofact '92 382
  Manufacturing Engineering
    350, 356, 654, 655
  Tool and manufacturing
    engineers handbook
    [ed. Bakerjian] 124,
    368–9

Robotics International 381,
  387
Robotics Today 378
Society of Mechanical Engineers
  see Amer. Soc. of
  Mechanical Engineers
Society of Mining, Metallurgy
  and Exploration
  (SME(AIME) [US]) 594,
  601, 610
  Minerals and Metallurgical
    Processing 596
  Mining Engineering 596, 610
  publications 597, 598, 599,
    613
Society of Mining Professors/
  Societät der Bergbaukunde
  610
Society of Naval Architects of
  Japan (SNAJ) 475
  Journal of SNAJ 476
Society of Naval Architects of
  Korea 475
Society of Naval Architects and
  Marine Engineers (SNAME
  [US]) 475
  Journal of Ship Production
    476
  Journal of Ship Research 476
  Marine Technology 476
  Notes on ship slamming 477
  publications 476–7
  Ship design bulletins 477
  STAR symposia 476
  Transactions 476
Society of Naval Engineers (SNE
  [US]) 475
Society of Petroleum Engineers
  [of AIME] 240
Society of Photo-Optical
  Engineers Proceedings 65
Society of Photo-Optical
  Instrumentation Engineers
  373
Society of Powder Technology
  [Japan] 359
Society for the Protection of
  Ancient Buildings (SPAB)
  523, 534, 551, 560, 561
Society of Tribologists and
  Lubrication Engineers [US]
  173
  Tribology Transactions 173–4,
    175
Society for Underwater
  Technology (SUT) 475
  Underwater Technology 476
sockets, electrical 16
Soderstrom, T. and Stoica, P.
  391
Sodha, M.S. et al. 242, 244
sodium (Na), liquid, as reactor
  coolant 260, 261
software 36, 183–4, 281, 432–3,
  452
  for CD-ROM 426, 427
  abstracts 113, 528
  ASCII format 168
  bearings 161tab, 173, 175–6
  book-accompanying 151–2,
    183–4, 455

CAD/CAM 158, 159, 160–1,
  161tab, 170–2tab, 281
  historical aspects 155
  chemical engineering 273, 281
  COBA 508
  construction 581–2
  control systems 398
  costing 582, 583
  databases on disk 426, 581–2
  databases of 113, 584
  data transfer 310–12
  defence 310–11
  directories of 587
  discussed 151–2, 183–4
  DXF format 168
  expert systems 312, 396
  fluid dynamics 294–6
  as information source 926
  Internet navigation 117–18
  journals 105, 169, 460
  for machines/mechanisms 151,
    160–1, 161tab, 163, 165,
    173, 175–6, 183–4, 432–3
  reliability 179
  materials 304–5
  mathematical 160, 184, 452,
    456, 460
  algebra 451, 452, 455
  for calculators 452
  finite elements 137, 139,
    148–9
  optimization 137, 147
  organizations 413
  patentability of 36
  on pocket calculators 452
  post-processing 137
  process control 366
  reliability 179
  robotics 379, 385
  safety-related 319, 324, 328
  Standards 139
  stress analysis 137, 147, 306–7
  switching systems 443
  thermal systems 203, 205,
    432–3
  transport 503, 508
  see also computing and under
    specific topics
Software Abstracts for Engineers
  528
SOFTWARE ABSTRACTS FOR
  ENGINEERS (SAFE) database
  113
software engineering, Internet
  resources 119
Soft Warehouse Inc. 452
Software packages for
  mechanism analysis and
  synthesis (pub. Spring
  Research) 169
Software in safety-related systems
  (pub. IEE/BCS) 319
Sohnel, O. and Garside, J. 274
soil see earth; ground
solar energy 210, 242–5
  conferences 243
  economic aspects 205, 242–3
  heat pumps 208
  organizations 225
Solar Energy 245
Solar Energy Research Institute
  (SERI [US]) 254

Solar Energy Research Institute
(SERI [US]) (contd)
Materials Branch, Semiannual
Report 245
SERI Journal 245
Solar Progress 245
Solar radiation data for the UK
(pub. Meteorological Office)
243
Solar Thermal Energy
Technology 245
soldering 361–2
solidification dynamics 364
solid mechanics
bulk solids **278**
organizations 138
and stress analysis 133, 138
see also powder etc.
solid state devices 417
see also circuits
solid state physics 100
Solid State and Superconductivity
Abstracts 100
Solvay [org'n] 305
solvents 323
solvent extraction, mineral
processing 599, 601
Solymar, L. 457
Somasundaran, P. 601
Somasundaran, P. and Moudgil,
B.M. 600
Someya, T. 175
Soni, A.H. et al. 167
Sonnenenergie und Warmepumpe
255
Sonnenenergie und Warmetechnik
245
Sonntag, R.E. and Van Wylen,
G. 202
Soong, T.T. 458
Sorensen, B. [renewable energy]
210
Sorenson, H.W. [control systems]
392
sound
and telephony 441
see otherwise noise
Sound Research Laboratories 493
Noise control in industry 181,
492
Sources of technical information
for ... masonry structures
(pub. Brit. Masonry Soc.)
540
South Africa
minerals/mining 593–4, 595,
596, 608–9, 610, 611, 612
Standards 27
South Africa Bureau of
Standards 27
Bulletin 27
South African Institution of
Mining and Metallurgy
(SAIMM) **594**, 610
Journal (JSAIMM) **596**
South African Mining, Coal,
Gold and Base Minerals
611
South Africa's mineral industry
(pub. S.African Minerals
Bureau) 612

South America
energy 232
minerals 596
patents 46
Standards 20
South Korea
marine technology 475
nuclear energy 261
Soviet Union, former/CIS
energy 233
manufacture 357
nuclear energy 261–2
Chernobyl accident 266
Ex-USSR Nuclear Society
265
reports 11, 228
see also Russia etc.
Sowden, A.M. 541, 555, 560
SPAB see Soc. for the Protection
of Ancient Buildings
Spacecraft thermal ... systems
(pub. ESA) 204
space engineering see aerospace
Spain
machines 161
minerals 596
nuclear energy 261
Standards 27
see also Spanish language
Spalding, D.B. and Cole, E.H.
202
Spanish language
dictionaries 154, 271
Standards 20
SPATE 147
SPC [1] (statistical process
control) 358, **366–7**, 369
SPC [2] (stored-program control)
444
Specification (pub. EMAP) 527
Specification phase ... (pub.
SEED) 333
specifications 15, 25, **83**, 155–7,
178, **302**, **578–9**
aerospace 23
components 163–4
construction 89, 578
Project Specification Code
578–9
databases 91, **230**, 368
data transfer 310
defence 23, 26, 91
design **155–7**, 331fig, 333
as grey literature 10
handbooks 124–5
manufacturing 124
materials 302
metals 125
National Building Specification
578
National Standard 578
PDS 333
and product information 82,
83, 85, 91–2
projects 578–9
robotics 384
standard elements 579
Standard for 178
see also Standards
speckle patterns, optical **135–6**,
146–7

spectra, radio 443, 444
Spectrum **416**, 427
speech, Internet 503
Speight, J.G. 274
Spiegler, K.S. and Laird, A.D.K.
244
Spindler & Hoyer UK Ltd 147
Spine 638
Spitzer, D.W. 165
spline functions 167–8
Spon, E. & F.N. Ltd [publishers]
527, 531
Spon's plant and equipment
guide 543
Spon's quarry guide ... 608
Spong, M.W. and Vidyasagar,
M. 394
Sporn, P. et al. 208
sports grounds 552
sports science 624, 640, 646
Sprackling, M.T. 205
Spring Design Manual (pub.
SAE) 164, 168–9
Spring Research and
Manufacturers Association,
Software packages for
mechanism analysis and
synthesis 169
spring rollers 293
springs 161tab, **168–9**, 341
Standards 169
SRIS see Brit. Library: Science
Reference and
Information ...
Stability of buildings (pub.
IStructE) 550
Stacey, W.M. 251
Stack, B. 614
Stagg, G.W. and El-Abiad, A.H.
432–3
Stahlbau 533
Stahlschlüssel [ed. Wegst] 302–3
Stallings, W. 445, 446
Stambuleanu, A. 237
Stanat, D.F. and McAllister, D.F.
457
Standard for the Exchange of
Product Data see STEP
Standard and guide ... for
curtain walling (pub.
CWCT) 564
standard industrial classification
(SIC) 85, 92
Standardiseringkommission i
Sverige (SIS) 27
Maanadens standard 27
STANDARDLINE database 530
Standard method of detailing ...
concrete (pub. IStructE/CS)
557
Standard performance testing of
wind energy systems (pub.
AWEA) 246
Standard periodical directory
(pub. Oxbridge) 72–3
Standards **15–29**, 89, 152, 472
benchmarks 139, 433
classification of 15, **18**, 19
Codes of Practice see Codes
of Practice
currency 18, 535

Index 757

Standards (contd)
  databases 22, 24, **113**, 230, 368
  de facto 29, 450
  Directives [of EC] **19–20**, **180–1, 183**
    see also under Europe
  European see regional and international below
  Guidelines 16
  historical aspects 17, **20–1**
  industry-led 16, 19–20
    company-devised **28–9**
    *EURONORM*s 20
    Manchester Steam Users' Ass'n 21
    US 25–6
  information centres 18, 24
  international **16–19**, 21–2, 450, **524**, 529
    equivalence 26
    harmonization 16, 17, **19–20**, 26, 324, 508
    ISONET 18
    NORMIMAGE 22
    PERINORM **22**, 152, **582**
    ITU-T 449
  journals 18, 19, 25, 28
  Laboratories **27–8**
  listed **22, 152**
    AFNOR catalogue...s 22, **23**
    Annual book of ASTM standards 25, 302, 314, 340
    BRITISH STANDARDS 90
    'BSI bestsellers' 22
    BSI standards catalogue **21**, 22, **152**, 154, 340, 367, 527
    Catalogue of American national standards (pub. ANSI) 25
    Catalogue of Canadian Standards Ass'n 27
    Catalogue of English translations (pub. AFNOR) 23
    Catalogue des ouvrages (pub. AFNOR) 22
    Catalogue des recueils (pub. AFNOR) 24
    DIN Katalog für technische Regeln 24
    DITR DATENBANK database 24, 583, 584
    EURONORM 20
    IEC catalogue 19
    IHS INTERNATIONAL STANDARDS AND SPECIFICATIONS **230**
    International standards index (pub. Woodhead) 26
    ISO catalogue 18
    JIS yearbook 27
    KWIC index of international standards 18
    non-print media 22, **113**
    Normes françaises (pub. AFNOR) 23

  NORMIMAGE 22
  PERINORM **22**, 152, **582**
  STANDARDLINE database 530
  Standards India 27
  STANDARDS INFODISK **113**
  STANDARDS AND SPECIFICATIONS DATABASE **113**, **230**, **368**
  *Technical guide*s [IEC/ISO] 19
  WORLDWIDE STANDARDS SERVICE 529
  national **20–7**, 302
    incorporation of international standards 19
    product information from 89
  regional 16, **19–20**
    CE Marks 19–20
    Common standards for enterprises (pub. EU) 20
    ENVs (Eur. pre-Standards) **549**
    ETSI 449
    Eurocodes **549**
    Standards for access to the EC markets (pub. CEN/CENELEC) 20
  Regulations see Regulations
  SI units 16, **28**
  specifications see specifications
    specificity 302
  standards engineers 28–9
    user-dedication 302
    see also British Standards Inst'n etc. and specific topics
  Standards for access to the EC markets (pub. CEN/CENELEC) 20
  Standards activities organizations in the United States (pub. NIST) 25
  Standard for safety for power supplies (pub. UL) 321
  Standards Association of Australia (SAA) 524
  Standards Council of Canada (SCC) 524
  standards engineers 28–9
  Standards India (pub. Bureau of Indian Standards) 27
  STANDARDS INFODISK database **113**
  Standards ... for instrumentation and control (pub. ISA) 279
  STANDARDS AND SPECIFICATIONS DATABASE **113**, **230**, **368**
  Standards and specifications information bulletin (pub. NSA) 25
  Standby vessels of the world 475
  Standing Advisory Committee on Trunk Road Assessment see SACTRA
  Standing Committee on Structural Safety (of IStructE), *Report* 550
  Stanek, V. 276, 278
  Stanford, H.W. 207

Stanford University [US] 113, 119, 656
STAR reports see *Scientific and Technical Aerospace* ...
STAR symposia [marine technology] 476
Stark, J. 652
Starkey, C.V. [*design*] 332
Starkie, D. [*transport*] 507
Starling, K.E. 273
Starr, M.R. *and* Palz, W. 244
State of the Art Reviews (pub. TRL) **499**, 545, 566
state space models 390, 391
Statgraphics [program] 452
Statistical dictionary (pub. Hungarian Central Statistical Office) 454
Statistical Graphics Corp'n 452
statistical process control (SPC) 358, **366–7**, 369
Statistical report on road accidents (pub. ECMT) 510
Statistical Science: a Review Journal 460
Statistical trends in transport (pub. ECMT/OECD) 510
statistics **126**, **451–66**, 510
  nature and role 451–2, 460
  abstracts/indexes of **113**, **460–1**
  books of/on 124–5, 126, 452, **457–8**
  dictionaries/encyclopaedias **453**, **454**
  directories 510
  handbooks 454
  commodities 75
  construction **572–4**
  construction management 569
  employment **574**
  energy-related **231–5**, 252
    coal 237
    oil 239
  hypothesis tests 451
  industry 75
  journals of/on 65, 73, 459, **460**
    Current Index ... 113, **460–1**
  manufacturing 371
  newspapers 73
  on organizations 113, **453**
  organizations for user groups 510
  probability **451**, **454**, **457–8**, **460**
  robotics 382
  software for **452**
  statistical process control (SPC) 358, **366–7**, 369
  tables 154–5, **452**, **454**
  trade 75
  transport 497, **509–11**
  users of, groups 510
  see also mathematics
Statistics of road traffic accidents in Europe (pub. ECE) 510
Statistics Users Council, *and* TSUG 510
status reports 4
Statute of Monopolies (1623) **32**, 36

758 Index

Statutory instruments: (dangerous goods)... (pub. HMSO) 477
steam 202, **235–7**
  boilers 21, 206
  nuclear reactors 259, 260
  and coal 590
  and electricity generation 428, 429
  engines 200
  journals 237
  organizations 21
  properties 235–7, 253–5
  Standards 206
  thermodynamic properties 202–3
  turbines *see* turbines
Steam tables... (pub. Electrical Research Ass'n) 235
steel 64, **538–40**, 552, **557–9**, 590, 593
  books, handbooks 302–3, 362, 534, 538, **539**, 558–9
  bridges 566–7
  cladding 564
  duplex 362
  fire hazard 555
  fixings 565
  historical aspects 538
  journals 64, 68, 532, 533
  loadings 552
  offshore structures 567
  organizations 20, **538–9**, 557–8
  seismic hazard 554
  stainless 558
  Standards 25, 26, **302–3**, **539**, 549, **558**
  steel industry, hydraulics 286, 288
  wire 302
Steel Construction Institute (SCI) **539**, 558
  publications 538, 540, 555, 558, 559, 566
  catalogue 527
  New Steel Construction 532
  P112: Interim guidance... topside structures... 567
Steel construction yearbook (pub. Telford) 534
Steele, D.J. [*mining machinery*] 614
Steele, R. [*telecommunications*] 446, 447
Steel Times 68
Steelwork design guide... (pub. SCI) 558
Steemers, T.C. 244
Stefanko, R. 613, 616
STEP (Standard for the Exchange of Product Data [*ISO 10303*]) 307, **309–10**, 314
Stephanopoulos, G. 279, 280
Stephens, A. [*abstracts directory*] 227
Stephens, J.H. *and* Ryder, C. [*energy*] 227
steplessly variable drives **165**
stereolithography 653

Stern, A.C. 491
Sterritt, R.M. *and* Lester, J.N. 486
Stewart, D.R. *et al.* [*mining*] 615
Stewart, R.D. *et al.* [*costing*] 337
STIC database host 422
stiction 290
STIMLINE database 602
STN [INTERNATIONAL] database host **231**
  CEBA 112
  citations 109
  COMPENDEX 110
  conference *Proceedings* 77, 110
  electrical/electronics 111, 422
  energy **231**, 267
  grey literature 10
  ICONDA 113
  INSPEC 111, 422
  journals 75
  mechanical engineering 111
  METADEX 112
  NTIS 109
  patents 59
  RAPRA ABSTRACTS 112
  science 109
  theses 78
stochastic processes **451**, **457–8**, **460**
stochastic control theory 390, 391, 393, 395–6
Stochastics and Stochastic Reports 460
Stocks, J. *et al.* 613
Stoecker, W.F. 204
Stoecker, W.F. *and* Jones, J.W. 207
Stoecker, W.F. *and* Stoecker, P.A. 205
Stokes, A. 164
Stolarski, T.A. 173
Stoliarov, D.E. *and* Kuzmin, I.A. 123
Stoll, H.W. 337
Stollard, P. *and* Johnston, L. 555
stone
  in constructions *see* masonry
  gemstones *see* gemstones
  historical aspects 607
  as mineral source *see* rock
Stone Federation of Great Britain 540, 559, 560
Stonham, T.J. 423
Stoorvogel, A.A. 398
Stopford, M. 282
Stopher, P.R. *et al.* 507
storage
  automated 385
  reservoirs 430
  safety 324
Storch, R.L. *et al.* 483
stored-program control (SPC) 444
Storrar, C.D. 614
Stott, B. 433
strain **134**
  BSSM **137**, 138, 141, 142, **144**, 145
  handbooks 141–2, 341
  journals 138, 139, 142, 145, 338

strain gauges **135**, 136, 146
surfaces 135
see otherwise stress analysis
Strain 138, **142**, 145
Strainstall Engineering Services Ltd 146
Stramler, J.H. 123
Strandh, S. 154
Strang, J. *and* Mackenzie-Wood, P. 617
Stranks, J.W., *RoSPA handbook...* 181
'Strategic directions for control' *(pub.* IEE Colloquium *Digest)* 398
Strategic Planning for Energy and the Environment 257
Strathclyde, Building Centre 520
Straub, H. 535
Streetman, B.G. 423
Strehlow, R.A. 237
Stremler, F.G. 443
Strengthening of historic... structures (pub. ICOMOS) 552
strength of materials 133
  see otherwise load; stress analysis; strain
stress, human 335
stress analysis/experimental mechanics **133–49**
  nature and role **133–4**
  books 133, **145–9**, 176, 319
  handbooks **141–2**, 341
  reports 142
  theses 142
  'chunky bodies' 137
  conferences 138, **139–41**, **144–5**
  crack-tip stresses 136
  dimensional aspects 137
  direct stress 133
  equipment/suppliers 145–9
  journals 138, 139, 141, **142–4**, 338
  normal stress 133
  organizations **137–41**
  residual stress **134**, 135, 136, 148
  shear stress 133
  software 137, 147, **306–7**
  techniques **133–7**
    experimental 133–4, **135–6**
    numerical 133–4, **136–7**, 148–9, 616
  types of 133, 134
  see also materials
Stresscoat [co.] 148
Stress Photonics [co.] 147
Strigle, R.F., Jr 274
Strike, J. 535
Strong, S.R. *and* Fahrni, G.R., Jr 176
Structural design standards... (pub. CIRIA) 527
Structural Design of Tall Buildings 532
Structural effects of alkali-silica reaction... (pub. IStructE) 557
structural engineering **547–68**

Index   759

structural engineering (contd)
  nature and role 519, **547–8**, 563
  building types **565–8**
  air-supported structures 568
  analysis of structures **133–49**, 327, 550
  finite elements 137
  appraisal 522, 537, 550, **551–2**, 557
  architectural aspects see architecture
  books
    bibliographies 552, 567
    directories 552
    handbooks **550**, 554, 559–60, 561
  bridges **566–7**
  building envelope **563–4**
  conferences 206, 551
  conservation aspects **551–3**
  design 547–68
  'design-and-build' 548
  energy services see electricity etc.
  existing structures 549, **551–3**, 557, 559, 560, 561, 564, 566–7
  fire engineering 554–6
  fixings 565
  foundations 122, 543, **564–5**
  historical aspects 93, 522, 527, **534–5**, 538, 549, **551–3**
  integrity 137, 140
  journals **142–4**, **531–3**, 551, 580
  loadings 327, 551, 552, **553–4**, 566
  management 550
  marine 469tab, 474
  materials **555–63**
  mining 238, 616
  offshore structures 240, 545, **567**
  organizations 549–50, 551, 552, 553, 556, 558, 559, 560, 561, 562, 563, 564, 565, 566, 585
  authorities 547
  building services 206
  quality 550
  failure hazard 316
  renovation **551–3**
  restoration work, archival sources 93–5
  safety 315, 319, **326–7**, 547–8, 549
  silos/tanks etc. **567**
  spatial structures 532, 566
  Standards/guidelines **548–9**, 553–4, 556, 558, 559–60, 561, 562, 564, 566
    Approved Documents 549, 550, 555, 561
    Building Regulations **549**, 553
    Eurocodes **549**, 553–4, 556, 558, 560, 561, 562
  stress see stress
  thin-walled structures 143
  towers/masts **567**

uncertainty analysis 327
  see also construction; building
Structural Engineering International (pub. IABSE) 532
Structural Engineering Review 532
Structural Engineer (pub. IStructE) **522**, **530**, 531, 553, 557, **580**
Structural failures ... (pub. IStructE) 550
Structural Safety 532
Structural Steel Advisory Service [of BSC] **524–5**
Structural Survey 532, 551
Structural use of lightweight aggregate concrete (pub. IStructE) 557
Structure 533
Structures and Buildings (ICE Proc.) 531
Studies in Applied Mathematics 459
Studies in Automation and Control 401
Study on design of steel buildings in earthquake zones (pub. ECCS) 554
Stull, D.R. et al. 273
Stultz, S.C. and Kitto, J.B. 237
sub-contracting see contractors
Sube, R. 267
Subject category descriptions (pub. NTIS) 6
submarines
  conferences 476
  nuclear-powered 261
subsea see underwater
Successful health and safety management (pub. HMSO) 477
Suess, M.J. 487, 490
Suess, M.J. and Huismans, J.W. 490
Suh, N.P. 331
Sukhatme, S.P. 210
sulphides, metal 591
Sumner, J.A. 208
sun see solar
Sundararajan, C. 196
Sun Hydraulics Corp'n 287
Sunley, J. and Bedding, B. 542
Sunworld 245
superconductivity
  abstracts 100
  encyclopaedia 312
  high-temperature 420
'superhighway' [information networks] 502
'supersheds' 566
supertrams 512
surface mining 613, **614–15**, 616
Surface Mining Conference 613
surfaces **173–7**
  marine 469tab
  mathematics of 454, **455**
  strain 135
  treatments 28, 310, 361, 369
  conferences 174, 358, 359
  journals 68, 357, 358

tribology 173–5, **176–7**
  see also finishing
surgery, robotics 378
Survey of building regulations worldwide (pub. BYGGDOK) 527
surveying 547, 551–2, 561
  conferences 500, **501**
  and construction 579
  and energy efficiency 253
  journals 67, 490, 532, 551, 579
  organizations 514, **515**
  quantity surveying 579
  and transport 500, 501, 514, 515
  and waste management 490
Survey Methods in Transport conference 500, **501**
Surveyor 67, 490
Susman, G.I. 652
SUT see Soc. for Underwater Technology
Sutherland, K.L. and Wark, I.W. [flotation] 599
Sutherland, R.J.M. 559
Sutphin, S.E. 208
Suzuki, K. et al. 298
Svarovsky, L. 274, 598, 600
Svoboda, J. 599
Sweden
  construction 527, 584
  energy conservation 253, 255
  materials 303
  minerals 596
  nuclear energy 261
  patents 45
  Standards 20, **27**, 113, 303
  timber 560
  transport 517
Swedish Council for Building Research 253, 255
Swedish language
  databases 584
  dictionaries 154
Swindells, D.J. and Hutchings, M. 561
Swinnerton-Dyer, P. 114
switches/switchgear **435**, **436**
  drives 434, 435
  handbooks 431
  journals 460
  packet switching 444–5
  programming 443
  PWM 436
  software 443
  stator switching 437
Switzerland
  construction 522–3
  design 162
  machinery 67
  Standards 22
  timber 560
SWOV [Inst. for Road Safety Research (Netherlands)] 516
symbols
  dictionaries **124**, 287
  robot control 379
  Standards 28, 287
symposia see conferences
Symposium on Robot Control (SYROCO) 381

760 Index

synectics 346
synfuels *see* fuels: synthetic
synopsis journals 70, **71**
SYROCO (Symposium on Robot Control) 381
system analysis *see under* systems
Système international [d'unités] *see* SI units
Système International d'Unités [*International Units System*] (*pub.* BIPM) 28
System for Information on Grey Literature in Europe *see* SIGLE
systems **389–90**
  autonomous 379
  controls **389–409**
  dynamics 389, 391, 395
  expert *see* expert systems
  fuzzy **457**
  integrated 366
  journals 378–9, 399–403, 415–17, 460
  linearity 389, 391, **392–4, 456**
  non-linear systems 391, 393, 394, **457, 460**
  quality 368
  reliability 368
  robust **390**, 393, 394–5, 396, **397**
  Standards 368
  subsystems 385, 432
  system analysis 115, 273–4, 279, 331–2, **390, 432–3**
  power engineering 432–3
  system design *see under* design applications
  system transfers 390, 393
  uncertainty 389, 390, **395**, 396, 397
  see also controls; automation; CAD/CAM *etc.*
Systems and Control Letters 401
Sze, S.M. 423

T

Tabb, P. 244
tables, data **124–5, 452,** 454
  see also statistics
Taft, C.K. *and* Twill, J.P. 298
Taggart, A.F. 590, 597
tagging, radio-control 385
Taguchi, G., design system 177, **179–80, 334,** 367
tailings 592
Taiwan, catalogues 90
Takahashi, K. *et al.* 298
Takahashi, K. *and* Konagai, M. 244
Takamura, S. *et al.* 443
Talavage, J. *and* Hannam, R.G. 363
Tallian, T.E. 165, 168, **177**
TAM *see* Traffic appraisal manual
Tamrock [co.], *Tamrock handbooks*
  ... *on surface drilling and blasting* 614, 615
  ... *of underground drilling* 614, 615

Tanaka, N. *et al.* 458
Taniguchi, N. *et al.* 363–4
tanks [containers] 324, **567**
Tanner, J.P. 361
Tannock, J.D.T. 366
tape
  databases on 108, 422
  digital audio (DAT) 108
  tape leasing **108,** 450
Tapley, B.D. *and* Poston, T.R., *Eshbach's handbook*... 124
tar 240, 241, 590
Tattersalll, G.H. 538
Taub, H. *and* Schilling, D.L. 443
taxation
  carbon 233
  energy 233, 234
taxis 511
Taylor, C.F. [*engines*] 237
Taylor, C.M. [*valve trains*] 168
Taylor, David W., Naval Ship Research and Development Center [US] 478
Taylor, G.J. [*wind energy*] 247
Taylor, H.P.J. [*concrete*] 538, 557
Taylor, M. *et al.* [*traffic*] **506,** 507
Taylor, R.H. [*alternative energy*] 210
Taywood Engineering Ltd 94
TBV [co.] 94
TC (Technical Committee, of ISO) 17
TCP/IP (Transmission Control Protocol/Internet Protocol) 118
TCSU (Traffic Control Systems Unit) **511**
Tebbutt, T.H.Y. 488
Technica Ltd, *Techniques for assessing... hazards* 317
*Technical data book*s (*pub.* API) 273
  ... *petroleum refining* 273
Technical Engineering Services (TES) **594**
Technical guides (*pub.* ISO/IEC) 19
Technical Indexes Ltd (TI/ti) **90,** 340, 528–9
  Engineering Design and Manufacturing Index **156–7,** 165, 370
PERINORM 582
  product information **90,** 96, 340
RIBA/TI database 581–2
  Standards 22
  and Technical Information Microfile 579
Technical Information Service [co.] 112
Technical Information Microfile 579
Technical Institute of Research [Japan] 478
technical packages *see* package libraries
Technical Reports in Computer Science 113

Technical Services and Research Executive (TSRE), *former* 594
technicians 624
Techni Measure [co.] 146
Techniques for assessing... hazards (*pub.* Technica) 317
Technische Informationsbibliothek *see* TIB
technological information 119, **122,** 574
technology, history of *see* engineering: history
Technology and Disability 643
Technology Exchange Ltd, *Technoshop '92* 157
Technology Reports Centre (TRC) 8
technology transfer 371, 415
technometrics 460
Technoshop '92 (*pub.* Technology Exchange) 157
TED [Tenders Electronic Daily] database **584**
telecommunications **441–50**
  nature and role 411, 412, **441–4**
  abstracts 100, 101, **421–2**
  analogue signals 412, **423, 441–2**
  bandwidths 115
  broadband 442
  books 423, 442
  handbooks 125, **443–8**
  IEE series **443–4**
  cables 441
  conferences **417, 450**
  costs/tariffs 442, 443
  databases **421–2,** 450
  data compression 115
  digital signals **412, 423, 441–2**
  economic aspects 443
  Internet resources 119
  journals 414, **415–17, 448–9,** 459
  local 443
  mobile equipment 442, **446**
  networks **441–2,** 443, **444,** 580
  broadband 442, 444, **445**
  data communication 412, 443, 444, 445–6, **445–6**
  see also networks: computer/data
  digital buses 125
  integrated (ISDN) **442, 445**
  intelligent (IN) 442
  journals 449, 460
  network management (NM) 442
  personal communications (PCN) 442
  POTS and PANS 444
  telephony **441–2,** 444
  open systems communication (OSI) 450
  optical fibres 412, **447–8**
  organizations 414, 448–50, **516**
  packet switching 444
  PCM 442
  radio **447**

telecommunications (contd)
  satellites 412, 443–4, **447**
  signals 444
  see also signal processing
  Standards 16, 22, 420–1,
    449–50
    'Blue Book' 449
  telematics 502, 508
  theoretical basis **442–3**
  traffic engineering 443
  see also electronics radio *etc.*
Telecommunications 449
telegraphy 443
telematics
  transport 502, 508, 516
  see also Minitel;
    telecommunications *etc.*
teleoperation/remote control 383,
  400
telephones/telephony **441–2**, **444**
  cellular 442, **446**, 449
  cordless 442
  exchanges 444
  journals 449
  mobile 412, **442**, **446**, 449
  modems 581, 583
  optical fibres 412
  POTS 444
  radio telephony 442, **446**, 449
Télésystèmes-Questel *see*
  QUESTEL
television
  early 411
  high definition (HDTV) 413
  as information source 497–8
  TV holography 136
Telford, Thomas [publishers]
  522, 527, **531**, 534, 556,
  558, 582
  Coastal structures and
    breakwaters 567
  Concrete yearbook 534
  NCE road construction . . .
    directory 534
  Thomas Telford Software 582
  Timber in excavations 545
  Underground directory 534
Telnet [Internet command] 118
temperature **199–201**
  control 202
  high-temperature materials 314
  measurement 16, 202
  Standards 16, 27–8
  see otherwise heat *etc.*
Tenders Electronics Daily *see*
  TED
Tensor 458
tensors **456**, **458**
TERA (total energy resource
  analysis), *TERA Analysis*
  235
terminals, very small aperture
  (VSAT) 444
terminology
  Standards 123, 154, 368
  see otherwise dictionaries *and*
    specific topics
Terminology of heating,
  ventilation, air-conditioning
  and ventilation (pub.
  ASHRAE) 253

Terry, G.J. 156, 181
Tesla, Nikola 434
TES (Technical Engineering
  Services) 594
testing 178–80, 360, 362, **432–3**
  automated 379
  benchmarks 433
  CHETAH programme 324
  databases 355
  electronics 420
  environmental engineering
    486–7
  journals 87, **357–8**, 379
  machines 355
  materials 355
  noise 368
  non-destructive 357–8, 360,
    362, 369
  organizations 25
  power engineering 432–3
  product information 82, 83,
    310
  sample results 433
  Standards 25, 179, 310, 368
  verification 433
  see also reliability: condition
    monitoring
tetrahedra 137
Texas Instruments [co.]
  data books 418, 419
  digital signal processors 419
  TTL data book 418
TEXTLINE database 71
Theocaris, P.S. 147
Theocaris, P.S. *and* Gdoutos,
  E.E. 145
Theodore, L. *et al.* 326
Theodore, L. *and* Buonicore, A.
  491
Theoretical Chemical
  Engineering 283
Theoretical Chemical
  Engineering Abstracts,
  former 283
Therapy Weekly 643
THERM [database] 314
Thermal Engineering 255
thermal loading 134
thermal reactors 260
thermal stress 137
thermal systems 174, 199–203,
  204–11
  nature and role **204–5**, 269–70
  theoretical basis **199–201**,
    203–5
  abstracts/indexes 205
  analysis 205
  books 235–7, 276–7
  bibliographies 208, 209
  dictionaries 201, 207, 253
  directories 206
  handbooks 205, 206, 207,
    208, 209, 233, 253–4,
    271–3
  conferences 204, 205–6,
    207–8, 209, 210, 254
  control systems 391
  cooling **207**
  current trends **208–10**
  databases 205
  efficiency **204**, 253–5

fans **208**
heat exchangers/recovery
  206–7
heating **206**
heat transfer **205–6**
historical aspects 206
journals 205, 207, 208, 209,
  237, **252–3**
nuclear 259–62
organizations 204, 205, 206,
  207, 208, 209, **210–11**,
  227
power generation **208–10**,
  428–30
process engineering **276–7**
pumps **208**
  heat pumps **207–8**
software 203, 205
Standards **206**, 208
storage systems **206–7**
see also thermodynamics;
  energy technology
thermionic valves 411, 412
thermochemistry 199, 202–3,
  272–3, 282
thermodynamics **199–222**, 269
Laws of **199–201**, 204
availability **201**, 204
books **202–3**, 235–7, 240, 269,
  272–3, 275–7
dictionaries **201**
chemical **199**, 275–6, 282
  biochemical 279
conferences 277
cyclic processes **200–1**
  Carnot cycles **201**, 203
  combined cycles 204
  Rankine cycles 203, 207
efficiency **200**, **203–4**, 205,
  277, **428**
entropy **201**
  see also entropy
equilibrium **199**, 272, 275
exergy **201**, **203–4**
exothermic systems 259–60
fluid mechanics **205–6**
heat transfer **205–6**
journals **205**, 282
mass transfer **199**, **273**, 274,
  275, 282
organizations **210–11**
power density **261**
processes 199–201, **203–5**,
  269–70
properties data **202–3**, 253–4
reversibility **201**, 204
solar energy 243
statistical 202
thermal systems *see* thermal
  systems
transport processes **273**
see also energy *and specific*
  topics
thermoelasticity **136**
thermoelastic stress analysis **136**,
  147
thermophysics 203, 205
see also thermodynamics *etc*
thesauri
  of databases 354, 421
  see otherwise dictionaries

Index

theses/dissertations 9, 11, 63, **77–9**
  directories 79
  as grey literature 10, 78
  information on 78–9, 112, 142, **530–1**
  reports 7
  reprints 530–1
Thiel, M.J. 560
Thin Walled Structures **143**
Thollot, P.A. 435
Thoma, J.U. [simulation/hydraulics] 287
Thomas, L.J [*mining*] 613
Thomas, R. [*traffic*] 507
Thomas, R. [*minerals*] 600
Thomas [co.]
  Thomas register of American manufacturers **84–5**, 340
  online 91, 92
Thomas Cook [co.], *Thomas Cook European and worldwide timetables* 510
Thomas Online [co.] 92
THOMAS NEW INDUSTRIAL PRODUCTS DATABASE 92
Thomas Skinner Directories [publishers], *Willing's press guide* 73
Thomas Telford Software 582
Thomen, J.R. 326
Thompson, J.R. [*safety*] 181
Thompson, P. [*global warming*] 256
Thompson, T.J. *and* Beckerley, J.G. [*nuclear energy*] 266
Thomson, I. [*documentation/EU*] 126
Three-Mile-Island (TMI), nuclear incident, *Report* [ed. Kemeny] 266
Threlkeld, J.L. 205
Thring, M.W. *and* Laithwaite, E.R. 332
Thumann, A. 256–7
thyristors 435, 436
TI/ti *see* Technical Indexes Ltd
TIB (Technische Informationsbibliothek) 10, 11
TIBQUICK service 11
tidal power **247–9**
  Severn Estuary project 248
TIG (tungsten inert gas) 362
Tijunelis, D. *and* McKee, K.E. 369
tiles, roofing 564
Tiller, J. *and* Creech, D.B. 244, 254
timber **541–2**, **560–1**
  bridges 566
  databases 308
  fire hazard 555–6
  fixings 565
  journals 68
  organizations 303, **541**, 560
  Standards **542**, 549, 555, **561**
  temporary works 545
Timber construction manual (*pub.* Amer. Inst. Timber Construction) 542

Timber in excavations (*pub.* Telford/TRADA) 545
Timberline [program] 582
Timber Research and Development Association (TRADA) 527, 528, **541**, 542, 560, 561
Timber in excavations 545
time scales, Standards 27–8
Timken Co. 176
Timoshenko, S.P. [*design/structures*] 133, 141, **550**, 553
tin
  bronze 589
  early uses 589
  organizations 303
  placer mining 615
  tips, mineral waste 592
Tissue Viability Society 630
Tizzard, A. 335
TK Solver [programs] 159, **160**, 165
TMI *see* Three-Mile-Island
TMS-AIME *see* Metallurgical Soc.
TN (CIRIA Technical Note) *see under* Construction Industry Research and Information Ass'n
TNO [Inst. of Spatial Organization (Netherlands)] 517
Todhunter, I. *and* Pearson, K. 553
Toft, B. *and* Reynolds, S. 318
Tokhi, M.O. *and* Leitch, R.R. 182
Tokuz, L.C. *and* Jones, J.R. 172
tolerances 310, 337, **354**
Tolley Publishing Co. Ltd,
  Tolley's health and safety . . . handbook [ed. Dewis/Murdoch] 181
Tomasi, W. 443, 447
Tomlinson, M.J. 565
Tompkins, W.G. 280
Tong, R.M. 396
Tool and manufacturing engineers handbook [ed. Bakerjian] 124, **368–9**
tools/tooling *see* machine tools
Topalian, A. 333
Top 3000 directories and annuals (*pub.* Dawson) 86, 125
Torafson, L.E. 156
Toronto, Kilborn Engineering [co.] 595
torque, hydraulic systems 290
Tortora, G.J. *and* Grabowski, S.R. 634, 635
total design 330, 331, 333
total energy resource analysis *see* TERA
Total Marine [co.], *Annual Report* 481
total quality management 366–7, 368, 371
Towell, J.E. *and* Sheppard, H.E. 124

towers **567**
towns
  transport 499–500, 502, 504–5, 506–7, 509
  traffic 495–6, 499
  waste management 490
  see also London; local authorities
  see also planning
Townsend, A. [*safety/maintenance*] 326
Townsend, D.P., *Dudley's gear handbook* 164
toxicity 322, 323
Toyota [co.] 363, 386
traction 174, 435, 436
TRADA *see* Timber Research and Development Ass'n
trade, statistics 75
trade associations *see under* organizations
Trade associations and professional bodies of the United Kingdom 370
trade directories *see under* directories
Trade directories of the world (*pub.* Croner) 86
trade fairs *see* exhibitions
Trade Indemnity plc 576
trade journals *see under* journals
trade literature **82–3**, **93–6**, **473**, **481**
  access to 66, 83, 87, 90, 93–5, 473, 481
  online 91
  archives 93–5
  catalogues *see* catalogues
  collections 93–5
  as grey literature 10
  house journals as 69
  and product information 82–3
  Standards, BS4940: . . . presentation of technical information . . . 81, 82
Trademark register of the US (*pub.* Trademark Register) 88
trademarks/trade names **88**
  directories **88**, 92, 100, 341
  journals 341
  licensing 41
  and product information 84, **88**, 92, 96
  USPTO 45
Trademarks (*pub.* BL) 121
TRADEMARKSCAN database 88
Trademarks Registry 88
trade names *see* trademarks
trade secrets 36
and patents 60
Trade and Technical Press,
  Valves, piping and pipelines handbook 285
traffic **495–517**
  abstracts/indexes 504–5
  books **506–7**, 545
  handbooks 508, 534
  conferences 500–1
  control systems/signing 495–6, 498–9, 500, 507, **508**, 511, 513, 515

Index 763

traffic (contd)
  current trends **495–6**
  databases 504–5
  government reports 499–500
  journals 497–9
  networks 502–3
  organizations **511–17**
  police and 511
  regulations 511
  Standards 507–8
  statistics 509–11
  see also transport
*Traffic appraisal manual (TAM;
  pub.* DoT) 500, **508**
Traffic Control Systems Unit
  (TCSU) **511**
Traffic Engineering and Control
  498, **499**, 502, 516
Traffic generation (pub.
  SACTRA) 500
Traffic in Great Britain (pub.
  DoT) 509
Traffic Quarterly, former 499
Traffic signs manual (pub. DoT)
  500
Traffic Technology International
  498, 499
Traffic in towns [Buchanan
  Report] 499
training/education, of engineers
  distance learning 420
  in-house 429
  journals 282, 338
  organizations 282, 371
  qualifications 371, 593, 609
  short courses 497, **502**
  videos 420
  architecture 586
  chemical engineering 270, 282
  construction 520, 521
  control systems 399
  design 329, 332, **334**, **336**, 338
    Design Council review 336
    OU courses 331, 334
    SEED publications see
      SEED
  electronics 413, 414, 415, 420
  environmental management
    491
  fluid mechanics 288
  instrumentation 415
  lasers 364
  management 329, 366
  manufacturing 364, 366, 371
  marine technology 479
  mathematics 461
  mineral processing 595–6
    mining 595, 610
  nuclear engineering 267
  power engineering 429
  quality 334, 371
  robotics 379
  transport 496–7, **502**
  welding 371
trains *see* railways
trams, supertrams 512
Transactions 64–5, 72, **476**
  *see otherwise under specific
    organizations*
TRANSDOC database **504–5**
transducers 166

transformers 431
transforms 390, 451, **457**
  Laplace 390, 457
TransIMM see Instn of Mining
  and Metallurgy:
  Transactions
transistors **411–12**, 419, 435
  MOSFETs 435
TRANSIT [list] **502**
translations 9
  abstracts 58, 101, 112
  conferences 359
  as grey literature 10
  journals 357
  patents 39, 58
  reports 7
  Standards 20, 21, 23, 24
  see also dictionaries:
    multilingual
  transmission, mechanical 164
    hydraulic 289, 292–4
    see also gear *etc.*
Transmission Control Protocol/
  Internet Protocol (TCP/IP)
  118
TRANSP-L [list] **502**
TRANSPORT CD-ROM database **505**,
  511
transport **495–517**
  nature and role **495–6**, 547
  abstracts/indexes **503–4**
  air *see* aerospace
  books **506–7**
    data books 509–11
    directories 510–11, 534
    handbooks 508
    official publications 497,
      507–8, 509–11
    timetables 510
  coal 238
  commercialization 511–12
  conferences 497, **500–1**, 505,
    506, 508, 513, 514
  congestion 495–6
  consultancies 511, **512**
  consultation procedures 508
  control systems *see under*
    traffic
  cycling 496
  databases 355, 503, **504–5**,
    511
  design 338, 500, 508, 511, 566
  economic aspects 498
  e-mail **502–3**, 511
  energy issues 255, 495
  environmental aspects 495,
    496, 498, 500, 507–8
  of goods
    hazardous substances 323,
      477
    Standards, *Conteneurs pour
      le transport des
      marchandises (pub.*
      AFNOR) 23
  statistics 509
  history of 498
  information systems 502, 508,
    516
  inquiries 499
  intelligent 503
  journals 68, 338, **497–9**, **503**,
    511, 513

  and land use **496**
    legislation 496, 507–8
    licensing 508
    local 495–6, 498
    minerals 592
    off-highway 615
    organizations 323, **497**,
      498–506, 507–10,
      511–17, 523
      international **516–17**
      official 326, **511–12**
      passenger 499, 509–10
      new approaches 495–6,
        511–12
    pedestrians 496
    planning **495–6**, 498–9, 502,
      505, 506–7, **511–12**, 513,
      515, 517
    policy **495–6**, 498–9, 500, 505,
      507, **511–12**
      European input 508
      privatization 502
    product information 498
    public 495–6, 498–9, 502, 507,
      509–11
    research 499, 500, 502,
      515–16
    review studies 498–9
    roads *see* roads
    safety 316, **317**, 323, 324, 477,
      496, 500, 513, 516, 517
      accidents 323, 324, 326,
        498, **509–10**
    sea *see* marine
    shipping **467–83**
    software 503, 508
    Standards 477, **507–8**
      harmonization measures **508**,
        516
    statistics 497, **509–11**
    surveying 500
    surveys 509–11
    telematics 502
    timetables 510
    training 496–7
    travel behaviour 500, 507, 509
    tribology 174
    see also roads; railways; traffic
  Transport 498
  TRANSPORT CD-ROM database **505**,
    511
  Transport: annual statistics (pub.
    EUROSTAT) 510
  *Transport accident reports/briefs
    (pub.* NTIS/USDOT) 510
  transportation *see* transport
  Transportation 499
  Transportation Association of
    Canada **517**
  Transportation Journal 499
  *Transportation Planning and
    Technology* 499
  Transportation Quarterly 499
  Transportation Research 499
  *Transportation Research
    Abstracts* 504
  Transportation Research Board
    (TRB [US]) **501**, **517**
    conference **500**, 501
    Highway capacity manual 500,
      506

764 Index

Transportation Research Board (TRB [US]) (contd)
Special Reports 506
and TRANSPORT CD-ROM 505
Transportation Research Abstracts 504
Transportation Research Record 499
TRB Publications 504
TRIS **503**, **504**, 505
Urban Transportation Abstracts 504
Transportation Research Forum [US] **517**
Transportation Research Information Service see TRIS
Transportation Research Record 499
Transportation Science 499
Transport and the environment [Royal Commission report (pub. HMSO)] 500
Transport of goods by road in Great Britain (pub. HMSO) 509
Transport Policy 499
transport processes **273**, 274
Transport Research Board (TRB) 504
 Highway capacity manual 500, **506**
Transport Research Laboratory (TRL) **512**, **515**, **523**
 Bus registration index 510
 Contractor Reports 545, 566
 Current Topics in Transport **503**, 504
 database searches 505
 and IRRD 505, 512
 library 505, 512
 privatization, proposed 512
 reports 512
 Research Reports 545, 566
 State of the Art Reviews **499**, 545, 566
 Supplementary reports 545
 TRL Headstart 504
 Transport Reviews **499**
Transport and Road Research Laboratory (TRRL), former **512**, **523**
 see otherwise Transport Research Laboratory
Transport Sectional List (pub. HMSO) 504
Transport statistics Great Britain (pub. HMSO) **509**
Transport statistics for London (pub. HMSO) 509
Transport Statistics User Group (TSUG) **510**
 Directory of sources in transport statistics **510–11**
Transport 2000 [org'n] **514**
 Vital travel statistics 510
TRANSPORT.UK [list] **502**
Transport and Water (ICE Proc.) 499, 531
Trantolo, D.J. and Wise, D.L. 238

travel behaviour 500, 507, 509
TRB
 TRB Publications 504
 see otherwise Transportation Research Bd
TRC (Technology Reports Centre) 8
Treichler, J.R. et al. 395
trenches/trenching 545
Treybal, R.E. 274
tribology 168, **173–7**
 books 154
 dictionaries 154
 conferences 168, **173–4**
 journals 173, **173–4**
 organizations 173, 175
 software 173
Tribology International 173, 175
Tribology Transactions 173–4, 175
TRICS (Trip Rate Information Computer System) [database] 511
trigonometry 451
Trip Rate Information Computer System (TRICS) 511
TRIS (Transportation Research Information Service), TRIS database **503**, **504**, 505
tritium, and fusion process 261
TRL
 TRL Current Topics in Transport **503**, 504
 TRL Headstart 504
 see otherwise Transport Research Laboratory
Troitsky, M.S. 566
Trott, A.R. 207
TRRL see Transport and Road Research Laboratory
Truscott, S.J. 590
Tsai, S.W. 313
TSRE see Technical Services and Research ...
TSUG see Transport Statistics User Group
Tsukiji, T. et al. 298
TTL data book (pub. Texas Instruments) 418
tubes
 and hydraulics 289–90
 see also pipes
Tuhtar 322
Tulsa, University [US]
 Petroleum Abstracts 228, 230
 TULSA database 230
Tuma, J.J. 453
tungsten inert gas (TIG) welding 362
tunnels 545
 mining **615**, **616**
turbines 200, 205, 208, **235–7**
 and electricity generation 252, 259, 260, **428–9**, 430
 gas 208, 209, **235–7**
 co-generation 209
 combined-cycle processes 429
 journals 87, 209, **235**
 steam 208, **235–7**, **428–9**
 turbo-alternators 259, 260

wind 246–7, **430**
Turbomachinery International 87
turbulence, hydraulics 294, 297
Turino, J.L. 649, 653
Turk, C. and Kirkman, J. 338
Turkey, minerals 594, 596
Turnbull, R.G.H. 493
Turner, W.C. 257
turning
 diamonds 357
 metals 363
Tutt, T. and Adler, D. 565
TV holography 136
Twelvetrees, W.N. 540
Twidell, J. 247
Twidell, J. and Weir, A.D. 210, 242
Twigg, D. and Voss, C.A. 652
Twort, A.C. et al. 489
Tyn, Myint-U. and Debnath, L. 456
tyres, Standards 20
Tzafestas, S.G. 393, 399

U
UAP (Universal Availability of Publications) 10
UCPTE see Union for the Co-ordination of ... Electric Power
UDC see Universal Decimal ...
UITP
 UITP Index 504
 UITP Revue, former 499
 see otherwise Internat. Union of Public Transport
UK see United Kingdom
UKAEA see Atomic Energy Authority
UK CONSTRUCTION AND CIVIL ENGINEERING INDEX 90
UK Oil Operators' Association (UKOOA) 479
UK trade names 92
UK workshop on heat pumps (pub. DoE) 208
UL see Underwriters Laboratories
Ullman, D.G. [design] 156, 312, 331
Ullman's encyclopedia of industrial chemistry (pub. VCH) 271
Ulrich, K.T. and Eppinger, S.D. 332
Ulrich's International Periodicals Directory (pub. R.R. Bowker) **73**, 74, 99, 227, 400
 online 73
ultra-precision engineering 359
ultrasound
 clinical uses 639
 ultrasonic processes 363
 Ultrasound in Medicine and Biology 639
Umana, Q.A. 233
UME see Internat. Conf. on Ultra-Precision ...
UMI see University Microfilms Internat.

Index 765

UMIST (University of Manchester Inst. of Science and Technology) 358–9
UN *see* United Nations Org'n
uncertainty analysis 327
UNCTAD database 503
Underground directory (pub. Telford) 534
underground engineering *see* mining *etc.*
underwater engineering 476, 479
  cables 481
  electricity transmission 432
  organizations 475
  product information 481
  robotics 378
  subsea installations 470*tab*
Underwater Technology (pub. SUT) 476
Underwater technology (pub. WEGEMT) 479
Underwriters Laboratories (UL) [co.], *Standards* 22, **321**
UNE (pub. Asociación Española...) 27
UNEP (UN Environment Programme) 242, 249, 493
Unesco (UN Educational, Scientific and Cultural Org'n), *and* energy **225**
ENERGY 225
International directory of new and renewable energy... 242
Unicorn Construction Services Ltd 94
UNIDO (UN Industrial Development Org'n) 249
Unimates [automated tools] 377
uninterruptible power supplies (UPS) 414
Union for the Co-ordination of the Production and Transport of Electric Power (UCPTE) 225
Union Internationale des Transports Publics *see* Internat. Union of Public Transport
Union of International Organizations, *Yearbook...* 126
UNIPEDE *see* Internat. Union of... Electrical Energy
United Kingdom (UK)
  Atomic Energy Authority *see* Atomic Energy Authority
  British Geological Survey (BGS) 602, 612
  Budgets 570
  Building Research Establishment *see* Building Research Establishment
  Central Statistical Office (CSO) 126, 572
  Civil Aviation Authority 421
  Commonwealth *see* Commonwealth
  and concurrent engineering 656, 657

Defence Research Agency (DRA) 8
Defence Research Information Centre (DRIC) 8–9, **23**
Dept of Energy (DoE) 205, 208, 210
  Combined heat and electrical power generation... 213–14
  Combined Heat and Power Group 213–14
  Compressed air and energy use 208
  conferences 207
  Development of alternative sources... 210
  District heating... 209
  District Heating Working Party 208–9
  Economic use of gas-fired boiler plant 206
  Efficient use of energy 205
  Energy Managers' Workshops 205
  Fuel Efficiency Booklets 206, 208
  Offshore installations... 567
  Recovery of waste heat... 206
  Renewable energy sources 210
  UK workshop on heat pumps... 208
  wave energy programme 248
Dept of the Environment (DoEnv)
  Better trade literature 96
  BRIX/FLAIR 230, 530, **583**
  Control of landfill gas 614
  Data co-ordination... in the construction industry 577–8
  Dept of the Environment publications [abstracts] 504
  Environmental assessment... 493
  Landfilling wastes 614
  Library Bulletin 503
  Marine dredging for sand and gravel 615
  Mineral planning guidance 614
  Planning policy guidance notes 500
  and transport 500, 507–8
  Waste management papers 490
Dept of Health, Medical Devices Directorate (MDD) 645
Dept of Scientific and Industrial Research, *former* 512
Dept of Trade and Industry (DTI)
  Condition Monitoring and Sound Assessment Club 178–9

and design 342
and electronics **415**, 421
and energy policy **225–6**, 234
Energy Trends 234
Joint Framework for Information Technology (JFIT) 415
Manufacturing into the late 1990s 657
New earnings survey **574**
reports 8
research 355
and Standards 22–3, 28, 421
Dept of Transport (DoT) 497, **499–500**, 509, **511–12**, **515**
  agencies **511–12**
  Annual vehicle census Great Britain 509
  bulletins 511
  COBA manuals 500, **508**
  Department of Transport Publications 504
  Design manual for roads and bridges 500, **508**, 545, 566–7
  Environmental evaluation 500
  guidance/circulars 507–8
  information provision **497**, 504, 507, 509, **511**
  inquiries, public 499
  library 511
  Library Bulletin 503
  Manual of contract documents for highway works 545
  Microprocessor-based traffic signal controller... 500
  National road maintenance condition survey 509
  New motor vehicle registrations Great Britain 509
  Quarterly road casualties Great Britain 509
  Quarterly transport statistics 509
  reports 499, 508, 509, 511
  Road Research Technical Papers 507
  Roads and traffic in urban areas **506**, 507
  Road traffic statistics Great Britain 509
  standing committees 493, 499, 508
  statistics **509–10**, 511
  Traffic appraisal manual (TAM) 500, 508
  Traffic generation 500
  Traffic in Great Britain 509
  Traffic signs manual 500
  Urban road appraisal 499
  Vehicle speeds in Great Britain 510
  deregulation **511–12**
  and design 342
  Economic and Social Research Council/ESRC 510, 511

766 Index

United Kingdom (UK) (*contd*)
electricity, privatization **429**
electronics 415, 421
energy agencies **225–6**
Energy Efficiency Office
  (EEO) 208, 210, 253
energy policies 208–9, 225–6,
  234, 242
  nuclear energy 261–2, 263,
    264, 266
Energy Technology Support
  Unit (ETSU) **226**, 243,
  245, 246, 247
Engineering and Physical
  Sciences Research
  Council (EPSRC) 415,
  515
English Heritage 552
environment 486–7, 490, 493
Government Publications
  [abstracts] 504
Her Majesty's Nuclear
  Installations Inspectorate
  (NII) **264**
Highlands and Islands
  Development Board 481
Highways Agency **511**
Home Office 511
House of Commons Select
  Committees 508
  ... on Science and
    Technology 210, 248
  ... on Transport 500, 508
House of Lords Committee on
  Science and Technology
  508
Institute of Energy *see* Inst. of
  Energy
Institute of Fuel, *former* 205
local authorities *see* local
  authorities
marine technology 475
Meteorological Office 243
metrology 28
Min. of Aviation, *former*, DTD
  Standards 23
Min. of Defence (MoD)
  preferred products 91, 421
  reports 8–9, 11
  and Standards 23, 421
Min. of Transport 512
monopolies 31–2, 36
Museums and Galleries
  Commission 552
National Economic
  Development Council/
  Office, *former* 344, 570,
  575
National Engineering
  Laboratory 139, 207, 211
National Industrial Fuel
  Efficiency Service 206,
  211, 236
National Physical Laboratory
  23, **28**, 147
National Radiological
  Protection Board **264**
National Weights and
  Measures Laboratory 23,
  28
Nautical Institute 475, 476

Naval Research Laboratory
  252
patents 31–61, 341
  compulsory licensing 42
  EPO patents 46
  Letters Patent 31–2
  Patents Act, 1977 32
  procedures **37–42**
  Statute of Monopolies
    (1623) **32**, 36
  see also Patent Office
physics 415
privatization 226, 429, 502,
  512, 594
Property Services Agency *see*
  Property Services Agency
publications **126**, 504
reports 5, **8–9**, 266, 499, 508,
  509, 511
research 8, 226, **515**
robotics 382
Science and Engineering
  Research Council (SERC)
  11, 78, 329, 468
Sound Research Laboratories
  181, 492, 493
Standards 20, **21–3**, 28, 29,
  113, 421
  Laboratories 28
  see also British Standards
    Institution
Standing Advisory Committee
  on Trunk Road
  Assessment (SACTRA)
  493, **499–500**, 508
statistics **126**, 232, 235,
  509–10, 511
Technology Reports Centre
  (TRC) 8
tidal energy projects 248
trade directories 85–6
Trademarks Registry 88
Traffic Control Systems Unit
  (TCSU) **511**
transport 493, 499–500, 507–8,
  511–12
Transport Research Laboratory
  see Transport Research
  Laboratory
United Kingdom Automation
  Council 403
United Kingdom Energy Statistics
  (*pub*. HMSO) 235
*United Kingdom minerals
  yearbook* (*pub*. BGS) 612
United Kingdom Oil Operators'
  Association (UKOOA) 479
*United Kingdom shipping
  industry: international
  earnings and expenditure*
  (*pub*. HMSO) 510
United Nations Educational,
  Scientific and Cultural Org'n
  see Unesco
United Nations Organization
  (UN[O])
  databases 584
  SCAN-A-BID **584**
  Development Business/
    Development Forum 584
  Economic Commission for

Europe (ECE) 237, 382,
  493, 510, **516**
energy 237, 249
Environment Programme
  (UNEP) 242, 249, 493
Industrial Development
  Organization (UNIDO)
  249
publications **510**
Annual bulletins [of Eur.
  statistics] 232, 237,
  252, 510
*Census of motor traffic on
  main international
  traffic arteries* 510
Energy statistics yearbook
  232
*Policies ... of
  environmental impact
  assessment* 493
*Post project analysis in
  environmental impact
  assessment* 493
*Statistics of road traffic
  accidents in Europe*
  510
*Workshop on biomass
  thermal processing ...*
  249
transport 510
WIPO 45
see also Unesco; World Health
  Org'n; Internat. Labour
  Office; Food and
  Agriculture Org'n
United States of America (US)
  abstracts/indexes 100, 102,
    230, 355
  aerospace 421
  American Bureau of Shipping
    477
  American Public Health
    Association 494
  Armed Services Technical
    Information Agency,
    former (Astia) 5
  biographical data 126
  Bureau of Mines (USBM) 594,
    608
    Information Circulars 608
    Mineral Commodity Profiles
      608
    Minerals Today 597
    Minerals yearbook 601, 608,
      612
    Reports on Investigations
      608
  catalogues 90
  Coast Guard, *CHRIS
    hazardous chemicals ...*
    323
  Committee on Scientific and
    Technical Information,
    former (COSATI) 6, 9
  and concurrent engineering
    655, 656, 657
  Concurrent Engineering
    Research Center (CERC)
    654, **656**
  conferences 76
  Congressional Information
    Bureau 251

Index

United States of America (US) (contd)
  Constitution 43
  Council on Tall Buildings and the Urban Habitat [US] 530
  Dept of Commerce 6, 230, 248
    Final... impact statement for... [OTEC]... 248, 256
    Office of Ocean Minerals and Energy 248, 256
    publications 266
  Dept of Defense (DOD)
    CALS software **310–11**, 314, 655
    and concurrent engineering 655
    DARPA 656
    DICE 656
    MIL-HDBKs 91, 307
    Navy 91
    Standards 26, **26–7**
    US Dept of Defense index of specifications and standards 26
  Dept of Energy (DOE) **226**, 241
    Annual energy review 232
    Annual Technical Report: Energy Materials 233
    Buildings Energy Technology 254
    Coal conversion... data book 237
    databases **112**, **229**, **230**
    EIA data index 232
    EIA data index (pub. USDOE) 232
    Energy data base... 8, 232
    Energy Materials Coordinating Committee 233–4
    Energy Research Abstracts (ERA) **7–8**, 227, 267
    Fusion Power Program... Report 251
    Geothermal Energy 246
    Hydrogen Energy Coordinating Committee, Annual Report 241
    Monthly Energy Review 232
    Nuclear Fuel Cycle 251
    and Nuclear Safety 267
    publications 232, 234, 251, 254, 266, 267
    Quarterly coal report 237
    reports **7–8**, 11, 237, 266
    Technical Information Center **226**
  Dept of the Interior (DOI)
    energy policy 226
    Geological Survey 113
    see also Bureau of Mines above
  Dept of Transport (DOT) 517
    Hazardous materials emergency... 326
    National transportation statistics 510
  Transport accidents reports/briefs 510
  US Dept of Transport national transportation statistics 510
  Educational Resource Information Center (ERIC) 11
  electricity 320, 430, 433
  Electric Power Research Institute see Electric Power Research Inst.
  electronics 91, 421
  energy agencies 226–7, 264, 266
  Energy Information Administration (EIA) **226**, 232
  Engineering Design Research Center 656
  environmental protection 323, 486, 488
  Environmental Protection Agency 323, 486
  failure rates 91
  Federal Communications Commission (FCC) 421
  Federal Council for Science and Technology 6, 9
  Federal Document Retrieval service 74, 400
  Federal Highway Administration 517
  fire hazard 321
  Food and Drug Administration 645
  Government-Industry Data Exchange Program (GIDEP) 91–2
  Government Printing Office (USGPO) 26, 237, 612
  Governor's Division of Energy... and National Resources 244, 254
  journals 72–3, 74
  Library of Congress 103
  manufacture, CIM 362
  marine technology 421, 475, 477, 478, 480
  materials 92
  metrology 28, 91
  NASA see Nat. Aeronautics and Space...
  National Center for Supercomputing Applications 118
  National Electrical Code 320, 321
  National Fire Protection Ass'n **321**, 322, 323, 324, **327**
  National Highway Traffic Safety Administration 517
  National Institute for Standards and Technology **25**, **28**, 304
  National Laboratories 116, 242, 251, 254
  National Maritime Research Center 480
  National Petroleum Council (NPC) 226
  National Repository for Concurrent Engineering 656
  National Research Council 650
  National Safety Council 315–16, 317
  National Science Foundation **342**, 359
  National Standards Ass'n **25**, 113, 230, 368
  National Technical Information Service (NTIS) see Nat. Technical Information Service
  Navy 91, 261
  newspapers 75
  nuclear energy 261–2, **264**
  Nuclear Regulatory Commission see Nuclear Regulatory Commission
  Reports 266
  Occupational Safety and Health Administration (OSHA) 320
  Office of Building and Community Systems 254
  Office of Energy Technology 237
  Office of Ocean Minerals and Energy 248, 256
  Office of Scientific Research and Development 4
  Office of Scientific and Technical Information 254, 267
  patents 33, 41, **43–5**, 53, 58–9, 341
  Polaroid v Kodak 43
  Patent and Trademark Office (USPTO) 44, 45, 53
  product information 90–1, 93
  Radio Technical Commissions for Aeronautics (RTCA) 421
  for Maritime Services 421
  reliability 91
  reports **4**, 5–8, 11, 355
  research 355
  Standards 17, 20, 22, **25–7**, 421
    ANSI 16, 20, **25**, 123, 449
    databases **113**, **230**
    Federal Supply Classification Codes 25
    Food and Drug Administration 645
    military specifications 22, **26–7**
    National Standards Ass'n (NSA) **25**, **113**, 230, 368
    safety 181
    statistics 510
    stress analysis **140–1**
  theses 11
  timber 560
  trade literature 90–1, 93
  trademarks 44, 88
  transport 500, 501, 510, 517
  Transportation Research Board see Transportation Research Bd

United States of America (US) (contd)
 Water Pollution Control Federation 488, 490
 wind power 430
 see also North America and specific topics
unit operations 269, **273–4**, 488
 books 271–2
units, Standard 27–8
Universal Availability of Publications (UAP) programme 10
Universal Decimal Classification system (UDC) 102, 121
Universal joint and drive shaft manual (pub. SAE) 195
Universal Technical Software Ltd 160
universities
 directories 126
 and electronic publishing 114, 117
 libraries 152
 Follett Report 120
 mining/minerals **595–6**
 networks
  BIDS 109
  Joint Academic Network (JANET) **109**, 505, 527, **584**
 research 77–9, **126**, 478–9, 515–16
 specialist groupings 479, 501, 515–16
Universities Transport Studies Group see UTSG
University Microfilms International [co.] (UMI) 78, 111, 426, 530–1
University Programs in Computer-Aided Engineering, Design and Manufacture conference (UPCAEDM) 359
UNO see United Nations Org'n
UPCAEDM see University Programs in Computer-Aided...
Upp Loy, E. and Daniel Industries Staff 276
upstream processes 278
UPS (uninterruptible power supplies) 414
UPT (Forschungsgemeinschaft Ultrapräzisionstechnie) 359
uranium **259–62**, 590
 enriched 260, 261
 isotopes 259–60
 leaching 599
 mining 609
Urban Abstracts and ACOMPLINE 505
Urban road appraisal (pub. SACTRA) 499
Urban Transportation Abstracts 504
Urological Research 639
Ursu, I. 251
US[A] see United States of America

USBM see United States: Bureau of Mines
US electronics industry directory (pub. Harris) 418
Usenet client-server 305, 502
user groups 479, 510
User guide on process integration... (pub. IChemE) 277
USGPO see United States: Government Printing Office
USPA database 58
USPTO see United States: Patent and Trademark Office
USSR Reports (pub. NTIS) 228
US VENDOR CATALOGUES/MASTER DIRECTORY [CD-ROM] 90
Utah, University 596
utilities
 privatized 226, 429
 process integration 204
 thermal systems 204–5
 see also electricity etc.
UTSG (Universities Transport Studies Group) 515
 Conference **500**, 501
 UTSG Current Research Listing 503
 UTSG Mailbox [list] **502**

V
Valentin, F.H.H. et al. 491
Valette, L. and Focquet, J.P. 209
Validated engineering data index (pub. ESDU) 142
value engineering 337
valves
 hydraulic 285, 286, 288, **291**, 292, 293–4, 296, 297–8
 control systems and 279, 286
 long transmission lines 292–4
 pneumatic 285
 servovalves 286, 290, 292–3, 297
 thermionic 411, 412
Valves, piping and pipelines handbook (pub. Trade and Technical) 285
VAMAS see Versailles Agreement...
Vanderberg, M. 564
Van der Schaaf, T.W. et al. 318
Vanderspek, P.G. 366, 385
Van Dulken, S. 61
Van Duuren, J. et al. 445
Van Hulle, F.J.L. et al. 247
Van Nostrand's scientific encyclopedia [ed. Considine/Considine] **122**
Van Vuren, T. and Van Vliet, D. 507
Van Wylen, G.J. and Sonntag, R.E. 202–3
Van Zyl, D.J.A. et al. 599
vapours see gases
Vas, P. 437
VCH [publishers], Ullman's encyclopedia of industrial chemistry 271

VDA see Verband der Automobilindustrie
VDE see Verband Deutscher Elektrotechniker
VDI/VDI–GP see Verein Deutscher Ingenieure
vectors 158, **456**, **458**
Vehicle Navigation and Information Systems Conference (VNIS) 500, **501**
vehicles
 AGVs 363, 385, 386, 470*tab*
 batteries 252
 conferences 500, 501
 design 338
 and electrical power 252
 heavy goods (HGVs) 509
 licensing 508
 safety 317
 space vehicles 176, 204
 speeds 510
 statistics 509–10
 underwater vehicles 470*tab*, 481
 see otherwise automotive; transport etc.
Vehicle speeds in Great Britain (pub. DoT) 510
Veilleux, R.F. 370
Vellinga, P. and Grubb, M. 233
veneers 303
Ventilation in coal mines (pub. NCB) 617
ventilation systems 204–5, 207, 253–5
 conferences 613, 617
 mining 610, 611, 613, **616–17**
 organizations 403
Verband der Automobilindustrie ([Automobile Industry Ass'n] VDA) 25
Verband Deutscher Elektrotechniker ([Ass'n of German Electrical Engineers] VDE) 24
 Guidelines 16
Verein Deutscher Ingenieure ([Ass'n of German Engineers] VDI) **523**
 Bauingenieur 532
 Gesellschaft Produktionstechnik (VDI–GP) 342, 651
 Guidelines 16
 Standards 421
Verlag TuV Rheinland, Safety-related computers... 319
Vernon, P.E. 332
Versailles Agreement on Advanced Materials and Standards (VAMAS) 304
very large scale integrated circuit (VLSI) 415, 417
very small aperture terminals (VSAT) 444
Vesilind, P.A. 488
Vest, C.M. 146
Veziroglu, T.N. et al. 241
Veziroglu, T.N. and Takahashi, P. 241

Index 769

vibration 184, 191, 193, 195, **492**
  analysis of 367
  handbooks 125, 184, 492
  hydraulic systems 292
  Internet resources 119
  journals 492
  marine technology 467, 469*tab*
video conferencing 342
video imaging *see* imaging
videos
  development 412
  safety 320
  training 420
  videodisk patented 51
Vidyasagar, M. 394
Vieira, C. *et al.* 474
Vienna, Convention on traffic signs 508
Vilenius, M.J. *et al.* 296
Villemeur, A. 319
Vincent, J.F.V. 635
Vine, E. *et al.* 257
Vine, E. and Crawley, D. 257
VINITI [Russian All-Union Inst. of Scientific and Technical Information] 101
Virginia Polytechnic Institute . . . [US], *and* electronic publishing 115
virtual libraries 119–20
viscosity 202, 235, 276
  viscoelasticity 134
  viscous force 290, 291, 298
vision, automated 363, 377, 382, 384, 385, 386
vitalism 625
Vital travel statistics (pub. Transport 2000) 510
VLSI (very large scale integrated circuit) 415, 417
VME (xxx) 393
VNIS *see* Vehicle Navigation and Information . . .
VNU Science Press, Topics in Transportation series 499
Vogwell, J. 197
volatility, *and* fire hazard 322
Volborth, A. 238
Vollman, T.E. *et al.* [*manufacturing*] 365
Vollmer, E., *Encyclopedia of hydraulics, soil* . . . 122
voltage **320**
  extra-high (EHV) 252, **429**
  high, direct current (HVDC) 432
  motors 435, 436
volume
  hydraulic fluids 289
  thermodynamic aspects 200–1
Von Kube H.L. *and* Steimle, F. 254
Vorres, K.S. 238
VROMDOC database 584
VSAT (very small aperture terminals) 444
Vutukuri, V.S. *and* Lama, R.D. 617
VVER reactors 261

W
Waddell, J.J. *and* Dobrowolski, J.A. 538
Wadden, R.A. *and* Scheff, P.A. 326
Wade, J.G. 445
Wahl, A.M. 168
Wainwright, S.A. *et al.* 635
WAIS (Wide Area Information Server) 117
Wait, J.V. *et al.* [*electronics*] 423
Wait, R. *and* Mitchell, A.R. [*finite elements*] 455
Wakerly, J.F. 423
Walas, S.M. 274, 276
Waldner, J.B. *and* Duffin, W.J. 363
Walford's guide to reference material [ed.Mullay/Schlicke] 121
Walker, D.J. *et al.* [*creativity*] 332
Walker, J.R. [*machining*] 363
Walker, P.M.B.
  Cambridge dictionary of science and technology 123
  Chambers materials science and technology dictionary 123
Walker, S.C. [*mining*] 621
walking machines 383
Wall, R.A. 93, 96, 339
Wallace, W.A. *et al.* [*bioengineering*] 648
Wallis, R.A. [*fans/ducts*] 208
walls 556, 560, 563, **563–4**
  cladding 557, 563–4
Walls, W.L. [*petroleum*] 322
Wall technology (pub. CIRIA) 563
Walsh, V. *et al.* [*design*] 332
Walshaw, A.C. [*thermodynamics*] 202
Walter, M.H. *and* Cox, R.F. 197
Walton, J.W. 332
Wan Changsen [*bearings*] 176
Wang, Q. *et al.* [*chain-drives/expert systems*] 164
WANO (World Ass'n of Nuclear Operators) **263**
WAN (wide area network) 109
Warland, E.G. 560
Warne, D.F. [*wind energy*] 247
Warnes, L.A.A. [*electronic materials*] 419
Warnick, C.C. *et al.* 248
Warren Spring Laboratory (WSL), *former* 494, 594, 595
Warrington, A.R. van C. 432
Warship '93 conference 476
Warwick, K. 393
Warwick University, Macroeconomic Modelling Bureau 575
W.A.S. [program] 176
washing, minerals 590
Wasserman, S.R. 122
Wasserwirtschaft 68
waste/waste treatment **488**, **489–90**, 493

air pollution 490–2
combustion 199
developing countries 488, 490
energy production 241–2
environmental impact 493
handbooks 534
hazardous wastes **489–90**
  radioactive waste 250, **260**
heat 206–7, **208–9**
industrial wastes 488, 489–92
journals 488, 490, 491
landfill 490, 614
mineral processing **591**, 592
odour **491–2**
solid wastes **489–90**
wastewater 485, 486, **488**, 491, 545
databases 485
Internet resources 119
mineral processes 592
safety 324
Waste Management 490
Waste management series (pub. DoEnv), Waste management papers 490
Waste Manager, The 490
Waste, recycling and environmental directory (pub. Telford) 534
Wastewater engineering (pub. Metcalf & Eddy) 545
Wastewater stabilization ponds . . . *(pub. WHO)* 488
watches, electronic 412
water
  databases 113
  energy from 210, **247–9**, **430**
  journals 249
  environmental protection *see* environmental engineering; water engineering
  and geothermal energy 245
  heavy (D_2O) 260, 261
  and hydraulic fluids **288–90**, 293
  and hydrogen fuels 241
  and machining processes 363
  and mineral processing 590
  and power generation 210, **247–9**, 255–7, **428**
  as neutron-moderator 260, 261
  as reactor coolant 260–1
  see also hydroelectric *etc.*
  properties 202, 235, 236, 253–5, 288–90
  sea *see* ocean
  steam engines 200, 237
  ultrasonic jet processes 363
  see also water engineering
Water Bulletin 489
Water directory (pub. Telford) 534
water engineering 269, **485–9**
  analysis **486–7**
  books **487**
  databases 113, **485**
  developing countries 488, 489
  ecology 487
  environmental aspects **486–9**, 493

770 Index

water engineering (*contd*)
  Internet resources 119
  journals 68, 113, 249, **487**, **488**, **489**
  safety 322, 324
  Standards 486–7
  thermodynamic aspects 205
  treatment **489**
  underwater engineering 378, 432, 470*tab*
  water energy 247–9, **430**
  waterfront structures 560, **567**
  wetlands 488
Water Engineering and Management 68, 489
Water and Environmental Management 487, 488, 489
Water Environment Research 487, 488
Water International 487
Waterloo Maple Software [co.] 452
Waterman, D.A. [*control systems*] 396
Waterman, N.A. *and* Ashby, M.F. [*materials*] 124, 313
Water Pollution Control Federation [US]
  Hazardous waste treatment processes 490
  Operation of... wastewater... plants 488
water power *see* water: energy from
Water Quality International 487, 488, 489
Water Record 489
Water Research 487, 488
WATER RESOURCES ABSTRACTS 113
Water Science and Technology 487, 488, 489
water treatment **489**
Water treatment handbook (*pub.* Intercept) 489
Water and Waste Treatment 488, 489
Water and Wastewater International 488, 489
Watson, J.P. [*roads*] 507
Watson, M.N. [*plastics*] 362
Watton, J. 288, 291, 293, 295, 298
Watton, J. *et al.* 296
Watton, J. *and* Salters, D.G. 296
Watton, J. *and* Tadmori, M.J. 293
Watton, J. *and* Xue, Y. 290
Waugh, W. 636
Wave energy (*pub.* IMechE) 248
wavelets [mathematics] 457
wave loading 567
wave power **247–9**
WCTR (World Conf. on Transport Research) **500**, **501**
WDK (Workshop Design–Konstruktion) 342, **651**
WDK Heurista [publishers] 162
weapons *see* defence
wear 173, 174
Wear 174, **175**

Weatherall, A. 363
Weaver, L.A. 324
'Web [WWW] virtual library' 119
Web *see* World Wide Web
Webster, F.V. *et al.* [*transport*] 507
Webster, F.V. *and* Cobbe, J. [*transport*] **506**, 507
Webster, J.G. [*biomedical engineering*] 635
WEC *see* World Energy Cncl
Weck, M. 360
Weck, M. *et al.* 361
Weedy, B.M. 427–8
WEGEMT *see* Western Eur. Graduate Education...
Wegst, C.W., *Stahlschlüssel* 302–3
weights, Standards 27–8
Weinberg, A.M. *and* Wigner, E.P. [*nuclear energy*] 266
Weinberg, F.J. [*combustion*] 237
Weiss, G. [*hazard/chemicals*] 323
Weiss, N.L. [*minerals*] 597, 600
Welch, S. [*telecommunications*] 443
WELDASEARCH database 341, 355
welding 361, **362**, **386**
  computer-aided 362, 386
  conferences 360
  databases 341, 355
  explosive 362
  hydraulics 292
  journals 68, 357
  organizations 138, **303**, **371**
  Standards 26
  training 371
Welding Institute **138**, **303**, 375, 387
  conferences 360
  WELDASEARCH database 341, 355
Welding International 357
Welding and Metal Fabrication 68
Wellburn, A. 491
Wells, H.G., *Time machine, The* 34
Wellstead, P.E. 391
Wellstead, P.E. *and* Zanker, P. 395
Wellstead, P.E. *and* Zarrop, M.B. 395
Welsh, R.J. [*bearings/design*] 175
Welty, J.R. *et al.* 277
Wennrich, P. 124
Wesselingh, J.A. *and* Krishna, R. 274
Wessex Group/Wessex Software UK Ltd 582, 583
WESSEX [program] **582**
West, M. *et al.* [*alternative energies*] 209, 210
West, R.E. *and* Kreith, F. [*thermal systems/solar energy*] 205
Western European Graduate Education in Marine Technology (WEGEMT) 479

Underwater technology 479
Westerterp, K.R. *et al.* 275
Westinghouse Electric Corp'n, *Westinghouse transmission and distribution reference book* 431
West Kentucky, University, Center for Coal Science 239
West Midlands Joint Data Team, GENERATE 511
Weston, J. [*robotics*] 386
Weston, W. *and* Lalor, M.J. [*fluid power*] 298
West Virginia, University, CERC 654, **656**, 657
WET-ICE [enabling technologies] conferences 656
Wexler, A. *et al.* 254
Whalley, P.B. [*thermodynamics*] 202, 206, 237
Whalley, R. [*control systems*] 394
What to Buy for Business 105
What's New... series (*pub.* Morgan Grampian) **87**
  ... in Design 157, 169, 339
  ... in Industry 67, 86, **87**, 339
  ... in Processing 87
What's On (*pub.* IChemE) 281
wheel-rims, Standards 20
Whelan, A. 365
Whelan, A. *and* Goff, J.P. 365
Which? 105
Whitaker, J. & Sons Ltd [publishers] 104
Whitaker's book list 104
Whitaker's books in print 104, 224
White, D.A. *and* Sofge, D.A. [*artificial intelligence*] 397
White, L. [*minerals*] 600
White, P.R. [*transport/planning*] 507
Whiteman, M.F. 338
Whitney, D.E. 337
Whittaker, B.N. *and* Frith, R.C. 545, 616
Whittle, R.T. 557
Whitworth, J. 20
WHO *see* World Health Org'n
'Whole Internet catalog' [list] 119
Who's who (*pub.* Black) 126
Who's who of British engineers (*pub.* Simon) 126
Who's who in science and engineering, 1994–1995 (*pub.*Marquis [US]) 126
Who's who in science in Europe (*pub.* Longman) 126
Whyte, I.L. [*construction*] 544
Wide Area Information Server (WAIS [client-server]) 117
wide area network (WAN) 109
  see also networks
Widrow, B. *and* Stearns, S.D. 395
Wignall, A. *et al.* 545
Wilbur, L.C. 242

Index 771

Wilby, C.B. 557
Wilcock, D.F. [bearings] 177
Wilcox, A.D. et al. [design/
    electrical] 331
Wildi, T. 427
Wiley, John, & Sons Ltd
    [publishers] 423, 527
    Kirk-Othmer encyclopaedia ...
    271
Wilkinson, B.W. and Barnes,
    R.W. [co-generation] 209
Wilkinson, C. [structures] 566
Williams, A.F. [solar energy]
    244
Williams, B.W. [power
    electronics] 436
Williams, D.F. [bioengineering]
    636
Williams, D.J. [robotics/systems]
    385
Williams, D.J. and Rogers
    P.[manufacturing cells] 385
Williams, F.A. [combustion] 237
Williams, G.B.A. [chimneys] 560
Williams, J.A. [tribology] 175
Williams, T.I. [technology
    history] 154
Williams, T.J. [control systems]
    394
Williamson, A.C. [electricity]
    208
Williamson, M. [aerospace] 123
Willing's press guide (pub.
    Thomas Skinner) 73
Wills, B.A. 597
Wilson, C.E. and Sadler, J.P.
    [machines] 159, 167,
    170–1tab
Wilson, H.W., Co. [publishers]
    110
    Applied Science and
        Technology Index 380
    Cumulative book index 103
Wilson, J.G. [water/biology] 487
WilsonDisc [co.] 110
WilsonLine database host 110
Wimpey, George, plc 525
Winburn, D.C. 320
wind
    energy, US 430
    energy from 210, 241, 242,
        246–7, 430–1
    abstracts 228
    databases 229
    journals 247, 431
    organizations 242
    wind farms 430
    and structures 554
    wind loads 553, 554
Wind Energy Abstracts 228
Wind Energy News 247
Wind Energy Weekly 247
Wind Engineering 431
Windkraft Journal 247
Window, A.L. 146
Window Industries 67
windows 562, 563–4
    journals 67
    organizations 562
Windpower Monthly 247
Wind turbines ... (pub. BWEA)
    246

Wings 68
Winston, P.H. 335, 397
Winter, D.A. 636
Winter concreting (pub. BCA)
    537
WIPO (World Intellectual
    Property Org'n) 45
wires 302, 364
wire wrapping 361
Wiring regulations/Guidance
    (pub. IEE) 320, 321
Wistrich, E. 507
Withers, D.J. 444
WME (World Mining Equipment)
    597, 611
Wohl, M. and Hendrickson, C.
    507
Wolff, J. 633
Wolfram Research (UK) Ltd 160,
    452
Wolfson, R. 251
Wong, H.Y. 233
wood
    energy from 242, 249–50, 256
    as material see timber
Wood, A.J. and Wollenberg, B.F.
    [electricity] 428
Wood, C. and Jones, C.
    [environment] 493
Wood, D. [design] 172tab
Wood, D.A., Assessing ... road
    schemes ... 493
Wood, D.A., Committee [road
    assessment] 493
Woodcock, C.R. and Mason, J.S.
    [bulk solids] 278
Woodcock, J.T. [mining/
    metallurgy] 600
Wood Energy Monthly Update
    250
Woodgate, R.W. 362
Woodhead Publishing [co.],
    International standards
    index: welding ... 26
Wood Information Sheets (pub.
    TRADA) 541
Woods Acoustics [co.], Woods
    practical guide to noise
    control [ed. Sharland] 492
Wood Technology 68
Woodward, M.E. 446
work equipment see equipment
work, human
    construction 578
    see otherwise ergonomics;
        manufacturing etc.
workplaces
    building sites 543, 578
    construction of 544
    Regulations 182
    safety 180–4, 315–24, 325–7
    Workplace Directive 183
    see also plant etc.
Works construction guides (pub.
    ICE) 544
Workshop on biomass thermal
    processing ... (pub.
    UNIDO) 249
Workshop Design–Konstruktion
    (WDK) 342, 651
Workshops on Enabling

Technologies
    Infrastructure ...
    (WET-ICE) 656
Workshop on VLSI Signal
    Processing 417
Work Study 358
Work study in mines (pub. NCB)
    614
work, thermodynamic 199–201,
    203, 223, 425, 428
World Aluminium Abstracts 57
    online 110
World Association of Nuclear
    Operators (WANO) 263
World Bank [International Bank
    for Reconstruction and
    Development] 516
    and energy 247, 248, 249, 256
    Internet data 503
    and safety 317
    and transport 516
World business directory (pub.
    Gale/World Trade Centers)
    84
World Business Publications,
    World's New Products 87
WORLDCAT database 103
World Coal, former 597
World Conference on Transport
    Research (WCTR) 500, 501
World databases in patents (pub.
    Bowker-Saur) 57
World directory of
    organizations ... (pub.
    Engineering Research
    Centres) 126
World Energy Conference 248
    Energy terminology ... 233
World Energy Council (WEC)
    224
    Energy for tomorrow's world
        233
World energy and nuclear
    directory 233
World Health Organization
    (WHO), publications 490
    Air quality guidelines ... 491
    European Series 491
    Wastewater stabilization
        ponds ... 488
World industrial robot statistics
    (pub. IFR) 382
World Intellectual Property
    Organization (WIPO) 45
World of learning (pub. Europa)
    126
World meetings (pub. Macmillan)
    76–7
World Meteorological
    Association 244
World mineral statistics (pub.
    BGS) 612
World Mining, former 597
World Mining Congress 612
World Mining Equipment (WME)
    597, 611
World nuclear industries
    handbook (pub. Nuclear
    Engineering Internat.) 261,
    267
World Oil 241

World Patent Information 61
WORLD PATENTS INDEX (WPI)
 [Derwent] database 55, 58,
 59, 380
World Petroleum Congress
 (WPC) 225
World road statistics (pub. IRF)
 510
World Ship Society, *Marine
 News* 481
World's new products (pub.
 World Business) 87
World Trade Centers
 Association, *World business
 directory* 84
World transport data (pub.
 IRTU) 510
*World Water and Environmental
 Engineer* 488, 489
*World-wide guide to equivalent
 irons and steels (pub.* ASM)
 26
*Worldwide pipelines and
 constructors' directory* 475
*Worldwide public transport
 timetables (pub.* ABC) 510
WORLDWIDE STANDARDS SERVICE
 database 529
World Wide Web (WWW)
 client-server 116, **117–18**,
 305, 341, **502–3**, 505
 HSE publications 182
 product information 312
 'Virtual Library' Internet list
 119
Wound Management, former 639
Wound and Skin Care 639
WPC (World Petroleum
 Congress) 225
WPI *see* WORLD PATENTS INDEX
wrench keys, Standards 16

Wright, C.J. 334
writing *see* communication: skills
Wrixon, G.T. *et al.* 242
WSL *see* Warren Spring
 Laboratory
WWW *see* World Wide Web
Wykes, K. 336

X

X.25 protocol 305, **445**
Xanalog Corp'n 295
 Xanalog program 295
Xray Associates [co.] 148
X-ray diffraction stress analysis
 136, 147–8

Y

yacht keel, patented 51
'Yahoo' Internet list 119, 312
Yang, S.J. *and* Ellison, A.J. 182
Yao, J.T.P. 319
Yariv, A. 448
Yaverbaum, L. 237
Yaws, C.L. 273
*Yearbook of international
 organizations, (pub.* Union
 of Internat. Org'ns) 126
yearbooks 122, **125–6**
 directories of 72, 73
Yeomans, D. 542
Yeomans, D.T. 561
Yoshida, T. 323, 324
Young, W.C., *Roark's formulas
 for stress and strain* 142
Yu, Jinghong 290
Yugoslavia, *former*
 design 162
 machines 161–2
 see also Croatia
Yuncu, H. *et al.* 244

Z

Zambia, minerals 596
ZDE [Zentralsstelle
 Dokumentation
 Electrotechnik] database
 101, 400
Zeid, I. 362
*Zeitschrift für Angewandte
 Mathematik und Mechanik*
 459
*Zeitschrift für Angewandte
 Mathematik und Physik* 459
Zeitschrift für Metallkunde 68
Zemansky, M.W. *et al.* 202
Zemansky, M.W. *and* Dittman,
 R.H. 202
Zemansky, M.W. *and* Van Ness,
 H.C. 202
Zentralsstelle Dokumentation ...
 see ZDE
Zeroth Law of thermodynamics
 199
Zhu, G. 168
Ziegler, J.G. *and* Nichols, N.B.
 392
 Ziegler-Nichols controls
  approaches 392–3
Zienkiewicz, O.C. *and* Taylor,
 R.L. 148
Zimmerman, K.H. *and* Powell,
 R.H. 208
zinc 593
 abstracts/indexes 57
 cladding/roofing 564
 organizations 564
Zinc Abstracts 57
Zinc Development Association
 564
Zolynski, B. 126
zoology 635, 640
Zunkel, A.D. *et al.* 601